Healthy Ageing and Longevity

Volume 14

Rapidly changing demographics worldwide towards increased proportion of the elderly in the population and increased life-expectancy have brought the issues, such as "why we grow old", "how we grow old", "how long can we live", "how to maintain health", "how to prevent and treat diseases in old age", "what are the future perspectives for healthy ageing and longevity" and so on, in the centre stage of scientific, social, political, and economic arena. Although the descriptive aspects of ageing are now well established at the level of species, populations, individuals, and within an individual at the tissue, cell and molecular levels, the implications of such detailed understanding with respect to the aim of achieving healthy ageing and longevity are ever-changing and challenging issues. This continuing success of gerontology, and especially of biogerontology, is attracting the attention of both the well established academicians and the younger generation of students and researchers in biology, medicine, bioinformatics, bioeconomy, sports science, and nutritional sciences, along with sociologists, psychologists, politicians, public health experts, and health-care industry including cosmeceutical-, food-, and lifestyle-industry. Books in this series will cover the topics related to the issues of healthy ageing and longevity. This series will provide not only the exhaustive reviews of the established body of knowledge, but also will give a critical evaluation of the ongoing research and development with respect to theoretical and evidence-based practical and ethical aspects of interventions towards maintaining, recovering and enhancing health and longevity.

More information about this series at http://www.springer.com/series/13277

Suresh I. S. Rattan · Gurcharan Kaur
Editors

Nutrition, Food and Diet in Ageing and Longevity

Springer

Editors
Suresh I. S. Rattan
Department of Molecular Biology
and Genetics
Aarhus University
Aarhus C, Denmark

Gurcharan Kaur
Department of Biotechnology
Guru Nanak Dev University
Amritsar, Punjab, India

ISSN 2199-9007 ISSN 2199-9015 (electronic)
Healthy Ageing and Longevity
ISBN 978-3-030-83016-8 ISBN 978-3-030-83017-5 (eBook)
https://doi.org/10.1007/978-3-030-83017-5

This Springer imprint is published by the registered company Springer Nature Switzerland AG
The registered company address is: Gewerbestrasse 11, 6330 Cham, Switzerland

Preface: Nutrition, Food and Diet

Nutrition, food and diet are the terms often used casually and uncritically as overlapping ones. Whereas nutrition generally refers to the macro- and micro-nutrients essential for health and survival, it is normally consumed as animal- or plant-based foods. Even more importantly, there are crucial cultural dimensions which transform nutritional components through various methods of cooking and preservation and then identify the resulting food as a specific diet. Moreover, in the background of all this lurks the evolutionary history of *Homo sapiens* from being hunter–gatherers to becoming agriculturists and to the present state of becoming the modern consumers of highly processed food products and overeating behaviours. This book is a multi-chapter compilation reviewing and discussing the present state of information and knowledge about those aspects.

The contents of the book are divided into five parts: (1) the first part "Nutritional components", explains the role of macro- and micro-nutrients including proteins, carbohydrates, fats and oils to pro- and prebiotics, and hormetins, along with an article on the evolutionary biology of nutrition and its implications in the modern lifestyle; (2) the second part "Food for health" discusses various types of food in relation to their claimed health benefits from animal- and plant-based food, fermented foods, anti-inflammatory foods, milk and other dairy products, nutraceuticals and functional foods, and foods for brain health; (3) the third part "Diet and culture" discusses the general principles of healthy dietary patterns, nutritional regulation of ageing and longevity, circadian rhythms and diet, dietary principles and practice according to the traditional Chinese medicine, Indian traditional diet, and ketogenic diet; (4) the fourth part "Nutritional and dietary interventions" critically reviews various social and clinical approaches through history, including fasting and calorie restriction, molecular mimetics of calorie restriction, nutrition-based senolytic compounds, natural and synthetic short peptides and specific herb-based chemicals, along with providing some practical and cautionary advice for having the optimal nutrition, and managing the problems related to obesity. Finally, in the fifth part, a concluding chapter by the editors recapitulates and summarizes the central messages and suggestions emerging from all other chapters included in this book.

Prevalence of malnutrition and under-nutrition increases in old age due to both physiological changes in appetite, digestion and metabolism, and other factors such as decreased physical activity, cognitive decline, and multiple co-morbidities. Thus, making right choices of food, based on the scientific information and within a cultural context, are the most effective strategies to enhance individual health and public health. Wholesome foods consumed with the hormetic principles of pleasure, moderation, and variety may be the mantra for healthy ageing and longevity. This book provides reviews and critical analyses of the past, present, and futuristic research on "Nutrition, Food and Diet" in the context of human health, well-being, ageing, age-related diseases, lifespan, and healthspan.

Suresh I. S. Rattan
Gurcharan Kaur

Contents

List of Contributors

Stephen A. Adefegha Functional Foods and Nutraceuticals Unit, Department of Biochemistry, Federal University of Technology, Akure, Akure, Nigeria

Ebenezer I. O. Ajayi Diabesity Complications and Other Neglected Infectious Diseases Group, Department of Biochemistry, Osun State University, Osogbo, Nigeria;
Membrane Biophysics and Nanotechnology Laboratories, Mercedes and Martin Ferreyra Institute of Medicine, IMMF-INIMEC-CONICET-UNC, Cordoba, Argentina

Laxmi Ananthanarayan Department of Food Engineering and Technology, Institute of Chemical Technology, Mumbai, India

Eleonora Baiamonte Geriatric Unit, Department of Internal Medicine and Geriatrics, University of Palermo, Palermo, Italy

Mario Barbagallo Geriatric Unit, Department of Internal Medicine and Geriatrics, University of Palermo, Palermo, Italy

Éric Le Bourg Centre de Recherches sur la Cognition Animale (CRCA), Centre de Biologie Intégrative (CBI Toulouse), Université de Toulouse, CNRS, UPS, Toulouse, France

Kaustuv Bhattacharya International Flavor & Fragrances, Brabrand, Denmark

Edward J. Calabrese Department of Environmental Health Sciences, Morrill I, N344, University of Massachusetts, Amherst, MA, USA

Vittorio Calabrese Department of Biomedical and Biotechnological Sciences, University of Catania, Torre Biologica, Catania, Italy

Shivani Chourasia Department of Applied Agriculture, Central University of Punjab, Bathinda, India

Roberto Crea Oliphenol LLC., Hayward, CA, USA

Lisette de Groot Division of Human Nutrition, Wageningen University, Wageningen, The Netherlands

Giuseppe Dionisio Department of Molecular Biology and Genetics, Research Center Flakkebjerg, Aarhus University, Slagelse, Denmark

Ligia J. Dominguez Geriatric Unit, Department of Internal Medicine and Geriatrics, University of Palermo, Palermo, Italy

Aderonke E. Fakayode Department of Biochemistry and Molecular Biology, Faculty of Science, Obafemi Awolowo University, Ile-Ife, Ile-Ife, Nigeria

Sukant Garg AIST-INDIA DAILAB, DBT-AIST International Center for Translational and Environmental Research (DAICENTER), National Institute of Advanced Industrial Science and Technology (AIST), Tsukuba, Japan

Donovan P. German Department of Ecology and Evolutionary Biology, University of California, Irvine, CA, USA

Ceren Gezer Department of Nutrition and Dietetics, Faculty of Health Sciences, Eastern Mediterranean University, Famagusta, North Cyprus via Mersin 10, Turkey

Jyoti S. Gokhale Department of Food Engineering and Technology, Institute of Chemical Technology, Mumbai, India

Valentina Greco Department of Biomedical and Biotechnological Sciences, University of Catania, Torre Biologica, Catania, Italy

Pol Grootswagers Division of Human Nutrition, Wageningen University, Wageningen, The Netherlands

Martina Guarrera Geriatric Unit, Department of Internal Medicine and Geriatrics, University of Palermo, Palermo, Italy

S. Heeba Agro Processing and Technology Division, CSIR-National Institute for Interdisciplinary Science and Technology, Trivandrum, Kerala, India

Dino Demirovic Holmquist Eurofins DQCI, LLC, Quincy Street, MN, USA

Anita Jagota Neurobiology and Molecular Chronobiology Laboratory, Department of Animal Biology, School of Life Sciences, University of Hyderabad, Hyderabad, India

Majeed Jamakhani Molecular and Cellular Epigenetics (GIGA) and Molecular Biology (GxABT), University of Liège, Liège, Belgium

Aayushi Kadam School of Biotechnology and Bioinformatics, D. Y. Patil Deemed to be University, Navi Mumbai, India

Deepak Kadam Anatek Services Pvt. Ltd., Mumbai, India

Sunil C. Kaul AIST-INDIA DAILAB, DBT-AIST International Center for Translational and Environmental Research (DAICENTER), National Institute of Advanced Industrial Science and Technology (AIST), Tsukuba, Japan

Amritpal Kaur Department of Food Science and Technology, Guru Nanak Dev University, Amritsar, Punjab, India

Gurcharan Kaur Department of Biotechnology, Guru Nanak Dev University, Amritsar, India

M. Sultan Khan Neurobiology and Molecular Chronobiology Laboratory, Department of Animal Biology, School of Life Sciences, University of Hyderabad, Hyderabad, India

Zeeshan Akhtar Khan Neurobiology and Molecular Chronobiology Laboratory, Department of Animal Biology, School of Life Sciences, University of Hyderabad, Hyderabad, India

Vladimir Kh. Khavinson Saint Petersburg Institute of Bioregulation and Gerontology, St. Petersburg, Russia;
Pavlov Institute of Physiology of the RAS, Saint-Petersburg, Russia

Mallika Khurana AIST-INDIA DAILAB, DBT-AIST International Center for Translational and Environmental Research (DAICENTER), National Institute of Advanced Industrial Science and Technology (AIST), Tsukuba, Japan

Marios Kyriazis National Gerontology Centre, Larnaca, Cyprus

Ignazio La Mantia Surgical Specialties Department, ENT Unit - General Surgery and Medical, University of Catania, Catania, Italy

S. S. Lele Department of Food Engineering and Technology, Institute of Chemical Technology, Mumbai, India;
Institute of Chemical Technology, Matunga, Mumbai, India

Marco Mensink Division of Human Nutrition, Wageningen University, Wageningen, The Netherlands

Olorunfemi R. Molehin Department of Biochemistry, Faculty of Science, Ekiti State University, Ado-Ekiti, Ado-Ekiti, Nigeria

Alexey Moskalev Institute of Biology of FRC “Komi Science Center”, Ural Division of RAS, Syktyvkar, Russia

Azza Silotry Naik Department of Biological Sciences, School of Natural Sciences, Faculty of Science and Engineering, University of Limerick, Limerick, Ireland

Jan O. Nehlin Department of Clinical Research, Copenhagen University Hospital Amager and Hvidovre-Copenhagen, Hvidovre, Denmark

P. Nisha Agro Processing and Technology Division, CSIR-National Institute for Interdisciplinary Science and Technology, Trivandrum, Kerala, India;
Academy of Scientific and Innovative Research (AcSIR), Ghaziabad, India

Gozde Okburan Department of Nutrition and Dietetics, Faculty of Health Sciences, Eastern Mediterranean University, Famagusta, North Cyprus via Mersin 10, Turkey

Johnson O. Oladele Biochemistry Unit, Department of Chemical Sciences, Kings University, Ode Omu, Osun, Nigeria

Maria Laura Ontario Department of Biomedical and Biotechnological Sciences, University of Catania, Torre Biologica, Catania, Italy

Angela Parisi Geriatric Unit, Department of Internal Medicine and Geriatrics, University of Palermo, Palermo, Italy

Francesco Rampulla Department of Biomedical and Biotechnological Sciences, University of Catania, Torre Biologica, Catania, Italy

Suresh I. S. Rattan Department of Molecular Biology and Genetics, Aarhus University, Aarhus, Denmark

Michael R. Rose Department of Ecology and Evolutionary Biology, University of California, Irvine, CA, USA

Yuliya Safarova Center for Life Sciences, National Laboratory Astana, Nazarbayev University, Nur-Sultan, Kazakhstan;
School of Engineering and Digital Sciences, Nazarbayev University, Nur-Sultan, Kazakhstan

Timur Saliev S.D. Asfendiyarov Kazakh National Medical University, Almaty, Kazakhstan

Olawumi A. Samuel Membrane Biophysics and Nanotechnology Laboratories, Mercedes and Martin Ferreyra Institute of Medicine, IMMF-INIMEC-CONICET-UNC, Cordoba, Argentina

Sebastiano Sciuto Department of Biomedical and Biotechnological Sciences, University of Catania, Torre Biologica, Catania, Italy

Maria Scuto Department of Biomedical and Biotechnological Sciences, University of Catania, Torre Biologica, Catania, Italy

Heena Shah K. J. Somaiya College of Science and Commerce, Mumbai, India

Sandeep Sharma Department of Comparative Biology and Experimental Medicine, University of Calgary, Calgary, AB, Canada

Sushree Abhidhatri Sharma Neurobiology and Molecular Chronobiology Laboratory, Department of Animal Biology, School of Life Sciences, University of Hyderabad, Hyderabad, India

Khetan Shevkani Department of Applied Agriculture, Central University of Punjab, Bathinda, India

Balwinder Singh P.G. Department of Biotechnology, Khalsa College, Amritsar, Punjab, India

Jatinder Pal Singh Department of Food Science and Technology, Guru Nanak Dev University, Amritsar, Punjab, India

Ilia Stambler Vetek (Seniority) Association—The Movement for Longevity and Quality of Life, Tel-Aviv, Israel

Federica Tagliaferri Geriatric Unit, Department of Internal Medicine and Geriatrics, University of Palermo, Palermo, Italy

Svetlana V. Trofimova Saint Petersburg Institute of Bioregulation and Gerontology, St. Petersburg, Russia

Angela Trovato Department of Biomedical and Biotechnological Sciences, University of Catania, Torre Biologica, Catania, Italy

Kanchanlata Tungare School of Biotechnology and Bioinformatics, D. Y. Patil Deemed to be University, Navi Mumbai, India

Bauyrzhan Umbayev Center for Life Sciences, National Laboratory Astana, Nazarbayev University, Nur-Sultan, Kazakhstan

Madhavi R. Vernekar School of Biotechnology and Bioinformatics, D.Y. Patil Deemed to be University, Navi, Mumbai, India

Renu Wadhwa AIST-INDIA DAILAB, DBT-AIST International Center for Translational and Environmental Research (DAICENTER), National Institute of Advanced Industrial Science and Technology (AIST), Tsukuba, Japan

Aislu Yermekova Center for Life Sciences, National Laboratory Astana, Nazarbayev University, Nur-Sultan, Kazakhstan

Yifang Zhang Shanghai East International Medical Center, Shanghai, China

Vincenzo Zimbone Department of Biomedical and Biotechnological Sciences, University of Catania, Torre Biologica, Catania, Italy

Božena Ćurko-Cofek Faculty of Medicine, Department of Physiology and Immunology, University of Rijeka, Rijeka, Croatia

Part I
Nutritional Components

Chapter 1
Dietary Proteins: Functions, Health Benefits and Healthy Aging

Khetan Shevkani and Shivani Chourasia

Abstract Dietary proteins are the source of indispensable, dispensable and functional amino acids essential for the synthesis of body proteins and the regulation of physiological processes. They also are beneficial in alleviating lifestyle and/or age-related health problems (e.g. loss of muscle mass and strength, obesity/sarcopenic obesity, dyslipidaemia, bone mineral loss and type-2 diabetes) because of their roles in the promotion of muscle protein synthesis, satiation, optimisation of growth factors, inhibition of inflammatory substances and regulation of major pathways of metabolism, though protein-quantity and quality are major determinants of nutritional and disease-preventing effects. The intake of high quality proteins in adequate amounts together with healthy lifestyle can contribute to healthy aging through maintenance of muscle mass and/or enhanced ability of recovering from diseases, while proteins from legumes/pulses and milk (e.g. whey proteins) may benefit elderly people by reducing the risk of coronary artery diseases, obesity, bone density loss, type-2 diabetes and associated morbidities/mortalities. This chapter discusses health benefits of increased intake of dietary proteins in elderly people and provides an overview of protein quality and methods for evaluating the same. The chapter also outlines functions of dietary proteins and compares animal and plant proteins for their quality and health benefitting effects.

Keywords Dietary proteins · Health benefits · Health-span · Healthy aging · Plant vs animal proteins

1.1 Introduction

Proteins are complex nitrogenous compounds that are made up of amino acids linked through peptide bonds. At the elemental level, they contain 50–55% carbon, 20–25% oxygen, 12–19% nitrogen, 6–7% hydrogen and 0.2–3.0% sulphur (Damodaran 2017). Proteins are considered as the most fundamental component of body. They are the second most abundant constituent in human body, next only to water. A healthy

K. Shevkani (✉) · S. Chourasia
Department of Applied Agriculture, Central University of Punjab, Bathinda 151401, India

S. I. S. Rattan and G. Kaur (eds.), *Nutrition, Food and Diet in Ageing and Longevity*, Healthy Ageing and Longevity 14, https://doi.org/10.1007/978-3-030-83017-5_1

man weighing 70 kg contains around 11 kg (approximately 16%) proteins in the body against around 10 kg (approximately 14%) in a woman of the same weight. About 45–50% of total body protein is present in muscles, 15–20% in bones, 10–15% in skin and the remaining portion in other tissues. Thousands different types of proteins occur in human body, which are involved in almost all metabolic and physiological processes. However, human body does not have mechanisms to store proteins. Therefore, a constant supply of a certain minimum amount of proteins through diet is essential for maintaining growth and other metabolic processes. Normal healthy adults are recommended to take 0.8 g quality proteins per kg body weight per day to meet protein requirements (Wolfe et al. 2017). Although body proteins differ considerably from proteins present in foods for chemical/biochemical and structural characteristics, dietary proteins provide amino acids and nitrogen that are utilized for synthesis of body proteins and other compounds vital for survival.

Milk, meat, fish, eggs, pulses/legumes and cereals are major sources of dietary proteins. These foods differ widely for protein content and characteristics. Protein content of selected foods are shown in Table 1.1. In general, animal foods (milk, egg, fish and meat) are considered as a source of high-quality proteins due to their 'completeness' in the context of amino acid composition and the absence of anti-nutritional constituents. However, plant foods, nowadays, are being promoted because of 1) the absence of cholesterol, 2) the presence of health-beneficial bioactive compounds (e.g. lectins, enzyme inhibitors, phytosterols, saponins, phenolic compounds, dietary fibre, etc.), and 3) their roles in ensuring sustainable agriculture and food security (Shevkani et al. 2015; Singh et al. 2017; Lonnie et al. 2018; Magrini et al. 2018; Bessada et al. 2019; Shevkani et al. 2021).

Dietary proteins also can be beneficial in alleviating lifestyle and age-related disorders/diseases and reducing morbidity/mortality in elderly people. Increased intake of dietary proteins was reported beneficial in the maintenance of muscle mass and the prevention of sarcopenia (Houstan et al. 2008; Zhang et al. 2020), obesity (Weigle et al. 2005; Millward et al. 2008; Zhou et al. 2014), bone demineralisation/osteoporosis (Thorpe et al. 2008; Shams-White et al. 2017; Groenendijk et al. 2019) and dyslipidaemia (Santesso et al. 2012; Hosomi et al. 2013). Also, a high protein diet caused about 28% reduction in mortality in elderly people of more than 65 years of age (Levine et al. 2014). This chapter discusses the health-benefitting effects of increased intake of dietary proteins in elderly people and provides an overview of protein quality and methods for evaluating the same. In addition, the chapter also outlines functions of dietary proteins and compares animal and plant proteins for their quality and health benefits.

Table 1.1 Protein content of selected foods

Food	Protein content (g/100 g)
Cereals and pseudocereals	
Wheat	8.0–18.0[a]
Brown rice	7.92–8.20[b]
Milled rice	7.46–8.02[b]
Maize	8.80[c]
Barley	10.94[c]
Ragi millet	7.16[c]
Sorghum	9.97[c]
Amaranth grains	12.5–15.2[d]
Quinoa	14.1–15.4[e]
Legumes/pulses	
Pinto bean	22.80[f]
Lima bean	23.92[f]
Small red bean	25.68[f]
Red kidney bean	25.60[f]
Black bean	25.37[f]
Navy bean	25.73[f]
Black eye bean	24.58[f]
Mung bean	27.10[f]
Lentils	28.05[f]
Chickpea	22.37[f]
Horse gram	21.73[c]
Urad bean	21.97[c]
Moth bean	19.75[c]
Pigeon pea/red gram	20.47[c]
Lupine	36.17[g]
Soybean	36.49[g]
Soybean, white	37.8[c]
Soybean, brown	35.58[c]
Soybean, split dehulled	37.5[h]
Lentils, split dehulled	27.6[h]
Mung bean, split dehulled	25.0[h]
Urad bean, split dehulled	31.4[h]

(continued)

Table 1.1 (continued)

Food	Protein content (g/100 g)
Chickpea, split dehulled	21.3[h]
Peas, split dehulled	31.1[h]
Cowpea, split dehulled	29.0[h]
Oilseeds	
Mustard seeds	19.51[c]
Groundnut	23.65[c]
Coconut kernel	7.27[c]
Sunflower seeds	23.53[c]
Safflower seeds	17.66[c]
Sesame seeds, whole, dried	17.73[g]
Flaxseed	19.35[g]
Tree nuts	
Almonds	18.41[c]
Cashew nuts	18.78[c]
Walnuts	14.92[c]
Pistachio nuts	23.35[c]
Fruits and vegetables	
Pomegranate, pulp	2.47[i]
Kinnow, pulp	1.76[i]
Mango, pulp	1.69[i]
Banana, pulp	1.82[i]
Jambolan	1.75[i]
Grapes	1.66[i]
Sapodila, pulp	1.54[i]
Beetroot	1.12[i]
Brinjal	1.68[i]
Carrot, orange	1.75[i]
Bitter gourd	1.64[i]
Mentha	3.30[i]
Carrot, black	1.75[i]
Spinach	1.71[i]

(continued)

Table 1.1 (continued)

Food	Protein content (g/100 g)
Potato	1.35–1.54[c]
Milk and Milk products	
Cow milk, whole	3.26[c]
Buffalo milk, whole	3.68[c]
Cheddar cheese	23.3[g]
Paneer	18.86[c]
Khoa	16.34[c]
Egg, meat and fish	
Egg	12.4[g]
Lamb, ground	20.33[g]
Chicken	16.07[g]
Ham, minced	16.28[g]
Fish, salmon, raw	20.5[g]
Fish, tuna, raw	24.4[g]
Fish, mackerel, raw	19.08[g]
Fish, porgy, raw	18.88[g]
Crab	10.23[c]
Oyster	9.51[c]
Tiger prawns, brown	14.85[c]
Lobster, brown	15.96[c]

References:
[a]Posner (2000).
[b]Sandhu et al. (2018).
[c]Longvah et al. (2017).
[d]Shevkani et al. (2014).
[e]Ghumman et al. (2021).
[f]Du et al. (2014).
[g]United States Department of Agriculture (available online at https://fdc.nal.usda.gov/).
[h]Shevkani et al. (2021).
[i]Singh et al. (2016).

1.2 Proteins as a Source of Amino Acids

Dietary proteins are the source of amino acids and nitrogen required in body for synthesis/formation of tissues/organs, enzymes, hormones and cell-specific physiologically important low molecular weight substances e.g. nitric oxide, glutathione, carnitine, carnosine, serotonin, melanin, melatonin, etc. (Blachier et al. 2011; Kong et al. 2012; Wu 2013). Although hundreds of amino acids occur in nature, only 20–21 are involved in synthesis of proteins in human body and in most other forms of life. Structurally, amino acids are composed of a carbon atom linked covalently with a hydrogen atom, an amine group, a carboxyl group and an alkyl group (also referred to as amino acid side chain). Different amino acids differ for the side chain, the characteristics of which decide physicochemical properties and reactivity of amino acid as well as properties of proteins. Based on the characteristics of side chains, amino acids are categorized in to acidic (aspartic acid and glutamic acid), basic (arginine, histidine and lysine), polar uncharged (serine, threonine, asparagine, glutamine and cysteine), aliphatic and nonpolar (alanine, isoleucine, leucine, valine, glycine, proline and methionine) and aromatic and nonpolar amino acids (phenylalanine, tyrosine and tryptophan), though a few (e.g. glycine, histidine and cysteine) do not fit perfectly into a particular group and their assignments to a particular group is not considered absolute (Nelson and Cox 2013; Ustunol 2015).

On the basis of nutritional properties, amino acids are grouped as essential (indispensable), non-essential (dispensable) and conditionally-essential (semi-essential/conditionally-indispensable) amino acids (Fig. 1.1). Although human body needs all 21 amino acids for proper functioning, growth and maintenance, only 9 have been classified as indispensable amino acids as these cannot be synthesized endogenously in human body. Some amino acids (arginine, cysteine, glutamine, glycine, proline and tyrosine) are conditionally-indispensable, the human body is unable to synthesize them endogenously under certain physiological conditions (e.g. infancy, pregnancy, lactation, injury, burn, infections or diseased conditions). Amino acids have also been classified as functional amino acids based on their specific functions in the body. Functional amino acids are defined as the amino acids that participate in regulation of key metabolic pathways involved in growth, development, lactation, reproduction and overall health of organisms (Wu 2010, 2013). Functional amino acids contribute to and/or play crucial roles in 1) intestine health, 2) foetal survival/development, 3) immunity, 4) neurological functions and synthesis of neurotransmitters, 5) secretion of insulin, 6) activation of taste receptors and 7) recovery from injury (Li et al. 2007; Brosnan and Brosnan 2013; Rezaei et al. 2013; Wu et al. 2017). For example, dietary glutamate, glutamine and aspartate serve as major metabolic fuels for enterocytes and small intestine (Rezaei et al. 2013). Glutamate, glutamine, arginine and glycine also can contribute to intestinal health through enhancing villus-height, width and surface area, mucosal protein synthesis, activity of antioxidant enzymes (e.g. catalase, superoxide dismutase and glutathione peroxidase) in small intestine and health of enterocytes (Jiao et al. 2015; Yi et al. 2018). In addition, glutamine is also involved in providing ATPs to macrophages and lymphocytes and serves as an exclusive source of energy for arteries in post-absorptive state (Li et al. 2007). Further, arginine and glutamine play important functions in placental metabolism and supplementation of these amino acids can help enhance foetal survival and growth (Wu et al. 2017). Selenocystein, a more recently discovered 21st proteinogenic amino acid, also performs important functions in human body. It is involved in the regulation of thyroid metabolism, removal of reactive oxygen species and protection of cells from oxidative damage, maintenance of cellular redox balance, regulation of signalling cascades, promotion of protein folding and maintenance of selenium homeostasis (Schmidt and Simonović 2012).

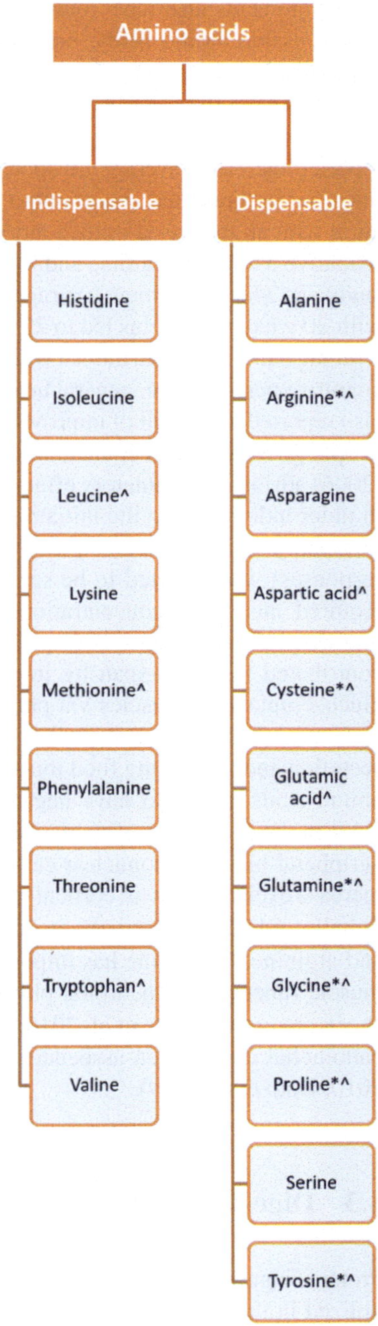

Fig. 1.1 Classification of amino acids. * Conditionally essential amino acids. ^ Functional amino acids (Wu 2013)

Amino acids also have roles in healthy aging and longevity. Amongst different indispensable amino acids, branched chain amino acids (leucine, isoleucine and valine) have contributory roles in healthy aging and longevity. They do not undergo metabolism in liver, hence get circulated almost immediately after absorption and become available to the body (Holeček 2018; Dato et al. 2019). D'Antona et al. (2010) reported increased average lifespan and improved performance/endurance for male mice fed on branched chain amino acids diets. The effect was observed in association with an increased mitochondrial biogenesis, sirtuin-1 expression and reduced oxidative damage in cardiac and skeletal muscles (D'Antona et al. 2010). In another study on 24 elderly men (average age 74.3 years), co-ingestion of 2.5 g leucine with 20 g casein proteins led to 22% higher muscle protein synthesis rate compared with ingestion of casein alone (Wall et al. 2013). Improvements in nutritional status, cognitive performance, general health, muscle mass, strength and performance were also reported as a result of improved mitochondria functions in malnourished elderly people given branched chain amino acids (Buondonno et al. 2020). Fujita and Volpi (2006) attributed stimulatory effect of these amino acids on muscle protein synthesis in older individuals to the initiation of mRNA translation. Also, mammalian target of rapamycin (mTOR) Complex 1 (a key pathway in amino acid induced anabolic responses) was reported to be sensitive for leucine, while muscles in older adults required increasing concentrations of this amino acids for maintaining anabolic responses through the mTOR pathway (Dillon 2013). Furthermore, leucine, also contributed to health-span by improving glucose metabolism through promoting glucose uptake in muscles via phosphatidylinositol 3-kinase and protein kinase C pathways (Nishitani et al. 2002) and reducing body weight by increasing leptin secretion and decreasing food intake (Valerio et al. 2011). However, branched chain amino acids may also have negative health implications owing to their involvement in the promotion of oxidative stress and mitochondrial dysfunction in human peripheral blood mononuclear cells (Zhenyukh et al. 2017) as well as the ability to increase excitotoxicity in cortical neurons in brain (Contrusciere et al. 2010). Conditionally indispensable amino acids contributing to health-span include glutamine and arginine. Glutamine has important roles in the maintenance of normal skeletal muscle function and neuronal physiology owing to its ability of controlling heat-shock responses (Leite et al. 2016), while arginine can contribute to reduced risk of endothelial dysfunction-associated cardiovascular risk with aging (Heffernan et al. 2010; Dato et al. 2019).

1.3 Digestion and Absorption of Proteins

Proteins ingested through food are required to be digested before being absorbed and utilized in the body. Digestion involves breakdown of protein molecules to constituting amino acids. Proteins in liquid foods pass through the mouth almost unaltered (Loveday 2019). Semisolid and solid foods, however, undergo physicochemical changes during mastication, though most of protein digestion occurs in stomach and small intestine. In the stomach, proteins get mixed with gastric acid and pepsinogen (inactive form of pepsin also called as zymogen). The acid in the stomach transforms pepsinogen to active form (pepsin) and makes proteins susceptible to pepsinolysis by partially denaturating (unravelling) them. Pepsin acts on acid denatured proteins and catalyse hydrolysis of peptide bond next to phenylalanine, tryptophan and tyrosine residues. The partially digested proteins/polypeptides in *chyme* (stomach contents) then enters the duodenum, where they trigger the release of hormone cholecystokinin. The hormone reaches pancreas through blood stream and makes the organ to release alkaline secretions with proteolytic enzymes including trypsin, chymotrypsin and pancreatin in their inactive forms that get activated under specific conditions. The enzymes in pancreatic secretions selectively hydrolyse bonds in polypeptides, shorter peptides or oligopeptides to liberate amino acids and small peptides. These amino acids and small peptides are absorbed by the absorptive cells of the small intestine which then break down small peptides (dipeptides and tripeptides) to amino acids and transfer them into blood capillaries. Eventually, absorbed amino acids are metabolized by the liver for synthesising body proteins, to be used as an energy source, converting to sugars or fats and/or releasing into the blood stream (Byrd-Bredbenner et al. 2009).

Human digestive system is very effective in achieving breakdown of proteins to amino acids and small peptides, though some proteins and peptides may resist complete digestion in the small intestine. Such proteins and peptides eventually reach the large intestine where gut microflora ferment them to short chain fatty acids (SCFA) or amines through deamination or decarboxylation, respectively (Fan et al. 2015; Rios-Covian et al. 2020). Acetates, propionates and butyrates are major SCFA produced in the gut. These SCFA exert beneficial effects on host through maintaining acidic pH in the gut, fuelling epithelial cells for mucin production, increasing colonic blood flow and preventing the growth of abnormal colonic cells (Rios-Covian et al. 2016). The SCFA that are not utilized in colon may exert beneficial effects against diet-induced obesity and hyperlipidaemia by interfering with carbohydrate and lipid metabolism after reaching liver via hepatic vein (Rios-Covian et al. 2016). In addition to acetates, propionates and butyrates, branched-SCFA e.g. isobutyric, isovaleric, and 2-methylbutyric acids are also produced in gut as a result of fermentation of amino acids (e.g. valine, leucine and isoleucine) by *Bacteroides* and *Clostridium* spp. (Aguirre et al. 2016; Rios-Covian et al. 2020). Therefore, the presence of these SCFA in stools is considered as the indication of the fermentation of proteins in gut. However, the fermentation of indigested proteins also may have some negative effect on gut health as some products with potential harmful effects on colon epithelium (e.g. biogenic amines, sulphides, ammonia, cresols, phenol, etc.) are also produced during fermentation (Nie et al. 2018; Rios-Covian et al. 2020).

1.4 Functions of Dietary Proteins

Proteins are involved in nearly each physiological processes in the body. They function as transport proteins (e.g. haemoglobin and ferritin that are involved in transport of oxygen and Fe, respectively), structural proteins (e.g. elastin and collagen function to support the body), contractile proteins (e.g. actin and myosin required for movement), enzymes (required for catalysing several biochemical reactions), hormones (required for bodily functions), receptors (e.g. proteins involved in nerve cell transmission) or transcription factors (Fig. 1.2). Generalized functions of dietary proteins in human body are outlined below.

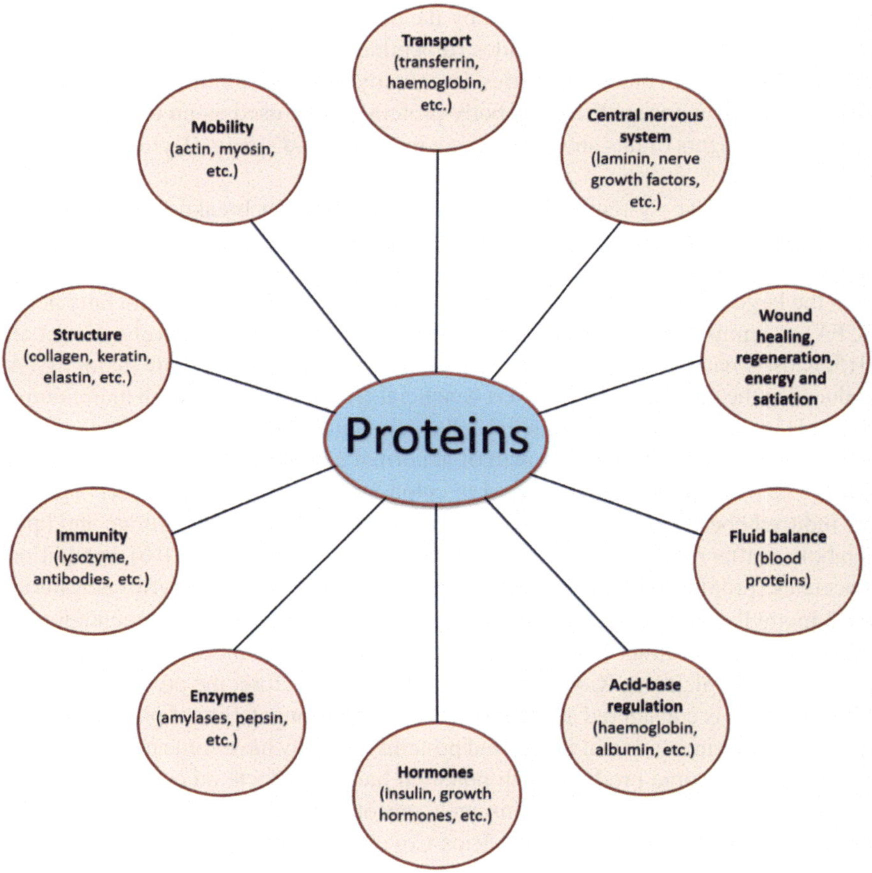

Fig. 1.2 Major functions of proteins in human body

1.4.1 Building Blocks

Dietary proteins are fundamentally the source of amino acids and nitrogen that serve as raw materials in the synthesis of body proteins required for formation/synthesis of vital organs, muscles, bones, biological fluids, hormones, enzymes, neurotransmitters, nucleic acids, etc. (Henley et al. 2010; Wu 2013; Wu et al. 2017). A large portion of proteins in animal body is involved in providing structural stability and mobility. These proteins include collagen, actin and myosin that are essential in muscle structure and functioning (e.g. contraction and relaxation). Collagen proteins also function to make a large portion of the organic structural matrix of bones. Dietary proteins provide amino acids for the synthesis of such structural and contractile proteins. As the human body is incapable of maintaining reserves of proteins, a constant supply of dietary protein is essential for maintaining tissue repairing/growth and other physiological functions. Insufficient intake of proteins especially during development and growth years can affect all tissues/organs of the body (Boye et al. 2012), while adverse protein deficiency in adults results in breakdown of lean tissues and muscle mass for providing amino acids and other nitrogenous precursors required for synthesis of compounds critical for survival e.g. insulin and haemoglobin (Hoffer 1994).

1.4.2 Biological Catalysts

Proteins are involved in biochemical reactions as biological catalysts or enzymes that accelerate the rate of reactions. All processes in nature require enzymes to occur at required rates, while metabolic pathways in cells are said to be determined by the type and amount of enzymes present (van Oort 2010). Enzymes are essential in signal transduction, cell regulation, mobility/movement, digestion, and so one (Hunter 1995; Berg et al. 2001). Their functioning in specific order is essential for creating metabolic pathways, whereas their absence will cause metabolism to neither progress nor be fast enough to serve the requirements of cells. For example, glucose can react directly with ATP to get phosphorylated during glycolysis. However, in the absence of enzymes, phosphorylation is insignificant, while the presence of enzyme hexokinase speeds up the phosphorylation of glucose to glucose-6-phosphate which gets finally converted to pyruvate in a series of reactions involving activities of a number of enzymes.

1.4.3 Immune Function

Proteins are key component in immune system. Amino acids play important roles in immune responses by regulating 1) the activation of T-lymphocytes, B-lymphocytes, natural killer cells and macrophages, 2) cellular redox state, gene expression and lymphocyte proliferation and 3) the production of antibodies, cytokines and other cytotoxic substances (Li et al. 2007). Glutamine serves as a major energy substrate for immunocytes (Wu 2013). During conditions of infection, amino acid requirement of immune cells increases rapidly to meet synthesis of immunological proteins (Le Floch et al. 2004). The deficiency of dietary proteins can make the body severely prone to infections and diseases owing to the unavailability/lesser availability of materials required by immune system, while dietary supplementation of specific amino acids (e.g. arginine, glutamine and cysteine) to individuals with malnutrition/infections could contribute to enhanced immunity and reduced morbidity/mortality (Li et al. 2007).

1.4.4 Transporting Nutrients

Proteins in body also function as carriers for transporting nutrients between different organs/tissues/cells (Byrd-Bredbenner et al. 2009). Haemoglobin and myoglobin (involved in transportation of oxygen from lungs to organs/tissues/cells) are the most common examples of transport proteins. Lipoproteins function to transport lipid molecules from the small intestine to various locations. Apart from these, important transfer proteins involved in nutrient metabolism in human body include ferritin (a protein that can store Fe and release the same in a controlled manner), transferrin (plasma protein responsible for transporting Fe in body), retinol-binding protein (a protein for carrying Vit-A) and ceruloplasmin (a carrier protein for copper).

1.4.5 Maintenance of Fluid Balance and Regulation of pH

Proteins are involved in maintenance of fluid balance in the body (Byrd-Bredbenner et al. 2009). Blood pressure in arteries makes blood to enter minute blood vessels (also called as capillary beds), from where blood fluids pass to extracellular spaces for providing nutrients to cells. As proteins in the blood (e.g. albumins) are too large to enter extracellular spaces, they stay there and contribute to the maintenance of fluid balance by attracting proper amount of fluid back from cellular spaces to the blood stream. Accumulation of excessive amount of fluids in tissues in the presence of lesser amounts of proteins in bloodstream caused by inadequate intake of dietary proteins results in swelling of tissues, a clinical condition called as *oedema*. Proteins in the membrane also contribute to the regulation of acid-base balance and maintenance of pH in human body by acting as buffers for the body (Byrd-Bredbenner et al. 2009).

1.4.6 Providing Glucose During Fasting or Starvation

Human body needs to maintain a certain minimum level of glucose in the blood stream to keep organs/tissues/cells functioning. During the period of fasting or starvation, body uses proteins for synthesizing glucose through gluconeogenesis. Body proteins provide most of the glucose (about 90%) needed during the first few days of fasting for functioning of body (Byrd-Bredbenner et al. 2009).

1.4.7 Satiation

Dietary proteins also contribute to the satiety and suppression of apatite. They provide greater satiety than carbohydrates and fats (Weigle et al. 2005). Protein induced satiety is attributed to the combined effects of elevated plasma amino acid and anorexigenic hormone levels, prolonged suppression of hunger hormone (ghrelin), increased diet-induced thermogenesis (increased dissipation of energy at rest) and ketogenic state (Drummen et al. 2018). Satiation effect of dietary proteins is discussed in detail in the Sect. 1.5.2 of the chapter.

1.4.8 Providing Energy

Similar as carbohydrates and lipids, proteins can provide energy to humans. The liver is able to break amino acids to the carbon skeleton, which can then be used for energy. *In vivo* oxidation of proteins to water and carbon dioxide releases 4.1 kcal/g, which is equal to starch (4.1 kcal/g) but lesser than that released from lipids (9.4 kcal/g) (Wu 2016). However, as these nutrients are not digested completely and because protein oxidation in the body is incomplete, the energy values of 4, 4 and 9 kcal/g, respectively are taken for dietary proteins, available carbohydrates (e.g. starches, sugars, etc.) and fats/oils (Wu 2016). Usually 10–20% of energy is derived from dietary proteins. However, amino acids are not preferred as a source of energy by body. They exhibit lower energetic efficiency than fatty acids and glucose (Wu 2013). Also, they burden kidneys and liver with excessive amount of processing and metabolism for being utilized as a source of energy (Byrd-Bredbenner et al. 2009).

1.5 Health Benefits and Healthy Aging

In addition of performing essential functions in metabolic and physiological processes, dietary proteins also contribute to disease prevention and healthy aging. High protein intake has been associated with reduced overall mortality and morbidity in elderly people owing to the positive effect of protein consumption towards reduced risk of age/lifestyle-related diseases including sarcopenia, cardiovascular diseases, osteoporosis, obesity and diabetes (Gaffney-Stomberg et al. 2009; Vikøren et al. 2013; Levine et al. 2014; Wu 2016). This section discusses health-benefitting effects of dietary proteins in elderly people.

1.5.1 Maintenance of Muscle Mass

Progressive muscle loss and associated reduction in lean body mass with aging is termed as sarcopenia. It is considered as an insidious process characterized by 3–8% reduction in lean muscle mass per decade after the age of 30 years (Paddon-Jones and Rasmussen 2009). Sarcopenia is one of the major causes of functional impairment/disability and mortality in elderly people (Houston et al. 2008). The condition is quite prevalent in elderly people affecting about 30% of population of over 60 years of age and more than half of those above 80 years (Baumgartner et al. 1998). Sarcopenia can develop because of a number of reasons including sedentary lifestyle, inadequate protein intake, impaired turnover rate of proteins, neurodegeneration, reduced anabolic hormone production, dysregulation of cytokines and inflammation (Gaffney-Stomberg et al. 2009; Paddon-jones and Rasmussen 2009; Fukagawa 2013). However, low protein intake is considered as a key factor in the age-related muscle loss as amino acids are a prerequisite for the synthesis of muscle proteins while many elderly people do not take dietary proteins in adequate amounts owing to several reasons including 1) high cost, 2) tooth decay and difficulties in chewing, 3) perceived intolerance to certain foods and 4) avoidance of animal foods to minimize cholesterol and lipids intake (Chernoff 2004; Houston et al. 2008). Moreover, reduction in anabolic signals for muscle protein synthesis with ageing further accelerates muscle loss and development of sarcopenia. Guillet et al. (2004) reported lower increase in muscle protein synthesis (0.023% per hour) with infusion of amino acids and insulin in older people than adult subjects (0.041% per hour).

Physical activity and intake of dietary proteins are main anabolic stimuli for synthesis of muscle proteins. Regular exercise with protein supplementation not only can reduce the progression of sarcopenia but also treat the same by increasing muscle-strength, size and mass (Campbell 2007; Strasser et al. 2018). However, as older adults require to take greater amount of proteins for achieving maximum stimulation of muscle protein synthesis than younger ones (Moore et al. 2015), they may be advised to take proteins in excess of the recommended dietary allowance (0.8–1.0 g proteins/kg body weight/day) for prevention of sarcopenia through maintenance of positive nitrogen balance and preservation of muscle mass (Campbell 2007; Gaffney-Stomberg et al. 2009; Kerstetter et al. 2015; Baum and Wolfe 2015; Lonnie et al. 2018). Genaro et al. (2015) while evaluating differences in protein intake in women with or without sarcopenia reported higher muscle mass for group consuming >1.2 g proteins/kg body weight/day. Rizzoli (2015) and Gaffney-Stomberg et al. (2009) also recommended protein intake of 1.0–1.2 g/kg body weight/day with repartition for preventing sarcopenia and maintaining nitrogen balance without having any harmful health effects. However, in addition to protein quantity, protein quality (amino acid composition and bioavailability) also is relevant in muscle protein synthesis and prevention of sarcopenia. Indispensable amino acids, particularly branched-chain amino acids, appear to be highly effective in stimulating muscle protein synthesis and enhancing muscle strength. The supplementation of branched-chain amino acids together with L-glutamine reversed muscle loss in total gastrectomized rats (Haba et al. 2019). Also, leucine supplementation improved condition of patients with muscle wasting diseases (Wandrag et al. 2015). In this regard, proteins from animal sources may be considered effective in stimulating muscle protein synthesis owing to high content of essential amino acids including leucine (Baum and Wolfe 2015; Shang et al. 2018). However, some recent studies highlight that proteins from plants (e.g. soybean) alone or in combination with animal proteins can be equally beneficial in minimizing muscle wasting and preventing sarcopenia. For example, Jarzaguet et al. (2018) reported increased anabolic response of skeletal muscles in aged rats with meals comprising of >25% soy/whey proteins. In another clinical study on bedridden patients, soybean proteins in diet were found to be superior to milk proteins (casein) in enhancing muscle strength, though casein supplementation resulted in greater enlargement of muscle volume (Hashimoto et al. 2015). However, more studies are required for evaluating the effectiveness of proteins from different plant sources in the prevention of sarcopenia.

1.5.2 Weight Management

Obesity is characterized by excessive deposition of fat in body. A person is said to be obese when exhibits body mass index (a measure of the healthy weight of an adult for her/his height, which is calculated by dividing body mass to squared height; BMI) value of $\geq$30 kg/m^2. Obesity prevails in all age groups and affects a significant portion of the population of the world. In 2016, about 13% of adults (11% men and 15% women of >18 years of age), while 6% of girls and 8% of boys of 5–19 years of age were obese (WHO 2020a). Moreover, its prevalence is increasing in many parts of the world including middle and low-income countries. In Africa, the number of overweight children has increased by approximately 24% since 2000, while almost 50% of children less than 5 years of age who were overweight or obese in 2019 lived in Asia (WHO 2020a).

Medically, obesity is classified as a complex condition that not only impairs physical ability but also can detrimentally affect longevity, health-span and quality of life by increasing the risk of many diseases including diabetes, cardiac diseases (including congestive heart failure, hypertension, atherosclerosis, etc.), non-alcoholic fatty liver disease, some types of cancers, psoriasis, inflammatory bowel disease, musculoskeletal disorders, chronic limb pains, osteoarthritis, Alzheimer's disease, dementia, impaired respiratory function, obstructive sleep apnoea and increased susceptibility to infections (Elia 2001; Coggon et al. 2001; Frasca and McElhaney

2019). In the elderly, obesity can lead to early onset of chronic diseases, respiratory tract infections, functional impairment and premature mortality (Lorenzo et al. 2006; Amarya et al. 2014; Frasca and McElhaney 2019). Causes of obesity include hereditary/genetics, hormonal imbalances, metabolic factors and life-style related conditions, though it is mostly because of life-style related factors that include chronic positive energy balance (regular intake of calories in excess of expenditure) and lack of physical activity (Paddon-Jones et al. 2008). The treatment of obesity requires a negative energy balance which may be achieved by indulging in physical activities and/or sticking to energy restricted diets. Further, increased satiety and suppressed apatite can also contribute to weight loss and obesity prevention. However, the weight loss strategies for overweight/obese elderly people are required to be considered carefully in order to achieve weight loss with minimum loss of lean muscles as a specific condition, termed as sarcopenic obesity (a medical condition characterized by an excessive fat deposition in combination with a detrimental loss of lean body mass), is more common in elderly people (Mathus-Vliegen 2012).

Protein intake has been associated with weight management/obesity prevention. Protein intake at levels of requirement through energy restricted diets contributes to weight loss, while increased intakes are effective in maintaining fat-free mass (Drummen et al. 2018). Soenen et al. (2013) reported reduction in body and fat-mass in overweight and obese men and women of 18–80 years of age with BMI >25 kg/m^2 given dietary proteins at and above the levels of requirements (0.8 g/kg body weight/day and 1.2 g/kg body weight/day, respectively). In comparison, the dietary protein intake below requirement levels could result in lesser weight loss and higher risk of weight regain (Acheson 2013). Weigle et al. (2005) reported positive outcomes of increased protein and reduced fat intake (from 15 to 30% and from 35 to 20% of energy, respectively) at constant carbohydrate intake on weight loss. Protein induced weight loss can be attributed to satiation and increased energy expenditure. Protein-rich diets were more satiating than that high in fats and carbohydrates. Marmonier et al. (2000) while investigating the effect of protein, fat or carbohydrates-rich snack (each providing 250 kcal) after *ad libitum* lunch on apatite (requests for next meal) observed the longest delay in requests for next meal (60 min) for protein-rich snack. High-carbohydrate and high-fat snacks, in comparison, delayed dinner requests by 35 and 25 min, respectively (Marmonier et al. 2000). The contribution of dietary proteins to satiety and weight loss has been attributed to their effect on hormones involved in apatite regulation, e.g. ghrelin, leptin, insulin, amylin, adiponectin, glucagon-like peptides, cholecystokinin, etc. (Lonnie et al. 2018), as well as to specific amino acids, e.g. tryptophan, tyrosine and/or histidine, which are used for synthesis of neurotransmitters (e.g. serotonin, dopamine, norepinephrine and histamine) involved in regulation of food intake and apatite (Keller 2011). Moreover, the satiation effect of a few food proteins (e.g. casein proteins of milk) also can be associated with acid-induced coagulation in stomach, which may inhibit gastric digestion by slowing diffusion of pepsin into coagulated proteins (Thévenot et al. 2017). This affects release of partially digested proteins/peptides into small intestine and contributes to appetite control by creating a prolonged feeling of fullness (Loveday 2019).

In addition, protein-induced weight loss also is attributed to high energy expenditure as protein-rich foods require greater amount of energy for digestion, absorption and metabolism (thermic effect of food; TEF) in comparison to carbohydrates and fats. The TEF value of protein-rich foods range from 20 to 30%, whereas carbohydrates and lipids-rich foods show TEF values of 5–10% and 0–3%, respectively (Byrd-Bredbenner et al. 2009). Increased energy expenditure for protein-rich foods is attributed to metabolic inefficiency of protein oxidation in comparison to glucose (more energy is required for producing ATP from amino acids than from glucose), gluconeogenesis and urea synthesis (Westerterp-Plantenga et al. 2009; Keller 2011). Furthermore, the energy expenditure of high protein intake has also been attributed to its effect on thyroid hormones, androgens, catecholamines and growth hormone (Mikkelsen et al. 2000).

Plant proteins also demonstrate effects similar to that of animal proteins in the context of the ability to increase satiety and delay hunger. Neacsu et al. (2014) while investigating appetite responses to high-protein weight-loss diets in obese men of 34–71 years of age reported statistically similar hunger, desire-to-eat and weight loss for meat and soybean-based diets. Veldhorst et al. (2009) compared the effect of casein, soybean and whey-protein breakfasts on appetite regulating hormones (glucagon-like peptide-1 and insulin). Although the strongest effect was shown by whey, mediating effects on these hormones were also observed for soybean proteins and casein (Veldhorst et al. 2009). In another report, Scully et al. (2017) also reported no significant differences in appetite and food intake for participants of 23–63 years of age with BMI of 19.3–38.9 kg/m^2 given buckwheat and fava bean based diets. In addition, legumes/pulses also have an edge over animal products for weight loss strategies in elderly people because of high protein to calorie ratio which is attributable to low lipids content (for most legumes/pulses with exception of dried seeds/kernels of soybean, lupine and groundnut, though lipids in these legumes are rich in health benefitting unsaturated fatty acids) and high content of unavailable complex carbohydrates including oligosaccharides, resistant starch and soluble and insoluble dietary fibres.

1.5.3 Cardiac Health

Cardiac diseases are responsible for approximately 18 million deaths annually. High levels of serum cholesterol, triacylglycerols, low-density lipoprotein cholesterol (LDL-C), high LDL-C to high-density lipoprotein cholesterol (HDL-C) ratio, smoking, stress/depression, obesity and alcohol abuse are major risk factors in cardiac diseases. Although the development of cardiac diseases is associated with many aspects of lifestyle, besides heredity, the most significant risk factors are related to food habits (Cam and de Mejia 2012). Dietary proteins are considered an important source of bioactive peptides that can contribute to lower the risk of cardiac diseases owing to anti-hypercholesterolemic, anti-hypertensive, anti-thrombotic and anti-inflammatory activities (Nasri 2017; Tapal et al. 2019). However, proteins from plant sources, in general, have an edge over animal proteins in the context of their effect on cardiac health. Virtanen et al. (2019) reported high mortality risk associated with ischaemic heart diseases amongst Finnish males of 42–60 years of age consuming greater amounts of animal proteins than plant proteins. Li et al. (2017) based on a meta-analysis highlighted reduction in blood LDL-C, non-HDL-C and apolipoprotein-B levels after substitution of animal proteins with plant proteins. Similarly, Bernstein et al. (2010) and Bernstein et al. (2012) while investigating the effect of protein sources on cardiac diseases also highlighted reduction in the risk of strokes and other coronary heart diseases with replacement of red meat as protein source with poultry, nuts, fish and low-fat milk products.

Amongst different plant proteins, legume proteins (particularly soybean) have been investigated extensively for their effects on cardiac health. Intake of soybean proteins was found to upregulate LDL receptors in the liver (Baum et al. 1998) and reduce levels of total cholesterol, LDL-C and triacylglycerol (Torres et al. 2006; Borodin et al. 2009). Sacks et al. (2006) based on 22 randomized trials recommended a minimum daily intake of 50 g of soybean proteins for achieving reduction in serum lipids levels. The lipid lowering effects of soybean proteins have been attributed to the presence of isoflavones (Anthony et al. 1996) and release of small peptides (<15 amino acids) after digestion/hydrolysis (Sirtori et al. 2007). In addition, the beneficial effect of soybean proteins on cardiac health may also be due to the lower ratio of leucine to arginine than animal proteins as high concentration of arginine in plasma was associated with hypocholesterolemic effect, whereas leucine acted as a cholesterol precursor having a strong insulinotropic effect (Chalvon-Demersay et al. 2017). Proteins from other legumes, e.g., lupine, chickpeas, cowpea and peas, also exert beneficial effects on cardiac health. In studies by Bettzieche et al. (2009) and Fontanari et al. (2012), lupine proteins were shown to increase plasma HDL-C level while reducing levels of triacylglycerol and LDL-C in hypercholesterolemic rats and hamsters. Similarly, the study by Frota et al. (2015) also reported 2.7% increase in HDL-C while 12% reduction in total cholesterol, 18.9% in LDL-C and 16% non-HDL-C in hypercholesterolemic subjects (30–70 years of age) with consumption of 25 g cowpea protein isolate per day. In relatively recent works, chickpea peptides lowered lipids level and reversed liver damage in hyperlipidaemic mice (Xue et al. 2018), while hydrolysed pea proteins inhibited lipid accumulation in adipose tissue cells (3T3-L1 cells) in a concentration dependent manner (Flores-Medellín et al. 2021).

Milk proteins (e.g. whey proteins) also exert beneficial effects on cardiac health owing to their ability to improve vascular functions and reduce LDL-C and total cholesterol. Ballard et al. (2013) reported improvements in vascular function in overweight individuals after ingestion of whey peptides, while Pal et al. (2010) showed reduction in triacylglycerol, total cholesterol and LDL-C levels by 22, 11 and 7%, respectively in overweight to obese subjects of 18–65 years of age and 25–40 kg/m^2 BMI with ingestion of whey protein isolate (27 g twice a day) for 12 weeks. The beneficial effects of milk proteins have been attributed to increased lipolysis, reduced lipid accretion, reduced absorption of dietary cholesterol/fatty acids and increased insulin resistance (Chen and Reimer 2009; Lillefosse et al. 2014; Fekete et al. 2016).

Proteins from fish also have hypocholesterolemic and hypolipidimic effects in animal models (Shukla et al. 2006; Kawabata et al. 2015; Drotningsvik et al. 2016). The effect of fish protein intake on blood lipid profile was investigated by Shukla et al. (2006). Rats fed on fish proteins showed lower levels of triacylglycerol and cholesterol while higher levels of liver LDL-receptors and sterol regulatory-binding proteins than casein fed rats. However, unlike legume proteins, fish proteins also decreased levels of HDL-C (Shukla et al. 2006). Reduced serum and liver cholesterol levels have also been reported for fish protein-fed rats by Hosomi et al. (2011) and Hosomi et al. (2013). The hypolipidimic effects of fish proteins were attributed to the inhibition of the absorption of bile acid in the small intestine and enhanced excretion of cholesterol in faeces as a result of increased expression of cholesterol 7α-hydroxylase and carnitine palmitoyltransferase-2 enzymes involved in lipid metabolism (Hosomi et al. 2011; Hosomi et al. 2013).

1.5.4 Bone Health

Proteins make up about half of the bone volume and about one-third of the bone weight (Heaney 2007) as they (mainly collagen and various other non-collagen proteins) are primarily responsible for the organic structural matrix of bones. The loss of bone mass/density with aging results in enhanced bone fragility and increased risks of fractures which not only causes people to become bedridden but also can cause life threatening complications in elderly individuals. Besides Ca, P and Vit-D, dietary proteins and physical activities are considered critical in achieving optimal bone mass and preventing bone loss in elderly people (Rizzoli 2014; Gaffney-Stomberg et al. 2014; Morris-Naumann and Wark 2015). However, high dietary protein intake has been shown to have positive as well as negative effects on bone health. For example, high protein intake was associated with demineralisation of bones and development of osteoporosis because of stimulation of urinary Ca excretion and acidification of blood (Heaney and Layman 2008; Wu 2016). Zwart et al. (2005) reported negative Ca balance and increased urinary excretion of this mineral in male subjects on diet high in essential amino acids. They attributed this response to the increased intake of sulphur containing amino acids that increase Ca excretion by increasing endogenous sulphuric acid production. On the other hand, many studies regarded dietary proteins as an essential nutrient in bone health owing to their involvement in 1) providing organic structural matrix, 2) optimising the level of growth factors (e.g. insulin-like growth factors which can increase bone mass by increasing osteoblasts activity), 3) increasing absorption of calcium in intestine, 4) stimulation of bone-mineralisation and 5) inhibition of bone resorption and inflammatory cytokines that could activate bone degradation (Fernandes et al. 2003; Heaney and Layman 2008; Millward et al. 2008; Gaffney-Stomberg et al. 2009; Hardy and Cooper 2009; Zhang et al. 2011; Gaffney-Stomberg et al. 2011; Kerstetter et al. 2015; Wolfe 2015; Shang et al. 2018). In addition, high protein diets have also been associated with higher Ca intake, which can, therefore, compensate for any moderate increase in urinary excretion of the mineral (Wu 2016). Shams-White et al. (2017) based on a meta-analysis evaluation concluded no adverse effect of high protein intake on bone health and highlighted that a higher protein intake may reduce bone mineral density loss in older adults in comparison to a lower protein intake. Similarly, Groenendijk et al. (2019) while performing a systematic review and meta-analysis on the effect of high/low protein intake on bone health in elderly people also highlighted a positive association of higher protein intakes with bone (femoral neck and hip) mineral density and reduced hip bone fracture risks. However, the effect of dietary protein intake on bone health relates with the intake of other nutrients. Dawson-Hughes et al. (2002) showed crucial roles of Ca and Vit-D on the impact of dietary protein on bone health as greater protein intake was associated with increased bone mineral density in older adults during a 3-year supplementation period of these nutrients, though such improvement required sufficient dietary Ca and Vit-D intake as relationships between protein intake and bone density were not observed for placebo group (Dawson-Hughes and Harris 2002).

1.5.5 Type-2 Diabetes

Diabetes is a chronic disease of metabolism characterized by high blood glucose levels occurring due to the lack of insulin production or ineffective use of insulin by cells (Roglic 2016). The disease affects about 8.5% population of the world and is considered as a major cause of kidney diseases, blindness, heart diseases and limb amputation (WHO 2020b). Obesity is one of the most important risk factors in type-2 diabetes. Moreover, reduced muscle mass as a result of sarcopenia/sarcopenic obesity may also result in reduced insulin sensitivity and decreased glucose uptake which in turn can increase the risk for hyperglycemia and insulin-resistance syndrome in elderly people with diabetes (Solerte et al. 2008). Healthy lifestyle (regular physical activity and consumption of healthy diets for maintenance of normal body weight) can, therefore, be useful in delaying or managing this chronic disease. However, the association of dietary protein intake with type-2 diabetes varies with proteins sources. Pal et al. (2010) reported favourable effects of whey protein intake towards reduced risk of diabetes in obese men and women of BMI 25–40 kg/m^2 and 18–65 years in age, while Sluijs et al. (2010) based on a prospective investigation highlighted increased risk of type-2 diabetes with increased intake of total and animal proteins. Based on a systematic review and meta-analysis study, Tian et al. (2017) also reported varying effect of protein source on the risk of diabetes. Red and processed meats increased risk of diabetes, soybean and milk had protective effect, while fish and egg intake had no association with decreased risk of diabetes (Tian et al. 2017). In comparison, Vikøren et al. (2013) in a randomised study on fish protein (3–6 g/day for 8 weeks) effect on glucose metabolism in adults of 20–70 years of age with BMI $\geq$27 kg/m^2 reported favourable effects of the protein supplementation on glucose metabolism (lower fasting and postprandial glucose levels in comparison to placebo).

1.6 Protein Quality and Its Evaluation

Protein quality is an index of how well proteins meet requirements of amino acids and physiological needs of organisms (Shivakumar et al. 2018). Protein quality depends on amino acid composition and digestibility. The presence of all nine indispensable amino acids in adequate amounts reflects completeness of proteins, while high digestibility indicates the bioavailability of those amino acids. Therefore, highly digestible proteins that provide indispensable amino acids in amounts equal to or greater than that required for growth and maintenance are said to be of high quality. As elderly people generally consume foods (including proteins) in lesser amounts than younger adults, protein quality becomes critical in meeting enhanced requirements of indispensable and functional amino acids. Positive associations between the consumption of foods with high protein quality and healthy aging have been reported (Mathus-Vliegen 2012; Paddon-Jones et al. 2015; Hidayat et al. 2018).

1.6.1 Evaluation of Protein Quality

Different methods, ranging from chemical and biochemical (enzymatic) to microbiological and biological in nature, have been developed and employed for evaluation of protein quality (Damodaran 2017). Chemical methods, e.g. *amino acid score* (AAS) also referred to as *chemical score* measure protein quality based on the amount of most limiting indispensable amino acid in test protein and the content of the same in a reference protein (usually egg white). It is calculated as:

$$\text{Amino acid score or chemical score} = \frac{\text{mg of limiting amino acid in 1 g of test protein}}{\text{mg of the same amino acid per g in reference protein}}$$

Enzymatic methods use proteolytic enzymes (pepsin, trypsin, chymotrypsin, peptidases and/or proteases) under specific conditions of the test to determine in vitro digestibility of proteins (Bodwell et al. 1980; Calsamiglia and Stern 1995). Microbiological methods, in comparison, measure quality of proteins by determining their ability to support growth of microorganisms, including bacteria (e.g. *Streptococcus zymogenes*, *S. faecalis*, *Leuconostoc mesenteroides* and *Clostridium perfringens*) and protozoa (e.g. *Tetrahymena pyriformis*) exhibiting amino acid requirements similar to humans (Ford 1981).

Chemical, enzymatic and microbiological methods have some advantages. Chemical methods are simple and less time consuming and involve comparison of amino acid composition of test and reference proteins. Enzymatic methods are specifically useful in comparing different proteins for digestibility, while microbiological assays can provide useful information on protein quality depending on the amino acid requirements of organisms. However, these methods are subjected to several drawbacks, e.g., these assays do not take into account the effect of toxins, antinutritional constituents, food additives and/or ingredients (e.g. common salt, nitrates, spices, etc.) affecting protein quality. Moreover, enzymatic and chemical assays provide information only about in vitro digestibility/bioavailability and amino acid composition, respectively, while microbiological assays are useful for foods with known composition (Satterlee et al. 1979; Pellett and Young 1980).

Biological methods, in comparison, measure protein quality based on nitrogen retention and the ability of proteins to support growth (weight gain) in test animals (Damodaran 2017). The protocols involve feeding test animals (usually rats) with a test diet containing proteins in limited amounts (10% on dry basis) for a specific period of time (9 days). Protein-free diet is given to the control group. The test diet is formulated to provide adequate amount of energy so that proteins can be utilized in the animal body to the maximum possible extent. Animal weight is noted daily and faeces and urine are collected for nitrogen/protein content determination (FAO/WHO 1991). The data obtained is used in different ways to express protein quality as *protein efficiency ratio*, *net protein ratio*, *true digestibility*, *biological value* and *net protein utilisation* (Damodaran 2017).

Protein efficiency ratio is the simplest and most commonly used expression which is defined as weight gained per gram of test protein consumed. It is expressed as:

$$\text{Protein efficiency ratio} = \frac{\text{Weight gain in test animal on test diet}}{\text{Amount of protein ingested}}$$

The *protein efficiency ratio* is directly related to protein quality as the gain in weight and growth achieved are dependent on the incorporation of dietary proteins in body tissue. However, this expression is criticized for not taking into account the dietary protein utilized for maintenance as only gain in weight is taken in the calculation. *Net protein ratio*, in comparison, takes into account the weight lost in protein-free group, hence provide information on the ability of protein to support both maintenance and growth. This expression is calculated as:

$$\text{Net protein ratio} = \frac{(\text{Weight gain}) - (\text{Weight loss for animals on protein free diet})}{\text{Amount of protein ingested}}$$

True digestibility and *biological value* determine protein quality by measuring nitrogen uptake and nitrogen lost by test animals. These expressions involve faecal nitrogen content determination to take in to account digestibility/nitrogen retention as well as the metabolic/endogenous nitrogen. Therefore, these expressions are the measure of the ability of dietary proteins to convert into body proteins. However, in comparison to *true digestibility*, *biological value* also involves analysis of urine for nitrogen content. *True digestibility* and *biological value* are calculated as:

$$\text{True digestibility}\,(\%) = \frac{\text{Protein ingested} - (\text{FN} - \text{FKN})}{\text{Protein ingested}} \times 100.$$

$$\text{Biological value} = \frac{\text{Protein ingested} - (\text{FN} - \text{FKN}) - (\text{UN} - \text{UKN})}{\text{Protein ingested} - (\text{FN} - \text{FKN})} \times 100.$$

where, FN and UN are nitrogen in faeces and urine of test animal, respectively, while FKN and UKN represent metabolic/endogenous nitrogen lost through faeces and urine, respectively of test animals on protein-free diet.

Another useful expression of protein quality is *net protein utilisation*. This expression provides the information on percentage of diary proteins retained in animal body. It is a product of *true digestibility* and *biological value* and is calculated as:

$$\text{Net protein utilisation} = \frac{\text{Protein ingested} - (\text{FN} - \text{FKN}) - (\text{UN} - \text{UKN})}{\text{Protein ingested}} \times 100$$

Protein digestibility corrected amino acid score (PDCAAS) and *digestible indispensable amino acid score* (DIAAS) are relatively recent developments in protein quality evaluation. PDCAAS is obtained as the product of the *true digestibility* and AAS as:

$$\text{PDCAAS} = \text{Amino acid score} \times \text{True digestibility}$$

The PDCAAS is one of the recommended methods of protein quality evaluation in human nutrition. This method has been widely adapted for protein quality evaluation (Schaafsma 2000), though it has also been criticised for some shortcomings. For example, 1) the expression does not provide information of individual indispensable amino acids bioavailability, 2) there may be overestimation of protein quality, specifically, of products containing known antinutritional constituents and 3) the method is considered inappropriate for regulatory uses as there is overestimation of quality for poorly digestible proteins supplemented with limiting amino acids (FAO 2013; Rutherfurd et al. 2015; Mathai et al. 2017). The DIAAS is an alternative to PDCAAS. In calculation of this score, the digestibility values of each indispensable amino acid is taken into calculation. The DIAAS is calculated as:

$$\text{DIAAS} = \frac{\text{mg of dietary digestible indispensable amino acid in 1 g dietary protein}}{\text{mg of the same dietary indispensable amino acid in 1 g of the reference protein}} \times 100$$

FAO now recommends DIAAS method for protein quality evaluation instead of PDCAAS (FAO 2013). Also, it is now considered necessary to determine the digestibility of individual indispensable amino acids at the end of the small intestine (*ileum*) in order to avoid any change in amino acid composition made by microorganisms in the large intestine (FAO 2013; Mathai et al. 2017) as the large microbial load and long transit time in the large intestine can either promote microbial synthesis of amino acids or their utilization by the microorganisms (Sauer et al. 1975; Trottier and Walker 2015). Moreover, growing pig is now recognised as more appropriate animal model for protein quality analysis because of its resemblance with humans for rate of protein synthesis, 2) physiology of stomach (glandular-type stomach lined with cardiac, gastric and pyloric mucosa) and small intestine, 3) topography of portal vein, mesenteric vessels and duodenum and 4) functionality of liver and pancreas (Kararli 1995; Swindle and Smith 1998; Deglaire et al. 2009; FAO 2013).

1.7 Animal vs Plant Proteins

Proteins from different sources differ for protein quality. Animal products (milk, egg, meat and fish) contribute about 35% of total proteins globally (Wu et al. 2014). They are generally considered as a source of high quality proteins because of the presence of all nine indispensable amino acids in amounts adequate or more than adequate for supporting growth (with the exception of gelatin/partially hydrolysed collagen that lacks tryptophan and contains threonine, isoleucine and methionine in low amounts) and high digestibility. Proteins of animal origin are highly digestible (94–97%) and show higher PDCAAS (1.0 against 0.91, 0.67, 0.57, and 0.45 for soybeans, peas, oats and wheat, respectively) than plant foods (FAO/WHO/UNU 1985; van Vliet et al. 2015; Damodaran 2017). However, as proteins do not occur in foods in isolation, the entire food matrix should to be taken into account in order to evaluate health benefits of a particular diet (Millward et al. 2008). While milk and fish are sources of high quality proteins, healthy lipids (e.g. omega-3 oils in fish or conjugated linoleic acids in milk), bioavailable haem-Fe and Vit-B12, the consumption of red and processed meats has been associated with increased risk of chronic diseases (e.g. coronary artery diseases, dyslipidaemia and some cancers) and all-cause mortality (Song et al. 2004; Vang et al. 2008; Bernstein et al. 2010; Chan et al. 2011; Allen et al. 2013; Mirzaei et al. 2014; Song et al. 2016; Sacks et al. 2017). The diets high in animal proteins also can increase the risk of type-2 diabetes (Sluijs et al. 2010; Tian et al. 2017). A high consumption of animal proteins brought about 75% increase in overall mortality and about four-fold increase in the risk of developing cancer in the individuals of 50–65 years in age (Levine et al. 2014).

Plant proteins (particularly those from staple foods viz. cereals and legumes/pulses), in comparison, are deficient in one or more essential amino acids. Histidine, isoleucine, leucine, phenylalanine, tryptophan and valine are present in adequate amounts in plant foods, though lysine, threonine, tryptophan and/or methionine occur as limiting amino acids (indispensable amino acids present in amounts less than that recommended for ensuring optimal growth and maintenance) in plant proteins. For example, lysine is the most limiting amino acid in all cereals followed by tryptophan in maize and threonine in most other cereals (Serna-Saldivar 2010). Legumes/pulses, in comparison, are rich in lysine but lack methionine and cysteine (Shevkani et al. 2019a; Ge et al. 2021). Therefore, individuals on diets containing either cereal or legume proteins face difficulties in maintaining health/growth, though a mixed diet containing proteins from both sources in adequate amounts can provide all amino acids for supporting maintenance and growth. In addition, plant-based foods also contain dietary fibre and phytochemicals, e.g. polyphenols, enzyme inhibitors, phytates/phytic acid, saponins, etc. (Shevkani and Singh 2015; Singh et al. 2017; Shevkani et al. 2019b), which, nowadays, are regarded as bioactive compounds owing to several beneficial effects on health, e.g., regulation of blood glucose level, improvement in lipid profile and reduced risk of some cancers and coronary artery diseases (Schlemmer et al. 2009; Singh et al. 2017). Moreover, plant protein sources (e.g. pulses/legumes, nuts, pseudocereals, etc.) also can contribute to reduced risks of lifestyle associated chronic diseases (e.g. diabetes, dyslipidaemia, hypertension and some cancers) owing to the presence of active constituents e.g. lectins/hemagglutinins, enzyme inhibitors, peptides and amino acids with antioxidative effects (Duranti 2006; Mendonca et al. 2009; Carbonaro et al. 2015; De Souza et al. 2017; Shevkani et al. 2019a; Singh et al. 2019; Tovar-Pérez et al. 2019). Song et al. (2016) highlighted association of high plant protein intake and replacement of animal protein sources with plant sources with reduced all-cause and cardiovascular-mortality. Ginter (2008) also reported lower ischemic heart disease mortality in vegetarians. This association was attributed to high antioxidant status, lower prevalence of obesity and low cholesterol and blood pressure levels. World Cancer Research Fund (2018) also recommends to increase consumption of a plant-based diets (high in whole grains, vegetables, fruits and legumes/pulses) while limiting the consumption of red meats (beef, veal, pork, lamb, mutton, horse, and goat) to 3 servings (each of 350–500 g) per week and that of processed meats (i.e. meats processed by salting, curing, fermentation, smoking, etc.) to very little amounts. Furthermore, a few relatively recent studies on the effect of plant-based diet on physical performance show little or no differences in endurance capacity and performance (Lynch et al. 2016; Craddock et al. 2016; Nebl et al. 2019). However, in spite of the nutritional and nutraceutical advantages of plant proteins, a complete shift to vegan/vegetarian diets may not be recommended for elderly people considering the risk of Vit-B12 deficiency, elevating homocysteine levels (Obersby et al. 2013; Lonnie et al. 2018) and the association of protein quality with reduced morbidity or mortality. A mixed diet, e.g. lacto-vegetarian diet, may be regarded adequate taking in account high indispensable amino acids to calorie ratio of dairy foods (particularly low and reduced-fat milk and milk products) and considering dairy products as an excellent source of vitamins and minerals (e.g. Ca, P and Vit-D that are essential for optimal bone health), bioactive proteins/peptides (e.g. lactoferrin), probiotics (specifically fermented milk products) and high quality proteins. Song et al. (2016) while comparing major protein sources associated dairy product intake with lower mortality than the consumption of processed meats and eggs. New dietary guidelines (e.g. The Eatwell Guide) also recommend to increase intake of plant foods (fruits, vegetables, pulses and whole grain cereals) and include low fat dairy products/alternatives (e.g. low fat milk, reduced-fat cheese, plain low-fat yoghurt, soy milk, etc.) in daily diet while reducing the intake of red meats, processed meats and foods high in fat, sugar and salt-content (Public Health England 2016).

1.8 Conclusion

Proteins are essential for survival and perform critical functions in the growth and maintenance of human body. Increased intake of dietary proteins can contribute to healthy aging and longevity by preventing/delaying chronic age-related diseases, enhancing ability of recovering from the diseases and reducing morbidity/mortality. Consumption of plant proteins should be encouraged because of the associated health-benefitting effects, though a mixed diet providing both animal and plant proteins in adequate amounts may be recommended for elderly people for ensuring balanced intake of amino acids and other essential nutrients.

Acknowledgements KS acknowledges Central University of Punjab for RSM grant and Dr. Amritpal Kaur, Department of Food Science and Technology, Guru Nanak Dev University, Amritsar for her valuable suggestion.

Compliance with Ethical Standards

Funding Authors acknowledge Central University of Punjab, Bathinda for Research Seed Money Grant.

Conflict of Interest All authors declare they have no conflict of interest.

References

Acheson KJ (2013) Diets for body weight control and health: the potential of changing the macronutrient composition. Eur J Clin Nutr 67(462):466

Aguirre M, Eck A, Koenen ME, Savelkoul PHM, Budding AE, Venema K (2016) Diet drives quick changes in the metabolic activity and composition of human gut microbiota in a validated in vitro gut model. Res Microbiol 167:114–125

Allen NE et al (2013) Macronutrient intake and risk of urothelial cell carcinoma in the European prospective investigation into cancer and nutrition. Int J Cancer 132:635–644

Amarya S, Singh K, Sabharwal M (2014) Health consequences of obesity in the elderly. J Clin Gerontol Geriatr 5:63–67

Anthony MS, Clarkson TB, Hughes CL, Jr, Morgan TM, Burke GS (1996) Soybean isoflavones improve cardiovascular risk factors without affecting the reproductive system of peripubertal rhesus monkeys. J Nutr 126:43–50

Ballard KD, Kupchak BR, Volk BM, Mah E, Shkreta A, Liptak C, Ptolemy AS, Kellogg MS, Seip RL, Maresh CM, Kraemer WJ, Volek JS (2013) Acute effects of ingestion of a novel whey-derived extract on vascular endothelial function in overweight, middle-aged men and women. Br J Nutr 109:882–893

Baum J, Wolfe R (2015) The link between dietary protein intake, skeletal muscle function and health in older adults. Healthcare 3:529–543

Baum JA, Teng H, Erdman JW Jr, Weigel RM, Klein BP, Persky VW, Freels S, Surya P, Bakhit RM, Ramos E, Potter SM (1998) Long-term intake of soy protein improves blood lipid profiles and increases mononuclear cell low-density-lipoprotein receptor messenger RNA in hypercholesterolemic, postmenopausal women. Am J Clin Nutr 68:545–551

Baumgartner RN, Koehler KM, Gallagher D, Romero L, Heymsfield SB, Ross RR, Garry PJ, Lindeman RD (1998) Epidemiology of sarcopenia among the elderly in New Mexico. Am J Epidemiol 147:755–763

Berg JS, Powell BC, Cheney RE (2001) A millennial myosin census. Mol Biol Cell 12:780–794

Bernstein AM, Sun Q, Hu FB, Stampfer MJ, Manson JE, Willet WC (2010) Major dietary protein sources and the risk of coronary heart disease in women. Circulation 122:876–883

Bernstein AM, Pan A, Rexrode KM, Stampfer M, Hu FB, Mozaffarian D, Willett WC (2012) Dietary protein sources and the risk of stroke in men and women. Stroke 43:637–644

Bessada SM, Barreira JC, Oliveira MBP (2019) Pulses and food security: dietary protein, digestibility, bioactive and functional properties. Trends Food Sci Technol 93:53–68

Bettzieche A, Brandsch C, Eder K, Stangl GI (2009) Lupin protein acts hypocholesterolemic and increases milk fat content in lactating rats by influencing the expression of genes involved in cholesterol homeostasis and triglyceride synthesis. Mol Nutr Food Res 53:1134–1142

Blachier F, Davila AM, Benamouzig R, Tome D (2011) Channelling of arginine in NO and polyamine pathways in colonocytes and consequences. Front Biosci 16:1331–1343

Bodwell CE, Satterlee LD, Hackler LR (1980) Protein digestibility of the same protein preparations by human and rat assays and by in vitro enzymic digestion methods. Am J Clin Nutr 33:677–686

Borodin EA, Menshikova IG, Dorovskikh VA, Feoktistova NA, Shtarberg MA, Yamamoto T, Takamatsu K, Mori H, Yamamoto S (2009) Effects of two-month consumption of 30 ga day of soy protein isolate or skimmed curd protein on blood lipid concentration in Russian adults with hyperlipidemia. J Nutr Sci Vitaminol 55:492–497

Boye J, Wijesinha-Bettoni R, Burlingame B (2012) Protein quality evaluation twenty years after the introduction of the protein digestibility corrected amino acid score method. Br J Nutr 108:S183–S211

Brosnan JT, Brosnan ME (2013) Glutamate: a truly functional amino acid. Amino Acids 45:413–418

Buondonno I, Sassi F, Carignano G, Dutto F, Ferreri C, Pili FG, Massaia M, Nisoli E, Ruocco C, Porrino P, Ravetta C, Riganti C, Isaiaa GC, D'Amelio P (2020) From mitochondria to healthy aging: the role of branched-chain amino acids treatment: MATeR a randomized study. Clin Nutr 39:2080–2091

Byrd-Bredbenner C, Moe G, Beshgetoor D, Berning J (2009) Wardlaw's Perspective in Nutrition. McGrawHill, USA

Calsamiglia S, Stern MD (1995) A three-step in vitro procedure for estimating intestinal digestion of protein in ruminants. J Anim Sci 73:1459–1465

Cam A, de Mejia EG (2012) Role of dietary proteins and peptides in cardiovascular disease. Mol Nutr Food Res 56:53–66

Campbell WW (2007) Synergistic use of higher-protein diets or nutritional supplements with resistance training to counter sarcopenia. Nutr Rev 65:416–422

Carbonaro M, Maselli P, Nucara A (2015) Structural aspects of legume proteins and nutraceutical properties. Food Res Int 76:19–30

Chalvon-Demersay T, Azzout-Marniche D, Arfsten J, Egli L, Gaudichon C, Karagounis LG, Tomé D (2017) A systematic review of the effects of plant compared with animal protein sources on features of metabolic syndrome. J Nutr 147:281–292

Chan DSM, Lau R, Aune D, Vieira R, Greenwood DC, Kampman E, Norat T (2011) Red and processed meat and colorectal cancer incidence: meta-analysis of prospective studies. PLoS ONE 6(6):e20456

Chen Q, Reimer RA (2009) Dairy protein and leucine alter GLP-1 release and mRNA of genes involved in intestinal lipid metabolism in vitro. Nutrition 25:340–349

Chernoff R (2004) Protein and older adults. J Am Coll Nutr 23:627S-630S

Coggon D, Reading I, Croft P, McLaren M, Barrett D, Cooper C (2001) Knee osteoarthritis and obesity. Int J Obes Relat Metab Disord 25:622–627

Contrusciere AV, Paradisi S, Matteucci A, Malchiodi-Albedi F (2010) Branched-chain amino acids induce neurotoxicity in rat cortical culture. Neurotox Res 17:392–398

Craddock JC, Probst YC, Peoples GE (2016) Vegetarian and omnivorous nutrition - comparing physical performance. Int J Sports Nutr Exerc Metab 26:212–220

Damodaran S (2017) Amino acids, peptides and proteins, 5th edn. In: Damodaran S, Parkin KL (eds) Fennema's food chemistry. CRC Press, Boca Raton, pp 235–356

D'Antona G, Ragni M, Cardile A, Tedesco L, Dossena M, Bruttini F, Caliaro F, Corsetti G, Bottinelli R, Carruba MO, Valerio A, Nisoli E (2010) Branched-chain amino acid supplementation promotes survival and supports cardiac and skeletal muscle mitochondrial biogenesis in middle-aged mice. Cell Metab 12:362–372

Dato S, Hoxha E, Crocco P, Iannone F, Passarino G, Rose G (2019) Amino acids and amino acid sensing: implication for aging and diseases. Biogerontology 20:17–31

Dawson-Hughes B, Harris SS (2002) Calcium intake influences the association of protein intake with rates of bone loss in elderly men and women. Am J Clin Nutr 75:773–779
Deglaire A, Bos C, Tomé D, Moughan PJ (2009) Ileal digestibility of dietary protein in the growing pig and adult human. Br J Nutr 102:1752–1759
Dillon EL (2013) Nutritionally essential amino acids and metabolic signaling in aging. Amino Acids 45:431–441
Drotningsvik A, Mjøs SA, Pampanin DM, Slizyte R, Carvajal A, Remman T, Hogoy I, Gudbrandsen OA (2016) Dietary fish protein hydrolysates containing bioactive motifs affect serum and adipose tissue fatty acid compositions, serum lipids, postprandial glucose regulation and growth in obese Zucker fa/fa rats. Br J Nutr 116:1336–1345
Drummen M, Tischmann L, Gatta-Cherifi B, Adam T, Westerterp-Plantenga M (2018) Dietary protein and energy balance in relation to obesity and co-morbidities. Front Endocrinol 9:443
Du SK, Jiang H, Yu X, Jane JL (2014) Physicochemical and functional properties of whole legume flour. LWT-Food Sci Technol 55:308–313
Durant M (2006) Grain legume proteins and nutraceutical properties. Fitoterapia 77:67–82
Elia M (2001) Obesity in the elderly. Obes Res 9:244S-248S
Fan P, Li L, Rezaei A, Eslamfam S, Che D, Ma X (2015) Metabolites of dietary protein and peptides by intestinal microbes and their impacts on gut. Curr Protein Pept Sci 16:646–654
FAO (2013) Dietary protein quality evaluation in human nutrition. Report of an FAO Expert Consultation. FAO Food and Nutrition Paper 92, Auckland, New Zealand, 31 March–2 April 2011
FAO/WHO (1991) Protein quality evaluation. Report of a Joint FAO/WHO expert consultation. FAO Food and Nutrition Paper 51, FAO, Geneva, Switzerland, pp 23–24
FAO/WHO/UNU (1985) Energy and protein requirements. Report of a joint FAO/WHO/UNU expert consultation, World Health Organization Technical Report Series 724, WHO, Geneva, Switzerland
Fekete ÁA, Givens DI, Lovegrove JA (2016) Can milk proteins be a useful tool in the management of cardiometabolic health? An updated review of human intervention trials. Proc Nutr Soc 75:328–341
Fernandes G, Lawrence R, Sun D (2003) Protective role of n-3 lipids and soy protein in osteoporosis. Prostaglandins Leukot Essent Fatty Acids 68:361–372
Flores-Medellín SA, Camacho-Ruiz RM, Guízar-González C, Rivera-Leon EA, Llamas-Covarrubias IM, Mojica L (2020) Protein hydrolysates and phenolic compounds from fermented black beans inhibit markers related to obesity and type-2 diabetes. Legume Sci 2020(1):17
Fontanari GG, Batistuti JP, da Cruz RJ, Saldiva PHN, Arêas JAG (2012) Cholesterol-lowering effect of whole lupin (*Lupinus albus*) seed and its protein isolate. Food Chem 132:1521–1526
Ford JE (1981) Microbiological methods for protein quality assessment. In: Bodwell CE, Adkins JS, Hopkins DT (eds) Protein quality in humans: assessment and in vitro estimation. AVI Publishing Corporation, Westport, pp 278–305
Frasca D, McElhaney J (2019) Influence of obesity on Pneumococcus infection risk in the elderly. Front Endocrinol 10:71
Frota K, dos Santos RD, Ribeiro VQ, Arêas JAG (2015) Cowpea protein reduces LDL-cholesterol and apolipoprotein B concentrations, but does not improve biomarkers of inflammation or endothelial dysfunction in adults with moderate hypercholesterolemia. Nutr Hosp 31:1611–1619
Fujita S, Volpi E (2006) Amino acids and muscle loss with aging. J Nutr 136:277S-280S
Fukagawa NK (2013) Protein and amino acid supplementation in older humans. Amino Acids 44:1493–1509
Gaffney-Stomberg E, Insogna KL, Rodriguez NR, Kerstetter JE (2009) Increasing dietary protein requirements in elderly people for optimal muscle and bone health. J Am Geriatr Soc 57:1073–1079
Gaffney-Stomberg E, Lutz LJ, Rood JC, Cable SJ, Pasiakos SM, Young AJ, McClung JP (2014) Calcium and vitamin D supplementation maintains parathyroid hormone and improves bone

density during initial military training: a randomized, double-blind, placebo controlled trial. Bone 68:46–56

Gaffney-Stomberg E, Cucchi C, Sun BH, Bataille A, Renfro L, Kerstetter J, Insogna K (2011) Dietary protein increases intestinal calcium absorption in part by increasing tight junction expression of Claudin-2. Exp Biol Meeting Abstracts 25:983.30–983.30

Ge J, Sun C-X, Mata A, Corke H, Gan R-Y, Fang Y (2021) Physicochemical and pH-dependent functional properties of proteins isolated from eight traditional Chinese beans. Food Hydrocolloids 112:106288

Genaro PDS, Pinheiro MDM, Szejnfeld VL, Martini LA (2015) Dietary protein intake in elderly women: association with muscle and bone mass. Nutr Clin Pract 30:283–289

Ghumman A, Mudgal S, Singh N, Ranjan B, Kaur A, Rana JC (2021) Physicochemical, functional and structural characteristics of grains, flour and protein isolates of Indian quinoa lines. Food Res Int 140:109982

Ginter E (2008) Vegetarian diets, chronic diseases and longevity. Bratislavske Lekarske Listy/bratislava Med J 109:463–466

Groenendijk I, den Boeft L, van Loon LJ, de Groot LC (2019) High versus low dietary protein intake and bone health in older adults: a systematic review and meta-analysis. Comput Struct Biotechnol J 17:1101–1112

Guillet C, Prod'homme M, Balage M, Gachon P, Giraudet C, Morin L, Grizard J, Boirie Y (2004) Impaired anabolic response of muscle protein synthesis is associated with S6K1 dysregulation in elderly humans. FASEB J 18:1586–1587

Haba Y, Fujimura T, Oyama K, Kinoshita J, Miyashita T, Fushida S, Harada S, Ohta T (2019) Effect of oral branched-chain amino acids and glutamine supplementation on skeletal muscle atrophy after total gastrectomy in rat model. J Surg Res 243:281–288

Hardy R, Cooper MS (2009) Bone loss in inflammatory disorders. J Endocrinol 201:309–320

Hashimoto R, Sakai A, Murayama M, Ochi A, Abe T, Hirasaka K, Ohno A, Teshima-Kondo S, Yanagawa H, Yasui N, Inatsugi M, Doi D, Takeda M, Mukai R, Terao J, Nikawa T (2015) Effects of dietary soy protein on skeletal muscle volume and strength in humans with various physical activities. J Med Invest 62:177–183

Heaney RP (2007) Effects of protein on the calcium economy. In: Burckhardt P, Heaney RP, Dawson-Hughes B (eds) Nutritional aspects of osteoporosis. Elsevier Inc., Lausanne, Switzerland, Amsterdam, Netherlands, pp 191–197

Heaney RP, Layman DK (2008) Amount and type of protein influences bone health. Am J Clin Nutr 87:1567–1570

Heffernan KS, Fahs CA, Ranadive SM, Patvardhan EA (2010) L-Arginine as a nutritional prophylaxis against vascular endothelial dysfunction with aging. J Cardiovasc Pharmacol Ther 15:17–23

Henley EC, Taylor JRN, Obukosia SD (2010) The importance of dietary protein in human health: combating protein deficiency in sub-Saharan Africa through transgenic biofortified sorghum. Adv Food Nutr Res 60:21–52

Hidayat K, Chen GC, Wang Y, Zhang Z, Dai X, Szeto IMY, Qin LQ (2018) Effects of milk proteins supplementation in older adults undergoing resistance training: a meta-analysis of randomized control trials. J Nutr Health Aging 22:237–245

Hoffer LJ (1994) Starvation. In: Shils ME, Olson JA, Shike M (eds) Modern Nutrition in Health and Disease, 8th edn. Lea and Febiger, Philadelphia, pp 927–949

Holeček M (2018) Branched-chain amino acids in health and disease: metabolism, alterations in blood plasma, and as supplements. Nutr Metab 15:33

Hosomi R, Fukunaga K, Arai H, Kanda S, Nishiyama T, Yoshida M (2011) Effect of simultaneous intake of fish protein and fish oil on cholesterol metabolism in rats fed high-cholesterol diets. Open Nutraceuticals J 4:12–19

Hosomi R, Fukunaga K, Arai H, Kanda S, Nishiyama T, Yoshida M (2013) Effect of combination of dietary fish protein and fish oil on lipid metabolism in rats. J Food Sci Technol 50:266–274

Houston DK, Nicklas BJ, Ding J, Harris TB, Tylavsky FA, Newman AB, Lee JS, Sahyoun NR, Visser M, Kritchevsky SB (2008) Dietary protein intake is associated with lean mass change in older, community-dwelling adults: the health, aging, and body composition (health ABC) study. Am J Clin Nutr 87:150–155
Hunter T (1995) Protein kinases and phosphatases: the yin and yang of protein phosphorylation and signaling. Cell 80:225–236
Jarzaguet M, Polakof S, David J, Migné C, Joubrel G, Efstathiou T, Rémond D, Mosoni L, Dardevet D (2018) A meal with mixed soy/whey proteins is as efficient as a whey meal in counteracting the age-related muscle anabolic resistance only if the protein content and leucine levels are increased. Food Funct 9:6526–6534
Jiao N, Wu Z, Ji Y, Wang B, Dai Z, Wu G (2015) L-Glutamate enhances barrier and antioxidative functions in intestinal porcine epithelial cells. J Nutr 145:2258–2264
Kararli TT (1995) Comparison of the gastrointestinal anatomy, physiology, and biochemistry of humans and commonly used laboratory animals. Biopharm Drug Dispos 16:351–380
Kawabata F, Mizushige T, Uozumi K, Hayamizu K, Han L, Tsuji T, Kishida T (2015) Fish protein intake induces fast-muscle hypertrophy and reduces liver lipids and serum glucose levels in rats. Biosci Biotechnol Biochem 79:109–116
Keller U (2011) Dietary proteins in obesity and in diabetes. Int J Vitam Nutr Res 81:125–133
Kerstetter JE, Bihuniak JD, Brindisi J, Sullivan RR, Mangano KM, Larocque S, Kotler BM, Simpson CA, Cusano AM, Stomberg EM, Kleppinger A, Reynolds J, Dziura J, Kenny AM, Insogna KL (2015) The effect of a whey protein supplement on bone mass in older Caucasian adults. J Clin Endocrinol Metab 100:2214–2222
Kong XF, Tan BE, Yin YL, Gao H, Li X, Jaeger LA, Bazer FW, Wu G (2012) L-Arginine stimulates the mTOR signaling pathway and protein synthesis in porcine trophectoderm cells. J Nutr Biochem 23:178–1183
Le Floch N, Melchior D, Obled C (2004) Modifications of protein and amino acid metabolism during inflammation and immune system activation. Livest Prod Sci 87:37–45
Leite JSM, Cruzat VF, Krause MS, de Bittencourt PIH (2016) Physiological regulation of the heat shock response by glutamine: implications for chronic low-grade inflammatory diseases in age-related conditions. Nutrire 41:1–17
Levine ME, Suarez JA, Brandhorst S, Balasubramanian P, Cheng CW, Madia F, Fontana L, Mirisola MG, Guevara-Aguirre J, Wan J et al (2014) Low protein intake is associated with a major reduction in IGF-1, cancer, and overall mortality in the 65 and younger but not older population. Cell Metab 19:407–417
Li P, Yin YL, Li D, Kim SW, Wu G (2007) Amino acids and immune function. Br J Nutr 98:237–252
Li SS, Mejia SB, Lytvyn L, Stewart SE, Viguiliouk E, Ha V, de Souza RJ, Leiter LA, Kendall CWC, Jenkins DJA, Sievenpiper JL (2017) Effect of plant protein on blood lipids: a systematic review and meta-analysis of randomized controlled trials. J Am Heart Assoc 6(12):e006659
Lillefosse HH, Clausen MR, Yde CC, Ditlev DB, Zhang X, Du ZY, Liaset B (2014) Urinary loss of tricarboxylic acid cycle intermediates as revealed by metabolomics studies: an underlying mechanism to reduce lipid accretion by whey protein ingestion? J Proteome Res 13:2560–2570
Longvah T, Anantan I, Bhaskarachary K, Venkaiah K (2017) Indian food composition tables, pp 2–58. National Institute of Nutrition, Indian Council of Medical Research, Hyderabad
Lonnie M, Hooker E, Brunstrom J, Corfe B, Green M, Watson A, Williams E, Stevenson E, Penson S, Johnstone A (2018) Protein for life: review of optimal protein intake, sustainable dietary sources and the effect on appetite in ageing adults. Nutrients 10(3):360
Lorenzo MD, Cameron C, Bruno V, Valuria DB, Carlo C (2006) In preparation for the international symposium on obesity in the elderly. Ageing Health 2:47–51
Loveday SM (2019) Food proteins: technological, nutritional, and sustainability attributes of traditional and emerging proteins. Ann Rev Food Sci Technol 10:311–339
Lynch H, Wharton C, Johnston C (2016) Cardiorespiratory fitness and peak torque differences between vegetarian and omnivore endurance athletes: a cross-sectional study. Nutrients 8(11):726

Magrini MB, Anton M, Chardigny JM, Duc G, Duru M, Jeuffroy MH, Meynard JM, Micard V, Walrand S (2018) Pulses for sustainability: breaking agriculture and food sectors out of lock-in. Front Sustain Food Syst 2:64

Marmonier C, Chapelot D, Louis-Sylvestre J (2000) Effects of macronutrient content and energy density of snacks consumed in a satiety state on the onset of the next meal. Appetite 34:161–168

Mathai JK, Liu Y, Stein HH (2017) Values for digestible indispensable amino acid scores (DIAAS) for some dairy and plant proteins may better describe protein quality than values calculated using the concept for protein digestibility-corrected amino acid scores (PDCAAS). Br J Nutr 117:490–499

Mathus-Vliegen E (2012) Obesity and the elderly. J Clin Gastroenterol 46:533–544

Mendonca S, Saldiva PH, Cruz RJ, Areas JA (2009) Amaranth protein presents cholesterol-lowering effect. Food Chem 116:738–742

Mikkelsen PB, Toubro S, Astrup A (2000) Effect of fat-reduced diets on 24-h energy expenditure: comparisons between animal protein, vegetable protein, and carbohydrate. Am J Clin Nutr 72:1135–1141

Millward DJ, Layman DK, Tomé D, Schaafsma G (2008) Protein quality assessment: impact of expanding understanding of protein and amino acid needs for optimal health. Am J Clin Nutr 87:1576S-1581S

Mirzaei H, Suarez JA, Longo VD (2014) Protein and amino acid restriction, aging and disease: from yeast to humans. TEM 25:558–566

Moore DR, Churchward-Venne TA, Witard O, Breen L, Burd NA, Tipton KD, Phillips SM (2015) Protein ingestion to stimulate myofibrillar protein synthesis requires greater relative protein intakes in healthy older versus younger men. J Gerontol Ser A Biomed Sci Med Sci 70:57–62

Morris-Naumann FL, Wark JD (2015) Exercise, nutrition, and bone health. In: Nutrition and bone health. Humana Press, New York, pp 543–560

Nasri M (2017) Protein hydrolysates and biopeptides: production, biological activities, and applications in foods and health benefits - a review. Adv Food Nutr Res 81:109–159

Neacsu M, Fyfe C, Horgan G, Johnstone AM (2014) Appetite control and biomarkers of satiety with vegetarian (soy) and meat-based high-protein diets for weight loss in obese men: a randomized crossover trial. Am J Clin Nutr 100:548–558

Nebl J, Haufe S, Eigendorf J, Wasserfurth P, Tegtbur U, Hahn A (2019) Exercise capacity of vegan, lacto-ovo-vegetarian and omnivorous recreational runners. J Int Soc Sports Nutr 16(1):23

Nelson DL, Cox MM (2013) Lehninger principles of biochemistry, 7th edn. Worth Publishers, New York

Nie Y, Luo F, Lin Q (2018) Dietary nutrition and gut microflora: A promising target for treating diseases. Trends Food Sci Technol 75:72–80

Nishitani S, Matsumura T, Fujitani S, Sonaka I, Miura Y, Yagasaki K (2002) Leucine promotes glucose uptake in skeletal muscles of rats. Biochem Biophys Res Commun 299:693–696

Obersby D, Chappell DC, Dunnett A, Tsiami AA (2013) Plasma total homocysteine status of vegetarians compared with omnivores: a systematic review and meta-analysis. Br J Nutr 109:785–794

Paddon-Jones D, Rasmussen BB (2009) Dietary protein recommendations and the prevention of sarcopenia: protein, amino acid metabolism and therapy. Curr Opin Clin Nutr Metab Care 12:86–90

Paddon-Jones D, Westman E, Mattes RD, Wolfe RR, Astrup A, Westerterp-Plantenga M (2008) Protein, weight management, and satiety. Am J Clin Nutr 87:1558S-1561S

Paddon-Jones D, Campbell WW, Jacques PF, Kritchevsky SB, Moore LL, Rodriguez NR, van Loon LJC (2015) Protein and healthy aging. Am J Clin Nutr 101:1339S-1345S

Pal S, Ellis V, Dhaliwal S (2010) Effects of whey protein isolate on body composition, lipids, insulin and glucose in overweight and obese individuals. Br J Nutr 104:716–723

Pellett PL, Young VR (1980) Nutritional evaluation of protein foods. The United Nations University, Japan

Posner ES (2000) Wheat. In: Kulp K, Ponte JG (eds) Handbook of cereal science and technology. Marcel Dekker Inc., New York, pp 1–29

Public Health England (2016) The Eatwell Guide, Crown Copyright. https://www.gov.uk/government/publications/the-eatwell-guide

Rezaei R, Wang W, Zhenlong Wu, Dai Z, Wang J, Guoyao Wu (2013) Biochemical and physiological bases for utilization of dietary amino acids by young pigs. J Anim Sci Biotechnol 4(1):7

Ríos-Covián D, Ruas-Madiedo P, Margolles A, Gueimonde M, De Los Reyes-gavilán CG, Salazar N (2016) Intestinal short chain fatty acids and their link with diet and human health. Front Microbiol 7:185

Rios-Covian D, González S, Nogacka AM, Arboleya S, Salazar N, Gueimonde M, de Los Reyes-Gavilán CG (2020) An overview on fecal branched short-chain fatty acids along human life and as related with body mass index: associated dietary and anthropometric factors. Front Microbiol 11:973

Rizzoli R (2014) Dairy products, yogurts, and bone health. Am J Clin Nutr 99:1256S-1262S

Rizzoli R (2015) Nutrition and sarcopenia. J Clin Densitom 18:483–487

Roglic G (2016) WHO Global report on diabetes: a summary. Int J Noncommunicable Dis 1:3–8

Rutherfurd SM, Fanning AC, Miller BJ, Moughan PJ (2015) Protein digestibility-corrected amino acid scores and digestible indispensable amino acid scores differentially describe protein quality in growing male rats. J Nutr 145:372–379

Sacks FM, Lichtenstein A, Van Horn L, Harris W, Kris-Etherton P, Winston M (2006) American Heart Association Nutrition Committee. Soy protein, isoflavones, and cardiovascular health: an American Heart Association Science Advisory for professionals from the Nutrition Committee. Circulation 113:1034–1044

Sacks FM, Lichtenstein AH, Wu JHY, Appel LJ, Creager MA, Kris-Etherton PM, Miller M, Rimm EB, Rudel LL, Robinson JG, Stone NJ, Van Horn LV (2017) Fats and cardiovascular disease: a presidential advisory from the American Heart Association. Circulation 136:e1–e23

Sandhu RS, Singh N, Kaler RSS, Kaur A, Shevkani K (2018) Effect of degree of milling on physicochemical, structural, pasting and cooking properties of short and long grain Indica rice cultivars. Food Chem 260:231–238

Santesso N, Akl EA, Bianchi M, Mente A, Mustafa R, Heels-Ansdell D, Schünemann HJ (2012) Effects of higher-versus lower-protein diets on health outcomes: a systematic review and meta-analysis. Eur J Clin Nutr 66:780–788

Satterlee LD, Marshall HF, Tennyson JM (1979) Measuring protein quality. J Am Oil Chem Soc 56:103

Sauer FD, Erfle JD, Mahadevan S (1975) Amino acid biosynthesis in mixed rumen cultures. Biochem J 50:357–372

Schaafsma G (2000) The protein digestibility-corrected amino acid score. J Nutr 130:1865S-1867S

Schincaglia R, Pimentel G, Mota J (2017) Nuts and human health outcomes: a systematic review. Nutrients 9(12):1311

Schlemmer U, Frølich W, Prieto RM, Grases F (2009) Phytate in foods and significance for humans: Food sources, intake, processing, bioavailability, protective role and analysis. Mol Nutr Food Res 53:S330–S375

Schmidt RL, Simonović M (2012) Synthesis and decoding of selenocysteine and human health. Croat Med J 53:535–550

Scully A, Neacsu M, Russell W, Vaughan N, Fyfe C, Hudson M, Taylor K, Johnstone AM (2017) Plant protein influence on appetite and food intake in healthy subjects. Proc Nutr Soc 76:E44

Serna-Salvidar SO (2010) Cereal grains properties, processing and nutritional attributes. CRC Press, Boca Raton, FL

Shams-White MM, Chung M, Du M, Fu Z, Insogna KL, Karlsen MC, LeBoff MS, Shapses SA, Sackey J, Wallace TC, Weaver CM (2017) Dietary protein and bone health: a systematic review and meta-analysis from the National Osteoporosis Foundation. Am J Clin Nutr 105:1528–1543

Shang N, Chaplot S, Wu J (2018) Food proteins for health and nutrition, chap 12. In: Proteins in food processing. Woodhead Publishing, pp 301–336

Shevkani K, Singh N (2015) Relationship between protein characteristics and film-forming properties of kidney bean, field pea and amaranth protein isolates. Int J Food Sci Technol 50:1033–1043

Shevkani K, Singh N, Kaur A, Rana JC (2014) Physicochemical, pasting, and functional properties of amaranth seed flours: effects of lipids removal. J Food Sci 79:C1271–C1277

Shevkani K, Singh N, Kaur A, Rana JC (2015) Structural and functional characterization of kidney bean and field pea protein isolates: a comparative study. Food Hydrocolloids 43:679–689

Shevkani K, Singh N, Rattan B, Singh JP, Kaur A, Singh B (2019) Effect of chickpea and spinach on extrusion behaviour of corn grit. J Food Sci Technol 56:2257–2266

Shevkani K, Singh N, Chen Y, Kaur A, Yu L (2019) Pulse proteins: secondary structure, functionality and applications. J Food Sci Technol 56:2787–2798

Shevkani K, Kaur M, Singh N (2021) Composition, pasting, functional, and microstructural properties of flours from different split dehulled pulses (*dhals*). J Food Process Preserv 45:e15485

Shivakumar N, Minocha S, Kurpad AV (2018) Protein quality and amino acid requirements in relation to needs in India. Indian J Med Res 148:557–568

Shukla A, Bettzieche A, Hirche F, Brandsch C, Stangl GI, Eder K (2006) Dietary fish protein alters blood lipid concentrations and hepatic genes involved in cholesterol homeostasis in the rat model. Br J Nutr 96:674–682

Singh JP, Kaur A, Shevkani K, Singh N (2016) Composition, bioactive compounds and antioxidant activity of common Indian fruits and vegetables. J Food Sci Technol 53:4056–4066

Singh B, Singh JP, Shevkani K, Singh N, Kaur A (2017) Bioactive constituents in pulses and their health benefits. J Food Sci Technol 54:858–870

Singh N, Singh P, Shevkani K, Virdi AS (2019) Amaranth: potential source for flour enrichment. In: Preedy VR, Watson RR (eds) Flour and breads and their fortification in health and disease prevention (second edition). Academic Press, Elsevier-UK, pp 123–135

Sirtori CR, Anderson J, Arnoldi A (2007) Nutritional and nutraceutical considerations for dyslipidemia. Futur Lipidol 2:313–339

Sluijs I, Beulens JW, Spijkerman AM, Grobbee DE, van der Schouw YT (2010) Dietary intake of total, animal, and vegetable protein and risk of type 2 diabetes in the European Prospective Investigation into Cancer and Nutrition (EPIC)-NL study. Diab Care 33:43–48

Soenen S, Martens EA, Hochstenbach-Waelen A, Lemmens SG, Westerterp-Plantenga MS (2013) Normal protein intake is required for body weight loss and weight maintenance, and elevated protein intake for additional preservation of resting energy expenditure and fat free mass. J Nutr 143:591–596

Solerte SB, Fioravanti M, Locatelli E, Bonacasa R, Zamboni M, Basso C, Mazzoleni A, Mansi V, Geroutis N, Gazzaruso C (2008) Improvement of blood glucose control and insulin sensitivity during a long-term (60 weeks) randomized study with amino acid dietary supplements in elderly subjects with type 2 diabetes mellitus. Am J Cardiol 101:82E-88E

Song M, Fung TT, Hu FB, Willett WC, Longo VD, Chan AT, Giovannucci EL (2016) Association of animal and plant protein intake with all-cause and cause-specific mortality. JAMA Intern Med 176:1453–1463

Song Y, Manson JE, Buring JE, Liu S (2004) A prospective study of red meat consumption and type 2 diabetes in middle-aged and elderly women: the women's health study. Diab Care 27:2108–2115

Strasser B, Volaklis K, Fuchs D, Burtsche M (2018) Role of dietary protein and muscular fitness on longevity and aging. Aging Dis 9:119–132

Swindle M, Smith AC (1998) Comparative anatomy and physiology of the pig. Scand J Lab Anim Sci 25:11–21

Tapal A, Vegarud GE, Sreedhara A, Tiku PK (2019) Nutraceutical protein isolate from pigeon pea (*Cajanus cajan*) milling waste by-product: Functional aspects and digestibility. Food Funct 10:2710–2719

Thévenot J, Cauty C, Legland D, Dupont D, Floury J (2017) Pepsin diffusion in dairy gels depends on casein concentration and microstructure. Food Chem 223:54–61

Thorpe MP, Jacobson EH, Layman DK, He X, Kris-Etherton PM, Evans EM (2008) A diet high in protein, dairy, and calcium attenuates bone loss over twelve months of weight loss and maintenance relative to a conventional high-carbohydrate diet in adults. J Nutr 138:1096–1100
Tian S, Qian Xu, Jiang R, Han T, Sun C, Na L (2017) Dietary protein consumption and the risk of type 2 diabetes: a systematic review and meta-analysis of cohort studies. Nutrients 9(9):982
Torres N, Torre-Villalvazo I, Tovar AR (2006) Regulation of lipid metabolism by soy protein and its implication in diseases mediated by lipid disorders. J Nutr Biochem 17:365–373
Tottier N, Walker R (2015) Nutritional aspects of proteins, chap 8. In: Ustunol Z (ed) Applied food protein chemistry. Wiley, Chichester
Tovar-Pérez EG, Lugo-Radillo A, Aguilera-Aguirre S (2019) Amaranth grain as a potential source of biologically active peptides: a review of their identification, production, bioactivity, and characterization. Food Rev Intl 35:221–245
Ustunol Z (2015) Amino acids, peptides and proteins, chap 3. In: Ustunol Z (ed) Applied food protein chemistry. Wiley, Chichester
Valerio A, D'Antona G, Nisoli E (2011) Branched-chain amino acids, mitochondrial biogenesis, and healthspan: an evolutionary perspective. Aging 3:464
van Oort, M (2010) Enzymes in food technology – introduction, chap 1. In: Whitehurst RJ, van Oort M (eds) Enzymes in food technology. Wiley-Blackwell, Chinchester, pp 1–17
van Vliet S, Burd NA, van Loon LJ (2015) The skeletal muscle anabolic response to plant-versus animal-based protein consumption. J Nutr 145:1981–1991
Vang A, Singh PN, Lee JW, Haddad EH, Brinegar CH (2008) Meats, processed meats, obesity, weight gain and occurrence of diabetes among adults: findings from adventist health studies. Ann Nutr Metab 52:96–104
Veldhorst MA, Nieuwenhuizen AG, Hochstenbach-Waelen A, van Vught AJ, Westerterp KR, Engelen MP, Brummer RJM, Deutz NEP, Westerterp-Plantenga MS (2009) Dose-dependent satiating effect of whey relative to casein or soy. Physiol Behav 96:675–682
Vikøren LA, Nygård OK, Lied E, Rostrup E, Gudbrandsen OA (2013) A randomised study on the effects of fish protein supplement on glucose tolerance, lipids and body composition in overweight adults. Br J Nutr 109:648–657
Virtanen HE, Voutilainen S, Koskinen TT, Mursu J, Kokko P, Ylilauri M, Tuomainen TP, Salonen JT, Virtanen JK (2019) Dietary proteins and protein sources and risk of death: the Kuopio Ischaemic heart disease risk factor study. Am J Clin Nutr 109:1462–1471
Wall BT, Hamer HM, de Lange A, Kiskini A, Groen BB, Senden JM, Gijsen AP, Verdijk LB, van Loon LJ (2013) Leucine co-ingestion improves postprandial muscle protein accretion in elderly men. Clin Nutr 32:412–419
Wandrag L, Brett SJ, Frost G, Hickson M (2015) Impact of supplementation with amino acids or their metabolites on muscle wasting in patients with critical illness or other muscle wasting illness: a systematic review. J Hum Nutr Diet 28:313–330
Weigle DS, Breen PA, Matthys CC, Callahan HS, Meeuws KE, Burden VR, Purnell JQ (2005) A high-protein diet induces sustained reductions in appetite, ad libitum caloric intake, and body weight despite compensatory changes in diurnal plasma leptin and ghrelin concentrations. Am J Clin Nutr 82:41–48
Westerterp-Plantenga MS, Nieuwenhuizen A, Tome D, Soenen S, Westerterp KR (2009) Dietary protein, weight loss, and weight maintenance. Annual Revision of Nutrition 29:21–41
WHO (2020a) Obesity and overweight. https://www.who.int/news-room/fact-sheets/detail/obesity-and-overweight. Accessed 03 Mar 2021
WHO (2020b) Diabetes. https://www.who.int/news-room/fact-sheets/detail/diabetes. Accessed 04 Mar 2020
Wolfe RR (2015) Update on protein intake: importance of milk proteins for health status of the elderly. Nutr Rev 73:41–47
Wolfe RR, Cifelli AM, Kostas G, Kim IY (2017) Optimizing protein intake in adults: interpretation and application of the recommended dietary allowance compared with the acceptable macronutrient distribution range. Adv Nutr 8:266–275

World Cancer Research Fund/American Institute for Cancer Research (2018). Diet, nutrition, physical activity and cancer: a global perspective. Continuous Update Project Expert Report
Wu G (2010) Functional amino acids in growth, reproduction and health. Adv Nutr 1:31–37
Wu G (2013) Functional amino acids in nutrition and health. Amino Acids 45:407–411
Wu G (2016) Dietary protein intake and human health. Food Funct 7:1251–1265
Wu G, Fanzo J, Miller DD, Pingali P, Post M, Steiner JL, Thalacker-Mercer AE (2014) Production and supply of high-quality food protein for human consumption: sustainability, challenges, and innovations. Ann N Y Acad Sci 1321:1–19
Wu G, Bazer FW, Johnson GA, Herring C, Seo H, Dai Z, Wang J, Wu Z, Wang X (2017) Functional amino acids in the development of the pig placenta. Mol Reprod Dev 84:870–882
Xue Z, Hou X, Yu W, Wen H, Zhang Q, Li D, Kou X (2018) Lipid metabolism potential and mechanism of CPe-III from chickpea (*Cicer arietinum* L.). Food Res Int 104:126–133
Yi D, Li B, Hou Y, Wang L, Zhao D, Chen H, Wu G (2018) Dietary supplementation with an amino acid blend enhances intestinal function in piglets. Amino Acids 50:1089–1100
Zhang Q, Wastney ME, Rosen CJ, Beamer WG, Weaver CM (2011) Insulin-like growth factor-1 increases bone calcium accumulation only during rapid growth in female rats. J Nutr 141:2010–2016
Zhang J, Yu Y, Wang J (2020) Protein Nutritional support: the classical and potential new mechanisms in the prevention and therapy of Sarcopenia. J Agric Food Chem 68:4098–4108
Zhenyukh O, Civantos E, Ruiz-Ortega M, Sanchez MS, Vazquez C, Peiro C, Egido J, Mas S (2017) High concentration of branched-chain amino acids promotes oxidative stress, inflammation and migration of human peripheral blood mononuclear cells via mTORC1 activation. Free Radical Biol Med 104:165–177
Zhou D, Lezmi S, Wang H, Davis J, Banz W, Chen H (2014) Fat accumulation in the liver of obese rats is alleviated by soy protein isolate through β-catenin signaling. Obesity 22:151–158

Chapter 2
Carbohydrates as Nutritional Components for Health and Longevity

Gozde Okburan and Ceren Gezer

Abstract Carbohydrates act as signaling molecules, energy sources, and structural components. The importance of carbohydrates in human life has been reinforced by understanding the tight and important relationship between a carbohydrate-rich diet and chronic metabolic diseases. High quality and quantity of carbohydrate intake increases 5'adenosine monophosphate-activated protein kinase (AMPK) and sirtuins while causing a reduction in rapamycin (mTOR) and insulin/IGF-1/GH levels. An increase in AMPK and sirtuins is thus a decrease in mTOR, and improved insulin/IGF-1/GH levels improve cellular processes related with aging such as mitochondrial biogenesis, autophagy, cellular metabolism, oxidative stress, genome maintenance, and protein synthesis. Along with all these consequences, it is assumed that, with the right type and amount of carbohydrates, human life is expected to be positively affected. Concerning the amount of carbohydrates and health consequences, both high and low carbohydrate intake were associated with increased mortality; thus, 50–60% of total energy intake is recommended. Also, carbohydrate type is related with non-communicable chronic disease risk and mortality. Consumption of added and free sugars, in particular, is related to obesity, diabetes, and cardiovascular diseases. Sufficient intake of fibre-rich sources such as non-starchy vegetables, whole fruits, whole grains, and legumes is generally recommended to reduce the alarming rate of chronic diseases and mortality. Therefore, 25–29 g/day dietary fibre and <10% of daily energy from free sugars intake is recommended. In conclusion, both quality and quantity of carbohydrates notably modifies health, aging, and longevity.

Keywords Aging · Carbohydrates · Dietary fibre · Health · Longevity · Sugar

---

G. Okburan (✉) · C. Gezer
Department of Nutrition and Dietetics, Faculty of Health Sciences, Eastern Mediterranean University, Famagusta, North Cyprus via Mersin 10, Turkey
e-mail: gozde.okburan@emu.edu.tr

C. Gezer
e-mail: ceren.gezer@emu.edu.tr

S. I. S. Rattan and G. Kaur (eds.), *Nutrition, Food and Diet in Ageing and Longevity*, Healthy Ageing and Longevity 14, https://doi.org/10.1007/978-3-030-83017-5_2

2.1 Introduction

Nutrition has critical long-term consequences for health. Dietary patterns and nutrition are among the vital factors affecting the development and progression of chronic diseases such as cardiovascular diseases, diabetes, and cancer. Since half of the deaths in the world are caused by chronic diseases, prevention, and management of these diseases are global priorities. At this point, it would be useful to underline how important a role nutrition plays in human life (Mckay and Mathers 2011). All organisms need to get nutrients from their environment in order to survive. These nutrients are identified as organic chemicals such as carbohydrates, proteins, lipids, vitamins, and inorganic substances such as minerals and water. These nutrients are essential for biological functions including metabolism, growth, and repair. All of the various activities and processes that take place in the body require energy. Carbohydrates, proteins, and fats are known as macronutrients, which provide energy and are measured in the form of calories (Shukla et al. 2017). Calorie restriction, that is, restriction of dietary energy intake without causing malnutrition, has been shown to improve the maintenance of the biological system and enhance lifespan (Kenyon 2010). The effects of nutrition and caloric restriction on human life expectancy has been recognized since the days of Hippocrates, who noted that excessive eating shortens lifespan (Hyung 2010). In more recent history, data presented by McCay et al. (1935) found, for the first time, that the restriction of calories without malnutrition prolonged mean and maximal lifespan in rats compared with ad libitum feeding. Since then, reduced food intake, known as calorie restriction, has become an established pattern in the area of aging studies. It is considered a robust and reproducible intervention regarding aging, delayed aging, and increased lifespan. A study indicating the relationship between energy restriction and prolongation of life showed that, compared to a study group on an ad-libitum diet (control group), members of a study group with a dietary energy restriction of approximately 30–50% of the typically consumed amount and supplementation of micronutrients, increased their life expectancy (Speakman and Mitchell 2011). Many studies have shown that reduced calorie intake by alteration of nutrient content, such as fat, carbohydrates, or amino acids, can have different impacts on longevity in animal models and in human studies (Guasch-Ferré et al. 2015; Martinez-Gonzalez and Martin-Calvo 2016; Kirkpatrick et al. 2019). Regarding the effect of energy density and macronutrient composition on aging and lifespan, The Geometric Framework for Nutrition concluded that, while keeping energy intake stable, reducing the ratio of protein to carbohydrates (without causing malnutrition) extends life (Lee et al. 2008; Solon-Biet et al. 2015a, b; Couteur et al. 2016a, b). Carbohydrates are organic compounds that contain carbon, hydrogen, and oxygen in a 1:2:1 ratio. Carbohydrates act as signaling molecules, energy sources, and structural components. The importance of carbohydrates in human life has been reinforced by understanding the closely linked and important relationship between a carbohydrate-rich diet and chronic metabolic diseases. It has been suggested that some types of carbohydrates directly affect lifespan through signaling pathways in vitro and in vivo in various organisms (Lee et al. 2015). Therefore, the potential influence of carbohydrates on human health, aging, and longevity will be discussed.

2.2 Potential Mechanisms and Roles for Dietary Carbohydrates in Human Aging

Carbohydrates are the most abundant organic molecules in nature and are also key molecules for cellular processes such as energy production. Several forms of carbohydrates are converted to glucose in order to be used as the primary source of energy in cells (Buyken et al. 2018). They also act as a stored form of energy in the body and serve as cell membrane components that mediate some forms of intercellular communication. Carbohydrates are classified according to the degree of polymerization, or number of saccharide units, in the carbohydrate molecule. Table 2.1 provides the various saccharide categories along with examples of commonly consumed carbohydrates. Glucose and fructose are the main monosaccharides are present in fruits and vegetables, while fructose is also used in processed foods. Sucrose is generally the most abundant disaccharide in Westernized diets, with foods providing naturally

Table 2.1 Classification of carbohydrates (Burke 2000)

Type	Examples
Monosaccharide (1 unit)	Glucose Fructose Galactose
Disaccharides (2 units)	Sucrose Lactose Maltose
Oligosaccharides (3–20 units)	Raffinose (3 units) Stachyose (4 units) Verbascose (5 units) Fructo-oligosaccharides Commercially derived glucose polymers/maltodextrins (5–15 units)
Polysaccharides (20–1000 units) Starch	Amylose Amylopectin
Non-starch polysaccharides	Cellulose Hemicellulose Pectins β-glucans Fructans Gums Mucilages Algal polysaccharides

occurring and/or added sources of this sugar, while lactose is provided primarily by dairy foods. Oligosaccharides form only a small amount of dietary carbohydrates intake and are found in legumes, while fructo-oligosaccharides are found in other vegetables. Starch provides plant storage for carbohydrates, and appears primarily in grains, legumes, and some vegetables and fruit. The non-starch polysaccharides (NSPs) include structural cell wall components (hemicellulose and cellulose, along with pectins) as well as storage polysaccharides, gums, and mucilages. These NSPs share the characteristic of being largely undigested in the small intestine, and together with lignin comprise dietary fibre (Burke 2000).

Studies have shown that glucose, which is a monosaccharide, affects lifespan in several model organisms (Lee et al. 2015). Generally, the influence of carbohydrates on lifespan may show either positive or negative effects, depending on the quality and quantity of the dietary carbohydrates (Ludwig et al. 2018). While some studies have shown that glucose-enriched diets decrease lifespan, other research has found that various other carbohydrates and carbohydrate metabolites, including trehalose, pyruvate, malate, fumarate, and N-acetylglucosamine (GlcNAc) increase longevity (Lee et al. 2015, 2017). Various mechanisms have been shown to identify the nutrient-sensing cellular pathways that link diet and aging. Low protein, high carbohydrate (LPHC) diets have been found to retard aging by mechanisms and pathways similar to calorie restricted diets (Couteur et al. 2016). The mechanisms which link diet and delayed aging have four main pathways. First, LPHC diets influence nutrient sensing pathways such as sirtuins (SIRTs) mechanistic target of mTOR, AMPK and insulin/insulin growth factor (IGF-1)/growth hormone (GH) (Simpson and Raubenheimer 2005). An LPHC diet increases AMPK and SIRTs while reducing mTOR and insulin/IGF-1/GH levels. Increased AMPK and SIRTs and reduced mTOR and insulin/IGF-1/GH improve cellular processes related with aging such as mitochondrial biogenesis, autophagy, cellular metabolism, oxidative stress, genome maintenance, and protein synthesis. Those pathways enable organisms to survive periods of food shortages by transferring resources from reproduction to survival and cellular maintenance (David et al. 2016).

On the other hand, as mentioned above, the type of carbohydrate in the diet plays a vital role regarding impact on health. Some studies have shown that glucose-enriched diets shorten lifespan via various pathways (Mortuza et al. 2013; Zhang et al. 2013a, b; Lee et al. 2017). In particular, human endothelial cells and fibroblasts are negatively affected by a high glucose diet (Lee et al. 2017). In human studies, it was determined that glucose-enriched diets increase some types of aging-related phenotypes such as increased levels of senescence-associated (SA) β-gal staining, decreased proliferation, irregular morphology, and increased reactive oxygen species (ROS). SIRTs have been shown to be associated with glucose-induced accelerated cellular aging. It has been shown that glucose-enriched diets suppress and down-regulate SIRTs (Mortuza et al. 2013; Zhang et al. 2013a, b). Also, it has been demonstrated that high dietary glucose activated p38 mitogen-activated protein kinase (MAPK); thus, this enzyme accelerates the senescence of human endothelial progenitor cells (EPCs). Additionally, with the presence of high glucose levels, cellular aging phenotypes such as SA β-gal staining levels increased while cellular proliferation reduced; thus, p38 MAPK is activated in EPCs. Restoration of aging phenotypes has been demonstrated when treated with p38 MAPK inhibitors. Thus, it is proposed that p38 MAPK mediates the pro-aging consequences of high glucose levels (Kuki et al. 2006). Another study, by Micó et al. (2017), also found that high glucose levels trigger insulin release, which then causes an increase in IGF-1. At that point, IGF-1 binds to its receptor and causes autophosphorylation to switch on, leading to the following activation of PI3K: PI3K phosphorylates and activates protein kinase B (AKT). Activated AKT phosphorylates and activates mTOR and inhibits FOXO. Thus, an increase in glucose levels affects this signaling cascade. Hereby, for optimal health, longevity, and aging, the amount and especially the type of dietary carbohydrates play a critical role (Ludwig et al. 2018).

2.3 Effects of Carbohydrates on Aging and Longevity

Carbohydrates, proteins, and fats, known as macronutrients, are sources of energy, yet they also have various other physiological functions (Carreiro et al. 2016). Recent studies show that the composition and balance of macronutrients have a greater effect on the extension of human life than does total energy intake (Mair et al. 2005; Solon-Biet et al. 2015a, b). U.S. Department of Health and Human Services and U.S. Department of Agriculture (2015) 2015–2020 Dietary Guidelines for Americans recently established a new approach to the recommended dietary intake of macronutrients. According to the 2015–2020 Dietary Guideline for Americans, the dietary reference intakes for carbohydrates, fats, and proteins are 45–65, 25–35, and 10–30% of calories, respectively (IOM 2015; Buyken et al. 2018). The Prospective Urban Rural Epidemiology (PURE) and the Atherosclerosis Risk in Communities (ARIC) studies, two large-scale and long-term follow up epidemiological studies, examined the effect of carbohydrates on mortality and lifespan. According to PURE study results, high carbohydrate intake is associated with a higher risk of mortality, whereas ARIC study results indicated a U-shaped association between carbohydrate intake and all-cause mortality, with 50–55% carbohydrate intake linked to the longest lifespan (Dehghan et al. 2017; Seidelmann et al. 2018). Another cohort study had parallel findings with the ARIC study, showing U-shaped association between carbohydrate intake and all-cause mortality along with finding that a 50–60% carbohydrate intake, which is accepted as a moderate intake of dietary carbohydrates, lowered all-cause mortality risk (Kwon et al. 2020). A meta-analysis comprising 17 cohort studies on dietary carbohydrate intake and mortality reported that low dietary carbohydrates is associated with an increased risk of mortality (Noto et al. 2013). Therefore, it can be suggested that both low carbohydrate intake of <40% and high carbohydrate intake of >70% both cause a rise in mortality risk compared to a moderate intake of 50–60%. In fact, these percentage ranges are dependent on dietary patterns in such a way that in low-carbohydrate diets, an increase in animal fat and protein intake is observed with the change in dietary pattern (Ludwig et al. 2018; Seidelmann et al. 2018). Specifically, it has been shown that a low carbohydrate diet causes an increase intake of animal fat and protein; thus, a decrease in carbohydrate intake leads to a decrease in vegetable protein and dietary fibre intake. Therefore, studies have drawn attention to the importance of the type of dietary fat and protein that should be consumed in a low-carbohydrate diet and concluded that while a low carbohydrate diet based on animal sources is associated with higher mortality risk, a vegetable-based low carbohydrate diet is associated with lower mortality risk (Ludwig et al. 2018; Seidelmann et al. 2018; Shamima et al. 2020). Some clinical studies have reported that short-term low-carbohydrate diets have a greater effect on weight loss than do low-fat diets. While it is thought that body weight loss may improve lifespan indirectly, in the long term, due to low compliance, the efficiency of this diet decreases (Hu 2015; Mancini et al. 2016; Mansoor et al. 2016). The DIETFITs randomized clinical study demonstrated that low-carbohydrate diets for weight loss did not provide a significant advantage over low-fat diets, yet suggested limiting sugar, refined grains, and processed foods in both diet types. Regarding the effect of carbohydrate type and amount on health outcomes, the study concluded that the type of carbohydrate is more important than the total amount of carbohydrate consumed (Gardner et al. 2018).

The Okinawan diet, shown to be among the healthiest eating models, emphasizes the importance of the amount of carbohydrates as well as the carbohydrate type on health outcomes. The longevity advantage of the Okinawan diet is primarily attributed to the healthy lifestyles of its adherents. That lifestyle includes a traditional diet which is energy-restricted but nutritionally dense, and especially rich in vitamins, minerals, and phytonutrients, many of which have nutraceutical potential (Willcox et al. 2014). In the traditional Okinawan diet, 85% of energy is provided by carbohydrates, 9% by protein, and 6% by total dietary fat (saturated fat 2%) (Willcox et al. 2007), whereas in the modern Okinawan diet, the energy derived from carbohydrates is about 58%, while 15% comes from protein and 28% from dietary fat (7% saturated fat). (Willcox et al. 2014). At this point, it is useful to know the principles and characteristics of the traditional Okinawan diet and to determine the dietary pattern of it. The typical characteristics of the Okinawan diet are as follows: low energy intake; high amount of vegetable consumption, especially root and green-yellow vegetables; high consumption of legumes, especially soy; reasonable consumption of fish; low consumption of meat and dairy products; low fat intake; mostly low glycemic index carbohydrates and high fibre foods; and moderate alcohol consumption (Willcox et al. 2014). It can be observed that the Okinawan diet has similar constructive effects on health as the Mediterranean diet; both diets underline the importance of moderate carbohydrate (50–60%) consumption of the type of carbohydrates consumed. Although the amount of carbohydrates is significant, the importance of the carbohydrate type on health outcomes should be emphasized (Willcox et al. 2014; Seidelmann et al. 2018; Ekmekcioglu 2020). This finding is supported by two meta-analyses indicating that the risk of all-cause mortality decreased with a Mediterranean-type diet comprising moderate carbohydrate consumption (50–60%) and high fibre intake via vegetables and fruits as well as whole grain products as carbohydrate sources (Bonaccio et al. 2018; Eleftheriou et al. 2018).

As a result, both low and high carbohydrate intake was associated with mortality risk, while the minimum risk of mortality was found to be linked with 50–60% carbohydrate intake. It has been reported that compared to a high carbohydrate diet, low-carbohydrate diets are associated with a higher risk of death due to the increased intake of animal-derived proteins and the accompanying increase in dietary fat sources. However, it has also been shown that protein and dietary fats are associated with lower mortality rates when obtained from plant sources such as vegetables, nuts, and whole grains. In this case, it is thought that the type of carbohydrates derived from staple foods significantly affects mortality rate (Virtanen et al. 2019; Chen et al. 2020).

2.4 Relation Between Carbohydrate Types, Health, and Longevity

2.4.1 Dietary Fibre

Nutritional interventions and dietary patterns have a very critical role in the management of aging. Dietary fibre is known as non-digestible complex carbohydrates. Many clinical and epidemiological studies have shown that dietary fibre, especially at higher intakes, reduces many age-related diseases and therefore the risk of mortality. Studies have confirmed that dietary fibre intake causes a decrease in inflammatory and oxidative stress parameters, and this improves health in the aging process and prolongs lifespan (Huang et al. 2015; Zhang et al. 2018; Reynolds et al. 2019). In fact, an insufficient amount of dietary fibre has been proven to speed up the aging process. A substantial intake of dietary fibre is an essential element of healthy aging (Abdolghaffari et al. 2020). Fibre increases the feeling of satiety through various mechanisms and also regulates carbohydrate and lipid metabolism. In the large intestine, fibre is almost completely broken down by the resident microbiota in a series of anaerobic reactions known as fermentation. The gut microbiota play an important role and improve human health, including protection against pathogens, enhancement of the immune system, vitamin synthesis, and metabolism of xenobiotics (Alasmar et al. 2019). Whole grain products, one of the greatest food sources of fibre, are made up of a variety of components including germ, endosperm, and bran. The bran and germ contain phytochemicals, while the endosperm contains a high amount of starch (Zhang et al. 2018; Reynolds et al. 2019). It is thought that whole grain products, with high fibre, mineral, and antioxidant content—which certainly advance human health—provide individuals a means to maintain a healthy body weight, improve metabolism, and suppress inflammation (Huang et al. 2015; Zhang et al. 2018).

A cohort study has indicated that a fibre-rich diet reduces all-cause mortality (Dominguez et al. 2019). In addition, some studies have shown that especially high-fibre diets have a potential to decrease the risk of cardiovascular disease, stroke, type 2 diabetes mellitus, and some type of cancers (Aune et al. 2012; Zhang et al. 2013a, b; Yao et al. 2014; Micha et al. 2017; Song et al. 2018). Mechanisms by which high fibre consumption may reduce the risk of these diseases include: reducing postprandial glucose responses (Gibb et al. 2015); prevention of obesity through an increased feeling of satiety (Emilien et al. 2017); decreasing intestinal absorption of cholesterol (Hartley et al. 2016); reducing inflammatory parameters such as C-reactive protein, interleukin-6, and tumor necrosis factor-alpha (Krishnamurthy et al. 2012; Jiao et al. 2015; Gibb et al. 2015; Monfort-Pires and Ferreira 2016; Emilien et al. 2017, Hartley et al. 2016). Recent years have brought emphasis on another mechanism, the fermentation of dietary fibre by gut microbiota. Due to the fermentation of dietary fibre, short chain fatty acids (SCFAs) are produced; thus, those SCFAs act as anti-inflammatory and immunomodulatory agents (Sonnenburg and Backhed 2016; Donovan 2017). Huang et al. (2015) also found higher dietary fibre intake to be associated with lower all-cause mortality and cardiovascular disease, cancer, and ischemic heart disease mortality. For each 10 g per day increase in dietary fibre intake, mortality rate was lowered by 11% for all-cause mortality, by 17% for cardiovascular disease, by 9% for cancer, by 20% for coronary heart disease, and by 34% for ischemic heart disease. Dominguez et al. (2019) showed an inverse relationship between total dietary fibre intake and overall mortality risks, determining that each additional intake of 5 g/1000 kcal of dietary fibre was related with a 9% reduction in all-cause mortality. They also examined the sources of dietary fibre on mortality risk as well as the impact of soluble and insoluble fibre on the risk of all-cause mortality. In terms of dietary fibre sources, while a significant inverse relation was observed with the fibre derived from vegetables, there was no significant relation found between fibre derived from fruits, legumes, cereals, or other sources. Comparing soluble fibre versus insoluble fibre, they showed inverse significant associations for both soluble and insoluble dietary fibre and all-cause mortality, with a slightly stronger relation for soluble fibre (Dominguez et al. 2019).

The systematic review and meta-analysis of prospective cohort and randomized controlled trials on carbohydrate quality and health which had the largest dataset indicated that dietary fibre intake exhibits 15–30% reduction in all-cause and cardiovascular-related mortality. Correspondingly, from the prospective cohort data, the incidence of coronary heart disease, stroke and mortality, type 2 diabetes, and colorectal cancer were also reduced by 15–30% when comparing the highest dietary fibre consumers with the lowest consumers. It has been demonstrated that important, positive health consequences were greatest when daily fibre consumption was between 25 and 29 g. The meta-analysis also reported that higher fibre intake led to even more positive results in protection from cardiovascular diseases, type 2 diabetes, and colorectal cancer. Dose–response data showed that above 30 g per day conferred further benefits (Reynolds et al. 2019). Kim and Je (2014) also indicated that total mortality risk was 23% lower in the highest category of dietary fibre intake (average 26.9 g/day) compared to lowest (15 g/day) intake. In addition, in terms of dietary sources of fibre, they specified that vegetable and, especially, cereal fibre were meaningfully linked with lower total mortality while fruit fibre did not show any effect.

2.4.2 Added and Free Sugar

Added sugars are sugars added to foods during processing or preparation. They include corn syrup, dextrose, fructose, glucose, sucrose, and high-fructose corn syrup. Free sugars include added sugars as well as sugars that occur naturally in juiced or pureed fruit and vegetables (Mela and Woolner 2018). The World Health Organisation (2015) recommends obtaining <10% of daily energy from free sugars intake. A dietary pattern which includes foods rich in sugar is associated with increased mortality (Kiefte-de Jung et al. 2014). A recent prospective cohort study conducted in the United Kingdom indicated a linear correlation between sugar intake and cardiovascular diseases, and mortality risk is lower with sugar intake accounting for 5–20% of energy compared to 35% (Ho et al. 2020). Another prospective cohort study among the Swedish population demonstrated a U-shaped correlation between added sugar intake and cardiovascular and cancer mortality, while added sugar intake at >20% of energy has been linked to a 30% increase in mortality. Added sugar sources including sugar-sweetened beverages are also related with increased mortality risk (Ramne et al. 2019). Moreover, in a United States cohort, sugar-sweetened beverages positively related with total, cardiovascular, and cancer mortality (Malik et al. 2019). A study among adults in the US demonstrated that cardiovascular disease mortality risk is increased 1.3-fold and 2.75-fold as a result of added sugar intake at 10–24.9% and ≥25% of energy, respectively (Yang et al. 2014). In addition, a retrospective cohort study observed that added sugar is most predictive of all-cause mortality (Shah et al. 2018). A meta-analysis of prospective cohort studies demonstrated a nonlinear association between total sugars, added sugars, fructose, and cardiovascular diseases mortality (Khan et al. 2019). Furthermore, it has been suggested that there is a positive association between added sugar intake, in particular through sugar-sweetened beverages, and obesity, type 2 diabetes, dyslipidemia, non-alcoholic fatty liver disease, and cardiovascular mortality (Haque et al. 2020). The reasons for these pathologies are explained by a term of carbotoxicity. Increased excess fructose intake is one of the featured reasons for carbotoxicity (Kroemer et al. 2018). Fructose is naturally contained in fruit, but excess intake is related with increased intake of high-fructose corn syrup. Excess fructose increases de novo lipogenesis and thus triglyceride and LDL levels in circulation. Also, it causes mitochondrial dysfunction, inflammation, and oxidative stress due to ATP depletion and increased uric acid production. Consequently, these mechanisms of action possibly lead to chronic disease such as obesity, diabetes, metabolic syndrome, gout, non-alcoholic fatty liver disease, and aging (Dornas et al. 2015; Hannou et al. 2018). To sum up, lowering added and free sugars intake can cause a decrease in non-communicable chronic diseases and mortality risk, and may retard aging by prevention of oxidative stress and inflammatory processes which induce aging mechanisms such as senescence and organelle dysfunctions.

2.5 Conclusion

In conclusion, a growing body of evidence and research shows that carbohydrate quality appears to have more important role in public health than the amount of carbohydrates consumed. Concerning the level of carbohydrate intake and health consequences, both high and low carbohydrate consumption were associated with increased mortality; thus, moderate (50–60% of total energy) intake is recommended. Also, carbohydrate type is related with non-communicable chronic disease risk and mortality. Consumption of added and free sugars is especially related to obesity, diabetes, and cardiovascular diseases. Sufficient intake of fibre-rich sources such as non-starchy vegetables, whole fruits, whole grains, and legumes generally works to reduce the alarming rate of chronic diseases and its derived mortality. Data suggests that both quality and quantity of carbohydrates notably modifies health, aging, and longevity.

Compliance with Ethical Standards

Conflict of Interest All authors declare they have no conflict of interest /or/ I have no conflict of interest.

References

Abdolghaffari AH, Farzaei MH, Lashgari NA, Roudsari NM, Haddadi NS, Singh AK et al (2020) Dietary fiber and aging. Nutrients and nutraceuticals for active and healthy ageing. Springer, Singapore, pp 111–145

Alasmar RM, Varadharajan K, Shanmugakonar M, Al-Naemi HA (2019) Gut microbiota and health: understanding the role of diet. Food Nutr Sci 10:1344–1373

Aune D, Chan DS, Greenwood DC, Vieira AR, Rosenblatt DA, Vieira R et al (2012) Dietary fiber and breast cancer risk: a systematic review and meta-analysis of prospective studies. Ann Oncol 23:1394–1402

Bonaccio M, Castelnuovo AD, Costanzo S, Gialluisi A, Persichillo M, Cerletti C et al (2018) Mediterranean diet and mortality in the elderly: a prospective cohort study and a meta-analysis. Br J Nutr 120:841–854

Buyken AE, Mela DJ, Dussort P, Johnson IT, Macdonald IA, Stowell JD et al (2018) Dietary carbohydrates: a review of international recommendations and the methods used to derive them. Eur J Clin Nutr 72:1625–1643

Burke LM (2000) Nutrition in sport. In: Maughan RJ (ed) Dietary carbohydrates, 1st edn. Blackwell Science, London, pp 73–84

Carreiro AL, Dhillon J, Gordon S, Higgins KA, Jacobs AG, McArthur BM et al (2016) The Macronutrients, appetite, and energy intake. Annu Rev Nutr 36:73–103

Chen Z, Glisic M, Song M, Aliahmad AM, Zhang X, Moumdjian AC et al (2020) Dietary protein intake and all cause and cause specifc mortality: results from the Rotterdam Study and a meta analysis of prospective cohort studies. Eur J Epidemiol 35:411–429

Couteur DG, Biet SS, Cogger VC, Mitchell SJ, Senior A, Cabo R et al (2016) The impact of low-protein high-carbohydrate diets on aging and lifespan. Cell Mol Life Sci 73:1237–1252

Couteur DG, Solon-Biet S, Cogger VC, Mitchell SJ, Senior A, Cabo R et al (2016) The impact of low-protein high-carbohydrate diets on aging and lifespan. Cell Mol Life Sci 73:1237–1252

Dehghan M, Mente A, Zhang X, Swaminathan S, Li W, Mohan V et al (2017) Associations of fats and carbohydrate intake with cardiovascular disease and mortality in 18 countries from five continents (PURE): a prospective cohort study. Lancet 390:2050–2062
Dominguez LJ, Rastrollo MB, Toledo E, Gea A, Fresán U, Barbagallo M et al (2019) Dietary fiber intake and mortality in a Mediterranean population: the "Seguimiento Universidad de Navarra" (SUN) project. Eur J Nutr 58:3009–3022
Donovan SM (2017) Introduction to the special focus issue on the impact of diet on gut microbiota composition and function and future opportunities for nutritional modulation of the gut microbiome to improve human health. Gut Microb 8:75–81
Dornas WC, de Lima WG, Pedrosa ML, Silva ME (2015) Health implications of high-fructose intake and current research. Adv Nutr 6:729–737
Ekmekcioglu C (2020) Nutrition and longevity – from mechanisms to uncertainties. Crit Rev Food Sci Nutr 60(18):3063–3082
Eleftheriou D, Benetou V, Trichopoulou A, La Vecchia C, Bamia C (2018) Mediterranean diet and its components in relation to all-cause mortality: meta-analysis. Br J Nutr 120(10):1081–1097
Emilien CH, Hsu WH, Hollis JH (2017) Effect of resistant wheat starch on subjective appetite and food intake in healthy adults. Nutrition 43(44):69–74
Gardner CD, Trepanowski JF, Del Gobbo LC, Michelle EH, Joseph R, John PA et al (2018) Effect of low-fat vs low-carbohydrate diet on 12-month weight loss in overweight adults and the association with genotype pattern or insulin secretion: the DIETFITS randomized clinical trial. JAMA 319:667–679
Gibb RD, McRorie JW Jr, Russell DA, Hasselblad V, D'Alessio DA (2015) Psyllium fiber improves glycemic control proportional to loss of glycemic control: a meta-analysis of data in euglycemic subjects, patients at risk of type 2 diabetes mellitus, and patients being treated for type 2 diabetes mellitus. Am J Clin Nutr 102:1604–1614
Guasch-Ferré M, Babio N, Martínez-Gonzalez MA, Corella D, Ros E et al (2015) Dietary fat intake and risk of cardiovascular disease and all-cause mortality in a population at high risk of cardiovascular disease. Am J Clin Nutr 102:1563–1573
Hannou SA, Haslam DE, McKeown NM, Herman MA (2018) Fructose metabolism and metabolic disease. J Clin Invest 128(2):545–555
Haque M, McKimm J, Sartelli M, Samad N, Haque SZ, Bakar MA (2020) A narrative review of the effects of sugar-sweetened beverages on human health: a key global health issue. J Popul Ther Clin Pharmacol 27(1):e76–e103
Hartley L, May MD, Loveman E, Colquitt JL, Rees K (2016) Dietary fibre for the primary prevention of cardiovascular disease. Cochrane Database Syst Rev 2016:CD011472
Ho FK, Gray SR, Welsh P, Peterman-Rochan F, Foster H, Waddell H et al (2020) Associations of fat and carbohydrate intake with cardiovascular disease and mortality: prospective cohort study of UK Biobank participants. BMJ 368:m688
Hu FB (2015) Effect of low-fat diet interventions versus other diet interventions on long-term weight change in adults: a systematic review and meta-analysis. Lancet Diabetes Endocrinol 3:968–979
Huang T, Xu M, Lee A, Cho S, Qi L (2015) Consumption of whole grains and cereal fiber and total and cause-specific mortality: prospective analysis of 367,442 individuals. BMC Med 13:59
Hyung WP (2010) Longevity, aging, and caloric restriction: Clive Maine McCay and the construction of a multidisciplinary research program. Hist Stud Nat Sci 40(1):79–124
Jiao J, Xu JY, Zhang W, Han S, Qin LQ (2015) Effect of dietary fiber on circulating C-reactive protein in overweight and obese adults: a meta-analysis of randomized controlled trials. Int J Food Sci Nutr 66:114–119
Kenyon CJ (2010) The genetics of ageing. Nature 464:504–512
Khan TA, Tayyiba M, Agarwal A, Mejia SB, de Souza RJ, Wolever TMS et al (2019) Relation of total sugars, sucrose, fructose, and added sugars with the risk of cardiovascular disease: a systematic review and dose-response meta-analysis of prospective cohort studies. Mayo Clin Proc 94(12):2399–2414

Kiefte-de Jung JC, Mathers JC, Franko OH (2014) Nutrition and healthy ageing: the key ingredients. Proc Nutr Soc 73:249–259

Kim Y, Je Y (2014) Dietary fiber intake and total mortality: a meta-analysis of prospective cohort studies. Am J Epidemiol 180(6):565–573

Kirkpatrick CF, Bolick JP, Kris-Etherton PM, Sikand G, Aspry KE, Soffer DE et al (2019) Review of current evidence and clinical recommendations on the effects of low-carbohydrate and very-low-carbohydrate (including ketogenic) diets for the management of body weight and other cardiometabolic risk factors: a scientific statement from the National Lipid Association Nutrition and Lifestyle Task Force. J Clin Lipidol 13:689–711

Krishnamurthy VM, Wei G, Baird BC, Murtaugh M, Chonchol MB, Raphael KL et al (2012) High dietary fiber intake is associated with decreased inflammation and all-cause mortality in patients with chronic kidney disease. Kidney Int 81:300–306

Kroemer G, Lopez-Otin C, Madeo F, de Cabo R (2018) Carbotoxicity-noxious effects of carbohydrates. Cell 175:605–614

Kuki S, Imanishi T, Kobayashi K, Matsuo Y, Obana M, Akasaka T (2006) Hyperglycemia accelerated endothelial progenitor cell senescence via the activation of p38 mitogen-activated protein kinase. Circ J 70(8):1076–1081

Kwon YJ, Lee HS, Park JY, Lee JW (2020) Associating intake proportion of carbohydrate, fat, and protein with all-cause mortality in Korean adults. Nutrients 12:3208

Lee D, Hwang W, Artan M, Jeong DE, Lee SJ (2015) Effects of nutritional components on aging. Aging Cell 14(1):8–16

Lee D, Son HG, Jung Y, Lee SJV (2017) The role of dietary carbohydrates in organismal aging. Cell Mol Life Sci 74:1793–1803

Lee KP, Simpson SJ, Clissold FJ, Brooks R, Ballard JWO, Taylor PW et al (2008) Lifespan and reproduction in Drosophila: new insights from nutritional geometry. Proc Natl Acad Sci 105:2498–2503

Ludwig D, Frank H, Luc T, Jennie BM (2018) Dietary carbohydrates: role of quality and quantity in chronic disease. BMJ (Clin Res Ed) 361:k2340

Mair W, Piper MDW, Partridge L (2005) Calories do not explain extension of life span by dietary restriction in Drosophila. PLoS Biol 3:e223

Malik VS, Pan A, Koning LD, Schernhammer E, Willet WC, Hu FB (2019) Long-term consumption of sugar-sweetened and artificially sweetened beverages and risk of mortality in US adults. Circulation 139:2113–2125

Mancini JG, Filion KB, Atallah R, Eisenberg MJ (2016) Systematic review of the Mediterranean diet for long-term weight loss. Am J Med 129:407–415

Mansoor N, Vinknes KJ, Veierød MB, Retterstøl K (2016) Effects of low-carbohydrate diets v. low-fat diets on body weight and cardiovascular risk factors: a meta-analysis of randomized controlled trials. Br J Nutr 115:466–479

Martinez-Gonzalez MA, Martin-Calvo N (2016) Mediterranean diet and life expectancy; beyond olive oil, fruits, and vegetables. Curr Opin Clin Nutr Metab Care 19:401–407

McCay CM, Crowell MF, Maynard LA (1935) The effect of retarded growth upon the length of life span and upon the ultimate body size. J Nutr 10:63–79

Mckay JA, Mathers JC (2011) Diet induced epigenetic changes and their implications for health. Acta Physiol 202(2):103–118

Mela DJ, Woolner EM (2018) Perspective: total, added, or free? What kind of sugars should we be talking about? Adv Nutr 9:63–69

Micha R, Shulkin ML, Penalvo JL, Khatibzadeh S, Singh GM, Rao M et al (2017) Etiologic effects and optimal intakes of foods and nutrients for risk of cardiovascular diseases and diabetes: systematic reviews and meta-analyses from the Nutrition and Chronic Diseases Expert Group. PLoS One 12(4):e0175149

Micó V, Berninches L, Tapia J, Daimiel L (2017) NutrimiRAging: micromanaging nutrient sensing pathways through nutrition to promote healthy aging. Int J Mol Sci 18:915

Monfort-Pires M, Ferreira SRG (2016) Modification in a single meal is sufficient to provoke benefits in inflammatory responses of individuals at low-to-moderate cardiometabolic risk. Clin Nutr 35:1242–1250

Mortuza R, Chen S, Feng B, Sen S, Chakrabarti S (2013) High glucose induced alteration of SIRTs in endothelial cells causes rapid aging in a p300 and FOXO regulated pathway. PLoS One 8(1):e54514

No authors listed (1989) The atherosclerosis risk in communities (ARIC) study: design and objectives. The ARIC investigators. Am J Epidemiol 129:687–702

Noto H, Goto A, Tsujimoto T, Noda M (2013) Low-carbohydrate diets and all-cause mortality: a systematic review and meta-analysis of observational studies. PLoS One 8:e55030

Ramne S, Dias JA, González-Padilla E, Olsson K, Lindahl B, Engström G et al (2019) Association between added sugar intake and mortality is nonlinear and dependent on sugar source in 2 Swedish population–based prospective cohorts. Am J Clin Nutr 109:411–423

Reynolds A, Mann J, Cummings J, Winter N, Mete E, Morenga LT (2019) Carbohydrate quality and human health: a series of systematic reviews and meta-analyses. Lancet 393:434–445

Seidelmann SB, Claggett B, Cheng S, Henglin M, Shah A, Steffen LM et al (2018) Dietary carbohydrate intake and mortality: a prospective cohort study and meta-analysis. Lancet Public Health 3:e419–e428

Shah NS, Leonard D, Finley CE, Rodriguez F, Sarraju A, Ce B et al (2018) Dietary patterns and long-term survival: a retrospective study of healthy primary care patients. Am J Med 131:48–55

Shamima A, Mizoue T, Nanri A, Goto A, Noda A, Sawada N et al (2020) Low carbohydrate diet and all cause and cause-specific mortality. Clin Nutr 40:2016–2024

Shukla AK, Kumar V, Sharma P, Choudhury B, Singh P, Kumar S (2017) Role of macronutrient in health. World J Pharm Res 6(3):373–381

Simpson SJ, Raubenheimer D (2005) Obesity: the protein leverage hypothesis. Obes Rev 6:133–142

Simpson SJ, Raubenheimer D (2012) The nature of nutrition: a unifying framework from animal adaptations to human obesity. Princeton University Press, Oxford

Solon-Biet SM, Mitchell SJ, De Cabo R, Raubenheimer D, Le Couteur DG, Simpson SJ (2015) Macronutrients and caloric intake in health and longevity. J Endocrinol 226:R17–R28

Solon-Biet SM, Walters KA, Simanainen UK, McMahon AC, Ruohonen K, Ballard JW et al (2015) Macronutrient balance, reproductive function, and lifespan in aging mice. Proc Natl Acad Sci 112:3481–3486

Song M, Wu K, Meyerhardt JA, Ogino S, Wang M, Fuchs CS et al (2018) Fiber intake and survival after colorectal cancer diagnosis. JAMA 4:71–79

Sonnenburg JL, Backhed F (2016) Diet-microbiota interactions as moderators of human metabolism. Nature 535:56–64

Speakman JR, Mitchell SE (2011) Caloric restriction . Mol Aspects Med 32:159–221

U.S. Department of Health and Human Services and U.S. Department of Agriculture 2015–2020 Dietary Guidelines for Americans. 8th Edition, December 2015. https://health.gov/our-work/food-nutrition/previous-dietary-guidelines/2015. Accessed 27 Jan 2021

Virtanen HEK, Voutilainen S, Koskinen TT, Mursu J, Kokko P, Ylilauri MPT et al (2019) Dietary proteins and protein sources and risk of death: the Kuopio Ischaemic heart disease risk factor study. Am J Clin Nutr 109(5):1462–1471

Willcox BJ, Willcox DC, Todoriki H, Fujiyoshi A, Yano K, He Q et al (2007) Caloric restriction, the traditional Okinawan diet, and healthy aging: the diet of the world's longest-lived people and its potential impact on morbidity and life span. Ann N Y Acad Sci 1114:434–455

Willcox DC, Scapagninid G, Willcox BJ (2014) Healthy aging diets other than the Mediterranean: a focus on the Okinawan diet. Mech Ageing Dev 136(137):148–162

World Health Organisation (2015) Guideline: sugars intake for adults and children. WHO, Geneva

Yang Q, Zhang Z, Gregg EW, Flander WD, Merritt R, Hu FB (2014) Added sugar intake and cardiovascular diseases mortality among US Adults. JAMA Intern Med 174(4):516–524

Yao B, Fang H, Xu W, Yan Y, Xu H, Liu Y et al (2014) Dietary fiber intake and risk of type 2 diabetes: a dose-response analysis of prospective studies. Eur J Nutr 29:79–88

Zhang B, Cui S, Bai X, Zhuo L, Sun X, Hong Q et al (2013) SIRT3 overexpression antagonizes high glucose accelerated cellular senescence in human diploid fibroblasts via the SIRT3-FOXO1 signaling pathway. Age 35(6):2237–2253
Zhang B, Zhao Q, Guo W, Bao W, Wang X (2018) Association of whole grain intake with all-cause, cardiovascular, and cancer mortality: a systematic review and dose–response meta-analysis from prospective cohort studies. Eur J Clin Nutr 72:57–65
Zhang Z, Xu G, Ma M, Yang J, Liu X (2013) Dietary fiber intake reduces risk for gastric cancer: a meta-analysis. Gastroenterology 145:113–120

Chapter 3
Fats and Oils for Health and Longevity

Kaustuv Bhattacharya and Suresh I. S. Rattan

Abstract Fats and oils are one of the three primary nutritional components, and they deliver essential fatty acids and lipids vitally important for the normal biological structure and functions. The physiological functions of lipids include storing and providing energy, acting as inter- and intra-cellular signaling molecules, dissolving some vitamins to make them bioavailable, and acting as the crucial structural components of the cell membranes. This article provides an overview of the chemical nature and composition of fats and oils, their biological functions, their metabolic pathways of synthesis and catabolism, and dietary guidelines for their consumption for a healthy and possibly a longer life.

Keywords Animal fats · Plant oils · Health · Signaling · Cell membrane · Lipidome · Vitamins

3.1 Introduction

Fats and oils are the third primary nutritional component after proteins and carbohydrates, and they deliver essential fatty acids (FA) vitally important for the normal biological structure and physiological functions. Fats have the highest caloric density among foodstuffs (9 kcal/g), and are also the solubilizers and carriers of vitamins A, D, E and K. Our intake of oils and fats is largely through cooking- and salad-oils; butter, margarines and other spreads; baked and fried products; dairy-products including milk, cheese, desserts, chocolate and sugar confectionery; and through culinary applications, such as mayonnaise and other dressings. Fats are also consumed as a part of the animal-based foods. All these sources make up a complex matrix

K. Bhattacharya
International Flavor & Fragrances, Brabrand, Denmark
e-mail: kaustuv.bhattacharya@iff.com

S. I. S. Rattan (✉)
Department of Molecular Biology and Genetics, Aarhus University, Aarhus, Denmark
e-mail: rattan@mbg.au.dk

S. I. S. Rattan and G. Kaur (eds.), *Nutrition, Food and Diet in Ageing and Longevity*, Healthy Ageing and Longevity 14, https://doi.org/10.1007/978-3-030-83017-5_3

of various visible and invisible oils and fats that end up in our body (Bhattacharya 2006).

Oils and fats contribute to both the textural-structural properties of the food, and organoleptic aspects including flavor, mouthfeel, palatability and appearance. Food additives such as emulsifiers and hydrocolloids play an ever-increasing role in providing the structural integrity in processed foods to deliver lubrication, enhance and stabilization of aeration, crumb structure etc., and also contribute to the shelf-life of such foods (Bhattacharya 2006).

The aim of this article is to provide an overview of the chemical nature and composition of fats and oils, their biological functions, their metabolic pathways of synthesis and catabolism, and dietary guidelines for their consumption for a healthy and possibly a longer life.

3.2 Terminology and Chemical Composition of Fats and Oils

Before describing the chemical nature and composition of fats and oils, it may be useful to clarify some terminology. FA are the basic unit of all fats and oils. Lipids are the compounds based on fatty acids or on closely related compounds, such as the corresponding alcohols or the sphingosine base, and are mainly, but not entirely, mixtures of triacylglycerols (often termed triglycerides). Collectively, these are known as oils or fats, depending on whether, at a room temperature of about 24 °C, they are liquid or solid, respectively. Often these terms are used interchangeably, and so one should be cautiously aware and precise about these.

The broad group of lipids can be subdivided into classes based on their chemical structure, for example diacylglycerols, phosphatidylethanolamines, ceramides and others. They can be further separated into individual molecular species based on their acyl groups. Fats and oils of animal and plant origin consist almost exclusively of triglycerides, which consist of a glycerol moiety with each hydroxyl group esterified to a fatty acid. Triacylglycerols are synthesized by enzyme systems, which determine that a centre of asymmetry is created about carbon-2 of the glycerol backbone, so they exist in enantiomeric forms, that is with different fatty acids in each position. The positions of the fatty acids in the glycerol backbone are denoted by sn-1 or sn-3, the two terminal positions and sn-2, the middle position. (The abbreviation 'sn' stands for 'stereospecific numbering'). The naturally occurring fatty acids are mainly straight-chain compounds containing an even number of carbon atoms, and can be grouped into three classes, (i) saturated; (ii) monounsaturated and polyunsaturated; and (iii) branched chain. Unsaturated fatty acids may contain one or more double or triple bonds and can be classified as monounsaturated, polyunsaturated, and acetylenic. FA composition and distribution of triacylglycerols can be altered via a process called interesterification for a specific ratio of n-3, n-6, n-9 and saturated FA (SFAs), which play an important role in lowering cholesterol and blood lipid levels leading (Akoh 2002).

3.2.1 Saturated Fatty Acids (SFA)

SFA are composed of straight-chain FA esterified to the glycerol backbone. Among the most common and commercially available vegetable oils, such as coconut oil, palm kernel oil and palm oil, the most common SFA are lauric (C12), myristic (C14), palmitic (C16) FA. Presence of stearic acid (C18) is quite low (often below 5%) in common vegetable oils and fats. Dairy fats contain C4 to C10 FA while longer chain SFA up to C38 are found in waxes (Bhattacharya and Rattan 2006).

3.2.2 Monounsaturated Fatty Acids (MUFA)

The most abundant MUFA in common liquid vegetable oils such as olive, canola, sunflower, peanut, and safflower is oleic acid, a FA with one unsaturation or one double bond at the 9th carbon atom from the methyl end. Oleic acid is the precursor of biosynthesis of omega-9 class of FA. There are also high oleic varieties of sunflower, canola and soybean oil developed via selective breeding and genetic modification techniques (Kristott 2003).

3.2.3 Polyunsaturated Fatty Acids (PUFA)

FA with two or more unsaturation or double bonds in the carbon chain are known as PUFA, and the two major PUFA families are linoleic acid (delta-9,12–18:2 omega-6), and alpha-linolenic acid (delta-9,12,15–18:3 omega-3). These two class of PUFA are considered as essential FA as they cannot be synthesized *in vivo* by humans, and so the dietary intake of linolenic and linoleic acids is commonly via canola, soybean, sunflower, flaxseed and corn oils. Once linoleic and linolenic acids are consumed, they undergo elongation and desaturation via enzymatic pathways into higher monologues such as arachidonic acid (AA, 20:4, omega-6), eicosapentaenoic acid (EPA, 20:5, omega-3) and docosahexaenoic acid (DHA, 22:6, omega-3). EPA and DHA are the most bioavailable forms of omega-3 for humans. (Murphy and Howe 2008).

3.2.4 Trans Fatty Acids (TFA)

Various health-detrimental effects of TFA have been extensively documented over the last two decades. The double-bond geometry of TFA is in the trans (E) configuration as the hydrogen atoms are placed on the opposite sides of the double bond (Hunter 1992; Hastert 1996). While dairy fats contain small amounts of naturally occurring

TFA, it is the industrially-produced TFA during the process of partial hydrogenation that has been studied the most. However, fully hydrogenated oils have less than 2% TFA and are often considered as zero trans. The use of partially hydrogenated oils have also been removed from food application in most parts of the world.

3.3 Physiological Functions of Fats and Oils

The physiological functions of lipids include storing energy, inter- and intra-cellular signaling, and being the structural components of the cell membranes, and are briefly discussed below. In addition, some vitamins, for example vitamins A, D, E and K, require fats for getting dissolved and becoming bioavailable as essential micronutrients stored in the liver and other fatty tissues. The major sources of energy storage in adipose tissue, both in animals and plants, are triglycerides. Lipases in the body regularly synthesize and break down the adipocytes or fat cells, and oxidation of FA provides high caloric content (about 9 kcal/g or 38 kJ/g), compared with 4 kcal/g or 17 kJ/g obtained from the breakdown of carbohydrates and proteins.

The role of FA in the composition of eukaryotic cell membranes is crucial. In animal cells, the intracellular components are physically separated from the extracellular environment by the plasma membrane. The main structural components of the cellular plasma membrane and the intracellular membranes of organelles are glycerophospholipids along with non-glyceride lipid components namely, sphingomyelin and sterols (Stryer et al 2007). It is often considered that the formation of lipids into protocell membranes was a key step in the origin of life (Segré et al 2001).

Another important physiological function of FA is the lipid-based signaling in cellular functioning (Wang 2004). Lipid signaling is initiated by the activation of G protein-coupled nuclear receptors, and several different lipid categories are identified as the signaling molecules acting as the messengers (Eyster 2007). Such signaling molecules include ceraminde-derived sphingosine-1-phosphate, which is involved in the regulation of calcium mobilization (Hinkovska-Galcheva et al 2008), cell growth, and apoptosis. Other examples of signaling molecules are diacylglycerols (DAG), phosphatidylinositol phosphates, prostaglandins and phosphatidylserines involved in the signaling for the phagocytosis of apoptotic cells or other broken components of cells (Saddoughi et al 2008).

3.4 Lipid Metabolism

Triglycerides, sterols and phospholipids from animals and plants are the main dietary sources. Lipid metabolism includes the production and the degradation of lipids inside the cells both for the production of energy and for the synthesis of structural and functional lipids. However, unlike long chain-PUFAs, very long chain PUFAs are generally not obtained from the ordinary dietary sources, and therefore need to be synthesized *in situ* from shorter FA precursors.

Lipids being hydrophobic molecules, require solubilization before being metabolised. In comparison, PUFA are more readily mobilized and oxidized than other fats. PUFA also have an influence on gene expression and appetite-controlling peptides, providing some protective value against obesity. The initial step of lipid metabolism is enzymatic hydrolysis of the triglycerides into its constituent FA in the digestive system. This is followed by the absorption of the FA into the epithelial cells of the intestinal wall where FA are packaged and transported to the rest of the body.

The processes of lipid digestion begin with lingual lipases starting the breakdown of dietary fats in the mouth. However, cholesterols from the food are not broken down by the lipases and generally remain intact until they reach the epithelium of the small intestine. The chemical digestion of fats continues by gastric lipases and the mechanical digestion, peristalsis, begins. In the small intestine, pancreatic lipases and bile salt-dependent lipases help breakdown the triglycerides along with further mechanical digestion. Triglycerides are finally converted into individual FA units and are absorbed into the small intestine's cells (Voet et al 2013).

Fat absorption is the next step in lipid metabolism. The triglycerides moieties FA and glycerol along with cholesterol, aggregate into colloidal structures (micelles) before diffusing across the membrane to enter the intestinal epithelial cells. Monoglycerides and FA resynthesize triglycerides in the cytosol of epithelial cells, forming clusters of bigger particles called chylomicrons which are amphipathic structures that transport digested lipids as they travel through the bloodstream to enter adipose and other tissues in the body (Jo et al 2016).

Transportation of lipids between organs in an aqueous environment is facilitated by lipoproteins, which are complexes of lipids with specific apoproteins. Such transportation process through blood is necessary because of the hydrophobic character of membrane lipids, triglycerides and cholesterols. The lipoprotein lipase breaks down the lipoproteins in the luminal surface of endothelial cells in capillaries, and triglycerides are released, which are then next split into FA and glycerol before entering the cells, and the remaining cholesterol travels through the blood to the liver. The glycerol is converted to glyceraldehyde 3-phosphate, and is subsequently oxidized for production of energy while long chain FA are converted to fatty acyl-CoA in order to pass across the mitochondria membrane (Feingold and Grunfeld 2000).

Triglyerides, membrane lipids and cholesterols can also be synthesized by the organisms through various pathways. Biosynthesis of glycerophospholipids and sphingolipids, the two main classes of membrane lipids occurs in the endoplasmic reticulum membrane. The first step is synthesis of sphingosine or glycerol as the backbone followed by the esterification of fatty acids to the backbone to make phosphatidic acid, which is further altered with the attachment of different hydrophilic head groups to the backbone (Gault et al 2010; Choe et al 2016). Triglyceride biosynthesis occurs in the cytosol via phosphatidic acid which acts as a precursor (Lok et al 1976). Cholesterol biosynthesis occurs in the cytosol of liver cells from acetyl-CoA through a multiple-step pathway known as isoprenoid pathway.

3.4.1 Sources of Lipids – Natural and Prepared Foods

Oils extracted from crops such as soybean, canola, sunflower, safflower, corn, palm, palm kernel, coconut, etc. are widely used in food items, including baked and fried products, margarines and spreads, chocolate and various other confectionery products, ice creams, salad dressings and mayonnaise. Individual oils may be used in these applications, or two or more oils may be blended in defined proportions with or without modification. Such wide array of oils and fats provide us with a diverse range of FA which fall into the categories as mentioned earlier.

Table 3.1 gives a comparative list of FA composition of common vegetable oils. Only by understanding the relative importance and the health-beneficial or health-damaging effects of various FA, one can choose among various sources according to one's needs and preferences.

3.5 Guidelines for Fat Intake

With the globalization of internet and social media, the way consumers obtain nutrition information has changed dramatically. This has also vastly contributed to conflicting information of uncertain and variable quality leading to confusion in many cases. One prominent example is the erroneous understanding and belief about dietary fats leading to its general avoidance.

For years, an emphasis of nutrition communication was to balance calorie intake and energy expenditure, and to decrease dietary fat intake. Reductions in total dietary fat were recommended to reduce saturated fat, trans fat with the overall aim of reducing calorie consumption. However, this resulted in an unintentional negative

Table 3.1 Typical major fatty acid composition of common vegetable oils

Plant source	C12:0 Lauric acid	C16:0 Palmitic acid	C18:0 Stearic acid	C18:1 Oleic acid	C18:2 Linoleic acid	C18:3 Linolenic acid
Coconut	46.5	9.2	2.9	6.9	1.7	
Palm oil	1	43.5	4.5	38.8	9.5	0.3
Sunflower		6	4.6	15.7	71.4	0.6
Canola		4.5	1.8	58	21	9.9
Cottonseed		22.9	2.5	17.5	54.5	0.5
Soybean		10.3	3.9	22.2	54.3	8.5
Olive		11	3	75	9.5	0.5
Safflower		6	4.6	15.7	71.4	0.6
Rice bran		20	2	42.5	31.5	
Corn		9.9	2	28.7	56.9	1.1

health consequences. Reduction of fat in industrially produced food often led to its substitution with refined carbohydrates and added sugars, which led to significant increases in total energy intake and obesity rates (Liu et al 2017). In addition, it led to avoidance of nutrient-dense foods rich in healthy unsaturated fats such as nuts, seeds, avocados and vegetable oils.

3.5.1 Current Recommendations for Dietary Fat Intake

Various national and international health organizations have put forward guidelines on dietary fat intake, as summarized in Tables 3.2 and 3.3. Dietary guidelines from the World Health Organization and other health organizations recommend a total fat intake between 20 and 35% of total calories. The minimum of 20% is to ensure adequate consumption of total energy, essential fatty acids, and fat-soluble vitamins and prevent atherogenic dyslipidemia which occurs with low-fat, high carbohydrate diets and increases risk of coronary heart disease. Table 3.2 shows the recommended percent of energy from fats as recommended by various organizations and institutions; and Table 3.3 lists the recommended levels of n-3 LC-PUFA as per various international organizations. These guidelines are as per 2017 (Liu et al 2017).

Table 3.2 Recommended percent of energy from fats and oils

Organization	Recommended percent of energy				
	Total	Saturated fats	Trans fats	n-6 PUFA	n-3 PUFA
World Health Organization	20–35%	< 10%	< 1%	2.5–9%	0.5–2%
Food and Nutrition Board, Institute of Medicine, WHERE?	20–35%	Limit	Limit	5–10%	0.6–1.2%
United States Department of Health and Human Services and United States Department of Agriculture		< 10%	Limit		
American Heart Association/American College of Cardiology		5–6%	Limit		

Adapted from (Liu et al 2017)

Table 3.3 Recommended n-3 LC-PUFA from various organizations

Organisation	n-3 LC-PUFA mg/day
UK Committee on Medical Aspects of Food Policy, 1994	200
British Nutrition Foundation, 1992	400–1000
American Heart Association, 2002	400–500
ISSFAL, 2004	500
North Atlantic Treaty Organization	800
Japanese Ministry of Health, Labor and Welfare, 1999	1600
France, CNERNA-CNRA; AFSSA, 2001	400–500
National Heart Foundation of Australia, 2008	500
Health Council of Netherlands, 2001	200

Adapted from (Murphy and Howe 2008)

3.6 Lipidome During Aging

More than 75% of the human metabolome is represented by lipids, and the complete set of all biological lipids is called lipidome (Ejsing et al 2009; Psychogios et al 2011). The plasma lipidome consists of thousands of lipids performing different functions and having different structures (Almeida et al 2021). Changes in dietary lipid intake during aging would influence the plasma or serum lipidome. For example, there is a correlation between the FA ingested in the diet and the phospholipids of the cell membranes and the TG of the adipose tissue and plasma (Abbott et al 2012). These studies also showed that whereas dietary SFA, MUFA and PUFA did not significantly influence membrane lipids, the composition of TG in adipose tissue and plasma appeared to be influenced in accordance with the dietary FA (Almeida et al 2021).

During ageing, there is a gradual decrease in the total intake of lipids. More specifically, there is an increase in the intake of SFA and a decrease in the intake of MUFA and PUFA (Almeida et al 2021). With regard to PUFAs, there appears to be a deficit in the intake of omega-3-PUFA, such as α-linoleic acid, eicosapentaenoic acid (EPA) and docosahexaenoic acid (DHA) in the elderly (Carrière et al 2007). This specific group of PUFA is extremely important as it has numerous benefits. For example, omega3-PUFA are associated with the maintenance of bone health and muscle tone, inhibition of TG synthesis in the liver, decrease of the inflammatory processes, and decrease in the cognitive performance decline associated with aging (Almeida et al 2021).v

Due to the central role that lipids play in the body, changes in their metabolic pathways can lead to the development of diseases, such as atherosclerosis, DM2, arterial hypertension, dyslipidemia, cardiovascular, and neurodegenerative diseases.

A comprehensive description of the age-related changes in the lipid profiles during normal ageing and pathological situations in old age is available in Almeida et al (2021). Currently, there are several dietary supplements marketed with the claims of increasing the plasma levels of omega3-PUFA (mainly EPA and DHA); and often, these omega3-PUFA used in supplements are derived from fish oils (Almeida et al 2021). However, to what extent such food supplements can slow down or reverse age-related changes in lipidome is presently not well demonstrated.

3.7 Conclusion

Our daily diet is made up of a complex mix of oils and fats from both vegetable, animal and marine sources. Fats and oils are essential for our health and survival at all ages, and no single category of FA can be labelled as good or bad in its entirety. Of course, any imbalance between the amounts of FA and the physiological requirements of the body can lead to health problems and several diseases. Therefore, a focus should be maintained on consumption of balanced intake between saturated, monounsaturated and polyunsaturated rich oils in accordance with the needs of the body at different stages of life. In future, more focus will be put on personalized diet that fits the individual blood lipid profiling (Almeida et al 2021).

In case of industrially-produced food products, the consumer has only a limited say in deciding the oils and fats used in the application. Selection of these oils and fats is based on their contribution to the texture, structure and organoleptic properties of the food item. Declaration on the nutritional labels and guideline recommendations as mentioned in the above tables could be useful. One should also consider proper protection of mono- and polyunsaturated oils against oxidation. Consumption of recommended calories from carbohydrates, proteins and oils and fats, regular physical activity, maintenance of proper food safety and personal hygiene all contribute to health and longevity.

Compliance with Ethical Standards

Conflict of Interest All authors declare they have no conflict of interest.

References

Abbott SK, Else PL, Atkins TA, Hulbert AJ (2012) Fatty acid composition of membrane bilayers: importance of diet polyunsaturated fat balance. Biochim Biophys Acta Biomembr 1818:1309–1317

Akoh CC (2002) Structured lipids. In: Akoh CC (ed) Food Lipids-Chemistry. Nutrition and Biotechnology, Marcel Dekker, Inc., New York, pp 877–908

Almeida I, Magalhães S, Nunes A (2021) Lipids: biomarkers of healthy aging. Biogerontology 22:273–296

Bhattacharya K (2006) Speciality oils and their applications in food. In: Gunstone F (ed) Modifying lipids for use in food. Woodhead Publishing, England, pp 539–566
Bhattacharya K, Rattan SIS (2006) Dietary fats and age-related diseases. In: Rattan SIS, Kassem M (eds) Prevention and Treatment of Age-related diseases. Springer, The Netherlands, pp 335–356
Carrière I, Delcourt C, Lacroux A et al (2007) Nutrient intake in an elderly population in southern France (POLANUT): deficiency in some vitamins, minerals and ω-3 PUFA. Int J Vitam Nutr Res 77:57–65
Choe SS, Huh JY, Hwang IJ, Kim JI, Kim JB (2016) Adipose tissue remodeling: its role in energy metabolism and metabolic disorders. Front Endocrinol 7:30
Ejsing CS, Sampaio JL, Surendranath V et al (2009) Global analysis of the yeast lipidome by quantitative shotgun mass spectrometry. Proc Natl Acad Sci USA 106:2136–2141
Eyster KM (2007) The membrane and lipids as integra participants in signal transduction: lipid signal transduction for the non-lipid biochemist. Adv Physiol Educ 31:5–16
Feingold KR, Grunfeld C (2000) Introduction to lipids and lipoproteins. In: De Groot LJ, Chrousos G, et al (eds) MDText.com, Inc., South Dartmouth (MA). NBK305896
Gault CR, Obeid LM, Hannun YA (2010) An overview of sphingolipid metabolism: from synthesis to breakdown. Adv Exp Med Biol 688:1–23
Hastert RC (1996) Hydrogenation. In: Hui YH (ed) Bailey's industrial oil & fat products, vol 3. John Wiley & Sons, New York, pp 212–300
Hinkovska-Galcheva V et al (2008) The role of sphingosine-1-phosphate and ceramide-1-phosphate in calcium homeostasis. Curr Opin Investig Drugs 9:1192–1205
Hunter JE (1992) Safety and health effects of isomeric fatty acids. In: Chow CK (ed) Fatty acids in foods and their health implications. Marcel Dekker, Inc., New York, pp 857–868
Jo Y, Okazaki H, Moon YA, Zhao T (2016) Regulation of lipid metabolism and beyond. Int J Endocrinol 2016:5415767
Kristott J (2003) High-oleic oils – how good they are for frying? Lipid Technol 15:29–32
Liu AG et al (2017) Dietary fat guidelines: a healthy approach to dietary fats: understanding the science and taking action to reduce consumer confusion. Nutr J 16:53
Lok CM, Ward JP, van Dorp DA (1976) The synthesis of chiral glycerides starting from D- and L-serine. Chem Phys Lipid 16(2):115–122
Murphy KJ, Howe RC (2008) Food sources and intake of Omega-3 fatty acids'. In: Watson R (ed) Fatty acids in health promotion and disease causation. AOCS Press, USA, pp 787–818
Psychogios N, Hau DD, Peng J, et al (2011) The human serum metabolome. PLoS One 6:1–23
Saddoughi SA, Song P, Ogretmen B (2008) Roles of bioactive sphingolipids in cancer biology and therapeutics. lipids in health and disease. Subcell Biochem 49:413–440
Segré D, Ben-Eli D, Deamer DW, Lancet D (2001) The lipid world. Orig Life Evol Biosph 31:119–145
Stryer L, Berg JM, Tymoczko JL (2007) Biochemistry, 6th edn. W.H. Freeman, San Francisco, pp 329–334
Voet D, Voet JG, Pratt CW (2013) Fundamentals of biochemistry: life at the molecular level, 4th edn. Wiley, Hoboken
Wang X (2004) Lipid signaling. Curr Opin Plant Biol 7:329–336

Chapter 4
Micronutrients in Ageing and Longevity

Božena Ćurko-Cofek

Abstract Ageing is a biological process that can be described as the accumulation of molecular damage to cells in response to stress. A tendency in modern society is to optimize longevity by trying to minimize a physical and mental decline and to decrease susceptibility to disease. There are some nutritional factors, including micronutrients, which can support this process of successful ageing. Micronutrients are vitamins (e.g., A, B group, C, D, E, folate) and minerals (e.g., copper, iron, magnesium, selenium, zinc), needed in small amounts but essential for healthy living. They have numerous significant functions in the organism as antioxidants, coenzymes, cofactors in metabolism, and genetic control. Therefore, micronutrients contribute to the normal functions of the immune, nervous, and endocrine system, protect the organism from oxidative stress, thus contributing to longevity and successful ageing. The dietary intake of micronutrients is usually insufficient in the elderly due to low income, reduced mobility, oral health problems, intestinal malabsorption, presence of chronic diseases, and changes in cognition. The result is an inadequate status of micronutrients which may contribute to suppressed immunity and consequent predisposition to infections, cognitive decline, neurodegeneration, development of the cardiovascular disease, disturbance of immune response and other health disorders.

Keywords Ageing · Immunosenescence · Inflammaging · Longevity · Micronutrients · Minerals · Neurocognitive disorders · Nutrition · Oxidative stress · Vitamins

4.1 Introduction

Term micronutrients is commonly used for vitamins and minerals needed in small amounts but vital for a healthy living. They have numerous functions in the organism as antioxidants, coenzymes, cofactors in metabolism, and genetic control. Among

---

B. Ćurko-Cofek (✉)
Faculty of Medicine, Department of Physiology and Immunology, University of Rijeka, B. Branchetta 20, 51000 Rijeka, Croatia
e-mail: bozena.curko.cofek@uniri.hr

S. I. S. Rattan and G. Kaur (eds.), *Nutrition, Food and Diet in Ageing and Longevity*, Healthy Ageing and Longevity 14, https://doi.org/10.1007/978-3-030-83017-5_4

the micronutrients, only vitamin D is produced in the body while the rest must be taken from the diet (Celep et al. 2017). The appropriate intake of the recommended amounts is important since micronutrients deficiency can have severe consequences for health.

The elderly population is particularly at risk of micronutrients deficiencies. They undergo many physiological changes which decrease their energy requirements. At the same time, the requirements for micronutrients are the same as for younger adults (Kehoe et al. 2019). There are also many other limiting factors in access and consummation of nutrient-reach food for the elderly. Some of them are low income, reduced mobility, oral health problems, and changes in taste, smell, and cognition (Marsman et al. 2018). Furthermore, in older age, most people have chronic diseases. The presence of disease can affect the need for calories and micronutrients (Marian and Sacks 2009). Medications used in the treatment of chronic diseases can be nutrient wasting and cause a decrease in the status of micronutrients essential for health (Marsman et al. 2018).

Micronutrient deficiency makes the older individuals a higher risk group in the body's immune response and increased susceptibility to infection (Bourke et al. 2016). Recent researches suggest that micronutrients participate in immune functions through several mechanisms. Those mechanisms are maintenance of physical barriers, production of antimicrobial proteins, and regulation of immune cells activity. Vitamins A, C, E, and the trace element zinc contribute to the enhancement of the skin barrier function. The vitamins A, B6, B12, C, D, E, folic acid, and the trace elements iron, zinc, copper, and selenium synergistically support the protective activities of the immune cells. All these micronutrients, except vitamin C and iron, are essential for antibody production. Therefore, the inadequate status of mentioned vitamins and trace elements may result in suppressed immunity and consequent predisposition to infections (Maggini et al. 2007).

Combined effects of ageing and insufficient intake of micronutrients contribute to cognitive decline, development of cardiovascular disease, disturbance of immune response and other health disorders. Uncontrolled use of dietary supplements can also have detrimental effects on health (Watson et al. 2018). Therefore, it is necessary to maintain an appropriate balance between the need for micronutrients and their intake, as shown in Table 4.1.

There is a tendency in modern society to optimize longevity by trying to minimize a physical and mental decline (Bowling and Dieppe 2005). This process of successful ageing depends on several factors. Some of them, such as genetic background, cannot be modified. In contrast, social, cultural, or lifestyle choices can be influenced (Porter et al. 2016). The ageing process can be described as the accumulation of molecular damage to cells in response to stress. Among other exogenous factors, there are some nutrition components that can modify the rate of damage (Sofi et al. 2008). For instance, the trace elements from the food can modulate the rate at which damage accumulates in the cells. Also, they influence metabolic pathways, such as oxidative and inflammatory processes, which alter during ageing (Meplan 2011). Therefore, we can say that dietary, with pharmacological and lifestyle interventions, may promote health and longevity.

Table 4.1 Estimated average requirements and tolerable upper intake level for selected micronutrients in the elderly (>65 years) (IOM 2006; IOM 2011)

Micronutrient	Estimated average requirements		Tolerable upper intake level
	males	females	
VITAMINS			
Vitamin A	625 μg/day	500 μg/day	3,000 μg/day
Vitamin B_1	1.0 mg/day	0.9 mg/day	ND*
Vitamin B_2	1.1 mg/day	0.9 mg/day	ND*
Vitamin B_6	1.4 mg/day	1.3 mg/day	100 mg/day
Vitamin B_{12}	2.0 μg/day		ND*
Folate	320 μg/day		1,000 μg/day
Vitamin C	75 mg/day	60 mg/day	2,000 mg/day
Vitamin D	10 μg/day (400 IU)		100 μg/day (4,000 IU)
Vitamin E	12 mg/day		1,000 mg/day
MINERALS			
Copper	700 μg/day		10,000 μg/day
Iron	6.0 mg/day	5.0 mg/day	45 mg/day
Magnesium	350 mg/day	265 mg/day	350 mg/day
Selenium	45 μg/day		400 μg/day
Zinc	9.4 mg/day	6.8 mg/day	40 mg/day

*ND = not determinable

4.2 Vitamins

Vitamins are organic substances that function as regulators in the body. They are divided into two groups: fat-soluble vitamins (vitamin A, D, E and K) and water-soluble vitamins (vitamin B1, B2, B6, B12, C, folic acid, etc.) (Celep et al. 2017).

4.2.1 Vitamin A

Vitamin A includes several fat-soluble substances like retinol, retinyl palmitate, and beta-carotene. There are two forms of vitamin A obtained through diet. From the animal sources, preformed vitamin A (retinol, retinyl ester) is derived while provitamin A (beta-carotenoid) is derived from plants. For the use in biologic processes, retinol and beta-carotenoid must be converted to biologically active forms, retinal and retinoic acid (Moise et al. 2007).

Vitamin A has a great impact on the ageing process due to its role in immune function and oxidative processes. Therefore, vitamin A deficiency is associated with a defective immune response to infection. The active form of vitamin A regulates immune cell differentiation and activates T cell responses. Recent researches are trying to enlighten the role of vitamin A in the enhanced T cell response in some diseases associated with ageing, such as cancer, infection, inflammation, and immune-mediated diseases (Raverdeau and Mills 2014).

The organ particularly susceptible to oxidative damage is an eye due to its exposure to light and high metabolism (Rasmussen and Johnson 2013). Age-related macular degeneration is one of the age-related degenerative diseases caused, among other factors, by high oxidative stress (Gorusupudi et al. 2017). It is the second most common cause of blindness after cataract in Europe (Bourne et al. 2018). Vitamin A is one of the most effective vitamins (together with vitamins C and E) for reducing the risk of macular degeneration. It plays an essential role in the human retinal pigment epithelial cells. Carotenoids lutein and zeaxanthin are concentrated in the macula and therefore known as macular pigments. They are the most potent antioxidants for the prevention and reduction of the risk of age-related macular degeneration. The human body is not able to synthesize lutein and zeaxanthin. Therefore, they must be obtained from the diet (Khoo et al. 2019).

The role of vitamin A in skin changes is also well studied. The ageing process promotes imbalance of collagen homeostasis resulting in the wrinkled appearance and atrophy of aged skin (Cole et al. 2018). The basis of this process is down-regulation of type I collagen accumulation and promotion of collagen degradation. Type I collagen is the major structural protein in the skin (Quan et al. 2011). Vitamin A and its metabolites promote new deposition of collagen and prevent its degradation by increasing type I procollagen and reducing matrix metalloproteinase-1 activity (Bielli et al. 2019).

Retinoids play a significant role in the development and normal functions of the human brain. Therefore, there is a great interest in potential therapeutic applications, especially for Alzheimer's disease (Das et al 2019). Alzheimer's disease is the most common neurodegenerative disease and the most common cause of dementia and loss of memory in old adults (Andreeva et al. 2017). Retinoids inhibit the expression of chemokines and neuroinflammatory cytokines in microglia and astrocytes, which are activated in Alzheimer's disease. Stimulation of retinoic acid receptors and retinoid X receptors slows down the accumulation of amyloids, reduces neurodegeneration, and thereby prevents pathogenesis of Alzheimer's disease (Das et al. 2019).

It is important to maintain the balance between the vitamin A intake and clearance from the organism since the elderly may have difficulty clearing it (Bolzetta et al. 2015). Some of the manifestations of chronic vitamin A excess are fatigue, hair loss, dry mucous membranes and skin, bone fractures, and abnormal liver function (Marian and Sacks 2009).

4.2.2 Vitamin B

B vitamins are a group of essential water-soluble vitamins which contribute to normal physiological and biochemical functioning of the body. B vitamins include B1 (thiamine), B2 (riboflavin), B3 (niacin), B5 (pantothenic acid), B6 (pyridoxamine), B7 (biotin), B9 (folate) and B12 (cobalamin). Vitamin B4 (choline) is not considered as part of the B vitamins complex (Mikkelsen and Apostolopoulos 2018).

B vitamins have significant roles throughout life, from childhood to old age (McNulty et al. 2019). As coenzymes, they participate in many enzymatic reactions and metabolic processes, play a crucial role in the methylation cycle, synthesis and repair of DNA and RNA, and maintenance of phospholipids. B vitamins are essential for the normal function of the immune and nervous system and for maintaining the cognitive functions (Mikkelsen and Apostolopoulos 2018).

Vitamin B deficiency within the ageing population can be caused by low intake (B2, B9), malabsorption (B12) or increased requirement with ageing (B6) (Porter et al. 2016). The ageing process affects absorption, transport, and metabolism of B vitamins within the body. Some other factors can also contribute to B vitamin deficiency such as interactions between drugs and nutrients, genetic disorders, and some medical conditions (Mikkelsen and Apostolopoulos 2018).

Several diseases are connected to vitamin B deficiency, such as pellagra, beriberi, and pernicious anaemia. Furthermore, vitamin B deficiency contributes to neurocognitive disorders (Mitchell et al. 2014), immune dysfunction, inflammation (Mikkelsen et al. 2017), liver damage, peripheral neuropathy, and anaemia (Mikkelsen et al 2016). In the older population, vitamin B deficiency is linked to cardiovascular disorders, cognitive dysfunction, osteoporosis, and methylation disorders (Porter et al. 2016).

4.2.3 Vitamin B1 (Thiamine)

Vitamin B1 is part of the coenzyme thiamine pyrophosphate (Mikkelsen and Apostolopoulos 2018). It has a crucial role in the normal function of the nervous system because it participates in the generation of nerve impulses and synthesis of neurotransmitters (Nemazannikova et al. 2017).

Deficiency of the vitamin B1 can cause neurological damage by the production of free radicals and increased oxidative stress in the brain and neuronal tissue (Liu et al. 2017). The result is axonal damage, inadequate myelin production, and glutamate-mediated excitotoxicity (Abdou and Hazell 2014). B1 deficiency also causes immune effects through the T-cell infiltration and increased production of pro-inflammatory cytokines, leading to neuroinflammation. Neuroinflammation affects mitochondrial function and increases oxidative stress which leads to endoplasmic reticulum stress

(Wang et al. 2017). All these factors are included in the pathogenesis of ageing-related diseases such as dementia, Alzheimer's disease, Parkinson's disease, and Huntington's disease (Mikkelsen and Apostolopoulos 2018).

4.2.4 Vitamin B2 (Riboflavin)

Vitamin B2 is a powerful antioxidant involved in numerous oxidation/reduction reactions in two coenzymatic forms, flavin mononucleotide and flavin adenine dinucleotide (Moore et al. 2018).

The free radical theory of ageing proposes aerobic metabolism as a cause of oxidative damage that accumulates in body cells and contributes to the ageing process (Wickens 2001). Vitamin B2 acts as an antioxidant by preventing the lipid peroxidation and attenuating reperfusion oxidative injury (Ashoori and Saedisomeolia 2014). Accordingly, riboflavin could be involved in prolonging the life span, but so far, scientific research did not establish a strong link between riboflavin and slowing the human ageing process.

Along with other B vitamins, riboflavin plays a role in slowing cognitive decline and possibly reducing the risk of depression in ageing (Moore et al. 2018). Therefore, B2 deficiency in older adults is linked to reduced cognitive outcome, depression, personality changes and distinct alterations within the central nervous system (Mikkelsen et al. 2016, 2016a).

4.2.5 Vitamin B6 (Pyridoxamine)

Vitamin B6 can be found in three forms: pyridoxine, pyridoxal and pyridoxamine. It has many essential functions within the endocrine, neurological and immune systems. A biologically active form of B6 is coenzyme pyridoxal-5 phosphate which, as a cofactor, aids the synthesis of the neurotransmitters serotonin, dopamine, epinephrine and GABA (Mikkelsen and Apostolopoulos 2018).

Pyridoxamine is an effective inhibitor of the formation of advanced glycation end-products. The pathological complications of diabetes are directly related to the effects of these nonenzymatic reactions between proteins and sugars (Ramis et al. 2019). The prevalence of diabetes is more than two times higher among elderly adults compared to middle age or young adults (Cowie et al. 2009), and the incidence in the older population is constantly growing (Narayan et al. 2006). A large portion of the diabetes-related cost involves treating diabetes-related complications (ADA 2013). Several studies (Degenhardt et al. 2002; Murakoshi et al. 2009; Stitt et al. 2002) have shown that pyridoxamine has therapeutic effects on various complications of diabetes. Vitamin B6 forms stable complexes with metal ions that catalyze the oxidative reactions taking place in the advanced stages of the protein glycation cascade (Adrover et al. 2008). It also reacts with reactive carbonyl compounds generated as by-products of protein glycation, thereby preventing further protein damage (Voziyan et al. 2002).

Since the advanced glycation end-products have been linked to the increased production of free radicals, they are implicated in causing tissue damage associated with ageing. Advanced glycation end-products can be deposited anywhere within the body and cause abnormal function of the organ or tissue, for example, cataract, arthritis, nephrosis and plaque formation within the vessel wall (Voziyan and Hudson 2005).

4.2.6 Vitamin B9 (Folate)

The term folates is commonly used for the entire family of compounds, naturally occurring and synthetic variants. Vitamin B9 (folate, folic acid) has a vital role in the normal functioning of the body. It is involved in immune response, brain function, DNA synthesis, and it is inevitable for cell division. Without folate, cells cannot divide and function (Craenen et al. 2020).

Folates play a key role in methylation reactions and DNA synthesis as one-carbon carrier/donor (Imbard et al. 2013). It has been known for years that folic acid food fortification and supplementation can reduce the prevalence of birth defects (Blom et al. 2006). Furthermore, it has been noticed that in healthy individuals plasma folate levels decrease while homocysteine levels increase with age, especially between the age of 40 and 90 years (Magnus et al. 2009). Increased homocysteine cytotoxicity has been conferred to have a role in ageing, neuronal plasticity, and neurodegenerative diseases such as Alzheimer's disease, Parkinson's disease, and Huntington's disease (Obeid and Herrmann 2006; Kalani et al. 2014).

Dementia and depression are considered to be major disorders of ageing (WHO 2016). Folate deficiency has been connected to cognitive decline and memory deficits. Apart from that, folate deficiency can affect the duration and clinical severity of depression. Also, it is associated with a weaker response to antidepressant medication (Reynolds 2006).

4.2.7 Vitamin B12 (Cobalamin)

Vitamin B12 is necessary for the normal function of the nervous system, nerve cell maintenance, cell proliferation and survival, and breakdown of fatty acids and amino acids (Mikkelsen et al 2016a).

Vitamin B12 (together with folate) is involved in the prevention of chronic diseases associated with ageing through the methylation of homocysteine. This process has a vital role in the prevention of amyloid and tau protein accumulation, which lead to cognitive decline (Watson et al. 2018).

An optimal vitamin B12 status does not depend only on adequate dietary intake, and deficiency of the vitamin B12 is not rare. An effective absorption is crucial, but it diminishes with age (Hughes et al. 2013). The absorption of food-bound B12 is difficult because of the decline in gastric acid secretion and the lack of intrinsic factor production in old age (Sawaengsri et al. 2016; Marsman et al. 2018). Deficiency leads to megaloblastic anaemia, demyelinating neurological symptoms, irreversible nerve damage and neuropathy (Sawaengsri et al. 2016). Furthermore, depression, mania, psychosis, and suicidal behaviours can develop (Petridou et al. 2016).

The risk of low B12 levels in elderly is increased by use of proton pump inhibitors and histamine H2 blockers. They inhibit gastric acid secretion (Maes et al. 2017) and interfere with the release of B12 from binding protein, thus reducing the absorption of vitamin B12 (Wong 2015). Since type-2 diabetes is frequent in the elderly, it should be kept in mind that Metformin (a medication used for the treatment of type-2 diabetes) reduces serum B12 levels and could worsen diabetes-associated neuropathy (Out et al. 2018).

Low vitamin B12 levels were found to increase blood pressure, contributing to cardiovascular diseases. Additionally, a decrease in vitamin B12 concentration can result in hyperhomocysteinemia, which could quadruplicate risk of stroke due to atrial fibrillation in elderly patients (Spence 2017). A recent study (Ao et al. 2019) reported that insufficiency of vitamin B12 and folate is a risk for decreased muscle strength, which increases the risk of falling and fractures.

Subclinical deficiency of vitamin B12 seems to be implicated in the development of several chronic age-related diseases. Evidence suggests that suboptimal levels of B12 can raise homocysteine which is associated with the risk of bone fractures (van Wijngaarden et al. 2013) and risk of cardiovascular diseases (Kehoe et al. 2019).

4.2.8 Vitamin C (Ascorbic Acid)

Vitamin C is involved in biological functions as a potent antioxidant and radical scavenger. It protects the cell against oxidative stress-mediated by reactive oxygen species (ROS) and free radicals. Vitamin C is considered the most effective antioxidant in the plasma because of its water solubility and the wide range of ROS that it can scavenge (Harrison 2012).

Since vitamin C acts as a cofactor of the monooxygenase and dioxygenase enzymes, it is required for the synthesis of several crucial biomolecules (Morelli et al. 2020). Vitamin C-dependent enzymatic reactions are involved in the biosynthesis of collagen and cellular procollagen secretion (Kehoe et al. 2019). Also, they participate in the biosynthesis of L-carnitine, norepinephrine, epinephrine, and other molecules. Vitamin C improves the absorption of the non-heme iron, thus increasing the bioavailability of iron (Sourabh et al. 2019).

The ability to transport ingested vitamin C from the intestines into the blood is limited by the saturable sodium-dependent vitamin C transporter (SVCT1) in the gut and thus very high intakes and the use of supplements are often erroneously considered to be of greater benefit than they really are (Harrison 2012). It has been reported that ageing, oxidative stress, and inflammatory factors cause changes in the expression of SVCT 1 and SVCT2, transporters expressed in liver, brain, heart, chondrocytes, and osteoblast (Patterson et al. 2021).

Experimental studies on animals and human pointed to the important role of vitamin C in the synthesis, remodelling, and maintenance of the dermal extracellular matrix. Environmental and intrinsic oxidative stress decreases the natural cutaneous antioxidative mechanisms and induces pro-ageing signalling pathways and the accumulation of structural and functional changes, thus inducing intrinsic ageing and photoageing. The hallmarks of ROS-induced skin ageing are wrinkle formation, decreased resilience, pigmentation changes, telangiectasia, dehydration of the skin (Crisan et al. 2015). Vitamin C, as a natural antioxidant, is effective in preventing and treating skin ageing. It stimulates the barrier function of the endothelial cells, protects keratinocytes from UV radiation, and shows photoprotective effects (Barbosa and Kalaaji 2014). Topically applied vitamin C contributes to the maintenance of the collagen quantity and density at the dermal level and is also involved in strengthening the collagen fibres (Crisan et al. 2015).

Vitamin C is essential for wound healing. It decreases the expression of pro-inflammatory mediators and enhances the expression of various wound healing mediators. Leukocytes, which are the major players in wound healing, actively accumulate vitamin C (Mohammed et al. 2016).

Severe vitamin C deficiency which leads to scurvy has become rare in most countries. Low levels of vitamin C are much more common. They have been associated with high blood pressure, endothelial dysfunction, heart disease, atherosclerosis, and stroke (Morelli et al. 2020). Long-term vitamin C deprivation, evidenced by low plasma levels in men with normal diets, was linked to nervousness, depression, and emotional lability (Harrison 2012). Furthermore, studies have shown that patients with conditions, such as diabetes, COPD, chronic hypertension, and viral-induced sepsis, have decreased levels of serum and plasma vitamin C (Patterson et al. 2021).

4.2.9 Vitamin D

Vitamin D is a nutrient, a pro-hormone and steroid hormone with an important role in calcium and bone metabolism. The biologically inactive form of vitamin D, vitamin D3, can be obtained from dietary sources or dermal synthesis. D3 is converted in the liver and kidneys to the active metabolite calcitriol (Hill et al. 2018).

Optimal vitamin D status is necessary for mineral homeostasis, bone health, and function of skeletal muscles (Cianferotti et al. 2017). Vitamin D is important for immune response because it stimulates phagocytosis by macrophages and protects immune cells against apoptosis (Watson et al. 2018). As for the current Covid-19 pandemic, some studies offer evidence that vitamin D status may influence the severity of response to Covid-19 and that the prevalence of vitamin D deficiency in Europe coincides with Covid-19 mortality (Laird et al. 2020). Through its interactions with different types of cells, vitamin D may have several ways to reduce the risk of acute respiratory tract infections and COVID-19: reducing the survival and replication of viruses, reducing the risk of inflammatory cytokine production, increasing angiotensin-converting enzyme-2 concentrations, and maintaining endothelial integrity (Mercola et al. 2020).

Vitamin D intake and absorption drastically decreases with age due to various environmental and biological factors (Boucher 2012). Reduced sunlight exposure, nutrient intake, fat absorption, and conversion to its active form (Watson et al. 2018) contribute to more common and severe hypovitaminosis D in older people (Boucher 2012). It has been reported that over 90% of older adults in most European countries have intakes below 10 μg/day, which is an average daily requirement for vitamin D (Kehoe et al. 2019). Impaired mobility and residential care often limit sun exposure and cause a decrease in the synthesis of vitamin D in the skin while a decline in renal function reduces vitamin D activation (de Jongh et al. 2017).

Regarding the ageing, vitamin D deficiency can be associated with impaired cognition, depression, cancer, and cardiovascular disease (Watson et al. 2018). For sure, the most significant impact of deficiency is on bone metabolism. For people over 70 years of age, the predominant bone health concerns are bone loss and the resulting osteoporotic fractures (Pfortmueller et al. 2014). Osteoporosis is a silent disease, and it is responsible for about 9 million fractures annually worldwide (Johnell and Kanis 2006). A meta-analysis conducted by the National Osteoporosis Foundation found that calcium plus vitamin D supplementation resulted in a statistically significant 15% reduced risk of total fractures and 30% reduced risk of hip fractures (Weaver et al. 2016). Supplementation may reduce the number of subsequent fractures, enhance muscular strength, and improve balance (Childs et al. 2016).

When taking vitamin D supplements, it must be careful not to cause an excess of the vitamin. Consequences of vitamin D overuse are acute kidney injury and pancreatitis, secondary to hypercalcemia and hypercalciuria (Razzaque 2018).

4.2.10 Vitamin E (Alpha-Tocopherol)

Vitamin E includes a group of eight structurally related antioxidants: four tocopherols and four tocotrienols, each designated as α, β, γ and δ. Among them, α-tocopherol is the most abundant and bioavailable antioxidant form of vitamin E in human tissues (Rigotti 2007).

Vitamin E has an important function as an antioxidant, protects cells from oxidative stress caused by ROS (La Fata et al. 2014). ROS are mainly produced in mitochondria and represent important regulators of cell signalling. At high concentrations, they are harmful and cause cellular damage contributing to the ageing process (Bratic and Larsson 2013). It is known that older adults, over 65 years of age, have less effective enzymatic antioxidant defence and compromised immune and inflammatory responses. This result in an increased risk of infectious and non-infectious chronic diseases in the elderly (Meydani et al. 2018).

The brain is highly susceptible to oxidative stress. For this reason, oxidative stress which increases during ageing is considered a major contributor to neurodegeneration. It is involved in the onset of pathological conditions typical of old age, such as Alzheimer's disease and dementia. Vitamin E could help in these conditions since its high plasma levels were repeatedly associated with better cognitive performance (La Fata et al. 2014). Due to the lack of appropriate treatment for Alzheimer's disease, vitamin E, because of its relative safety and low cost, could be a nutritional compound to promote healthy brain ageing and delay Alzheimer's disease-related functional decline (La Fata et al. 2014).

Vitamin E also has important non-antioxidant functions. The immunomodulatory role of vitamin E includes lymphocyte proliferation, prostaglandin E2 production, gene transcription, translation, cell membrane, and signal transduction (Zingg and Azzi 2004). Therefore, some of the age-related dysregulations of the immune and inflammatory responses have shown improvement due to vitamin E supplementation (Wu and Meydani 2008).

4.3 Minerals

Minerals are essential nutrients divided into two major groups. Major or macrominerals (calcium, phosphorus, sodium, potassium, magnesium, and chloride) are present in the body at levels greater than 0.01% and required in amounts greater than 100 mg/day. Therefore, their function is both structural and regulatory. Trace or micro-minerals (iron, cobalt, chromium, copper, fluoride, iodine, manganese, selenium, zinc, and molybdenum) are present in the body at levels less than 0.01% and required in amounts less than 50 mg/day. Trace minerals have primarily regulatory role in metabolic and immune functions (Celep et al. 2017).

Trace elements are involved in numerous important processes altered during ageing, such as immune function, oxidative stress, insulin sensitivity, and cognitive function. These elements are vital to enzymatic activity, free radical scavenging, modulation of oxidative damage, DNA repair capacity, and protein functions (Meplan 2011). As with other nutrients, conditions that are common in the elderly, such as malnutrition, interaction with medications, and reduced digestion and absorption, can contribute to reduced intake of the trace elements. Low or inadequate intake of trace elements can affect the ageing process and has been linked to disease development in the older age (Roussel 2002).

4.3.1 *Copper*

Copper (Cu) is an essential trace element to human health because of its involvement in a wide range of biological functions, especially in a central nervous system. It is an integral component of various enzymes, such as cytochrome C oxidase, lysyl oxidase, superoxide dismutase, dopamine β-oxidase, catalase (Zheng and Monnot 2012). Reduction in the Cu concentration affects the activity of these enzymes and consequently decreases the capacity of the organism to reduce oxidative damage or enhance repair capacity (Mocchegiani et al. 2014).

Copper deficiency has also been linked to the appearance of osteoporosis. Lysyl oxidase is a copper-dependent enzyme that catalyzes the formation of aldehydes from lysine residues in collagen and elastin precursors. In the case of lower Cu intake, the reduction of lysyl oxidase activity occurs and leads to bone mass loss, called osteopenia of copper deficiency (Arredondo et al. 2018).

As an unbound metal ion, Cu participates in the metabolism of neurotransmitters and nerve myelination (Zheng and Monnot 2012). Free Cu ions are present in an organism in a very low concentration to prevent the possibility of inducing highly reactive free radicals (Zoroddu et al. 2019). Due to its chemical reactivity, both a deficiency and an excess of Cu can be detrimental to the central nervous system (Zheng and Monnot 2012). An increase of the circulating Cu which is not bound to ceruloplasmin can contribute to the development of neurodegenerative diseases in the ageing population, such as Alzheimer's disease, Parkinson's disease, amyotrophic lateral sclerosis, dementia with a vascular origin, spongiform encephalitis (Creutzfeld-Jakob disease), and Huntington's disease (Squitti et al. 2014; Cerpa et al. 2005).

4.3.2 *Iron*

Iron (Fe) is an essential trace element inevitable for many biological functions such as oxygen transport, DNA synthesis, and mitochondrial oxidation (Ćurko-Cofek et al. 2017). Therefore, it has an important role during ageing maintaining the immune and antioxidant function and cognitive abilities (Mocchegiani et al. 2012). However, iron homeostatic mechanisms change during physiological ageing and create the basis for iron deficiency or increased body iron stores (Fairweather-Tait et al. 2014).

Anaemia and iron deficiency are two of the most prevalent disorders worldwide. According to the WHO, anaemia affects more than 25% of the population, while iron deficiency is even more prevalent. Chronic inflammation has an important role in the impairment of iron status (Dao and Meydani 2013) and its effects can be further enhanced by malnutrition (Fairweather-Tait et al. 2014). Both, chronic inflammation and malnutrition, are often present in the elderly, making this population susceptible to the development of iron deficiency. Iron deficiency in older adults has been associated with a decline in physical performance, cognitive impairment, increased susceptibility to falling and frailty, and mortality (Price et al. 2011). In contrast, iron overload is mainly associated with pathological conditions, but it could also be relevant during ageing (Grubić Kezele and Ćurko-Cofek 2020). In older adults, body iron levels can be elevated due to consumption of iron supplements or vitamin C, which enhances non-heme iron absorption (Fleming et al. 2002). Previous researches associate high body iron with coronary heart disease (Hunnicutt et al. 2013), type 2 diabetes (Bao et al. 2012), and risk of cognitive impairment (Penke et al 2012).

During ageing, iron accumulates in different brain regions, creating an imbalance between ROS production and antioxidant defence. Oxidative damage caused by the accumulation of iron in the brain increases the susceptibility of the aged brain to disease and could be the reason why ageing is the major risk factor for neurodegenerative diseases (Ashraf et al. 2018).

4.3.3 Magnesium

Magnesium (Mg) has great physiological importance as a cofactor for numerous biological processes. It participates in oxidative phosphorylation, energy production, protein synthesis, glycolysis, nucleic acid synthesis and stability (Dominguez et al. 2021). Through its role in the active transport of ions across cell membranes, Mg modulates neuron excitability, muscle contraction, and normal heart rhythm (Barbagallo and Dominguez 2013). Furthermore, Mg is involved in immune response participating in signalling pathways that regulate the development, homeostasis, and activation of immune cells (Tam et al. 2003).

Total body Mg and Mg in the intracellular compartment tend to decrease with age (Barbagallo et al. 2009). According to the published data, 73% of older men and 41% of older women in the Western world have inadequate Mg intake (ter Borg et al. 2015). Other frequent causes of Mg deficits in the elderly are reduced Mg intestinal absorption and bone stores, and the excess urinary loss (Barbagallo et al. 2009). Low Mg intake and low Mg body levels are associated with chronic conditions usual in the elderly, such as high blood pressure, type-2 diabetes, osteoporosis (Marsman et al. 2018), and low-grade chronic inflammation. This type of chronic inflammation involves several tissues and organs, which are frequently associated with multiple chronic diseases, and that has been named "inflammaging" (Franceschi and Campisi 2014). The fact that chronic Mg deficiency is frequent in old age (Barbagallo and Dominguez 2013) could be significant in a time of COVID-19 pandemic. Namely, Mg deficiency creates an appropriate microenvironment for the virus to promote thromboembolism (Iotti et al 2020), the main feature of COVID-19.

In contrast, use of Mg supplementation reduces plasma C-reactive protein concentrations (Simental-Mendia et al 2017) and significantly lower blood pressure in those with insulin resistance, prediabetes, and other chronic diseases (Dibaba et al 2017). Also, it has been shown that Mg supplementation improves glucose metabolism and insulin sensitivity in type-2 diabetes (Gommers et al 2016).

The consequences of Mg imbalance in the elderly may contribute to the ageing process itself. Namely, Mg is required to maintain intracellular genomic stability, and it is an essential cofactor in almost all enzymatic systems involved in DNA processing (Hartwig 2001). DNA is continuously damaged either by environmental mutagens or by endogenous processes. Mg has a significant role in removing DNA damage caused by any of these causes (Barbagallo and Dominguez 2010). The data from research (Killilea and Maier 2008) have shown that Mg deficiency may accelerate cellular senescence in the cultured human fibroblasts with a loss of replicative capacity and accelerated expression of senescence-associated biomarkers.

4.3.4 Zinc

Zinc (Zn) is involved in many homeostatic mechanisms as a structural and regulatory catalyst ion for many enzymes, proteins, and signal transcription factors, including cell proliferation, genome stability, and the immune system efficiency (Cabrera 2015). The immune system is specially affected by the ageing process, and therefore the term immunosenescence is often used in this context. Immunosenescence is characterized by a progressive abnormal regulation of immune responses, both innate and adaptive. The result is low-grade inflammation, susceptibility to infections (Alonso and De la Fuente 2011), and lower efficacy of vaccines (Lang et al 2011). There are remarkable similarities in the immunological changes during ageing and Zn deficiency. In the case of Zn content decrease, immune cells show a decline in function. There is a decrease in monocyte cytotoxicity, reduced phagocytosis in neutrophils, increased apoptosis in B cells, deterioration of T cells functions, but also an increase in autoreactivity (Ibs and Rink 2003). Zn deficiency shows the closest link to ageing and immunosenescence through the oxidative inflammatory ageing (oxi-inflamm-ageing) process. The oxi-inflamm-ageing theory associates oxidative stress with ageing effects, particularly on the nervous, endocrine, and immune cells, which have a regulatory systems function (De la Fuente and Miguel 2009). Zn deficiency increases oxidative stress and causes the generation of inflammatory cytokines, such as IL-1β, IL-2, IL-6 and TNF-α (Cabrera 2015). Together, immunosenescence and inflamm-ageing contribute to most of the diseases of the elderly, such as infections, cancer, autoimmune disorders, and chronic inflammatory diseases. However, recent data suggest that these two processes are not only detrimental but also adaptive and remodelling, and therefore may be needed for extended survival/longevity (Fulop et al 2018). Since many studies confirm a decline of Zn levels with age, bearing in mind the important role of Zn in immunity, oral Zn supplementation has the potential to improve immunity and downregulate chronic inflammatory responses in the elderly (Haase and Rink 2009).

4.3.5 Selenium

Selenium (Se) is a trace element essential for human health. However, the safe range of exposure to Se is generally narrow, and both Se deficiency and excess can be harmful to human health (Garcia-Esquinas et al 2021). Se is an essential component of several major metabolic pathways, including thyroid hormone metabolism, antioxidant defence systems, and immune function. The biological functions of Se are exerted by selenoproteins in which Se is incorporated in the form of the amino acid selenocysteine (Kryukov et al 2003). Selenoproteins have antioxidant effects and are involved in regulating antioxidant activities. Reactive oxygen species are initial factors in ageing and ageing-related diseases. Therefore, Se alleviates ROS-mediated processes, such as inflammation or DNA damage (Cai et al 2019). Skeletal muscle is one of the major sites of Se storage (30–45% of the total pool), and several selenoproteins are involved in muscular function. Other potential health benefits of Se status in older populations are reduced risk of immune dysfunction, cognitive decline, cardiovascular disease, certain tumours, and overall mortality (Garcia-Esquinas et al 2021). However, there are still many contradictions regarding the role of Se in longevity and ageing-related diseases, as well as about Se supplementation. In the case of Se excess, Se is non-specifically incorporated into proteins other than selenoproteins, changing protein structures and thereby affecting their function (Cai et al 2019).

4.4 Conclusion

Micronutrients are vital for human health, especially in the ageing process. They support the function of immune, nervous, and cardiovascular system, protect cells from oxidative stress and enable call division, reduce cognitive decline. The best way for the elderly to maintain the quality of life and prevent disease is a healthy diet and a healthy lifestyle. Diet should be rich in fruits, vegetables, whole grains, and lean proteins and within micronutrients DRI values recommended by IOM. However, many factors in the older population influence dietary intake and cause undernutrition. In that case, oral supplements are needed but should be used with caution. Potential drug-nutrient interaction and excess intake should be avoided. Therefore, the best way is to adjust the intake to the actual needs of the organism determined by nutrition assessment.

Compliance with Ethical Standards

Conflict of Interest The author declares no conflict of interest.

References

Abdou E, Hazell AS (2014) Thiamine deficiency: an update of pathophysiologic mechanisms and future therapeutic considerations. Neurochem Res 40:353–361. https://doi.org/10.1007/s11064-014-1430-z

Adrover M, Vilanova B, Frau J et al (2008) The pyridoxamine action on amadori compounds: a reexamination of its scavenging capacity and chelating effect. Bioorg Med Chem 16:5557–5569

Alonso P, De la Fuente M (2011) Role of the immune system in aging and longevity. Curr Aging Sci 4(2):78–100

American Diabetes Association (ADA) (2013) Economic costs of diabetes in the U.S. in 2012. Diabetes care 36(4), 1033–1046. https://doi.org/10.2337/dc12-2625

Andreeva TV, Lukiw WJ, Rogaev EI (2017) Biological basis for amyloidogenesis in Alzheimer's disease. Biochemistry (Mosc) 82:122–139

Ao M, Inuiya N, Ohta J et al (2019) Relationship between homocysteine, folate, vitamin B12 and physical performance in the institutionalized elderly. J Nutr Sci Vitaminol 65:1–7. https://doi.org/10.3177/jnsv.65.1

Arredondo M, Gonzales M, Latorre M (2018) Copper. In: Malavolta M, Mocchegiani (eds) Trace elements and minerals in health and longevity, pp 35–62. Springer, Zurich. https://doi.org/10.1007/978-3-030-03742-0_2

Ashoori M, Saedisomeolia A (2014) Riboflavin (vitamin B2) and oxidative stress: a review. Br J Nutr 111:1985–1991. https://doi.org/10.1017/S0007114514000178

Ashraf A, Clark M, So P-W (2018) The aging of iron man. Front Aging Neurosci 10:65

Bao W, Rong Y, Rong S et al (2012) Dietary iron intake, body iron stores, and the risk of type 2 diabetes: a systematic review and meta-analysis. BMC Med 10, 119

Barbagallo M, Belvedere M, Dominguez LJ (2009) Magnesium homeostasis and aging. Magnes Res 22(4):235–246. https://doi.org/10.1684/mrh.2009.0187

Barbagallo M, Dominguez LJ (2010) Magnesium and aging. Curr Pharm Des 16(7):832–839. https://doi.org/10.2174/138161210790883679

Barbagallo M, Dominguez LJ (2013) Magnesium metabolism in type 2 diabetes mellitus. Encycl Metalloproteins 458:1277–1281

Barbosa NS, Kalaaji AN (2014) CAM use in dermatology. Is there a potential role for honey, green tea, and vitamin C? Complement Ther Clin Pract 20(1):11–15

Bielli A, Scioli MG, D'Amico F et al (2019) Cellular retinoic acid binding protein-II expression and its potential role in skin aging. Aging 11(6):1619–1631

Blom HJ, Shaw GM, den Heijer M et al (2006) Neural tube defects and folate: case far from closed. Nat Rev Neurosci 7(9):724–731. https://doi.org/10.1038/nrn1986

Bolzetta F, Veronese N, De Rui M et al (2015) Are the recommended dietary allowances for vitamins appropriate for elderly people? J Acad Nutr Diet 115(11):1789–1797

Boucher BJ (2012) The problems of vitamin D insufficiency in older people. Aging Dis 3(4):313–329

Bourke CD, Berkley JA, Prendergast AJ (2016) Immune dysfunction as a cause and consequence of malnutrition. Trends Immunol 37:386–398

Bourne RR, Jonas JB, Bron AM et al (2018) Prevalence and causes of vision loss in high-income countries and in Eastern and Central Europe in 2015: magnitude, temporal trends and projections. Br J Ophthalmol 102:575–585. https://doi.org/10.1136/bjophthalmol-2017-311258

Bowling A, Dieppe P (2005) What is successful ageing and who should define it? BMJ 331:1548–1551. https://doi.org/10.1136/bmj.331.7531.1548

Bratic A, Larsson NG (2013) The role of mitochondria in aging. J Clin Investig 123:951–957

Cabrera AJR (2015) Zinc, aging, and immunosenescence: an overview. Pathobiol Aging Age Relat Dis 5:25592. https://doi.org/10.3402/pba.v5.25592

Cai Z, Zhang J, Li H (2019) Selenium, aging and aging-related diseases. Aging Clin Exp Res 31(8):1035–1047. https://doi.org/10.1007/s40520-018-1086-7

Celep GS, Kaynar P, Rastmanesh R (2017) Biochemical functions of micronutrients. Adv Obes Weight Manag Control 6(2):43–45. https://doi.org/10.15406/aowmc.2017.06.00147

Cerpa W, Varela-Nallar L, Reyes AE et al (2005) Is there a role for copper in neurodegenerative disease? Mol Aspects Med 26, 405–420

Childs BR, Andres BA, Vallier HA (2016) Economic benefit of calcium and vitamin D supplementation: does it outweigh the cost of nonunions? J Orthop Trauma 30(8):e285–e288

Cianferotti L, Parri S, Gronchi G et al (2017) The use of cholecalciferol in patients with hip fracture. Clin Cases Miner Bone Metab 14(1):48–53

Cole MA, Quan T, Voorhees JJ et al (2018) Extracellular matrix regulation of fibroblast function: redefining our perspective on skin aging. J Cell Commun Signal 12:35–43. https://doi.org/10.1007/s12079-018-0459-1

Cowie CC, Rust KF, Ford ES et al (2009) Full accounting of diabetes and pre-diabetes in the U. S. population in 1988–1994 and 2005–2006. Diabetes Care 32(2), 287–294

Craenen K, Verslegers M, Baatout S et al (2020) An appraisal of folates as key factors in cognition and ageing-related diseases. Crit Rev Food Sci Nutr 60(5):722–739. https://doi.org/10.1080/10408398.2018.1549017

Crisan D, Roman I, Crisan M et al (2015) The role of vitamin C in pushing back the boundaries of skin aging: an ultrasonographic approach. Clin Cosmet Investig Dermatol 8:463–470. https://doi.org/10.2147/CCID.S84903

Ćurko-Cofek B, Grubić Kezele T, Barac-Latas V (2017) Hepcidin and metallothioneins as molecular base for sex-dependent differences in clinical course of experimental autoimmune encephalomyelitis in chronic iron overload. Med Hypotheses 107, 51–54. https://doi.org/10.1016/j.mehy.2017.07.022

Dao MC, Meydani SN (2013) Iron biology, immunology, aging, and obesity: four fields connected by the small peptide hormone hepcidin. Adv Nutr 4(6):602–617

Das BC, Dasgupta S, Ray SK (2019) Potential therapeutic roles of retinoids for prevention of neuroinflammation and neurodegeneration in Alzheimer's disease. Neural Regen Res 14(11):1880–1892. https://doi.org/10.4103/1673-5374.259604

Degenhardt TP, Alderson NL, Arrington DD et al (2002) Pyridoxamine inhibits early renal disease and dyslipidemia in the streptozotocin-diabetic rat. Kidney Int 61:939–950

de Jongh RT, van Schoor NM, Lips P (2017) Changes in vitamin D endocrinology during aging in adults. Mol Cell Endocrinol 453:144–150

De la Fuente M, Miquel J (2009) An update of the oxidation-inflammation theory of aging: the involvement of the immune system in oxi-inflam-aging. Curr Pharm Des 15(26):3003–3026

Dibaba DT, Xun P, Song Y et al (2017) The effect of magnesium supplementation on blood pressure in individuals with insulin resistance, prediabetes, or noncommunicable chronic diseases: a meta-analysis of randomized controlled trials. Am J Clin Nutr 106(3):921–929

Dominguez LJ, Veronese N, Guerrero-Romero F et al (2021) Magnesium in infectious diseases in older people. Nutrients 13:180. https://doi.org/10.3390/nu13010180

Fairweather-Tait SJ, Wawer AA, Gillings R, Jennings A et al (2014) Iron status in the elderly. Mech Ageing Dev 136–137:22–28

Fleming DJ, Tucker KL, Jacques PF et al (2002) Dietary factors associated with the risk of high iron stores in the elderly framingham heart study cohort. Am J Clin Nutr 76:1375–1384

Franceschi C, Campisi J (2014) Chronic inflammation (inflammaging) and its potential contribution to age-associated diseases. J Gerontol Ser A Biomed Sci Med Sci 69:S4–S9

Fulop T, Larbi A, Dupuis G et al (2018) Immunosenescence and inflamm-aging as two sides of the same coin: friends or foes? Front Immunol 8:1960. https://doi.org/10.3389/fimmu.2017.01960

Garcia-Esquinas E, Carrasco-Rios M, Ortola R et al (2021) Selenium and impaired physical function in US and Spanish older adults. Redox Biol 38, 101819.https://doi.org/10.1016/j.redox.2020.101819

Gommers LM, Hoenderop JG, Bindels RJ et al (2016) Hypomagnesemia in type 2 diabetes: a vicious circle? Diabetes 65(1):3–13

Gorusupudi A, Nelson K, Bernstein PS (2017) The age-related eye disease 2 study: micronutrients in the treatment of macular degeneration. Adv Nutr 8:40–53. https://doi.org/10.3945/an.116.013177
Grubić Kezele T, Ćurko-Cofek B (2020) Age-related changes and sex-related di erences in brain iron metabolism. Nutrients 12, 2601.https://doi.org/10.3390/nu12092601
Haase H, Rink L (2009) The immune system and the impact of zinc during aging. Immun Ageing 6:9. https://doi.org/10.1186/1742-4933-6-9
Harrison FE (2012) A critical review of vitamin C for the prevention of age-related cognitive decline and Alzheimer's disease. J Alzheimers Dis 29(4):711–726. https://doi.org/10.3233/JAD-2012-111853
Hartwig A (2001) Role of magnesium in genomic stability. Mutat Res 475:113–121
Hill TR, Granic A, Aspray TJ (2018) Vitamin D and ageing. Subcell Biochem 90:191–220. https://doi.org/10.1007/978-981-13-2835-0_8
Hughes CF, Ward M, Hoey L et al (2013) Vitamin B12 and ageing: current issues and interaction with folate. Ann Clin Biochem 50:315–329
Hunnicutt J, He K, Xun P (2013) Dietary iron intake and body iron stores are associated with risk of coronary heart disease in a meta-analysis of prospective cohort studies. J Nutr 144:359–366
Ibs RH, Rink L (2003) Zinc-altered immune function. J Nutr 133(5 Suppl 1), 1452–1465
Imbard A, Benoist JF, Blom HJ (2013) Neural tube defects, folic acid and methylation. J Environ Res Public Health 10(9):4352–4389. https://doi.org/10.3390/ijerph10094352
Institute of Medicine (IOM) (2006) Dietary Reference Intakes: The essential guide to nutrient requirements. Washington (DC): National Academies Press (US). http://www.nap.edu/catalog/11537.html. Accessed 3 Jan 2021
Institute of Medicine (IOM) (2011) Dietary Reference Intakes for Calcium and Vitamin D. Washington (DC): National Academies Press (US). https://www.ncbi.nlm.nih.gov/books/NBK56070/. https://doi.org/10.17226/13050
Iotti S, Wolf F, Mazur A et al (2020) The COVID-19 pandemic: is there a role for magnesium? Hypotheses and Perspectives. Magnes Res 33(2):21–27. https://doi.org/10.1684/mrh.2020.0465
Johnell O, Kanis JA (2006) An estimate of the worldwide prevalence and disability associated with osteoporotic fractures. Osteoporos Int 17:1726
Kalani APK, Kamat S, Givvimani K et al (2014) Nutri-epigenetics ameliorates blood-brain barrier damage and neurodegeneration in hyperhomocysteinemia: role of folic acid. J Mol Neurosci 52(2):202–215. https://doi.org/10.1007/s12031-013-0122-5
Kehoe L, Walton J, Flynn A (2019) Nutritional challenges for older adults in Europe: current status and future directions. Proc Nutr Soc 78(2):221–233. https://doi.org/10.1017/S0029665118002744
Khoo HE, Ng HS, Yap W-S et al (2019) Nutrients for prevention of macular degeneration and eye-related diseases. Antioxidants 8(4):85. https://doi.org/10.3390/antiox8040085
Killilea DW, Maier JAM (2008) A connection between magnesium deficiency and aging: new insights from cellular studies. Magnes Res 21(2):77–82
Kryukov GV, Castellano S, Novoselov SV et al (2003) Characterization of Mammalian Selenoproteomes. Science 300(5624):1439–1443
La Fata G, Weber P, Mohajeri MH (2014) Effects of vitamin E on cognitive performance during ageing and in Alzheimer's disease. Nutrients 6:5453–5472. https://doi.org/10.3390/nu6125453
Laird E, Rhodes J, Kenny RA (2020) Vitamin D and inflammation: potential implications for severity of covid-19. Ir Med J 113(5):81
Lang PO, Govind S, Michel JP et al (2011) Immunosenescence: implications for vaccination programs in adults. Maturitas 68(4):322–330
Liu D, Ke Z, Luo J (2017) Thiamine deficiency and neurodegeneration: the interplay among oxidative stress, endoplasmic reticulum stress, and autophagy. Mol Neurobiol 54:5440–5448. https://doi.org/10.1007/s12035-016-0079-9
Maes ML, Fixen DR, Linnebur SA (2017) Adverse effects of proton-pump inhibitor use in older adults: a review of the evidence. Ther Adv Drug Saf 8(9):273–297

Maggini S, Wintergerst ES, Beveridge S et al (2007) Selected vitamins and trace elements support immune function by strengthening epithelial barriers and cellular and humoral immune responses. Br J Nutr 98(Suppl 1), 29–35

Magnus EM, Bache-Wiig JE, Aanderson TR et al (2009) Folate and vitamin B12 (cobalamin) blood levels in elderly persons in geriatric homes. Scand J Haematol 28(4):360–366. https://doi.org/10.1111/j.1600-0609.1982.tb00539.x

Marian M, Sacks G (2009) Micronutrients and older adults. Nutr Clin Pract 24:179–195

Marsman D, Belsky DW, Gregori D et al (2018) Healthy ageing: the natural consequences of good nutrition – a conference report. Eur J Nutr 57(Suppl 2), 15–34. https://doi.org/10.1007/s00394-018-1723-0

McNulty H, Ward M, Hoey L et al (2019) Addressing optimal folate and related B-vitamin status through the lifecycle: health impacts and challenges. Proc Nutr Soc 78(3):449–462. https://doi.org/10.1017/S0029665119000661

Meplan C (2011) Trace elements and ageing, a genomic perspective using selenium as an example. J Trace Elem Med Biol 25:S11–S16. https://doi.org/10.1016/j.jtemb.2010.10.002

Mercola J, Grant WB, Wagner CL (2020) Evidence regarding vitamin D and risk of COVID-19 and its severity. Nutrients 12(11):3361. https://doi.org/10.3390/nu12113361

Meydani SN, Lewis ED, Wu D (2018) Perspective: should vitamin E recommendations for older adults be increased? Adv Nutr 9(5):533–543. https://doi.org/10.1093/advances/nmy035

Mikkelsen K, Apostolopoulos V (2018) B vitamins and ageing. Subcell Biochem 90:451–470. https://doi.org/10.1007/978-981-13-2835-0_15

Mikkelsen K, Stojanovska L, Apostolopoulos V (2016) The effects of vitamin B in depression. Curr Med Chem 23:4317–4337

Mikkelsen K, Stojanovska L, Tangalakis K et al (2016a) Cognitive decline: a vitamin B perspective. Maturitas 93:108–113. https://doi.org/10.1016/j.maturitas.2016.08.001

Mikkelsen K, Stojanovska L, Prakash M et al (2017) The effects of vitamin B on the immune/cytokine network and their involvement in depression. Maturitas 96:58–71. https://doi.org/10.1016/j.maturitas.2016.11.012

Mitchell ES, Conus N, Kaput J (2014) B vitamin polymorphisms and behaviour: evidence of associations with neurodevelopment, depression, schizophrenia, bipolar disorder and cognitive decline. Neurosci Biobehav Rev 47C:307–320. https://doi.org/10.1016/j.neubiorev.2014.08.006

Mocchegiani E, Costarelli L, Giacconi R et al (2012) Micronutrient (Zn, Cu, Fe)–gene interactions in ageing and inflammatory age-related diseases: implications for treatments. Ageing Res Rev 11(2):297–319

Mocchegiani E, Costarelli L, Giacconi R et al (2014) Micronutrient-gene interactions related to inflammatory/immune response and antioxidant activity in ageing and inflammation. A Syst Rev Mech Ageing Dev 137:29–49

Mohammed BM, Fisher BJ, Kraskauskas D et al (2016) Vitamin C promotes wound healing through novel pleiotropic mechanisms. Int Wound J 13:572–584. https://doi.org/10.1111/iwj.12484

Moise AR, Noy N, Palczewski K et al (2007) Delivery of retinoid-based therapies to target tissues. Biochemistry 46(15):4449–4458

Moore K, Hughes CF, Ward M et al (2018) Diet, nutrition and the ageing brain: current evidence and new directions. Proc Nutr Soc 77:152–163

Morelli MB, Gambardella J, Castellanos V et al (2020) Vitamin C and cardiovascular disease: an update. Antioxidants 9:1227. https://doi.org/10.3390/antiox9121227

Murakoshi M, Tanimoto M, Gohda T et al (2009) Pleiotropic effect of pyridoxamine on diabetic complications via CD36 expression in KK-Ay/Ta mice. Diabetes Res Clin Pract 83:183–189

Narayan KM, Boyle JP, Geiss LS et al (2006) Impact of recent increase in incidence on future diabetes burden: U. S., 2005–2050. Diabetes Care 29(9), 2114–2116

Nemazannikova N, Mikkelsen K, Stojanovska L et al (2017) Is there a link between vitamin B and multiple sclerosis? Med Chem 14:170–180. https://doi.org/10.2174/1573406413666170906123857

Obeid R, Herrmann W (2006) Mechanisms of homocysteine neurotoxicity in neurodegenerative diseases with special reference to dementia. FEBS Lett 580(13):2994–3005. https://doi.org/10.1016/j.febslet.2006.04.088

Out M, Kooy A, Lehert P et al (2018) Long-term treatment with metformin in type 2 diabetes and methylmalonic acid: post hoc analysis of a randomized controlled 4.3 year trial. J Diabetes Complicat 32(2), 171–178

Patterson T, Isales CM, Fulzele S (2021) Low level of vitamin C and dysregulation of vitamin C transporter might be involved in the severity of COVID-19 Infection. Aging Dis 12(1):14–26. https://doi.org/10.14336/AD.2020.0918

Penke L, Valdés Hernandéz MC, Maniega SM et al (2012) Brain iron deposits are associated with general cognitive ability and cognitive aging. Neurobiol Aging 33, 510–517

Petridou ET, Kousoulis AA, Michelakos T et al (2016) Folate and B12 serum levels in association with depression in the aged: a systematic review and metaanalysis. Aging Ment Health 20(9):965–973

Pfortmueller CA, Lindner G, Exadaktylos AK (2014) Reducing fall risk in the elderly: risk factors and fall prevention, a systematic review. Minerva Med 105(4):275–281

Porter K, Hoey L, Hughes CF et al (2016) Causes, consequences and public health implications of low B-vitamin status in ageing. Nutrients 8(11):725. https://doi.org/10.3390/nu8110725

Price EA, Mehra R, Holmes TH et al (2011) Anemia in older persons: etiology and evaluation. Blood Cells, Mol Dis 46:159–165

Quan T, Qin Z, Shao Y et al (2011) Retinoids suppress cysteine-rich protein 61 (CCN1), a negative regulator of collagen homeostasis, in skin equivalent cultures and aged human skin in vivo. Exp Dermatol 20:572–576. https://doi.org/10.1111/j.1600-0625.2011.01278.x

Ramis R, Ortega-Castro J, Caballero C et al (2019) How does pyridoxamine inhibit the formation of advanced glycation end products? the role of its primary antioxidant activity. Antioxidants 8(9):344. https://doi.org/10.3390/antiox8090344

Rasmussen HM, Johnson EJ (2013) Nutrients for the aging eye. Clin Interv Aging 8:741–748. https://doi.org/10.2147/CIA.S45399

Raverdeau M, Mills KHG (2014) Modulation of T cell and innate immune responses by retinoic acid. J Immunol 192(7):2953

Razzaque MS (2018) Can adverse effects of excessive vitamin D supplementation occur without developing hypervitaminosis D? J Steroid Biochem Mol Biol 180:81–86. https://doi.org/10.1016/j.jsbmb.2017.07.006

Reynolds E (2006) Vitamin B12, folic acid, and the nervous system. Lancet Neurol 5:949–960

Rigotti A (2007) Absorption, transport, and tissue delivery of vitamin E. Mol Aspects Med 28:423–436

Roussel AM (2002) Trace element deficiencies and supplementations in the elderly. Trace Elem Man Animal 10:409–416

Sawaengsri H, Bergethon PR, Qiu WQ et al (2016) Transcobalamin 776C→G polymorphism is associated with peripheral neuropathy in elderly individuals with high folate intake. Am J Clin Nutr 104(6):1665–1670

Simental-Mendia LE, Sahebkar A, Rodriguez-Moran M et al (2017) Effect of magnesium supplementation on plasma C-reactive protein concentrations: a systematic review and meta-analysis of randomized controlled trials. Curr Pharm Des 23(31), 4678–4686

Sofi F, Cesari F, Abbate R et al (2008) Adherence to Mediterranean diet and health status: meta-analysis. BMJ 337:a1344. https://doi.org/10.1136/bmj.a1344

Sourabh S, Bhatia P, Jain R (2019) Favourable improvement in haematological parameters in response to oral iron and vitamin C combination in children with iron refractory iron deficiency anemia (IRIDA) phenotype. Blood Cells Mol Dis 75:26–29

Spence JD (2017) Increased coagulation with aging: importance of homocysteine and vitamin B12. Circ J 81(2):268

Squitti R, Simonelli I, Ventriglia M et al (2014) Meta-analysis of serum non-ceruloplasmin copper in Alzheimer's disease. J Alzheimers Dis 38:809–822

Stitt A, Gardiner TA, Alderson NL et al (2002) The AGE inhibitor pyridoxamine inhibits development of retinopathy in experimental diabetes. Diabetes 51:2826–2832

Tam M, Gómez S, GonzalezGross M et al (2003) Possible roles of magnesium on the immune system. Eur J Clin Nutr 57:1193–1197

ter Borg S, Verlaan S, Hemsworth J et al (2015) Micronutrient intakes and potential inadequacies of community-dwelling older adults: a systematic review. Br J Nutr 113:1195–1206

van Wijngaarden JP, Doets EL, Szczecińska A et al (2013) Vitamin B(12), folate, homocysteine, and bone health in adults and elderly people: a systematic review with meta-analyses. J Nutr Metab 486186. https://doi.org/10.1155/2013/486186

Voziyan PA, Metz TO, Baynes JW et al (2002) A post-amadori inhibitor pyridoxamine also inhibits chemical modification of proteins by scavenging carbonyl intermediates of carbohydrate and lipid degradation. J Biol Chem 277:3397–3403

Voziyan PA, Hudson BG (2005) Pyridoxamine: the many virtues of a maillard reaction inhibitor. Ann N Y Acad Sci 1043:807–816. https://doi.org/10.1196/annals.1333.093

Wang X, Xu M, Frank JA et al (2017) Thiamine deficiency induces endoplasmic reticulum stress and oxidative stress in human neurons derived from induced pluripotent stem cells. Toxicol Appl Pharmacol 320:26–31. https://doi.org/10.1016/j.taap.2017.02.009

Watson J, Lee M, Garcia-Casal MN (2018) Consequences of inadequate intakes of vitamin A, vitamin B12, vitamin D, calcium, iron, and folate in older persons. CurrGeriatr Rep 7(2), 103–113. https://doi.org/10.1007/s13670-018-0241-5

Weaver CM, Alexander DD, Boushey CJ et al (2016) Calcium plus vitamin D supplementation and risk of fractures: an updated meta-analysis from the national osteoporosis foundation. Osteoporosis Int 27:367–376

Wickens AP (2001) Ageing and the free radical theory. Respir Physiol 128:379–391

Wong CW (2015) Vitamin B12 deficiency in the elderly: is it worth screening? Hong Kong Med J 21:155–164

World Health Organisation (2016) Mental health and older adults. WHO Available at: http://www.who.int/mediacentre/factsheets/fs381/en/. Accessed 15 Dec 2020

Wu D, Meydani SN (2008) Age-associated changes in immune and inflammatory responses: impact of vitamin E intervention. J Leukoc Biol 84(4):900–914

Zheng W, Monnot AD (2012) Regulation of brain iron and copper homeostasis by brain barrier systems: implication in neurodegenerative diseases. Pharmacol Ther 133(2):177–188. https://doi.org/10.1016/j.pharmthera.2011.10.006

Zingg JM, Azzi A (2004) Non-antioxidant activities of vitamin E. Curr Med Chem 11(9):1113–1133

Zoroddu MA, Aaseth J, Crisponi G et al (2019) The essential metals for humans: a brief overview. J Inorg Biochem 195:120–129. https://doi.org/10.1016/j.jinorgbio.2019.03.013

Chapter 5
Probiotics and Prebiotics in Healthy Ageing

Aayushi Kadam, Deepak Kadam, Kanchanlata Tungare, and Heena Shah

Abstract Probiotics and prebiotics have been reported to be useful in maintaining health and ameliorating various disease conditions. This chapter provides insights into the beneficial role of various probiotics and prebiotics along with their probable mechanisms of action. Several pieces of evidence from clinical trials and *in-vitro* studies have been presented and discussed in this chapter specifically in relation to ageing, longevity and general well-being of an individual. Probiotics have been reported to significantly enhance the skin elasticity, moisture content and gloss with a reduction in wrinkle depth which are all aging related consequences. These biotics increase antioxidants levels with decreased hair loss, skin ulcers and age-related inflammation. It promotes the number of mitochondria in cell and extends the lifespan. Also, they display a beneficial role in the prevention of abdominal pain, diarrhoea, diabetes mellitus, infant colic, irritable bowel syndrome, *C. difficile* infection and ulcerative colitis. On the other hand, prebiotics is known for elevating the antioxidant enzyme levels thereby minimizing harmful reactive radicals and increase immuno-regulatory cytokines which may further impart double protection when given with the right combination of probiotics. The results from several studies reflect the potential of these biotics in various therapeutic interventions, albeit it is also recommended to the researcher's community to plan and execute active surveillance to understand other side effects associated with the usage of these biotics.

Keywords Probiotics · Prebiotics · Ageing · Health benefits · Mechanism · Safety issues

A. Kadam (✉) · K. Tungare
School of Biotechnology and Bioinformatics, D. Y. Patil Deemed to be University, Plot no. 50, Sector-15, CBD Belapur, Navi Mumbai, India

D. Kadam
Anatek Services Pvt. Ltd., C.S.T. Road, Kalina, Santacruz (E), Mumbai 400 098, India

H. Shah
K. J. Somaiya College of Science and Commerce, Vidyavihar, Mumbai 400077, India

S. I. S. Rattan and G. Kaur (eds.), *Nutrition, Food and Diet in Ageing and Longevity*, Healthy Ageing and Longevity 14, https://doi.org/10.1007/978-3-030-83017-5_5

5.1 Introduction

The microbiome comprises trillions of bacteria, viruses, and fungi in/on our body (Brody 2020; Wang et al. 2017). The microbes found in the cecum of our large intestine are referred to as the gut microbiome (Davani-Davari et al. 2019). This gut microbiome is home to over 100 trillion good micro-organisms that include 1000 species and more than 9000 strain of each species. The gut is also considered as the largest immune organ in the human body and a vital part of the endogenous host defense system. It is the home of more than 80% of the antibody-producing cells which is a little brain within the human body. Many of the gut microbiomes are distinct pathogenic and beneficial bacterial species (Brody 2020). This gut microbiota imparts many benefits to the host, through various physiological functions such as strengthening of gut integrity or shaping of the intestinal epithelium (Natividad and Verdu 2013; Thursby and Juge 2017), harvesting energy (den Besten et al. 2013; Thursby and Juge 2017), protecting against pathogens (Bäumler and Sperandio 2016; Thursby and Juge 2017), modulating the metabolic phenotype, regulating epithelial development (Wang et al. 2017), and influencing innate immunity (Gensollen et al. 2016; Thursby and Juge 2017; Wang et al. 2017). The gut microbiota starts colonizing right from birth and keeps on altering till the end of life. The modulation of gut microbiota from time to time depends upon the ages, sexes, races, diets of the host, illness, antibiotic treatments, environment, individual genetics, stress, hygiene sanitation practice followed, etc. (Arrieta et al. 2014; Thursby and Juge 2017; Wang et al. 2017).

Symbiotic relation of human and normal gut microbiota positively metabolizes non-digestible compounds, supplies essential nutrients, maintains energy homeostasis, prevents the colonization by opportunistic pathogens, and contributes to the formation of intestinal architecture enriching the long healthy lifespan (Round and Mazmanian 2009). These non-digestible foods are primarily dietary fibers such as cellulose and hemicellulose, which are commonly found in vegetables and can be digested by a specific species of *Bacteroides* (Larsbrink et al. 2014). Other non-digestible soluble fibers, for example fructooligosaccharides and galactooligosaccharides are utilized by beneficial microbes, such as *Lactobacillus* and *Bifidobacterium* (Goh and Klaenhammer 2015). As an end product, this gut microbiota produces short-chain fatty acids (SCFAs), such as acetic acids, propionic acids, and butyric acids (Duncan et al. 2009). These SCFAs are used as an energy source to the host intestinal epithelium, as well as are absorbed in the colon to serve various roles in regulating gut motility, inflammation, glucose homeostasis, and energy harvesting (Cani et al. 2013). Studies have been done to show the role of the gut microbiota in lipid and protein homeostasis, microbial synthesis of essential nutrients and vitamins such as folates, vitamin K, biotin, riboflavin (B2), cobalamin (B12), and possibly other B vitamins (Morowitz et al. 2011). Moreover, this gut-microbiota encourages the normal development of the humoral, cellular mucosal, innate immune systems of the host (Cebra 1999; Thaiss et al. 2016). Also gut microbiota leads to the normal development of gut-associated lymphoid tissue and antibody production (Round and

Mazmanian 2009) and inhibits the dendritic cells mediated T-helper cell (Th17) anti-inflammatory pathway (Magrone and Jirillo 2013).

However, with the modern lifestyle, there are changes in food habits, lack of physical activity and excess stress which has led to the fluctuations in the population of gut microbiota that culminates in various immune-mediated diseases. Moreover many life events experienced at an early age disrupt the microbiota which may result in the development of diseases later in life (Arrieta et al. 2014; Penders et al. 2007). Too much of gut bacteria will ferment more fiber into excess fatty acids which may get deposited in the liver and lead to "metabolic syndrome" that often leads to various conditions such as type 2 diabetes, heart diseases, and obesity. Reduction in the population count of anti-inflammatory gut bacteria may lead to inflammatory bowel diseases, including Crohn's disease and ulcerative colitis. On the contrary, when the population of inflammatory bacteria increases it may cause rheumatoid arthritis. Disorders of the CNS example—anxiety, depression, and autism spectrum disorder are also linked by the ecosystem of gut bacteria. Dysbiosis, i.e., imbalances in the ecosystem of gut bacteria may be the reason for the condition of colon cancer, chronic fatigue syndrome, etc. (Menees and Chey 2018; Nagy-Szakal et al. 2017).

In this chapter, we will discuss the various probiotic strains and prebiotics (non-digestible fibers) and their mechanisms of action as well as their health benefits specifically in relevance to aging and longevity. Various risks and safety-related issues related to the use of these biotics will also be discussed.

5.2 Probiotics

A healthy individual has 10 times more the number of gut microbiota in comparison to the number of cells present in the body. Four dominant phyla i.e. *Firmicutes, Bacteridetes, Proteobacteria* and *Actinobacteria* encompass different organisms of different genus and species. The phyla of *Firmicutes* include all organisms belonging to the genus *Lactobacilli, Staphylococcus* and *Clostridium* whereas *Proteobacteria* encompasses most of the pathogen like *Enterobacteria, Salmonella, Escherichia* and *Shigella* (Stojanov et al. 2020).The phylum called *Actinobacteria* includes *Bifidobacteria.* Many of these microbiotas are strategically associated with the gut epithelial lining called Gut Associated Lymphoid Tissue (GALT) which further serves as a habitat for 70% of all immunological active cells and also remains in constant communication with other immunologically active cells of the intestine (Belkaid and Hand 2014; Jandhyala et al. 2015).

There are specific probiotic strains known to display enhanced functionality, specifically *Lactobacillus* and *Bifidobacteria* and hence popularly called psychobiotics for their potential therapeutic benefits (Sarkar et al. 2016). A vast majority of probiotic bacteria belongs to the genus *Bifidobacteria* or *Lactobacilli* and within the genus, many bacterial species display different probiotic activities with a difference in the rate of survival and response. *Lactobacillus* are very popularly used as probiotics and are part of the lactic acid bacteria (LAB) family that converts hexose sugars to lactic acid in the intestine thereby producing an acidic environment which inhibits the growth of several harmful bacterial species (Fayol-Messaoudi et al. 2005; Florou-Paneri et al. 2013). Further LAB family includes *Lactobacillus*, *Lactococcus*, *Enterococcus*, *Oenococcus*, *Pediococcus*, *Streptococcus* and *Leuconostoc* species and several other strains of the *Bifidobacterium* that are known for their resistive mechanisms to bile salts (Fijan 2014; Ruiz et al. 2013). Similarly a strain called *L. acidophilus* is resistant to bile acid and possess strong antimicrobial effect on other intestinal pathogens including fecal *E. coli* strains (Plaza-Diaz et al. 2019). *L. acidophilus* mainly ferments the non-digestible carbohydrate (include galactose, mannose, trehalose, saccharose and esculin), convert it into organic acids (lactic acid and acetic acid) and produce antibiotic substances (Lactocidin, Acidophilin, Acidolin, Lactocin B) (Nagpal et al. 2012). All these organisms together impart probiotic effect and influence the physiology of the individual.

5.2.1 Role of Probiotics in Aging, Longevity and Well Being

The microbiota and their metabolites in the gastrointestinal tract (GIT) system play a critical role in modulating gut-associated immune systems (Magrone and Jirillo 2013). There is a need for anti-aging and stress-reducing probiotics in all age groups. Particularly, in the elderly population due to gradual deterioration of their anatomy and physiological functions that leads to an imbalance in their gut microbiota ecosystem. This change in microbial composition mainly contributes to metabolic and inflammatory diseases such as irritable bowel disease, diabetes, cardiovascular disease, celiac diseases, food allergies, rheumatoid arthritis and colorectal cancer (Geier et al. 2006; Nagpal et al. 2018).

Moreover, the aging gut also contributes to the over-expression of proinflammatory cytokine IL-6, which significantly affects the function of the intestinal barrier and mucosal immune system (Nagpal et al. 2018). Further decrease in mucus secretion by these healthy bacteria causes changes in intestinal permeability which leads to the development of celiac disease, colorectal cancer, inflammatory bowel disease and even systemic as well as CNS disorders (Kho and Lal 2018). However, a healthy lifestyle with a customized nutritional diet including probiotics can protect against several age-related chronic diseases in elderly people (Landete et al. 2017).

The majority of probiotics inhabited in GIT are anaerobic and their mutualistic behavior possesses wild ranging metabolic activity in the maintenance of physiological functions, including intestinal homeostasis, digestion of complex carbohydrates, protection against pathogens, synthesis of essential nutrients and vitamins, and stimulation of the immune system (Gorbach 1996; Judkins et al. 2020). For example, many probiotics in the lower intestine tract ferment wide a variety of dietary fibers in food to produce SCFAsandother metabolites including acetate, propionate and butyrate that have a distinct role in promoting gut health. SCFA is also involved in controlling anorexigenic hormones by signaling to the gut via free fatty acid receptors (Lu et al. 2018). Acetate is mainly metabolized by the peripheral tissues and propionate is gluconeogenic. Butyrate is the major energy source for the colonic epithelium that is significantly reduced in elderly people (Parada Venegas et al. 2019). The species called *F. Prausnitzii* and *Roseburia* notably involves in butyryl CoA: acetate CoA transferase route for butyrate formation (Shinohara et al. 2019). Moreover, butyrate can be generated from the LAB family in the colon from the lactate as a precursor. For example, *Bifidobacterium* species in combination with *A. hadrus* and *Eubacteriumhallii* form butyrate in the host colon that confers several health benefits (Rivière et al. 2016). Studies also reveal that the supplementation of probiotics in the elderly population, with or without specific diet composition can improve the functionality of microbiota (Landete et al. 2017). In addition to this, a diet rich in phytoestrogens has benefited the aging population since they are pro-estrogenic and antioxidant in nature (Rietjens et al. 2017).

Lactobacillus species like *L. acidophilus*, *L. fermentum*, *L. reuteri* and *B. Bifidum* have been reported to up-regulate transforming growth factor beta (TGF-β), peroxisome proliferator activated receptor gamma (PPAR-γ) along with down-regulation of interleukin-1 (IL-1), interleukin-8 (IL-8) and tumor necrosis factor-α (TNF-α) in older patients with Parkinson's disease when supplemented for 12 weeks (Borzabadi et al. 2018). Probiotics supplementation has also been beneficial in older patients with osteoporosis. Treatment with Kefir fermented milk constituting *Leuconostoc* and *Lactobacillus* for 6 months was found to increase bone formation with elevated bone mineral density at the femoral neck and hip region (de Oliveira Leite et al. 2013; Tu et al. 2015). Further reports have suggested that probiotics can modulate serum hs-CRP levels, pain symptoms and may improve quality of life in patients with a history of knee osteoarthritis (Lei et al. 2017). Also in the same context, *L. casei* strain resulted in minimized pain and early recovery in wrist flexion and grip strength of elderly patients with distal radius fracture (Lei et al. 2016). Reports also point out the beneficial aspect of probiotics in relieving constipation issues. Bifidobacteria in yogurt has been reported to improve stool frequency (Tanaka and Shimosaka 1982). A commercially available *L. rhamnosus* LC705 and *Propionobacter freundreichii* mixture have been investigated to increase in defecation frequency of elderly subjects by 24% (Ouwehand et al. 2002). On the other hand, *B. subtitlis* which is an active ingredient of "natto" is proposed to be contributing to the long and healthy longevity of the Japanese population. At recommended doses *B. subtilis* in human food may decrease the rate of aging and stamp out disease because of the downregulation of insulin/IGF-1 signaling with enhancement of innate immunity (Ayala et al. 2017). More studies on various anti-aging properties of probiotics are represented in Table 5.1.

The probiotics in the gut ecosystem promise two major benefits that include immunomodulatory activity that alleviate many age-related pathologies as well as the formation of bioactive metabolites from dietary compounds (Hemarajata and Versalovic 2013). The immune and protective functions of the microbiota are mediated by different mechanisms. One of them is competition with the pathogen in the lumen and by enhancing the mucosal barrier (Bron et al. 2017; Plaza-Diaz et al. 2019). Microbiota stimulates the epithelium to secrete mucin and strengthen their tight junctions between cells (Takiishi et al. 2017). There are micro-organisms that

Table 5.1 Studies conducted on anti-ageing properties of probiotics

Sr.no	Strains	Models	Duration of exposure	Research finding	Reference
1	*Lactobacillus plantarum*	Humans	12 weeks	Significant enhancement in skin elasticity, reduction in the wrinkle depth associated with the improvement in skin moisture content and gloss	(Lee et al. 2015)
2	*Lactococcuslactis*	Mice strain (Senescence accelerated)	15 weeks	Elevated alpha interferon levels associated with minimised aging related skin thinning and expression of muscle degeneration gene. Increased expression levels of tight junction genes. Treated mice displayed considerably reduced senescence score	(Tsuji et al. 2018)
3	*Lactobacillus salivarius*	*C. elegans*	–	Increased antioxidant status of the model thereby extending the lifespan	(Zhao et al. 2013)
4	*Bacillus licheniformis*	*C. elegans*	–	Extended lifespan of the model organisms and proposed to be associated with the serotonin pathway	(Park et al. 2015)
5	*Lactococcuslactis*	Mice model (Senescence accelerated)	5 weeks	Extended lifespan associated with less consequences of senescence through activation of plasmacytoid dendritic cells	(Sugimura et al. 2018)

(continued)

Table 5.1 (continued)

Sr.no	Strains	Models	Duration of exposure	Research finding	Reference
6	*Lactococcuslactis*	(Senescence accelerated)	2–5 mo	Decreased hair loss, skin ulcers and number of *Staphylococcus* spp along with improved bone density	(Kimoto-Nira et al. 2007)
7	*Lactobacillus gasseri*	*C. elegans*	–	Elevated gene expression of skn-1 and numbers of mitochondria	(Nakagawa et al. 2016)
8	*B. bifidum, L. acidophilus* and *Ba. Coagulans*	Human Volunteers	–	Increasing saccharolytic fermentation and decreased inflammation associated with aging	(Liu et al. 2016)
9	*Lactobacillus* interventions alone or incombination with *Bifidobacterium, Bacillus coagulans*	364 healthy elderly subjects	3 to 12 weeks	Significantly increased NK cell activity	(Gui et al. 2020)
10	Human-origin probiotic cocktail containing 5 *Lactobacillus* and 5 *Enterococcus* strains	Mice model	–	Reduced leaky gut by increasing tight junctions, which in turn reduced inflammation. The action was attributed to increase bile salt hydrolase activity, which in turn increased taurine abundance in the gut that stimulated tight junctions and suppressed gut leakiness	(Ahmadi et al. 2020)

induce the immune system to secrete antibodies, specifically IgA. Whereas, other classes of micro-organisms modulates the cellular immune response in the gut by stimulating both Th1 and Th2 cell types (Yan and Polk 2011).

The composition of microbiota in the gut may affect brain function in adults, thereby having an impact on stress, anxiety, depression, and cognition. Evidences for the effect of the brain on the gut microbiome can be found in studies documenting that parental, early-life and psychological stress changes the composition of the gut microbiota (Mohajeri et al. 2018). On the other hand, GIT physiology depends on the function of the gut including its motility, its sensation, diet, microbiota and immune function. Function of GITmicrobiota primarily proves beneficial in digestion and metabolic activities. Many of these bacteria synthesize vitamins along with the fermentation of non-digestible carbohydrates that reduce the pH of the gut and also metabolize carcinogens (Gorbach 1996; Judkins et al. 2020; Rowland 2000).

Furthermore, immune development, modulation and enhancement of gut barrier function are major components of gut microbiota. *Bifidobacterialactis* (strain Bp-12) was the first bacteria to achieve GRAS status from the US-FDA for use in infants from birth. The use of this *B. lactis* as oral probiotics by infants has a decade of safety record (Pham et al. 2017). There have been systematic reviews and clinical trials showing a group of babies growing adequately without having any severe side effects and considered safe by regulatory agencies (Sanders et al. 2010). There are reports on babies who have colic also expresses dysbiosis. Populations of children with colic are reported to harbor less *Lactobacilli* or have less diversity of bacteria in their gut ecosystem than babies who do not have colic. This unhealthy ecosystem has more coliform bacteria like *E. coli* which leads to an increase in gut inflammatory markers (Pham et al. 2017). Many reports explain the relationship between these symptoms of crying in colic and an abnormal gut ecosystem. Supplementation of *L. reuteri* increases the amount of *Lactobacilli* in the stools of babies and decrease of the presence of *E. coli* in the gut thereby improving the symptom of infantile colic along with the reduced frequency of functional regurgitation (Chau et al. 2015; Garofoli et al. 2014; Indrio et al. 2015; Savino et al. 2010). Also, *L. reuteri* supplementation significantly decreased the episodes and duration of diarrhea along with decreased respiratory tract infection in children compare with placebo (Gutierrez-Castrellon et al. 2014).

5.2.2 Incorporating Probiotics into Foods

The incorporation of probiotics in foods has decades of history and is added into several dairy and fermented foods to improve their structural and sensorial functionality. Although the techniques have matured over a period of time but their mechanism of action remains the same. These probiotics can be incorporated in fresh, refrigerated dairy products and a broader range of supplementation as an ingredient. However, the selection of compatible probiotic strain and incorporating into foods and most importantly keeping them alive throughout shelf life is a challenge for food biotechnologists. Besides fermented food products, one has to ensure that the food matrix will support probiotic growth. Furthermore, it is also necessary to ensure that the safety of incorporated probiotics does not adversely impact health as well as taste and texture of food during new product development.

These probiotic ingredients in food mainly are selected from *Lactobacillus* and *Bifidobacterium* genus due to their predominant inhabitance in human GI microbiota. A wild range of species in this genus has been used in food supplementation. Notably, some strains of those species have exhibited healthy probiotic attributes. For example, strain like *Bifidobacterium infantis, B. adolescentis, B. animalis subsp animalis, B. animalis subsp lactis, B. bifidum, B. longum, B. Breve* etc. are demonstrated to be effective probiotics in literature (O'Callaghan and van Sinderen 2016).

Phytoestrogens such as coumestans, ellagitannins, lignans, and isoflavones are similar to endogenous estrogen and have both anti-estrogenic and estrogenic effects. They are present in plants or foods derived from plants such as soya, cereals, vegetables, fruit etc. Phytoestrogens protect against various age-related chronic diseases such as cardiovascular and bone diseases, cancers, and cognitive function (Landete et al. 2017). These health benefits derived from phytoestrogens consumption are attributed to bioactive metabolites generated by gut bacteria (Bolca et al. 2013). Thus, the intake of a diet rich in isoflavones, lignans and ellagitannins in combination with selected probiotic bacteria may lead to the production of equol, enterolignans, and urolithins in the gut, respectively (Gaya et al. 2017; Romo-Vaquero et al. 2015; Shimada et al. 2010). This combination of bioactive-rich food with probiotics should be looked upon in amelioration, mitigation and prevention of aging-related pathologies. Nowadays, different bacteria such as *Butyribacterium methylotrophicum, Eubacterium callanderi*, and *Peptostreptococcus productus* and the strains *Eubacterium limosum, Ruminococcus productus, Clostridium scindens, Peptostreptococcus productus* SECO-Mt75m3, and *Eggerthellalenta* SECO-Mt75m2 have been involved are being used in the production of enterolignans which further protect from age-related diseases (Landeteetal 2017). Albeit direct anti-aging probiotic formulations are still to be explored, extensive research is being conducted to fortify various food products with probiotics that influence the health status, nutritional levels and well-being of a consumer.

Bifidobacteriumanimalis subsp. *lactis* which are included in LAB family are also used in the fermentation of milk due to their proteolytic activity. These probiotics are included in the dairy product for enhancing the sensorial property along with their proteolytic activity. Besides *Bifidobacterium* strain also induces immunoglobulin production that improved the nutritional value of food by assimilation of substrates not metabolized by the host (Maldonado Galdeano et al. 2019). Thus these probiotics with different functionality are used in combination to yield better results. For example *Lactobacillus delbrueckii* ssp and *Bulgaricus species* are used in combination for acid formation and the production of aroma substances such as acetaldehyde are popular (Chen et al. 2017). Similarly, for the production of yogurt optimum combination of *L. acidophilus, Bifidobacterium lactis* and *S. thermophiles* are used. The strain *S. thermophiles* also show a symbiotic relationship with *L. bulgaricus* in the yoghurt production. During yogurt production, there is increase in the acidity of the media and oxygen consumption which is favorable for the growth of *L. bulgaricus* that further forms valine, an essential growth component for *S. thermophilus*. In addition, these strains are also used in various starter cultures to produce fermented dairy products and cheese. Strain *L. casei* is also used to enhance the sensorial property of traditional dairy products such as kefir and cheese (Horiuchi and Sasaki 2012). Studies have also shown the use of *L. rhamnosus* GG strain in dairy probiotic products marketed for infant formulations. This strain is indigenous to human intestinal flora and thus it has resistance to low pH values with superior adherence ability to the gastrointestinal tract. Another genus called *Enterococcus* are also present in a higher amount in dairy products and have been demonstrated to exhibit widespread technological properties owing to the production ability of bacteriocin (Banwo et al. 2013). Studies have also shown effective use of *E. faecium and L. gasseri* in diarrhea treatment and thus can serve as a possible alternative to the usage of antibiotics (Margreiter et al. 2006). As a probiotic, *E. faecium* have been used for reducing the absorption of cholesterol from the digestive system whereas *L. gasseri's* probiotic activity is attributed to its reducing fecal mutagenic enzyme (Kumar et al. 2012).

5.3 Prebiotics

In 1995, Gibson and Roberfroid defined prebiotics as "a non-digestible food ingredient that provides beneficial effects to host by selectively stimulating the growth and/or activity of one or a limited number of bacteria in the colon" (Carlson et al. 2018). In other words, prebiotics is food source which remains undigested by host enzymes, but in the large intestine these are used by microbiota for nourishment. These substances should be selectively used by the host microorganisms and fermented by one or a few colonic bacteria that are beneficial. Also, it has to confer a health benefit which is a critical part because the health benefit has to be measurable and therefore confirmable. For example, a prebiotic called fructooligosaccharide (FOS) is used as a growth substrate for bifidobacteria and some other colonic bacteria as well (Rossi et al. 2005). FOS has been shown to improve calcium absorption, therefore has measurable health benefits (Whisner and Castillo 2018). Similarly, other carbohydrates with low digestibility are also being tested for health benefits which include mono-oligosaccharide (MOS), pectooligosaccharide (POS), galactooligosaccharide (GOS), and xylooligosaccharide (XOS) (Belorkar and Gupta 2016). Several other prebiotics such as inulin, oligofructose, lactulose, human milk oligosaccharide, arabinoxylan, resistant starch, polyphenol, etc. are being studied for their modes of action and health benefits. Recently clinical trial has been conducted using arabinoxylan oligosaccharide (AXOS) as an oral supplementation that reveals the increased total bacterial populations and fecal butyrate concentrations (Sanchez et al. 2009). The natural food sources of the prebiotics are wheat, onions, bananas, honey, garlic, berries, legumes, beans, peas, oats, jerusalem artichokes, asparagus, dandelion, apple skin, chicory root, and leeks. In contrast to fibers, such as cellulose, pectins, and xylans, which promote the growth of many microorganisms in the gut, prebiotics such as fructooligosaccharides and galactooligosaccharides mainly stimulate the proliferation of *Lactobacillus* and *Bifidobacterium* ("Probiotics and Prebiotics | World Gastroenterology Organisation" 2018; Quigley 2019). Fructooligosaccharides are one of the most common prebiotics, whereas other non-digestible carbohydrates like non-starch polysaccharides, plant wall polysaccharides, and pectins, are not necessarily prebiotic agents, but are termed as dietary fibers. Thus all prebiotics are not fiber and all fibers are not prebiotics. The only thing common in both prebiotics and fibers is that both cannot be digested in the human small intestine and are fermented by the gut microbiota (Floch 2014).

5.3.1 Mechanism of Prebiotic Action

Understanding the mechanism of prebiotic action is very critical as probiotic bacteria use this prebiotics for their growth by fermenting them and releasing other metabolites as by-products. There are different routes for product formation which involve bacterial cross-feeding where the intermediate products are substrates for other bacteria (Blaak et al. 2020).

Production of SCFA by gut microorganisms from prebiotics and other substrates facilitates direct utilization of these prebiotics or complex carbohydrates that generate butyrate or propionate as beneficial SCFA. Further degradation of these complex compounds by short-chain fatty acid-producing bacteria or non-fermentative microorganisms can yield a different product. For example, Bifidobacteria can produce intermediate products such as lactate or even the short-chain fatty acid acetate that are then used by gut bacteria (Blaak et al. 2020). Moreover the antioxidant action of inulin-type fructans on colon mucosa and contractility is also reported. Inulin, through SCFA, can act as a scavenger of reactive oxygen species (ROS) and appears to resist cooking or digestion. Inulin can modulate responses to pathogenic bacterial lipopolysaccharide and protect the gut from inflammatory processes. This mode of inulin action is probably a defense mechanism against ROS by up-regulating colonic mucosal detoxification enzymes like glutathione S transferase thereby restoring the level of some important proteins involved in intestinal smooth muscle contraction (Guarino et al. 2020). However, the exact mechanisms by which inulin acts on intestinal muscle functions to exert direct and/or indirect response to colonic mucosa are not well understood. The various *in-vitro* effects of inulin-type prebiotics are documented not only to stimulate the antioxidant enzymes and scavenge reactive radicals but also to prevent lipid peroxidation in the stomach, replace vitamin C as dietary supplements and inhibit degradation of ascorbate (Busserolles et al. 2003; Kanner and Lapidot 2001; Miene et al. 2011; Phillips et al. 1995). β-GOS is also reported to exhibit immune-modulating function by increasing the immunoregulatory cytokine IL-10, with a significant reduction of IL-1β expression levels (Vulevic et al. 2015). GOS mixture is documented to increase the blood level of interleukin 8 (IL-8) and C-reactive protein and to improve Natural Killer (NK) cell activity as well (Vulevic et al. 2015). Another report indicates that supplementation with GOS in mice improved lipid metabolism with significant enrichment of mouse microbiota (Cheng et al. 2018). LBA also has anti-inflammatory properties, and in a study on mice, it was demonstrated that its administration is associated with control of obesity and associated metabolic parameters. Lactulose on the other hand is reported to increase *Bifidobacterium* count (Bouhnik et al. 2004), but not Lactobacilli with low production of SCFAs. 5 g/day dose extends the correct balance among the microbial population and SCFAs production, while 10 g/day decreases butyrate production and increases acetate content (Bothe et al. 2017). XOS and soybean oligosaccharides increases the population of *Bifidobacteria*, *Lactobacilli*, butyrate fecal concentration and inhibit clostridium growth (Lecerf et al. 2012; Lin et al. 2016; Mäkeläinen et al. 2010). Polyphenols are also known to increase the growth of *Lactobacilli*, *Bifidobacterium*, *F. Prausnitzii* and reduce *Clostridium* growth (Okubo et al. 1992; Tzounis et al. 2011). It also inhibits pro-inflammatory mediators cyclooxygenase-2 (COX2), IL-6, TNF-α, Nuclear Factor kB (NFkB) and Vascular-Endothelial Growth Factor. Polyphenols have also been reported to reduce serum triacylglycerol and C- reactive protein (Guarino et al. 2020). Thus dietary intake of prebiotics seems to have a positive modulatory effect on intestinal microbiota by not only promoting the growth of good intestinal bacteria, but also by producing metabolites that are potentially protective of gut functionality.

5.3.2 The Health Benefits of Prebiotics

The health benefits of prebiotics need to be validated in controlled studies on the target subjects. This beneficial effect(s) are mainly categorized into local effects (gut targeted) or systemic effects (whole-body). There are three immediate local effects that prebiotics can have. First, prebiotics like dietary fiber contribute to fecal bulking and also increased transit rate through the colon that protecting the gut or the colonic epithelial cells from any toxic compounds that might have been ingested in the diet. Increased transit rate make sure that any such toxins move through the large intestine more quickly again protecting the gut cells. Second, bacterial fermentation of the prebiotic causes a lowering of the pH. This helps to inhibit the growth of pathogenic bacteria which donot generally grow in an acidic pH. Moreover it also improves calcium solubility and therefore uptake of calcium and influences bone health (Slavin 2013). SCFAs like butyrate, acetate and propionate are important energy sources for the gut epithelial cells which keep the gut healthy and help gut cell turnover (Parada Venegas et al. 2019). So these local effects have very specific health benefits. Furthermore, the European food safety authority (EFSA) has also given a positive opinion for the consumption of Native chicory inulin that increase stool frequency (Micka et al. 2017). Probiotics like GOS and FOS supplementation in infant formula reveals the increase in Bifidobacteria number (Vandenplas et al. 2014). Also, consumption of FOS and inulin resulted in increased in calcium absorption that improved bone health in adolescents and menopausal women (Whisner and Castillo 2018). The systemic effects of prebiotic fermentation by the gut microorganism have been studied during clinical trials and have revealed metabolic and immunomodulatory functions (Hemarajata and Versalovic 2013). The effects are accomplished through fermentation by the gut bacteria and then through uptake of metabolites or interaction with the host. Food ingredients particularly diet as prebiotics are fermented by the gut microbiota into the SCFA that is used as an energy source by the epithelial cells. SCFA enters the liver through the bloodstream where most of it gets involved in gluconeogenesis (den Besten et al. 2013). Moreover, this SCFA also reaches other organs like brain, muscle and adipose tissue. SCFA may stimulate a particular effect through interaction with receptors present in these organs. For example G-protein-coupled receptors (GPR41/43) present in epithelial cells interact with SCFA and lead to the secretion of hormones PYY and GLP-1 which then reach the brain and lead to satiety (Koh et al. 2016). Similarly in adipose and muscle tissue SCFA lowers inflammation (Vinolo et al. 2011). A randomized double-blind trial suggests a fiber-containing yogurt sweetened with lactitol as a natural means of treating chronic constipation in elderly hospitalized subjects (Rajala et al. 1988). GOS relieves constipation in few but not all elderly people by ensuring an easy defecation process (Teuri and Korpela 1998). They suggested different people have different responses upon GOS ingestion. Another trial to investigate the repercussion of lactose or inulin on the bowel habits of constipated elderly patients revealed that inulin served as a better laxative effect than lactose and reduced functional constipation with mild discomfort (Kleessen et al. 1997). Furthermore, the consumption of oral supplementation with FOS and inulin for 12–13 weeks is reported to improve physical function, nutritional status, quality of life, as well as frailty degree (Jayanama and Theou 2020). Inulin with vitamin D is also reported to increase physical function, nutritional status and quality of life in a multicentric prospective observational study (Abizanda et al. 2015). Inulin with FOS enhanced handgrip strength and modulated Barthel index, body mass index along with frailty phenotype (Kleessen et al. 1997; Theou et al. 2019). The beneficial effect of GOS in increasing *Bifidobacteria, Lactobacillus, Enterococcus* with decrease in pathogenic organisms is also well established (Vulevic et al. 2015). This article also reports increased IL-10 levels with a decrease in IL-6, IL-1β, and TNF-α post GOS ingestion in a double-blind, cross-over, randomized controlled trial.

5.4 Risk and Safety Issues

Although probiotics have been used safely over a period of hundred years in the food and dairy industry, the safety outcomes have not yet been effectively reported during clinical trials. According to Marteau 2001, "the zero risk does not exist, and that acceptance of the concept that probiotics may not only have positive effects but potentially also side effects is important." A report by the Agency for Healthcare Research and Quality (AHRQ) on the safety of probiotics extended comprehensive literature on 622 organisms from 6 genera viz., *Bifidobacterium, Saccharomyces, Lactobacillus, Streptococcus, Enterobacillus and Bacillus* but did not provide conclusive evidence of risk and rather insinuated that the literature is not sufficiently equipped to claim the safety nature of the probiotics with assurance (Hempel et al. 2011). Clinical trials in hospitalized children, hospitalized adults and immunocompromised subjects using various strains have revealed no toxic effects. Likewise, trials conducted in pregnant women, premature neonates, elderly people and patients with inflammatory bowel disease also showed no harmful repercussion of the probiotics (Doron and Snydman 2015). A report by World's Health Organisation (WHO) in 2002 suggested that probiotics may have four types of typical side effects like a) Systemic infections b) Excessive immune stimulation, c) detrimental metabolic activities and d) Transfer of genes FAO/WHO (2002). There exists a plethora of literature suggesting the side effects of the probiotic strains such as *S. boulardii, L. rhamnosus, B. subtilis, S. pyogenes, K. pneumonia, L. acidophilus, B. infantis, S. thermophilus, L. bulgaricus* during clinical trial. Some reports have revealed the occurrence of systemic infections caused by *S. boulardii* in subjects receiving treatment against fungemia and have reported complication such as fever spike, septic shock, contamination of central venous catheter, massive colonization by yeast and transmission of infection (Cesaro et al. 2000; Hennequin et al. 2000; Lherm et al. 2002; Muñoz et al. 2005; Perapoch et al. 2000). Munoz et al. 2005 have recommended that probiotics can prove to be critical specifically in immunosuppressed patients. A child with the short gut syndrome and a young man were reported with an incidence of bacteremia and endocarditis along with septic arthritis respectively owing to treatment with *L. rhamnosus* (De Groote et al. 2005; Presterl et al. 2001).On the other hand, nosocomial bacteremia and distinct septicemic episodes were recorded in subjects given with oral preparation of *B. Subtilis* (Oggioni et al. 1998; Richard et al. 1988). The organism caused severe immunodeficiency and persisted in the intestinal tract of 73 years old male with chronic lymphocytic leukemia (Oggioni et al. 1998). *Lactobacilli* strain was also reported to cause a systemic infection like bacteremia by Land et al. 2005.

In other reports, *S. pyogenes, L. rhamnosus, K. pneumonia* and other bacterial probiotic cocktail caused immunostimulation in monocytes, monocytes derived immature dendritic cells and bone marrow derived dendritic cells. *L. rhamnosus* caused moderate increase in the expression of cell-surface co-stimulatory molecules and chemokine response, whereas, *S. pyogenes* strongly induced maturation of monocyte derived dendritic cells (Veckman et al. 2004). *K. pneumoniae* induced Th1 immune responses via dendritic cells and differential response to various bacterial strains exist owing to differential modulation of dendritic cells (Braat et al. 2004). Probiotic cocktail was reported to modulate dendritic cell surface phenotype and cytokine release in granulocyte–macrophage (Drakes et al. 2004). A provoked D-lactic acidosis case was reported in a Chinese boy with short bowel syndrome upon probiotic (*L. acidophilus* and *B. infantis*) supplemention (KU et al. 2006). Furthermore, 61 pediatric patients when supplemented with *L. rhamnosus*, the zilch effect was observed in reducing the incidence of nosocomial infections and instead promoted the infection (Honeycutt et al. 2007). *L. acidophilus, B. lactis, S. thermophilus*and*L.bulgaricus* with oligofructose favorably altered the microbial composition of the upper gastrointestinal tract but had no effect on intestinal permeability (Jain et al. 2004).

Lactic acid bacteria possess plasmids containing genes conferring resistance to various antibiotics such as tetracycline, erythromycin, macrolide, chloramphenicol or lincosamide streptomycin, and streptogrammin. There are some reports that pediococcus and leuconostoc species can accept broad host range antibiotic resistance plasmids from lactococcus species. Conjugation transfer from enterococci to lactobacilli and lactococci can occur in the gut of animals as well as in vitro; however, the transfer to lactobacilli is quite rare (Doron and Snydman 2015). Lateral gene transfer between probiotic organisms to other organisms in the gut or other sites is possible though no clinical evidence of transfer of antimicrobial resistance has ever has been reported. This is particularly important to investigate as probiotics arecommonly used to rejuvenate the good microflora of gut post/during antibiotic treatment. Furthermore application of mono-strains or multi-strains need to be evaluated carefully for their synergistic modulatory role.

On the other hand, prebiotics are known to cause significant change in the gut microbiota composition, treat chronic constipation and facilitate easy defecation in elderly people. Prebiotics posse's better laxative potential and can improve nutritional status, physical function and quality of life of elderly subjects. Prebiotics are also reported to improve stool quality (pH, frequency and consistency) in children as well (Bozzi Cionci et al. 2018). It also reduced the risk of gastroenteritis and improves the general well-being of a person. Not many reports are available with respect to effect of prebiotics in elderly aged groups and thus more studies need to be undertaken for component characterization, functional characterization, product formulation and safety assessment using double blind, randomised-controlled human clinical trials.

5.5 Research Trends, Research Gaps and Future Perspective

Classical probiotics have shown promising effects on human gut microflora but there is always an urge to develop better strains followed by the improved selection process. Previous studies have lead to the possible next-generation probiotic strains like *Clostridium* clusters IV, XIVa and XVIII, *F. prausnitzii*, *Akkermansiamuciniphila*, *Bacteroidesuniformis*, *Bacteroidesfragilis*and *Eubacteriumhallii*. The next-generation probiotics were evaluated in preclinical trials and yielded positive outcomes of possessing modulatory roles in inflammatory and metabolic disorders. Extensive clinical trials can give information on the effective use of these next generation microbes in mono-strains or multi-strains based formulations against many age-related diseases. In addition, new techniques are required for the development of new probiotic products containing strains of human origin (El Hage et al. 2017). Other than next-generation microbes, postbiotics and paraprobiotics are the upcoming horizons in the field. Postbiotics and paraprobiotics are cell constituents, metabolic by-products and non-viable microbial cells, respectively that contribute in health improvement (Nataraj et al. 2020). They are made from many probiotic strains and postbiotics by different inactivating methods. Various postbiotics include metabolic by products of living probiotic organisms such as vitamins, cell free supernatants, bio-surfactants, phenols (urolithins, equol, enterolactone, valerolactones, 8-prenylnaringenin and enterodiol) flavonoids (norathyriol, daidzein, desaminotyrosine, equoldaidzein), terpenoids (paeoniflorin, genipin, paeonimetabolin I, II, III and paeoni lactone glycosides) (Cortés-Martín et al. 2020; Wang et al. 2019). On the other hand, parabiotic include ruptured components of probiotic cells such as teichoic acids, muropeptides, pili, fimbriae, flagella, exopolysaccaride etc. (Chung et al. 2019; Shenderov 2013).

The use of such molecules in therapeutic studies provides an upper edge in understanding the molecular mechanisms of each purified cell component as using probiotics may yield confusing outcomes owing to complex bacterial structure. They have been linked to harbor immunomodulatory, anti-inflammatory, anti-hypertensive, hypocholesterolemic, anti-obesogenic, anti-proliferative, and antioxidant activities (Vallejo-Cordoba et al. 2020). Postbiotic preparations have also received patents as 1) anti-tumour agents, 2) bio-therapeutics for immunomodulation specific claims and feed additives for monogastric animals (Nataraj et al. 2020). These reports suggest the excellent potential of these molecules to boost host health by mitigation and prevention of the diseased condition. But their mechanism of action and elated signal transduction pathways are still unexplored and may be extensively researched by using of metatranscriptomic, metabolomic and metaproteomic approaches that may contribute in understanding their mechanism of action. Further their potential applications in the pharmaceutical and food industry can be revealed. Also, advancement in modern techniques and methods can lead to the production of bioengineered novel recombinant probiotics (Vallejo-Cordoba et al. 2020). More number of clinical trials authenticating the claims of these bioactive molecules may prove their direct therapeutic implications. Trials on subjects with low immune competence can unfold the tolerance status of these bioactive molecules by immunocompromised subjects. Stability related studies of para and postbiotics in in vitro and in vivo digestive conditions can prove beneficial in further exploring their health benefits.

There are still some important unanswered questions related to probiotics that need the urgent attention from researchers such as scientific validation of all claimed benefits by the definition of probiotics? There are very meagre reported meta-analyses, systematic reviews in comparison to the clinical trials conducted which reflects the inadequate comparison of these trials. Predominantly, the efficacy of probiotics is investigated widely in gastrointestinal diseases like antibiotic associated diarrhea, *Clostridium difficile* diarrhea, inflammatory bowel disease and necrotizing enterocolitis. Also, another major area of research is probiotic induced allergy and atopy. Despite of so much research, there are issues with understanding the efficacy of probiotics due to poor quality clinical trials, abysmal clinical trial reports and sufficiently evaluated safety reports. Moreover, Agency for Healthcare Research and Quality (2011) report complained about the inadequate literature to substantiate whether probiotics consumptions are safe, but still probiotics are being consumed by millions of people on daily basis further raising the concern.

5.6 Conclusion

Aging leads to several pathologies that may be directly or indirectly be associated with the imbalances of the gut microbiota and associated immune system. Moreover, intestine is considered to be the prime organ to enhance and improve the quality of life in age-related senescence process. These beneficial organisms residing in the gut may impart a powerful ameliorative effect in the prevention of age-associated health deterioration by its immunomodulatory activity. Reduction in proinflammatory status and age-related pathologies can be mitigated by adopting a healthy lifestyle along with a customized diet for elderly people. Probiotics possess excellent potential in preserving the integrity of the gut barrier and in evading infection. Prebiotics rich diet can facilitate the probiotics to exhibit their important function and may lead to the generation of equol, enterolignans, and urolithins, which are considered protective against chronic diseases related to aging. Although investigating the toxicological/ safety aspect of probiotic and prebiotic is a pressing priority, their applications in treatment, prevention and amelioration of the diseases seem to be gaining popularity and have been reported to be beneficial by many researchers. The limited data expressed in the risk and safety section should not dishearten the investigator to promote the usage of good bacteria in different food products. Moreover, it is recommended to the researcher community to plan and execute active surveillance to understand infections and other side effects associated with the usage of these biotics. Also, it is important to investigate whether the probiotic usage is appropriate in subjects with low immune competence, short bowel syndrome, cardiac valve disease and central venous catheters.

Compliance with Ethical Standards

Conflict of Interest All authors declare that they have no conflict of interest.

References

Probiotics and Prebiotics | World Gastroenterology Organisation [WWW Document] (2018). https://www.worldgastroenterology.org/guidelines/global-guidelines/probiotics-and-prebiotics. Accessed 4 Jun 2021

Abizanda P, López M.D, García VP, Estrella J. de Dios da Silva González, Á, Vilardell NB, Torres KA (2015) Effects of an oral nutritional supplementation plus physical exercise intervention on the physical function, nutritional status, and quality of life in frail institutionalized older adults: the activnes study. J Am Med Dir Assoc 16:439.e9–439.e16

Ahmadi S, Wang S, Nagpal R, Bo W, Jain S, Razazan A, Mishra SP, Zhu X, Wang Z, Kavanagh K, Yadav H (2020) A human-origin probiotic cocktail ameliorates aging-related leaky gut and inflammation via modulating the microbiota/taurine/tight junction axis. JCI Insight 5(9):e132055

Arrieta M-C, Stiemsma LT, Amenyogbe N, Brown EM, Finlay B (2014) The intestinal microbiome in early life: health and disease. Front Immunol 5:427

Ayala F, Bauman C, Cogliati S, Lenini C, Bartolini M, Grau R (2017) Microbial flora, probiotics, Bacillus subtilis and the search for a long and healthy human longevity. Microb Cell 4(4):133–136

Banwo K, Sanni A, Tan H (2013) Technological properties and probiotic potential of Enterococcus faecium strains isolated from cow milk. J Appl Microbiol 114:229–241

Bäumler AJ, Sperandio V (2016) Interactions between the microbiota and pathogenic bacteria in the gut. Nature 535:85–93

Belkaid Y, Hand T (2014) Role of the Microbiota in Immunity and inflammation. Cell 157:121–141

Belorkar SA, Gupta AK (2016) Oligosaccharides: a boon from nature's desk. AMB Express 6:82

Blaak EE, Canfora EE, Theis S, Frost G, Groen AK, Mithieux G, Nauta A, Scott K, Stahl B, van Harsselaar J, van Tol R, Vaughan EE, Verbeke K (2020) Short chain fatty acids in human gut and metabolic health. Benef Microbes 11:411–455

Bolca S, Van de Wiele T, Possemiers S (2013) Gut metabotypes govern health effects of dietary polyphenols. Curr Opin Biotechnol 24:220–225

Borzabadi S, Oryan S, Eidi A, Aghadavod E, Daneshvar Kakhaki R, Tamtaji OR, Taghizadeh M, Asemi Z (2018) The effects of probiotic supplementation on gene expression related to inflammation, insulin and lipid in patients with Parkinson's Disease: a randomized, double-blind, placebocontrolled trial. Arch Iran Med 21:289–295

Bothe MK, Maathuis AJH, Bellmann S, van der Vossen JMBM, Berressem D, Koehler A, Schwejda-Guettes S, Gaigg B, Kuchinka-Koch A, Stover JF (2017) Dose-dependent prebiotic effect of lactulose in a computer-controlled in vitro model of the human large intestine. Nutrients 9(7):767

Bouhnik Y, Attar A, Joly FA, Riottot M, Dyard F, Flourié B (2004) Lactulose ingestion increases faecal bifidobacterial counts: a randomised double-blind study in healthy humans. Eur J Clin Nutr 58:462–466

Bozzi Cionci N, Baffoni L, Gaggìa F, Di Gioia D (2018) Therapeutic microbiology: the role of bifidobacterium breve as food supplement for the prevention/treatment of paediatric diseases. Nutrients 10(11):1723

Braat H, de Jong EC, van den Brande JMH, Kapsenberg ML, Peppelenbosch MP, van Tol EAF, van Deventer SJH (2004) Dichotomy between Lactobacillus rhamnosus and Klebsiella pneumoniae on dendritic cell phenotype and function. J Mol Med (berl) 82:197–205

Brody H (2020) The gut microbiome. Nature 577:S5

Bron PA, Kleerebezem M, Brummer R-J, Cani PD, Mercenier A, MacDonald TT, Garcia-Ródenas CL, Wells JM (2017) Can probiotics modulate human disease by impacting intestinal barrier function? Br J Nutr 117:93–107

Busserolles J, Gueux E, Rock E, Demigné C, Mazur A, Rayssiguier Y (2003) Oligofructose protects against the hypertriglyceridemic and pro-oxidative effects of a high fructose diet in rats. J Nutr 133:1903–1908

Cani PD, Everard A, Duparc T (2013) Gut microbiota, enteroendocrine functions and metabolism. Curr Opin Pharmacol 13:935–940

Carlson JL, Erickson JM, Lloyd BB, Slavin JL (2018) Health effects and sources of prebiotic dietary fiber. Curr Dev Nutr 2(3):nzy005

Cebra JJ (1999) Influences of microbiota on intestinal immune system development. Am J Clin Nutr 69:1046S-1051S

Cesaro S, Chinello P, Rossi L, Zanesco L (2000) Saccharomyces cerevisiae fungemia in a neutropenic patient treated with Saccharomyces Boulardii. Support Care Cancer 8:504–505

Chau K, Lau E, Greenberg S, Jacobson S, Yazdani-Brojeni P, Verma N, Koren G (2015) Probiotics for infantile colic: a randomized, double-blind, placebo-controlled trial investigating Lactobacillus reuteri DSM 17938. J Pediatr 166:74–78

Chen C, Zhao S, Hao G, Yu H, Tian H, Zhao G (2017) Role of lactic acid bacteria on the yogurt flavour: a review. Int J Food Prop 20:S316–S330

Cheng W, Lu J, Lin W, Wei X, Li H, Zhao X, Jiang A, Yuan J (2018) Effects of a galacto-oligosaccharide-rich diet on fecal microbiota and metabolite profiles in mice. Food Funct 9:1612–1620

Chung I-C, OuYang C-N, Yuan S-N, Lin H-C, Huang K-Y, Wu P-S, Liu C-Y, Tsai K-J, Loi L-K, Chen Y-J, Chung A-K, Ojcius DM, Chang Y-S, Chen L-C (2019) Pretreatment with a heat-killed probiotic modulates the NLRP3 inflammasome and attenuates colitis-associated colorectal cancer in mice. Nutrients 11(3):516

Cortés-Martín A, Selma MV, Tomás-Barberán FA, González-Sarrías A, Espín JC (2020) Where to look into the puzzle of polyphenols and health? The postbiotics and gut microbiota associated with human metabotypes. Mol Nutr Food Res 64:e1900952

Davani-Davari D, Negahdaripour M, Karimzadeh I, Seifan M, Mohkam M, Masoumi SJ, Berenjian A, Ghasemi Y (2019) Prebiotics: definition, types, sources, mechanisms, and clinical applications. Foods 8(3):92

De Groote MA, Frank DN, Dowell E, Glode MP, Pace NR (2005) Lactobacillus Rhamnosus GG bacteremia associated with probiotic use in a child with short gut syndrome. Pediatr Infect Dis J 24:278–280

de Oliveira Leite AM, Miguel MAL, Peixoto RS, Rosado AS, Silva JT, Paschoalin VMF (2013) Microbiological, technological and therapeutic properties of kefir: a natural probiotic beverage. Braz J Microbiol 44:341–349

den Besten G, van Eunen K, Groen AK, Venema K, Reijngoud D-J, Bakker BM (2013) The role of short-chain fatty acids in the interplay between diet, gut microbiota, and host energy metabolism. J Lipid Res 54:2325–2340

Doron S, Snydman DR (2015) Risk and safety of probiotics. Clin Infect Dis 60:S129–S134

Drakes M, Blanchard T, Czinn S (2004) Bacterial probiotic modulation of dendritic cells. Infect Immun 72:3299–3309

Duncan SH, Louis P, Thomson JM, Flint HJ (2009) The role of pH in determining the species composition of the human colonic microbiota. Environ Microbiol 11:2112–2122

El Hage R, Hernandez-Sanabria E, Van de Wiele T (2017) Emerging trends in "smart probiotics": functional consideration for the development of novel health and industrial applications. Front Microbiol 8:1889

Fayol-Messaoudi D, Berger CN, Coconnier-Polter M-H, Liévin-Le Moal V, Servin AL (2005) pH-, lactic acid-, and non-lactic acid-dependent activities of probiotic lactobacilli against Salmonella Enterica Serovar Typhimurium. Appl Environ Microbiol 71:6008–6013

Fijan S (2014) Microorganisms with claimed probiotic properties: an overview of recent literature. Int J Environ Res Public Health 11:4745–4767

Floch MH (2014) Probiotics and prebiotics. Gastroenterol Hepatol (NY) 10:680–681

Florou-Paneri P, Christaki E, Bonos E (2013) Lactic acid bacteria as source of functional ingredients. In: Lactic acid bacteria - r&d for food, health and livestock purposes

Garofoli F, Civardi E, Indrio F, Mazzucchelli I, Angelini M, Tinelli C, Stronati M (2014) The early administration of Lactobacillus reuteri DSM 17938 controls regurgitation episodes in full-term breastfed infants. Int J Food Sci Nutr 65:646–648

Gaya P, Peirotén Á, Medina M, Landete JM (2017) Bifidobacterium adolescentis INIA P784: the first probiotic bacterium capable of producing enterodiol from lignan extracts. J Funct Foods 29:269–274

Geier MS, Butler RN, Howarth GS (2006) Probiotics, prebiotics and synbiotics: a role in chemoprevention for colorectal cancer? Cancer Biol Ther 5:1265–1269

Gensollen T, Iyer SS, Kasper DL, Blumberg RS (2016) How colonization by microbiota in early life shapes the immune system. Science 352:539–544

Goh YJ, Klaenhammer TR (2015) Genetic mechanisms of prebiotic oligosaccharide metabolism in probiotic microbes. Annu Rev Food Sci Technol 6:137–156

Gorbach SL (1996) Microbiology of the gastrointestinal tract, chap. 95. In: Baron S (ed) Medical microbiology. University of Texas medical branch at Galveston, Galveston, TX. PMID: 21413258

Guarino MPL, Altomare A, Emerenziani S, Di Rosa C, Ribolsi M, Balestrieri P, Iovino P, Rocchi G, Cicala M (2020) Mechanisms of action of prebiotics and their effects on gastro-intestinal disorders in adults. Nutrients 12(4):1037

Gui Q, Wang A, Zhao X, Huang S, Tan Z, Xiao C, Yang Y (2020) Effects of probiotic supplementation on natural killer cell function in healthy elderly individuals: a meta-analysis of randomized controlled trials. Eur J Clin Nutr 74:1630–1637

Gutierrez-Castrellon P, Lopez-Velazquez G, Diaz-Garcia L, Jimenez-Gutierrez C, Mancilla-Ramirez J, Estevez-Jimenez J, Parra M (2014) Diarrhea in preschool children and Lactobacillus reuteri: a randomized controlled trial. Pediatrics 133:e904-909

Hemarajata P, Versalovic J (2013) Effects of probiotics on gut microbiota: mechanisms of intestinal immunomodulation and neuromodulation. Therap Adv Gastroenterol 6:39–51

Hempel S, Newberry S, Ruelaz A, Wang Z, Miles JNV, Suttorp MJ, Johnsen B, Shanman R, Slusser W, Fu N, Smith A, Roth B, Polak J, Motala A, Perry T, Shekelle PG (2011) Safety of probiotics used to reduce risk and prevent or treat disease. Evid Rep Technol Assess (full Rep) 200:1–645

Hennequin C, Kauffmann-Lacroix C, Jobert A, Viard JP, Ricour C, Jacquemin JL, Berche P (2000) Possible role of catheters in Saccharomyces Boulardii Fungemia. EJCMID 19:16–20

Honeycutt TCB, El Khashab M, Wardrop RM, McNeal-Trice K, Honeycutt ALB, Christy CG, Mistry K, Harris BD, Meliones JN, Kocis KC (2007) Probiotic administration and the incidence of nosocomial infection in pediatric intensive care: a randomized placebo-controlled trial. Pediatr Crit Care Med 8(5):452–458

Horiuchi H, Sasaki Y (2012) Short communication: effect of oxygen on symbiosis between Lactobacillus bulgaricus and Streptococcus thermophilus. J Dairy Sci 95:2904–2909

Indrio F, Di Mauro A, Di Mauro A, Riezzo G, Panza R, Cavallo L, Francavilla R (2015) Prevention of functional gastrointestinal disorders in neonates: clinical and socioeconomic impact. Benef Microbes 6:195–198

Jain PK, McNaught CE, Anderson ADG, MacFie J, Mitchell CJ (2004) Influence of synbiotic containing Lactobacillus acidophilus La5, Bifidobacterium lactis Bb 12, Streptococcus thermophilus, Lactobacillus bulgaricus and oligofructose on gut barrier function and sepsis in critically ill patients: a randomised controlled trial. Clin Nutr 23:467–475

Jandhyala SM, Talukdar R, Subramanyam C, Vuyyuru H, Sasikala M, Reddy DN (2015) Role of the normal gut microbiota. World J Gastroenterol 21:8787–8803

Jayanama K, Theou O (2020) Effects of Probiotics and Prebiotics on Frailty and Ageing: A Narrative Review. Curr Clin Pharmacol 15:183–192

Judkins TC, Archer DL, Kramer DC, Solch RJ (2020) Probiotics, nutrition, and the small intestine. Curr Gastroenterol Rep 22:2

Kanner J, Lapidot T (2001) The stomach as a bioreactor: dietary lipid peroxidation in the gastric fluid and the effects of plant-derived antioxidants. Free Radic Biol Med 31:1388–1395

Kho ZY, Lal SK (2018) The human gut microbiome – a potential controller of wellness and disease. Front Microbiol 9:1835

Kimoto-Nira H, Suzuki C, Kobayashi M, Sasaki K, Kurisaki J-I, Mizumachi K (2007) Anti-ageing effect of a lactococcal strain: analysis using senescence-accelerated mice. Br J Nutr 98:1178–1186

Kleessen B, Sykura B, Zunft HJ, Blaut M (1997) Effects of inulin and lactose on fecal microflora, microbial activity, and bowel habit in elderly constipated persons. Am J Clin Nutr 65:1397–1402

Koh A, De Vadder F, Kovatcheva-Datchary P, Bäckhed F (2016) From dietary fiber to host physiology: short-chain fatty acids as key bacterial metabolites. Cell 165:1332–1345

Ku W, Lau D, Huen K (2006) Probiotics provoked d-lactic acidosis in short bowel syndrome: case report and literature review. Hong Kong J Paediatr 11:246–254

Kumar M, Nagpal R, Kumar R, Hemalatha R, Verma V, Kumar A, Chakraborty C, Singh B, Marotta F, Jain S, Yadav H (2012) Cholesterol-lowering probiotics as potential biotherapeutics for metabolic diseases. Exp Diabetes Res 2012:902917

Land MH, Rouster-Stevens K, Woods CR, Cannon ML, Cnota J, Shetty AK (2005) Lactobacillus sepsis associated with probiotic therapy. Pediatrics 115:178–181

Landete JM, Gaya P, Rodríguez E, Langa S, Peirotén Á, Medina M, Arqués JL (2017) Probiotic bacteria for healthier aging: immunomodulation and metabolism of phytoestrogens. Biomed Res Int 2017:5939818

Larsbrink J, Rogers TE, Hemsworth GR, McKee LS, Tauzin AS, Spadiut O, Klinter S, Pudlo NA, Urs K, Koropatkin NM, Creagh AL, Haynes CA, Kelly AG, Cederholm SN, Davies GJ, Martens EC, Brumer H (2014) A discrete genetic locus confers xyloglucan metabolism in select human gut Bacteroidetes. Nature 506:498–502

Lecerf J-M, Dépeint F, Clerc E, Dugenet Y, Niamba CN, Rhazi L, Cayzeele A, Abdelnour G, Jaruga A, Younes H, Jacobs H, Lambrey G, Abdelnour AM, Pouillart PR (2012) Xylo-oligosaccharide (XOS) in combination with inulin modulates both the intestinal environment and immune status in healthy subjects, while XOS alone only shows prebiotic properties. Br J Nutr 108:1847–1858

Lee DE, Huh C-S, Ra J, Choi I-D, Jeong J-W, Kim S-H, Ryu JH, Seo YK, Koh JS, Lee J-H, Sim J-H, Ahn Y-T (2015) Clinical evidence of effects of Lactobacillus plantarum HY7714 on skin aging: a randomized, double blind, placebo-controlled study. J Microbiol Biotechnol 25:2160–2168

Lei M, Hua L-M, Wang D-W (2016) The effect of probiotic treatment on elderly patients with distal radius fracture: a prospective double-blind, placebo-controlled randomised clinical trial. Benef Microbes 7:631–637

Lei M, Guo C, Wang D, Zhang C, Hua L (2017) The effect of probiotic Lactobacillus casei Shirota on knee osteoarthritis: a randomised double-blind, placebo-controlled clinical trial. Benef Microbes 8:697–703

Lherm T, Monet C, Nougière B, Soulier M, Larbi D, Le Gall C, Caen D, Malbrunot C (2002) Seven cases of fungemia with Saccharomyces boulardii in critically ill patients. Intensive Care Med 28:797–801

Lin S-H, Chou L-M, Chien Y-W, Chang J-S, Lin C-I (2016) Prebiotic effects of Xylooligosaccharides on the improvement of microbiota balance in human subjects. Gastroenterol Res Pract 2016:e5789232

Liu Y, Gibson GR, Walton GE (2016) An in vitro approach to study effects of prebiotics and probiotics on the faecal microbiota and selected immune parameters relevant to the elderly. PLoS ONE 11:e0162604

Lu VB, Gribble FM, Reimann F (2018) Free fatty acid receptors in enteroendocrine cells. Endocrinology 159:2826–2835

Magrone T, Jirillo E (2013) The interplay between the gut immune system and microbiota in health and disease: nutraceutical intervention for restoring intestinal homeostasis. Curr Pharm Des 19:1329–1342

Mäkeläinen H, Saarinen M, Stowell J, Rautonen N, Ouwehand AC (2010) Xylo-oligosaccharides and lactitol promote the growth of Bifidobacterium lactis and Lactobacillus species in pure cultures. Benef Microbes 1:139–148

Maldonado Galdeano C, Cazorla SI, Lemme Dumit JM, Vélez E, Perdigón G (2019) Beneficial effects of probiotic consumption on the immune system. ANM 74:115–124

Margreiter M, Ludl K, Phleps W, Kaehler ST (2006) Therapeutic value of a Lactobacillus gasseri and Bifidobacterium longum fixed bacterium combination in acute diarrhea: a randomized, double-blind, controlled clinical trial. Int J Clin Pharmacol Ther 44:207–215

Marteau P (2001) Safety aspects of probiotic products. Näringsforskning 45:22–24

Menees S, Chey W (2018) The gut microbiome and irritable bowel syndrome. F1000Res 7:1029

Micka A, Siepelmeyer A, Holz A, Theis S, Schön C (2017) Effect of consumption of chicory inulin on bowel function in healthy subjects with constipation: a randomized, double-blind, placebo-controlled trial. Int J Food Sci Nutr 68:82–89

Miene C, Weise A, Glei M (2011) Impact of polyphenol metabolites produced by colonic microbiota on expression of COX-2 and GSTT2 in human colon cells (LT97). Nutr Cancer 63:653–662

Mohajeri MH, La Fata G, Steinert RE, Weber P (2018) Relationship between the gut microbiome and brain function. Nutr Rev 76:481–496

Morowitz MJ, Carlisle E, Alverdy JC (2011) Contributions of intestinal bacteria to nutrition and metabolism in the critically Ill. Surg Clin North Am 91:771–785

Muñoz P, Bouza E, Cuenca-Estrella M, Eiros JM, Pérez MJ, Sánchez-Somolinos M, Rincón C, Hortal J, Peláez T (2005) Saccharomyces cerevisiae fungemia: an emerging infectious disease. Clin Infect Dis 40:1625–1634

Nagpal R, Kumar A, Kumar M, Behare PV, Jain S, Yadav H (2012) Probiotics, their health benefits and applications for developing healthier foods: a review. FEMS Microbiol Lett 334:1–15

Nagpal R, Mainali R, Ahmadi S, Wang S, Singh R, Kavanagh K, Kitzman DW, Kushugulova A, Marotta F, Yadav H (2018) Gut microbiome and aging: physiological and mechanistic insights. Nutr Healthy Aging 4(4):267–285
Nagy-Szakal D, Williams BL, Mishra N, Che X, Lee B, Bateman L, Klimas NG, Komaroff AL, Levine S, Montoya JG, Peterson DL, Ramanan D, Jain K, Eddy ML, Hornig M, Lipkin WI (2017) Fecal metagenomic profiles in subgroups of patients with myalgic encephalomyelitis/chronic fatigue syndrome. Microbiome 5:44
Nakagawa H, Shiozaki T, Kobatake E, Hosoya T, Moriya T, Sakai F, Taru H, Miyazaki T (2016) Effects and mechanisms of prolongevity induced by Lactobacillus gasseri SBT2055 in Caenorhabditis elegans. Aging Cell 15:227–236
Nataraj BH, Ali SA, Behare PV, Yadav H (2020) Postbiotics-parabiotics: the new horizons in microbial biotherapy and functional foods. Microb Cell Fact 19:168
Natividad JMM, Verdu EF (2013) Modulation of intestinal barrier by intestinal microbiota: pathological and therapeutic implications. Pharmacol Res 69:42–51
O'Callaghan A, van Sinderen D (2016) Bifidobacteria and their role as members of the human gut microbiota. Front Microbiol 7:925
Oggioni MR, Pozzi G, Valensin PE, Galieni P, Bigazzi C (1998) Recurrent septicemia in an immunocompromised patient due to probiotic strains of Bacillus subtilis. J Clin Microbiol 36:325–326
Okubo T, Ishihara N, Oura A, Serit M, Kim M, Yamamoto T, Mitsuoka T (1992) In vivo effects of tea polyphenol intake on human intestinal microflora and metabolism. Biosci Biotechnol Biochem 56:588–591
Ouwehand AC, Lagström H, Suomalainen T, Salminen S (2002) Effect of probiotics on constipation, fecal azoreductase activity and fecal mucin content in the elderly. Ann Nutr Metab 46:159–162
Parada Venegas D, De la Fuente MK, Landskron G, González MJ, Quera R, Dijkstra G, Harmsen HJM, Faber KN, Hermoso MA (2019) Short chain fatty acids (SCFAs)-mediated gut epithelial and immune regulation and its relevance for inflammatory bowel diseases. Front Immunol 10:277
Park MR, Oh S, Son SJ, Park D-J, Oh S, Kim SH, Jeong D-Y, Oh NS, Lee Y, Song M, Kim Y (2015) Bacillus licheniformis Isolated from traditional korean food resources enhances the longevity of caenorhabditis elegans through serotonin signaling. J Agric Food Chem 63:10227–10233
Penders J, Stobberingh EE, van den Brandt PA, Thijs C (2007) The role of the intestinal microbiota in the development of atopic disorders. Allergy 62:1223–1236
Perapoch J, Planes AM, Querol A, López V, Martínez-Bendayán I, Tormo R, Fernández F, Peguero G, Salcedo S (2000) Fungemia with Saccharomyces cerevisiae in two newborns, only one of whom had been treated with ultra-levura. Eur J Clin Microbiol Infect Dis 19:468–470
Pham V, Lacroix C, Braegger C, Chassard C (2017) Lactate-utilizing community is associated with gut microbiota dysbiosis in colicky infants. Sci Rep 7(1):11176
Phillips J, Muir JG, Birkett A, Lu ZX, Jones GP, O'Dea K, Young GP (1995) Effect of resistant starch on fecal bulk and fermentation-dependent events in humans. Am J Clin Nutr 62:121–130
Plaza-Diaz J, Ruiz-Ojeda FJ, Gil-Campos M, Gil A (2019) Mechanisms of action of probiotics. Adv Nutr 10:S49–S66
Presterl E, Kneifel W, Mayer HK, Zehetgruber M, Makristathis A, Graninger W (2001) Endocarditis by Lactobacillus rhamnosus due to yogurt ingestion? Scand J Infect Dis 33:710–714
Quigley EMM (2019) Prebiotics and probiotics in digestive health. Clin Gastroenterol Hepatol 17:333–344
Rajala SA, Salminen SJ, Seppänen JH, Vapaatalo H (1988) Treatment of chronic constipation with lactitol sweetened yoghurt supplemented with guar gum and wheat bran in elderly hospital in-patients. Compr Gerontol A 2:83–86
Richard V, Van der Auwera P, Snoeck R, Daneau D, Meunier F (1988) Nosocomial bacteremia caused by Bacillus species. Eur J Clin Microbiol Infect Dis 7:783–785
Rietjens IMCM, Louisse J, Beekmann K (2017) The potential health effects of dietary phytoestrogens. Br J Pharmacol 174:1263–1280

Rivière A, Selak M, Lantin D, Leroy F, De Vuyst L (2016) Bifidobacteria and butyrate-producing colon bacteria: importance and strategies for their stimulation in the human gut. Front Microbiol 7:979
Romo-Vaquero M, García-Villalba R, González-Sarrías A, Beltrán D, Tomás-Barberán FA, Espín JC, Selma MV (2015) Interindividual variability in the human metabolism of ellagic acid: contribution of gordonibacter to urolithin production. J Funct Foods 17:785–791
Rossi M, Corradini C, Amaretti A, Nicolini M, Pompei A, Zanoni S, Matteuzzi D (2005) Fermentation of fructooligosaccharides and inulin by Bifidobacteria: a comparative study of pure and fecal cultures. Appl Environ Microbiol 71:6150–6158
Round JL, Mazmanian SK (2009) The gut microbiota shapes intestinal immune responses during health and disease. Nat Rev Immunol 9:313–323
Rowland RH, Ian R (2000) Metabolic activities of the gut microflora in relation to cancer. Microb Ecol Health Dis 12:179–185
Ruiz L, Margolles A, Sánchez B (2013) Bile resistance mechanisms in Lactobacillus and Bifidobacterium. Front Microbiol 4:396
Sanchez JI, Marzorati M, Grootaert C, Baran M, Van Craeyveld V, Courtin CM, Broekaert WF, Delcour JA, Verstraete W, Van de Wiele T (2009) Arabinoxylan-oligosaccharides (AXOS) affect the protein/carbohydrate fermentation balance and microbial population dynamics of the simulator of human intestinal microbial ecosystem. Microb Biotechnol 2:101–113
Sanders ME, Akkermans LM, Haller D, Hammerman C, Heimbach J, Hörmannsperger G, Huys G, Levy DD, Lutgendorff F, Mack D, Phothirath P, Solano-Aguilar G, Vaughan E (2010) Safety assessment of probiotics for human use. Gut Microbes 1:164–185
Sarkar A, Lehto SM, Harty S, Dinan TG, Cryan JF, Burnet PWJ (2016) Psychobiotics and the manipulation of bacteria–gut–brain signals. Trends Neurosci 39:763–781
Savino F, Cordisco L, Tarasco V, Palumeri E, Calabrese R, Oggero R, Roos S, Matteuzzi D (2010) Lactobacillus reuteri DSM 17938 in infantile colic: a randomized, double-blind, placebo-controlled trial. Pediatrics 126:e526–e533
Shenderov BA (2013) Metabiotics: novel idea or natural development of probiotic conception. Microb Ecol Health Dis 24. https://doi.org/10.3402/mehd.v24i0.20399
Shimada Y, Yasuda S, Takahashi M, Hayashi T, Miyazawa N, Sato I, Abiru Y, Uchiyama S, Hishigaki H (2010) Cloning and expression of a novel NADP (H)-dependent daidzein reductase, an enzyme involved in the metabolism of daidzein, from equol-producing Lactococcus strain 20-92. Appl Environ Microbiol 76:5892–5901
Shinohara R, Sasaki K, Inoue J, Hoshi N, Fukuda I, Sasaki D, Kondo A, Osawa R (2019) Butyryl-CoA:acetate CoA-transferase gene associated with the genus Roseburia is decreased in the gut microbiota of Japanese patients with ulcerative colitis. Biosci Microbiota Food Health 38:159–163
Slavin J (2013) Fiber and prebiotics: mechanisms and health benefits. Nutrients 5:1417–1435
Stojanov S, Berlec A, Štrukelj B (2020) The Influence of probiotics on the Firmicutes/Bacteroidetes ratio in the treatment of obesity and inflammatory bowel disease. Microorganisms 8(11):1715
Sugimura T, Jounai K, Ohshio K, Suzuki H, Kirisako T, Sugihara Y, Fujiwara D (2018) Long-term administration of pDC-stimulative Lactococcus lactis strain decelerates senescence and prolongs the lifespan of mice. Int Immunopharmacol 58:166–172
Takiishi T, Fenero CIM, Câmara NOS (2017) Intestinal barrier and gut microbiota: Shaping our immune responses throughout life. Tissue Barriers 5(4):e1373208
Tanaka R, Shimosaka K (1982) Investigation of the stool frequency in elderly who are bed ridden and its improvements by ingesting bifidus yogurt. Nihon Ronen Igakkai Zasshi 19:577–582
Teuri U, Korpela R (1998) Galacto-oligosaccharides relieve constipation in elderly people. Ann Nutr Metab 42:319–327
Thaiss CA, Zmora N, Levy M, Elinav E (2016) The microbiome and innate immunity. Nature 535:65–74
Theou O, Jayanama K, Fernández-Garrido J, Buigues C, Pruimboom L, Hoogland AJ, Navarro-Martínez R, Rockwood K, Cauli O (2019) Can a prebiotic formulation reduce frailty levels in older people? J Frailty Aging 8:48–52

Thursby E, Juge N (2017) Introduction to the human gut microbiota. Biochem J 474:1823–1836
Tsuji R, Komano Y, Ohshio K, Ishii N, Kanauchi O (2018) Long-term administration of pDC stimulative lactic acid bacteria, Lactococcus lactis strain Plasma, prevents immune-senescence and decelerates individual senescence. Exp Gerontol 111:10–16
Tu M-Y, Chen H-L, Tung Y-T, Kao C-C, Hu F-C, Chen C-M (2015) Short-Term effects of kefir-fermented milk consumption on bone mineral density and bone metabolism in a randomized clinical trial of osteoporotic patients. PLoS ONE 10:e0144231
Tzounis X, Rodriguez-Mateos A, Vulevic J, Gibson GR, Kwik-Uribe C, Spencer JPE (2011) Prebiotic evaluation of cocoa-derived flavanols in healthy humans by using a randomized, controlled, double-blind, crossover intervention study. Am J Clin Nutr 93:62–72
Vallejo-Cordoba B, Castro-López C, García HS, González-Córdova AF, Hernández-Mendoza A (2020) Postbiotics and paraprobiotics: a review of current evidence and emerging trends, chap. 1. In: da Cruz AG, Prudencio ES, Esmerino EA, da Silva MC (eds) Advances in Food and Nutrition Research, Probiotic and Prebiotics in Foods: Challenges. Academic Press, Innovations and Advances, pp 1–34
Vandenplas Y, Greef ED, Veereman G (2014) Prebiotics in infant formula. Gut Microbes 5:681–687
Veckman V, Miettinen M, Pirhonen J, Sirén J, Matikainen S, Julkunen I (2004) Streptococcus pyogenes and Lactobacillus rhamnosus differentially induce maturation and production of Th1-type cytokines and chemokines in human monocyte-derived dendritic cells. J Leukoc Biol 75:764–771
Vinolo MAR, Rodrigues HG, Nachbar RT, Curi R (2011) Regulation of inflammation by short chain fatty acids. Nutrients 3:858–876
Vulevic J, Juric A, Walton GE, Claus SP, Tzortzis G, Toward RE, Gibson GR (2015) Influence of galacto-oligosaccharide mixture (B-GOS) on gut microbiota, immune parameters and metabonomics in elderly persons. Br J Nutr 114:586–595
Wang B, Yao M, Lv L, Ling Z, Li L (2017) The human microbiota in health and disease. Engineering 3:71–82
Wang Y, Qin S, Jia J, Huang L, Li F, Jin F, Ren Z, Wang Y (2019) Intestinal microbiota-associated metabolites: crucial factors in the effectiveness of herbal medicines and diet therapies. Front Physiol 10:1343
Whisner CM, Castillo LF (2018) Prebiotics, bone and mineral metabolism. Calcif Tissue Int 102:443–479
Yan F, Polk DB (2011) Probiotics and immune health. Curr Opin Gastroenterol 27:496–501
Zhao Y, Zhao L, Zheng X, Fu T, Guo H, Ren F (2013) Lactobacillus salivarius strain FDB89 induced longevity in Caenorhabditis elegans by dietary restriction. J Microbiol 51:183–188

Chapter 6
Nutritional Hormetins in Ageing and Longevity

Suresh I. S. Rattan

Abstract Unlike proteins, carbohydrates and fats as the main macromolecules necessary for biological structure and functions, numerous other chemicals present in the food have a paradoxical nature. Various dietary components in the food are actually toxins which induce molecular damages in a biphasic dose response. A successful handling of the damage, through a complex series of biochemical pathways of maintenance and repair can result in achieving health beneficial effects in the body. This is the phenomenon of mild stress-induced hormesis, and the conditions causing hormesis are known as hormetins. Food components, which induce one or more pathways of stress response in cells and organisms, are potential hormetins which can have health beneficial, anti-ageing and longevity promoting effects. Various dietary components present in spices, herbs, and numerous medicinal plants, and some vitamins, trace elements, minerals and others are such hormetins. These nutritional hormetins have almost no nutritional value in terms of providing proteins, carbohydrates and fats, but have pleiotropic biological effects in maintaining, recovering, and enhancing health and longevity.

Keywords Stress · Lifespan · Homeostasis · Homeodynamics · Health · Poisons · Toxins

6.1 Introduction

Hormetins are the conditions—physical, chemical, biological and psychological—which can bring about health beneficial effects by inducing stress responses in a living system (Rattan 2017). Exercise is the most well-known example as a physical hormetin, which initially induces the production of free radicals, acids and cell distortions, but eventually leads to numerous health benefits (Pingitore et al. 2015).

S. I. S. Rattan (✉)
Department of Molecular Biology and Genetics, Aarhus University, 8000 Aarhus C, Denmark
e-mail: rattan@mbg.au.dk

S. I. S. Rattan and G. Kaur (eds.), *Nutrition, Food and Diet in Ageing and Longevity*,
Healthy Ageing and Longevity 14, https://doi.org/10.1007/978-3-030-83017-5_6

This phenomenon of stress-induced potential health benefits is known as physiological hormesis (Calabrese et al. 2007, Rattan and Le Bourg 2014, Rattan and Kyriazis 2019).

Nutritional hormetins are the natural compounds present in the food, which actually induce stress, but can elicit a range of biological effects, such as antioxidative, anti-inflammatory, prebiotic, antibiotic, and metabolic regulatory effects. Almost all spices, herbs and various other so-called medicinal plants come under this category of being nutritional hormetins, without providing any significant amounts of the usual nutritional components—proteins, carbohydrates and lipids. This article aims to give a brief overview of the phenomenon, principles and mechanisms of hormesis, and to discuss some examples of natural and synthetic nutritional hormetins, especially with respect to their effects on health, ageing and longevity.

6.2 Hormesis and Its Molecular Basis

Hormesis is the non-linear stress response often described using various terms addressing the shape of the dose–response curve as biphasic, bimodal, bitonic, U-shaped, inverted-U-shaped, J-shaped, nonmonotonic, functional antagonism and stimulatory inhibitory curve (Calabrese and Baldwin 2001b). Furthermore, terms such as adaptive response, preconditioning, autoprotection, heteroprotection, and paradoxical effects have also been used to describe the nature of the biological dose response patterns (Calabrese and Baldwin 2001a). However, the term hormesis, which was coined originally in 1943 (Southam and Ehrlich 1943), has now been widely adopted as a result of a collective article written by more than 50 leading researchers from a variety of disciplines, including biomedical gerontology (Calabrese et al. 2007). Further refinements of the term hormesis were also made in that article, such as conditioning hormesis, post-exposure conditioning hormesis and physiological hormesis, dealing with specific aspects of the biological consequences of stress.

The key conceptual features of hormesis are the disruption of homeostasis, the modest overcompensation during counter-action, the reestablishment of homeodynamics and the adaptive nature of the process (Rattan and Le Bourg 2014, Rattan and Kyriazis 2019). The ultimate consequences of stress can be both harmful and beneficial depending both on the intensity, duration and frequency of the stress, and on the physiological price paid in terms of energy utilisation and other metabolic disturbances. But the most important aspect of stress response is that it is not monotonic with respect to the dose of the stressor, and is almost always characterized by a nonlinear biphasic relationship (Calabrese and Mattson 2017). Another frequent observation in studies of hormesis is that a single hormetic agent can improve the overall physiological characteristics such as enhanced immunity, improved cognition, and increased tolerance to other stresses, in a holistic process of biological amplification of health beneficial effects (Rattan 2008).

Physiological hormesis is the phenomenon in which low doses of a stressor, which is generally toxic at high doses, could be beneficial for the system. The terms "stressor" and "stress" are defined as any physical, chemical or biological factors that disrupt the homeostasis of a biological system and initiate a series of events in order to counteract, adapt and survive (Rattan 2008, Giudice et al. 2018). This intrinsic and dynamic ability of all living systems to respond, to counteract and to adapt to the external and internal sources of disturbance is known as homeodynamic space (Rattan 2020a). A wide range of molecular, cellular and physiological pathways are the basis of this homeodynamic ability, and these range from multiple pathways of nuclear and mitochondrial DNA repair to free radical counteracting mechanisms, protein turnover and repair, detoxification mechanisms, and other processes including immune- and stress-responses (Rattan 2019). All these processes involve numerous genes whose products and their interactions give rise to a "homeodynamic space" or the "buffering capacity", which is the ultimate determinant of an individual's chance and ability to survive and maintain a healthy state (Rattan 2020a).

A critical component of the homeodynamic space of living systems is the capacity to respond to stress, which also provides the molecular basis to the phenomenon of hormesis. Table 6.1 gives a list of the main molecular stress responses (SR), some examples of commons stressors, and various immediate and late molecular sensors and effectors, which are integral to the organismic property of homeodynamics and the phenomenon of hormesis (Bhattacharya and Rattan 2019, Calabrese and Kozumbo 2021).

Each of these SR pathways is well understood in terms of its molecular biology, and several excellent review articles can be found in the published literature and online sources. Briefly, and as discussed in detail previously (Bhattacharya and Rattan 2019), autophagy response is the lysosome-mediated and chaperone-mediated sequestering of damaged membranes and organelles, which is a SR induced during nutritional limitation, starvation, and hypoxia (Ryter and Choi 2013, Filomeni et al. 2015). Heat shock response (HSR) is a universal and primordial stress response to protein denaturation, achieved by the activation of the HS transcription factors, and followed by the preferential synthesis of several heat shock proteins (HSP) (Verbeke et al. 2001, Ciocca and Caderwood 2005, Leak 2014). Similarly, accumulation of misfolded proteins in the endoplasmic reticulum (ER) leads to the so-called ER stress response, also known as the unfolded protein response (UPR), resulting in the synthesis, activation and translocation of several chaperones (Lee 2001, Lin et al. 2007, Yoshida 2007, Banhegyi et al. 2007). Other primary SR include hypoxia-induced SR via hypoxia-inducible factor, DNA damage response involving several DNA repair genes, inflammatory response by the activation of NFkB, and energy and nutritional deficiency SR involving AMPK, mTOR, sirtuins and the deacetylation of histones and other proteins in response to reduced levels of metabolic energy (Bhattacharya and Rattan 2019).

In the context of nutritional hormetins, a widely studied pathway through which cells respond to oxidative stress is the Nrf2-mediated regulation of transcription of antioxidant genes (Calabrese and Kozumbo 2021). The main regulator of this specific antioxidant phenotype is the transcription factor Nrf2 which regulates the basal and inducible expression of numerous detoxifying and antioxidant genes (Motohashi and Yamamoto 2004, Singh et al. 2010). Under normal conditions, Nrf2 is held in the cytoplasm by the specific inhibitory protein Keap1. Oxidative modification of cysteine residues of KEAP1 induces conformational changes and a loss of Nrf-2 binding, allowing Nrf2 to translocate to the nucleus where it heterodimerizes with specific co-factors, and leads to the transcription of various genes through the regulatory regions of antioxidant response elements (AREs). Some of the genes activated by stress-induced activation of Nrf2 are heme oxygenas-1 (HO-1), NAD(P)H-quinone oxidoreductase-1 (NQO1), and glutathione S-transferases (GSTs) (Bhattacharya and Rattan 2019).

Although there may be some overlap, generally the initial induction of SR is quite stressor-specific. The specificity of the response is mostly determined by the nature of the initial damage induced by the stressor and the variety of downstream effectors involved (Bhattacharya and Rattan 2019). For example, cytoplasmic induction of protein denaturation by heat, heavy metals and antibiotics will initiate HSR by inducing the synthesis of HSP followed by the activation of proteasome-mediated protein degradation. But, unfolded proteins in the ER will induce UPR, and will initiate the induction of synthesis of a totally different set of proteins and their

Table 6.1 Main molecular level stress response pathways as the basis of hormesis

Stress response *(arranged in alphabetical order)*	Some common stressors	Molecular sensors and effectors
Autophagic response	Nutritional limitation, hypoxia, damaged organelles	Autophagosomes, lysosomes
DNA damage response (DDR)	Radiation, reactive oxygen species	DNA damage sensors (ATM, ATR), p53, DNA repair proteins
Energy-deficiency response	Energy depletion, metabolic imbalance	Sirtuins
Heat shock response (HSR)	Heat, exercise, hypergravity, heavy metals, natural and synthetic small molecules, antibiotics	Heat shock transcription factors, heat shock proteins, chaperones, proteasome
Inflammatory response	Pathogens, allergens, damaged macromolecules	Nuclear factor-κB transcription factors, cytokines, nitric oxide synthase
Oxidative stress response	Oxidants, free radicals, reactive oxygen species	Transcription factors (Nrf2, FOXO), hemeoxygenase, antioxidative enzymes (SOD, catalase, glutathione etc.)
Unfolded protein response (UPR)	Unfolded and misfolded proteins in endoplasmic reticulum (ER) and mitochondria, cytokines	ER-chaperones, mitochondrial-chaperonins, proteasome

downstream effectors. Similarly, whereas oxidative damage to proteins will generally initiate Nrf2-mediated antioxidant response, damage to DNA by free radicals or other agents will result in the activation of DNA repair enzymes. In the same vein, whereas nutritional deprivation and low energy levels will activate autophagy and FOXO-sirtuin pathways, infections and antigenic challenge will generally initiate pro-inflammatory NFκB response (Bhattacharya and Rattan 2019).

However, often the source (stressor) of activation of SR cannot be easily identified, and may involve more than one stressor and their effectors. Examples of such SR include early inflammatory SR and neuroendocrinal SR, which lead to the synthesis and release of interleukins and corticoid hormones, respectively. Similarly, pathways involving NF-κB, Nrf2, FOXO, sirtuins and heme-oxygenase (HO) activation may involve more than one type of stressors and stress signals, including pro-oxidants, free radicals, reactive oxygen species (ROS), and nutritional components (Bhattacharya and Rattan 2019). Following the initial activation of one or more primary SR, higher order (cellular, organ level and body level) responses are manifested, which include apoptosis, inflammation, and hyper-adrenocorticism etc. (Giudice et al. 2018).

Therefore, although the exact nature of the initial molecular damage caused by a compound may not be easily identified, activation of one or more primary SR pathways is a good indicator of the potential nature of a compound as a hormetin.

6.3 Hormetins in Food

Nutritional hormetins are the hormesis-inducing chemicals present in the food, which bring about any health beneficial effects by initially causing stress (damage) in a biphasic dose response. As discussed in the earlier section, damage-induced activation of one or multiple pathways of SR can lead to potentially health beneficial effects by removing, repairing or adapting to the causative damages. Often the biological end point, such as anti-oxidative effects observed after exposure to a food component, is mistakenly interpreted as if the original chemical compound was an anti-oxidant. In reality, however, such compounds are not direct anti-oxidants, but are hormetins (Ali and Rattan 2006, Rattan 2017), which are sometimes also referred to as adaptogens (Wiegant et al. 2009, Panossian et al. 2021) or indirect anti-oxidants (Gutteridge and Halliwell 2010, Sadowska-Bartosz and Bartosz 2014)}. Various dietary components present in spices, herbs, and numerous medicinal plants, and some vitamins, trace elements, minerals and others are actually hormetins (Franco et al. 2019).

As an example of a prevalent misinterpretation of the biochemical nature of a compound as being antioxidant is the induction of Nrf2-mediated oxidative SR which leads to antioxidative biological effects (Calabrese and Kozumbo 2021). Activation of Nrf2 transcription factor follows the electrophilic modification/damage of its inhibitor protein Keap1, which then leads to the accumulation, heterodimerization, nuclear translocation and DNA binding of Nrf2 at the antioxidant response element (Balstad et al. 2011). This binding results in the downstream expression of a large number of the so-called anti-oxidant genes, such as haeme-oxygenease-1

(HO-1), superoxide dismutase, glutathione, and catalase, giving rise to a biologically antioxidative result. Some well-known phytochemicals, which strongly induce Nrf2-mediated SR, include curcumin, quercetin, genistein and eugenol and food extracts, such as coffee, turmeric, rosemary, broccoli, thyme, clove and oregano (Balstad et al. 2011, Lima et al. 2011, Calabrese and Kozumbo 2021. In other cases, such as alpha lipoic acid and coenzyme Q10, it is their pro-oxidant activity in producing hydrogen peroxide, which induces defensive responses {Linnane, 2007 #6132). Similarly, certain mimetics of superoxide dismutase also appear to work through hormetic pathways by inducing oxidative SR (Melov et al. 2000, Bayne and Sohal 2002, Keany et al. 2004). None of these compounds should be called as anti-oxidants.

Another SR pathway, which has been studied in detail, and can be the basis for identifying nutritional hormetins is the HSR. Induction of proteotoxic stress, such as protein misfolding and denaturation initiate HSR, leads to the activation of the processes of protein repair, refolding, and selective degradation of abnormal proteins cleaning up the debris and an overall improvement in the structure and function of the cells (Verbeke et al. 2001, Ciocca and Caderwood 2005, Leak 2014). Various phytochemicals and nutritional components have been shown to induce HSR and have health beneficial effects. These potential hormetins include phenolic acids, polyphenols, flavonoids, ferulic acid, geranylgeranyl, rosmarinic acid, kinetin, zinc and the extracts of tea, dark chocolate, saffron and spinach (Ambra et al. 2004, Son et al. 2008, Berge et al. 2008, Barone et al. 2009, Sonneborn 2010, Wieten et al. 2010).

Components of various medicinal plants used frequently in the traditional Chinese medicine (TCM) and in the Indian Ayurvedic system of medicine appear to be achieved through hormetic pathways. For example, celastrols and paeoniflorin present in some medicinal herbs used in TCM, have cytoprotective effects and induce HSP in human cells (Westerheide et al. 2004, Yan et al. 2004). Other pathways of SR, which are involved in initiating hormetic effects of nutritional components are the NFkB, FOXO, sirtuins, DNA repair response and autophagy pathways. Resveratrol and some other mimetics of calorie restriction work by the induction of one or more of these SR pathways (Longo 2009, Sonneborn 2010).

With respect to ageing and longevity, Table 6.2 provides some examples of the purified chemical entities from various food sources, which have shown stress-induced health beneficial hormetic effects. Most of these studies have used isolated active compounds and tested them individually for their effects on survival, ageing and lifespan, using experimental model systems, such as the fruitfly *Drosophila,* the nematode *Caenorhabditis elegans,* rodents rats and mice, and human cells in culture. Some extension of lifespan and healthspan, along with slowing down of the emergence of various ageing markers, is a common feature of all such studies, some of which are cited in the Table 6.2. What makes these food components qualify as nutritional hormetins is the fact that all these purified compounds from food sources induce one or more SR described in the earlier section.

As regards the chemical nature of these nutritional hormetins affecting ageing, lifespan and healthspan, the major chemical categories of identified so far are polyphenols, flavonoids, and terpenoids. Additionally, vitamins, minerals, trace

Table 6.2 Examples of food-based hormetin molecules reported for their effects on ageing and lifespan of experimental model systems

Main chemical category	Sub-chemical category	Active compounds	Natural food sources	Ageing- and longevity-related references
Polyphenols and flavonoids	Curcuminoids	Curcumin	Turmeric, mustard	(Lee et al. 2010, Lima et al. 2011, Zhu et al. 2015)
	Catechins	Epigallocatechin gallate (EGCG)	Green tea, black tea, grapes	(Kumar et al. 2019, Unno 2016, Lopez et al. 2014)
	Hydroxyflavanol	Fisetin	Apples, grapes, strawberries	(Zhu et al., 2017, Yousefzadeh et al., 2018)
	Stilbenoids	Resveratrol	Grapes, peanuts, cocoa, blueberries	(Valenzano et al. 2006, Li et al. 2017)
	Tannin	Gallotanin, digallic acid, urolithin	Grapes, Pomegranate	(Saul et al. 2010, Ryu et al. 2016)
	Anthocyanins	Cyanidin	Black berry, black rice, red cabbage, purple corn	(Choi et al. 2010, Buonocore et al. 2012)
Terpenoids	Mono-, di-, tri-tetra-terpenes	β-caryophyllene, camphor, ursolic acid, ginsenoside, carotene, lycopene, lutein	Pine, rosemary, coriander, kale, spinatch, asparagus, broccoli, pepper, lettuce, apricot	(Pant et al. 2014, Proshkina et al. 2020)
Other chemical categories	Vitamins	Vitamin-D Vitamin-C	Red meat, oily fish, egg yolk Citrus fruits, kiwi, papaya	(Chen et al. 2019, Sadowska-Bartosz and Bartosz 2014, Burger et al. 2017)
	Minerals and trace-elements	Zinc Selenium, Lithium	Shellfish, meat, legumes Grains, mustard, kelp, pistachio	(Mocchegiani et al. 2006, Putics et al. 2008, Cabrera 2015, Sonneborn 2010, McColl et al. 2008, Castillo-Quan et al. 2016)

(continued)

Table 6.2 (continued)

Main chemical category	Sub-chemical category	Active compounds	Natural food sources	Ageing- and longevity-related references
	DNA, RNA, protein components	Caffeine	Coffee beans, cocoa nuts, guarana berries	(Paganini-Hill et al. 2007, Wanke et al. 2008, Brunquell et al. 2018)
		Cytokinins—kinetin, zeatin	Coconut milk, maize	(Rattan and Clark 1994, Rattan and Sodagam 2005, Kadlecova et al. 2018)
		Dipeptide carnosine	Muscle tissues of beef, pork, turkey	(McFarland and Holliday 1994, Hipkiss et al. 2001, Calabrese et al. 2011)

elements, and modified bases of nucleic acids are other chemical categories of nutritional hormetins in the context of ageing and longevity (Table 6.2).

Although the exact nature of the initial molecular damage caused by such hormetins may not be easily identified, an activation of one or more SR is a good indicator of the immediate and primary action of the compound. However, induction of a specific SR pathway as the first response (immediate response) does not rule out the induction of one or more other SR pathways later on (delayed response). A complete and successful SR for effective homeodynamics and for the maintenance of the homeodynamic space includes both immediate and delayed SR. Furthermore, as argued previously (Demirovic and Rattan 2013), it is important that all SR pathways are analysed simultaneously and a complete stress response profile is established under a given condition, such as age-, health- and disease status, and during and after exposure to single or multiple hormetins. Being able to map the kinetics and amplitude of different SR, and their effects on each other, can form the basis to evaluate the health status of an individual and to develop effective means of aging modulators and maintainers of homeodynamic space.

6.4 Hormetins as Nutrition, Food and Diet

It seems that non-nutritional foods (in the sense that they do not provide any significant amounts of proteins, carbohydrates and fats) have always been an integral component of the dietary habits of human beings. Such food categories include mainly spices, herbs and other plants claimed to have various medicinal and other beneficial properties, and enhance the flavor and taste of the food. Different cultural practices have given rise to a fascinatingly wide variety of food preparation and consumption methods, along with various health-associated claims. But there is no universally agreed notion of the best food for health and longevity. The mythology of food and the cultural practices are full of stories and claims, which may or may not be scientifically verified *(see the chapter by Ilias Stambler in this book).*

Modern scientific analyses of numerous food sources and non-food sources have identified individual active molecules which account for their specific biological effects. Such discoveries have also led to the successful development of some very effective drugs, such as aspirin, atropine, digoxin, colchicine, morphine, quinine and many others used in therapy against many diseases. Nutritional hormetins are a novel category of "drugs for health", which can stimulate body's intrinsic abilities for maintenance, repair, tolerance and adaptation towards strengthening its homeodynamic space and survival (Rattan 2017).

In practice, nutritional hormetins in food are already being used for thousands of years in the dietary patterns of different societies with different food cultures and habits; and there are numerous folk-remedies and recommendations based on their experiences (Gerber et al. 1999, Rahman 2003). There are also epidemiological data showing the correlation between certain food habits and the occurrence of various diseases in different populations (Martucci et al. 2017, Chopan and Littenberg 2017). However, modern scientific research, generally by using experimental model systems, has identified the active compounds in some of those foods, and has provided mechanistic explanations and information about the concentrations of those compounds required to gain health benefits.

An important issue to emerge from such studies is that, owing to the serious issues of low bio-availability of active compounds in vivo, it is almost impossible to achieve those relatively high doses from normal dietary composition and consumption of food. That is why there is a great incentive to develop and market single- or multiple-hormetin products in the form of nutraceuticals, cosmeceuticals and food supplement products (Rattan et al. 2013). Unfortunately, in the absence of any rigorous tests performed on human beings, these commercial products are often based on naïve extrapolations from model organisms, anecdotal evidence, overhyped claims and false promises (Rattan 2020b).

6.5 Challenges and Unresolved Issues

We now understand that several things that we eat as food in our diet, and which are considered to be generally health beneficial, are actually damage-causing stressors, termed as nutritional hormetins. Plants generally produce them as toxic metabolites for self-defense (Pallauf et al. 2016, Butt et al. 2018). It is basically with trial-and-error over long periods of cultural practices that humans have developed various methods of cooking, mixing and preserving for safe consumption of such toxins on a regular basis (Harrison and Bartels 2006). Modern scientific and analytical approaches have opened up the possibilities of identifying and using the most active chemicals in these hormetins in the form of nutritional supplements, functional foods and nutraceuticals.

However, several important issues remain to be resolved in order to develop nutritional hormetin-based health care product and recommendations.

1. Determining the interactive and pleiotropic effects of multiple hormetins. This is because hormetins in diet are almost always consumed as a part and combination of the original food sources, such as curcumin in turmeric, capsaicin in chilli peppers, rosemarinic acid in rosemary, resveratrol in grapes or wine and so on where numerous other chemicals are also present.
2. Establishing the regimens for the consumption of hormetins in terms of the quantities, timing and frequency.
3. Adjusting the levels of hormetins to account for age-related and general health-related changes in the sensitivity to stress-inducing compounds.

Finally, it may be useful to recount that not all components of the food are good or bad in a simple and straight forward manner. Unlike the proteins, carbohydrates and fats as the major macromolecules necessary for cell structure and physiological functions, numerous other chemicals and small molecules have a paradoxical nature. This applies especially to the plant metabolites which are generally toxic and off-putting for most animals, but we humans have discovered and invented ways of using those food sources safely. Nutritional hormetins are an integral part of the nutrition, food and diets all around the world with its cultural variations and myths.

Compliance with Ethical Standards

Conflict of Interest The author declares having no conflict of interest.

References

Ali RE, Rattan SIS (2006) Curcumin's biphasic hormetic response on proteasome activity and heatshock protein synthesis in human keratinocytes. Ann NY Acad Sci 1067:394–399

Ambra R, Mocchegiani E, Giacconi R, Canali R, Rinna A, Malavolta M, Virgili F (2004) Characterization of the hsp70 response in lymphoblasts from aged and centenarian subjects and differential effects of in vitro zinc supplementation. Exp Gerontol 39:1475–1484

Balstad TR, Carlsen H, Myhrstad MC, Kolberg M, Reiersen H, Gilen L, Blomhoff R (2011) Coffee, broccoli and spices are strong inducers of electrophile response element-dependent transcription in vitro and in vivo—studies in electrophile response element transgenic mice. Mol Nutr Food Res 55:185–197

Banhegyi G, Baumeister P, Benedetti A, Dong D, Fu Y, Lee AS, Yang W (2007) Endoplasmic reticulum stress. Ann N Y Acad Sci 1113:58–71

Barone E, Calabrese V, Mancuso C (2009) Ferulic acid and its therapeutic potential as a hormetin for age-related diseases. Biogerontology 10:97–108

Bayne AV, Sohal RS (2002) Effects of superoxide dismutase/catalase mimetics on life span and oxidative stress resistance in the housefly Musca domestica. Free Rad Biol Med 32:1229–1234

Berge U, Kristensen P, Rattan SIS (2008) Hormetic modulation of differentiation of normal human epidermal keratinocytes undergoing replicative senescence in vitro. Exp Gerontol 43:658–662
Bhattacharya S, Rattan SIS (2019) Primary stress response pathways for pre-conditioning and physiological hormesis. In: Rattan SIS, Kyriazis M (eds) The science of hormesis in health and longevity. Academic Press, UK, pp 35–54
Brunquell J, Morris S, Snyder A, Westerheide SD (2018) Coffee extract and caffeine enhance the heat shock response and promote proteostasis in an HSF-1-dependent manner in Caenorhabditis elegans. Cell Stress Chaperones 23:65–75
Buonocore D, Lazzeretti A, Tocabens P, Nobile V, Cestone E, Santin G, Marzatico F (2012) Resveratrol-procyanidin blend: nutraceutical and antiaging efficacy evaluated in a placebocontrolled, double-blind study. Clin Cosmet Investig Dermatol 5:159–165
Burger MG, Steinitz A, Geurts J, Pippenger BE, Schaefer DJ, Martin I, Pelttari K (2017) Ascorbic acid attenuates senescence of human osteoarthritic osteoblasts. Int J Mol Sci 18:2517
Butt J, Mir ZA, Tariq H, Arshad S (2018) Poison-drug enigma. EC Pharmacol Toxicol 6(7):542–558
Cabrera AJ (2015) Zinc, aging, and immunosenescence: an overview. Pathobiol Aging Age Relat Dis 5:25592
Calabrese E, Mattson M (2017) How does hormesis impact biology, toxicology and medicine? NPJ Aging Mech Dis 3:13
Calabrese EJ, Bachmann KA, Bailer AJ, Bolger PM, Borak J, Cai L, Mattson MP (2007) Biological stress response terminology: Integrating the concepts of adaptive response and preconditioning stress within a hormetic dose-response framework. Toxicol Appl Pharmacol 222:122–128
Calabrese EJ, Baldwin LA (2001) Hormesis: a generalizable and unifying hypothesis. Crit Rev Toxicol 31:353–424
Calabrese EJ, Baldwin LA (2001) Hormesis: U-shaped dose responses and their centrality in toxicology. Trends Pharmacol Sci 22:285–291
Calabrese EJ, Kozumbo WJ (2021) The hormetic dose-response mechanism: Nrf2 activation. Pharmacol Res 167:105526
Calabrese V, Cornelius C, Dinkova-Kostova AT, Iavicoli I, Di Paola R, Koverech A, Calabrese EJ (2011) Cellular stress responses, hormetic phytochemicals and vitagenes in aging and longevity. Biochim Biophys Acta 1822:753–783
Castillo-Quan JI, Li L, Kinghorn KJ, Ivanov DK, Tain LS, Slack C, Partridge L (2016) Lithium promotes longevity through Gsk3/Nrf2-dependent hormesis. Cell Rep 15:1–13
Chen L, Yang, R, Qiao W, Zhang W, Chen J, Mao L, Miao D (2019) 1,25-Dihydroxyvitamin D exerts an antiaging role by activation of Nrf2-antioxidant signaling and inactivation of p16/p53-senescence signaling. Aging Cell 18:e12951
Choi MJ, Kim BK, Park KY, Yokozawa T, Song YO, Cho EJ (2010) Anti-aging effects of cyanidin under a stress-induced premature senescence cellular system. Biol Pharm Bull 33:421–6
Chopan M, Littenberg B (2017) The association of hot red chili pepper consumption and mortality: a large population-based cohort study. PLoS One 12:e0169876
Ciocca DR, Caderwood SK (2005) Heat shock proteins in cancer: diagnostic, prognostic, predictive, and treatment implications. Cell Stress Chaperones 10:86–103
Demirovic D, Rattan SIS (2013) Establishing cellular stress response profiles as biomarkers of homeodynamics, health and hormesis. Exp Gerontol 48:94–98
Filomeni G, De Zio D, Cecconi F (2015) Oxidative stress and autophagy: the clash between damage and metabolic needs. Cell Death Differ 22:377–388
Franco R, Navarro G, Martinez-Pinilla E (2019) Hormetic and mitochondria-related mechanisms of antioxidant action of phytochemicals. Antioxidants (Basel) 8:373
Gerber LM, Williams GC, Gray SJ (1999) The nutrient-toxin dosage continuum in human evolution and modern health. Q Rev Biol 74:273–89
Giudice MD, Buck CL, Chaby L, Gormally BM, Taff CC, Thawley CJ, Wada H (2018) What is stress? A systems perspective. Integr Comp Biol 58:1019-1032
Gutteridge JM, Halliwell B (2010) Antioxidants: molecules, medicines, and myths. Biochem Biophys Res Commun 393:561–564

Harrison AP, Bartels EM (2006) A modern appraisal of ancient Etruscan herbal practices. Am J Pharmacol Toxicol 1:21–24
Hipkiss AR, Brownson C, Carrier MJ (2001) Carnosine, the anti-ageing, anti-oxidant dipeptide, may react with protein carbonyl groups. Mech Ageing Dev 122:1431–1445
Kadlecova A, Jirsa T, Novak O, Kammenga J, Strnad M, Voller J (2018) Natural plant hormones cytokinins increase stress resistance and longevity of caenorhabditis elegans. Biogerontology 19:109–120
Keany M, Matthijssens F, Sharpe M, Vanfleteren J, Gems D (2004) Superoxide dismutase mimetics elevate superoxide dismutase activity in vivo but do not retard aging in the nematode Caenorhabditis elegans. Free Rad Biol Med 37:239–250
Kumar R, Sharma A, Kumari A, Gulati A, Padwad Y, Sharma R (2019) Epigallocatechin gallate suppresses premature senescence of preadipocytes by inhibition of PI3K/Akt/mTOR pathway and induces senescent cell death by regulation of Bax/Bcl-2 pathway. Biogerontology 20:171–189
Leak RK (2014) Heat shock proteins in neurodegenerative disorders and aging. J Cell Commun Signal 8:293–310
Lee A (2001) The glucose-regulated proteins: stress induction and clinical applications. Trends Biochem Sci 26:504–510
Lee KS, Lee BS, Semnani S, Avanesian A, Um CY, Jeon HJ, Jafari M (2010) Curcumin extends life span, improves health span, and modulates the expression of age-associated aging genes in drosophila melanogaster. Rejuvenat Res 13:561–570
Li YR, Li S, Lin CC (2019) Effect of resveratrol and pterostilbene on aging and longevity. Biofactors 44:69-82
Lima CF, Pereira-Wilson C, Rattan SI (2011) Curcumin induces heme oxygenase-1 in normal human skin fibroblasts through redox signaling: relevance for anti-aging intervention. Mol Nutr Food Res 55:430–442
Lin JH, Li H, Yasumura D, Cohen HR, Zhang C, Panning B, Walter P (2007) IRE1 signaling affects cell fate during the unfolded protein response. Science 318:944–949
Longo VD (2009) Linking sirtuins, IGF-I signaling, and starvation. Exp Gerontol 44:70–74
Lopez T, Schriner SE, Okora M, Lu D, Chiang BT, Huey J, Jafari M (2014) Green tea polyphenols extend the lifespan of male drosophila melanogaster while impairing reproductive fitness. J Med Food 17:1–8
Martucci M, Ostan R, Biondi F, Bellavista E, Fabbri C, Bertarelli C, Santoro A (2017) Mediterranean diet and inflammaging within the hormesis paradigm. Nutr Rev 75:442–455
McColl G, Killilea DW, Hubbard AE, Vantipalli MC, Melov S, Lithgow GJ (2008) Pharmacogenetic analysis of lithium-induced delayed aging in Caenorhabditis elegans. J Biol Chem 283:350–7
McFarland GA, Holliday R (1994) Retardation of the senescence of cultured human diploid fibroblasts by carnosine. Exp Cell Res 212:167–175
Melov S, Ravenscroft J, Malik S, Gill MS, Walker DW, Clayton PE, Lithgow GJ (2000) Extension of life-span with superoxide dismutase/catalase mimetics. Science 289:1567–1569
Mocchegiani E, Rink L, Blasco MA (2006) Zinc and ageing (ZINCAGE project). Biogerontology 7:305–306
Motohashi H, Yamamoto M (2004) Nrf2-Keap1 defines a physiologically important stress response mechanism. Trends Mol Med 10:549–557
Paganini-Hill A, Kawas CH, Corrada MM (2007) Non-alcoholic beverage and caffeine consumption and mortality: the Leisure World Cohort study. Prev Med 44:305–310
Pallauf K, Duckstein N, Rimbach G (2017) A literature review of flavonoids and lifespan in model organisms. Proc Nutr Soc 76:145-162
Panossian AG, Efferth T, Shikov AN, Pozharitskaya ON, Kuchta K, Mukherjee PK, Wagner H (2021) Evolution of the adaptogenic concept from traditional use to medical systems: pharmacology of stress- and aging-related diseases. Med Res Rev 41:630–703
Pant A, Mishra V, Saikia SK, Shukla V, Asthana J, Akhoon BA, Pandey R (2014) Beta-caryophyllene modulates expression of stress response genes and mediates longevity in Caenorhabditis elegans. Exp Gerontol 57:81–95

Pingitore A, Lima GP, Mastorci F, Quinones A, Iervasi G, Vassalle C (2015) Exercise and oxidative stress: potential effects of antioxidant dietary strategies in sports. Nutrition 31:916–922

Proshkina E, Plyusnin S, Babak T, Lashmanova E, Maganova F, Koval L, Moskalev A (2020) Terpenoids as Potential Geroprotectors. Antioxidants (Basel) 9:529

Putics A, Vodros D, Malavolta M, Mocchegiani E, Csermely P, Soti C (2008) Zinc supplementation boosts the stress response in the elderly: Hsp70 status is linked to zinc availability in peripheral lymphocytes. Exp Gerontol 43:452–461

Rahman K (2003) Garlic and aging: new insights into an old remedy. Ageing Res Rev 2:39–56

Rattan SIS (2008) Hormesis in aging. Ageing Res Rev 7:63–78

Rattan SIS (2017) Hormetins as drugs for healthy aging. In: Vaiserman M (ed) Anti-aging drugs: from basic rsearch to clinical practice. The Royal Society of Chemistry, UK, pp 170–180

Rattan SIS (2020) Homeostasis, homeodynamics and aging. Encycl Biomed Gerontol 2:238–241

Rattan SIS (2020a) Biological health and homeodynamic space. In: Sholl J, Rattan SIS (eds) Expaning health across the sciences. Springer, Switzerland, pp. 43–51

Rattan SIS (2020b) Naive extrapolations, overhyped claims and empty promises in ageing research and interventions need avoidance. Biogerontology 21:415–421

Rattan SIS, Clark BFC (1994) Kinetin delays the onset of ageing characteristics in human fibroblasts. Biochem Biophys Res Commun 201:665–672

Rattan SIS, Kryzch V, Schnebert S, Perrier E, Carine Nizard C (2013) Hormesis-based anti-aging products: a case study of a novel cosmetic. Dose Response 11:99–108

Rattan SIS, Kyriazis M (eds) (2019) The science of hormesis in health and longevity. Academic Press, UK

Rattan SIS, Le Bourg E (eds) (2014) Hormesis in health and disease. CRC Press, Boca Raton

Rattan SIS, Sodagam L (2005) Gerontomodulatory and youth-preserving effects of zeatin on human skin fibroblasts undergoing aging in vitro. Rejuven Res 8:46–57

Ryter SW, Choi AMK (2013) Autophagy: an integral component of the mammalian stress response. J Biochem Pharmacol Res 1:176–188

Ryu D, Mouchiroud L, Andreux PA, Katsyuba E, Moullan N, Nicolet-Dit-Felix AA, Auwerx J (2016) Urolithin a induces mitophagy and prolongs lifespan in C. elegans and increases muscle function in rodents. Nat Med 22:879–888

Sadowska-Bartosz I, Bartosz G (2014) Effect of antioxidants supplementation on aging and longevity. Biomed Res Int 2014:404680

Saul N, Pietsch K, Menzel R, Sturzenbaum SR, Steinberg CE (2010) The longevity effect of tannic acid in caenorhabditis elegans: disposable soma meets hormesis. J Gerontol A Biol Sci Med Sci 65:626–635

Singh S, Vrishni S, Singh BK, Rahman I, Kakkar P (2010) Nrf2-ARE stress response mechanism: a control point in oxidative stress-mediated dysfunctions and chronic inflammatory diseases. Free Radic Res 44:1267–1288

Son TG, Camandola S, Mattson MP (2008) Hormetic dietary phytochemicals. Neuromolecular Med 10:236–246

Sonneborn JS (2010) Mimetics of hormetic agents: stress-resistance triggers. Dose Response 8:97–121

Southam CM, Ehrlich J (1943) Effects of extracts of western red-cedar heartwood on certain wood-decaying fungi in culture. Phytopathology 33:517–524

Unno K (2016) Prevention of brain aging by green tea components: Role of catechins and theanine. J Phys Fitness Sports Med 5:117–122

Valenzano DR, Terzibasi E, Genade T, Cattaneo A, Domenici L, Cellerino A (2006) Resveratrol prolongs lifespan and retards the onset of age-related markers in a short-lived vertebrate. Curr Biol 16:296–300

Verbeke P, Fonager J, Clark BFC, Rattan SIS (2001) Heat shock response and ageing: mechanisms and applications. Cell Biol Int 25:845–857

Wanke V, Cameroni E, Uotila A, Piccolis M, Urban J, Loewith R, De Virgilio C (2008) Caffeine extends yeast lifespan by targeting TORC1. Mol Microbiol 69:277–285

Westerheide SD, Bosman JD, Mbadugha BNA, Kawahara TLA, Matsumoto G, Kim S, Morimoto RI (2004) Celastrols as inducers of the heat shock response and cytoprotection. J Biol Chem 279:56053–56060
Wiegant FA, Surinova S, Ytsma E, Langelaar-Makkinje M, Wikman G, Post JA (2009) Plant adaptogens increase lifespan and stress resistance in C. elegans. Biogerontology 10:27–42
Wieten L, van der Zee R, Goedemans R, Sijtsma J, Serafini M, Lubsen NH, Broere F (2010) Hsp70 expression and induction as a readout for detection of immune modulatory components in food. Cell Stress Chaperones 15:25–37
Yan D, Saito K, Ohmi Y, Fujie N, Ohtsuka K (2004) Paeoniflorin, a novel heat shock protein-inducing compound. Cell Stress Chaperones 9:378–389
Yoshida H (2007) ER stress and diseases. FEBS J 274:630–658
Yousefzadeh MJ, Zhu Y, McGowan SJ, Angelini L, Fuhrmann-Stroissnigg H, Xu M, Niedernhofer LJ (2018) Fisetin is a senotherapeutic that extends health and lifespan. EBioMedicine 36:18–28
Zhu W, Wu Y, Meng YF, Wang JY, Xu M, Tao JJ, Lu J (2015) Effect of curcumin on aging retinal pigment epithelial cells. Drug Des Devel Ther 9:5337–5344
Zhu Y, Doornebal EJ, Pirtskhalava T, Giorgadze N, Wentworth M Fuhrmann-Stroissnigg H, Kirkland JL (2017) New agents that target senescent cells: the flavone, fisetin, and the BCL-XL inhibitors, A1331852 and A1155463. Aging (Albany NY) 9:955-963

Chapter 7
Notes Toward an Evolutionary Biology of Nutrition

Donovan P. German and Michael R. Rose

Abstract One of the most important foci for adaptation among animal species is their consumption, digestion, and utilization of food. Despite this centrality, the topic has not been a major focus for evolutionary biology, with a few salient exceptions. Here we raise fundamental evolutionary questions involving the biology of animal nutrition.

The most important limitation on the evolutionary biology of nutrition is the problem of specialization. No animal is entirely omnivorous. Instead, all animal species specialize on the consumption of a finite range of foods. This central limitation is then more subtly modulated by a series of secondary constraints that we delineate here: (i) long-term evolutionary history of prior nutritional adaptations; (ii) ecological fluctuations producing rapid changes in the favored features of nutritional adaptation; (iii) problems of reduced population size and inbreeding degrading nutritional adaptation; and (iv) falling forces of natural selection with adult age, which limit the degree to which later-life features effective adaptation to current diets.

We apply these evolutionary considerations to three different contexts for research that are affected by the evolutionary biology of nutrition. (1) Laboratory research with recently caught wild animals, or their immediate descendants. Such research has revealed something of the profound diversity of animal adaptations to the predicaments of nutrition. (2) Laboratory research using long-cultured model organisms. This type of experimentation inherently involves a spectrum of basic evolutionary issues, because long-cultured organisms inevitably undergo laboratory evolution. (3) Clinical and related research on human nutrition and its associated health consequences. This is the case of greatest medical interest, but it too is entangled with profound evolutionary considerations, such as interactions between our evolutionary history and the range of alternative foods that humans now consume.

Overall, we believe that our discussion raises deep evolutionary issues that have been neglected too often in scientific studies of nutrition ranging from animal husbandry to experimental physiology to the epidemiology of human chronic disease.

D. P. German (✉) · M. R. Rose
Department of Ecology and Evolutionary Biology, University of California, Irvine, CA 92617, USA
e-mail: dgerman@uci.edu

S. I. S. Rattan and G. Kaur (eds.), *Nutrition, Food and Diet in Ageing and Longevity*, Healthy Ageing and Longevity 14, https://doi.org/10.1007/978-3-030-83017-5_7

Keywords Evolution · Aging · Longevity · Ecology · Food · Culture

7.1 Introduction

All animals must eat in order to survive, leaving aside the rare animals that have photosynthetic commensals. But the evolution of adaptations that enable animals to live off diverse nutrients is a subject that has too often been neglected, outside of the optimal foraging literature (MacArthur and Pianka 1966; Pyke 1984) and the few species cultured for human consumption (e.g., Van Soest 1994).

It needs to be said at the outset that there are frequently problems with the use of evolutionary reasoning by scientists or physicians who did not receive extensive training in the field of evolutionary biology. Evolution by natural selection is a powerful force. It has produced extensive diversity among the animals of our planet, and no doubt still greater diversity across all the planets that have animal species. But that evident power often blinds those who aren't evolutionary biologists to the profound limitations on what evolution by natural selection can achieve. And such blindness has been a long-standing problem with respect to thinking about the evolutionary biology of nutrition.

Let's begin with a hypothetical Omnivorous Darwinian Beast that can devour any kind of substrate that contains nutrients. In effect, if there is fuel that can generate calories in a bomb-calorimetry assay, then that Omnivorous Darwinian Beast ("ODB") can extract a large proportion of those calories. Furthermore, assume that the ODB can synthesize all of the proteins, lipids, and other hydrocarbons required for its metabolism from a minimal diet that contains water, a carbon source, a nitrogen source, and trace minerals, from sulfur to iron. That is to say, this ODB has no need of "vitamins" and faces no limitation with respect to the kinds of nutrients it can use as catabolic or anabolic fuel.

The nutrition of such an ODB would be of relatively little evolutionary interest. It can eat anything and thereby survive, reproduce, and live in excellent health, leaving aside infectious disease, predators, parasites, and the general scourge of aging. Its diet is not of particular interest from the standpoint of evolutionary thinking.

But that is not the way actual animal species ever evolve. Evolving species of animals face a long list of evolutionary constraints and contexts which preclude achieving such an ODB, and this is true of any organism from bacteria to animals to plants (Allison et al. 2014). Here we provide a sketch of some of the more important evolutionary constraints that would prevent such perfect omnivory. Then we continue with three important applications of our evolutionary reasoning: (i) the appropriate design and interpretation of nutritional experiments with laboratory animals; (ii) the design and interpretation of nutritional experiments with recently caught wild animals; and (iii) the appropriate invocation of evolutionary reasoning in discussions of optimal human nutrition.

7.2 The Core Problem: Specialization

The core problem for the evolutionary biology of animal nutrition is the heterogeneity of potential foods. Nutrition can be derived from microbial organisms, such as the commensal bacteria to be found in many animal guts, or from environmental microbes (e.g., detritus), which is common in terrestrial and aquatic systems; fermented human foods would represent an extension of this mode of feeding. Nutrition can be obtained from plant matter, or it can be obtained from the tissues of other animals. In rare cases, important nutrients can be obtained from mineral sources, such as rock salts.

Within each of these broad categories, the mechanical properties of the sources of nutrition vary widely. Some microbes and mineral sources may provide nutrients that can be directly absorbed through gut walls. But many microbial species have cell walls made of peptidoglycan or chitin that require catalytic or mechanical destruction, before the nutrients that they contain can be absorbed. Plant tissues vary from soft, in the case of some fruit, to hardened, such as bark and the shells of nuts. The latter often require the application of considerable mechanical force, fostering the evolution of mandibles of some type and even manual destruction by animals with limbs. The efficiency with which animals digest foods declines with the amount of refractory material (e.g., fiber; Karasov and Douglas 2013) present in the food (Fig. 7.1). Animals that are used for food require discovery, capture, subjugation, and devouring. The adaptations that their predators have evolved to achieve those ends are endlessly diverse, as are the adaptations that such prey have evolved to evade their destruction. The natural history of acquiring and consuming prey is vast.

Even when nutritional material is in the gut of an animal, further adaptations are required to extract nutrients from the gut contents, including the detoxification of noxious substances present in the food. Large proteins must be broken down into oligopeptides, dipeptides, and amino acids that can be absorbed. Fats must be broken down into fatty acids and their glycerol or phosphate backbones that a particular animal can absorb. Simple sugars, like those in nectar, can usually be absorbed

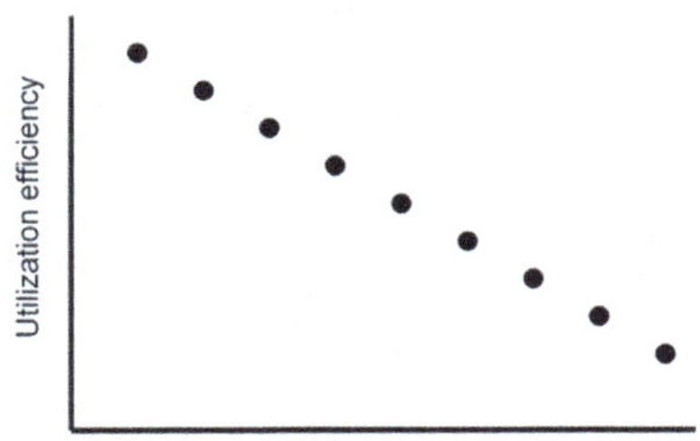

Fig. 7.1 As a general rule, utilization efficiency of a food, whether on the scale of energy, or nutrients absorbed, declines with increasing amount of refractory material in food. Refractory material usually refers to plant insoluble fiber (e.g., cellulose), but can also refer to anything that dilutes or makes soluble nutrients more inaccessible to digestive enzymes in the gut. After Karasov and Douglas (2013)

directly, but sugar polymers like starch and cellulose require specific enzymes to be broken down into their absorbable, monomeric units.

The diversity of foods needing to be digested has resulted in a range of digestive system morphologies and functions (Fig. 7.2). Some animals require the aid of microbes to break down polymers (e.g., cellulose), or detoxify secondary compounds, and they do this by providing habitat within the alimentary canal (Moran et al. 2019) (or other body cavities, like the gills; Waterbury et al. 1983; Altamia et al. 2020) where microbes can establish and flourish within that system. Some animals do this in the foregut (e.g., ruminants, hoatzin), others do that in the hindgut (e.g., rats, termites). Some animals have acidic stomachs, like most mammals, reptiles, and about 75% of fish species, where hydrochloric acid aids in the initial digestive process, usually of protein, but thousands of taxa lack this specialization. All animals have some absorptive tissue, which is usually in a tubular form, like an intestine (Stevens and Hume 1995). A tube gives the highest surface area for absorption within a body cavity

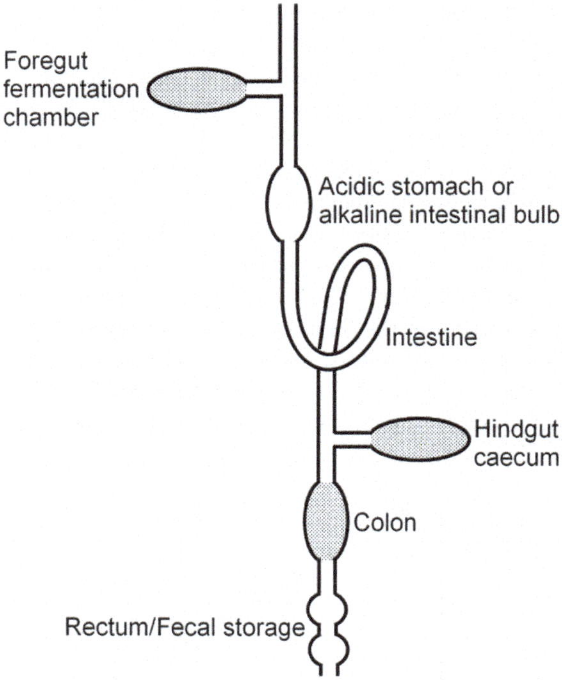

Fig. 7.2 Basic design of a gut. Most animals have some form of an intestine, but vary as to whether they possess other compartments such as crop, forestomach, stomach, caecum, and large intestine/colon. As a general rule, catalytic enzymatic reactions occur in the intestine, whereas microbial fermentation can occur in the forestomach, caecum, and large intestine/colon (shown with stippled areas). Foregut fermentation is known to occur in four major clades of mammals and in at least one avian species (the hoatzin). Hindgut fermentation, either in the caecum or large intestine/colon, occurs in many clades of mammals, birds, reptiles and fishes, and some insects (e.g., Termites). Modified from Karasov and Douglas (2013)

(particularly with the folding patterns of the absorptive tissues and cell membranes; Karasov and Hume 1997).

How an animal digests its food generally falls into a "Rate vs Yield" continuum (Clements and Raubenheimer 2006; German et al. 2015; Fig. 7.3). On one end, there are rate maximizers that have high intake, rapid transit of material through the gut, and an endogenous digestive process that mostly extracts soluble nutrients

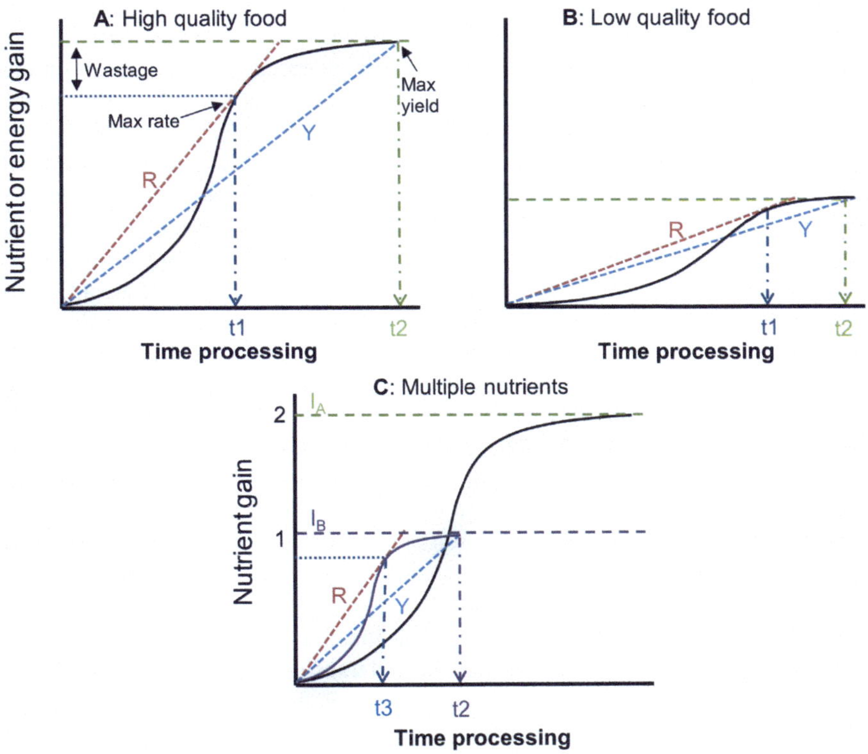

Fig. 7.3 Cumulative nutrient or energy gained by an animal as a function of time spent processing a meal. **a**: For high quality food; **b**: for low quality food. The green horizontal dashed line represents the extractable nutrient ingested in the meal. The slope of the red line "R" represents the maximum rate at which the nutrient or energy can be absorbed from the meal. This is achieved by evacuating the contents at time 1 (t1) and taking a new meal, although a portion of the nutrient or energy consumed is lost in the feces ("wastage"). Maximum yield (blue line "Y") is attained by extending processing time to time 2 (t2), however, this is done at the cost of reduced digestive rate. For the low quality food, long retention is unavoidable for maximum digestive rate. **c**: For two ingested nutrients, I_A and I_B. Nutrients IA and IB are contained in the food in a 2:1 ratio, and thus, twice the amount of nutrient IA is consumed as I_B. The fish is capable of assimilating the nutrients in a reciprocal 1:2 ratio, however, if they void the contents at time 3 (t3), since this is the point at which the rate of gain of nutrient IB is twice that of I_A. Lines and times needed for maximum rate (R) and maximum efficiency (Y) provided. Reproduced from Clements KD and Raubenheimer D (2006) Feeding and nutrition. In: Evans DH and Claiborne JB (eds) *The Physiology of Fishes*, 3rd Edition, pp. 47–82. Boca Raton, FL: CRC

from ingested foods. These animals defecate the ingesta from the digestive system before refractory material is digested, resulting in "wastage", or nutrients that were not acquired. Because these animals eat frequently, they will just take another meal and make up for wastage through more intake. Transit times in these animals are fast because of the high intake, and thus, they tend to have less reliance on microbial digestion within their alimentary canals, with some rate-maximizing animals lacking a demonstrable gut microbiome (Hammer et al. 2017). On the other hand, yield maximizers eat less, hold food for a longer period of time in their guts, and digest more of the available nutrients or energy. A yield-maximizing digestive strategy can involve enteric microbes since it provides the time investment needed by microbes, resulting in higher digestive efficiency, even of refractory material that was indigestible to the animal's endogenous digestive process (Fig. 7.3). Even at the elevated temperatures experienced in a mammalian gut, yield-maximizing herbivores reliant on microbial degradation will have transit times of ingesta through the whole system of greater than 20 h, whereas more rate-maximizing taxa that are reliant on endogenous digestive processes to digest more soluble nutrients will have shorter transit times (e.g., <10 h; Hofmann 1989). Animals eating more animal material will fall on the yield-maximizing side of this continuum, but they tend to have a higher-quality diet than herbivores adopting this strategy. Of course, many animals have diverse diets and can slide along the rate vs yield continuum depending on what was ingested, but they are limited in how far they can go in either direction by their gut morphology (Fig. 7.2) and transit times (Fig. 7.3). Moreover, when one dissects ingesta into specific nutrient categories, interesting patterns can emerge where an animal can appear to be a rate maximizer for a common nutrient (e.g., cellulose), but a yield maximizer for a targeted nutrient (e.g., protein; Fig. 7.3C). Hence, defining an animal as a rate or yield maximizer can vary depending on the time and scale at which it is examined.

Based on the discussion thus far, it is clear that digestion is a complex process and to understand digestive efficiency, intake and transit time must be considered. In addition to intake and transit time of material through the gut, nutrients (i.e., carbohydrates, proteins, lipids, nucleic acids) are usually acquired through the catalytic breakdown by digestive enzymes, and the efficiency of those enzymes will vary on several levels. Following Karasov and Hume (1997), and Karasov and Douglas (2013), the following proportions are helpful to put gut function in the context of the variables we have been covering thus far:

$$\text{Digestive Efficiency} \propto \frac{\text{Enzyme Activities}}{\text{Substrate Concentration}} \propto \text{Time} \propto \frac{\text{Gut Size}}{\text{Digesta Transit Rate}}$$

The digestive efficiency of a given food is proportional to the amount and efficiency of digestive enzymes present as a function of the substrate concentrations for those enzymes. Thus, if a substrate is abundant in a given food type, then enzyme activities (e.g., amylase activity) against that substrate (e.g., starch) must be proportionally elevated to digest the available substrate. This hypothesis, known broadly as the Adaptive Modulation Hypothesis (Karasov 1992; Karasov and Martínez del Rio 2007), is widely upheld for starch concentrations in the diet and the digestive enzyme amylase, in a broad range of animals. There is a rich literature examining the genetic underpinnings of this match between enzyme and substrate (e.g., Perry et al. 2007; Axelsson et al. 2013; Heras et al. 2020), showing changes in gene copy number and gene expression in different taxa.

More limiting, essential nutrients, like protein and lipid, can also invoke higher enzyme activities to acquire these nutrients from diets low in them. When animals ramp up digestive enzyme activities against more limiting nutrients, it is called "Nutrient Balancing" (Clissold et al. 2010; Heras et al. 2020). There are good examples showing elevated lipase activities in fishes consuming high-fiber, low lipid diets (German et al. 2004; Leigh et al. 2018; Heras et al. 2020), which has resulted in expanded gene copy numbers of lipases in the genome (Heras et al. 2020). Consistent with the rate vs yield continuum, digesta held for extended periods of time in the gut are more efficiently digested, and time is proportionally related to gut size as a function of transit rates of digesta through the gut. Hence, if an animal has rapid transit of material through the alimentary canal, they must have a larger gut to allow for enough time to digest anything. Indeed, one of the hallmarks of digestive tract morphology is that animals with lower quality diets have higher intake, and larger, and longer guts (Fig. 7.4).

Once nutrients have been absorbed from the gut, further processing, usually in a liver or hepatic-like organ, is required, and this may be where the downstream limitations of dietary specialization manifest (Merkin et al. 2012; Betancor et al. 2018; Jin et al. 2018; Herrera et al. 2021). In mammals, and now in fishes, it is clear that hepatic gene expression and metabolism is somewhat set by an animal's genetics, including epigenetics, at the population or species level. Hepatic gene expression tends to not change dramatically with dietary fluctuations, instead reflecting the genetic background of the organism (Herrera et al. 2021). The gut, on the other hand, is more variable in gene expression, enzyme activity, and size in response to dietary perturbations in the same individual (Sullam et al. 2015; Leigh et al. 2018; Herrera et al. 2021). No animal can simply use, transport, or store any and all nutrient molecules with equal ease. Some nutrients, such as glucose, can be used immediately for routine metabolic processes. But overly high initial spikes even in glucose levels can cause metabolic problems, such as glycation, as well as disruptions of the osmolarity of body fluids, which not all species can tolerate. A variety of polypeptides could disrupt "hormonal" signaling among the animal's organs. Indeed, a "leaky gut" developing from lesions in the human intestine caused by autoimmune problems that stem from a disruptive enteric microbiome, can lead to the absorption of compounds (e.g., microbial peptides) that normally do not end up in the blood stream. The metabolic outcomes can manifest in severe, systemic autoimmune problems like Rheumatoid Arthritis or Multiple Sclerosis (Brandl et al. 2021). High levels of absorbed fats could produce mechanical strain for circulatory organs, like hearts, across both open and closed circulatory systems. Dyslipidemia is also the true precursor to insulin resistance in humans (Krssak et al. 1999; Montell et al. 2001; Basciano et al. 2005; Dekker et al. 2010).

Finally, the functional allocation of nutrients across organ systems and functions is not uniform. In a tradeoff known as "Capital vs Income" breeding, income breeders

Fig. 7.4 Trimmed phylogeny and photos of fishes (with their unraveled guts beneath them) in the family Stichaeidae, showing a gradient in the amount of algal material in the diet and gut length relative to body size. Increasing the amount of refractory material in the diet leads to higher intake, leading to more rapid transit of digesta through the gut, which in turn requires a longer gut to maintain digestive efficiency (German 2011; Leigh et al. 2018). From top to bottom: *Cebidichthys violaceus* (herbivore), *Xiphister mucosus* (herbivore), X. *atropurpureus* (omnivore), *Phytichthys chirus* (omnivore), *Anoplarchus purpurescens* (carnivore). After German et al. (2014, 2015). The slope of positive allometry of gut size relative to body size follows the gradient in algal material in the diet (German et al. 2004; German et al. 2014), and that allometric pattern is maintained even when the fishes are raised on a high-protein, animal-based diet in the laboratory (German and Horn 2006). Such a relationship between gut size and diet is inherent in nearly all animal clades examined (Stevens and Hume 1995)

(e.g., many insect taxa) use almost all of their metabolically available nutrients for immediate reproduction, whether synthesizing gametes, provisioning zygotes, or even caring for their young. Capital breeders (e.g., some marine mammals), on the other hand, build up extensive stores of nutrients for subsequent reproduction or for later survival under conditions of restricted food availability (Davis et al. 2016).

Thus it is inevitable that no animal has or will ever evolve to be an ODB. All animals specialize nutritionally, and this specialization can happen in the digestive system itself (including the mouth, which can vary in shape, opening mechanism, or dentition), or in downstream nutrient processing (Karasov and Martínez del Rio 2007). Animals seek different foods, behaviorally. They attack and devour different foods, equipped with specific structures and behaviors for prey subjugation and consumption. Even herbivores are confronted with plant adaptations, from thorns to woody structures to poisons, that make their consumption of plants as food give rise to natural selection for particular types of foraging and consumption, including mechanical processing, detoxification, and catalytic processing.

As is apparent from the discussion above, there is a limited set of nutritional adaptations that will always be beneficial, and therefore consistently and generally fostered by natural selection. There has to be some flexibility in gut function to accommodate fluctuations in resource availability, but there can be constraints built in (e.g., mouth shape or size, gut morphology, available digestive enzymes, enteric microbial diversity) based on the animal's phylogenetic history that can limit what they can efficiently digest, absorb, and metabolize. Nutritional adaptations may involve genetic trade-offs, or "antagonistic pleiotropy," that will lead to the evolution of different nutritional adaptations among animal populations exploiting different sources of food. Nutritional specialization is thus an inevitable product of evolution by natural selection acting on animal species that confront different environments with different body plans and physiologies. Humans can complicate the understanding of digestive adaptations within primates since we cook our food, which is essentially predigesting it with heat, chemicals, or microbial fermentations. Thus, there are limitations to what can be inferred about human biology by studying other animals.

The core problem of animal specialization with respect to nutritional adaptation then interacts with a variety of general evolutionary constraints on adaptation that affect most, if not all, animal species. Among these constraints are evolutionary history, ecological fluctuation, effective population size, and declining forces of natural selection with age. We deal with each in turn.

7.3 Evolutionary History

Evolution by natural selection is not a globally optimizing process, contrary to the suppositions of some applied mathematicians who dabbled in evolutionary theory during the 1970s and 1980s (e.g. Gadgil and Bossert 1970). Instead, adaptation is an historically contingent process that is subject to constraints which prevent the achievement of optimal outcomes (Gould and Lewontin 1979; Rose and Lauder 1996).

As such, the functional biology of nutrition is conditioned by the prior evolutionary history of each and every animal species (e.g., German et al. 2004; Herrel et al. 2008; Wehrle et al. 2020). Furthermore, recent research in the genomic foundations of adaptation has revealed that animal evolution is dominated by the fine-tuning of extant genetic variation, rather than the wholesale introduction of novel mutations that then undergo substitution (Burke et al. 2010; Burke 2012; Phillips et al. 2016; Graves et al. 2017). For instance, the expansion in gene copy number in the salivary amylase gene (*amy1*) in humans from agrarian backgrounds (Perry et al. 2007), and in pancreatic amylase (*amy2*) in dogs in comparison to wolves (Axelsson et al. 2013) led to advantageous increases in amylolytic activity to accommodate starchy diets. Thus we should expect that the adaptations involved in the biology of nutrition involve tinkering with the same basic functional elements (e.g., Heras et al. 2020), even when such adaptations involve many-fold changes to physiological characters (vid. Rose et al. 2004, 2005). An excellent example in the context of digestive physiology is the wood-eating catfishes in the family Loricariidae. Although one might expect these fishes consuming a wood diet would develop the adaptations (e.g., expanded hindgut that houses cellulolytic microbes; Fig. 7.2) seen in well-known wood-eating, and digesting, animals like termites or beavers, the wood-eating catfishes maintain a rather long, thin-walled intestine and a rate-maximizing digestive strategy consistent with other loricariid catfishes (German 2009; German and Bittong 2009; Lujan et al. 2011; McCauley et al. 2020). Thus, wood-eating catfishes are detritivores that digest the soluble nutrients in their detrital diet and virtually none of the cellulose (German 2009; German and Bittong 2009; German and Miles 2010; Lujan et al. 2011). Furthermore, wood-eating has evolved three separate times in that family (Lujan et al. 2017), each time with the same outcome (poor wood digestibility), and the only "adaptive" feature being the repeated evolution of thick, cupped teeth that are excellent for scraping wood. Thus, some parts of the digestive system (i.e., the mouth) showed changes, but others did not (German 2009; German and Bittong 2009), suggesting that selection can indeed act on some aspects of nutritional biology, but loricariids are otherwise rate-maximizing in their approach to intake and digestion which limits what nutrients they can efficiently digest and assimilate.

In the same way that cetacean respiration uses lungs rather than gills, we should expect that the evolutionary biology of digestive systems within sizeable taxonomic groups will nonetheless feature variations on common adaptations for feeding, digestion, and the processing of nutrients within intermediary metabolism. Excellent examples of this abound within entire families of fishes where the gastric stomach has been lost (Wilson and Castro 2011; Castro et al. 2014), yet a gut that is essentially only an intestine can digest a wide array of food items ranging from detritus, to plankton, to animal material, to algae. Absent gut commensals, we should not expect to find animals which can evolve to consume radically novel foods, with one salient exception. That exception would be humans that have learned how to use extra-somatic tools for foraging and processing foods; we will address this exception later in this article.

7.4 Ecological Fluctuations

Within the context of limited scope for immediate adaptation, evolving animal species nonetheless show remarkable capacities to track environmental change. The most famous example is the evolution of industrial melanism among multiple Lepidopteran species over more than a century of coal burning in Europe. Less widely appreciated, the introduction of environmental regulations in the 1950s and 1960s led to the reverse evolution of melanic alleles across these same species, resulting in dramatic declines in dark coloration among European Lepidoptera. Natural selection acting on standing genetic variation readily tracks shifting selection regimes, due to the relative rarity of selective sweeps that proceed to fixation (e.g. Burke et al. 2016; Graves et al. 2017).

Rutledge et al. (2020, 2021) have recently published evidence showing rapid adaptation to banana-molasses medium among fruit flies cultured on that food for hundreds of generations, despite an evolutionary history that did not feature use of bananas for nutrition. These same fruit flies have never been exposed to orange-based medium, and show measureable deficits in adaptation to that novel diet. [We will return to this example of dietary adaptation later.]

Yet most animals will face seasonal and other climatic variation in the availability of foods that are within their capacity to ingest and digest. To the extent to which such variation is considerable, then evolution by natural selection may do a poor job of adapting species like small, short-lived insects to such variable diets. Natural selection may have not tracked such ecological fluctuations well in specific insect lineages, or led to different generations actually specializing on different resources.

On the other hand, larger and longer-lived animal species, such as most mammals and many birds, may be selected to accommodate the range of foods that arise over the course of lifespans that stretch across multiple years. In this way, such species may develop the ability to adjust their nutritional physiology in a manner that accommodates considerable ecological variation in food supplies. For example, birds that experience considerable seasonal fluxes in available foods (e.g., fruits vs seeds), or migrate into different habitats with different resources, do indeed show digestive flexibility (Karasov and Martinez del Rio 2007). A well-cited example with yellow-rumped warblers shows that it takes greater than six days for the digestive physiology of the bird to transition from a fruit diet to a seed diet (Fig. 7.5). In fishes, it can take as few as four weeks for a gut to show incredible plasticity to dietary

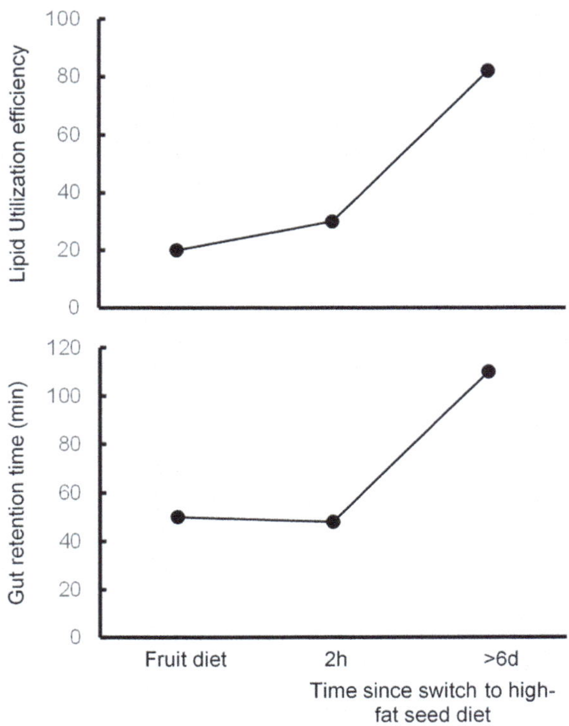

Fig. 7.5 When yellow-rumped warblers habituated to a sugary fruit-based diet, were transferred to a high-fat seed diet, lipid digestive efficiency (**Top**) and gut retention time (**Bottom**) had to be adjusted to the new diet. It took greater than six days for the gut to adjust to the new intake of lipid-rich seeds. Modified from Karasov and Douglas (2013)

shifts, but again, it is the liver that doesn't necessarily adjust to the influx of different nutrient classes (Herrera et al. 2021).

Nonetheless, given the commonplace pattern of seasonal and episodic variation in climate, and thus food availability, most species should have their adaptations to the consumption and digestion of varying foods impaired *relative to* their ability to adapt to the consumption of consistently available foods (Leigh et al. 2018). Such impairment can arise at many functional levels, from feeding behavior to dentition to digestion (Figs. 7.2, 7.3) to metabolic use of varied foods. Again, this is a corollary of the core principle that trade-offs prevent the evolution of an ODB, that can eat anything equally well.

7.5 Effective Population Size and Inbreeding

One of the chief findings of theoretical and experimental work on the evolutionary genetic machinery of adaptation is the importance of effective population size ("N_e") (Lynch and Walsh 1998). As N_e decreases, the response to selection demonstrably attenuates. This is not usually a problem for abundant species evolving in the wild, but it is a considerable difficulty for less abundant species in the wild, as well as virtually all laboratory stocks, which are kept at much smaller N_e compared to their wild ancestors.

This fundamental limitation on the power of natural selection will prevent adaptation to novel foods in animal populations that are at lower population sizes. Note that this problem is expected to afflict an animal population (i) that has recently colonized a new habitat, (ii) that has undergone population declines as a result of climatic and other ecological changes reducing availability of its favored foods, or (iii) that has been sampled from nature for rearing in a laboratory setting. These challenges are notably common in the face of ecological disruptions arising from anthropogenic change, deliberate or inadvertent.

7.6 Declining Forces of Natural Selection with Adult Age

Another fundamental problem facing the evolution of nutritional adaptations is the declining forces of natural selection with adult age in animal populations with age-structure (Hamilton 1966; Charlesworth 1980). Such declines will produce age-dependent deterioration in nutritional adaptations even on a long-standing diet. In addition, recent major changes of diet will foster rapid evolution of nutritional adaptations at early ages, but not later ages (Rutledge et al. 2020, 2021). Below we discuss this limitation to nutritional adaptation in the context of both laboratory experiments and human nutrition.

7.6.1 Application One: Physiological Studies of Recently Caught Wild Animals

Nutritional studies on wild-caught animals go back centuries. The most detailed studies come from those animals that we raise for consumption: a handful of ruminants, pigs, chickens, and a limited number of mostly carnivorous fishes. Most studies of animals cultured for human consumption focus on getting the animals to grow the quickest, producing the most flesh for us to eat. However, this has led to detailed studies of digestion and metabolism, the development of chemical methods to examine the digestibility of feedstuffs (Van Soest 1994), and what happens with animal flesh once the animal is killed. The physiological genomic approaches using different lines of these cultured animals have revealed how, through artificial selection, we have generated animals that digest the grain-based foods we feed them with varying efficiency. The main exception being cattle consuming large amounts of corn-based feeds requiring the use of antibiotics. The antibiotics serve two purposes: they keep the animals from falling ill despite eating foods that aren't optimal for foregut fermenters (i.e., they should be eating grasses directly), and the antibiotics cause the animals to grow faster. It now takes about 15 months to grow a calf to a size fit for slaughter, whereas just seventy years ago it took nearly six years (Pollan 2006). Of course, just because they grow faster, doesn't mean they are healthier, and in fact, the animals are obese, leading to the meat being more marbled with fat than in the past (i.e., there is a double novelty here: rapid growth of the animal, and a human food source that is fattier than in the past). The mechanism of faster growth in antibiotic-treated animals remains incompletely understood, but may be linked to a shift in the enteric microbiome towards an "obese" microbiome.

Although there was a literature on nutritional adaptations prior to the 1980's (again, largely focused on animals raised for human consumption, or in zoo settings), in terms of western science, evolutionary nutritional physiology really emerged from the laboratory of Dr. Jared Diamond at University of California, Los Angeles. Although Dr. Diamond is broadly known for bird and island biogeography, and more recently, his popular science and societal writing, his training and laboratory studies focused on nutritional physiology and biochemistry. Studies out of the Diamond lab examined nutrient transport rates across the intestinal epithelium, and rates of enzymatic hydrolysis of various substrates in many different taxa in accordance with their natural diets. A core tenet of the work is the "Krogh Principle", which states that "for such a large number of problems there will be some animal of choice, or a few such animals, on which it can be most conveniently studied" (Krebs and Krebs 1980). That is, for a given question on a given animal system, there will be a species or line of animal that allows a researcher to most efficiently answer the question at hand. Indeed, the sheer number of vertebrate animal models (e.g., mammals, birds, amphibians, fish) used in the Diamond lab supports this contention. However, the diverse nature of the work itself established nutritional physiology in evolutionary and comparative contexts outside of the biomedical field. Although Optimal Foraging Theory (MacArthur and Pianka 1966; Pyke 1984) had certainly set the stage for thinking about what animals eat and the consequences of this on an ecological scale, Diamond's work (and that of his lab colleagues) showed that the gut itself was central to the entire process and that it, therefore, must be considered in studies of nutritional ecology in an evolutionary context (Karasov and Diamond 1988). The list of students and postdoctoral researchers who emerged from the Diamond Laboratory is impressive, and each made impacts on the field. Two in particular, Dr. William Karasov (University of Wisconsin; cited extensively in this chapter), and Dr. Stephen Secor (University of Alabama), have published hundreds of studies with their trainees and colleagues on how the gut responds to dietary perturbations (summarized in Karasov and Martínez del Rio 2007), feeding frequency (Secor and Diamond 2000), or how animals may be adapted to use specific resources (e.g., various forms of carbohydrates; Brun et al. 2020).

Core tenets emerging from the Diamond laboratory include the concepts of "evolutionary match" [rooted in Symmorphosis (Weibel et al. 1991) and including the concept of "Reserve Capacity"], and the "Adaptive Modulation Hypothesis". Within these hypotheses, it is argued that processes, like biochemical pathways, require the expression of proteins (enzymes, transporters) the synthesis of which costs ATP, and if embedded in a membrane, take up valuable space. Hence, a cost–benefit analysis suggests that pathway proteins should not be expressed at levels far exceeding what is needed by the system, nor should they be under-expressed so as to form bottle necks. Diamond (2002) argues that, typically, reserve capacities of 2-3X what is needed (in some cases up to 10X) are within the norm and allow an animal to handle larger fluxes in nutrients through pathways than are typically experienced, even titling an article "enough, but not too much" (Diamond and Hammond 1992). Several other studies have followed up on this in metabolic pathways, generally finding support (e.g., Salvador and Savageau 2006), and theoretical examinations suggest redundancy in regulation of pathway robustness (Frank 2007). However, this view is not without its critiques (Dudley and Gans 1991), especially relating to structural components, like bones or tendons. But, for pathways like digestion and absorption of individual nutrients, where specific enzymes and transporters are needed for each step of the process, reserve capacities appear to make sense, and led to the "Adaptive Modulation Hypothesis" (AMH; Karasov 1992; Karasov and Martínez del Rio 2007).

As originally postulated, and aligning the nutrient supply organ with Optimal Foraging Theory, the AMH suggests that rates of nutrient hydrolysis at the intestinal brushborder and absorption of nutrients like glucose, should match with ingested loads of the nutrient over evolutionary time scales (Karasov 1992). Indeed, numerous studies in a wide range of wild-caught animal taxa have found support for the AMH, particularly relating to carbohydrate digestion (e.g., Diamond and Hammond 1992; Perry et al. 2007; Karasov and Martínez del Rio, 2007; Kohl et al. 2011; German et al. 2010; Axelsson et al. 2013; German et al. 2015; Heras et al. 2020; Brun et al. 2020). Although there is some support for protein digestion following similar patterns in some animals (Schondube et al. 2001), this pattern is not universal (Buddington et al. 1987) because all animals need protein, whereas carnivores consume low amounts of carbohydrates and may have metabolism that is geared for gluconeogenesis from amino acids, or others a ketotic metabolism tied to enteric fermentations (Willmott et al. 2005; Heras et al. 2020). Another aspect of the AMH is that limiting micronutrients, like vitamins and minerals, may have the opposite pattern: when these essential nutrients are in short supply, an animal may upregulate intestinal transporters for them to ensure scavenging of these nutrients from their dilute diet. These patterns are also generally supported (Karasov and Martínez del Rio 2007). Recent work in herbivorous fishes suggests that this may also apply to lipid because essential fatty acids must be obtained from a diet that is already low in lipid, and fiber binds to fat, making it less available than in animal material. Hence, it is common to observe elevated lipolytic activities in the guts of herbivorous fishes, which also show expanded gene copy number for lipase genes (German et al. 2004; Leigh et al. 2018; Heras et al. 2020). Most other specializations largely stay in accord with the Rate vs Yield continuum discussed above (Fig. 7.3).

Because of the Krogh principle, studies of nutritional physiology of wild-caught animals have relied upon the comparative method to infer specializations. Many studies are on single species and qualitative comparisons are made with other organisms. Modern work in a phylogenetic context, utilizing closely related species with different dietary affinities, have revealed specializations by exposing animals to dietary perturbations in a laboratory setting, and observing what changes and what doesn't, or making use of natural clines. Such studies have revealed that digestive enzyme activities and nutrient transport rates of carbohydrates do indeed support the AMH broadly. A deeper dive into the molecular underpinnings of common patterns is warranted, but already it has been shown that not all increases in amylase activity, for example, are underlain by changes in amylase gene copy number (German et al. 2016; Heras et al. 2020). In other words, there is more than one way to skin a cat to achieve the same physiological phenotype. Nevertheless, laboratory studies of wild-caught animals have indeed led to a formulation of the general principles of digestion, and when put in the context of chemical reactor theory (Penry and Jumars 1987), actually provides great benefit to biomedical sciences in addition to comparative and evolutionary physiology. Rodent models in digestion and metabolism have continually failed to be totally utilitarian, leaving researchers looking for additional models, which can be found broadly in the animal kingdom. Animal populations reared in the laboratory can prove useful in this regard.

7.6.2 Application Two: Nutritional Research with Laboratory Domesticated Populations

A great deal of research on nutrition is conducted using animals stocks that have been maintained in the laboratory for a number of generations. These species range from rodents, such as *Mus musculus*, to insects, such as *Drosophila melanogaster*. The previously described difficulties create considerable problems for the use of such laboratory animals for studying the evolutionary biology of nutrition. We will discuss the importance of each in turn.

Evolutionary History: Almost all laboratory studies of nutrition which use established laboratory stocks are conducted on small animals, no bigger than dogs or cats. The laboratory rodents that are commonly studied afford an inherently limited window into the evolutionary biology of nutrition, even among mammals. In this regard, rodents may be particularly poor models for human nutritional biology, with a glaring difference being that rodents have an intact uricase gene (and enzyme) that makes their metabolism of fructose categorically different from higher primates (including humans), which lack an expressed and active uricase enzyme (Lanaspa et al. 2012; Kratzer et al. 2014). Thus, the elevated consumption of processed, fructose-rich foods that result in considerable hepatic uric acid production (and downstream effects) in humans produce problems that are **not** completely analogous with how rodents metabolize fructose in their livers (discussed in more detail below).

Among insects, the most commonly used species are those that can survive and reproduce on foods which can be readily supplied in labs, such as fruit flies and *Tribolium* species. Insect parasitoids, for example, feed off the living guts of other insects, but are rarely cultured in the laboratory. Despite that, they are among the most abundant of all insect species.

Thus our most penetrating and extensive studies of the biology of nutrition are limited by the evolutionary histories of the kinds of species that are most commonly studied. Groups with radically different kinds of evolutionary histories are not known in comparable detail, and thus, we are in need of more lines of laboratory animals evolved on different diets to study closely how an animal's digestive system and metabolism change to meet new nutritional needs.

Ecological Change: Every species that is brought into laboratories and cultured therein is undergoing a major ecological change from its first generation of lab maintenance. Natural selection will then set about adapting such laboratory populations to their new lab environment, especially with respect to the type(s) of food that are provided to them (e.g., Leigh et al. 2018).

Thus it is mistake to believe that such laboratory stocks provide a reliable window into the nutritional biology of the species in general. Research projects using laboratory populations of any species are necessarily a study of how that species responds to the evolutionary predicament of consuming often a single laboratory diet, among other features of the laboratory environment.

In effect, every experiment conducted with long-cultured laboratory stocks is an experiment in the field that is known as "experimental evolution" (Garland and Rose, 2009), whether the experimenter is aware of it or not. And if there is any type of selection which is likely to be fast acting, it is selection to adapt a population to a novel laboratory diet (Leigh et al 2018). Such selection will not usually be focused chiefly on juvenile survival, if the food being supplied is sufficient for the completion of development. Instead, the more likely targets of selection are adult longevity and reproduction (Fig. 7.6). That in turn is fraught with possibilities for age-dependent heterogeneity, as we will discuss below.

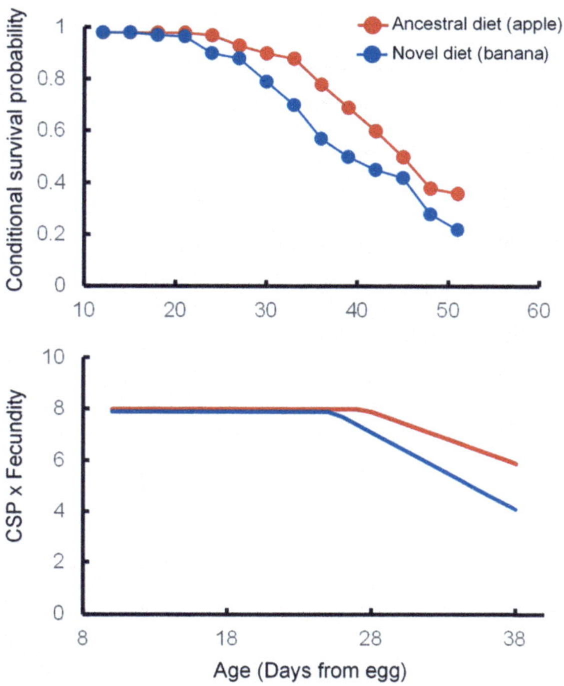

Fig. 7.6 **Top**: Conditional survival probability (CSP) as a function of adult age for female flies fed an ancestral (apple) diet, which they had been exposed to for many thousands of generations, or a novel (banana) diet they had been exposed to for about 1000 generations. Each point is pooled across five replicate lines of flies evolved on the different diets, and each is pooled across 3 days. The CSP is significantly higher in the apple-reared flies than the banana-reared flies for all points after day 12. Summarized from Rutledge et al. (2020). **Bottom**: CSP x fecundity, as a measure of female viability, as a function of adult age for the same female flies as the top graph. The "break day" at which the females began to decline, as well as the slope of CSP x fecundity over time, differed significantly among the two diets, with the flies on the banana diet declining sooner, and more rapidly, than those consuming the apple diet. From Rutledge et al. (2021)

Reduced N_e and inbreeding: Another inadvertent effect of laboratory cultivation will be greatly reduced N_e. And in some fields, such as population genetics, experimenters will actively inbreed laboratory stocks. An extreme example of this is the creation of "isofemale lines" from generation after generation of laboratory culture using only the progeny of sibling matings. Such heavily inbred lines are almost completely homozygous.

Whether moderate or severe, inbreeding degrades adaptation. Inbred stocks will have impaired nutritional adaptations on *any* diet. Furthermore, the ability of inbred laboratory strains to adapt to the novel laboratory diet will be specifically degraded.

In addition to this predictable "inbreeding depression" of functional characters from inbreeding and the failure of inbred lines to adapt to novel laboratory conditions, a collection of inbred lines will randomly diverge from their wild ancestors (Santos

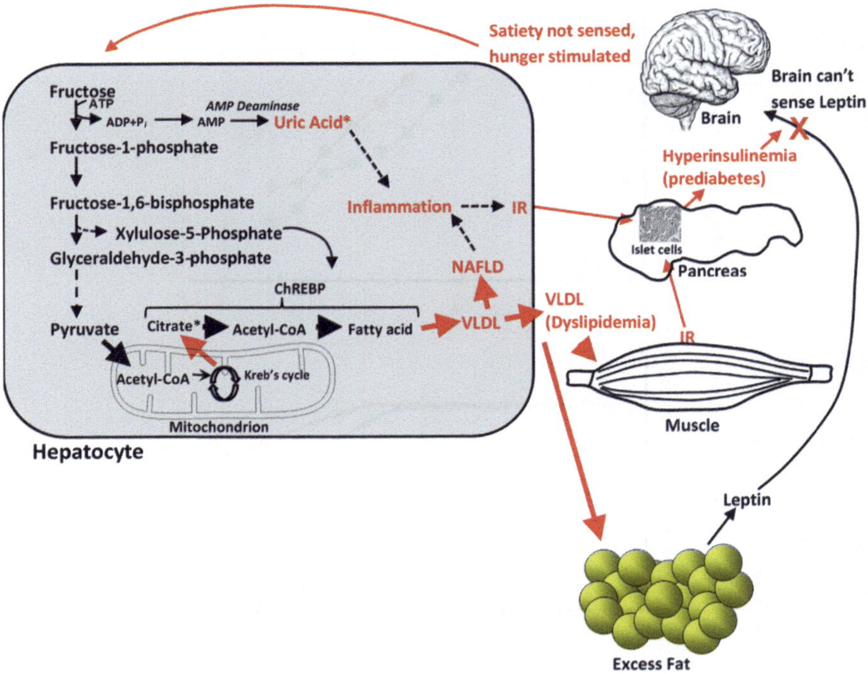

Fig. 7.7 Fructose metabolism in hepatocytes. Because 95% of all assimilated fructose is metabolized in the liver, an excessive carbon load is processed via glycolytic pathways, but starting with Fructose Kinase. This first step is consumptive of ATP and leads to significant uric acid production (Johnson et al. 2009; Lanaspa et al. 2012), which itself leads to oxidative stress in the cell. Although most of the fructose eventually enters the mitochondrion for entry into the Kreb's Cycle (thick black arrow), some F-1,6-BP is converted to Xylolose-5-Phosphate, which, through a cascade, activates Carbohydrate Responsive Element Binding Protein (ChREBP), which activates the fat synthesis cascade (Kabashima et al. 2003). Combined with the effects of uric acid (indicated by *, and detailed in Figure 7.8), this ensures a large amount of citrate exits the mitochondrion, resulting in excess fat production from fructose carbon. The resulting excess very low density lipoprotein (VLDL) concentrations in the blood leads to dyslipidemia, and insulin resistance (IR) by muscles (Krssak et al. 1999; Montell et al. 2001; Basciano et al. 2005; Dekker et al. 2010). Excess lipid returning to the liver leads to Non-Alcoholic Fatty Liver Disease (NAFLD; Ouyang et al. 2008; Dekker et al. 2010; Zámbó et al. 2013; Jegatheesan and De Bandt 2017; Jensen et al. 2018), and the associated inflammation leads to IR (and additional oxidative stress) in the liver (Rutledge and Adeli 2007; Dekker et al. 2010; Lim et al. 2010). With IR by many tissues, the pancreas over-produces insulin, leading to hyperinsulinemia (Lustig 2006; Shanik et al. 2008); this blocks the sensing of leptin at the brain (Kellerer et al. 2001; Shapiro et al. 2011), thus not allowing leptin to stimulate the satiety center and leaving the hunger center activated (Friedman and Halaas 1998; Ahima and Flier 2000; Isganatis and Lustig 2005). Leptin resistance leads to overeating, typically of energy dense foods, especially fructose. This repeating loop leads Type two diabetes to (T2D) (Isganatis and Lustig 2005). **Thick** arrows indicate large carbon loads. **Red** arrows and labels indicate metabolic abnormalities caused by excess fructose consumption. Dashed lines indicate multiple steps represented by that one line. Modified from Crummett (2020).

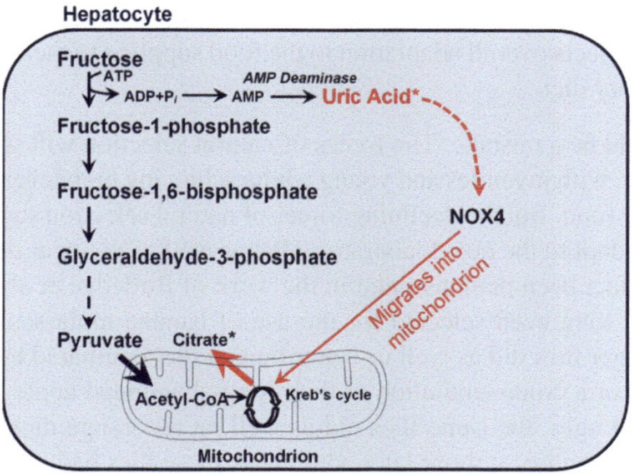

Fig. 7.8 The action of uric acid within a hepatocyte. Large amounts of uric acid are generated by excessive fructose metabolism and humans lack a working uricase enzyme (due to pseudogenization of the uricase gene; Kratzer et al. 2014) to degrade that uric acid. Excess uric acid causes oxidative stress in the cell, which causes the enzyme NADPH Oxidase 4 (NOX4) to migrate from its normal location in the cytoplasm, into the mitochondrion. Once in the mitochondrion, NOX4 is a powerful inhibitor of the Kreb's Cycle enzyme aconitase, which produces isocitrate from citrate. With aconitase inhibited, citrate piles up in the mitochondrion, and thus, leaks out in significant quantities (thick arrow) into the cytosol for use in fat production (see Fig. 7.7). Thus, other elements of fructose metabolism [carbon load at the liver (thick arrow into mitochondrion), xylulose-5-phosphate production upregulating fat synthesis enzymes] aid in fat production, but uric acid ensures that the starting point of fat synthesis (citrate) is in the cytosol in large amounts. Metabolic abnormalities indicated by redarrows and text. Based on Lanaspa et al. (2012).

et al. 2013). Thus, not only are inbred strains degraded, they are also randomly divergent. Worse material for the study of any type of functional biology is hard to imagine, excepting perhaps inbred lines that have been subjected to genome-wide mutagenesis using radiation or mutagenic compounds.

Yet such inbred and degraded organisms are the commonplace subjects of research on such phenomena as dietary restriction and the effects of dietary supplements (e.g. Weindruch and Walford 1988). For example, it is now known that while some lab rodent strains respond to dietary restriction with increased lifespan, other rodent strains do not. That is to say, the "reaction norm" that determines the response of lifespan to varying dietary intake is dependent on the genetic background of particular rodent lab strains. Reaction norms are examples of phenotypic plasticity, and the evolution of phenotypic plasticity is a long-standing and important subject within evolutionary biology. Such evolutionary phenomena should be taken into account in studies of the functional biology of nutrition among laboratory organisms.

Age-Dependent Adaptation: In laboratory populations that have not been deliberately inbred, but have been domesticated over as few as forty generations, we can expect considerable adaptation to laboratory conditions (Matos et al. 2000). It might then be supposed that the evolutionary biology of nutrition in such laboratory populations therefore reflects overall adaptation to the food supplied to them, again, usually a single laboratory diet.

But that would be a mistake. The forces of natural selection will scale adaptation according to age, with juveniles and young adults achieving high levels of adaptation to the laboratory food. But the declining forces of natural selection suggest that older adults will not adapt to the novel laboratory diet as well as younger organisms. This problem has in fact been demonstrated in the work of Rutledge et al. (2020, 2021). In flies that had long been selected for the use of banana-molasses medium, they found that younger flies did as well or better on that diet compared to either a novel orange medium or a crude emulation of their long-abandoned apple diet (Fig. 7.6). But at later adult ages, the same flies did as well on an orange diet as they did on the banana. Furthermore, at those later adult ages, these flies had *enhanced* survival and reproduction on lab medium that crudely emulated their long-abandoned apple diet. This is a signal example of the degree to which the design and interpretation of laboratory research on the evolutionary biology of nutrition should be informed by knowledge of the specific constraints that evolution places on the functional biology of nutrition.

7.6.3 Application Three: Nutritional Research with Human Patients

Human populations have a number of advantages with respect to the study of their nutritional biology, at least from an evolutionary perspective. Here we will again organize our discussion in terms of the same four evolutionary considerations that we have used hitherto.

Evolutionary History and Ecological Change: Humans are a relatively good target for research on the evolutionary biology of their nutrition. Above all, physical anthropologists have been extensively studying the evolutionary history of the patterns of food acquisition, processing, and consumption of our species. We have a fairly good idea of our food consumption patterns over the last million years: omnivory with extensive food processing, including the cooking of both animal tissues and otherwise toxic starchy foods. Different human populations vary with respect to the mix of animal and plant tissues used for nutrition, but extreme carnivory or herbivory were apparently rare before the advent of agriculture, up to 20,000 years ago.

Most importantly, it is well-known that most of our recently ancestral populations underwent a major change in food acquisition and consumption with the Neolithic agricultural revolution. Initially, it is clear that this dietary change had a substantially deleterious impact on human health, as shown in the skeletal remains from populations that underwent that change. But there is significant evidence for biological adaptation to the agricultural diet among Eurasian populations, with a continuum of adaptation from those populations to those of Australasia, the most recently agricultural large collection of human populations (e.g., Perry et al. 2007; Enattah et al. 2008). [There are also some minor non-agricultural populations that are peripheral to regions where people practiced agriculture, such as the Inuit of North America and the indigenous tribes of the Amazonian and Congolese rainforests.]

Finally, it is generally understood that the last few centuries have seen the widespread consumption of largely novel foods, from refined sugar to heavily processed "junk foods", to marbled meats, as mentioned above. Thus much recent nutritional physiology reflects a virtual absence of natural selection fostering adaptation to such novel "industrial" foods.

Perhaps one of the clearest examples is fructose metabolism in the human liver. Ancestral human diets had sucrose in fruits, some plant parts (e.g., beet root), and in honey. The first examples are encased in fibrous cell walls, and only available seasonally in many locations, the latter defended by bees (Lustig 2012). Thus, sucrose was somewhat difficult to obtain, even after mastication. With the advent of food processing, food processors realized that humans were drawn to sugar, and in fact, could become addicted to it. Thus, they added it, in large quantities, to processed foods, first as granulated sugar, then as high-fructose corn syrup after the Farm Bill of 1973 led to a glut of corn available for processing. Sugar consumption really took off after the 1982 "low fat guidelines" went into effect. These guidelines by the US congress suggested that Americans limit fat consumption to 30% of their daily caloric intake. This led food processors to make an array of new, "low-fat" or "non-fat" foods. Of course, fat tastes good, so when it is removed from food, the food doesn't taste good. This led to the addition of more sugar in processed foods to make it more palatable, and thus begun the obesity and type 2 diabetes epidemic (Lustig 2012).

In humans, 95% of fructose is metabolized in the liver. This leads to many problems, but the main issue is that if greater than about 10% of daily caloric intake is composed of sucrose, the fructose overloads the liver with a large carbon load that must be metabolized (Fig. 7.7). The way that fructose is brought into hepatocytes and metabolized there is categorically different than for glucose for many reasons. First, the phosphorylation of fructose is consumptive of ATP with little ATP given back in reward, and this process ends up generating Uric Acid (Fig. 7.7). Second, the activation of carbohydrate response element binding protein ensures that fat synthesis pathways are upregulated to use the excess citrate that will eventually be produced in, and leave, the mitochondria. The fat synthesis pathway leads to excess fat production from the fructose carbon, dyslipidemia, and eventually insulin resistance (Fig. 7.7). If it goes on long enough, insulin resistance can lead to extensive liver inflammation, non-alcoholic fatty liver disease, and type 2 diabetes. Prediabetes (hyperinsulinemia) can lead to leptin resistance, which leads to further overeating, and thus, reinforcement of the entire pathway. Hence, over consumption of sugar across hundreds or thousands of meals, leads to the metabolic syndrome and type 2 diabetes plaguing society, and it all starts with over consumption of fructose (Fig. 7.7), not glucose, per se (Stanhope et al. 2009).

Tied in with this is uric acid (Fig. 7.8). As mentioned above, uric acid is produced in large amounts in hepatic fructose metabolism (Lanaspa et al. 2012), and humans lack a working uricase enzyme to break down this excess uric acid (Kratzer et al. 2014). Beyond gout, uric acid leads to many problems on its own, including hypertension by inhibiting nitric oxide synthase, and to massive oxidative stress in hepatic mitochondria (Fig. 7.8). Essentially, elevated hepatic uric acid leads to a depolarization of the mitochondrial membrane potential, allows a cytoplasmic enzyme (NADPH oxidase 4) to enter the mitochondria, where it inhibits aconitase. Aconitase is the enzyme in the Citric Acid Cycle that converts citrate to isocitrate. Thus, this causes citrate to pile up and leave the mitochondria to enter fat synthesis, which is already upregulated by the fructose metabolic pathway (Fig. 7.8). Hence, excess uric acid ensures that fructose carbon will end up as fat. The lack of a working uricase enzyme in higher primates sets us apart from other mammals, such as monkeys, rodents, pigs, or other animal models (Kratzer et al. 2014). Therefore, there is no excellent model for human liver metabolism other than perhaps isolated human liver cells, which pose their own problems by not being part of the larger physiological organism.

Effective Population Size and Inbreeding: Another considerable advantage in the study of the functional biology of human nutrition is that we remain a "wild" population from the standpoint of population size and inbreeding depression. Our populations are not undergoing deliberate or inadvertent reductions in effective population size, so that inbreeding depression is not a common problem in studies of human nutrition.

Age-Dependent Forces of Natural Selection: The trickiest aspect of the functional biology of human nutrition is the interaction between our evolutionary and ecological history, on one hand, and the evolutionary biology of our aging, on the other hand. That is because there are effectively three different evolutionary forces acting on us simultaneously, each corresponding to different epochs in our evolutionary history, and to different phases of our life histories.

(i) Our ancient evolutionary history underpins our lifelong nutritional physiology, even to very late ages. For at least a million years, humans have been selected for adaptations to a relatively abundant and easily digested diet. With the invention of an extensive and efficient repertoire of tools for hunting, foraging, and food-processing more than a million years ago, our dietary intake was transformed. This is reflected in multiple unusual features of our nutritional anatomy and physiology.

Human dentition features teeth that are no longer used to attack other animals, or even conspecifics. We also lack thick enamel, continued adult growth of teeth, or continued replacement of lost teeth during adulthood, unlike some other mammalian species for which teeth play a large role in the mechanical crushing or tearing of food material. Evidently, this reflects dental evolution in the context of millions of years of tool use for hunting, gathering, and food processing, from spears to digging implements to pounding with stones. In a sense, our primary feeding "implements" are hand-held tools, not teeth. Because we "pre-masticate" our foods, we don't spend up to 80% of our time chewing, as our ancestors did.

The human gut is also different from that of many other mammals. The human stomach does not produce abundant acid to kill potential parasites, as in obligate carnivores. Indeed, the evolution of the stomach in elasmobranchs was a novel, vertebrate innovation, and it has changed much in many animal lineages, including its loss multiple times (Wilson and Castro 2011; perhaps most recently in humans undergoing gastric bypass surgery). As mentioned earlier, cooking is an entirely human endeavor. Although many animals, from insects, to snails, to fishes, farm foods for consumption, none of them deliberately cook their foods with heat, chemicals, or fermentations, before consuming them. Hence, humans are unique in that they "predigest" foods by cooking them. One of the main outcomes of cooking is the denaturation of proteins, which have led to a stomach pH (2) that is not as acidic as strict carnivores. Beyond the denaturation of protein, cooking and grinding our foods also makes fiber less of a barrier to digestion, meaning we are less reliant on enteric microbes to degrade fibrous material than our hominid ancestors or closely related apes. Thus, human intestines lack a large caecum for culturing commensal microbes to provide nutrients, particularly calories from fiber degradation. Instead, the human gastrointestinal tract seems to have evolved toward the absorption of nutrients that have been made relatively accessible through the use of cooking and other types of food preparation.

In a treatise on the evolution of cooking, Wrangham (2009) argues that cooking with fire was essential in human evolution. Once we started cooking foods, particularly animal material, our braincase increased in size, and tools became more sophisticated. One of the things that fire does is that it causes the crosslinking of proteins with glucose in what is known as the "Maillard" or browning reaction. The browning reaction releases glutamate, which is what is sensed by our umami taste buds. Hence, in many ways, we have evolved to be drawn to glutamate-rich foods, and cooking makes that more available with the browning reaction. This is also not lost on food processors, who add monosodium glutamate to foods to make them hyperpalatable. Moreover, in addition to denaturing proteins, cooking with heat sterilizes food, making a highly acidic stomach of true carnivores not as necessary for humans since the microbial load ingested with our food is reduced through cooking.

With so much of human digestion achieved extra-somatically, and a long-ancestral diet that featured omnivory, it is a common assumption that humans are essentially "pan-omnivorous," with a capacity to derive nutrition for almost any food that is not toxic. But we believe that this is an important mistake. There is no reason to expect that any human is well-adapted to the consumption of foods that are fundamentally novel, relative to our evolutionary history. No human population has evolved with sustained, high levels of sucrose consumption. No human population has evolved with sustained, high levels of consumption of low-fat, high-starch, high-salt, chemically-laden, fiberless junk foods, such as those which can be found in the middle aisles of most Western supermarkets. While human bodies do absorb abundant calories from such foods, these foods are nonetheless not foods that we are well-adapted to. Thus it is not surprising from these evolutionary considerations that the "Standard American Diet" ("SAD") is associated with higher levels of chronic illness, compared to the more traditionally agricultural "Mediterranean Diet." Of course, low-fiber diets are linked with an altered large intestine microbiome, which may have downstream health effects, including autoimmune disorders. With the recommendation of 25–60 g of fiber a day, most Americans only get 10–15 g. A recent Course-based Undergraduate Research Experience (CURE) course at UC Irvine showed how a high-fiber diet intervention shifted the microbiome of the participating students (Oliver et al. 2021).

(ii) With the Neolithic agricultural revolution, humans have been strongly selected for biological adaptations to diets based chiefly on the consumption of large grass species. Most humans now descend from populations that have chiefly fed on foods obtained through agricultural cultivation of large grass species: grains, rice, and corn, chiefly. Some Old World populations have also consumed foods derived from the milk of herd animals, chiefly ungulates. This was a major "ecological" change for the action of natural selection on human nutritional biology. And such natural selection has been sustained for hundreds of ancestral generations for much of the present human species.

Such sustained natural selection will necessarily have significant effects on human nutritional adaptation. Our long-ancestral diets generally featured omnivory, with an exceptional diversity of foods consumed by hunter-gatherer populations. By contrast, agricultural diets for the majority of "civilized" human populations lack diversity. In addition, both grass species and ungulate milk feature an abundance of pro-inflammatory agents. Grass species yield seeds that contain a great deal of omega-6 fatty acids. Ungulate milk, particularly in animals fed corn-based diets, is laden with short polypeptides, and branched amino acids, that are known to stimulate the adaptive immune system of calves. Humans who lack agricultural ancestry are well known to suffer a variety of chronic inflammatory disorders when they are subjected to agricultural diets (Lindeberg et al. 2007). By contrast, individuals who have relatively more ancestry from areas that feature milk-derived foods show longer maintenance of lactase production (Lindeberg et al. 2007; Enattah et al. 2008). Overall, it is apparent that agricultural human populations have become biologically adapted to the sustained consumption of agricultural foods, from bread to cheese to boiled rice. At least at early ages.

(iii) Later-life reversion to pre-agricultural nutritional biology. The evolutionary theory of ageing is fundamentally predicated on some degree of age-specificity in genetic effects on life history across biological ages (Rose 1991). Without such age-specificity, ageing is not expected to evolve.

The same principle in turn applies to the immediate evolution of species in response to a major ecological change. As shown by Rutledge et al. (2020, 2021), adaptation to a sustained change in diet produces strong dietary adaptation at early ages, but weaker adaptation at later ages (Fig. 7.6). In the case of human populations with agricultural ancestry, we can therefore expect that at later ages adaptations to agricultural diets will weaken. Eventually, given continued survival to later ages, the better diet may be one which emulates a diet that is no longer based chiefly on grass species or the milk of ungulates. That is to say, at late enough ages, reversion to a "Paleolithic Diet" may yield improved chronic health, with possibly reduced risk of chronic disorders like type 2 diabetes and some types of cardiovascular disease.

Note, however, that this is not an argument in favor of universal adoption of such diets. Children and young adults from human populations that have long consumed agricultural foods should be relatively well-adapted to such foods.

7.7 Conclusion

Theodosius Dobzhansky famously said, "Nothing in biology makes sense except in the light of evolution." We believe that this is certainly true with respect to nutritional biology, where evolutionary history, ecological change, inbreeding, and age will all have species-specific effects that will need to be taken into account in scientific research on the topic. The field of nutritional physiology, particularly in evolutionary and ecological contexts (i.e., outside of culturing animals for human consumption), is in its infancy. Much work remains to be done to tease apart how each facet of the inner tube of life truly feeds the fire of metabolism that is life itself. One thing is for sure, though, we cannot throw away the gut anymore. It deserves our full attention.

Compliance with Ethical Standards

Conflict of Interest The authors declare having no conflict of interest.

References

Ahima RS, Flier JS (2000) Leptin. Annu Rev Physiol 62:413–437

Allison SD, Chacon SS, German DP (2014) Substrate concentration constraints on microbial decomposition. Soil Biol Biochem 79:43–49

Altamia MA, Lin Z, Trindade-Silva AE, Uy ID, Shipway JR, Wilke DV, Concepcion GP, Distel DL, Schmidt EW, Haygood MG, Haygood MG (2020) Secondary Metabolism in the Gill Microbiota of Shipworms (Teredinidae) as Revealed by Comparison of Metagenomes and Nearly Complete Symbiont Genomes. mSystems 5(3):e00261–00220

Axelsson E et al (2013) The genomic signature of dog domestication reveals adaptation to a starch-rich diet. Nature 495:360–364

Basciano H, Federico L, Adeli K (2005) Fructose, insulin resistance, and metabolic dyslipidemia. Nutr Metab 2(1):5

Betancor MB, Olsen RE, Marandel L, Skulstad OF, Madaro A, Tocher DR, Panserat S (2018). Impact of dietary carbohydrate/protein ratio on hepatic metabolism in land-locked atlantic salmon (Salmo salar L.). Frontiers Physiol 9(1751)

Brandl C, Bucci L, Schett G, Zaiss MM (2021) Crossing the barriers: revisiting the gut feeling in rheumatoid arthritis. Europ J Immunol, in Press. https://doi.org/10.1002/eji.202048876

Brun A, Mendez-Aranda D, Magallanes ME, Karasov WH, Martínez del Rio C, Baldwin MW, Caviedes-Vidal E (2020) Duplications and functional convergence of intestinal carbohydrate-digesting enzymes. Mol Biol Evol 37(6):1657–1666

Buddington RK, Chen JW, Diamond J (1987) Genetic and phenotypic adaptation of intestinal nutrient transport to diet in fish. J Physiol 393:261–281

Burke MK (2012) How does adaptation sweep through the genome? Insights from long-term selection experiments. Proc Roy Soc B Biol Sci 279:5029–5038

Burke MK, Barter TT, Cabral LG, Kezos JN, Phillips MA, Rutledge GA, Phung KH, Chen RH, Nguyen HD, Mueller LD, Rose MR (2016) Rapid divergence and convergence of life-history in experimentally evolved Drosophila melanogaster. Evolution 70:2085–2098
Burke MK, Dunham JP, Shahrestani P, Thornton KR, Rose MR, Long AD (2010) Genome-wide analysis of a long-term evolution experiment with Drosophila. Nature 467(7315):587–590
Castro LFC, Gonçalves O, Mazan S, Tay BH, Venkatesh B, Wilson JM (2014) Recurrent gene loss correlates with the evolution of stomach phenotypes in gnathostome history. Proc Roy Soci Lond B: Biological Sci 281(1775):20132669
Charlesworth B (1980) Evolution in age-structured populations. Cambridge University Press, Cambridge
Clements KD, Raubenheimer D (2006) Feeding and nutrition. In: Evans DH (ed) The physiology of fishes. CRC Press, Boca Raton, pp 47–82
Clissold FJ, Tedder BJ, Conigrave AD, Simpson SJ (2010) The gastrointestinal tract as a nutrient-balancing organ. Proc Roy Soc B: Biological Sci 277(1688)
Crummett LT (2020) Industry's role in the metabolic disease pandemic. In Patterson A, Read I (eds.) The Shapes of Epidemics and Global Disease. Cambridge Scholars Publishing. Tyne, England
Davis RB, Javoiš J, Kaasik A, Õunap E, Tammaru T (2016) An ordination of life histories using morphological proxies: capital vs. income breeding in insects. Ecology 97(8):2112–2124
Dekker MJ, Su Q, Baker C, Rutledge AC, Adeli K (2010) Fructose: a highly lipogenic nutrient implicated in insulin resistance, hepatic steatosis, and the metabolic syndrome. Am J Physiol-Endocrinol Metabolism 299(5):E685–E694
Diamond J (2002) Quantitative evolutionary design. J Physiol 542(Pt 2):337–345
Diamond J, Hammond K (1992) The matches, achieved by natural selection, between biological capacities and their natural loads. Experientia 48:551–557
Dudley R, Gans C (1991) A critique of symmorphosis and optimality models in physiology. Physiol Zool 64(3):627–637
Enattah NS (2008) Independent introduction of two lactase-persistence alleles into human populations reflects different history of adaptation to milk culture. Am J Hum Genet 82(1):57–72
Frank SA (2007) Maladaptation and the paradox of robustness in evolution. PLoS ONE 2(10)
Friedman JM, Halaas JL (1998) Leptin and the regulation of body weight in mammals. Nature 395(6704):763–770
Gadgil M, Bossert WH (1970) Life historical consequences of natural selection. Am Nat 104(935):1–24
German DP (2009) Inside the guts of wood-eating catfishes: can they digest wood? J Comp Physiol B 179:1011–1023
German DP, Bittong RA (2009) Digestive enzyme activities and gastrointestinal fermentation in wood-eating catfishes. J Comp Physiol B 179:1025–1042
German DP, Foti DM, Heras J, Amerkhanian H, Lockwood BL (2016) Elevated gene copy number does not always explain elevated amylase activities in fishes. Physiol Biochem Zool 89:277–293
German DP, Horn MH, Gawlicka A (2004) Digestive enzyme activities in herbivorous and carnivorous prickleback fishes (Teleostei: Stichaeidae): ontogenetic, dietary, and phylogenetic effects. Physiol Biochem Zool 77:789–804
German DP, Miles RD (2010) Stable carbon and nitrogen incorporation in blood and fin tissue of the catfish *Pterygoplichthys disjunctivus* (Siluriformes, Loricariidae). Environ Biol Fishes 89:117–133
German DP, Nagle BC, Villeda JM, Ruiz AM, Thomson AW, Contreras-Balderas S, Evans DH (2010) Evolution of herbivory in a carnivorous clade of minnows (Teleostei: Cyprinidae): effects on gut size and digestive physiology. Physiol Biochem Zool 83:1–18
German DP, Sung A, Jhaveri PK, Agnihotri A (2015) More than one way to be an herbivore: convergent evolution of herbivory using different digestive strategies in prickleback fishes (family Stichaeidae). Zoology 118:161–170
Gould SJ, Lewontin RC (1979) The Spandrels of San Marco and the Panglossian paradigm: a critique of the adaptationist paradigm. Proc R Soc Lond B Biol Sci 205:581–598

Graves JL, Jr et al (2017) Genomics of parallel experimental evolution in drosophila. Mol Biol Evol 34(4):831–842

Hamilton WD (1966) The moulding of senescence by natural selection. J Theor Biol 12(1):12–45

Hammer TJ, Janzen DH, Hallwachs W, Jaffe SP Fierer N (2017) Caterpillars lack a resident gut microbiome. Proc Nat Acad Sci 114:9641–9646

Heras J, Chakraborty M, Emerson JJ, German DP (2020) Genomic and biochemical evidence of dietary adaptation in a marine herbivorous fish

Herrel A et al (2008) Rapid large-scale evolutionary divergence in morphology and performance associated with exploitation of a different dietary resource. Proc Natl Acad Sci 105:4792–4795

Herrera MJ, Heras J, German DP (2021) Comparative transcriptomics reveal tissue level specialization towards diet in prickleback fishes. J Comparative Physiol B, (In Revision)

Hofmann R (1989) Evolutionary steps of ecophysiological adaptation and diversification of ruminants - a comparative view of thier digestive system. Oecologia 78:443–457

Isganaitis E, Lustig RH (2005) Fast food, central nervous system insulin resistance, and obesity. Arterioscler Thromb Vasc Biol 25(12):2451–2462

Jegatheesan P, De Bandt JP (2017) Fructose and NAFLD: the multifaceted aspects of fructose metabolism. Nutrients 9(3)

Jensen T et al (2018) Fructose and sugar: A major mediator of non-alcoholic fatty liver disease. J Hepatol 68(5)

Jin Y et al (2018) A systemic study of lipid metabolism regulation in salmon fingerlings and early juveniles fed plant oil. Br J Nutr 120(6):653–664

Johnson RJ et al (2009) Lessons from comparative physiology: could uric acid represent a physiologic alarm signal gone awry in western society? J Comparative Physiol B 179(1), 67–76

Kabashima T, Kawaguchi T, Wadzinski BE, Uyeda K (2003) Xylulose 5-phosphate mediates glucose-induced lipogenesis by xylulose 5-phosphate-activated protein phosphatase in rat liver. Proc Natl Acad Sci 100(9):5107–5112

Karasov WH (1992) Tests of the adaptive modulation hypothesis for dietary control of intestinal nutrient transport. Am J Physiol 263:R496–R502

Karasov WH, Diamond J (1988) Interplay between physiology and ecology in digestion. Bioscience 38(9):602–611

Karasov WH, Douglas AE (2013) Comparative digestive physiology. Comprehensive. Physiol 3:741–783

Karasov WH, Hume ID (1997) Vertebrate gastrointestinal system. In: Dantzler WH (ed) Handbook of Comparative Physiology, vol 1. American Physiological Society, Bethseda, pp 407–480

Karasov WH, Martínez del Rio C (2007) Physiological ecology: how animals process energy, nutrients, and toxins. Princeton University Press, Princeton, NJ

Kellerer M et al (2001) Insulin inhibits leptin receptor signalling in HEK293 cells at the level of janus kinase-2: a potential mechanism for hyperinsulinaemia-associated leptin resistance. Diabetologia 44(9):1125–1132

Kohl KD, Brzek P, Caviedes-Vidal E, Karasov WH (2011) Pancreatic and intestinal carbohydrases are matched to dietary starch level in wild passerine birds. Physiol Biochem Zool 84:195–203

Kratzer JT (2014) Evolutionary history and metabolic insights of ancient mammalian uricases. Proc Natl Acad Sci 111(10):3763–3768

Krebs HA, Krebs JR (1980) The "August Krogh principle". Comparative biochemistry and physiology. B Comparative Biochem 67:379–380

Krssak M et al (1999) Intramyocellular lipid concentrations are correlated with insulin sensitivity in humans: a 1H NMR spectroscopy study. Diabetologia 42(1):113–116

Lanaspa MA et al (2012) Uric acid induces hepatic steatosis by generation of mitochondrial oxidative stress: POTENTIAL ROLE IN FRUCTOSE-DEPENDENT AND -INDEPENDENT FATTY LIVER*. J Biol Chem 287(48):40732–40744

Leigh SC, Nguyen-Phuc BQ, German DP (2018) The effects of protein and fiber content on gut structure and function in zebrafish (*Danio rerio*). J Comp Physiol B 188(2):237–253

Lim JS, Mietus-Snyder M, Valente A, Schwarz JM, Lustig RH (2010) The role of fructose in the pathogenesis of NAFLD and the metabolic syndrome. Nat Rev Gastroenterol Hepatol 7(5):251–264
Lindeberg S, Jönsson T, Granfeldt Y, Borgstrand E, Soffman J, Sjöström K, Ahrén B (2007) A Palaeolithic diet improves glucose tolerance more than a Mediterranean-like diet in individuals with ischaemic heart disease. Diabetologia 50(9):1795–1807
Lujan NK, Cramer CA, Covain R, Fisch-Muller S, López-Fernández H (2017) Multilocus molecular phylogeny of the ornamental wood-eating catfishes (Siluriformes, Loricariidae, Panaqolus and Panaque) reveals undescribed diversity and parapatric clades. Mol Phylogenet Evol 109:321–336
Lujan NK, German DP, Winemiller KO (2011) Do wood grazing fishes partition their niche? Morphological and isotopic evidence for trophic segregation in Neotropical Loricariidae. Funct Ecol 25:1327–1338
Lustig RH (2006) Childhood obesity: behavioral aberration or biochemical drive? Reinterpreting the First Law of Thermodynamics. Nat Clin Pract Endocrinol Metab 2(8):447–458
Lustig RH (2012) Fat Chance: Beating the odds against sugar, processed food, obesity, and disease. Avery, New York
Lynch M, Walsh B (1998) Genetics and analysis of quantitative traits. Sinauer Associates, Sunderland, Mass
Matos M, Rose MR, Rocha Pite MT, Rego C, Avelar T (2000) Adaptation to the laboratory environment in *Drosophila subobscura*. J Evol Biol 13:9–19
MacArthur R, Pianka E (1966) On optimal use of a patchy environment. Am Nat 100(916):603–609
McCauley M, German DP, Lujan NK, Jackson CR (2020) Gut microbiomes of sympatric Amazonian wood-eating catfishes (Loricariidae) reflect host identity but little role for in wood digestion. Ecol Evol 10:7117–7128
Merkin J, Russell C, Chen P, Burge CB (2012) Evolutionary dynamics of gene and isoform regulation in mammalian tissues. Science 338:1593–1599
Montell E et al (2001) DAG accumulation from saturated fatty acids desensitizes insulin stimulation of glucose uptake in muscle cells. Am J Physiol Endocrinol Metab 280(2):E229–E237
Moran NA, Ochman H, Hammer TJ (2019). Evolutionary and ecological consequences of gut microbial communities. Annual Rev Ecol Evol Systematics 50:20.21–20.25
Oliver A et al (2021) High-fiber, whole-food dietary intervention alters the human gut microbiome but not fecal short-chain fatty acids. mSystems 6(2)
Ouyang X et al (2008) Fructose consumption as a risk factor for non-alcoholic fatty liver disease. J Hepatol 48(6):993–999
Penry DL, Jumars PA (1987) Modeling animal guts as chemical reactors. Am Nat 129(1):69–96
Perry GH et al (2007) Diet and the evolution of human amylase gene copy number variation. Nature Genetics 39:1256–1260
Phillips MA et al (2016) Genome-wide analysis of long-term evolutionary domestication in *Drosophila melanogaster*. Sci Rep 6(1):39281
Pollan M (2006) The Omnivore's Dilemma: a natural history of four meals. Penguin Press, New York
Pyke G (1984) Optimal foraging theory: a critical review. Ann Rev Ecol Systematics 15:523–575
Rose MR (1991) Evolutionary biology of aging. Oxford University Press, New York
Rose MR, Lauder GV (1996) Adaptation. Academic Press, San Diego
Rose MR, Passananti HB, Matos M (eds) (2004) Methuselah flies
Rose MR, Passananti HB, Chippindale AK, Phelan JP, Matos M, Teotónio H, Mueller LD (2005) The effects of evolution are local: Evidence from experimental evolution in Drosophila. Integr Comp Biol 45:486–491
Rutledge AC, Adeli K (2007) Fructose and the metabolic syndrome: pathophysiology and molecular mechanisms. Nutr Rev 65(6 Pt 2):S13–23
Rutledge GA et al (2020) Hamiltonian patterns of age-dependent adaptation to novel environments. PLoS ONE, 15(10), e0240132. https://doi.org/10.1371/journal.pone.0240132

Rutledge GA, Phang HJ, Le MN, Bui L, Rose MR, Mueller LD, Jafari M (2021) Diet and botanical supplementation: combination therapy for healthspan improvement? Rejuvenation Research, in Press. https://doi.org/10.1089/rej.2020.2361

Salvador A, Savageau MA (2006) Evolution of enzymes in a series is driven by dissimilar functional demands. Proc Natl Acad Sci 103(7):2226–2231

Santos J, Pascual M, Simoes P, Fragata I, Rose MR, Matos M (2013) Fast evolutionary genetic differentiation during experimental colonizations. J Genet 92:183–194

Schondube J, Herrera L, Martinez Del Rio C (2001) Diet and the evolution of digestion and renal function in phyllostomid bats. Zoology 104:59–73

Secor SM, Diamond J (2000) Evolution of regulatory responses to feeding in snakes. Physiol Biochem Zool 73(2):123–141

Shanik MH, Xu Y, Skrha J, Dankner R, Zick Y, Roth J (2008) Insulin resistance and hyperinsulinemia: is hyperinsulinemia the cart or the horse? Diabetes Care 31(Suppl 2):S262–268

Shapiro A, Tümer N, Gao Y, Cheng KY, Scarpace PJ (2011) Prevention and reversal of diet-induced leptin resistance with a sugar-free diet despite high fat content. Br J Nutr 106(3):390–397

Stanhope KL et al (2009) Consuming fructose-sweetened, not glucose-sweetened, beverages increases visceral adiposity and lipids and decreases insulin sensitivity in overweight/obese humans. J Clin Investigation 119(5):1322–1334

Stevens CE, Hume ID (1995) Comparative physiology of the vertebrate digestive system, 2nd edn. Press Syndicate of the University of Cambridge, Melbourne, Australia

Sullam KE, Dalton CM, Russell JA, Kilham SS, El-Sabaawi R, German DP, Flecker AS (2015) Changes in digestive traits and body nutritional composition accommodate a trophic niche shift in Trinidadian guppies. Oecologia 177(1):245–257

Van Soest P (1994) Nutrional ecology of the ruminant. Cornell University Press, Ithica, NY

Waterbury JB, Calloway CB, Turner RD (1983) A cellulolytic, nitrogen-fixing bacterium cultured from the gland of Deshayes in shipworms (Bivalvia: Teredinidae). Science 221:1401–1403

Wehrle BA et al (2020) Rapid dietary shift in *Podarcis siculus* resulted in localized changes in gut function. Physiol Biochem Zool 93:396–415

Weibel E, Taylor C, Hoppeler H (1991) The concept of symmorphosis: a testable hypothesis of structure-function relationship. Proc Natl Acad Sci 88:10357–10361

Weindruch R, Walford RL (1988) The Retardation of Aging and Disease by Dietary Restriction. Springfield. Charles C. Thomas, Illinois

Willmott ME, Clements KD, Wells RMG (2005) The influence of diet and gastrointestinal fermentation on key enzymes of substrate utilization in Marine teleost fishes. J Exp Mar Biol Ecol 317:97–108

Wilson JM, Castro LFC (2011) Morphological diversity of the gastrointestinal tract in fishes. In: Grosell M, Farrell AP, Brauner CJ (eds) The multifunctional gut of fish, vol 30. Elsevier, San Diego, pp 1–55

Wrangham RW (2009) Catching fire: how cooking made us human. Basic Books, Philadelphia

Zámbó V, Simon-Szabó L, Szelényi P, Kereszturi E, Bánhegyi G, Csala M (2013) Lipotoxicity in the liver. World J Hepatol 5(10):550–557

Part II
Food for Health

Chapter 8
Animal- and Plant-Based Food for Health and Longevity

Azza Silotry Naik, Majeed Jamakhani, Madhavi R. Vernekar, and S. S. Lele

Abstract Epidemiological studies have shown certain diets such as Mediterranean diet, Okinawa diet, New Nordic diet and vegetarian diets to promote longevity. Omega-3 fatty acids, bioactive peptides, polyphenols and antioxidants are some of the implicated molecules derived from both plant and animal food that provide protection against cellular ageing. The role of gut microbiome and its interaction with dietary components has provided novel ways to understand the process of ageing. Beyond the inherent biochemical composition of plants and animals consumed, the methods employed for farming and agriculture (such as organic farming) as well as food preparation techniques (such as frying, grilling, steaming) have a major effect on the health benefits derived from the dietary source. In conclusion, health promoting effect of animal and plant based diets depend on obtaining recommended daily allowances of health promoting ingredients from a portion and minimizing intake of harmful components such as saturated fats, nitrites, nitrates, anti-nutrients amongst others. A well balanced personalized diet catering to an individual's age, genetic predisposition to allergies and diseases and microbiome needs to be followed for longevity. Novel food processing technologies such as 3-D printing that incorporates both plant & animal ingredients will inch us closer to fabricating ideal functional food products aimed at promoting longevity in a sustainable way.

Keywords Food bioactives · Anti-ageing · Dietary patterns · Microbiome · Food preparation · Agricultural practise · Food safety

---

A. S. Naik (✉)
Department of Biological Sciences, School of Natural Sciences, Faculty of Science and Engineering, University of Limerick, V94 T9PX, Limerick, Ireland

M. Jamakhani
Molecular and Cellular Epigenetics (GIGA) and Molecular Biology (GxABT), University of Liège, 4000 Liège, Belgium

M. R. Vernekar
School of Biotechnology and Bioinformatics, D.Y. Patil Deemed to be University, Navi, Mumbai, India

S. S. Lele
Institute of Chemical Technology, Matunga, Mumbai 400019, India

S. I. S. Rattan and G. Kaur (eds.), *Nutrition, Food and Diet in Ageing and Longevity*, Healthy Ageing and Longevity 14, https://doi.org/10.1007/978-3-030-83017-5_8

8.1 Introduction

Diet plays a major role in directly affecting longevity by altering biochemical constitution of cells and metabolic pathways or indirectly through interactions with the genome. Plant and animal based food and food-derived biomolecules are known to promote longevity. Animal and plant based dietary options are complex and preferences are classified as vegan, ovo-lacto-vegetarian, pesco-vegetarian or omnivore/non-vegetarian. The choice of plant and animal based diet is often based on geographical location, availability, cost, religious views, traditional practises, health status and age. Often vegetarian diets are cereal and legume based and in this chapter we will discuss that the benefits derived from vegetarian diets are also accounted to fresh fruits and vegetables. However availability and cost influence the addition of fresh produce to the so called vegetarian diets. Animal diets are often enjoyed for their taste and satisfaction and despite them being calorie laden and concentrated in saturated fats, are often a source of essential amino acids, minerals and vitamins.

The role of diet on longevity is studied based on the effect of consumption of dietary components on certain biomarkers of health and longevity. Most studies focus on markers implicated in diseases such as cardiovascular, cancer, neurodegenerative and common metabolic diseases such as diabetes known to reduce life span (Barnes 2015). However minor ailments and resulting low quality of life have an additive effect on ageing. Plant based diets are known to be low in saturated fats thereby playing a critical role in reduction of cardiovascular diseases in people following these diets. Common ailments such as scurvy can easily be prevented by including vitamin-C rich plant based food products. On the flip side, vegan diets are known to de devoid of vitamin B-12, deficiency of which is associated with stroke, Alzheimer's disease, vascular dementia, Parkinson's disease and cognitive impairments. Ageing is often accompanied with lowered level of digestive enzymes and higher incidences of food intolerances. Thus transformation in dietary preferences from animal to plant or vice versa as a result of age or sickness/deficiency is not uncommon. Lactose intolerant individuals seldom derive benefits from dairy products and have to switch to plant based alternatives just as individuals allergic to nuts such as cashews or almonds cannot consume plant based dairy alternatives. In general, ageing is accompanied with changes in microbiome along with reduced intestinal uptake of nutrients thus affecting dietary preferences across age groups (Vemuri et al. 2018).

The focus of this chapter is to compare composition of animal- and plant-based food and its effects on health, gut microbiome and ultimately human life span. It will also review common food preparation methods and agricultural practises and their role in altering biocomposition of select food items. The chapter concludes by enlisting novel alternatives available such as 3-D printed food for the health conscious consumers.

8.2 Epidemiological Studies and Dietary Preferences

An ideal diet should include safe, palatable, healthy and tasty food which can provide health, mental satiety and longevity. Several reports show that people in various parts of the world consume diets that enable them to live longer, e.g. Mediterranean diet and Japanese diet. Mediterranean diet involve abundant plant-based food ingredients such as fruits, vegetables, grains, beans, nuts, and olive oil seeds as the primary source of fat. A study revealed that, adherence to Mediterranean diet improved the health status of individuals with decrease in mortality (9%), mortality from cardiovascular diseases (9%), incidence of or mortality from cancer (6%), and incidence of Parkinson's disease and Alzheimer's disease (13%) (Sofi et al. 2008). Likewise, longevity factor is also observed in people living in Japanese island Okinawa. Health statistics show that Okinawan's have the longest documented life expectancy at 85.1 years for women and 77.2 years for men. Centenarians in Okinawa number about 35 per 100,000 people compared with 5 to 10 among the same population in the United States. There exists a variety of dietary patterns observed worldwide which are generally based on availability of specific food (plants and animals) and is largely governed by the geography, climate and environmental factors. Table 8.1 gives on overview of different diets and their benefits.

8.3 Factors Affecting Bio-composition of Plant and Animal Food Sources

For comparison of plant and animal diet and its effect on longevity it is imperative to study the bio-composition of the dietary source. It has been widely reported that the environment and use of specific agricultural and farming practises can affect the produce either favourably or unfavourably. Environmental factors such as water, air and soil can significantly affect the nutritional quality, safety and sensorial traits of the food. Air, water and soil are continuously monitored and regulatory limits for residues and pollutants are well established by agencies such as environmental protection agency (http://www.epa.ie/pubs/reports/indicators/SoE_Report_2016.pdf). Despite the limits, anthropogenic activities lead to pollutants such as gases (NH3, SO2, CO, C6H6, volatile phenols), suspended particles (PM10 and PM2.5, Pb, Ni, Cd, polyaromatic hydrocarbons) and ionizing radiation in air or chemicals (fertilizers & pesticides/industrial dyes/heavy metals) in water & soil that can be toxic to animals or plants (Gheorghe and Ion 2011). Toxic effects range from cancers in farm animal to presence of residues in animals and plants which when consumed by humans can be detrimental to their health and longevity. Agricultural practice followed for cultivation and rearing of animals has a direct impact on the composition of the plant and animal products. Thus grass fed cows produce nutritionally superior milk and the products produced from such milk rank high in palatability as well as bio-actives. The nutritional quality of sweet cream butter with respect to carotenoids and PUFAs

Table 8.1 Different dietary patterns followed world wide

Type of diet	Composition	Constituents according to the order of intake	Benefits	Reference
Mediterranean diet	60–65% of the calories as carbohydrates 10% proteins 25–30% fats 7–10% saturated fatty acids upto 20% MUFA 7% PUFA	Daily physical activity and exercise (at least 30 min per day) Water (1.5–2 L per day) Vegetables, fruits, extra virgin olive oil, whole meal grains and legumes, spices and aromatic herbs, olives, nuts and oil seeds Moderate intake of fish, poultry and eggs Occasionally red meat, processed food, sugary and fatty foods	High consumption of unsaturated fats, foods rich in fiber, antioxidants and anti-inflammatory compounds. These slow down ageing and decrease the risk of various chronic disease Significant reduction in Significantly reduced overall mortality rate, healthy longevity achieved	Martinez-Gonzalez and Martín-Calvo 2016; Sofi et al. 2008
Okinawa diet	60–65% of total calories as carbohydrates 20–25% fats Saturated fats 7% 10–15% proteins	Regular physical activity Drinking plenty of water or tea Staple carbohydrate sweet potato Vegetable based (raw and cooked vegetables, soups, sea weed), Tofu Jasmine Tea Small servings of fish, noodles, lean meat	Reduces risk of many chronic disease leading to healthy longevity	Willcox et al. 2009

(continued)

Table 8.1 (continued)

Type of diet	Composition	Constituents according to the order of intake	Benefits	Reference
New Nordic diet	45–60% of total calories as carbohydrates 33% fats 10–20% MUFA 5–10% PUFA Less than 10% saturated fatty acids Vitamin D supplementation	Local organic foods such as fruits, vegetables (wild aromatic herbs, cruciferous vegetables, green leafy vegetables and edible plant roots), mushroom, legumes Whole grain (rye, oat, barley), potatoes Nuts and seeds Skimmed dairy products, fish (Salmon, Sardines, Mackerel, Herring), Eggs, white meat, Canola Oil Rare occasions local red meat, fat dairy products Processed food and products rich in sugar	Significant reduction in overall mortality, help to attain a healthy longevity	Adamsson et al. 2012; Tetens et al. 2012

(continued)

Table 8.1 (continued)

Type of diet	Composition	Constituents according to the order of intake	Benefits	Reference
Vegetarian diet	60%–70% of total calories as carbohydrates 16% proteins 10–14% unsaturated fats Less than 5% saturated fat	Regular physical activity Adequate water and fluids intake (alcohol and sugar free) High consumption of variety of plant and unrefined foods Whole grains, legumes, vegetables and fruits, unsaturated fats as nuts, seeds and vegetable oil dairy products, eggs and sugars	Reduces the incidence of various chronic disease leading to healthy longevity	Marsh et al. 2012

Reference: Caprara 2018

is superior for grass fed cows as compared to total mixed ration diet as demonstrated through Raman spectroscopy by researchers in Ireland (Gómez-Mascaraque et al. 2020). On the poultry front, the preference for free range eggs over caged eggs is on a rise. It has been demonstrated that caged hens undergo tremendous stress that leads to production of stress hormones such as corticosterone in the hen which leads to lower quality off-springs (Bulmer and Gil 2008; Kim et al. 2014). Several studies have debated the nutritional superiority of caged versus range eggs. Few studies have demonstrated higher levels of pigments in yolk of free range eggs and lower total lipid content while other studies have shown higher microbial count on egg shells from free range systems (Molnár and Szollosi 2020). Nutritional composition of eggs and microbial contaminants will effect consumer health and hence selection of the egg type needs consideration. Similarly hormone induced- milking practises have detrimental effect on humans. Massive sales of organic produce demonstrate preference of consumers towards residue free product.

Mass cultivation using pesticides and fertilizers not only affect the farming community but the consumers at large. High percentage of cancers and other long term ailments from exposure to pesticides affect human life directly and needs cautious approach. Finally PUFA rich fish has been promoted as part of several longevity promoting diets such as Japanese or Eskimo diets (Sugano and

Hirahara 2000; López and Ortega 2003). An important criteria here is farmed fish versus wild caught. Often fish such as farmed catfish or tilapia have been condemned for their high content of pro-inflammatory omega-6 content (Weaver et al. 2008). On the other hand wild caught omega-3 rich mackerel, sardines and tuna have a positive anti-inflammatory effect. European food safety authority (EFSA) dictates caution on fish consumption by pregnant women as it is often a source of heavy metals and thus despite its health benefits fish consumption needs to be regulated (https://www.efsa.europa.eu/en/news/efsa-provides-advice-safety-and-nutritionalcontribution-wild-and-farmed-fish). A growing trend in aquaculture worth mentioning is that of multitrophic aquaculture. Integrated salmon-seaweed culture leads to utilization of salmon metabolic waste for the growth of seaweed (Fossberg et al. 2018). Such systems often result in high protein seaweed which are a rich source of branched amino acids such as leucine, directly implicated in muscle growth and prevention of sarcopenia in ageing population (Martínez-Arnau et al. 2020). Thus agricultural practices followed and residue limits as determined through good labelling practices and regulatory approvals help consumers select health promoting dietary components while rejecting detrimental plant and animal food products.

8.4 Plant and Animal Food for Longevity

Though several epidemiological studies, dietary pattern analysis as well as controlled feeding/intervention studies have explored the ever-lasting debate on plant versus animal food, no conclusive merit can be given to one over the other. This is mainly because any diet or intervention study starts with a hypothesis around a single food item but in reality daily dietary recalls or interventions need to consider the macro and micronutrient substitutions in the overall diet as well as the interaction of food matrix with the nutrient under study. Studies that have implicated carbohydrates as bad focus on refined sugars rather than complex carbohydrates, while the condemned fat molecules are generally saturated fats and not polyunsaturated fats. Similarly studies that compare animal and plant protein sources and their effect on metabolic diseases often disregard the overall macromolecular composition and the type of animal protein. For example the choice of a good protein source is not based on animal or plant origin, rather it should provide the recommended daily allowance (RDA) of protein which is 0.8 g protein/kg body weight of adults along with suppling RDA for each essential amino acid (*Dietary Reference Intakes for Energy, Carbohydrate, Fiber, Fat, Fatty Acids, Cholesterol, Protein, and Amino Acids* 2005). The next section will attempt to decipher the differential effect of animal and plant dietary components on longevity.

Plant Food

Fruits and vegetables provide a powerful boost of vitamins, minerals, and other nutrients. High intake of fruits and vegetables have been associated with reduced risk of cognitive impairment and dementia and lower incidences of variety of chronic diseases such as cardiovascular disease, cancer etc. Studies reveal that many of the phytonutrients in fruits and vegetables can both protect against and repair damage to the body's cells. E.g. Antioxidant quercetin found in onions and apples help in preserving brain cells. Anthocyanins found in berries, grapes, pomegranate, red onions, beans help to improve cholesterol and blood sugar levels and helps preventing heart diseases (Cassidy et al. 2011). Dietary carotenoids such as lutein, zeaxanthin found commonly in green leafy vegetables such as spinach, lettuce, pistachio nuts reduce the risk and progression of age-related macular degeneration (AMD), leading cause of blindness in old people (Eisenhauer et al. 2017). Soy protein has been implicated in reduction of cholesterol mainly due to the presence of phytosterols in soy bean (Lin et al. 2004). Many other fibre rich plant protein sources have demonstrated lipid binding & faecal elimination leading to cardio protective functionalities. Table 8.2 enumerates variety antioxidants available from plant and animal food and their benefits.

Bone health is also an important factor responsible for healthy active life. Micronutrients such as vitamin K, folate, magnesium, potassium, calcium and antioxidants such as vitamin C and carotenoids present in fruits and vegetables are useful for healthy bones. Consumption of bone protecting foods such as broccoli, cabbage, okra, prunes throughout life help to improve the indices of bone health (Higgs et al. 2017). Along with fruits and vegetables, legumes also have been linked with longevity food cultures such as the Japanese (soy, tofu, natto, miso), the Swedes (brown beans, peas), and the Mediterranean people (lentils, chickpeas, white beans). According to 'Food Habits in Later Life' (FHILL) a cross-cultural study conducted by International Union of Nutritional Sciences (IUNS) and the World Health Organization (WHO) reveal that for every 20 g increase in daily legumes intake there was an 8% reduction in the risk of death. Thus suggesting that higher legume intake is the most protective dietary predictor of survival amongst the elderly, regardless of their ethnicity (Darmadi-Blackberry et al. 2004).

Animal Food

Dairy products are rich in minerals (calcium, potassium, and magnesium), protein (casein and whey), and vitamins (riboflavin and vitamin B-12). In dietary guidelines, dairy products, including full or specifically low fat milk, are recommended as sources of protein and calcium for bone health, and in some diets as prevention against hypertension. However studies show that the proposed benefits of milk for bone health are not clear and controversial. A cohort studies in men and women carried out by Michaëlsson et al. 2014 showed that higher milk intake is associated with higher mortality. However this was not observed with fermented dairy products such as yoghurt and cheese. Recent study by Tucker 2019, also reported that consumption of high fat milk increased biological aging. Accordingly latest Dietary Guidelines for Americans (2015–2020) recommend consumption of low-fat milk as part of a healthy diet instead of high fat milk.

Eggs are rich source for high biological value protein, as well as vitamins and minerals such as folic acid, vitamin B12, vitamins E and D, selenium, choline, zinc, etc. About 50% of the fat in the egg is MUFA. However, because of the high saturated

Table 8.2 Antiaging antioxidants from plant and animal food

Antiaging biomolecules (antioxidants)	Major dietary sources	Benefits	References
β-carotene	**Plant food:** Sweet potatoes, Carrots, Spinach, Cantaloupe, Red Bell Peppers, Apricots, Peas, Pumpkin, Broccoli	Effective protection against cancer, heart disease, Alzheimer's disease	Vance et al. 2013
Chlorogenic acids	**Plant food:** Coffee beans, pears, apples	Anti-diabetic, Anti-carcinogenic, Anti-inflammatory and Anti-bacterial effects	Tajik et al. 2017
Caffeic acid	**Plant food:** Coffee beans, tea, red grapes	Anticancer activity, Hepatoprotective, antidiabetic, cardioprotective, antiviral activity	Espíndola et al. 2019
Cocoa	**Plant food:** Cocoa beans	Antioxidant, reduce risk of diabetes, protect skin from oxidative damage, protect nerves from inflammation	Katz et al. 2011
Kaempferol (Flavanoid)	**Plant food:** Ginko biloba, grape fruit, tea, broccoli, berries	Antioxidant, Reduces risk of cancer,	Chen and Chen 2013
α-Lipoic acid	**Plant food:** Spinach, broccoli, tomatoes, brussel sprouts **Animal food**: Meats particularly organs Heart, liver, Kidney	Antioxidant, Antidiabetic, obesity, Cancer, Alzheimer disease	Salehi et al. 2019
Lycopene	**Plant food:** Red tomatoes, Watermelon, grapefruit, Guava	Antioxidant, Protective effect on cardiovascular system, positive effect on neurodegenerative diseases, including Alzheimer's and Parkinson's	Przybylska 2020
Lutein and zeaxanthin	**Plant food:** Kale, spinach, broccoli, peas, lettuce **Animal food:** Egg yolk	Essential components for eye health, Reduces risk of age-related macular degeneration (AMD) and cataracts	Abdel-Aal 2013

(continued)

Table 8.2 (continued)

Antiaging biomolecules (antioxidants)	Major dietary sources	Benefits	References
Proanthocyanidines	**Plant food:** Grape seeds, Red grapes, Black grapes, Cranberry, Strawberry, Blue berry	Anti-oxidant, Anti-microbial, Anti-obesity, Anti-diabetic, Anti-neurodegenerative, Anti-osteoarthritis, Anti-cancer, and cardio- and eye-protective properties	Unusan 2020
Querectin (Flavonoid)	**Plant food:** Kales, Berries, Apple, Onions, Red grapes, Cherries, Broccoli	Anti-oxidant, Anti-inflammatory agent, Antiviral activity, cardiovascular properties and anticancer properties	Kumar et al. 2017
Selenium	**Plant food:** Brazil nuts, mushroom, Spinach, Brown rice **Animal food**: Sea foods, Meat, dairy products	Antioxidant reduces risk of cancer	Tinggi 2008
Vitamin A (Retinoids)	**Plant food:** Carrots, Sweet potatoes, Spinach, Broccoli, Pumpkin **Animal food:** Liver (Beef, Pork, chicken), Cheese, Eggs, Oily fish,Milk, Yoghurt	Eye health, Antixerophthalmic activity, Reduces risk of cancer	Akram et al. 2011
Vitamin C (Ascorbic acid)	**Plant food**: Citrus fruits, Strawberries, Kiwi, spinach, cabbage, garlic, cranberry	Effective protection against cancer	Block 1991, Roomi et al. 2018
Vitamin E (alpha tocopherol)	**Plant food:** Vegetable oils, legumes	Reduces risk of colon cancer	Bostick et al. 1993; Roomi et al. 2018

(continued)

Table 8.2 (continued)

Antiaging biomolecules (antioxidants)	Major dietary sources	Benefits	References
Zinc (trace element)	**Plant food:** legumes, nuts, oats, spinach, avocado, mushrooms, peas, garlic, pomegranate **Animal food:** Meats such as oysters, Angus beef, crab legs, lobster and roast chicken, Liver, egg,	Essential micronutrient for metabolism, Effective treatment for Wilson disease	Saper and Rash 2009

fat content (about 3 g/100 g) and cholesterol content (about 200–300 mg/100 g) frequent consumption of eggs was restricted based on the assumption that high dietary cholesterol is associated with high blood cholesterol levels and cardiovascular disease (Kritchevsky and Kritchevsky 2000). But studies revealed that dietary cholesterol in general and cholesterol in eggs in particular have limited effects on the blood cholesterol level and on cardiovascular disease (Jones 2009).

Fish is a rich source of high-quality proteins, n-3 polyunsaturated fatty acids (PUFAs), and other nutrients, such as minerals, trace elements, and vitamins (Hosomi et al. 2012). Epidemiologic evidence shows relationship between vitamin D deficiency and various age related and chronic diseases such as autoimmune disease, cancer, cardiovascular disease, depression, dementia, infectious diseases, musculoskeletal and bone health. A diet high in oily fish such as salmon, mackeral, sardines, tuna help in preventing vitamin D deficiency. Omega-3 [(n − 3)] fatty acids are known to be associated to healthy aging throughout life. Eicosapentaenoic acid (EPA) and docosahexaenoic acid (DHA) are n-3 polyunsaturated fatty acids (n-3 PUFAs) found abundantly in animal food e.g. fish, fish oil supplements, while alpha-linolenic acid (ALA) another omega -3 fatty acid is found in plant food such as nuts and seeds. These fatty acids show a wide array of health benefits in prevention of cardiovascular disease, cancer and metabolic syndrome (Sugawara et al. 2013). Of these, EPA and DHA are essential for proper foetal development and healthy aging (Dunstan et al. 2007). However, some studies have reported contradictory results. Tsuduki et al. 2011 demonstrated that administration of fish oil containing high levels of EPA and DHA to senescence-accelerated mice P8 (SAMP8) shortened their lifespan significantly. However further studies on *C. elegans* showed that large amounts of fish oil significantly reduced the lifespan of *C. elegans* but an appropriate amount of fish oil extended their lifespan significantly (Sugawara et al. 2013). In line with this Champigny et al. 2018 also observed increase in lifespan of Drosophilia fed with diets supplemented with EPA and DHA. Recently a cohort study revealed that higher levels of circulating omega-3 from seafood, especially EPA, is associated with healthy aging (Lai et al. 2018). Further research needs to be carried out in humans to assess various clinical outcomes.

Meat which includes unprocessed red meat (such as beef, pork and lamb), unprocessed white meat (such as chicken, turkey and poultry) are among the best dietary sources of high quality protein. Red meat contains essential micronutrients such as iron, zinc, selenium, potassium, and range of B vitamins including thiamin, niacin, riboflavin, pantothenic acid, and vitamins B6 and B12. For those who consume little or no oily fish, meat also helps in contributing long-chain omega-3 polyunsaturated fatty acids. Large number of studies have shown that consumption of red and processed meat is associated with increase in total mortality, whereas low meat intake is associated with greater longevity (Singh et al. 2003; Virtanen et al. 2019). Meat rich in branched chain amino acids that promote muscle synthesis are particularly beneficial to ageing populations showing signs of sarcopenia while vegan diets low in essential amino acids such as lysine, isoleucine, tryptophan need dairy or animal protein supplementation. However current research shows conflicting results not in favour of dietary advice to limiting unprocessed red meat (Micha et al. 2010; Binnie et al. 2014). Fresh lean cuts differ dramatically in their composition when compared to processed cured meat. Processed meat contains 400% more sodium and 50% more nitrates than unprocessed meat hence a critical risk to health. A systematic review has concluded that processed red meat is associated with 42% greater risk of CHD but the key word being 'processed' and not just red meat (Micha et al. 2010). The ill effects of red meat are often attributed to saturated fat and cholesterol or to advanced glycation and lipid oxidation products. Lean meat rich in branched amino acids and bio-actives has been accorded health benefits in contrast to sodium and nitrates laden cured meat. Having said that recent studies on dry cured ham as a source of bioactive peptides has completely spun the narrative (Gallego et al. 2019).

Thus the studies show that both animal and plant sources have molecules of health benefits but unless we have data of purified food fractions with known molecular compositions and inter-molecular interactions, the results need cautious interpretations. Nutritional experts and related organizations, such as the U.S. Department of Agriculture and U.S. Department of Health and Human Services of the Spanish Society of Community Nutrition (SENC), recommend an increased consumption of fruits and vegetables, grains, legumes, low-fat dairy products, lean meats and fish, especially fatty fish species that are high in n-3 PUFA for healthy life. Thus we can say consumption of minimally processed plant and animal food is equally important in providing health benefits in individual. Further, bio-accessibility and bioavailability of biomolecules from these foods have to be studied which plays a crucial role to exert beneficial health effect.

8.5 Effect of Plant and Animal Source Diet on Health Through Gut Microbiota

Human digestive system sequentially involves physical, biochemical and fermentation processes. Although the first few steps vary in a narrow range for most human beings, the fermentation in large intestine could be quite diverse in different subjects. Scientific research has established the importance of interplay between consumed food, gut microbiota and health. In fact, gut microbiota plays a central role in between food and human health. A schematic representation shown in Fig. 8.1 showcases different dietary sources and their effect on gut microbiome and health.

Human gut contains trillions of microorganisms with a variety of species showing dominance of *Actinobacteria* and *Proteobacteria* phyla in neonates whereas *Firmicutes* and *Bacteroidetes* in healthy adult (Eckburg et al. 2005). Major functions of gut microbiota are defending body from pathogens, modifying absorption of nutrients and producing vitamins. According to some studies there are similarities in the gut microbiota of healthy individuals but these are highly influenced by external factors like environment and diet (Conlon and Bird 2015) and a balanced microbiota is important for human longevity.

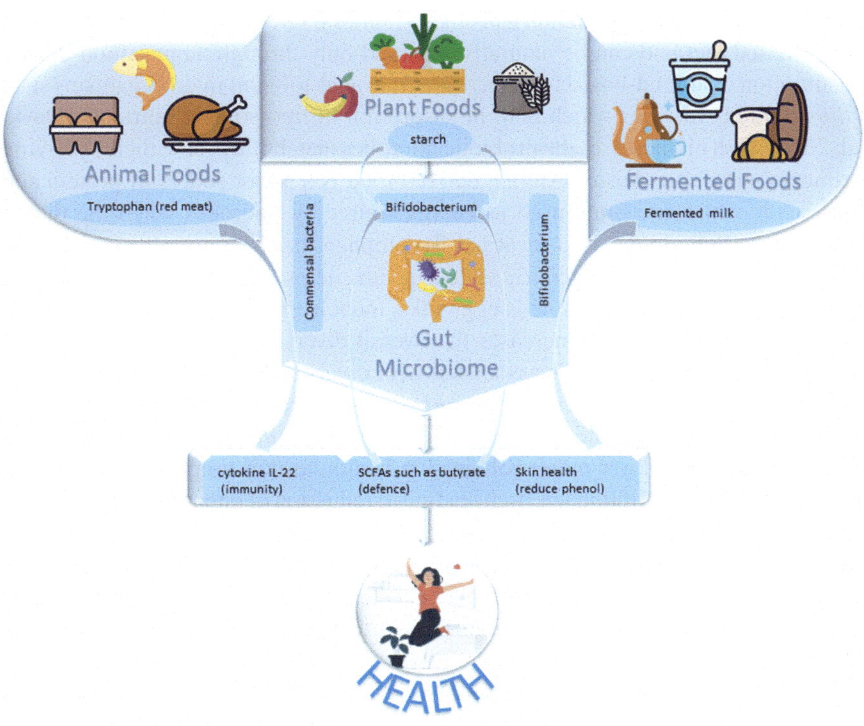

Fig. 8.1 Schematic highlights interplay of gut microbiome with plant and animal foods and their effect on health

Many researchers have attempted to decipher the effect of plant based diet on gut microbiome and human health. Broccoli has glucosinolates which reduces *Lactobacillus, Enterococcus and Clostridium perfringens* which helps in reduction of colonic inflammation (Raskin et al. 2002). Martinez showed diets rich in brown rice and barley increase the count of *Eubacterium rectale* which in turn increases the insulin response (Martínez et al. 2013). Cereal grains like sorghum, oat, rice, corn, wheat and barley rich in starch and xylan are digested by *Bifidobacterium, Ruminococcus* and *Roseburia* to produce essential defence elements SCFAs such as butyrate, acetate and propionate (Scott et al. 2013). Butyrate reduces propensity towards cardiovascular diseases and cancers (Wong et al. 2006). Another important disorder to focus on is obesity, known to impact quality of life. Dietary patterns effect composition of gut microbiome which in turn have profound effect on underlying mechanism that leads to obesity. Diet rich in high fibre and low fat such as spinach, tomato, seeds, broccoli, artichokes etc. helps to increase *Bifidobacteria* and *Lactobacilli* which tend to found at higher levels in healthy individuals compared to obese individuals. In contrary high fat and low fibre foods like processed juices, fried chips increases Firmicutes/Bacteroidetes which are directly linked to enhanced occurrence of obesity (Anhê et al. 2015). It was found that diet rich with cranberry extract increases *Akkermansia* species that help in controlling obesity.

Animal based food also plays an important role in gut microbiota composition and health. Some of the important nutrients like Vitamin B12 & D3 and DHA are absent in plant based diet and can be naturally obtained only through animal-food intake. Consumption of animal-based food increases bile acid-tolerant microorganisms such as *Bacteroides, Alistipes* which help in absorption of lipo-soluble nutrients (David et al. 2014). Fish oil increases the probiotic microbes metabolic profile thereby having detrimental effect on obesity. Animal derived food is a major source of protein and amino acids which are often used as energy source by microbes when the diet is limiting in carbohydrates. Red meat has Tryptophan that is converted to indole by commensal bacteria *Lactobacillus,* which in turn helps in production of immune cytokine IL-22. Another example is egg yolks, meat liver and fish that are known to be rich in vitamin D, which increases the overall diversity of gut microbiome and it was found that good health is associated with diverse microbiota. Scientists have found that active vitamin D is critical in maintaining increased levels of Bacteroidetes thereby promoting health and reducing risk for cardiovascular diseases and diabetes.

Fermented foods play a major role in skin health, as consumption of fermented milk helps to increase Bifidobacterium leading to reduction of phenol and its ill effects on skin (Miyazaki et al. 2014). Fermented foods also help in increased production of SCFAs such as butyrate. Kefir is a probiotic product that has kefir grains made of beneficial yeast and bacteria that promote bone health. Tempeh is another fermented food made of soybeans rich in protein known to reduce risk of heart disease by lowering LDL level. Daily consumption of yoghurt rich in live *lactobacilli* is a regular dietary habit in Indian subcontinent. Further, particularly in southern part of India known for its hot tropical climate (unsuitable for growth of lactobacilli), daily consumption of gourd family vegetables is common and Sreenivas and Lele (2013) have reported these vegetables to possess prebiotic properties especially ash gourd (*Benincasa Hispida*). Palamthodi and Lele (2014) have shown nutraceutical effects of some important gourds that are regularly consumed in India. Food additives such as non-caloric sweeteners alter gut microbiota by changing *Bacteroidetes* to *Firmicutes* and there by increases the chance of glucose intolerance and insulin resistance which has indirect effect of diabetes and obesity (Suez et al. 2014). Thus elimination of dietary components that have adverse effect on healthy gut microbiome and incorporation of probiotic and prebiotic plant and animal derived fermented food will promote health and longevity.

8.6 Healthy Food Preparation Practises and the Associated Risk/Benefits

While discussing the benefits and risks of various food groups, people often neglect the major role food preparation methods play in the final nutritional quality of a culinary preparation. Raw food in the form of vegetables, unpasteurized milk, fish (sushi) is consumed by many but consumption of cooked packaged food far exceeds raw consumption. Since the discovery of fire by the caveman and progress in the field of food processing and culinary sciences, novel processed products and innovative recipes have evolved. With advancement of science and communications the world is becoming flat and many food preparations are now used across the globe and geographical barriers have disappeared. Many consumers seek minimally processed or raw food. However, despite sushi and sashimi offering health benefits of PUFAs, pathogenic bacteria and parasites present in raw fish pose a potent health risk to connoisseurs (Muscolino et al. 2014). Consumption of raw vegetables if grown under unhygienic conditions of soil and water have risk of *E.coli* infections. Unpasteurized milk is known for bio-actives, enzymes and for promoting growth of lactobacillus in gut however needs to be carefully monitored for microbial loads and food safety precautions (Butler et al. 2020). Cooking methods such as blanching, steaming, roasting or using microwave to cook food is often associated with reduction in antinutrients like tannins, saponins, oxalates while preserving natural pigments and nutrients in the food. For example microwave cooking is associated with increased levels of Sulforaphane, an anti-cancer bioactive produced by hydrolysis of glucoraphanin in broccoli when compared to raw counterpart (Lu et al. 2020). Similarly sprouting of seeds and fermentation of plant and animal food products is often association with increased digestibility and enhanced bioactivities. Sprouts (lentils, flaxseed) and microgreens (kale, amaranth) have been labelled as superfoods that showcase several bioactivities including anti-diabetic and anticholinergic activity (Wojdylo et al. 2007). Fermented cabbage or 'kimchi', a plant based product has been reported to have anti-cancer, anti-oxidant, anti-obesity and anti-ageing properties amongst others (Park et al. 2014). On the other hand, bioactive peptides derived from *lactobacillus plantarum* fermented milk have been reported to possess the anti-inflammatory, antihaemolytic, antioxidant, antimutagenic, and antimicrobial activities (Aguilar-Toalá et al. 2017).

Deep frying of food at high temperatures is often associated with formation of acrylamide (Yang et al. 2016). Thus vegetables consumed in the form of fried fritters have high content of cancer causing acrylamide and beats the purpose of incorporating health promoting vegetables in the daily diet. Similarly mashed potatoes as the preferred carbohydrate source are better than a portion of potato chips despite the starting ingredient being the same humble potato. Meat products have several essential nutrients and are a source of essential amino acids, however the cooking method plays a major role in the projected health benefits. Barbecued meat is a source of nitrosamines while cured meat is rich in nitrates and nitrites all three implicated as cancer causing agents (Song et al. 2015). Acrylamide, nitrosamines, nitrites and anti-nutrients impact health negatively thereby affecting life span of the consumers.

Food preparation is often accompanied with utilization of add on ingredients to enhance taste, preserve food as well as aid in cooking. Spices, enzymes and extracts have been used for the mentioned purposes with some additional health promoting effects. Red meat is often tenderized using enzymes such as papain, bromelain, ficin to reduce cooking times and improve eating qualities. Often as a result of tenderization, proteins in the meat are hydrolysed into bioactive peptides with health benefits such as (ACE) inhibitory activity and antioxidant, lipid-lowering, antithrombotic and antimicrobial properties that can be incorporated in functional foods (Wu et al. 2015). Natural fermented sauces such as soy sauce or fish sauce are a source of many such bio-actives peptides that are often heat resistant and impart health benefits to the consumers. Spices form a major group of ingredients often added to food preparations in many Asian cuisines. Spices are preservative in nature due to anti-microbial molecules but also have several active ingredients that are known to be health promoting. Curcumin in turmeric, crocin and safranal in saffron, allicin in garlic, eugenol in cloves, carnosic acid in rosemary, piperine in black pepper, thymoquinone in cumin, cineole in cardamom, gingerols in ginger are few bio-actives worth mentioning for their innumerable health effects (Embuscado 2019). Spices are used in vegetarian and non-vegetarian preparations including savoury and sweet dishes. Thus the molecular composition of plant and animal food, the cooking method and the composition of added ingredients determine the final quality and benefit of the food items as shown in Table 8.3. Food labelling takes consideration of these factors and hence the list of ingredients as well as percent composition of essential nutrients is listed on packaged food products.

Table 8.3 Effect of food preparation method on molecular composition and food safety

Food preparation method/ingredient	Benefit/risk	Food group	Reference
Raw	PUFA rich while risk of *Anisaki* parasite	Animal- Fish based sushi	(Muscolino et al. 2014)
Boiling/steaming/blanching	Reduction in anti-nutrients such as oxalates	Plant - Produce such as spinach	(Chai and Liebman 2005)
Microwave	Enhanced anticancer bioactive -Sulphoraphane	Plant -Produce broccoli	(Lu et al. 2020)
Fermentation	Enhanced bioactivities such as anti-ageing	Plant- Kimchi from cabbage Animal- fermented milk	(Park et al. 2014) (Aguilar-Toalá et al. 2017)
Sprouting	Antidiabetic & anticholinergic bioactivities	Plant-alfalfa & flaxseed	(Mattioli et al. 2019)
Deep frying	Cancer causing acrylamide formation	Plant- Starchy food such as potato chips	(Yang et al. 2016)
Barbecuing	Cancer causing Nitrosamines	Animal- protein rich meat	(Song et al. 2015)
Curing	Cancer causing Nitrites High salt content- risk of Hypertension	Animal-meat sausages/Salamis/Mortadella	(Song et al. 2015)
Enzymes	Tenderization and formation of hydrolysates & bioactive peptides	Animal- red meat and fish	(Wu et al. 2015) (Naik et.al. 2020)
Spices	Antimicrobials and bio-actives	Both animal & plant	(Embuscado 2019)

8.7 Towards Future Through Novel Food Processing Technologies

Cultured meat, fermentation, textured analogues, 3-D printed meat are some examples where processing tries to combine multiple health benefits. There is a need for developing hybrid food products that have blended ingredients from plant and animal sources. Novel food processing technologies such as 3-D printing, high moisture extrusion, fermentation and cell culture are being employed for manufacturing meat analogues or animal free cultured meat. Isolated purified ingredients from animal and plant source are used, for instance plant derived proteins or polysaccharides as scaffolds for culturing meat (Ben-Arye and Levenberg 2019) or blended proteins such as soy, pea, oats and others for manufacturing extruded meat analogues or texturized vegetable protein (De Angelis et al. 2020). Another example would be 3-D printing nutritionally superior meat products derived from high quality protein ingredients derived from in vitro cell culture, meat by-products/waste, insects and plants devoid of saturated fats while demonstrating enhanced sensorial traits (Ramachandraiah 2021). Specialised textured meat is aimed at geriatric population to overcome their problems associated with consumption of tough meat and tenderization procedures aid in supplementing the diets of old people(Dick et al. 2019). Most of the processing technologies require ingredients that support the techno-functional properties of the meat product. Plant derived hydrocolloids provide functionality such as emulsion stability, viscosity, and gelation to ingredients. Meat analogues also require pigmentation or inks that paint the meat red or pink and provide the meat like aesthetics. Seaweed and fungi derived pigments or vegetable derived inks such as beetroot ink can be employed to manufacture faux meat. Quorn a fungi-derived mycoprotein rich in essential amino acids and fibre has been reported to support post exercise recovery better than milk protein (Monteyne et al. 2020). Thus, in our search for more sustainable and nutritionally superior protein sources, the trend shifts towards incorporation of low carbon food ingredients such as algae, fungi and insects for manufacturing functional food products aimed at promoting longevity.

8.8 Concluding Remarks

Food, life style, environmental factors, genetic constitution, gut microbiome are some of the important parameters that have significant impact on human longevity – a long healthy life. Food is one of the most prominent factors that could be easily managed for good health. Human digestion system works like a series of bioreactors and quality, quantity and frequency of meals in a day impacts the performance. Digestion and retrieval of nutrients from food is altered with age as a result of changes in expression of digestive enzymes as well as microbiome composition. Research shows that both plant and animal food can have beneficial as well as adverse health effects. Based on epidemiological studies, dietary interventions, in both in vitro and in vivo studies have reported that several plant and animal food products such as fruits, vegetables, dairy products, fresh fish and meat retard ageing through action of antioxidants, minerals, proteins, peptides, polyunsaturated fatty acids and other functional molecules. To obtain maximum benefit and minimise adverse effects, recommended daily allowances, food preparation practises and farming and agricultural methods need to be revisited before indulging in anti-ageing diets. Sudden drastic change in diet by inclusion of newer foods or exclusion of regular foods, may lead to developing allergies and perturbed microbiota. Interactions between bio-molecules such as dietary fat and beta-carotene or phytates and minerals resulting in synergistic or antagonistic effects requires more exploration. Extreme caution needs to be followed in extending benefits derived from whole food such as 'catechins in tea' to purified concentrated supplements such as green tea extract (GTE) that may alter metabolic pathways to a much greater extent.

Compliance with Ethical Standards

Conflict of Interest All authors declare they have no conflict of interest.

References

Abdel-Aal E, Akhtar H, Zaheer K, Ali R (2013) Dietary sources of lutein and zeaxanthin carotenoids and their role in eye health. Nutrients 5(4):1169–1185. https://doi.org/10.3390/nu5041169

Adamsson V, Reumark A, Cederholm T, Vessby B, Riserus U, Johansson G (2012) What is a healthy Nordic diet? Foods and nutrients in the NORDIET study. Food Nutr Res 56:18189. https://doi.org/10.3402/fnr.v56i0.18189

Aguilar-Toalá JE et al (2017) Assessment of multifunctional activity of bioactive peptides derived from fermented milk by specific Lactobacillus plantarum strains. J Dairy Sci US 100(1):65–75. https://doi.org/10.3168/jds.2016-11846

Akram M, Akhtar N, Asif HM, Shah PA, Saeed T, Arshad M, Malik NS (2011) Vitamin A: a review article. J Med Plant Res 5(20):4977–4979

Anhê FF et al (2015) A polyphenol-rich cranberry extract protects from diet-induced obesity, insulin resistance and intestinal inflammation in association with increased Akkermansia spp. population in the gut microbiota of mice. Gut 64(6):872–883. https://doi.org/10.1136/gutjnl-2014-307142

Barnes PJ (2015) Mechanisms of development of multimorbidity in the elderly. Eur Respir J 45(3):790–806. https://doi.org/10.1183/09031936.00229714

Ben-Arye T, Levenberg S (2019) Tissue engineering for clean meat production. Front Sustain Food Syst 3:46

Binnie MA, Barlow K, Johnson V, Harrison C (2014) Red meats: time for a paradigm shift in dietary advice. Meat Sci 98:445–451. https://doi.org/10.1016/j.meatsci.2014.06.024

Block G (1991) Vitamin C and cancer prevention: the epidemiologic evidence. Am J Clin Nutr 53:270S-282S

Bostick RM, Potter JD, McKenzie DR, Sellers TA, Kushi LH et al (1993) Reduced risk of colon cancer with high intake of vitamin E: the Iowa health study. Cancer Res 53:4230–4237

Bulmer E, Gil D (2008) Chronic stress in battery hens: measuring corticosterone in laying hen eggs. Int J Poult Sci 7(9):880–883

Butler MI et al (2020) Recipe for a healthy gut: intake of unpasteurised milk is associated with increased lactobacillus abundance in the human gut microbiome. Nutrients 12(5). https://doi.org/10.3390/nu12051468

Caprara G (2018) Diet and longevity: the effects of traditional eating habits on human lifespan extension. Mediterr J Nutr Metab 11(3):261–294. https://doi.org/10.3233/MNM-180225

Cassidy A, O'Reilly ÉJ, Kay C et al (2011) Habitual intake of flavonoid subclasses and incident hypertension in adults. Am J Clin Nutr 93(2):338–347. https://doi.org/10.3945/ajcn.110.006783

Chai W, Liebman M (2005) Effect of different cooking methods on vegetable oxalate content. J Agric Food Chem Am Chem Soc 53(8):3027–3030. https://doi.org/10.1021/jf048128d

Champigny CM, Cormier RPJ, Simard CJ, St-Coeur PD, Fortin S, Pichaud N (2018) Omega-3 monoacylglyceride effects on longevity, mitochondrial metabolism and oxidative stress: insights from drosophila melanogaster. Mar Drugs 16(11):453. https://doi.org/10.3390/md16110453

Chen AY, Chen YC (2013) A review of the dietary flavonoid, kaempferol on human health and cancer chemoprevention. Food Chem 138(4):2099–2107. https://doi.org/10.1016/j.foodchem.2012.11.139

Conlon MA, Bird AR (2015) The impact of diet and lifestyle on gut microbiota and human health. Nutrients. MDPI AG (1). https://doi.org/10.3390/nu7010017

Darmadi-Blackberry I, Wahlqvist ML, Kouris-Blazos A, Steen B, Lukito W, Horie Y et al (2004) Legumes: the most important dietary predictor of survival in older people of different ethnicities. Asia Pacific J Clin Nutr. 13(2):217–220

David LA et al (2014) Diet rapidly and reproducibly alters the human gut microbiome. Nature 505(7484):559–563. https://doi.org/10.1038/nature12820

De Angelis D et al (2020) Physicochemical and sensorial evaluation of meat analogues produced from dry-fractionated pea and oat proteins. Foods 9(12):1754. https://doi.org/10.3390/foods9121754

Dick A, Bhandari B, Prakash S (2019) 3D printing of meat. Meat Sci 153:35–44. https://doi.org/10.1016/j.meatsci.2019.03.005

Lupton JR, Brooks JA, Butte NF, Caballero B, Flatt JP, Fried SK (2005) Dietary reference intakes for energy, carbohydrate, fiber, fat, fatty acids, cholesterol, protein, and amino acids. The National Academies Press, Washington, DChttps://doi.org/10.17226/10490

Dunstan JA, Mitoulas LR, Dixon G, Doherty DA, Hartmann PE, Simmer K, Prescott SL (2007) The effects of fish oil supplementation in pregnancy on breast milk fatty acid composition over the course of lactation: a randomized controlled trial. Pediatr Res 62(6):689–694. https://doi.org/10.1203/PDR.0b013e318159a93a

Eckburg PB et al (2005) Microbiology: diversity of the human intestinal microbial flora. Science 308(5728):1635–1638. https://doi.org/10.1126/science.1110591

Eisenhauer B, Natoli S, Liew G, Flood VM (2017) Lutein and Zeaxanthin—food sources, bioavailability and dietary variety in age-related macular degeneration protection. Nutrients 9(2):120. https://doi.org/10.3390/nu9020120

Embuscado ME (2019) Bioactives from culinary spices and herbs: a review. J Food Bioactives 6(0 SE-Review). https://doi.org/10.31665/JFB.2019.6186

Espíndola KM et al (2019) Chemical and pharmacological aspects of caffeic acid and its activity in hepatocarcinoma. Front Oncol 9:541. https://doi.org/10.3389/fonc.2019.00541

Fossberg J et al (2018) The potential for upscaling kelp (Saccharina latissima) cultivation in salmon-driven integrated multi-trophic aquaculture (IMTA). Front Marine Sci 418. https://www.frontiersin.org/article/10.3389/fmars.2018.00418

Gallego M, Mora L, Toldrá F (2019) The relevance of dipeptides and tripeptides in the bioactivity and taste of dry-cured ham. Food Prod Process Nutr 1(1):2. https://doi.org/10.1186/s43014-019-0002-7

Gheorghe IF, Ion B (2011) The effects of air pollutants on vegetation and the role of vegetation in reducing atmospheric pollution. In: Khallaf MK (ed) The impact of air pollution on health, economy, environment and agricultural sources

Gómez-Mascaraque LG et al (2020) Raman spectroscopy: a rapid method to assess the effects of pasture feeding on the nutritional quality of butter. J Dairy Sci 103(10):8721–8731. https://doi.org/10.3168/jds.2020-18716

Higgs J, Derbyshire E, Styles K (2017) Nutrition and osteoporosis prevention for the orthopaedic surgeon: a wholefoods approach. EFORT Open Rev 2(6):300–308. https://doi.org/10.1302/2058-5241.2.160079

Hosomi R, Yoshida M, Fukunaga K (2012) Seafood consumption and components for health. Global J Health Sci 4(3):72–86. https://doi.org/10.5539/gjhs.v4n3p72

Jones PJ (2009) Dietary cholesterol and the risk of cardiovascular disease in patients: a review of the Harvard Egg Study and other data. Int J Clin Pract Suppl 63:28–36. https://doi.org/10.1111/j.1742-1241.2009.02155.x

Sreenivas KM, Lele SS (2013) Prebiotic activity of gourd family vegetable fibres using in vitro fermentation. Food Biosci 1:26–30

Katz DL, Doughty K, Ali A (2011) Cocoa and chocolate in human health and disease. Antioxid Redox Signal 10:2779–2811. https://doi.org/10.1089/ars.2010.3697

Kim Y-H et al (2014) Effects of dietary corticosterone on yolk colors and eggshell quality in laying hens. Asian-Australas J Anim Sci. 28(6):840–846. Asian-Australasian Association of Animal Production Societies (AAAP) and Korean Society of Animal Science and Technology (KSAST). https://doi.org/10.5713/ajas.14.0849

Kritchevsky SB, Kritchevsky D (2000) Egg consumption and coronary heart disease: an epidemiologic overview. J Am Coll Nutr 19:549S-555S. https://doi.org/10.1080/07315724.2000.10718979

Kumar R, Vijayalakshmi S, Nadanasabapathi S (2017) Health benefits of quercetin. Defence Life Sci J 2:142. https://doi.org/10.14429/dlsj.2.11359

Lai HT et al (2018) Serial circulating omega 3 polyunsaturated fatty acids and healthy ageing among older adults in the Cardiovascular Health Study: prospective cohort study. BMJ 363:k4067. https://doi.org/10.1136/bmj.k4067

Lin Y et al (2004) Soy protein enhances the cholesterol-lowering effect of plant sterol esters in cholesterol-fed hamsters. J Nutr 134(1):143–148. https://doi.org/10.1093/jn/134.1.143

López PM, Ortega RM (2003) Omega-3 fatty acids in the prevention and control of cardiovascular disease. Eur J Clin Nutr 57(1):S22–S25. https://doi.org/10.1038/sj.ejcn.1601805

Lu Y, Pang X, Yang T (2020) Microwave cooking increases sulforaphane level in broccoli. Food Sci Nutr 8(4):2052–2058. https://doi.org/10.1002/fsn3.1493

Marsh K, Zeuschner C, Saunders A (2012) Health implications of a vegetarian diet: a review. Am J Lifestyle Med 10(10):1–18. https://doi.org/10.1177/1559827611425762

Martínez I et al (2013) Gut microbiome composition is linked to whole grain-induced immunological improvements. ISME J 7(2):269–280. https://doi.org/10.1038/ismej.2012.104

Martínez-Arnau FM et al (2020) Effects of leucine administration in sarcopenia: a randomized and placebo-controlled clinical trial. Nutrients. https://doi.org/10.3390/nu12040932

Martinez-Gonzalez MA, Martín-Calvo N (2016) Mediterranean diet and life expectancy; beyond olive oil, fruits and vegetables. Curr Opin Clin Nutr Metab Care 19(6):401–407. https://doi.org/10.1097/MCO.0000000000000316

Mattioli S et al (2019) Effect of heat- and freeze-drying treatments on phytochemical content and fatty acid profile of alfalfa and flax sprouts. J Sci Food Agric 99(8):4029–4035. https://doi.org/10.1002/jsfa.9630

Micha R, Wallace SK, Mozaffarian D (2010) Red and processed meat consumption and risk of incident coronary heart disease, stroke, and diabetes mellitus: a systematic review and meta-analysis. Circulation 121(21):2271–2283. https://doi.org/10.1161/CIRCULATIONAHA.109.924977

Michaëlsson K, Wolk A, Langenskiöld S, Basu S, Lemming EW, Melhus H, Byberg L (2014) Milk intake and risk of mortality and fractures in women and men: cohort studies. BMJ 349:1–15. https://doi.org/10.1136/bmj.g6015

Miyazaki K et al (2014) Bifidobacterium fermented milk and galacto-oligosaccharides lead to improved skin health by decreasing phenols production by gut microbiota. Benef Microbes 5(2):121–128. https://doi.org/10.3920/BM2012.0066

Molnár S, Szollosi L (2020) Sustainability and quality aspects of different table egg production systems: a literature review. Sustain (Switz) 12(19). https://doi.org/10.3390/SU12197884

Monteyne AJ et al (2020) Mycoprotein ingestion stimulates protein synthesis rates to a greater extent than milk protein in rested and exercised skeletal muscle of healthy young men: a randomized controlled trial. Am J Clin Nutr 112(2):318–333. https://doi.org/10.1093/ajcn/nqaa092

Muscolino D et al (2014) Hygienic-sanitary evaluation of sushi and sashimi sold in Messina and Catania, Italy. Ital J Food Saf 3(2 SE-Original Articles). https://doi.org/10.4081/ijfs.2014.1701

Naik AS, Mora L, Hayes M (2020) Characterisation of seasonal mytilus edulis by-products and generation of bioactive hydrolysates. Appl Sci 10(19):6892. https://doi.org/10.3390/app10196892

Park K-Y et al (2014) Health benefits of kimchi (Korean fermented vegetables) as a probiotic food. J Med Food US 17(1):6–20. https://doi.org/10.1089/jmf.2013.3083

Przybylska S (2020) Lycopene – a bioactive carotenoid offering multiple health benefits: a review. Int J Food Sci Technol 55:11–32. https://doi.org/10.1111/ijfs.14260

Ramachandraiah K (2021) Potential development of sustainable 3d-printed meat analogues: a review. Sustain (Switz) 13(2):1–20. https://doi.org/10.3390/su13020938

Raskin I et al (2002) Plants and human health in the twenty-first century. Trends Biotechnol 20(12):522–531. https://doi.org/10.1016/s0167-7799(02)02080-2

Roomi MW, Niedzwiecki A, Rath M (2018) Scientific evaluation of dietary factors in cancer. J Nutri Med Diet Care 4:029. https://doi.org/10.23937/2572-3278.1510029

Palamthodi S, Lele SS (2014) Nutraceutical applications of gourd family vegetables: Benincasahispida, Lagenaria siceraria and Momordica charantia. Biomed Prev Nutr 4(1):15–21

Salehi B et al (2019) Insights on the use of α-lipoic acid for therapeutic purposes. Biomolecules 9(8):356. https://doi.org/10.3390/biom9080356

Saper RB, Rash R (2009) Zinc: an essential micronutrient. Am Fam Phys 79(9):768–772

Scott KP et al (2013) The influence of diet on the gut microbiota. Pharmacol Res (1). https://doi.org/10.1016/j.phrs.2012.10.020

Singh PN, Sabate J, Fraser GE (2003) Does low meat consumption increase life expectancy in humans? Am J Clin Nutr 78(3):526S-532S. https://doi.org/10.1093/ajcn/78.3.526S

Sociedad Española de Nutrición Comunitaria (SENC) (2011) Objetivos nutricionales para la población española. Rev Esp Nutr Comunitaria 4:178–199

Sofi F, Cesari F, Abbate R, Gensini GF, Casini A (2008) Adherence to Mediterranean diet and health status: meta-analysis. BMJ 337:a1344. https://doi.org/10.1136/bmj.a1344

Song P, Wu L, Guan W (2015) Dietary nitrates, nitrites, and nitrosamines intake and the risk of gastric cancer: a meta-analysis. Nutrients MDPI 7(12):9872–9895. https://doi.org/10.3390/nu7125505

Suez J et al (2014) Artificial sweeteners induce glucose intolerance by altering the gut microbiota. Nature 514(7521):181–186. https://doi.org/10.1038/nature13793

Sugano M, Hirahara F (2000) Polyunsaturated fatty acids in the food chain in Japan. Am J Clin Nutr US 71(1 Suppl):189S–196S. https://doi.org/10.1093/ajcn/71.1.189S

Sugawara S, Honma T, Ito J, Kijima R, Tsuduki T (2013) Fish oil changes the lifespan of Caenorhabditis elegans via lipid peroxidation. J Clin Biochem Nutr 52(2):139–145. https://doi.org/10.3164/jcbn.12-88

Tajik N, Tajik M, Mack I, Paul E (2017) The potential effects of chlorogenic acid, the main phenolic components in coffee, on health: a comprehensive review of the literature. Eur J Nutr 56:2215–2244. https://doi.org/10.1007/s00394-017-1379-1

Tetens IPAN et al (2012) Nordic Nutrition Recommendations. Integrating Nutrition and Physical Activity

Tinggi U (2008) Selenium: its role as antioxidant in human health. Environ Health Prev Med 13(2):102–108. https://doi.org/10.1007/s12199-007-0019-4

Tsuduki T, Honma T, Nakagawa K, Ikeda I, Miyazawa T (2011) Long-term intake of fish oil increases oxidative stress and decreases lifespan in senescence accelerated mice. Nutrition 27:334–337. https://doi.org/10.18632/aging.101021

Tucker LA (2019) Milk fat intake and telomere length in U.S. women and men: the role of the milk fat fraction oxidative medicine and cellular longevity. Article ID 1574021. https://doi.org/10.1155/2019/1574021

U.S. Department of Agriculture and U.S. Department of Health and Human Services (2010) Dietary Guidelines for Americans, 7th edn. U.S. Government Printing Office, Washington, DC, USA

Unusan N (2020) Proanthocyanidins in grape seeds: an updated review of their health benefits and potential uses in the food industry. J Funct Foods 67:103861. https://doi.org/10.1016/j.jff.2020.103861

Vance TM, Su J, Fontham ET, Koo SI, Chun OK (2013) Dietary antioxidants and prostate cancer: a review. Nutr Cancer 65:793–801

Vemuri R et al (2018) Gut microbial changes, interactions, and their implications on human lifecycle: an ageing perspective. In: Li W-J (ed) Hindawi, p 4178607. BioMed Research International. https://doi.org/10.1155/2018/4178607

Virtanen HEK et al (2019) Dietary proteins and protein sources and risk of death: the Kuopio Ischaemic Heart Disease Risk Factor Study. Am J Clin Nutr US 109(5):1462–1471. https://doi.org/10.1093/ajcn/nqz025

Weaver KL et al (2008) The content of favorable and unfavorable polyunsaturated fatty acids found in commonly eaten fish. J Am Diet Assoc US 108(7):1178–1185. https://doi.org/10.1016/j.jada.2008.04.023

Willcox DC, Bradley J, Willcox MD, Todoriki H, Makoto Suzuki MD (2009) The Okinawan diet: health implications of a low-calorie, nutrient-dense, antioxidant-rich dietary pattern low in glycemic load. J Am Coll Nutr 28(4):500S–516S. https://doi.org/10.1080/07315724.2009.10718117

Wojdylo A, Oszmianski J, Czemerys R (2007) Antioxidant activity and phenolic compounds in 32 selected herbs. Food Chem 105(3):940–949.https://doi.org/10.1016/j.foodchem.2007.04.038

Wong JMW et al (2006) Colonic health: fermentation and short chain fatty acids. J Clin Gastroenterol 235–243. https://doi.org/10.1097/00004836-200603000-00015

Wu J et al (2015) Bioactive peptides from meat proteins as functional food components. In: Functional polymers in food science, pp 181–208. Wiley Online Books. https://doi.org/10.1002/9781119108580.ch9

Yang Y, Achaerandio I, Pujolà M (2016) Influence of the frying process and potato cultivar on acrylamide formation in French fries. Food Control 62:216–223. https://doi.org/10.1016/j.foodcont.2015.10.028

Chapter 9
Fermented Foods in Aging and Longevity

S. Heeba and P. Nisha

Abstract Fermented foods have been in the human diet for thousands of years because of their nutritive, preservative, and organoleptic properties. Recently, fermented foods gained more attention among people due to the functional microbes, which possess antioxidant, antimicrobial, probiotics, peptide production properties etc., thereby modulating specific actions in the human body related to healthy aging and longevity. Growing evidence suggests that fermented foods change the metabolic activities in the gut, acquire new characteristics, and increase benefit in age-associated changes. Gut microbiota and the brain maintain bidirectional communication and modulate behavior via the "microbiota-gut-brain axis" in the human body. A series of relevant benefits associated with fermented foods in healthy aging and longevity have been reported that includes preventing allergic reactions, gastrointestinal disorders, diabetes, cardiovascular disease, cancer, obesity, improving metabolic function and neuroendocrine effects, etc. Therefore, the ultimate way for longevity may be obtained by healthy lifestyle habits that comprise a healthy diet inclusive of various fermented foods. This chapter focuses on reviews and critical analysis of past, current, and future research of fermented foods in aging and longevity.

Keywords Aging and longevity · Fermented foods · Gut microbiota · Probiotics · Short-chain fatty acids

9.1 Introduction

Fermented foods are evolved from an ancient food processing and preservation technique known as fermentation. Around one-third of the human population's diet consists of foods that are fermented and are consumed by people of various cultures

S. Heeba · P. Nisha (✉)
Agro Processing and Technology Division, CSIR-National Institute for Interdisciplinary Science and Technology, Trivandrum 695019, Kerala, India
e-mail: pnisha@niist.res.in

P. Nisha
Academy of Scientific and Innovative Research (AcSIR), Ghaziabad 201002, India

S. I. S. Rattan and G. Kaur (eds.), *Nutrition, Food and Diet in Ageing and Longevity*, Healthy Ageing and Longevity 14, https://doi.org/10.1007/978-3-030-83017-5_9

and lifestyles around the world (Carvalho et al. 2018). Existing scientific data show that functional microorganisms convert the chemical constituents of raw materials of food to beneficial metabolites during fermentation. Thereby, fermented foods promote the bio-availability of nutrients, improve organoleptic properties, impart bio-preservation and food safety, eliminate toxic components, develop antimicrobial and antioxidant compounds, induces probiotic activities, and thus modulate specific actions in the human body (Tamang et al. 2016). Fermented foods have gained attention, and the term probiotics was introduced in 1907, after assessing the importance of fermented milk on longevity by the Nobel Prize Laureate Elie Metchnikoff (Anukam and Reid 2007). Various studies reported that the consumption of fermented foods prevents cardiovascular diseases, cancer, gastrointestinal disorders, allergic reactions, diabetes and obesity (Tamang et al. 2016). Emerging evidence has shown a link between nutrition, diet, probiotics, and fermented foods for the human body's vital functions from the immune system to mental stability (Kim et al. 2016).

Aging refers to gradual changes in the dynamics of biological, behavioral, environmental, and social processes. It involves a gradual decline in the immune system, resulting in an imbalance between anti-inflammatory and pro-inflammatory activity (Das et al. 2020). Fermented foods enrich microbes residing in the gut known as gut microbiota, which is essential for health and helps to maintain many brain and immune functions. Emerging evidence suggests that the gut micro biota plays a major role in age-associated diseases (Badal et al. 2020). Therefore, the best way for longevity may be to adopt healthy lifestyle habits comprised of healthy diet and fermented foods. The present chapter provides a comprehensive review on the available literature the regarding the role of fermented foods in aging and longevity.

9.2 Microbes Involved in Food Fermentation

Several microorganisms are present in foods. The majority of them are associated with spoilage of food and toxin production, but a group of microorganisms are involved in enhancing food quality, nutrition, and also act as a natural preservative (Ray and Joshi 2014). Bacteria, yeasts, and mould are the common microbes involved in fermentation. Lactic acid bacteria like lactobacillus, streptococcus, oenococcus, pediococcus, and acetobacter species present in the food oxidize alcohol into acetic acid useful for vinegar production. Beneficial yeast like *S.cerevisiae* plays an important role in bread making and alcohol production. Schizosaccharomyces boulderi and *S. Pombe*areyeasttypes present in millet responsible for fermented beverage production (Joshi et al. 2011). Moulds like Penicillium species in some food products help in flavour development, such as cheese processing (Battcock and Azam-Ali 1998). Fermentation takes place in two ways. "Wild ferments" fermentation occurs naturally in the raw material or due to the environment, such as kimchi, sauerkraut. "Culture-dependent ferments" happen when a starter culture is added externally, for example, kefir. Another method is back slopping, in which some amount of already fermented batch is added to the raw material, for example, vinegar (Parker et al. 2018).

9.3 Mechanism Behind Fermented Foods in Health and Longevity

The gut microbiota plays a significant role in the human body by communicating between the gut and brain and modulate the behaviour via the "gut-brain axis" through hormonal, neural, and immune mediators (Cryan and Dinan 2012). Short-chain fatty acids (SCFAs) such as butyrate, propionate, acetate, etc., formed by fermentation of prebiotics by gut microbiota, inhibit the growth of many harmful microbes by lowering pH. It also promotes mineral absorption, metabolism of glucose and lipids, stimulate proliferation and differentiation of intestinal enterocytes. In addition to this, SCFAs were reported to have positive impact on central nervous system functions (Rowland et al. 2018). The SCFAs stimulate the secretion of GLP-1, which inhibits glucagon secretion, modulates insulin secretion, improves insulin sensitivity and decreases hepatic gluconeogenesis. Probiotics form intestinal barrier, which inhibits absorption of antigens to prevent damage to β cells. Strong evidence is available in literature about the gut microbiota being at the core of several age-associated changes (Bana and Cabreiro 2019).

During the fermentation process, the phytochemicals present in food are altered by the resident gut microbiota, which further inhibit the hypothalamic–pituitary–adrenal axis and regulate neurotransmitters activity and thereby maintain mental health (Cryan and Dinan 2012). Gamma-aminobutyric acid (GABA) is naturally produced by microbes, plants, and mammals, mostly present in a wide variety of foods. During fermentation, L-glutamate undergoes decarboxylation using the enzyme bacterial glutamate decarboxylase (GAD) and produce GABA. Increasing GABA content significantly inhibits excitatory neurotransmitters in the nervous system and promotes many physiological functions of the body (Gan et al. 2017). A pictorial representation of the mechanism behind fermented foods in the human body are presented in Fig. 9.1.

A weak immune system is one of the signs of aging. The antioxidant defence system of the body also becomes weak with aging due to which there is an increased accumulation of reactive oxygen species in the body leading to DNA damage. Fermentation metabolites present in many of the fermented foods (e.g. SCFAs) as well as the phytochemicals (e.g. wines) are reported to enhance the expression of immune players such as cytokines (interleukins- IL-6, IL-10, IL-12,and IL-1 β) and immunoglobulins (IgA, and IgG) thus help in health and longevity (Das et al. 2020). The production of antihypertensive peptides through proteolytic action during fermentation is reported to delay aging by modulating microflora in the gut thus improving immune response (Hsieh et al. 2015).

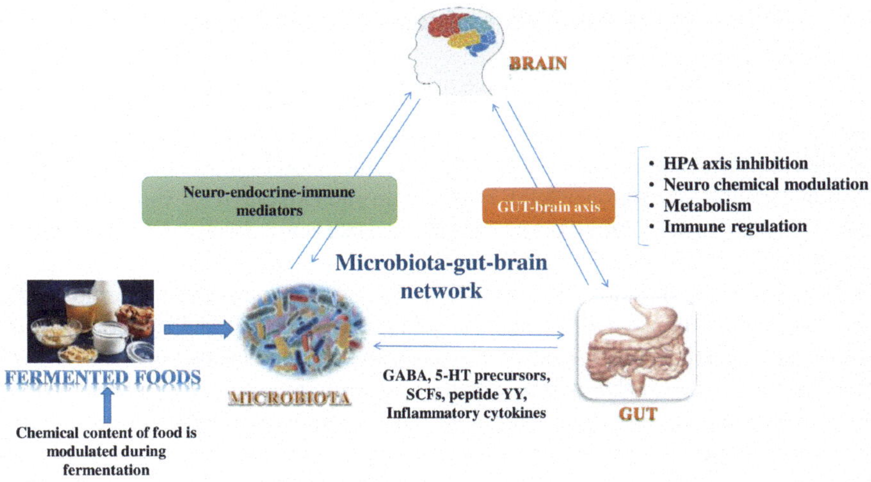

Fig. 9.1 A pictorial representation of the mechanism behind fermented foods in the human body

9.4 Types of Fermented Foods

Wide varieties of fermented foods are produced using fermentation technology that has changed from ancient practices and empirical science to industrial processes and life science-driven technologies. Fermented foods can be broadly divided into dairy-based fermented foods, meat-based fermented foods, fruit and vegetable-based fermented foods, cereal/pulse-based fermented foods, and miscellaneous fermented foods. Some of the health benefits of different types of fermented foods are summarized in the Table 9.1.

9.4.1 Dairy-Based Fermented Foods

Fermented dairy foods produced by lactic acid bacteria (LAB) by milk fermentation show a broad area of health benefits for longevity, such as the production of antihypertensive peptides through proteolytic action, modulate microflora in the gut, improve immune response thus delaying aging (Hsieh et al. 2015).

Yogurt is a coagulated milk product formed by fermentation using starter culture of *Lactobacillus bulgaricus* and *Streptococcus thermophilus*. Lactic acid fermentation results in folate production, increased protein content and digestibility, conjugated linoleic acid content, and calcium absorption. This further helps to restore gut microbiota and stimulates immune functions that decline during the process of aging (Eales et al. 2015). On the other hand, controversy still exists to prove the health benefits of yogurt through clinical trials.

Table 9.1 Some of the health benefits of different types of fermented foods are summarized (slightly modified and reproduced with permission from Das et al. 2020)

Fermented foods	Origin/study conducted region	Study type	Dose and duration	Active compounds	Health claims	Reference
Fermented milk	Japan	RPCT	200 ml of FM/day for 6 weeks	Probiotics (Lb. acidophilus L-92)	Anti-allergic activity	(Ishida et al. 2005)
Fermented milk	China	RPCT	100 g of FM for 2 weeks	Probiotics (B. lactis DN-173010)	Anti-constipation activity in adults	(Yang et al. 2008)
Fermented milk	Japan	RPCT	1.0 × 1011 cells per day for 12 weeks	Probiotics (Lb. casei strain Shirota)	Reduced upper respiratory tract infections	(Shida et al. 2017)
Dadih	Japan	In vivo (Mice model)	Enriched diet (g/100 g):12 days feeding period	Lactococcus lactis subsp. lactis IS-29862	Reduced hypocholesterolemic effect in rats	(Pato et al. 2004)
Vinegar	Japan	RPCT	500 ml daily of a beverage containing either 15 and 30 ml of vinegar for 12 weeks	Probiotics	Reduces body weight, body fat mass, and serum triglyceride levels in human	(Kondo et al. 2009)
Fermented red brown rice	Malaysia	In-vitro (plasmid DNA)	–	Phytochemical	Protect DNA from oxidative damage	(Kong et al. 2015)

(continued)

Table 9.1 (continued)

Fermented foods	Origin/study conducted region	Study type	Dose and duration	Active compounds	Health claims	Reference
Nuruk (alcoholic beverage)	Korean	In -vitro (H(2)O(2)-treated HepG2 cells)	–	Aspergillus kwachii Rhizopus oryzae KSD-815	Inhibitory Effects of Ethanol Extracts from Nuruk on Oxidative Stress, Melanogenesis, and Photo-Aging	(Lee et al. 2012)
Fermented fish oil	Korea	In- vivo (mouse model)	100 mg/kg FFO for 31 days	–	Suppresses Allergic skin inflammation	(Han et al. 2012)
Koumiss (fermented mare's milk)	China	In -vitro (intestinal endothelial cells)	–	Lactobacillus helveticus NS8	Anti-inflammatory, immunomodulatory properties	(Rong et al. 2015)

RPCT: Randomized place bo-controlled trail.

Increasing evidence of the benefits of fermented milk is reported through in vivo and in vitro studies. Shida et al. (2017) established the potential benefits of fermented foods and showed beneficial effects against infections in the upper respiratory tract. Similarly, Yang et al. (2008) reported improvement in constipation in adult Chinese women after consumption of fermented milk with probiotic strain yogurt starters (Streptococcus thermophilus, Lb. delbrueckii subsp. bulgaricus; 1.2 × 109 CFU per day) and Bifidobacterium lactis DN-173010 1.25 × 1010 CFU per day). Ishida et al. (2005) reported improved condition of a group of people in Japan with cedar pollen allergy and ocular symptoms medication scores after the consumption of Lb. acidophilus L-92 mediated fermented milk. Some other studies have also established the positive role of fermented milk against cardiovascular diseases, treatment of metabolic syndrome, inflammatory bowel disease (IBD), and modulating effects of adipocytes (Saez-Lara et al. 2015).

Mor kuzhambhu, a traditional recipe prepared in South Indian food, showed an excellent antibacterial response against pathogens L.monocytogenes and Salmonella typhi, which causes foodborne illness (Satish et al. 2013). Dadih, fermented milk from buffalo, is kept for 48 h in bamboo tubes and covered by banana leaves for fermentation to obtain yogurt-like texture in Indonesia. It was reported that Lc. lactis subsp. lactis IS 10,285 cultured Dadih reduces cholesterol levels in hypercholesterolemic rats due to its ability to colonize in the gut. However, Lc. lactissubsp.lactis IS 2986 did not normalize cholesterol levels in *invivo* studies because of its inactivity in the gut (Pato et al. 2004). Hence the benefits of specific strains need to be explored for promoting health and longevity. Studies on cheese showed that it is enriched with conjugated linoleic acids (CLA) which protects against atherosclerotic conditions and many other diseases (Hur et al. 2017). Koumiss, a raw mare's milk product from central and west Asia demonstrated a positive effect on endocrine glands, immune, nervous, cardiovascular systems, kidneys, liver, avitaminosis, gastric, and intestinal diseases (Rong et al. 2015).

9.4.2 Meat-Based Fermented Foods

LAB play a crucial function in the fermentation of meat by producing bacteriocins and adjusting optimum pH as well as protecting from the development of spoilage and pathogenic microorganisms, thereby providing safety, shelf life, and stability to fermented meat products (Dincer and Kivanc 2012).

LAB strains present in fermented fishes such as Tungtap, Hentak, and Ngari show better antibacterial effects, presence of proteolytic enzymes, and absence of biogenic amines (Satish et al. 2013). Sucuk, is a traditional Turkish fermented meat product prepared by adding a starter culture to the meat. Different combinations of microbes were used during sucuk processing, which provided many physicochemical, microbial, and biochemical properties in addition to its organoleptic properties Thapa et al. (2004). Several reports in literature provide evidence that meat fermented products as a probiotic starter culture carrier is increased due to its activities such as

ACE-inhibition and inhibition of fatty acid oxidation and proteolytic action, and it also prevents the growth of spoilage microbes that produce biogenic amines peptide (Ichimura et al. 2003). In contrast, some nutritionists criticize its high fat, biogenic amines, salt content, and possibilities of infection (Kjeldgaard et al. 2012). Deniz et al. (2016) Reported that Pastırma, a traditional semi-fermented meat product most popularly consumed by Turkish people, exhibits ACE inhibitory and anti-oxidant properties, which further helpto slow down the process of aging.It is reported that eicosapentaenoic acid (EPA) and docosahexaenoic acid (DHA) are present in higher amounts in fermented products such as fermented fish oil as compared to natural fish oil (Han et al. 2012). Ichimura et al. (2003) reported that fermented fish sauce has the potential to stimulate insulin secretion, which combats diabetes by bringing down hypertension by producing ACE-inhibitory peptides. Some fermented fish oil has a large amount of Omega-3 polyunsaturated fatty acids, which can prevent an allergic reaction in the human body (Han et al. 2012).

9.4.3 Cereal/Pulse-Based Fermented Foods

Short-chain fatty acids are produced in the large intestine during Cereals and legumes fermentation, forming an acidic environment in the gut that further multiplies lactic microbiota (Roopashri and Vardaraj 2009). Some anti aging-associated characteristics of fermented foods are reported in Table 9.2. Wide varieties of fermented cereals and legumes are consumed all over the globe every day, but their benefits for healthy aging and longevity are still not clear.

In most of the southern part of India, fermented cereal foods such as dosa, koozhu, pazhaiya soru (Fermented rice), idli, and ambali are taken regularly by locals (Nisha et al. 2005). Moreover, these foods are reported to have a higher amount of probiotics (Satish et al. 2013). Pal et al. (2005) reported that LAB isolated from dosa inhibited pathogens such as Listeria monocytogenes, Staphylococcus aureus, Pseudomonas aeruginosa, Aeromonas hydrophila, Bacillus cereus, V. parahaemolyticus, and the antibacterial activity was attributed to protein bacteriocins. Kefir, fermented milk produced from kefir grains during acid-alcoholic fermentation, was shown to have anti-hypertensive, hypocholesterolemic, anti-bacterial, anti-inflammatory anti-mutagenic, anti-allergenic, anti carcinogenic, anti-diabetic, anti-oxidant activity, and also probiotic effects in the immune system (Kesenka et al. 2013). Fermented legumes demonstrate anti-diabetic properties bymodifying the enzymes such as amylase, glucosidase, and acetylcholinesterase, during fermentation (Ademiluyi et al. 2015). Higher amount of ferulic acid was produced by fermented rice bran as compared to normal rice bran, which further improves health by reducing oxidative stress (Kim and Han 2011). Intake of fermented red rice is reported to reduce the level of cholesterol and oxidative stress-induced DNA damage (Kong et al. 2015). Ekinci et al. (2016) reported that Shalgam juice, a non-alcoholic beverage prepared by fermentation, is rich in probiotics, anti-oxidant, anti-proliferative properties, and also has a protective effect on colon cancer.

Table 9.2 Some anti aging-associated characteristics of fermented foods are reported in (Slightly modified and reproduced with permission from Das et al. 2020)

Fermented food	Experimental models	Biological Activities	References
Red-wine polyphenols	Lymphomonocytes Human B cells	Promoted the expression of IL-12, IL-6, IL-1β, IL-10, IgA, and IgG	(Magrone et al. 2008)
Rice wine	Human fibroblasts Human keratinocytes	Promoted the type I procollagen. Reduced the UV-induced MMP-1 and TNF-α. Increased antioxidant activity, and laminin-5 expression	(Seo et al. 2009)
Fermented rice bran	In vitro	Increases strongly the expression of adiponectin and PPAR-g. prevents production of reactive oxygen species, increases GLUT 4 associated with glucose transport and insulin sensitivity	(Kim and Han 2011)
Nuruk extracts	Human hepatoma	HepG2 Free radical scavenging activity. Increased expression of SOD in H2O2-treated HepG2 cells	(Lee et al. 2012)
Fermented milk	Adult human	Anti-constipation activity in adults	(Yang et al. 2008)

Beer has plenty of antioxidants and has a significant role in disease prevention by its anti-oxidative, vasodilatory, anti-mutagenic, probiotic, probiotic and anti-diabetic properties (SanJose et al. 2017). But the excessive consumption of beer may be harmful and leads to severe health issues (De Gaetano et al. 2016). During fermentation of sourdough, ACE-inhibitory peptides were produced, which further reduced the glycemic index and increased mineral availability in bread (Novotni et al. 2012). Intake of red wine in a particular amount also plays a role in the immune system and acts as an anti-aging agent (Magrone et al. 2008). Seo et al. (2009) reported that wine prepared from rice showed higher intracellular antioxidant capacity in human fibroblasts and keratinocytes. As observed from the literature, red.

yeast rice (produced by fermenting yeast over the rice) has cholesterol lowering effect in people affected with mildly hypercholesterolemic patients (Cicero et al. 2017). Nuruk, an alcoholic beverage prepared in Korea with several starters, possesses ABT Scavenging activity and produces Superoxide dismutase activity in

H_2O_2-treatedHepG2 cells (Lee et al. 2012). Rats fed with Chung kookjang, a Korean soybean fermented paste, has higher antioxidant potential (Kwak et al. 2007). GABA content of edible seeds such as soybean, faba bean, wheat, millet, rye, and quinoa is reported to increase during lactic acid fermentation (Gan et al. 2017). GABA plays a critical role in the regulation of blood pressure and many other physiological functions of the body.

9.4.4 Fruit and Vegetable-Based Fermented Foods

Fruits and vegetables are rich in naturally occurring bioactive compounds that have been identified having anti-aging effects through many in in-vivo and in-vitro studies. The microorganism in fermented fruits and vegetables help to prevent some diseases such as diarrhoea and cirrhosis and, also antioxidants may help to eliminate harmful free radicals that play a significant role in degenerative diseases (Swain et al. 2014).

Kimchi, fermented vegetables contain some species such as Leuconostoc, Weissella, and Lactobacillus during fermentation. Some health benefits are reported which including anti mutagenic and anticancer activities, constipation prevention, improvement of the immune system functions, and aging process (Patra et al. 2016). The effects of kimchi's anti-aging properties have been reported to reduce HT-29 human colon carcinoma cells by increasing the protein production and mRNA factors that could promote cell apoptosis and arrest the carcinoma cells during the cell cycle (Yu et al. 2020). Fermented olives, a rich source of probiotics, is reported to prevent cancer, allergic reaction, modulate immune system, and aid digestion (Peres et al. 2017). Sauerkraut (fermented vegetables) contains a high amount of vitamins, phenolic compounds, and minerals like calcium, iron, potassium, and phosphorus, and is reported to possess positive effect against carcinogens, inflammation, and reactive oxygen production (Penaset al. 2017). Acetic acid formed during vinegar production can inhibit body fat and hepatic lipid accumulation without changing skeletal muscle weight and food intake (Kondo et al. 2009).

9.4.5 Miscellaneous Fermented Products

Kombucha, a fermented tea product from northeast China promotes liver function, prevent cardiovascular diseases, various types of cancers and stimulates immune system which may be correlated with the healthy aging of this population (Jayabalan et al. 2014). Pulque, a Mexican alcoholic fermented beverage made from the cactus plant, was found to be rich in amino acids, vitamin B, and iron content during fermentation and also reported to have many health benefits. Gari, chickwanghe, fufu, lafun, agbelima, attieke, and kivunde are some of the fermented food products prepared from cassava roots in African countries, sochu an alcoholic beverage produced in

Japan from fermented sweet potatoes, are also reported for their functional properties (Ray and Sivakumar 2009).

9.5 Risk of Fermented Foods

Naturally fermented foods if not sterilized properly can have the risk of presence of pathogenic microbes. Production of toxic and harmful substances during fermentation is another risk factor associated with fermented foods. Biogenic amines are also a significant risk associated with fermented foods. However, consumption of fermented foods with an elevated amount of biogenic amines (>100 mg/kg) alters the detoxification system and are a cause of concern for health (Alvarez and Arribas 2014). Ethyl carbamate (urethane) is formed in fermented foods mostly in fermented alcoholic beverages and is considered as a possible cancer causing substance and has probable health risk for people regularly consuming alcoholic beverages (Gowd et al. 2018).

9.6 Future Prospects of Fermented Foods

Emerging technology for food processing may help to modify substrate, alter the fermentation process, apply particular stress to microbial load for fine tuning their beneficial effects to enhance the production of a wide variety of fermented products in the future. The use of innovative technologies for food fermentation at the industrial level is still new. Some researchers are focusing on developing new culture strains to improve the health benefits of fermented foods during fermentation. Advances in metabolic engineering, synthetic biology, and other genetic engineering tools have given hope to alter the microbial genome and thus modify some desirable properties of the starter cultures during fermentation (Terefe and Augustin 2020). In this advanced era, with the keen interest of food processing technologists and microbiologists, the scope for developing new products through fermentation technology with confirmed health benefits are tremendous that may help to combat aging and associated health problems.

9.7 Conclusion

Nowadays, increasing awareness about fermented foods attracts consumers to include more fermented food products in their daily diet. This chapter's primary intent was to review the mechanisms of potential benefits of fermented food for healthy aging and longevity. Consumption of fermented foods in the diet along with moderate physical activities may be beneficial for longevity as well as better quality of life. Although the anti-aging characteristic of fermented foods have been evaluate during different *invivo* and *invitro* models but further studies need to be planned at the clinical level on a wide variety of fermented foods to ascertain their health benefits in humans.

Compliance with Ethical Standards

Conflict of Interest All authors declare they have no conflict of interest.

References

Ademiluyi AO, Oboh G, Boligon AA, Athayde ML (2015) Dietary supplementation with fermented legumes modulate hyperglycemia and acetylcholinesterase activities in Streptozotocin-induced diabetes. Pathophysiology 22(4):195–220

Alvarez MA, Moreno-Arribas MV (2014) The problem of biogenic amines in fermented foods and the use of potential biogenic amine-degrading microorganisms as a solution. Trends Food Sci Technol 39(2):146–155

Anukam KC, Reid G (2007) Probiotics: 100 years (1907-2007) after Elie Metchnikoff.'s observation. In: Mendez-Vilas A (Ed) Communicating current research and educational topics and trends in applied microbiology, pp 466–474

Badal VD, Vaccariello ED, Murray ER, Yu KE, Knight R, Jeste DV, Nguyen TT (2020) The gut microbiome, aging, and longevity: a systematic review. Nutrients 12(12):1–25

Bana B, Cabreiro F (2019) The microbiome and aging. Ann Rev Genet 53(1):239–261

Battcock M, Azam-Ali S (1998) Fermented fruits and vegetables. A global perspective. FAO Agricultural Service

Carvalho MDN, Costa EM, Silva S, Pimentel L, Fernandes TH, Estevez PM (2018) Fermented foods and beverages in human diet and their influence on gut microbiota and health. Fermentation 4(4):1–13

Cicero AFG, et al (2017) Effect of a short-term dietary supplementation with phytosterols, red yeast rice or both on lipid pattern in moderately hypercholesterolemic subjects : clinical trial. Nutrition and Metabolism, 1–7

Cryan JF, Dinan TG (2012) Mind-altering microorganisms: the impact of the gut microbiota on brain and behaviour. Nat Rev Neurosci 13:701–712

Das G et al (2020) Traditional fermented foods with anti-aging effect: A concentric review. Food Research International 134:109269

De Gaetano G et al (2016) Effects of moderate beer consumption on health and disease: a consensus document. Nutr Metab Cardiovasc Dis 26(6):443–467

Deniz E, Mora L, Aristoy MC, Candogan K, Toldra F (2016) Free amino acids and bioactive peptides profile of Pastirma during its processing. Food Res Int 89(1):194–201

Dincer E, Kivanc M (2012) 'Characterization of lactic acid bacteria from Turkish Pastirma. Ann Microbiol 62(3):1155–1163

Eales J et al (2015) Is consuming yoghurt associated with weight management outcomes? Results from a systematic review. Int J Obes 40(5):731–746

Ekinci FY et al (2016) Characterization of chemical, biological, and antiproliferative properties of fermented black carrot juice, Shalgam. Eur Food Res Technol 242(8):1355–1368

Gan R, Li H, Gunaratne A, Sui Z, Corke H (2017) Effects of fermented edible seeds and their products on human health: bioactive components and bioactivities. Compr Rev Food Sci Food Saf 16:489–531

Gowda V, Sua H, Karlovskyb P, Chena W (2018) Ethyl carbamate: an emerging food and environmental toxicant. Food Chem 248:312–321

Han SC et al (2012) Fermented fish oil suppresses T helper 1/2 cell response in a mouse model of atopic dermatitis via generation of CD4+CD25+Foxp3+ T cells. BMC Immunol 13:1–12

Hsieh C, et al (2015) Milk proteins, peptides, and oligosaccharides : effects against the 21st century disorders. BioMed Res Int, 146840

Hur SJ, Kim HS, Bahk YY, Park Y (2017) Overview of conjugated linoleic acid formation and accumulation in animal products. Livest Sci 195:105–111

Ichimura T, Hu J, Aita DQUA (2003) Angiotensin I-converting enzyme inhibitory activity and insulin secretion stimulative activity of fermented fish sauce. J Biosci Bioeng 96(5):496–499

Ishida Y et al (2005) Effect of milk fermented with Lactobacillus acidophilus strain L-92 on symptoms of Japanese cedar pollen allergy: a randomized placebo-controlled trial. Biosci Biotechnol Biochem 69(9):1652–1660

Jayabalan R, Malbasa RV, Loncar ES, Vitas JS, Sathishkumar M (2014) A review on kombucha tea-microbiology, composition, fermentation, beneficial effects, toxicity, and tea fungus. Compr Rev Food Sci Food Saf 13(4):538–550

Joshi VK, Thakur NS, Bhat A, Grag C (2011) Wine and brandy: a perspective. In: Joshi VK (ed.) Handbook of enology, vol 1. Asia. Tech Publishers, Inc. New Delhi, pp 3–45

Kesenkas H, Yerlikaya O, Ozer E (2013) A functional milk beverage: Kefir. Agro Food Ind Hi Tech 24:53–55

Kim B, Hong VM, Yang J, Hyun H, Im JJ (2016) A review of fermented foods with beneficial effects on brain and cognitive function. Prev Nutr Food Sci 21:297–309

Kim D, Han GD (2011) Ameliorating effects of fermented rice bran extract on oxidative stress induced by high glucose and hydrogen peroxide in 3T3-L1 adipocytes. Plant Foods Hum Nutr 66(3):285–290

Kjeldgaard J, Cohn MT, Casey PG, Hill C, Ingmer H (2012) Residual antibiotics disrupt meat fermentation and increase risk of infection. mBio, 3(5):e00190-12

Kondo T, Kishi M, Fushimi T, Ugajin S, Kaga T (2009) Vinegar intake reduces body weight, body fat mass, and serum triglyceride levels in obese Japanese subjects. Biosci Biotechnol Biochem 73(8):1837–1843

Kong EL, Lee BK, Michelle Ginjom I, Nissom PM (2015) DNA damage inhibitory effect and phytochemicals of fermented red brown rice extract. Asian Pacific J Trop Dis 5(9):732–736

Kwak CS, Lee MS, Park SC (2007) Higher antioxidant properties of Chungkookjang, a fermented soybean paste, may be due to increased aglycone and malonylglycoside isoflavone during fermentation. Nutr Res 27(11):719–727

Lee SJ, Cho SW, Kwon YY, Kwon HS, Shin WC (2012) Inhibitory effects of ethanol extracts from Nuruk on oxidative stress, melanogenesis, and photo-aging. Mycobiology 40(2):117–123

Magrone T, Tafaro A, Jirillo F, Amati L, Jirillo E, Covelli V (2008) Elicitation of immune responsiveness against antigenic challenge in age- related diseases: effects of red wine polyphenols. Curr Pharm Des 14(26):2749–2757

Nisha P, Laxmi A, Rekha SS (2005) Effect of stabilizers on stabilization of idli (traditional south Indian food) batter during storage. Food Hydrocolloids 19:179–186

Novotni D, Cukelj N, Smerdel B, Bituh M, Dujmic F, Curic D (2012) Glycemic index and firming kinetics of partially baked frozen gluten-free bread with sourdough. J Cereal Sci 52(2):120–125

Pal V, Jamuna M, Jeevaratnam K (2005) Isolation and characterization of bacteriocin producing lactic acid bacteria from a south indian special dosa (appam) batter. J Cult Collect 4:53–60

Parker M, Zobrist S, Donahue C, Edick C, Mansen K, Hassan ZNM, Kort R (2018) Naturally fermented milk from northern Senegal: bacterial community composition and probiotic enrichment with Lactobacillus rhamnosus. Front Microbiol 9:2218

Pato U, Surono IS, Koesnandar H, A. (2004) Hypocholesterolemic effect of indigenous dadih lactic acid bacteria by deconjugation of bile salts. Asian Australas J Anim Sci 17(12):1741–1745

Patra JK, Das G, Paramithiotis S, Shin HS (2016) Kimchi and other widely consumed traditional fermented foods of Korea: a review. Front Microbiol 7:1493

Penas E, Villaluenga CM, Frias J (2017) Sauerkraut: production, composition, and health benefits. In: Fermented foods in health and disease prevention, technology and engineering, vol 24. Academic press, pp 557–576
Peres CM, Peres C, Xavier MF (2017) Role of natural fermented olives in health and disease. Fermen Foods Health Dis Prev 22:517–542
Ray RC, Joshi V (2014) Fermented foods: past, present and future. In: Microorganisms and fermentation of traditional foods, 1st edn, CRC press, pp 1–36
Ray RC, Sivakumar PS (2009) Traditional and novel fermented foods and beverages from tropical root and tuber crops: review. Int J Food Sci Technol 44:1073–1087
Rong, J, et al (2015) Probiotic and anti-inflammatory attributes of an isolate Lactobacillus helveticus NS8 from Mongolian fermented koumiss. BMC Microbiol 15(196)
Roopashri AN, Varadaraj MC (2009) Molecular characterization of native isolates of lactic acid bacteria, bifidobacteria and yeasts for beneficial attributes. Appl Microbiol Biotechnol 83:1115–1126
Rowland I et al (2018) 'Gut microbiota functions: metabolism of nutrients and other food components. Eur J Nutr 57(1):1–24
Saez-Lara MJ, Gomez LC, Plaza DJ, Gil A (2015) The role of probiotic lactic acid bacteria and bifidobacteria in the prevention and treatment of inflammatory bowel disease and other related diseases: a systematic review of randomized human clinical trials. BioMed Res Int 505878
SanJose MLG, Rodriguez PM, Valls-Belles V (2017) Beer and its role in human health. Ferm Foods Health Dis Prev 365–384.
Satish KR, Kanmani P, Yuvaraj N, Paari KA, Pattukumar V, Arul V (2013) Traditional Indian fermented foods: a rich source of lactic acid bacteria. Int J Food Sci Nutr 64:415–428
Seo MY et al (2009) Anti-aging effect of rice wine in cultured human fibroblasts and keratinocytes. J Biosci Bioeng 107(3):266–271
Shida K et al (2017) Daily intake of fermented milk with Lactobacillus casei strain Shirota reduces the incidence and duration of upper respiratory tract infections in healthy middle-aged office workers. Eur J Nutr 56(1):45–53
Swain MR, Anandharaj M, Ray RC, Parveen RR (2014) Fermented fruits and vegetables of Asia: a potential source of probiotics. Biotech Res Int 250424
Tamang JP, Shin DH, Jung SJ, Chae SW (2016) Functional properties of microorganisms in fermented foods. Front Microbiol 7:1–13
Terefe NS, Augustin MA (2020) Fermentation for tailoring the technological and health related functionality of food products. Crit Rev Food Sci Nutr 60(17):2887–2913
Thapa N, Pal J, Tamang JP (2004) Microbial diversity in ngari, hentak and tungtap, fermented fish products of North-East India. World J Microb Biotechnol 20:599–607
Yang YX et al (2008) Effect of a fermented milk containing Bifidobacterium lactis DN-173010 on Chinese constipated women. World J Gastroenterol 14(40):6237–6243
Yu T, Park ES, Zhao X, Yi RK, Park KY (2020) Lower Mg and S contents in solar salt used in kimchi enhances the taste and anticancer effects on HT-29 colon carcinoma cells. RSC Adv 10(9):5351–5360

Chapter 10
Milk and Other Dairy Product Trends in Health and Longevity

Dino Demirovic Holmquist

Abstract In recent years, the dairy industry is undergoing a disruption in consumer trends and efforts for a sustainable production. Being the only species consuming milk and dairy products post infancy and from other species, this has been an important nutritional source. Today, in the western society, the traditional consumption of dairy products is being challenged by a vast number of alternatives, which is forcing the dairy industry to innovate and bring new products into market. Some of the new trends have a focus on functionality, health, and availability. Here we will go through the basic nutritional elements of milk and dairy products, and cover some of the new trends and their effects on human health.

Keywords Milk · Dairy · Proteins · A2 milk

10.1 Introduction

Mammals are characterized by the presence of mammary glands which in females produce milk for feeding their young offspring. Humans seem to be the only species who consume milk and other dairy products after infancy, and the only species that consume milk and dairy products from other animals on a regular basis. Due to the high nutritional value of dairy products, these have been an optimal primary food source in many developing countries. In recent years, however, the consumption of milk has declined globally, which might be due to the myths and fallacies, or due to other trends in food consumption. Here we will go through the components in milk and dairy products, and their relevance in human health together with recent trends in dairy foods.

D. D. Holmquist (✉)
Eurofins DQCI, LLC, Mounds View, Quincy Street, MN 5205, USA

S. I. S. Rattan and G. Kaur (eds.), *Nutrition, Food and Diet in Ageing and Longevity*, Healthy Ageing and Longevity 14, https://doi.org/10.1007/978-3-030-83017-5_10

10.2 Nutrients in Dairy Products

Dairy products mainly consist of three macronutrients, fat, protein, and lactose, plus some other micronutrients and minerals such as calcium, potassium, magnesium and zinc. Furthermore, most dairy products have a significant concentration of vitamins D and B12. Bovine milk contains approximately 4.6% lactose, 4.2% fat, 3.4% protein, 0.8% minerals and 0.1% vitamins. This composition of milk undergoes changes depending on various factors such as breed, feeding, seasonality, days in milk, lactation stage, cow- and herd-managements etc. (Timlin et al. 2021).

The lipids in milkfat are present in microglobules as an oil and water emulsion consisting of mostly triglycerides (98%), cholesterol (less than 0.5%), phospholipids (about 1%) and free fatty acids (about 0.1%). It has often been reported that dairy products with a high fat content have a negative effect on longevity and health, while the opposite might be true with lower fat dairy products (Tucker 2019). However, other studies show that there is no increased mortality with intake of dairy products with higher fat content (Dehghan et al. 2018), suggesting that other dietary factors also play a major role in longevity and health.

10.2.1 Proteins in Dairy

Dairy proteins primarily include caseins (which is a conglomeration of alpha caseins 1 and 2, beta casein, and kappa casein), alpha-lactalbumin, beta-lactoglobulin, and bovine serum albumin. The beneficial properties of dairy proteins are often attributable to their bioactive peptides released during digestion or food processing (McGregor and Poppitt 2013). Studies have shown several bovine peptides having biologically active properties, including osteopontin, which have shown promising results in infant nutrition and has been shown to attenuate colitis and alcohol-induced liver injury (Christensen and Sorensen 2016). Furthermore, bovine milk contains a great number of peptides which either during food processing or in the gastrointestinal tract can be degraded into peptides that possesses antihypertensive effects (Siltari et al. 2019) and some of them in a dose-dependent matter (Leeuw et al. 2008). More than 200 dairy peptides have shown to have beneficial health effects on human health which therefore attract the interest of researchers as a health promoting functional food (Patil et al. 2017).

10.2.2 Dairy Sugars

Milk carbohydrates are predominantly lactose with trace amounts of monosaccharides and oligosaccharides. Lactose is a disaccharide of glucose and galactose. The pasteurization of the milk and dairy products usually does not have a significant effect on the sugars, however ultra-high temperature (UHT) pasteurization can cause Maillard reactions which can produce undesirable flavors and colors (Stojanskova et al. 2017). The majority (almost 65%) of the world's population have some sort of lactose malabsorption or intolerance, especially in the African and Asian regions (Scrimshaw and Murray 1988). This has caused an increasing demand for low-lactose or lactose-free dairy products.

10.3 New Product Development

While the dairy industry for a long time have had the principle of not manipulating the raw product as far as possible, and only subjecting it to pasteurization for the removal of pathogenic microorganisms, and homogenization for product stability. There is usually no further treatment of milk, although some countries may require addition of certain vitamins, such as in United States all fluid pasteurized milk must be fortified with vitamin D at a minimum of 400 International Units (IU) (FDA Pasteurized Milk Ordinance 2017), whereas fortification with 2000 IU vitamin A has been mandatory for lower-fat fluid milk and optional for whole milk (Yeh et al. 2017). This is constantly pressing the dairy industry to innovate the products through other means, and one of the new approaches is to manipulate fatty acid composition of the raw milk by both genetic breeding of the dairy cows and by manipulating their feeding rations and composition (Schroeder et al. 2003). Recent studies have shown that the major groups of fatty acids (saturated, mono-unsaturated, and poly-unsaturated) can be manipulated by up to 30% by feed alone (Holmquist et al. 2019). This is already leading to new product development in various countries, including Denmark, Norway, UK and Switzerland, which enables new flavors and textures of cheeses and butter products.

Today the dairy industry faces new challenges by new consumer trends with increasing sales of the so-called non-dairy alternatives, including beverages made from soy, almonds, oat etc. These products are made in a more sustainable production environment, which appeals to increasing number of consumers. However, despite some countries having the designation "milk", these products have rarely the same nutritional benefits as of the milk from mammals (Walsh and Gunn 2020). There is no significant difference in the consumption of these non-dairy alternatives when it comes to countries where the designation "milk" is allowed (such as USA) and where marketing of the word "milk" is prohibited (such as the European Union), according to data from Information Resources Inc. (IRI), Chicago, US 2020.

10.3.1 Low Fat – High Protein

One of the most recent and successful dairy trends has been the introduction of low-fat, high-protein products. Since the early 2000's the high protein products have mainly been targeted to serious heavy bodybuilding athletes and the old persons. The latter is due to the fact that the older persons are at high risk of insufficient protein ingestion. Older individuals therefore need to ingest a greater quantity of protein to maintain muscle function (Landi et al. 2016). More recent years, however, have seen an increase in high protein content products, such as various ready to drink products and yoghurts, which appeal to the general consumer. These products also tend to have lower amounts of total fat and usually no added sugars.

Recently, a new category of dairy products with A2 beta-caseins have emerged by the genetic profiling of the dairy herds. Approximately 5–10,000 years ago a point mutation occurred in some European dairy cattle introducing a new beta-casein type, the A1. The difference between A1 and A2 beta-caseins is that A2 beta-casein has a proline amino acid on the 67th position of the 209 amino acid long protein, whereas the A2 beta-casein has a histidine amino acid on this position. Once milk or milk products are consumed, the action of the digestive enzymes in the gut on A1 beta-caseins releases the bioactive opioid peptide beta-casomoprohin-7 (BCM-7), which seems to significantly increase gastrointestinal transit time, production of dipeptidyl peptidase-4 and the inflammatory marker myeloperoxidase (Pal et al. 2015). In humans a double-blind, randomized study has shown that participants who consumed A1 beta-casein type cows' milk had a statistically significant higher Bristol stool values, as compared to users of A2 beta-casein type of cow's milk (Ho et al. 2014).

The most common assumption for milk intolerance is the insufficient lactase enzyme activity; however, the National Institute of Health made a consensus statement in 2010 on lactose intolerance stating that "many who self-reported lactose intolerance shows no evidence of lactose malabsorption. Thus, the cause of their gastrointestinal symptoms is unlikely to be related to lactose" (Suchy et al. 2010). Therefore, some milk intolerance appears to be linked to the presence of BCM-7 after digestion of A1 beta-caseins, though more research has to be completed on these cases.

Other functions of dairy proteins have been linked to various bioactive peptides within the amino acid sequences of native milk proteins. Once bioactive peptides are liberated, they exhibit a wide variety of physiological functions in the human body such as gastrointestinal, cardiovascular, immune, endocrine and nervous system. Usually, the bioactive proteins are inactive within the sequence of the native or parent protein, however they can be released by the hydrolysis in the digestive system or by proteolytic microorganisms, or by proteolytic enzymes derived from microorganisms or plants (Korhonen and Pihlanto 2007).

10.4 Minerals, Vitamins and Antioxidants

Milk contains a large concentration of non-nutritional components, such as minerals, vitamins, and antioxidants. The latter have a role in the protection of the oxidation of the milk, and may also have antioxidant effects in the milk-producing cells and for the udder. The major antioxidants in milk are the vitamins A and E and the mineral selenium (Lindmark-Mansson and Akesson 2000).

Bovine milk has a calcium concentration of about 1 g/l, and daily intake of milk and dairy products has a central role in securing calcium intake in humans. Sufficient calcium intake is essential for healthy bones and teeth, and it also prevents hypertension, decrease the risks for colon and breast cancers and reduces the risk of kidney stones (Insel et al. 2004).

Another noticeable non-nutritional component is selenium, which has a role in immune- and antioxidant system and in DNA synthesis and repair (Dodig and Cepelak 2004). Milk also contains large amounts of iodine (about 60 μg iodine/l,), which is sufficient for up to 50% of the recommended daily intake for adults. Iodine plays a vital role in thyroids hormones, which regulates body metabolic rate, temperature regulation, reproduction and growth. Milk furthermore has a good concentration of magnesium, containing about 100 mg/l. Deficiency in magnesium has been linked to atherosclerosis, as studies have shown that deficiency may cause oxidative stress (Hans et al. 2002). Milk is also a good source of zinc, which has several functions in the human body including in DNA repair, cell growth and replication, gene expression, protein and lipid metabolism etc. (Insel et al. 2004).

As mentioned above, milk also has increased levels of vitamins, among which is Vitamin E, which is associated with decreased risk for cancer and coronary heart disease (Hayes et al. 2001). Other vitamins in milk includes vitamin A, which plays a role in vision and growth in children; folates which has been associated with the growth of neural networks; riboflavin which has antioxidant effects; and vitamin B12 which plays a vital role in homocysteine metabolism.

10.5 Conclusion

Milk and dairy products have had a great impact in our society and is still playing a major role in human nutrition in developing countries. In developed countries, consumer trends are moving to many alternatives to classical bovine (or other animals') milk, usually coming from a plant, nut, or bean source. With emerging companies such as the Israeli and US-based ImaginDairy and Perfect Day, there is another movement which is not looking to change existing milk and dairy products, but rather developing dairy products from ingredients produced by microorganisms in a laboratory. However, this new technology still needs both scalability for a larger production, and a general broad consumer acceptance.

Compliance with Ethical Standards

Conflict of Interest I have no conflict of interest.

References

Christensen B, Sorensen ES (2016) Structure, function and nutritional potential of milk osteopontin. Int Dairy J 57:1–6
Dehghan M, Mente A et al (2018) Association of dairy intake with cardiovascular disease and mortality in 21 countries from five continents (PURE): a prospective cohort study. Lancet 392:2288–2297
Dodig S, Cepelak I (2004) The facts and controverses about selenium. Acta Pharm 54:261–276
Hans CP, Chaudhary DP, Bansal DD (2002) Magnesium deficiency increases oxidative stress in rats. Indian J Exp Biol 40:1275–1279
Hayes K, Pronczuk A, Perlman D (2001) Vitamin E in fortified cow milk uniquely enriches human plasma lipoproteins. Am J Clin Nutr 74:211–218
Ho S et al (2014) Comparative effects of A1 versus A2 beta-casein on gastrointestinal measures: a blinded randomised cross-over pilot study. Eur J Clin Nutr 68:994–1000
Holmquist DD et al (2019) Where will new technologies take milk recording? Large scale screening of the Danish dairy cattle population for their fatty acid profile. In: ICAR 2019 Proceedings
Insel P, Turner RE, Ross D (2004) Nutrition. Second Ed. American dietetic association, Jones and Bartlett, USA
Korhonen H, Pihlanto A (2007) Bioactive peptides from food proteins. In: Hui YH (ed) Handbook of food products manufacturing. John Wiley & Sons, Inc, pp 5–37
Landi F, Calvanni R et al (2016) Protein intake and muscle health in old age: from biological plausibility to clinical evidence. Nutrients 8(5):295
Leeuw PW et al (2008) Dose-dependent lowering of blood pressure by dairy peptides in mildly hypertensive subjects. Blood Pressure 18:1–2
Lindmark-Mansson H, Akesson B (2000) Antioxidative factors in milk. Br J Nutr 84: S103–S110
McGregor RA, Poppitt SA (2013) Milk protein for improved metabolic health: a review of evidence. Nutr Metab (Lond) 3 10(1):46. https://doi.org/10.1186/1743-7075-10-46
Pal S et al (2015) Milk intolerance, Beta-Casein and Lactose. Nutrients 7(9):7285–7297
Patil P et al (2017) Milk-derived Bioactive Peptides: Health Benefits, Dairy Engineering, pp 209–228
Schroeder GF et al (2003) Milk fatty acid composition of cows fed a total mixed ration or pasture plus concentrates replacing corn with fat. J Dairy Sci 86(10):3237–3248
Scrimshaw NS, Murray EB (1988) The acceptability of milk and milk products in populations with a high prevalence of lactose intolerance. Am J Clin Nutr 48:1079–1159
Siltari A, Vapaatalo H, Korpela R (2019) Milk and milk-derived peptides combat against hypertension and vascular dysfunction: a review Int. J Food Sci Tech 54:1920–1929
Stojanskova S et al (2017) Maillard reaction and lactose structural changes during milk processing. Chem Res J 2(6):139–145
Suchy FJ et al (2010) NIH consensus development conference statement: Lactose intolerance and health. NIH Consens State Sci Statements 27:1–27
Timlin M, Tobin JT et al (2021) The impact of seasonality in pasture-based production systems in milk composition and functionality. Foods 10(3):607
Tucker LA (2019) Milk fat intake and telomere length in U.S. women and men: the role of the milk fat fraction. Oxidative Med Cellular Longevity, Oct. 28, 2019
Walsh MC, Gunn C (2020) Non-dairy milk substitutes: are they of adequate nutritional composition? In: Milk and Dairy Foods – Their functionality in Human Health and Disease, pp 347–369
Yeh EB, Barbano DM, Drake M (2017) Vitamin Fortification of Fluid Milk. J Food Sci 82(4):856–864

Chapter 11
Anti-inflammatory Foods in Ageing and Longevity

Ceren Gezer

Abstract Inflammageing underlies ageing- and age-related chronic diseases, while age-related chronic diseases also underpin inflammageing and ageing. The featured inflammageing mechanisms are the inflammasome; DNA damage including telomere shortening; accumulated cellular senescence; immunosenesence; increased synthesis of proinflammatory miRNAs through activation of pathways such as NF-κB, mTOR, and sirtuins; dysbiosis of the gut microbiota; and meta-inflammation. It is observed that fruits, vegetables, olive oil, fish oil, whole grains, legumes, nuts, flavonoid-rich green tea, carotenoids, omega-3 fatty acids, fibre, and pre- and pro-biotics can inhibit these mechanisms and promote the prevention of chronic diseases such as diabetes, cardiovascular diseases, cancer, and neurodegenerative diseases that underpin inflammageing and ageing. However, the studies are mostly in vitro and in vivo animal model studies. Thus, there is a need for more prospective and clinical studies on anti-inflammatory foods in ageing and longevity. Moreover, it is important to consider that explaining the relation between diet, low-grade chronic inflammation, and ageing not only depends on a single food component in the concept of anti-inflammatory food, but also dietary pattern. Even though there are scarce human studies on benefits of these potential anti-inflammatory effects in ageing and longevity, adopting a Mediterranean dietary pattern and recommended consumption amounts can be suggested.

Keywords Ageing · Food · Inflammation · Longevity · Phytochemical · Mediterranean diet

11.1 Introduction

Inflammation is a response of tissue to infection agents such as pathogens and toxins in the immune system. In acute inflammation, the aim is to provide physiological homeostasis to destroy pathogens and induce repair mechanisms of tissue. During

C. Gezer (✉)
Department of Nutrition and Dietetics, Faculty of Health Sciences, Eastern Mediterranean University, Famagusta, North Cyprus via Mersin 10, Turkey
e-mail: ceren.gezer@emu.edu.tr

S. I. S. Rattan and G. Kaur (eds.), *Nutrition, Food and Diet in Ageing and Longevity*, Healthy Ageing and Longevity 14, https://doi.org/10.1007/978-3-030-83017-5_11

this process of acute inflammation, heat, swelling, redness, and pain are distinguished (Fougère et al. 2017). Different from this acute response is chronic inflammation, which is a low-grade, sustained, and systemic response of tissue that causes degeneration via inflammatory markers such as inflammatory cytokines (interleukin (IL)-6, IL-1, and tumour necrosis factor (TNF)-α), chemokines (monocyte chemoattractant protein-1 (MCP-1)), cell adhesion molecules (VCAM-1, ICAM-1), and acute phase proteins (C-reactive protein).

Ageing, diet, smoking, etc., have been identified as chronic pro-inflammatory factors. It has been shown that chronic inflammation underlies ageing- and age-related diseases such as cancer, diabetes, obesity, and atherosclerosis (Guarner and Rubio-Ruiz 2015). Thus, ageing is characterized by increased levels of inflammatory markers, and this age-related inflammation is termed inflammageing (Sanada et al. 2018; Calder et al. 2017).

There are many mechanisms that have a role in inflammageing. Some of the mechanisms related to inflammageing are oxidative stress, glycation, immunosenesence, mitochondrial dysfunction, chronic infections, epigenetic and hormonal changes, and diet (Fougère et al. 2017; Ekmekcioğlu 2020). The balance between these inflammageing sources and sources of anti-inflammageing factors such as healthy nutrition is important to restrain age-related diseases. With respect to diet, dietary pattern is assumed that have positive effects on chronic inflammation in age-related diseases (Monti et al. 2017).

It has been shown that the food contents of fibre, flavonoids, carotenoids, and omega-3 fatty acids have anti-inflammatory effects, while simple carbohydrates, and saturated and trans fatty acids have pro-inflammatory effects (Galland 2010). Thus, foods rich in anti-inflammatory components (whole grains, fruits, vegetables, legumes, nuts, and tea) can be called anti-inflammatory foods. The Mediterranean Diet is a dietary pattern composed of a high amount of the anti-inflammatory foods and a low amount of foods poor in these pro-inflammatory components. This diet is related to lower chronic inflammation (Calder et al. 2011). Therefore, explaining the relation between diet, low-grade chronic inflammation, and ageing not only depends on single food content in the concept of anti-inflammatory food, but also dietary pattern. In this chapter, the aim is to underline featured foods with anti-inflammatory effects in ageing and longevity.

11.2 Anti-inflammatory Dietary Pattern, Ageing, and Longevity

Inflammation has been assessed with inflammatory cytokines, chemokines, cell adhesion molecules, and acute phase proteins as biomarkers. As an acute inflammation response in infection, microbial cell membranes and nucleic acid are recognized by pathogen-associated molecular patterns (PAMPs) such as toll-like receptors (TLRs) and retinoic acid-inducible gene-1-like receptors (RLRs). As a chronic inflammation response in inflammatory diseases, substances called danger-associated molecular patterns (DAMPs) are secreted. However, it is indicated that sometimes, it is very hard to differentiate these responses as either acute or chronic (Kourtgen and Bauer 2018).

Even though there are various biomarkers to assess acute and chronic inflammation, only a few of them are acceptable in clinical and epidemiological studies that are related to diseases and diet as a factor relevant to chronic inflammation. In this context, the mostly assessed pro-inflammatory cytokines are TNF-α, IL-1, and IL-6, and the anti-inflammatory cytokines are IL-4, IL-10, IL-13, and transforming growth factor (TGF)-β. Actually, there is not a certain line to define the cytokines as pro- and anti- inflammatory, and its depends on the local environment of released cytokines, cytokine receptor density, synergistic and competing factors, and tissue response. However, most commonly, TNF-α, IL-1, and IL-6 are found as pro-inflammatory.

Besides cytokines, other commonly assessed biomarkers are chemokines, and MCP-1/CCL2 is the most featured one. It is responsible for modulating migration and infiltration of monocytes and macrophages. Cell adhesion molecules are responsible for modulating migration of leucocytes, and VCAM-1 is one of the key types. Both MCP-1 and VCAM-1 have key roles in the basic inflammation process in endothelial cells. Also, acute-phase proteins are commonly assessed for inflammation, and lipopolysaccharide-binding protein (LBP), fibrinogen, serum amyloid A, ceruloplasmin, and CRP are the most used ones. CRP is used to evaluate the level of infection and the development and risk of chronic inflammatory diseases such as cancer (Wu and Schauss 2012; Casas et al. 2014). According to meta-analysis results, healthy dietary patterns are negatively associated with CRP levels in adults (Neale et al. 2016).

Overall, concerning dietary pattern or food components, the most used key biomarkers are included in a posteriori methods in research. While these a posteriori results are important to assess the relation between diet and inflammation, a priori methods using diet scores are also important (Calder et al. 2011). There are some indices that are used to evaluate dietary patterns and the risk of chronic diseases, such as the Healthy Eating Index, Dietary Approaches to Stop Hypertension Index, Diet Inflammatory Index, and Mediterranean Diet Scores (Calle and Andersen 2019; Marcason 2010). It has been indicated that the Diet Inflammatory Index is more advantageous than other indices to determine risk of inflammatory diseases related to diet (Wirth et al. 2016). On the other hand, Mediterranean Diet Scores have also been indicated as good indices to observe relations with inflammatory biomarkers to evaluate inflammatory disease risk (Casas et al. 2014).

In general, the studies on inflammatory biomarkers and dietary patterns are observational, and the common points of these different indices are that they are mostly based on fruits and vegetables (Barbaresko et al. 2013). The results mostly depend on observational short-term studies, and dietary patterns are more effective to evaluate diet-related inflammatory biomarkers and predict disease risk from a broader perspective (Corley et al. 2015). There are many cofactors that effect inflammatory biomarkers, and it is hard to identify mechanisms of dietary patterns and inflammation. Studies on inflammation and dietary patterns point out foods and food components that are potentially anti-inflammatory (Calle and Andersen 2019). Therefore, to understand the anti-inflammatory mechanisms of diet, intervention studies based on specific food are important. Thus, in a holistic view dietary pattern, food and food components are combined with each other, and all of them are related to ageing and lifespan. This chapter focuses on food components and inflammatory pathways to explain effects on ageing and lifespan.

11.3 Anti-inflammatory Foods, Ageing, and Longevity

Since older people have increased levels of cytokines and chemokines, it can be said that systemic low-grade chronic inflammation mostly rises with age. This inflammation type involves most of the tissues (such as adipose and muscle tissues), organs (such as the brain and liver), systems (such as the immune system), and ecosystems (such as the gut microbiota). The circadian rhythm, chronic stress, xenobiotics, diet, dysbiosis, obesity, chronic infections, and physical inactivity trigger systemic low-grade chronic inflammation. Eventually, chronic low-grade inflammation causes tissue damage and hence metabolic syndrome, type 2 diabetes, cardiovascular diseases, cancer, neurogenerative diseases, sarcopenia, osteoporosis, and immunosenesence (Monti et al. 2017; Furman et al. 2019). Thus, inflammageing underlies ageing- and age-related chronic diseases, while age-related chronic diseases underpin inflammageing and ageing (Franceschi et al. 2018).

There are several molecular mechanisms related to inflammageing: (i) oxidative stress, dysfunction of mitochondria, and endoplasmic reticulum stress cause increased production of DAMPs that activate inflammasome and NF-κB. (ii) DNA damage, including telomere shortening, occurs through oxidative stress and other factors and triggers secretion of pro-inflammatory compounds. (iii) Cellular senescence increases with age, accumulated senescence cells acquire senescence-associated secretory phenotype (SASP), and senescent cells drag along neighbour cells to senescence through SASP. (iv) In immunosenesence, age-related changes occur in innate and adaptive immune cells, which is also accepted as a characteristic of inflammageing as chronic activation of the innate immune system. (v) Age-related increase production of galactosylated immunoglobulin (Ig-G-GO) occurs, and (vi) there is increased synthesis of proinflammatory miRNAs through activation of pathways such as NF-κB, mTOR, sirtuins, and TGF-β pathways. (vii) There is also dysbiosis of the gut microbiota, and viii) meta-inflammation mediated by excess nutrients (particularly glucose and fatty acids) and energy trigger stress in pancreas, liver, muscle, and adipose tissue, resulting with secretion of cytokines, chemokines, and adipokines (Cevenini et al. 2013; Monti et al. 2017; Calder et al. 2017). Overall, the immune system, adipose tissue, liver, muscle, and gut microbiota contribute to inflammageing and meta-inflammation (circulating molecular mediator) and thus multi-system inflammation (cardiovascular diseases, obesity, type 2 diabetes, cancer, neurodegenerative diseases, etc.) (Cevenini et al. 2013; Monti et al. 2017).

A healthy dietary pattern such as the Mediterranean diet can prevent and mitigate inflammatory diseases by anti-inflammatory effects. This diet is characterized by high consumption of fruits, vegetables, whole grains, legumes, nuts, olive oil, and fish oil, so the intake of bioactive components such as phytochemicals, vitamins, and fatty acids is high (Estruch 2010). These components play a potential a key role in suppression of molecular mechanisms of inflammageing.

11.3.1 Fruits and Vegetables

Fruits and vegetables are indispensable parts of a healthy dietary pattern since not only are they rich in micronutrients, but they are also rich in various phytochemicals, such as phenolics and carotenoids. Each fruit and vegetable differs in phenolic composition, so consuming various types of fruits and vegetables is important while aiming to reach suggested healthy consumption amounts. Flavonoids are the most common phenolic compounds found in fruits and vegetables that and are relevant to inflammation (Zhang et al. 2015; Oz and Kafkas 2017).

Nuclear factor (NF)-κβ and activator protein-1 (AP-1) are commonly used molecular targets to assess inflammation and in phytochemical treatment studies. NF-κβ has a pivotal role in inflammatory response. The phosphorylation status of these factors is regulated by mitogen-activated protein kinase (MAPK) cascades. MAPK involves subgroups such as p38, extracellular signal-regulated kinase (ERK), and C-jun N-terminal kinase (JNK) signalling pathways and activates NF-κβ and AP-1 as an inflammatory response. In addition, lipopolysaccharide (LPS) is a PAMP that binds to TLR4 and hence activates NF-κβ and AP-1 mediated by MAPKs. Also, it is known that TNF-α, IL-6, and CRP activate NF-κβ. NF-κβ is also responsible for the expression of enzymes such as cyclooxygenase (COX)-2 and the inducible nitric oxide synthase (iNOS), as well as chemokines and adhesion molecules (Chung et al. 2021).

Resveratrol is the most common stilbene compound abundantly found in grapes, berries, peanuts, and wine. Resveratrol exhibits anti-inflammatory effects via suppression of NF-κB, JAK-STAT, and AP-1 signalling pathways, which cause the production of inflammatory mediators such as IL-1β, TNF-α, IL-6, IL-8, and NO in vitro and in vivo. Also, resveratrol suppresses COX and lipoxygenase (LOX) enzyme activities in the synthesis of leucotriens, eicosonoids such as thromboxanes, and prostanoids such as prostaglandins (Csiszar 2011; Ma et al. 2015; de Sá Coutinho et al. 2018; Banez et al. 2020). In addition, in association with these mechanisms, resveratrol exerts neuro- and cardioprotection effects, prevents cancer, and ameliorates the ageing process. These effects of resveratrol are supported in clinical trials, but there is a need for more studies (Banez et al. 2020).

Flavones, flavonols, and flavanones are the three flavonoid subgroups. Apigenin and luteolin are the most common flavones. Apigenin occurs in apple, orange, grapefruit, chamomile, celery, parsley, and onion. Apigenin inhibits activation of NF-κB, which is a transcription factor that has a crucial role in iNOS and COX-2 expression and mitigates IL-1β, TNF-α, IL-6, and inflammatory cell infiltration in vitro and in vivo (Ai et al. 2017; Dang et al. 2018; Xie et al. 2019). It is suggested that a diet rich in apigenin attenuates LPS-induced inflammation via regulation of gene expression in vivo by inhibiting IL-1β production via caspase-1 activation, inhibiting IL-1β and IL-6 production via inhibition of ERK1/2, and inhibiting TNF-α-induced activation of NF-kB (Wang et al. 2014; Zhang et al. 2014; Arango et al. 2015).

Apigenin also reduces LPS-induced inflammation in vitro by suppressing NO and prostaglandin production by inhibiting iNOS and COX-2, respectively, as well as caspase-3 activity, which is a central modulator of apoptosis. Thus, apigenin can reduce the risk of chronic diseases such as diabetes, cardiovascular diseases, cancer, and Alzheimer's disease (Duarte et al. 2013; Choi et al. 2014). In addition, both apigenin and luteolin inhibit TNFα-induced expression of CCL2/MCP-1 and CXCL1/KC (Funakoshi-Tago et al. 2011). Luteolin occurs in many vegetables, such as artichoke, broccoli, pepper, thyme, turnip, cucumber, and celery. Luteolin inhibits IL-1β and TNF-α expression by regulating transcription factors such as STAT3, NF-κB, and AP-1 in vitro and in vivo. Therefore, it has been suggested that luteolin may reduce risk of inflammatory diseases such as neurodegenerative diseases and cancer (Nabavi et al. 2015; Hayasaka et al. 2018; Aziz et al. 2018).

Quercetin is the most common flavonol and is found in apples, onions, berries, and capers. It decreases CRP, SAA, fibrinogen, TNF-α, IL-1β, IL-6, and NO in vitro and in vivo. Quercetin inhibits TNF-α by inactivation of ERK, c-Jun, and NF-κB, as well as the activation of peroxisome proliferator-activated receptor gamma (PPARγ) (Kleemann et al. 2011; Li et al. 2016). Also, quercetin inhibits COX-2 activity (Lesjak et al. 2018). Thus, similar to apigenin and luteolin, quercetin may also reduce the risk of chronic inflammatory diseases.

Hesperidin and naringenin are the most common flavonones and are abundantly found in citrus fruits (Yi et al. 2017). It is indicated that hesperidin inhibits NF-κB and thus iNOS and COX-2 activities in vitro and in vivo (Parhiz et al. 2015; Tejada et al. 2018). In addition, hesperidin suppresses the production of IL-2, IL-4, IL-10, NF-κB/p65, COX-2, IL-1β, TNF-α, MMP-3, and MMP-9, while it stimulates production of IL-10 in chondrocytes and chondroitin mesenchymal stem cells in vitro (Xiao et al. 2018a, b; Tsai et al. 2019). Another flavanone, naringenin, either activates SIRT1 enzyme and prevents senescence in vitro or reduces TNF-α and IL-6 in vivo in myocardial cells and tissues (Testai et al. 2020). Therefore, flavonones can prevent cardiovascular diseases, diabetes, and cancer (Yi et al. 2017).

Anthocyanins are another subgroup of flavonoids and are mostly found in berries, where they result in unique colours. The abundant types of anthocyanins are cyanidin, pelargonidin, peonidin, delphinidin, petunidin, and malvidin. Similar to other flavonoids, anthocyanins also exhibit anti-inflammatory effects by reducing iNOS and COX-2 activity, as well as modulate NF-κβ and MAPK signalling cascades, thus reducing TNF-α, IL-1β, IL-6, and NO production in vitro and in vivo (Vendrame and Klimis-Zacas 2015). Despite this, there is a need for more evidence on those mechanisms of action of anthocyanins to support their use in epidemiological and clinical trials (Joseph et al. 2014). Furthermore, it has been indicated that anthocyanins can also modulate the gut microbiota, and all of the other mechanisms of action provide protection from chronic inflammatory diseases (Morais et al. 2016; Ma et al. 2018). Overall, flavonoids have anti-inflammatory effects and play a role in preventing chronic diseases, such as cardiovascular diseases and cancer (Griffithd et al. 2016).

Lycopene, β-carotene, and lutein are carotenoids that are more commonly present in foods. Lycopene mainly occurs in tomatoes and tomato products. Lycopene inhibits LPS-induced NO, IL-6, TNF-α, and secretory phospholipase A2 production by suppressing ERK, p38MAPK, NF-κB, and high-mobility group box 1 (HMGB1) activation in vitro and in vivo (Feng et al. 2010; Lee et al. 2012; Marcotorchino et al. 2012; He et al. 2015). Lycopene also inhibits LPS-induced COX-2 and iNOS activities in human keratinocyte cells (Kim et al. 2014). Lycopene also suppresses high-fat-diet-induced inflammation via reduction of NF-κB/p65 (Fenni et al. 2017). Lycopene reduces SAA levels in overweight individuals (McEneny et al. 2013), and lycopene and lutein together attenuate VCAM-1, ICAM-1, and TNF-α-induced leukocyte adhesion and the NF-κB signalling pathway in human endothelial cells (Armoza et al. 2013).

Lutein is mainly found in corn and tangerines and attenuates IL-6 production and COX-2 activity. It also suppresses AP-1 activation by the inhibition of JNK/p38 in human keratinocyte cells (Arscott 2013; Oh et al. 2013). Lutein also suppresses NF-κB activation by inhibition of JNK/p38 and Akt in LPS-activated BV-2 microglia cells (Wu et al. 2015). In addition, lutein inhibits NF-κB and COX-2 activation and reduces IL-1β production in retinal ischemic/hypoxic injury in vitro and in vivo (Li et al. 2012). Lutein suppresses activation of NF-κB, inhibits COX-2 activity, and decreases IL-6, TNF-α, and IL-1β in primary chondrocyte cells (Qiao et al. 2018). Thus, lutein can reduce the risk of inflammatory diseases such as osteoarthritis, age-related macular degeneration, cardiovascular diseases, and neurodegenerative diseases (Kijlstra et al. 2012). β-carotene present in carrots (Arscott 2013).

LPS-induced NF-κB, JAK2/STAT3, and JNK/p38 MAPK activation is suppressed by β-carotene in macrophages (Li et al. 2018). β-carotene inhibits virus-induced NO, IL-1b, IL-6, and MCP-1 production and NF-κB, JNK/p38, and ERK activation in macrophages (Lin et al. 2012). Both β-carotene and lycopene reduce the inflammatory response in vitro (Di Tomo et al. 2012; Kawata et al. 2018). To sum up, even though there are contrary results about the mechanisms of action related to anti-inflammatory effects of carotenoids, they can lower the risk of cardiovascular diseases, type 2 diabetes, dementia, and cancer (Ciccone et al. 2013; Kaulmann and Bohn 2014; Honarvar et al. 2017).

In a comprehensive manner, phytochemicals in fruits and vegetables ameliorate the inflammatory response in the short and long term (Joseph et al. 2016). These effects on inflammageing mainly occur by mitigating NF-κB. Also, phytochemicals attenuate cellular senescence and immunosenesence in immune cells (Sharma and Padwad 2020). Studies have been mostly in vitro and in vivo animal studies, and there is a need for epidemiological and clinical studies to support these effects in humans and improving dietary recommendations.

11.3.2 Olive Oil

Olive oil is one of the crucial components of the Mediterranean diet responsible for positive health effects (Virruso et al. 2014). Olive oil is obtained solely from the olive tree fruit (*Olea europaea* L.). A number of processes are used, and virgin olive oil is obtained exclusively by specific mechanical processes that do not alter the oil. The virgin olive oils are classified as extra virgin olive oil (EVOO), virgin, and lampante according to the degree of acidity. All of these virgin oils mainly contain monounsaturated fatty acids, particularly oleic acid, as well as tocopherols, tocotrienols, b-carotene, phytosterols, flavonoids, and phenolic compounds, such as oleuropein, hydroxytyrosol, and oleocanthal (Ghanbari et al. 2012; Souza et al. 2017).

It has been suggested that phenolic compounds found in olive oil, such as hydroxytyrosol, tyrosol, and oleocanthal, can mitigate NF-kB and relevant signalling cascades, suppress eicosonoid synthesis, and enzyme activities i*n vitro* and in vivo (Souza et al. 2017). A systematic review and meta-analysis of randomized controlled trials on regular dietary intake of olive oil indicated that olive oil is associated with decreased levels of CRP, IL-6, and TNF-α (Fernandes et al. 2020). Oleic acid is abundantly found in olive oil, but it has higher concentrations in seed oils such as sunflower, rapeseed, and soybean oil. However, olive oil contains specific phenolics that are not present in seed oils. The specific phenolic compounds have anti-inflammatory effects in vitro and in *vivo,* hence reducing the risk of chronic disease development (Cicarelle et al. 2012).

It has been indicated that EVOO and its phenolic compounds inhibit NF-kB and STATs and modulate JAK/STATs, ERK/MAPKs, JNK, and AKT pathways in immune-mediated inflammatory diseases such as rheumatoid arthritis, inflammatory bowel disease, multiple sclerosis, and psoriasis (Santangelo et al. 2017). In addition, EVOO reduces NO and PGE2 by suppression of iNOS and COX-2 expression. It also inhibits MAPK and NF-kB and thus TNFα and IL-6 production in LPS or IFNγ-stimulated murine macrophages (Cárdeno et al. 2014a, b). Moreover, the olive oil polyphenol hydroxytyrosol reduces PGE2 due to inhibition of COX-2 activity in human monocytes (Rosignoli et al. 2013). Hydroxytyrosol also inhibits NF-kB and thus TNF-α (Killeen et al. 2014).

Another olive oil polyphenol oleocanthal inhibits COX-1, COX-2, and iNOS activities, as well as tau-tau interaction in neuron cells (Lucas et al. 2011; Cicerale et al. 2012). Thus, according to these in vitro an in vivo study results, virgin olive oils can prevent cardiovascular disease, cancer, diabetes, degenerative joint diseases, and neurodegenerative diseases, which is basically related to their hydroxytyrosol and oleocanthal content. However, clinical trials are scarce (Parkinson and Cicerale 2016).

11.3.3 Fish Oil

α-linolenic acid is a short-chain plant form of omega-3 fatty acid, and eicosapentaenoic acid (EPA) and docosahexaenoic (DHA) are long-chain fatty acids (FAs) that are the marine forms of omega-3 fatty acids. These fatty acids occur in sea foods in lean and oily fish. Some sea foods, oily fish, and livers of some lean fish include much more marine omega-3 fatty acids (Calder 2010). Omega-3 fatty acids inhibit pro-inflammatory leucocyte chemotaxis, adhesion molecule expression, and leucocyte-endothelial adhesive interactions. In particular, EPA upregulates 3 series of prostoglandins and leukotriens that are low-proinflammatory compounds compared to 2 series of prostoglandins and 4 series of leucotriens upregulated by AA (Calder 2012).

Obesity is seen as an inflammatory disease due to TNF-α and IL-6 increasing and adiponectin decreasing in adipocytes. In addition, CRP increases from hepatocytes mediated by IL-6 (Ellulu et al. 2015). Omaga-3 PUFAs suppress NF-kB and thus CRP and IL-6 production. Also, omega-3 FAs up-regulate PPARγ, which induces fat cell differentiation and maturation. Therefore, omega-3 FAs promote adipogenesis and a healthy expansion of adipose tissue, while omega-6 FAs have reverse effects on expansion of adipose tissue.

The free fatty acid receptor (FFAR) family has a role in both energy homeostasis and the inflammatory response in different cell types. Macrophage infiltration into adipose tissue is decreased via omega-3 PUFAs, which activate FFAR4/G protein-coupled receptor 120 (Jayarathne et al. 2017; Albracht-Schulte et al. 2018). In addition, resolving D series and resolving E series are synthesized from DHA and EPA, respectively, which have anti-inflammatory effects due to inhibiting inflammatory cytokine synthesis and clearance of inflammatory cells related to cardiovascular health (Oppedisano et al. 2020).

Many of these effects are related to changes in fatty acid composition of cell membranes. The changes can modify membrane fluidity and cell signalling, evoke gene expression changes, and change the pattern of lipid mediator production. For instance, richness in $n - 6$ fatty acid AA composition of cell membranes leads to an inflammatory response (Calder 2012). Thus, the omega-6/omega-3 balance is important for inflammatory processes and thus health maintenance.

It has been indicated that if the omega-6/omega-3 ratio is 3–5:1, mortality is decreased (Candela et al. 2011). In addition, it has been observed that omega-3 FAs attenuates cellular senescence of immune cells via telomere shortening and inhibition of SASP. Also, it ameliorates harmful effects of the cellular and humoral immune response during immunosenesence (Sharma and Padwad 2020).

Overall, in clinical studies, dietary omega-3 fatty acids have been related to lower inflammatory biomarkers such as CRP, TNF-α, IL-6, and IL-1, as well as endothelial activation in cardiovascular disease, diabetes, chronic renal disease, Alzheimer's disease, sepsis, acute pancreatitis, and cancer (Rangel-Huerta et al. 2012; Mocellin et al. 2016; Layè et al. 2017; Natto et al. 2019). In addition, marine $n - 3$ fatty acids were tested in animal models of rheumatoid arthritis (RA), inflammatory bowel disease (IBD) and asthma. They showed positive effects in clinical trials on RA, but not in IBD and asthma (Calder 2015; Akbar et al. 2017).

11.3.4 Whole Grains

The most consumed whole grain foods are whole wheat, whole oats, whole barley, whole rye and their flours, bulgur, brown rice, amaranth, sorghum, and their products such as bread, granola, etc. It is reported that whole grains lower CRP levels in epidemiological studies (Lefevre and Jonnalagadda 2012). Moreover, two meta-analyses of randomized control trials reported that whole grain consumption is negatively related to high-sensitivity-CRP (hs-CRP), CRP, TNF-α, IL-6, and IL-1β levels (Xu et al. 2018; Hajihashemi and Haghighatdoost 2018).

In this context, two main components of whole grains are featured: dietary fibre and phytochemicals since there is robust scientific evidence about them. However, they have primary effects on consumer acceptability related to qualities such as colour, texture of the whole-grain product, and their synergistic effects (Awika et al. 2018). Furthermore, the anti-inflammatory mechanisms of action depend on β-glukan (a dietary fibre), as well as alkylresorcinols and avenanthramides (phytochemicals). These show anti-inflammatory effects by modulating the gut microbiota. While β-glukan is fermented into short-chain fatty acids, alkylresorcinols and avenanthramides are fermented into phenolic acids such as cinnamic acid, cafeic acid, and ferulic acid by the gut microbiota. These microbial-derived metabolites inhibit LPS, IFN-γ, and reactive oxygen species-induced NF-κB expression and hence attenuate TNF-α, IL-1β, IL-6, and CRP (Sang et al. 2020). Thus, bioactive components of whole grains have anti-inflammatory activities that could reduce the risk of chronic diseases such as diabetes, cardiovascular disease, cancer, and dementia (Lee et al. 2015).

11.3.5 Legumes

Soybeans, kidney beans, lentils, and cowpeas are common legumes, which are rich in fibre and protein (Rebello et al. 2014). Legumes contain various bioactive compounds, such as phenolics, peptides, and saponins, which exhibit anti-inflammatory effects (Serventi and Dsouza 2020). Legume seed hulls are rich in phenolics, and it is indicated that lentil hulls inhibit 15-LOX, COX-1, and COX-2 activities (Boudjo et al. 2013). Legume proteins, particularly soybean and bean proteins, are digested into bioactive peptides that can modify NO, PGE2 (and thus iNOS), and COX- 2 activities, as well as cytokines and chemokines in vitro and in vivo (Reyes-Díaz et al. 2019). It is also observed that legume saponins, particularly soy saponins, attenuate NO, TNF-α, PGE2, and MCP-1, suppress iNOS and COX-2 activities, and mitigate NF-κB activation (Zhu et al. 2018).

A meta-analysis of randomized clinical trials demonstrated that decreased CRP and hs-CRP levels are associated with non-soy legume consumption, although more clinical studies are needed to clarified the effects of non-soy legume consumption on inflammatory markers (Salehi-Abargouei et al. 2015). To conclude, legumes contain various bioactive compounds that display anti-inflammatory effects, thus suggesting they could prevent inflammatory diseases. However, there is a need for more studies on the anti-inflammatory impact of legumes (not only clinical but also animal and molecular study models).

11.3.6 Nuts

The common tree nuts are almonds, Brazil nuts, cashews, hazelnuts, pecans, pistachios, and walnuts. Almonds, hazelnuts, pine nuts, pistachios, and walnuts are mostly involved in the Mediterranean diet. They are rich in unsaturated fatty acids, protein, fibre, tocopherols, phytosterols, and polyphenols (Ros 2015). The phytochemical content of tree nuts can vary considerably by nut type, and they have been associated with anti-inflammatory impact (Bolling et al. 2011). It has been demonstrated that increased nut consumption is related to decreased inflammatory disease mortality (Gopinath et al. 2011). In addition, an epidemiologic study, the National Health and Nutrition Examination Survey (NHANES), found that increased nut and seed consumption is related to increased telomere length, thus decreasing aging and cell senescence (Tucker 2017).

On the other hand, a meta-analysis of randomized controlled trials suggests that nut consumption significantly decreases leptin while having no significant effect on CRP, IL6, adiponectin, IL10, and TNF-a (Mazidi et al. 2016). Similarly, another meta-analysis of randomized controlled trials also suggests that nut consumption significantly reduced ICAM-1 levels but had no effect on other inflammatory markers such as CRP, IL-6 TNF-α, E-selectin, and VCAM-1 (Xiao et al. 2018a, b). Another meta-analysis of randomized controlled trials suggests favourable effects of nut consumption on flow-mediated dilation as a measure of endothelial function (Neale et al. 2017). The results of these three meta-analyses about non-significant changes in inflammatory biomarkers consider that there is a need for more randomized controlled trials on the relationship of nut consumption and inflammation.

11.3.7 Green Tea

Green tea has anti-inflammatory effects from the main bioactive component, catechins, which are in the flavonoid subgroup of flavan-3-ols. The four main catechins in green tea are (-)-epicatechin (EC), (-)-epigallocatechin (EGC), (-)-epicatechin-3-gallate (ECG), and (-)-epigallocatechin-3-gallate (EGCG). The most abundant catechin is EGCG (59%), followed by EGC ($\approx$19%), ECG ($\approx$14%), and EC ($\approx$6%).

Green tea catechins exert anti-inflammatory effects by increased IL-10 and decreased IL-1β, TNF-α, IL-6, IL-8, interferon gamma (INF-γ), CRP, matrix metalloproteinases (MMPs), ICAM-1, VCAM-1, and E-selection in vitro and in vivo (Reyaert 2017). In addition, green tea catechins also inhibit COX-2 expression. EGCG is the most effective green tea catechin and suppresses NF-κB/p38 expression as well (Riegsecker et al. 2013; Ohishi et al. 2016; Fechtner et al. 2017). High glucose increases VCAM-1 expression and thus induces inflammation, but EGCG mitigates PKC and NF-κB signalling in human umbilical-vein endothelial cells (Yang et al. 2013). EGCG also attenuates PCB-126-induced endothelial cell inflammation via suppression of NF-κB/p65, IL-6, CRP, ICAM-1, VCAM-1, and IL-1α/β production (Liu et al. 2016).

Similarly, EGCG suppresses infrasound-induced increases in NF-κB/p65 and inhibits IL-1β, IL-6, IL-18, and TNF-α in microglia (Cai et al. 2014). Moreover, EGCG attenuates TNF-a, MCP-1, ICAM-1, NO, VEGF, and MMP-2 production and NF-kB and MAPK signalling pathways in LPS-stimulated in L02 hepatocytes (Liu et al. 2014). Furthermore, it affects the accumulation of senescent cells related to ageing and age-related diseases. Activation of SIRT3 delays senescence and SASP-induced inflammation. EGCG possesses anti-inflammatory effects by activating SIRT3 and reducing IL-6 in 3T3-L1 preadipocytes (Lilja et al. 2020). It has been reported that EGCG's anti-inflammatory effects mostly depend on in vitro and in vivo animal studies (Lu and Yen 2015). However, green tea can demonstrate the prevention of RA, osteoarthritis, cardiovascular diseases, cancer, neurodegenerative disease, and metabolic syndrome in association with the main component, EGCG (Afzal et al. 2015).

11.3.8 Prebiotics and Probiotics

Firmicutes, Bacteroidetes, Proteobacteria, and Actinobacteria are the most abundant phyla of gut microbiota in humans. Gram-positive probiotic bacteria, lactic acid bacteria, and *Bifidobacteria* are in Firmicutes and Actinobacteria, respectively. In contrast, the phyla Bacteroidetes and Proteobacteria include Gram-negative bacteria that have LPS induce pro-inflammatory macrophage activity and hence cause infection or diseases under certain conditions (Wang et al. 2020). Dysbiosis is generally related to impairment of the gut barrier function, so LPS moves through leaky tight junctions into circulation. Also, LPS activates NF-κβ and AP-1 due to binding to TLR4. As a result of dysbiosis, types of secondary bile acids that are converted from primary bile acids in gut are also changed. Therefore, the pro-inflammatory farnesoid X receptor (FXR) signalling pathway is activated in enterocytes and adipocytes.

In addition, it has been suggested that IL-6, TNF-α, and CRP levels are linked with specific gut microbial species (Bander et al. 2020). A meta-analysis of randomized clinical trials demonstrated that CRP and NO levels are lowered by probiotic supplementation (Tabrizi et al. 2019). Another meta-analysis of randomized clinical trials indicated that hs-CRP, TNF-a, IL-6, IL-12, and IL-4 levels decrease as a result of probiotic supplementation (Milajerdi et al. 2020). Moreover, it has been reported that probiotics ameliorate the immune response during immunosenesence and mitigate cellular senescence in immune cells (Sharma and Padwad 2020).

Acetate, propionate, and butyrate are short-chain fatty acids that are known as products of the gut microbiota. Dietary fibre and resistant starches are prebiotics fermented by gut microbiota. Primarily, butyrate is the most effective SCFA associated with the inflammatory process. There are several anti-inflammatory mechanisms of action of butyrate. Firstly, butyrate minimizes transfer of LPS from the gut to circulation. Secondly, it binds to G-protein-coupled receptor 43 (GPR43), an SCFA receptor that regulates inflammatory signals and is expressed in macrophages. Thirdly, butyrate inhibits NF-kB, IFN-γ, and PPAR-γ (Lescheid 2014; Wang et al. 2020).

It is suggested that butyrate is mainly a product of Ruminococcaceae, Eubacterium, Clostridia, and Firmicutes (Ohira et al. 2017). It has been observed that altered gut microbiota (in other words, dysbiosis) is related to systemic inflammation and thus diabetes, obesity, cardiovascular diseases, cancer, and inflammatory bowel diseases (Boulangé et al. 2016; Lescheid, 2014; Wang et al. 2020). Prebiotics and probiotics ameliorate or prevent these diseases via the mechanisms of action mentioned. However, standardization of the methodology and biomarkers are needed to clarify the link between prebiotics, probiotics, the gut microbiota, inflammation, and morbidity.

11.4 Conclusion

In conclusion, fruits, vegetables, olive oil, fish oil, whole grains, legumes, nuts, flavonoid-rich green tea, carotenoids, omega-3 fatty acids, and fibre demonstrate anti-inflammatory effects. Basically, they inhibit TNF-α, IL-1, and IL-6 production by suppressing NF-κB, and they inhibit iNOS and COX activity. Also, they exhibit anti-inflammatory effects via modification of the gut microbiota, reduced accelerated senescence of cells, and reduced immunosenesence. According to these mechanisms, these foods promote the prevention of chronic diseases such as diabetes, cardiovascular diseases, cancer, and neurodegenerative diseases, which underpin inflammageing and ageing. However, the studies have mostly been in vitro and animal model studies. Thus, there is a need for more prospective and clinical studies on anti-inflammatory food components and food. Moreover, it is important to consider the dietary pattern to evaluate anti-inflammatory effects of foods from a wider perspective. Even though there are scarce human studies on benefits of these potential anti-inflammatory effects in ageing and longevity, adopting a Mediterranean dietary pattern and recommended consumption amounts can be suggested.

Compliance with Ethical Standards

Conflict of Interest All authors declare they have no conflict of interest /or/ I have no conflict of interest.

References

Afzal M, Safer A, Menon M (2015) Green tea polyphenols and their potential role in health and disease. Inflammopharmacology 23:151–161. https://doi.org/10.1007/s10787-015-0236-1

Ai X, Qin Y, Liu H et al (2017) Apigenin inhibits colonic inflammation and tumorigenesis by suppressing STAT3-NF-κB signaling. Oncotarget 8:100216–100226. https://doi.org/10.18632/oncotarget.22145

Akbar U, Yang M, Kurian D et al (2017) Omega-3 fatty acids in rheumatic diseases. J Clin Rheumatol 23:330–339. https://doi.org/10.1097/rhu.0000000000000563

Albracht-Schulte K, Kalupahana N, Ramalingam L et al (2018) Omega-3 fatty acids in obesity and metabolic syndrome: a mechanistic update. J Nutr Biochem 58:1–16. https://doi.org/10.1016/j.jnutbio.2018.02.012

Arango D, Mayra Diosa-Toro M (2015) Rojas-Hernandez LS (2015) Dietary apigenin reduces LPS-induced expression of miR-155 restoring immune balance during inflammation. Mol Nutr Food Res 59:763–772. https://doi.org/10.1002/mnfr.201400705

Armoza A, Haim Y, Basiri A et al (2013) Tomato extract and the carotenoids lycopene and lutein improve endothelial function and attenuate inflammatory NF-κB signaling in endothelial cells. J Hypertens 31:521–529. https://doi.org/10.1097/hjh.0b013e32835c1d01

Arscott SA (2013) Food sources of carotenoids. In: Tanumihardjo SA (ed) Carotenoids and human health, 1st edn. Springer, New York, pp 3–19

Awika J, Rose D, Simsek S (2018) Complementary effects of cereal and pulse polyphenols and dietary fiber on chronic inflammation and gut health. Food Funct 9:1389–1409. https://doi.org/10.1039/c7fo02011b

Aziz N, Kim MY, Cho JY (2018) Anti-inflammatory effects of luteolin: a review of in vitro, in vivo, and in silico studies. J Ethnopharmacol 225:342–358. https://doi.org/10.1016/j.jep.2018.05.019

Banez M, Geluz M, Chandra A et al (2020) A systemic review on the antioxidant and anti-inflammatory effects of resveratrol, curcumin, and dietary nitric oxide supplementation on human cardiovascular health. Nutr Res 78:11–26. https://doi.org/10.1016/j.nutres.2020.03.002

Bander ZA, Nitert M, Mousa A et al (2020) The gut microbiota and inflammation: an overview. Int J Environ Res Public Health 17:7618. https://doi.org/10.3390/ijerph17207618

Barbaresko J, Koch M, Schulze MB (2013) Dietary pattern analysis and biomarkers of low-grade inflammation: a systematic literature review. Nutr Rev 71(8):511–527. https://doi.org/10.1111/nure.12035

Bolling B, Chen C, McKay D et al (2011) Tree nut phytochemicals: composition, antioxidant capacity, bioactivity, impact factors. A systematic review of almonds, Brazils, cashews, hazelnuts, macadamias, pecans, pine nuts, pistachios and walnuts. Nutr Res Rev 24:244–275. https://doi.org/10.1017/s095442241100014x

Boudjou S, Oomah B, Zaidi F et al (2013) Phenolics content and antioxidant and anti-inflammatory activities of legume fractions. Food Chem 138:1543–1550. https://doi.org/10.1016/j.foodchem.2012.11.108

Boulangé C, Neves A, Chilloux J et al (2016) Impact of the gut microbiota on inflammation, obesity, and metabolic disease. Genome Med 8:42. https://doi.org/10.1186/s13073-016-0303-2

Cai J, Jing D, Shi M et al (2014) Epigallocatechin gallate (EGCG) attenuates infrasound-induced neuronal impairment by inhibiting microglia-mediated inflammation. J Nutr Biochem 25:716–725. https://doi.org/10.1016/j.jnutbio.2014.02.012

Calder PC (2010) Omega-3 fatty acids and inflammatory processes. Nutrients 2:355–374. https://doi.org/10.3390/nu2030355

Calder PC (2015) Marine omega-3 fatty acids and inflammatory processes: effects, mechanisms and clinical relevance. Biochim Biophys Acta Mol Cell Biol Lipids 1851:469–484. https://doi.org/10.1016/j.bbalip.2014.08.010

Calder PC (2012) Omega-3 polyunsaturated fatty acids and inflammatory processes: nutrition or pharmacology? Br J Clin Pharmacol 75:645–662. https://doi.org/10.1111/j.1365-2125.2012.04374.x

Calder PC, Ahluwalia N, Brouns F et al (2011) Dietary factors and low-grade inflammation in relation to overweight and obesity. Br J Nutr 106:S5-78. https://doi.org/10.1017/S0007114511005460

Calder PC, Bosco N, Bourdet-Sicard R et al (2017) Health relevance of the modification of low grade inflammation in ageing (inflammageing) and the role of nutrition. Ageing Res Rev 40:95–119. https://doi.org/10.1016/j.arr.2017.09.001

Candela CG, Lopez LMB, Kohen VL (2011) Importance of a balanced omega 6/omega 3 ratio for the maintenance of health nutritional recommendations. Nutr Hosp 26(2):323–329. https://doi.org/10.3305/nh.2011.26.2.5117

Cárdeno A, Sánchez-Hidalgo M, Aparicio-Soto M et al (2014a) Extra virgin olive oil polyphenolic extracts downregulate inflammatory responses in LPS-activated murine peritoneal macrophages suppressing NFκB and MAPK signalling pathways. Food Funct 5:1270–1277. https://doi.org/10.1039/c4fo00014e

Cárdeno A, Sanchez-Hidalgo M, Aparicio-Soto M, Alarcón-de-la-Lastra C (2014b) Unsaponifiable fraction from extra virgin olive oil inhibits the inflammatory response in LPS-activated murine macrophages. Food Chem 147:117–123. https://doi.org/10.1016/j.foodchem.2013.09.117

Calle MC, Andersen CJ (2019) Assessment of dietary patterns represents a potential, yet variable, measure of inflammatory status: a review and update. Dis Markers https://doi.org/10.1155/2019/3102870

Casas R, Emilio S, Ramon Estruch R (2014) The immune protective effect of the mediterranean diet against chronic low-grade inflammatory diseases. Endocr Metab Immune Disord Drug Targets 14:245–254

Cevenini E, Monti D, Franceschi C (2013) Inflamm-ageing. Curr Opin Clin Nutr Metab Care 16(1):14–20. https://doi.org/10.1097/MCO.0b013e32835ada13

Choi JS, Islam N, Ali Y et al (2014) Effects of C-glycosylation on anti-diabetic, anti-Alzheimer's disease and anti-inflammatory potential of apigenin. Food Chem Toxicol 64:27–33. https://doi.org/10.1016/j.fct.2013.11.020

Chung H, Kim D, Lee E et al (2021) Redefining chronic inflammation in aging and age-related diseases: proposal of the senoinflammation concept. Aging Dis 10(2):367–382. https://doi.org/10.14336/AD.2018.0324

Cicerale S, Lucas LJ, Keast RSJ (2012) Oleocanthal: a naturally occurring anti-inflammatory agent in virgin olive oil. In: Boskou D (ed) Olive oil-constituents, quality, health properties and bioconversions, 1st edn. IntechOpen, Rijeka, pp 357–374

Ciccone M, Cortese F, Gesualdo M et al (2013) Dietary intake of carotenoids and their antioxidant and anti-inflammatory effects in cardiovascular care. Mediators Inflamm 2013:1–11. https://doi.org/10.1155/2013/782137

Corley J, Kyle JAM, Starr JM (2015) Dietary factors and biomarkers of systemic inflammation in older people: the Lothian Birth Cohort 1936. Br J Nutr 114(7):1188–1198. https://doi.org/10.1017/S000711451500210X

Csiszar A (2011) Anti-inflammatory effects of resveratrol: possible role in prevention of age-related cardiovascular disease. Ann N Y Acad Sci 1215:117–122. https://doi.org/10.1111/j.1749-6632.2010.05848.x

Dang Y, Li Z, Wei Q et al (2018) Protective effect of apigenin on acrylonitrile-induced inflammation and apoptosis in testicular cells via the NF-κB pathway in rats. Inflammation 41:1448–1459. https://doi.org/10.1007/s10753-018-0791-x

de Sá Coutinho D, Pacheco M, Frozza R, Bernardi A (2018) Anti-inflammatory effects of resveratrol: mechanistic insights. Int J Mol Sci 19:1812. https://doi.org/10.3390/ijms19061812

Di Tomo P, Canali R, Ciavardelli D et al (2012) β-Carotene and lycopene affect endothelial response to TNF-α reducing nitro-oxidative stress and interaction with monocytes. Mol Nutr Food Res 56:217–227. https://doi.org/10.1002/mnfr.201100500

Duarte S, Arango D, Parihar A et al (2013) Apigenin protects endothelial cells from lipopolysaccharide (LPS)-induced inflammation by decreasing caspase-3 activation and modulating mitochondrial function. Int J Mol Sci 14:17664–17679. https://doi.org/10.3390/ijms140917664

Ekmekcioğlu C (2020) Nutrition and longevity – from mechanisms to uncertainties. Crit Rev Food Sci Nutr 60(18):3063–3082. https://doi.org/10.1080/10408398.2019.1676698

Ellulu M, Khaza'ai H, Abed Y et al (2015) Role of fish oil in human health and possible mechanism to reduce the inflammation. Inflammopharmacology 23:79–89. https://doi.org/10.1007/s10787-015-0228-1

Estruch R (2010) Anti-inflammatory effects of the Mediterranean diet: the experience of the PREDIMED study. Proc Nutr Soc 69:333–340. https://doi.org/10.1017/S0029665110001539

Fechtner S, Singh A, Chourasia M et al (2017) Molecular insights into the differences in anti-inflammatory activities of green tea catechins on IL-1β signaling in rheumatoid arthritis synovial fibroblasts. Toxicol Appl Pharmacol 329:112–120. https://doi.org/10.1016/j.taap.2017.05.016

Feng D, Ling WH, Duan RD (2010) Lycopene suppresses LPS-induced NO and IL-6 production by inhibiting the activation of ERK, p38MAPK, and NF-κB in macrophages. Inflamm Res 59:115–121. https://doi.org/10.1007/s00011-009-0077-8

Fenni S, Hammou H, Astier J et al (2017) Lycopene and tomato powder supplementation similarly inhibit high-fat diet induced obesity, inflammatory response, and associated metabolic disorders. Mol Nutr Food Res 61:1601083. https://doi.org/10.1002/mnfr.201601083

Fernandes J, Fialho M, Santos R et al (2020) Is olive oil good for you? A systematic review and meta-analysis on anti-inflammatory benefits from regular dietary intake. Nutrition 69:110559. https://doi.org/10.1016/j.nut.2019.110559

Funakoshi-Tago M, Nakamura K, Tago K et al (2011) Anti-inflammatory activity of structurally related flavonoids, apigenin, luteolin and fisetin. Int Immunopharmacol 11:1150–1159. https://doi.org/10.1016/j.intimp.2011.03.012

Furman D, Campisi J, Verdin E et al (2019) Chronic inflammation in the etiology of disease across the lifespan. Nat Med 25(12):1822–1832. https://doi.org/10.1038/s41591-019-0675-0

Fougère B, Boulanger B, Nourhashémi F et al (2017) Chronic inflammation: accelerator of biological aging. J Gerontol A Biol Sci Med Sci 72(9):1218–1225. https://doi.org/10.1093/gerona/glw240

Franceschi C, Garagnani P, Parini P et al (2018) Inflammaging: a new immune–metabolic viewpoint for age-related diseases. Nat Rev Endocrinol 14(10):576–590. https://doi.org/10.1038/s41574-018-0059-4

Galland L (2010) Diet and inflammation. Nutr Clin Pract 25:634–640. https://doi.org/10.1177/0884533610385703

Ghanbari R, Anwar F, Alkharfy K et al (2012) Valuable nutrients and functional bioactives in different parts of olive (Olea europaea L.)-a review. Int J Mol Sci 13:3291–3340. https://doi.org/10.3390/ijms13033291

Gopinath B, Buyken A, Flood V et al (2011) Consumption of polyunsaturated fatty acids, fish, and nuts and risk of inflammatory disease mortality. Am J Clin Nutr 93:1073–1079. https://doi.org/10.3945/ajcn.110.009977

Griffithd K, Aggarwal BB, Singh RB et al (2016) Food antioxidants and their anti-inflammatory properties: a potential role in cardiovascular diseases and cancer prevention. Diseases 4(3):28. https://doi.org/10.3390/diseases4030028

Guarner V, Rubio-Ruiz ME (2015) Low-grade systemic inflammation connects aging, metabolic syndrome and cardiovascular disease. In: Yashin AI, Jazwinski SM (eds) Aging and health – a systems biology perspective. Basel, Karger. Interdiscipl Top Gerontol vol 40, pp 99–106

Hajihashemi P, Haghighatdoost F (2018) Effects of whole-grain consumption on selected biomarkers of systematic inflammation: a systematic review and meta-analysis of randomized controlled trials. J Am Coll Nutr 38:275–285. https://doi.org/10.1080/07315724.2018.1490935

Hayasaka N, Shimizu N, Komoda T et al (2018) Absorption and metabolism of luteolin in rats and humans in relation to in vitro anti-inflammatory effects. J Agric Food Chem 66:11320–11329. https://doi.org/10.1021/acs.jafc.8b03273

He Q, Zhou W, Xiong C et al (2015) Lycopene attenuates inflammation and apoptosis in post-myocardial infarction remodeling by inhibiting the nuclear factor-κB signaling pathway. Mol Med Rep 11:374–378. https://doi.org/10.3892/mmr.2014.2676

Honarvar NM, Saedisomeolia A, Abdolahi M et al (2017) Molecular anti-inflammatory mechanisms of retinoids and carotenoids in Alzheimer's disease: a review of current evidence. J Mol Neurosci 61:289–304. https://doi.org/10.1007/s12031-016-0857-x

Jayarathne S, Koboziev I, Park OH et al (2017) Anti-inflammatory and anti-obesity properties of food bioactive components: effects on adipose tissue. Prev Nutr Food Sci 22(4):251–262. https://doi.org/10.3746/pnf.2017.22.4.251

Joseph SV, Edirisinghe I, Burton-Freeman FM (2014) Berries: anti-inflammatory effects in humans. J Agric Food Chem 62:3886–3903. https://doi.org/10.1021/jf4044056

Joseph S, Edirisinghe I, Burton-Freeman B (2016) Fruit polyphenols: a review of anti-inflammatory effects in humans. Crit Rev Food Sci Nutr 56:419–444. https://doi.org/10.1080/10408398.2013.767221

Kaulmann A, Bohn T (2014) Carotenoids, inflammation, and oxidative stress—implications of cellular signaling pathways and relation to chronic disease prevention. Nutr Res 34:907–929. https://doi.org/10.1016/j.nutres.2014.07.010

Kawata A, Murakami Y, Suzuki S (2018) Anti-inflammatory activity of β-carotene, lycopene and tri-n-butylborane, a scavenger of reactive oxygen species. In Vivo 32(2):255–264. https://doi.org/10.21873/invivo.11232

Kijlstra A, Tian Y, Kelly E et al (2012) Lutein: more than just a filter for blue light. Prog Retin Eye Res 31:303–315. https://doi.org/10.1016/j.preteyeres.2012.03.002

Killeen M, Linder M, Pontoniere P et al (2014) NF-κβ signaling and chronic inflammatory diseases: exploring the potential of natural products to drive new therapeutic opportunities. Drug Discover Today 19:373–378. https://doi.org/10.1016/j.drudis.2013.11.002

Kim C, Park M, Kim S et al (2014) Antioxidant capacity and anti-inflammatory activity of lycopene in watermelon. Int J Food SciTechnol 49:2083–2091. https://doi.org/10.1111/ijfs.12517

Kleemann R, Verschuren L, Morrison M et al (2011) Anti-inflammatory, anti-proliferative and anti-atherosclerotic effects of quercetin in human in vitro and in vivo models. Atherosclerosis 218:44–52. https://doi.org/10.1016/j.atherosclerosis.2011.04.023

Kourtgen A, Bauer M (2018) Biomarkers in inflammation. In: Cavaillon JM, Mervyn Singer M (eds) Inflammation: from molecular and cellular mechanisms to the clinic, 1st edn. Wiley-VCH, Weinheim, pp 1541–1566

Layé S, Nadjar A, Joffre C et al (2017) Anti-inflammatory effects of omega-3 fatty acids in the brain: physiological mechanisms and relevance to pharmacology. Pharmacol Rev 70:12–38. https://doi.org/10.1124/pr.117.014092

Lee Y, Han S, Song B et al (2015) Bioactives in commonly consumed cereal grains: implications for oxidative stress and inflammation. J Med Food 18:1179–1186. https://doi.org/10.1089/jmf.2014.3394

Lee W, Ku SK, Bae JW et al (2012) Inhibitory effects of lycopene on HMGB1-mediated pro-inflammatory responses in both cellular and animal models. Food Chem Toxicol 50:1826–1833. https://doi.org/10.1016/j.fct.2012.03.003

Lefevre M, Jonnalagadda S (2012) Effect of whole grains on markers of subclinical inflammation. Nutr Rev 70:387–396. https://doi.org/10.1111/j.1753-4887.2012.00487.x

Lescheid DW (2014) Probiotics as regulator of inflammation: a review. Funct Food Health Dis 4(7):299–311. https://doi.org/10.31989/ffhd.v4i7.2

Lesjak M, Beara I, Simin N et al (2018) Antioxidant and anti-inflammatory activities of quercetin and its derivatives. J Funct Foods 40:68–75. https://doi.org/10.1016/j.jff.2017.10.047

Lilja S, Oldenburg J, Pointner A et al (2020) Epigallocatechin gallate effectively affects senescence and anti-SASP via SIRT3 in 3T3-L1 preadipocytes in comparison with other bioactive substances. Oxid Med Cell Longev 2020:4793125. https://doi.org/10.1155/2020/4793125

Lin H, Chang T, Yang D et al (2012) Regulation of virus-induced inflammatory response by β-carotene in RAW264.7 cells. Food Chem 134:2169–2175. https://doi.org/10.1016/j.foodchem.2012.04.024

Li R, Hong P, Zheng X (2018) β-carotene attenuates lipopolysaccharide-induced inflammation via inhibition of the NF-κB, JAK2/STAT3 and JNK/p38 MAPK signaling pathways in macrophages. Anim Sci J 90:140–148. https://doi.org/10.1111/asj.13108

Li S, Fung F, Fu Z et al (2012) Anti-inflammatory effects of lutein in retinal ischemic/hypoxic injury: in vivo and in vitro studies. Invest Ophthamol Vis Sci 53:5976. https://doi.org/10.1167/iovs.12-10007

Liu D, Perkins J, Hennig B (2016) EGCG prevents PCB-126-induced endothelial cell inflammation via epigenetic modifications of NF-κB target genes in human endothelial cells. J Nutr Biochem 28:164–170. https://doi.org/10.1016/j.jnutbio.2015.10.003

Liu Q, Qian Y, Chen F et al (2014) EGCG attenuates pro-inflammatory cytokines and chemokines production in LPS-stimulated L02 hepatocyte. Acta Biochim Biophys Sin 46:31–39. https://doi.org/10.1093/abbs/gmt128

Li Y, Yao J, Han C et al (2016) Quercetin, inflammation and immunity. Nutrients 8:167. https://doi.org/10.3390/nu8030167

Lu CC, Yen GC (2015) Antioxidant and anti-inflammatory activity of functional foods. Curr Opin Food Sci 2:1–8. https://doi.org/10.1016/j.cofs.2014.11.002

Lucas L, Russell A, Keast R (2011) Molecular mechanisms of inflammation. anti-inflammatory benefits of virgin olive oil and the phenolic compound oleocanthal. Curr Pharm Des 17:754–768. https://doi.org/10.2174/138161211795428911

Ma C, Wang Y, Dong L et al (2015) Anti-inflammatory effect of resveratrol through the suppression of NF- B and JAK/STAT signaling pathways. Acta Biochim Biophys Sin 47:207–213. https://doi.org/10.1093/abbs/gmu135

Ma C, Sun Z, Zeng Y et al (2018) Molecular mechanism and health role of functional ingredients in blueberry for chronic disease in human beings. Int J Mol Sci 19:2785. https://doi.org/10.3390/ijms19092785

Marcason W (2010) J Am Diet Assoc 110(11):1780. https://doi.org/10.1016/j.jada.2010.09.024

Marcotorchino J, Romier B, Gouranton E et al (2012) Lycopene attenuates LPS-induced TNF-α secretion in macrophages and inflammatory markers in adipocytes exposed to macrophage-conditioned media. Mol Nutr Food Res 56(5):725–732. https://doi.org/10.1002/mnfr.201100623

Mazidi M, Rezaie P, Ferns G et al (2016) Impact of different types of tree nut, peanut, and soy nut consumption on serum C-reactive protein (CRP). Medicine 95:e5165. https://doi.org/10.1097/md.0000000000005165

McEneny J, Wade L, Young I et al (2013) Lycopene intervention reduces inflammation and improves HDL functionality in moderately overweight middle-aged individuals. J Nutr Biochem 24:163–168. https://doi.org/10.1016/j.jnutbio.2012.03.015

Milajerdi A, Mousavi S, Sadeghi A et al (2020) The effect of probiotics on inflammatory biomarkers: a meta-analysis of randomized clinical trials. Eur J Nutr 59:633–649. https://doi.org/10.1007/s00394-019-01931-8

Mocellin M, Camargo C, Nunes E et al (2016) A systematic review and meta-analysis of the n-3 polyunsaturated fatty acids effects on inflammatory markers in colorectal cancer. Clin Nutr 35:359–369. https://doi.org/10.1016/j.clnu.2015.04.013

Monti D, Ostan R, Borelli V et al (2017) Inflammaging and omics in human longevity. Mech Ageing Dev 165(Pt B):129–138. https://doi.org/10.1016/j.mad.2016.12.008

Morais CA, de Rosso VV, Estadella D et al (2016) Anthocyanins as inflammatory modulators and the role of the gut microbiota. J Nutr Biochem 33:1–7. https://doi.org/10.1016/j.jnutbio.2015.11.008

Nabavi SF, Braidy N, Gortzi O et al (2015) Luteolin as an anti-inflammatory and neuroprotective agent: a brief review. Brain Res Bull 119:1–11. https://doi.org/10.1016/j.brainresbull.2015.09.002
Natto Z, Yaghmoor W, Alshaeri H et al (2019) Omega-3 fatty acids effects on inflammatory biomarkers and lipid profiles among diabetic and cardiovascular disease patients: a systematic review and meta-analysis. Sci Rep 9:18867. https://doi.org/10.1038/s41598-019-54535-x
Neale EP, Batterham MJ, Tapsell LC (2016) Consumption of a healthy dietary pattern results in significant reductions in C-reactive protein levels in adults: a meta-analysis. Nutr Res 36:391–401. https://doi.org/10.1016/j.nutres.2016.02.009
Neale E, Tapsell L, Guan V et al (2017) The effect of nut consumption on markers of inflammation and endothelial function: a systematic review and meta-analysis of randomised controlled trials. BMJ Open 7:e016863. https://doi.org/10.1136/bmjopen-2017-016863
Oh J, Kim J, Park J et al (2013) Radical scavenging activity-based and AP-1-targeted anti-inflammatory effects of lutein in macrophage-like and skin keratinocytic cells. Mediators Inflamm 2013:787042. https://doi.org/10.1155/2013/787042
Ohira H, Tsutsui W, Fujioka Y (2017) Are short chain fatty acids in gut microbiota defensive players for inflammation and atherosclerosis? J Atheroscler Thromb 24:660–672. https://doi.org/10.5551/jat.rv17006
Ohishi T, Goto S, Monira P et al (2016) Anti-inflammatory action of green tea. Antiinflamm Antiallergy Agents Med Chem 15:74–90. https://doi.org/10.2174/1871523015666160915154443
Oppedisano F, Macrì R, Gliozzi M et al (2020) The anti-inflammatory and antioxidant properties of n-3 PUFAs: their role in cardiovascular protection. Biomedicines 8:306. https://doi.org/10.3390/biomedicines8090306
Oz AT, Kafkas E (2017) Phytochemicals in fruits and vegetables. In: Waisundara VY, Shiomi N (eds) Superfood and functional food - an overview of their processing and utilization, 1st edn. IntechOpen, Rijeka, pp 175–184
Parhiz H, Roohbakhsh A, Sultani F et al (2015) Antioxidant and anti-inflammatory properties of the citrus flavonoids hesperidin and hesperetin: an updated review of their molecular mechanisms and experimental models. Phytother Res 29(3):323–331. https://doi.org/10.1002/ptr.5256
Parkinson L, Cicerale S (2016) The health benefiting mechanisms of virgin olive oil phenolic compounds. Molecules 21:1734. https://doi.org/10.3390/molecules21121734
Rangel-Huerta O, Aguilera C, Mesa M et al (2012) Omega-3 long-chain polyunsaturated fatty acids supplementation on inflammatory biomakers: a systematic review of randomised clinical trials. Br J Nutr 107:S159–S170. https://doi.org/10.1017/s0007114512001559
Rebello C, Greenway F, Finley J (2014) A review of the nutritional value of legumes and their effects on obesity and its related co-morbidities. Obes Rev 15:392–407. https://doi.org/10.1111/obr.12144
Reyes-Díaz A, Del-Toro-Sánchez C, Rodríguez-Figueroa J et al (2019) Legume proteins as a promising source of anti-inflammatory peptides. Curr Protein Pept Sci 20:1204–1217. https://doi.org/10.2174/1389203720666190430110647
Reygaert W (2017) An update on the health benefits of green tea. Beverages 3:6. https://doi.org/10.3390/beverages3010006
Riegsecker S, Wiczynski D, Kaplan M et al (2013) Potential benefits of green tea polyphenol EGCG in the prevention and treatment of vascular inflammation in rheumatoid arthritis. Life Sci 93:307–312. https://doi.org/10.1016/j.lfs.2013.07.006
Ros E (2015) Contribution of nuts to the Mediterranean diet. In: Preedy VR, Watson RR (eds) The Mediterranean diet an evidence based approach, 1st edn. Academic Press, Massachusetts, pp 175–184
Rosignoli P, Fuccelli R, Fabiani R et al (2013) Effect of olive oil phenols on the production of inflammatory mediators in freshly isolated human monocytes. J Nutr Biochem 24:1513–1519. https://doi.org/10.1016/j.jnutbio.2012.12.011
Qiao Y, Jiang P, Gao Y (2018) Lutein prevents osteoarthritis through Nrf2 activation and down-regulation of inflammation. Arch Med Sci 14:617–624. https://doi.org/10.5114/aoms.2016.59871

Salehi-Abargouei A, Saraf-Bank S, Bellissimo N et al (2015) Effects of non-soy legume consumption on C-reactive protein: a systematic review and meta-analysis. Nutrition 31:631–639. https://doi.org/10.1016/j.nut.2014.10.018
Sanada F, Taniyama Y, Muratsu J et al (2018) Source of chronic inflammation in aging. Front Cardiovasc Med 5:12. https://doi.org/10.3389/fcvm.2018.00012
Sang S, Idehen E, Zhao Y et al (2020) Emerging science on whole grain intake and inflammation. Nutr Rev 78:21–28. https://doi.org/10.1093/nutrit/nuz079
Santangelo C, Varì R, Scazzocchio B et al (2017) Anti-inflammatory activity of extra virgin olive oil polyphenols: which role in the prevention and treatment of immune-mediated inflammatory diseases? Endocr Metab Immune Disord Drug Targets 18:36–50. https://doi.org/10.2174/1871530317666171114114321
Serventi L, Dsouza L (2020) Bioactives in legumes. In: Serventi L (ed) Upcycling legume water: from wastewater to food ingredients, 1st edn. Springer, New York, pp 139–153
Sharma R, Padwad Y (2020) Nutraceuticals-based immunotherapeutic concepts and opportunities for the mitigation of cellular senescence and aging: a narrative review. Ageing Res Rev 63:101141. https://doi.org/10.1016/j.arr.2020.101141
Souza P, Marcadenti A, Portal V (2017) Effects of olive oil phenolic compounds on inflammation in the prevention and treatment of coronary artery disease. Nutrients 9:1087. https://doi.org/10.3390/nu9101087
Tabrizi R, Ostadmohammadi V, Akbari M et al (2019) The effects of probiotic supplementation on clinical symptom, weight loss, glycemic control, lipid and hormonal profiles, biomarkers of inflammation, and oxidative stress in women with polycystic ovary syndrome: a systematic review and meta-analysis of randomized controlled trials. Probiotics Antimicrob Proteins. https://doi.org/10.1007/s12602-019-09559-0
Tejada S, Pinya S, Martorell M et al (2018) Potential anti-inflammatory effects of hesperidin from the genus *citrus*. Curr Med Chem 25:4929–4945. https://doi.org/10.2174/0929867324666170718104412
Testai L, Piragine E, Piano I et al (2020) The citrus flavonoid naringenin protects the myocardium from ageing-dependent dysfunction: potential role of SIRT1. Oxid Med Cell Longev 2020:4650207. https://doi.org/10.1155/2020/4650207
Tsai YF, Chen YR, Chen JP et al (2019) Effect of hesperidin on anti-inflammation and cellular antioxidant capacity in hydrogen peroxide-stimulated human articular chondrocytes. Process Biochem 85:175–184. https://doi.org/10.1016/j.procbio.2019.07.014
Tucker LA (2017) Consumption of nuts and seeds and telomere length in 5,582 men and women of the national health and nutrition examination survey (NHANES). J Nutr Health Aging 21(3):233–240. https://doi.org/10.1007/s12603-017-0876-5
Vendrame S, Klimis-Zacas D (2015) Anti-inflammatory effect of anthocyanins via modulation of nuclear factor-κB and mitogen-activated protein kinase signaling cascades. Nutr Rev 73(6):348–358. https://doi.org/10.1093/nutrit/nuu066
Virruso C, Accardi G, Colonna-Romano G et al (2014) Nutraceutical properties of extra-virgin olive oil: a natural remedy for age-related disease? Rejuvenation Res 17:217–220. https://doi.org/10.1089/rej.2013.1532
Wang J, Chen W, Wang Y (2020) The relationship between gut microbiota and inflammatory diseases: the role of macrophages. Front Microbiol 11:1065. https://doi.org/10.3389/fmicb.2020.01065
Wang J, Liu Y, Xiao L et al (2014) Anti-inflammatory effects of apigenin in lipopolysaccharide-induced inflammatory in acute lung injury by suppressing COX-2 and NF-kB pathway. Inflammation 37:2085–2090. https://doi.org/10.1007/s10753-014-9942-x
Wu W, Li Y, Wu Y et al (2015) Lutein suppresses inflammatory responses through Nrf2 activation and NF-κB inactivation in lipopolysaccharide-stimulated BV-2 microglia. Mol Nutr Food Res 59:1663–1673. https://doi.org/10.1002/mnfr.201500109
Wu X, Schauss AG (2012) Mitigation of inflammation with foods. J Agric Food Chem 60:6703–6717. https://doi.org/10.1021/jf3007008

Xie Y, Liang D, Wu Q et al (2019) A system-level investigation into the mechanisms of apigenin against inflammation. Nat Prod Commun 14:1934578X1987860. https://doi.org/10.1177/1934578x19878600

Xiao S, Liu W, Bi J et al (2018) Anti-inflammatory effect of hesperidin enhances chondrogenesis of human mesenchymal stem cells for cartilage tissue repair. J Inflamm 15:14. https://doi.org/10.1186/s12950-018-0190-y

Xiao Y, Xia J, Ke Y et al (2018) Effects of nut consumption on selected inflammatory markers: a systematic review and meta-analysis of randomized controlled trials. Nutrition 54:129–143. https://doi.org/10.1016/j.nut.2018.02.017

Xu Y, Wan Q, Feng J et al (2018) Whole grain diet reduces systemic inflammation. Medicine 97:e12995. https://doi.org/10.1097/md.0000000000012995

Yang J, Han Y, Chen C et al (2013) EGCG attenuates high glucose-induced endothelial cell inflammation by suppression of PKC and NF-κB signaling in human umbilical vein endothelial cells. Life Sci 92:589–597. https://doi.org/10.1016/j.lfs.2013.01.025

Yi L, Ma S, Ren D (2017) Phytochemistry and bioactivity of Citrus flavonoids: a focus on antioxidant, anti-inflammatory, anticancer and cardiovascular protection activities. Phytochem Rev 16:479–511. https://doi.org/10.1007/s11101-017-9497-1

Zhang Y, Gan R, Li S et al (2015) Antioxidant phytochemicals for the prevention and treatment of chronic diseases. Molecules 20:21138–21156. https://doi.org/10.3390/molecules201219753

Zhang X, Wang G, Gurley EC, Zhou H (2014) Flavonoid apigenin inhibits lipopolysaccharide-induced inflammatory response through multiple mechanisms in macrophages. PLoS ONE 9(9):e107072. https://doi.org/10.1371/journal.pone.0107072

Zhu F, Du B, Xu B (2018) Anti-inflammatory effects of phytochemicals from fruits, vegetables, and food legumes: a review. Crit Rev Food Sci Nutr 58(8):1260–1270. https://doi.org/10.1080/10408398.2016.1251390

Wirth MD, Hébert JR, Shivappa N, Hand GA, Hurley TG, Drenowatz C, McMahon D, Shook RP, Blair SN (2016) Anti-inflammatory Dietary Inflammatory Index scores are associated with healthier scores on other dietary indices. Nutr Res 36(3):214–219. https://doi.org/10.1016/j.nutres.2015.11.009

Chapter 12
Nutraceuticals and Functional Foods in Aging and Aging-Associated Diseases

Jatinder Pal Singh, Balwinder Singh, and Amritpal Kaur

Abstract Aging results in a gradual decline in the physiological functions of the body and is often accompanied by aging-associated diseases, which may have a great impact on the quality of life. Allopathic medicines are prescribed to alleviate the symptoms, but their use is associated with plenty of side effects and high cost. Recently, nutraceuticals and functional foods market has boomed and caught attention of the general public owing to their medicinal properties and health benefits, especially in safeguarding and treating chronic diseases. Nutraceuticals can be classified into phenolic compounds, carotenoids, organosulfur compounds, polyunsaturated fatty acids, minerals and vitamins that are useful in decelerating aging process and preventing aging-associated diseases. Nutraceuticals act mainly by boosting the immune system against cellular oxidation, improving the balance of gut flora and lowering blood cholesterol levels. Moreover, they may enhance health span by preventing/slowing down age-associated diseases such as Parkinson's disease, Alzheimer's disease, cardiovascular diseases, type 2 diabetes mellitus, osteoarthritis and cancer.

Keywords Nutraceuticals · Functional foods · Aging · Phenolic compounds · Carotenoids · Polyunsaturated fatty acids · Aging · Parkinsons disease · Alzheimers disease · Cardiovascular disease · Type 2 diabetes · Osteoarthritis · Cancer

12.1 Introduction

Nutraceuticals can be defined as foods or any of their part which can supply medicinal or health benefits, mostly safeguarding and treating diseases (Chiu et al. 2018; Vaclavik et al. 2021), so this term incorporates fortified/functional foods and their

J. P. Singh · A. Kaur (✉)
Department of Food Science and Technology, Guru Nanak Dev University, Amritsar 143005, Punjab, India

B. Singh
P.G. Department of Biotechnology, Khalsa College, Amritsar 143002, Punjab, India

S. I. S. Rattan and G. Kaur (eds.), *Nutrition, Food and Diet in Ageing and Longevity*, Healthy Ageing and Longevity 14, https://doi.org/10.1007/978-3-030-83017-5_12

nutrients (Ghosh et al. 2012). Nutraceuticals include a range of products from isolated nutrients (such as ω-3 fatty acids, β-carotene, gingerol, glucosamine, curcumin, ginseng, lycopene and folic acid), health supplements, herbal products or natural bioactive foods (such as garlic/green tea/soybeans) (Dominguez Diaz et al. 2020). Functional foods maybe described as a category of nutraceuticals as they also provide health benefits apart from nutrition. These foods are categorized as fortified, enriched, altered and enhanced products. Popular functional foods include margarine with incorporated phytosterols, yoghurt containing beneficial bacteria, fortified vitamins, mineral and fiber rich fruit juices (Longoria-García et al. 2018). Nutraceuticals and functional foods may be used to tackle the growing prevalence of chronic diseases, especially aging-associated diseases worldwide (Tang 2020).

Aging can be defined as a natural biological process that is associated with a gradual decline in the physiological functions and an increased vulnerability to diseases and mortality. The major aging-associated diseases include neurological disorders such as Parkinson's and Alzheimer's disease, cardiovascular disease such as atherosclerosis, type 2 diabetes, osteoarthritis and cancer. The conventional therapies for aging-associated diseases usually have many side effects and are costly as well. Since these diseases have been associated with unhealthy dietary habits, foods which impart positive health benefits besides their nutritional values may be an answer to prevent the onset of these diseases.

The rising interest in healthy diets has significantly impacted the direction of research for product development in the areas of nutraceuticals, functional foods and nanotechnology. Therefore, many pharmaceutical and nutritional products-based corporations have brought nutraceuticals and functional foods into the market owing to their property of not only having a nutritional value but also their therapeutic potential (Vaclavik et al. 2021). The health promoting effects of nutraceuticals and functional foods have been comprehensively described by many recent studies in literature (Grumezescu 2016; McWilliams 2017; Rajasekaran 2017; Pandareesh et al. 2018). Taking leads from the aforementioned knowledge, the purpose of this review article is to emphasize the role of nutraceuticals and functional foods in decelerating the aging process and preventing the incidence of aging-associated diseases.

12.2 Classes of Nutraceuticals/Functional Foods, Extraction and Their Delivery

Nutraceuticals and functional foods are categorized according to their chemical constituents into phenolic compounds, carotenoids, organosulfur compounds, polyunsaturated fatty acids, minerals and vitamins that are useful in preventing and alleviating aging-associated diseases. The major nutraceuticals, their sources and health promoting as well as anti-aging effects have been presented in Table 12.1. Phenolic compounds, carotenoids and organosulfur compounds are also termed as phytochemicals/phytonutrients because they are primarily synthesized in plants.

Phenolic compounds are the nutraceuticals and bioactive components of the diet which play an important role in good health and are further categorized as phenolic acids, flavonoids, tannins and lignins (Singh et al. 2016). The important phenolic compounds that act as nutraceuticals are presented in Fig. 12.1. Quercetin, naringenin, apigenin, kaempferol, genistein, cyanidin, epigallocatechin-3-gallate and

Table 12.1 Major nutraceuticals, their sources, and health-promoting as well as anti-aging activities

Category of Nutraceutical	Individual Nutraceutical	Health-promoting and anti-aging functions	Primary sources
Phenolic compounds	Curcumin	Reduces the incidence of neuroinflammation; lowers the levels of C-reactive protein; functions as an antioxidant, chemopreventive, chemosensitizer chondroprotective and anti-inflammatory agent	Turmeric rhizome and curry
	Epigallocatechin gallate	Mimics the anti-aging properties of metformin and rapamycin; has a neuroprotective role; reduces UVR-induced DNA damage and lowers ROS as well as free radical production.	Green tea, white, oolong, and black teas, blackberries
	Resveratrol	Restricts mitochondrial dysfunction and apoptosis; increases the intracellular free-radical scavenger glutathione amounts	Peanuts, pistachios, and grapes
	Genistein	Cardioprotective effects; protects from diabetes and cancer	Soybean and soy-based foods
Carotenoids	Lycopene	Relieves oxidative stress, increases NADH dehydrogenase and superoxide dismutase activity; lowers the number of low-density lipoproteins as well as pro-inflammatory cytokines	Tomatoes, watermelon and grapefruits
	β-Carotene	Lowers free radicals and singlet oxygen-induced lipid peroxidation; scavenges ROS to protect against oxidative stress and decreases UV exposures	Leafy vegetables, carrots, and sweet potatoes.
	Lutein and zeaxanthin	Reduces free radical damage and decreases exposure to UV damage	Green leafy vegetables and egg yolks

(continued)

Table 12.1 (continued)

Category of Nutraceutical	Individual Nutraceutical	Health-promoting and anti-aging functions	Primary sources
Polyunsaturated fatty acids	Eicosapentaenoic acid, α-Linolenic acid and docosahexaenoic acid	Reduce the amounts of triglycerides in the body; anti-inflammatory effects; enhance mitochondrial membrane lipids; lower calcium release and pyruvate dehydrogenase enzyme activity	Salmon and trout fish, flaxseed oil, canola oil and nuts
Minerals	Copper, manganese, and iron	Enzyme co-factors for antioxidant enzymes such as catalase and superoxide dismutase	Nuts, beans and leafy vegetables, red meat
	Zinc	Decreases oxidative damage and inflammation	Lean meat, poultry; eggs, beans, and nuts
	Selenium	Removes lipid hydroperoxides; supports DNA synthesis as well as repair	Meat and Seafood
Vitamins	Vitamin C	Free radical scavenger; stimulates the metabolism of chondrocytes, collagen, and proteoglycan synthesis	Citrus fruits, peppers, and broccoli
	Vitamin E	Has free radical scavenging properties; stabilizes cell membranes, reduces the number of apoptotic cells	Wheat germ oil, sunflower seeds, almonds, and peanuts

catechin are flavonoids (the largest class of phenolic compounds), while caffeic acid, curcumin and resveratrol are non-flavonoid phenolic compounds. Phenolic compounds can regulate the activity of many enzymes and cell receptors, such as in case of cancer they may inhibit enzymes which promote cell division or upregulate enzymes causing cell apoptosis (Rajasekaran 2017). Functional foods such as fruits, vegetables, herbs, nuts, tea and coffee contain adequate levels of these compounds (Singh et al. 2020).

Carotenoids are a vast group of hydrophobic pigments that are naturally produced inplants, algae, bacteria, and fungi. These compounds are carotenes or xanthophylls depending on their chemical structures (Rodriguez-Amaya 2015). The structures of major carotenoids acting as nutraceuticals is illustrated in Fig. 12.2. Lycopene, α-carotene and β-carotene are carotenes, while lutein, zeaxanthin, β-cryptoxanthin, astaxanthin and fucoxanthin are xanthophylls. One of the most important carotenoid pigments is lycopene (red in color) and has been recognized to have potent activities against aging-associated diseases (Chaudhary et al. 2018).

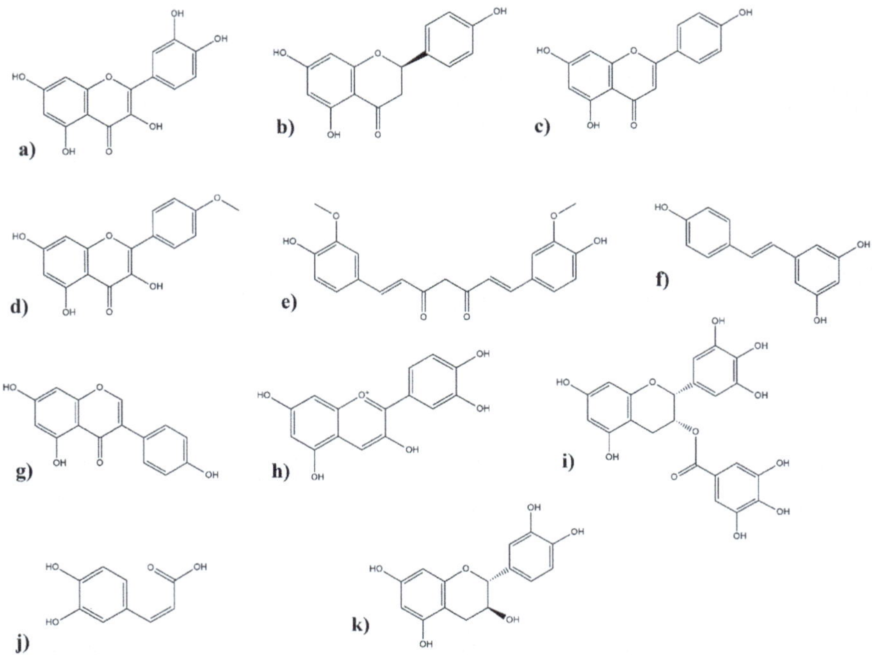

Fig. 12.1 Chemical structures of phenolic compounds having nutraceutical properties. The compounds shown are quercetin (a); naringenin (b); apigenin (c); kaempferol (d); curcumin (e); resveratrol (f); genistein (g); cyanidin (h); epigallocatechin-3-gallate (i); caffeic acid (j) and catechin (k)

Polyunsaturated fatty acids (ω-3 and ω-6 fatty acids) are an essential class of nutraceuticals important for human beings (Falinska et al. 2012). The structures of main polyunsaturated fatty acids are depicted in Fig. 12.3. Among these, linoleic acid, γ-linolenic acid, dihomo γ-linolenic acid, arachidonic acid are ω-6 fatty acids, while α-linolenic acid, eicosapentaenoic acid, docosapentaenoic acid and docosahexaenoic acid are ω-3 fatty acids. ω-3 and ω-6 fatty acids are processed by the identical enzymes and compete with each other for the enzyme active sites. Both ω-3 and ω-6 fatty acids produce compounds called eicosanoids during metabolism. Eicosanoids have been known to have anti-inflammatory roles (Saini and Keum 2018).

Organosulfur compounds are organic compounds containing sulfur in their chemical structures. These compounds are primarily present in cruciferous vegetables (especially broccoli, cauliflower, garlic, and onion). They can be further classified asallylic sulfur compoundssuch as aliin and allicin (present in garlic and onions), glucosinolates, isothiocyanates and indoles (Dwivedi et al. 2020).

Minerals can also act as nutraceuticals and are classified as macrominerals and trace minerals. Macrominerals are required in larger amounts than trace minerals and include calcium, phosphorus, magnesium, sodium, potassium, chloride and sulfur. On the other hand, trace minerals include iron, manganese, copper, iodine, zinc,

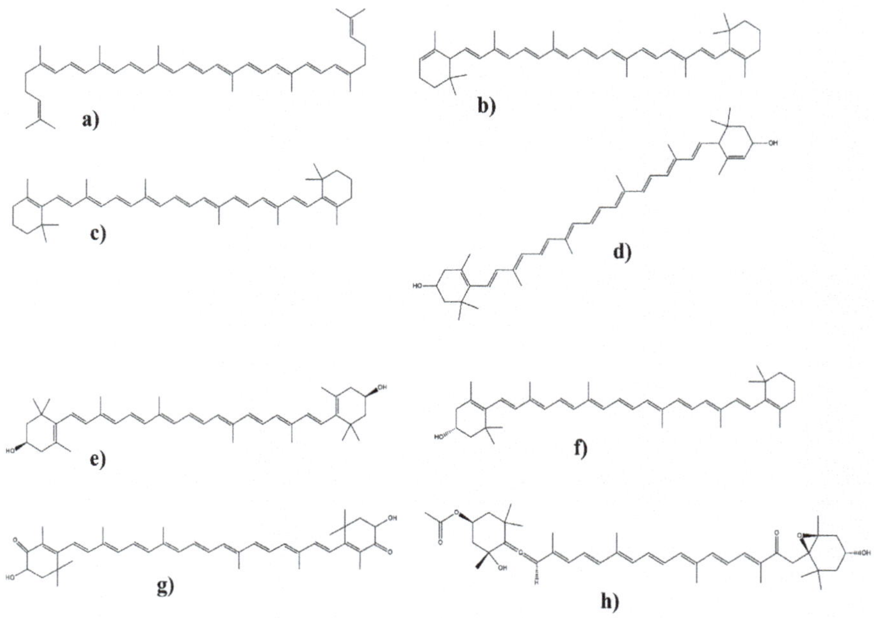

Fig. 12.2 Chemical structures of carotenoids having nutraceutical properties. The compounds shown are lycopene (a); α-carotene (b); β-carotene (c); lutein (d); zeaxanthin (e); β-cryptoxanthin (f); astaxanthin (g) and fucoxanthin (h)

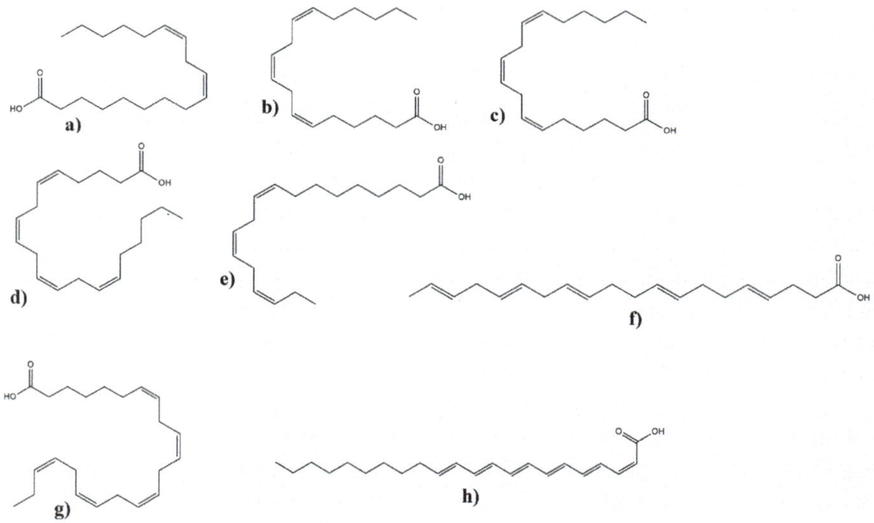

Fig. 12.3 Chemical structures of polyunsaturated fatty acids, linoleic acid (a); γ-linolenic acid (b); dihomo γ-linolenic acid (c); arachidonic acid (d); α-linolenic acid (e); eicosapentaenoic acid (f); docosapentaenoic acid (g) and docosahexaenoic acid (h)

cobalt, fluoride and selenium. Vitaminsare another class of nutraceuticals that have a complex organic structure and are required in small levels for a good human health. Fat soluble vitamins (A, D, E, and K) and water-soluble vitamins are the two different classes of vitamins. Water soluble vitamins include vitamin B and C having very important roles in the human body.

Apart from the aforementioned classes of nutraceuticals and functional foods, probiotics and prebiotics can be recognized as functional foods. Probiotics are health-promoting bacteria present in dairy products such as yogurt. *Lactobacillus acidophilus, L. bulgaricus, Bifidobacteriumlactis,* and *B. subtilis* are common examples of probiotic bacteria (Parker and Pace 2016). On the other hand, prebiotics (such as fructooligosaccharide, inulin, and honey) are ingredients that increase the growth of health-promoting bacteria.

Nutraceuticals are extracted from living sources by various methods and techniques. These include Soxhlet extraction (traditional method requiring good amounts of solvent and is time consuming), sonication assisted extraction (use of ultrasound extractor), supercritical fluid extraction and microwave assisted extraction (Grumezescu 2016). After extraction, nutraceuticals are characterized by different analytical techniques such as high-performance liquid chromatography, mass spectrometry, nuclear magnetic resonance and atomic spectroscopy.

Nutraceuticals often have low dispersibility in water and bioavailability (amount of nutraceutical that reaches the systemic circulation after consumption) (Singh et al. 2018; Tang 2020). Therefore, to reduce these limitation various delivery methods have been used recently. These include phospholipid-based delivery vehicles, liposome-based carrier system, and emulsion-based delivery system (such as microemulsions, nano emulsions, and double emulsions) (Grumezescu 2016). Nanoencapsulation is an emerging technique for not only improving the water dispersibility of nutraceuticals but also their bioavailability (Acevedo-Fani et al. 2017). Nano complexation (complexation of a nutraceutical with proteins at nanoscale) can be helpful for poorly soluble compounds such as curcumin (Singh 2007), while nano emulsions can also play an important role for improving solubility (Zhang et al. 2020). Apart from these, other latest developments for better delivery and bioavailability of nutraceuticals involve electrospinning and electro spraying techniques (do Evangelho et al. 2019; Coelho et al. 2020).

12.3 Current State of Knowledge About Nutraceuticals/Functional Foods

Nutraceuticals and functional foods can have a plant or animal origin, and the current research is primarily focused on understanding the mechanism of action, safety concerns and efficacy. Interestingly, nutraceuticals can be a boon for the treatment of long-term diseases in patients which do not qualify for a treatment by pharmaceuticals. Moreover, these substances may impact the health in a complicated and interactive manner. The current research on nutraceuticals has highlighted the need of newer processing methods and techniques (such as high-pressure processing and pulsed electric field processing) for nutraceuticals, so that they can not only have good microstructures but also acceptable sensory characteristics. Apart from this, newer packaging methods for nutraceuticals include the use of nanomaterials that can enhance their safety and efficacy without compromising bioavailability. Nutraceuticals and functional foods are suitable candidates for personalized nutrition which aims to provide health benefitting diets to people according to their genetic traits (especially for the prevention of aging-associated diseases) (Corzo et al. 2020). However, it must be realized that only nutrition without regular physical exercise cannot guarantee a long and robust life. Additionally, it needs to emphasized that nutraceuticals and functional foods are not substitutes for a healthy diet. With their increased advertising, many elderly people have started using health supplements which have nutraceutical claims but sometimes these supplements are deficient in what the labels claim. A healthy diet itself contains these nutraceuticals in balanced amounts, so the idea of healthy nutrition should never be neglected.

12.4 Role of Nutraceuticals/Functional Foods in Delaying the Aging Process

At present, the major goal of gerontology research is to discover natural chemicals that can modulate aging process. These compounds can be helpful either in suppression of senescent cells or their clearance. Among these, molecules having antioxidant or anti-inflammatory activities have received considerable interest in the recent years (Gurău et al. 2018). The use of nutraceuticals and functional foods in the diet has been considered as a promising approach against the aging process, especially in alleviating the impairment of body functions (Singh et al. 2021). However, as of now nutraceutical supplementation to reverse the aging process is very challenging, owing to the dosage as well as timing optimization and different individual responses. It has been widely recognized that a diet rich in fruits and vegetables, and whole grains can be effective because of the presence of adequate amounts of nutraceuticals.

Ferrari (2004) has comprehensively highlighted the roles of nutraceuticals and functional foods in delaying the progression of aging. He reported that these compounds can act as mitochondrial membrane stabilizers and enhancers, have metal chelating and antioxidant properties, and induce apoptosis of senescent cells. Apart from these, some nutraceuticals such as epigallocatechin gallate, allantoin and ginsenoside can mimic the anti-aging properties of metformin and rapamycin (anti-aging drugs) (Aliper et al. 2017). Moreover, nutraceuticals can decelerate aging process by improving the balance of gut flora and lowering blood cholesterol levels. The molecular mechanisms of their action have been reported to be by modulating the gene expression, modifying patterns of DNA methylation and acting as epigenetic modifiers (Lee et al. 2014). It has been validated that free radical clearance can be effectively enhanced by using nutraceuticals at adequate levels (Pisoschi and Pop 2015).

There have been many reports of nutraceuticals and functional foods in delaying the process of aging. Phytochemicals, probiotic bacteria and ω-3 fatty acids have been reported to have anti-cellular senescence capacity in the immune cells which affect aging (O'Shea et al. 2009). Flavonoids, along with other phenolic compounds, can protect against oxidation of polyunsaturated fatty acids present in the membranes and circumvent mitochondrial membrane disruptions (Pisoschi and Pop 2015). In terms of molecular events, foods rich in phenolic compounds can effectively modulate the activity of some enzymes such as DNA methyltransferase and histone deacetylases, which may be responsible for delaying the aging process (Gurău et al. 2018). ω-3 fatty acids have been documented to enhance mitochondrial membrane lipids, lower calcium release and pyruvate dehydrogenase enzyme activity (O'Shea et al, 2009). Nutraceuticals that function as antioxidants can regulate the mitochondrial functioning and lower the release of cytochrome c for apoptosis (Adachi and Ishii 2002). Nutraceutical microelements such as copper, manganese, and iron function as enzyme co-factors for antioxidant enzymes such as catalase and superoxide dismutase. This activity is responsible for clearance of free radicals responsible for early aging.

12.5 Nutraceuticals and Functional Foods Role in Reducing Aging-Associated Diseases

It is well established that aging is responsible for the start and progression of several diseases, that mainly include neurological disorders such as Parkinson's and Alzheimer's disease, cardiovascular disease such as atherosclerosis, type 2 diabetes mellitus diabetes, osteoarthritis and cancer (Hou et al. 2019; Yang et al. 2020). Aging progresses with malfunctioning in nutrient signaling, protein machinery and mitochondria and leads to cell senescence (Saraswat and Rizvi 2017). Neurological disorders in the elderly people cause changes in both the structural and biochemical functions of the brain. Owing to these changes, patients demonstrate many symptoms such as paralysis, muscle weakness, improper coordination, seizures, pain and loss of consciousness (Fonseca-Santos and Chorilli 2020). Other diseases such as atherosclerosis, type 2 diabetes mellitus diabetes, osteoarthritis and cancer also result from the physiological changes during aging. The roles of nutraceuticals and functional foods in reducing the incidence of aging-associated diseases is as follows:

12.5.1 Parkinson's Disease

Parkinson's disease is a neurodegenerative disease that affects around 1–2% of the world population (Khan et al. 2019). The usual treatment of Parkinson's disease involves providing dopamine replacement therapy. The combination medication of levodopa and carbidopa is generally used for the patients suffering from this disease. Nutraceuticals have recently gained a lot of attention because of their therapeutic properties, so an integrative use of nutraceuticals and functional foods along with the standard therapy of medication has been proven to be effective in many cases of Parkinson's disease (Lama et al. 2020). This addition can have an important role in improving the quality of life of the suffering patients. Nutraceuticals and functional foods generally are helpful in this disease by targeting and weakening various pathogenic events which include neuroinflammation, mitochondrial dysfunction, apoptosis and oxidative stress. Oxidative stress in Parkinson's disease manifests by forming α-synuclein aggregates and lowering amounts of neuromelanin (Knörle2018; Ludtmann et al. 2018). The biggest role played by nutraceuticals is to restore mitochondrial homeostasis by reducing oxidative stress and correct faulty electron transport chain as well as mitochondrial dynamics.

Many nutraceuticals have been investigated to alleviate the symptoms of Parkinson's disease. Among phenolic compounds, epigallocatechin-3-gallate (a major phenolic compound in tea plant) has shown a neuroprotective role because of its ability to cross the blood–brain barrier. Based on having a catechol-like structure, it has been described as a good radical scavenger and chelator of iron ions (Morgan and Grundmann 2017). Moreover, it also encourages the proper folding of α-synuclein monomers into stable oligomers in a concentration-dependent manner (Šneideris et al. 2015). Another phenolic compounds under investigation include curcumin and resveratrol. Curcumin reduces the incidence of nuclear factor κ-B (a family of inducible transcription factors) mediated neuroinflammation and targets Toll-like receptor-4 (known to modulate immune responses as well as stimulate synthesis of inflammatory chemokines and cytokines) (Zhu et al. 2014). Resveratrol has been reported to restrict mitochondrial dysfunction and apoptosis in nigrostriatal cells by acting through the protein kinase B/glycogen synthase kinase-3β pathway (Zeng et al. 2017). In a study by Ho and Pasinetti (2010), grape seed polyphenol extract was reported to be helpful in the reducing the incidence of Parkinson's disease that involve misfolded proteins. This was due to its ability to interfere with the development of misfolded-protein aggregates generated from Aβ peptides.

Lycopene has been documented to relieve oxidative stress, increase NADH dehydrogenase and superoxide dismutase activity in the corpus striatum and lower malondialdehyde levels (Prema et al. 2015). In addition, it also enhances the amounts of other antioxidant enzymes, such as catalase and glutathione peroxidase in animal as well as cellular models of Parkinson's disease (Kaur et al. 2011; Paul et al. 2020). Coenzyme Q10 prevents1-methyl-4-phenyl-1,2,3,6-tetrahydropyridine (MPTP)-induced neurotoxicity and blocks the electron transfer between complex 1 and other complexes of the electron transport chain (Cleren et al. 2008). Niacin (vitamin B3) is a precursor of NAD–NADH and is needed for producing dopamine and may be helpful for managing Parkinson's disease. Its effects include reduction in inflammation through NIARC1-associated mechanisms, enhancing the ratio of NAD/NADH, reestablishing complex 1 functioning and boosting dopamine synthesis in the striatum via the supply of NADPH to mitochondria (Bjørklund et al. 2020). Ginsenosides restrict astrogliosis as well as microgliosis and reduce the synthesis of proinflammatory cytokines in substantia nigra pars compacta (Fu et al. 2015). Probiotics can be helpful in restoring gut dysbiosis taking place during Parkinson's disease, improving gastrointestinal functioning, decreasing the enteric nervous system neuroinflammation and gut leakiness (Castelli et al.2020). In a recent study by Ma et al. (2020), peptides present in sesame cake reduced α-synuclein aggregation and reduced MPP+ -induced dopaminergic neuron degeneration.

12.5.2 Alzheimer's Disease

Alzheimer's disease is a neurological disorder which leads to loss of hippocampal and cortical neurons and clinically exhibits impairment of the cognitive capabilities along with aphasia, disorientation, and disinhibition (Scheltens et al. 2016). Nutraceuticals and functional foods have a role to play in reducing the risk of Alzheimer's disease. Phenolic compounds have been reported to be effective against this disease and act as antioxidants (Singh et al. 2016). Curcumin has been reported to lessen the assemblage of Aβpeptides in the neural tissue and its associated inflammation. He et al. (2016) reported that oral in take of curcumin can reduce the deposition and oligomerization of Aβ peptide, together with phosphorylation of the tau protein. In addition, curcumin can bind with metal ions, primarily Cu (II) and Zn (II), that are common in the central nervous system (especially in the synapse regions), adding to its neuroprotective activity (Sadhukhan et al. 2018). Genistein has the potential to reduce the activity of DNA topoisomerase as well as tyrosine protein kinase (Sadhukhan et al. 2018). Moreover, it has specificity for epidermal growth factor receptor, which is a growth regulatory transmembrane protein binding certain protein such as epidermal growth factor and transforming growth factor-β. Many phenolic moieties in the molecular structure of genistein leads to its good antioxidant activity. Resveratrol activates reactive protein kinase C that might stimulate α-secretase and result in lowering of Aβ synthesis, thereby increasing the intracellular free-radical scavenger glutathione amounts and lowering malondialdehyde and acetylcholinesterase levels (Han et al. 2004; Fonseca-Santos and Chorilli 2020). Carotenoids such as lycopene can stop the activities of the secretases (responsible for the processing of amyloid precursor protein) and glycogen synthase kinase 3β, which leads to neuroprotection in Alzheimer's disease (Paul et al. 2020). Fatty acids such as docosahexaenoic acid intake enhances the cognitive function and reduces the incidence of dementia (Cunnane et al. 2009). Additionally, eicosapentaenoic acid also has been reported to have a good role in the normal brain development (Cunnane et al. 2009).

12.5.3 Cardiovascular Disease

Cardiovascular disease is one major cause of death worldwide, becoming a significant healthcare problem in the developed countries (Sacks et al. 2017). The underlying cause of cardiovascular disease is atherosclerosis (a long-term inflammatory ailment of medium and large arteries). Taking this into consideration, prevention of atherosclerosis is the key target for lessening the risk of developing cardiovascular diseases. Nutraceuticals and functional foods can have a significant role in controlling atherosclerosis, especially during the promotion stage when circulating monocytes penetrate vascular endothelial cells and later differentiate into macrophages and foam cells. Among these compounds, catechin class of flavanols present in cocoa as well as green tea have been well researched and are well known for their antioxidant activities along with capacity to reduce inflammation by reducing release of cytokines and chemokines from the activated endothelial cells (Mangels and Mohler 2017). Curcumin has been reported to lower the levels of C-reactive protein (produced by the liver during inflammation and is the indicator of atherosclerosis) not only in animal models but also in humans (Sahebkar 2014). Carotenoid pigment lycopene has been documented to lower the amounts of low-density lipoproteins as well as pro-inflammatory cytokines in humans (Cheng et al. 2017; Chaudhary et al. 2018). Apart from these, phytosterols have also been reported to reduce atherosclerosisin mouse models and humans by lowering blood plasma low-density lipoprotein levels (Rocha et al. 2016). Fatty acid eicosapentaenoic acid is known to reduce the amounts of serum triglycerides without enhancing the amounts of low-density lipoproteins in the body (Ballantyne et al. 2012). Yan et al. (2020) have reported that the nutraceutical Coenzyme Q10 decreased atherosclerosis in apolipoprotein E-deficient mice and enhanced the efflux of cholesterol in humans.

12.5.4 Type 2 Diabetes

The conventional treatment of type 2 diabetes often involves many strategies which can be diet control, exercise, administration of insulin and drugs. Apart from these strategies, nutraceuticals and functional foods are generally utilized as an adjuvant/alternative therapy, in combination with relevant changes in the lifestyle. Many nutraceuticals and functional foods are known to ameliorate the symptoms of diabetes. Epigallocatechin-3-gallate (a catechin present in good levels in green tea) has significant antioxidant and anti-inflammatory activities (Singh et al. 2018). The administration of epigallocatechin-3-gallate on streptozotocin-induced diabetic rats has been reported to reduce glucose as well as lipid amounts, along with enhancement in insulin concentration. Epigallocatechin-3-gallate has also been reported to reduce troponin T level, lactate dehydrogenase, and aspartate aminotransferase enzyme activities in the serum (Othman et al. 2017). Monounsaturated fatty acids and polyunsaturated fatty acids (as components of olive oil and nuts) have been documented to improve glucose metabolism, insulin sensitivity and blood lipids in patients suffering from this disease (Mirabelli et al. 2020). The supplementation of chromium (as chromium histidinate) in the diet lowered the amounts of glucose, glycosylated hemoglobin, and total cholesterolin streptozotocin-induced diabetic rats (Ulas et al. 2015). Similarly, administration with zinc (a constituent of insulin and insulin receptors), decreased oxidative damage and inflammation in the kidneys of streptozotocin-induced diabetic rats (Tang et al. 2010). Garg (2016) reported that cinnamonas a supplement had a good role to play in managing type 2 diabetes.

12.5.5 *Osteoarthritis*

Osteoarthritis is the most prevalent form of arthritis that accounts for about half of the total cases, with other ones being rheumatoid arthritis, gout, and lupus (Wang et al. 2018). The conventional medical therapy includes oral and topical non-steroidal anti-inflammatory drugs but these can be toxic to the body producing rashes, increasing bleeding risk and can affect kidney functioning. Nutraceuticals and functional foods can be effective in reducing osteoarthritis pain. Epigallocatechin gallate has an anti-inflammatory activity on osteoarthritis chondrocytes by lowering the synthesis of important inflammatory mediators such as inducible nitric oxide synthase and prostaglandin-endoperoxide synthase 2 (Henrotin et al. 2011). Curcumin has been reported to play the role of an antioxidant, chondroprotective and anti-inflammatory agent (Kuptniratsaikul et al. 2009). Within a month of trials, the patients who were administered turmeric extracts (at 2 g per day dose) had a significant reduction in pain in comparison to those who took ibuprofen. In addition, their pain on level walking as well on stairs decreased over time. Procyanidins, particularly Procyanidin B2, have exhibited suppression of articular cartilage vascular endothelial growth factor (a signaling protein which induces new blood vessel growth), a key mediator of osteoarthritis pathogenesis as well as pain (Wang et al. 2018). Fatty acids such aseicosapentaenoic acid and docosahexaenoic acid have demonstrated anti-inflammatory effects (Henrotin et al. 2011). In case of animal models, vitamin C stimulated the metabolism of chondrocytes, collagen, and proteoglycan synthesis, while vitamin E was reported to have free radical scavenging properties (Lopez 2012). Functional foods such as green tea and fish oil have been conventionally recognized for their pain management activities in osteoarthritis patients (Schell et al. 2017). Moreover, functional foods are often enriched with natural active ingredients that can reduce pain and inflammation.

12.5.6 *Cancer*

Chemotherapy although is mainstream therapy for cancer but its usage has failed to achieve its maximum therapeutic potency. Because of the presence of acquired and intrinsic chemoresistance. There are many side effects of chemotherapy such as reduced immunity, loss of digestive tract lining along with hair loss (Nair et al. 2020). Nutraceuticals and functional foods can be helpful for the cancer patients especially for the management of adverse effects of chemotherapy.

Phenolic compounds such as curcumin has been reported as a potential chemo preventive and chemosensitizer agent among nutraceutical compounds because of its high efficacy (Mao et al. 2018). Puliyappadamba et al. (2015) have comprehensively reported the antiproliferative and pro-apoptotic activity of curcumin in lung cancer. Resveratrol has been reported to reduce inflammation, oxidative stress and multi-step tumorigenesis. It has been documented to be a chemo preventive and chemo sensitizing agent against breast cancer (Nair et al. 2020). Genistein has the property of being effective in lowering the growth of cancer cells without any side effects. Quercetin has been reported to enhance immune system response against tumor growth by increasing mitochondrial membrane permeabilization. Another nutraceutical named Polyphenon E (a green tea catechin mixture) has been successful in clinical trials against cancer cells proliferation (Zhang et al. 2004). Apart from phenolic compounds, lycopene has been effective against cancer by suppressing the phosphorylation of tumor suppressor p53 and stopping tumor cell division (Nair et al. 2020).

12.6 Conclusion

There is a conclusive evidence that the consumption of nutraceuticals and functional foods have health benefits and might keep aging-associated diseases at bay. However, more extensive research needs to be done in order to comprehensively demonstrate whether nutraceuticals and functional foods can successfully delay aging in humans. Moreover, the present limitations of nutraceuticals such as slow metabolism, poor solubility and reduced bioavailability needs more consideration. Despite of the present limitations, the future of nutraceuticals of both plant and animal origin holds exciting opportunities for the food and pharmaceutical industry to create novel products. There is a need to establish efficacy and safety parameters which comply with the regulatory requirements for maintaining consumer confidence.

Acknowledgements Authors are thankful to Science and Engineering Research Board, University Grants Commission and Council of Scientific and Industrial Research, New Delhi (CSIR Project No. 38(1419)/16/EMR-II) for providing financial assistance.

Compliance with Ethical Standards

Funding This study was funded by Science and Engineering Research Board, University Grants Commission and Council of Scientific and Industrial Research, New Delhi (CSIR Project No. 38(1419)/16/EMR-II).

Conflict of Interest All authors declare they have no conflict of interest.

References

Acevedo-Fani A, Soliva-Fortuny R, Martín-Belloso O (2017) Nanostructured emulsions and nanolaminates for delivery of active ingredients: improving food safety and functionality. Trends Food Sci Technol 60:12–22

Adachi M, Ishii H (2002) Role of mitochondria in alcoholic liver injury. Free Radical Biol Med 32(6):487–491

Aliper A, Jellen L, Cortese F, Artemov A, Karpinsky-Semper D, Moskalev A, Zhavoronkov A (2017) Towards natural mimetics of metformin and rapamycin. Aging (albany NY) 9(11):2245

Ballantyne CM, Bays HE, Kastelein JJ, Stein E, Isaacsohn JL, Braeckman RA, Soni PN (2012) Efficacy and safety of eicosapentaenoic acid ethyl ester (AMR101) therapy in statin-treated patients with persistent high triglycerides (from the ANCHOR study). Am J Cardiol 110(7):984–992

Bjørklund G, Dadar M, Anderson G, Chirumbolo S, Maes M (2020) Preventive treatments to slow substantia nigra damage and Parkinson's disease progression: a critical perspective review. Pharmacol Res 105065

Castelli V, d'Angelo M, Quintiliani M, Benedetti E, Cifone MG, Cimini A (2020) The emerging role of probiotics in neurodegenerative diseases: new hope for Parkinson's disease? Neural Regeneration Res 16(4):628

Chaudhary P, Sharma A, Singh B, Nagpal AK (2018) Bioactivities of phytochemicals present in tomato. J Food Sci Technol 55(8):2833–2849

Cheng HM, Koutsidis G, Lodge JK, Ashor A, Siervo M, Lara J (2017) Tomato and lycopene supplementation and cardiovascular risk factors: a systematic review and meta-analysis. Atherosclerosis 257:100–108

Chiu HF, Shen YC, Venkatakrishnan K, Wang CK (2018) Popular functional foods and nutraceuticals with lipid lowering activity and in relation to cardiovascular disease, dyslipidemia, and related complications: an overview. J Food Bioactives 2:16–27

Cleren C, Yang L, Lorenzo B, Calingasan NY, Schomer A, Sireci A, Beal MF (2008) Therapeutic effects of coenzyme Q10 (CoQ10) and reduced CoQ10 in the MPTP model of Parkinsonism. J Neurochem 104(6):1613–1621

Coelho SC, Estevinho BN, Rocha F (2020) Encapsulation in food industry with emerging electrohydrodynamic techniques: electrospinning and Electrospraying-a review. Food Chem 127850

Corzo L, Fernández-Novoa L, Carrera I, Martínez O, Rodríguez S, Alejo R, Cacabelos R (2020) Nutrition, health, and disease: role of selected marine and vegetal nutraceuticals. Nutrients 12(3):747

Cunnane SC, Plourde M, Pifferi F, Bégin M, Féart C, Barberger-Gateau P (2009) Fish, docosahexaenoic acid and Alzheimer's disease. Prog Lipid Res 48(5):239–256

do Evangelho JA et al (2019) Thermal and irradiation resistance of folic acid encapsulated in zein ultrafine fibers or nanocapsules produced by electrospinning and electrospraying. Food Res Int 124:137–146

Dominguez Diaz L, Fernández-Ruiz V, Cámara M (2020) The frontier between nutrition and pharma: the international regulatory framework of functional foods, food supplements and nutraceuticals. Crit Rev Food Sci Nutr 60(10):1738–1746

Durazzo A, Lucarini M, Santini A (2020) Nutraceuticals in human health. Foods 9:370

Dwivedi A, Singh S, Kumar S, Mittal PC (2020) Organosulfur phytochemicals against metabolic and neurodegenerative diseases: benefits and risks. In: Phytochemicals as Lead Compounds for New Drug Discovery, pp 179–194. Elsevier

Falinska AM, Bascoul-Colombo C, Guschina IA, Good M, Harwood JL (2012) The role of n-3 dietary polyunsaturated fatty acids in brain function and ameliorating Alzheimer's disease: opportunities for biotechnology in the development of nutraceuticals. Biocatal Agric Biotechnol 1(2):159–166

Ferrari CK (2004) Functional foods, herbs and nutraceuticals: towards biochemical mechanisms of healthy aging. Biogerontology 5(5):275–289

Fonseca-Santos B, Chorilli M (2020) The uses of resveratrol for neurological diseases treatment and insights for nanotechnology based-drug delivery systems. Int J Pharmaceutics 119832

Fu W, Zhuang W, Zhou S, Wang X (2015) Plant-derived neuroprotective agents in Parkinson's disease. Am J Transl Res 7(7):1189

Garg RC (2016) Nutraceuticals in Glucose Balance and Diabetes. In: Nutraceuticals, pp 145–160. Academic Press

Ghosh D, Das S, Bagchi D, Smarta RB (eds) (2012) Innovation in Healthy and Functional Foods. CRC press, Boca Raton

Grumezescu A (2016) Nutraceuticals. Academic Press

Guru F, Baldoni S, Prattichizzo F, Espinosa E, Amenta F, Procopio AD, Olivieri F (2018) Anti-senescence compounds: a potential nutraceutical approach to healthy aging. Ageing Res Rev 46:14–31

Han YS, Zheng WH, Bastianetto S, Chabot JG, Quirion R (2004) Neuroprotective effects of resveratrol against β-amyloid-induced neurotoxicity in rat hippocampal neurons: involvement of protein kinase C. Br J Pharmacol 141(6):997

He GL, Luo Z, Yang J, Shen TT, Chen Y, Yang XS (2016) Curcumin ameliorates the reduction effect of PGE2 on fibrillar β-amyloid peptide (1–42)-induced microglial phagocytosis through the inhibition of EP2-PKA signaling in N9 microglial cells. PLoS One 11(1):e0147721

Henrotin Y, Lambert C, Couchourel D, Ripoll C, Chiotelli E (2011) Nutraceuticals: do they represent a new era in the management of osteoarthritis? –A narrative review from the lessons taken with five products. Osteoarthritis Cartilage 19(1):1–21

Ho L, Pasinetti GM (2010) Polyphenolic compounds for treating neurodegenerative disorders involving protein misfolding. Expert Rev Proteomics 7(4):579–589

Hou Y, Dan X, Babbar M, Wei Y, Hasselbalch SG, Croteau DL, Bohr VA (2019) Ageing as a risk factor for neurodegenerative disease. Nat Rev Neurol 15(10):565–581

Kaur H, Chauhan S, Sandhir R (2011) Protective effect of lycopene on oxidative stress and cognitive decline in rotenone induced model of Parkinson's disease. Neurochem Res 36(8):1435–1443

Khan AU, Akram M, Daniyal M, Zainab R (2019) Awareness and current knowledge of Parkinson's disease: a neurodegenerative disorder. Int J Neurosci 129(1):55–93

Knörle R (2018) Neuromelanin in Parkinson's disease: from Fenton reaction to calcium signaling. Neurotox Res 33(2):515–522

Kuptniratsaikul V, Thanakhumtorn S, Chinswangwatanakul P, Wattanamongkonsil L, Thamlikitkul V (2009) Efficacy and safety of Curcuma domestica extracts in patients with knee osteoarthritis. J Alternative Complementary Med 15(8):891–897

Lama A et al (2020) Nutraceuticals: an integrative approach to starve Parkinson's disease. Brain, Behavior, Immunity-Health 2:100037

Lee JH, Lim H, Shehzad O, Kim YS, Kim HP (2014) Ginsenosides from Korean red ginseng inhibit matrix metalloproteinase-13 expression in articular chondrocytes and prevent cartilage degradation. Eur J Pharmacol 724:145–151

Longoria-García S, Cruz-Hernández MA, Flores-Verástegui MIM, Contreras-Esquivel JC, Montañez-Sáenz JC, Belmares-Cerda RE (2018) Potential functional bakery products as delivery systems for prebiotics and probiotics health enhancers. J Food Sci Technol 55(3):833–845

Lopez HL (2012) Nutritional interventions to prevent and treat osteoarthritis. Part II: focus on micronutrients and supportive nutraceuticals. PM&R 4(5):S155–S168

Ludtmann MH, Angelova PR, Horrocks MH, Choi ML, Rodrigues M, Baev AY, Al-Menhali AS (2018) α-synuclein oligomers interact with ATP synthase and open the permeability transition pore in Parkinson's disease. Nat Commun 9(1):1–16

Ma X, Li J, Cui X, Li C, Wang Z (2020) Dietary supplementation with peptides from sesame cake alleviates Parkinson's associated pathologies in Caenorhabditis elegans. J Functional Foods 65:103737

Mangels DR, Mohler ER III (2017) Catechins as potential mediators of cardiovascular health. Arterioscler Thromb Vasc Biol 37(5):757–763

Mao XY, Jin MZ, Chen JF, Zhou HH, Jin WL (2018) Live or let die: neuroprotective and anti-cancer effects of nutraceutical antioxidants. Pharmacol Ther 183:137–151

McWilliams M (2017) Foods: Experimental Perspectives. Pearson Education, Inc., London

Mirabelli M, Chiefari E, Arcidiacono B, Corigliano DM, Brunetti FS, Maggisano V, Brunetti A (2020) Mediterranean diet nutrients to turn the tide against insulin resistance and related diseases. Nutrients 12(4):1066

Morgan LA, Grundmann O (2017) Preclinical and potential applications of common western herbal supplements as complementary treatment in Parkinson's disease. J Dietary Suppl 14(4):453–466

Nair HH, Alex VV, Anto RJ (2020). Significance of nutraceuticals in cancer therapy. In: Evolutionary Diversity as a Source for Anticancer Molecules, pp 309–321. Academic Press

O'Shea KM, Khairallah RJ, Sparagna GC, Xu W, Hecker PA, Robillard-Frayne I, Stanley WC (2009) Dietary ω-3 fatty acids alter cardiac mitochondrial phospholipid composition and delay Ca2+-induced permeability transition. J Mol Cell Cardiol 47(6):819–827

Othman AI, El-Sawi MR, El-Missiry MA, Abukhalil MH (2017) Epigallocatechin-3-gallate protects against diabetic cardiomyopathy through modulating the cardiometabolic risk factors, oxidative stress, inflammation, cell death and fibrosis in streptozotocin-nicotinamide-induced diabetic rats. Biomed Pharmacother 94:362–373

Pandareesh MD et al (2018) Nutrition and nutraceuticals in neuroinflammatory and brain metabolic stress: implications for neurodegenerative disorders. CNS & Neurological Disorders-Drug Targets (Formerly Current Drug Targets-CNS & Neurological Disorders) 17(9):680–688

Parker R, Pace M (2016) Introduction to Food Science and Food Systems. Cengage Learning, Inc.

Paul R et al (2020) Lycopene-a pleiotropic neuroprotective nutraceutical: deciphering its therapeutic potentials in broad spectrum neurological disorders. Neurochemistry Int 140:104823

Pisoschi AM, Pop A (2015) The role of antioxidants in the chemistry of oxidative stress: a review. Eur J Med Chem 97:55–74

Prema A, Janakiraman U, Manivasagam T, Thenmozhi AJ (2015) Neuroprotective effect of lycopene against MPTP induced experimental Parkinson's disease in mice. Neurosci Lett 599:12–19

Puliyappadamba VT, Thulasidasan AKT, Vijayakurup V, Antony J, Bava SV, Anwar S, Anto RJ (2015) Curcumin inhibits B [a] PDE-induced procarcinogenic signals in lung cancer cells, and curbs B [a] P-induced mutagenesis and lung carcinogenesis. BioFactors 41(6):431–442

Rajasekaran A (2017) Nutraceuticals. Comprehensive Med Chem III(1):107–134

Rocha VZ, Ras RT, Gagliardi AC, Mangili LC, Trautwein EA, Santos RD (2016) Effects of phytosterols on markers of inflammation: a systematic review and meta-analysis. Atherosclerosis 248:76–83

Rodriguez-Amaya DB (2015). Food Carotenoids: Chemistry, Biology and Technology. John Wiley & Sons, Hoboken

Sacks FM, et al (2017) Dietary fats and cardiovascular disease: a presidential advisory from the American heart association. Circulation 136(3):e1–e23

Sadhukhan P, Saha S, Dutta S, Mahalanobish S, Sil PC (2018) Nutraceuticals: an emerging therapeutic approach against the pathogenesis of Alzheimer's disease. Pharmacol Res 129:100–114

Sahebkar A (2014) A systematic review and meta-analysis of randomized controlled trials investigating the effects of curcumin on blood lipid levels. Clin Nutr 33(3):406–414

Saini RK, Keum YS (2018) Omega-3 and omega-6 polyunsaturated fatty acids: dietary sources, metabolism, and significance—a review. Life Sci 203:255–267

Saraswat K, Rizvi SI (2017) Novel strategies for anti-aging drug discovery. Expert Opin Drug Discov 12(9):955–966

Schell J, Scofield RH, Barrett JR, Kurien BT, Betts N, Lyons TJ, Basu A (2017) Strawberries improve pain and inflammation in obese adults with radiographic evidence of knee osteoarthritis. Nutrients 9(9):949

Scheltens P, Blennow K, Breteler MM, de Strooper B, Frisoni GB, Salloway S, Van der Flier WM (2016) Alzheimer's disease. Lancet 388:505–517

Serviddio G, Davide Romano A, Cassano T, Bellanti F, Altomare E, Vendemiale G (2011) Principles and therapeutic relevance for targeting mitochondria in aging and neurodegenerative diseases. Curr Pharm Des 17(20):2036–2055

Singh S (2007) From exotic spice to modern drug? Cell 130(5):765–768

Singh JP, Kaur A, Singh N, Nim L, Shevkani K, Kaur H, Arora DS (2016) In vitro antioxidant and antimicrobial properties of jambolan (Syzygiumcumini) fruit polyphenols. LWT-Food Sci Technol 65:1025–1030

Singh B, Singh JP, Kaur A, Singh N (2016) Bioactive compounds in banana and their associated health benefits–a review. Food Chem 206:1–11

Singh B, Singh JP, Kaur A, Singh N (2018) Phenolic compounds as beneficial phytochemicals in pomegranate (Punica granatum L.) peel: a review. Food Chem 261:75–86

Singh B, Singh JP, Kaur A, Singh N (2020) Phenolic composition, antioxidant potential and health benefits of citrus peel. Food Res Int 109114

Singh P, Sivanandam TM, Konar A, Thakur MK (2021) Role of nutraceuticals in cognition during aging and related disorders. Neurochem Int 143:104928

Šneideris T, Baranauskienė L, Cannon JG, Rutkienė R, Meškys R, Smirnovas V (2015) Looking for a generic inhibitor of amyloid-like fibril formation among flavone derivatives. Peer J 3:e1271

Tang CH (2020) Nanocomplexation of proteins with curcumin: from interaction to nanoencapsulation (a review). Food Hydrocolloids 106106

Tang Y, Yang Q, Lu J, Zhang X, Suen D, Tan Y, Li X (2010) Zinc supplementation partially prevents renal pathological changes in diabetic rats. J Nutritional Biochem 21(3):237–246

Ulas M, Orhan C, Tuzcu M, Ozercan IH, Sahin N, Gencoglu H, Sahin K (2015) Anti-diabetic potential of chromium histidinate in diabetic retinopathy rats. BMC Complementary Alternative Med 15(1):16

Vaclavik VA, Christian EW, Campbell T (2021) Essentials of Food Science. Springer, Cham, pp 137–175

Wang A, Leong DJ, Cardoso L, Sun HB (2018) Nutraceuticals and osteoarthritis pain. Pharmacol Ther 187:167–179

Yang C et al (2020) A natural product solution to aging and aging-associated diseases. Pharmacol Therapeutics 107673

Zeng W, Zhang W, Lu F, Gao L, Gao G (2017) Resveratrol attenuates MPP+-induced mitochondrial dysfunction and cell apoptosis via AKT/GSK-3β pathway in SN4741 cells. Neurosci Lett 637:50–56

Zhang Q, Wei D, Liu J (2004) In vivo reversal of doxorubicin resistance by (–)-epigallocatechin gallate in a solid human carcinoma xenograft. Cancer Lett 208(2):179–186

Zhang R, Zhang Z, McClements DJ (2020) Nanoemulsions: an emerging platform for increasing the efficacy of nutraceuticals in foods. Colloids Surfaces B: Biointerfaces 194:111202

Zhu HT, Bian C, Yuan JC, Chu WH, Xiang X, Chen F, Lin JK (2014) Curcumin attenuates acute inflammatory injury by inhibiting the TLR4/MyD88/NF-κB signaling pathway in experimental traumatic brain injury. J Neuroinflammation 11(1):59

Chapter 13
Food for Brain Health

Vittorio Calabrese, Angela Trovato, Maria Scuto, Maria Laura Ontario, Francesco Rampulla, Vincenzo Zimbone, Ignazio La Mantia, Valentina Greco, Sebastiano Sciuto, Roberto Crea, Edward J. Calabrese, and Giuseppe Dionisio

Abstract Exploring molecular mechanisms of aging and determinants of lifespan will help reduce age-related morbidity, thus facilitating healthy brain aging. Recently, it has been demonstrated that nutritional polyphenols, the main constituents of the Mediterranean diet, maintain redox balance and neuroprotection through the activation of hormetic vitagene pathway. Mitochondria play pivotal roles in the

V. Calabrese and A. Trovato—Equally contributed to this manuscript

V. Calabrese (✉) · A. Trovato (✉) · M. Scuto · M. L. Ontario · F. Rampulla · V. Zimbone · V. Greco · S. Sciuto
Department of Biomedical and Biotechnological Sciences, University of Catania, Torre Biologica, Via Santa Sofia 97, 95125 Catania, Italy
e-mail: calabres@unict.it

A. Trovato
e-mail: trovato@unict.it

M. L. Ontario
e-mail: marialaura.ontario@ontariosrl.it

F. Rampulla
e-mail: francescorampulla1985@virgilio.it

V. Greco
e-mail: vgreco@unict.it

S. Sciuto
e-mail: ssciuto@unict.it

I. La Mantia
Surgical Specialties Department, ENT Unit - General Surgery and Medical, University of Catania, via Santa Sofia 87, 95125 Catania, Italy
e-mail: ilamantia@unict.it

R. Crea
Oliphenol LLC., 26225 Eden Landing Road, Suite C, Hayward, CA 94545, USA
e-mail: robertocrea@oliphenol.com

E. J. Calabrese
Department of Environmental Health Sciences, Morrill I, N344, University of Massachusetts, Amherst, MA 01003, USA
e-mail: edwardc@schoolph.umass.edu

S. I. S. Rattan and G. Kaur (eds.), *Nutrition, Food and Diet in Ageing and Longevity*, Healthy Ageing and Longevity 14, https://doi.org/10.1007/978-3-030-83017-5_13

mechanisms of cellular aging and lifespan extension, although further studies are required concerning optimal bioenergetic mechanisms promoting aerobic energy production and the underlying detrimental effects of reactive oxygen species (ROS) by-production with the interplayed nutrition and caloric intake modulatory effects. Consistently, ROS acting as sensors of intracellular nutrients and energy state regulate functional mitochondrial state. Interestingly, increasing evidence reports a functional crosstalk between ROS production by mitochondria and longevity pathways modulating lifespan across species thus ensuring healthy aging. Nrf2-dependent pathways of cellular stress response with their target antioxidant *vitagenes* are emerging as powerful systems capable to preserve redox homeostasis under environmental and metabolic stresses. *Vitagenes* encode redox longevity genes induced by oxidative damage including heat shock family (Hsp) Hsp32, Hsp70, glutathione, thioredoxin and sirtuin protein systems. During aging process, a gradual decline of the heat shock response occurs and this may prevent repair of protein damage. Therefore, there is a growing interest by scientific community in developing of novel preventive andpharmacological agents capable of inducing stress responses at the minimum dose within the broad frame of hormesis as therapeutic strategy in patients suffering from chronic degenerative diseases. The specialattention of this paper is focusedonpotential neuroprotective mechanisms of nutrition, in particular dietary polyphenols involved in the activation of *vitagenes* resulting in improved intracellular antioxidant defense systems against ROS damage leading to degeneration and death with considerable impact on brain health and longevity processes.

Keywords Vitagenes · Heat shock proteins · Nutritional polyphenols · Brain aging · Cellular stress response

13.1 Introduction

With the development of novel technologies and pharmacological interventions during last decades synergistic approaches based on the integration between Chinese and Western Medicine have emerged to preserve human brain health. The scientific community, recently, is spending consistent resources to prevent and combat aging and chronic degenerative diseases including Alzheimer disease (AD), Parkinson disease (PD) and cancer. Despite the best efforts of biomedical research resulting into promising outcomes by the sequencing of the human chromosomal single nucleotide polymorphisms (SNPs), and the identification of regional clusters of chromosomal SNPs (the HapMap) the underlying molecular mechanismsand genetic causes of the age-related diseases remain still elusive, while their incidence and morbidity either are continually increasing (Calabrese et al. 2006a; 2006b; 2006c; 2006d; Lodi et al.

G. Dionisio
Department of Molecular Biology and Genetics, Research Center Flakkebjerg, Aarhus University, Forsoegsvej 1, 4200 Slagelse, Denmark

2006; Scapagnini et al. 2006; Poon et al. 2006; Abdulet al. 2006; Perluigi et al. 2006; Mancuso et al. 2006a).

Aging is a highly complex process that encloses several vulnerabilities including environmental stressors, epigenetic alterations, mitochondrial deregulation and chronic inflammation. Interestingly, genetic and mitochondrial biology are being recognized as powerful candidate for expanding anatomical and mendelian paradigms to contribute to both accelerated and decelerated cellular or organismal aging (Mancuso et al. 2006b; Calabrese et al. 2007a, 2007b; Piroddi et al. 2007).The evolution of life requiresthe interplay between structure andmitochondrial energy and low environmental stressors as well. The endosymbiotic theoryexplains the origin of the eukaryotic cell from a glycolytic motile cell, which gave rise to the nucleus-cytosol, and an oxidative α-proteobacterium, which evolved into the mitochondrion about 2 billion years ago. Hence, these ancestral symbiotic mitochondrial events generate new insights on the evolutionary origin of aging. The discordance between cellular quiescence and mitochondrial proliferation results in cellular oxidative stress, ultimately leading to a gradual decline in host cell performance and age-related pathology. Therefore, aging evolved from a prominent conflict between maintenance of a normal quiescentstate and the evolutionarily conserved proliferative program driving the life cycle of mitochondria. In this light, mitochondrial energy is involvedfor lifeprocesses (Calabrese et al. 2007c, 2007d, 2007e, 2008a, 2008i; Athanasiou et al. 2007; Mancuso et al. 2007, 2008). Notably, in response to various environmental stressors, the cell employs oxidative phosphorylation and increases the flow of electrons through mitochondrial electron transport system (ETS) to generate a proton motive force and the extra ATP energy needed to defend and repair the cell from ROS facilitating healthy aging (Calabrese et al. 2011b). Emerging research focused its attention on dietary interventions or healthy nutrients targeting the mechanisms of brain aging to prevent neurodegenerative diseases (Flanagan et al. 2020; Miquel et al. 2018). According to the nutritional approach, existing evidence indicates that some nutrients or food ingredients, in particular specific vitamins, flavonoids and long chain ω-3 fatty acids have a potential to beneficially affect brain function (Samieri et al. 2018; Scarmeas et al. 2018).

It is noteworthy that an unhealthy lifestyle or poor nutrient intake can accelerate epigenetic aging. In contrast, healthy lifestyle and appropriate dietary supplementation delays biological age and the onset of chronic cognitive diseases (Quach et al. 2017). Therefore, there is now an increasing interest in how harmful lifestyle factors can be modified by bioactive nutrients or foods from the Mediterranean diet (MD) to prevent cognitive decline and generally improve health over the lifespan. One of the most recentrandomized studies conducted in Spain were the PREDIMED (Prevención con Dieta Mediterránea) studies investigating the impact of MD on human health (Estruch et al. 2018). Particularly, MD with extra-virgin olive oil was associated with higher scores of global brain performance following the 6.5 years nutritional intervention (Martinez-Lapiscina et al. 2013). Likewise, the NU-AGE study ("New dietary strategies addressing the specific needs of elderly population for healthy ageing in Europe") based on Mediterranean-style diet pattern improved brain health and quality of life by counteracting the deleterious effects of chronic inflammation and age-related cognitive decline in a total of 1250 older adults (aged 65–79) (Santoro et al. 2014). These promising outcomes are encouraged from a recent ongoing trial "MedEx-UK" designed to test the efficacy of MD intervention and physical activity as strategies to facilitate healthy brain ageing and reduce dementia risk in older adults (Shannon et al. 2021). Accordingly, higher adherence to MD over the year-long intervention was associated with statistically significant improvements to prevent the progression of cognitive decline and dementia compared to older adults with lower adherence (Marseglia et al. 2018). Notwithstanding, brain health benefits of this diet, further studies are needed to confirm the application and feasibility of the MD across the different European regions (Berendsen et al. 2018).

Currently, a great number of *in vitro* studies showed the capacity of nutritional polyphenols to modulate not only the expression of genes but also miRNA, proteins and other regulators of cell function, such as cell stress response pathways inducing epigenetic modifications to achieve healthy ageing (Claude et al. 2014; Vauzour et al. 2017). Moreover, experimental studies on animal models demonstrated that genetic and epigenetic manipulations as well asantioxidant polyphenols and caloric restriction (CR) improve brain health and extend maximal lifespan within the context of hormetic dose response.Specifically, CR appears to extend lifespan by reducing reactive oxygen species (ROS)-mediated oxidative damage. Indeed, low levels of ROS are positively implicated in the induction of many sensitive signal transduction cascades, thataretightly regulated by a complex network of cellular stress response pathways. Benefits of hormetic stimuli, such as the antioxidant supplementation and energy restriction lead to low ROS formation that escape from the mitochondrial electron transport chain and act as sensing signaling molecules to promote longevity signaling pathways, including *vitagenes* which are genes involved in preserving cellular redox homeostasis during stressful conditions. *Vitagenes* encode for heat shock proteins (Hsp) Hsp32, Hsp70, γ-glutamylcysteine synthetase, thioredoxin and sirtuins as biomarkers for stress adaptation, cross-tolerance and biological resilience underlying hormesis or preconditioning (Calabrese et al. 2011b). Importantly, antioxidant polyphenolsshowed to be neuroprotective through the activation of hormetic Nrf2-dependent pathways, including *vitagenes*. The hormesis concept asserts that low doses of a great diversity of environmental challenges (i.e. caloric restriction and antioxidant polyphenols) at a long-standing affect animal and human health prolonging lifespan and represent potential preventive and therapeutic strategies for chronic degenerative disorders (Calabrese et al. 2011a). Given the significance of cytoprotective properties of Nrf2-vitagene pathwaythere is now a strong interest in discovering and developing novel preventive and pharmacological agents capable of inducing hormetic stress response. In this review we discuss the most current and up to date understanding of the potential signaling mechanisms by caloric restriction, as well antioxidant polyphenols that activating *vitagenes* can enhance intracellular defensive systems involved in bioenergetic and cellular stress homeostasis with consequent impact on brain health and longevity processes (Calabrese et al. 2008b, 2008c, 2008d, 2008e, 2008f, 2008g; Di Renzo et al. 2008).

13.2 Hydroxytyrosol/hidrox® A Prototypical Polyphenolic Agent for Brain Health

13.2.1 Hydroxytyrosol/Hidrox®

Emerging research has recently focused on brain health benefits of the major olive oil polyphenols, particularly hydroxytyrosol (HT) and hydroxytyrosol (HT)-rich aqueous olive pulp extract (Hidrox®), that at low doses exert multiple preventive and pharmacological activities such as anti-inflammatory (Bitler et al. 2005), antioxidant (Bellumori et al. 2019), anti-aging (Di Rosa et al. 2020) and anti-proliferative effects in vitro and in vivo (Luo et al. 2013; Imran et al. 2018) Fig. 13.1. Aging is currently recognized as a multifactorial process involving diverse and complex

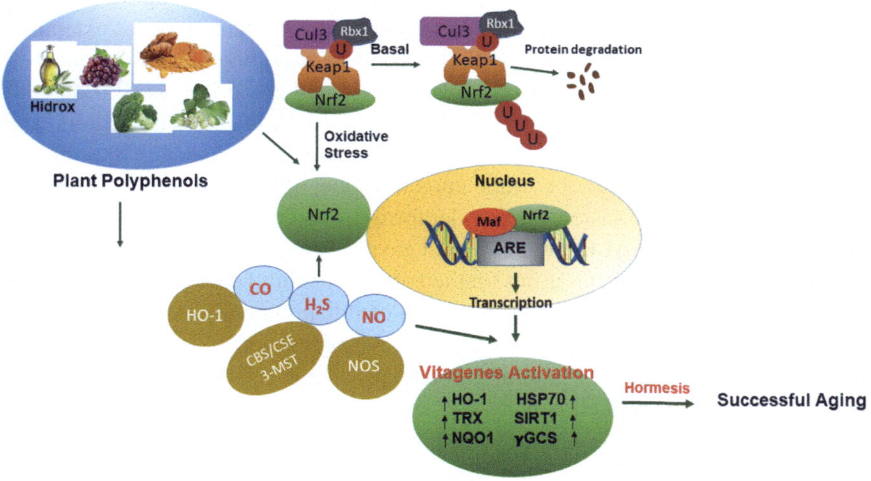

Fig. 13.1 Modulation of Nrf2-vitagene pathway by antioxidant polyphenols for brain health during aging.Under physiological conditions, Nrf2 is bound to its repressor Keap1 and is restricted to the cytosol where it undergoes ubiquitination and proteasomal degradation via association with the Cul3-Rbx1 based E3/ubiquitin ligase complex. Under stress condition, Nrf2 is released from Keap1 and it is translocated into the nucleus where it binds to the phase 2 of ARE in heterodimeric combination with Maf transcription factor in the DNA promoter region. Dysregulation of autophagy flux leading to p62 activation, an autophagy adaptor protein, which accumulates and sequesters Keap1 into autophagosomes for degradation. Consequently, Keap1 can no longer bind Nrf2, inducing an upregulation of Nrf2 pathway and related target genes. Moreover, antioxidant polyphenols are small molecules that at low dose reverse oxidative stress and inflammation by activating Nrf2 nuclear translocation and transcription (phase 2) of *vitagenes*. The upregulation of vitagene pathway such as HO-1, Hsp70, Trx, Sirt1, NQO1, γ-GCS induces neuroprotection andsuccessful aging. Nuclear factor-erythroid 2 p45-related factor 2 (Nrf2), Kelch-like ECH-associated protein 1 (Keap1), antioxidant response element (ARE), heme-oxygenase 1 (HO-1), heat shock protein 70 (Hsp70), thioredoxin (Trx), sirtuin 1 (Sirt1), NAD(P)H: quinone oxidoreductase 1 (NQO1), γ-glutamylcysteine synthetase (γ-GCS)

alterations including mitochondrial dysfunction, epigenetic alterations, deregulated nutrient sensing. Recent exhaustive evidence in the literature indicated as olive oil polyphenols significantly modulate the aging process, so tightly connected to oxidative stress and chronic inflammation. Thus, notable attention is paid to the anti-inflammatory and brain health effects of Hidrox® produced by harvesting it from an olive oil production by-product, the olive vegetation water (OVW or olive juice), and normally discarded as wastewater with an excellent safety profile, high bioavailability, tissue distribution and multiple mechanisms (Soni et al. 2006). Among the phenolic compounds, hydroxytyrosol (50–70%) represents the main constituent of the pulp extract, while other polyphenols present include oleuropein (5–10%), tyrosol (0.3%), oleuropein aglycone and gallic acid. The process developed by CreAgri, Inc is both solvent-free and environmentally friendly, employs citric acid (1%) for the mild acidic hydrolysis of the naturally occurring hydroxytyrosol esters and produces HT concentration ranging from 20 g/kg to 50 g/kg olive dry matter (Soni et al. 2006). Preclinical studies showed that HT efficiently exerts radical scavenging activities in vitro and significantly increases the expression of antioxidant enzymes extending lifespan in in vivo models of aging (Visioli et al. 1998; Rietjens et al. 2007; Luo et al. 2019; Calabrese et al. 2016). Of relevance, Bitler and coworkers (2005) demonstrated the anti-inflammatory effects of OVW obtained during olive oil milling, and containing 25 mg HT/g solid, (HT is > 45% of the total polyphenols). Other phytochemicals present in Hidrox® include apigenin, luteolin and elenolic acid. In this study, Hidrox® was administered in powder form (freeze dried OVW) to lipopolysaccharide (LPS)-treated BALB/c mice by oral gavage. Therefore, OVW and HT were tested at two different concentrations for their ability to inhibit the production of TNF-α, a primary cytokine involved in progression of inflammatory process. It has been observed that OVW at a dose of 125 mg/mouse (500 mg/kg) reduced serum TNF-alpha levels by 95%. Also, in the human monocyte cell line (THP-1) OVW suppressed LPS-induced TNF-α production by 50% at a concentration of 0.5 g/L (equivalent to approximately 0.03 g/L simple and polyphenols). OVW did not cause toxic effects in all of the experiments investigated. When OVW was combined with glucosamine, a component of proteoglycans and glycoproteins, their synergistic effects decreased the inducible nitric oxide synthase production in macrophage cells and serum TNF-α levels in the LPS-treated mice (Bliter et al. 2005). Interestingly, purified HT, in comparison, did not induce any reduction of TNF-α, in any of these experiments. Given the important role of HT in inflammatory signaling, the protective mechanism of HT includes, lastly the activation of the NF-κβ signaling cascades occurring in aging and chronic inflammatory diseases (Killen et al. 2014).

The beneficial effects of Hidrox® were confirmed by an in vitro study conducted by researchers at DSM (Basel) in 2011 (Richard et al. 2011). They used fractionated OVW to identify its anti-inflammatory constituents of the natural formulation. The fractions obtained by preparative HPLC were tested on macrophages (RAW264.7 cells) stimulated with LPS to evaluate their effect on the inhibition of inflammatory mediators (i.e. NO and PGE2, cytokines, interleukins and chemokines) in a hormetic dose–response manner, thereby supporting that HT exerts its anti-inflammatory effects, at least in part, mediated via the NF-kB cellular pathway.

The authors, however, indicated the potential synergistic/complementary action of unknown compounds present in the OVW. In addition, other in vitro studies on endothelial cells (Scoditti et al. 2014) and human monocytic cell line (Zhang et al. 2009) have highlighted the ability of HT to downregulate NF-kB activation and its translocation into the nucleus. Furthermore, animal studies have shown the safety profile of Hidrox®, since given by oral gavage to rats did not appear induce reproductive and developmental toxicity at the maximum dose of 2000 mg/kg/day (Christian et al. 2004). Consistent with the substantial studies above reported and based upon to the high safety profile of the OVW formulation (GRAS certification was obtained by CreAgri in 2004), human studies demonstrated that HT, in the form of freeze dry powder, orally administered in 4 capsules/day to 90 patients affected by osteo and rheumatoid arthritis induced a potent anti-inflammatory activity manifested as reduction of swelling, pain and improved daily motilities with an associated reduction of two inflammation biomarkers, CRP and homocysteine (Bitler et al. 2007). Other clinical trials tested hydroxytyrosol, in its natural environment (OVW, Olivenol™ or Hidrox®), on 100 healthy volunteers in Malaysia (Visioli et al. 2009) and 50 patients in Japan (Numano et al. 2007). In the first study, a single dose (2 ml OWV, 2.5 mg HT) administered orally significantly increased total (reduced and oxidized) glutathione antioxidant levels in plasma. Although the authors did not observe any difference of total antioxidant capacity in the blood. The second study conducted in Japan included patients with high cholesterol levels and confirmed the protective effect of olive polyphenols and HT on lipid profile. Doses as low as 5 mg HT/day produced a remarkable reduction of high MW cholesterol (VDL fraction 3, 4 and 5) while HDL cholesterol concentration was unchanged or improved (Visioli et al. 2009). Surprisingly, in this study the same HT dosage of 5–10 mg/daily carried out important improvements in 5 patients with skin disorders (psoriasis purulenta, atopic, allergic and seborrheic dermatitis), this could explain a potential protective effect of HT on the immune system (Numano et al. 2007). Consistent with this line of evidence, our research group published recent in vivo studies with olive polyphenols, in particular, Hidrox® that exerts higher anti-inflammatory and neuroprotective effects than the same amount of pure HT, improving overall of healthspan and longevity as well as stress resistance in a rotenone-stressed PD model of C. elegans and rodents (Brunetti et al. 2020; Di Rosa et al. 2020; Siracusa et al. 2020).

Additionally, recent double-blinded randomized preclinical studies indicated the neuroprotective effect of HT-enriched diet, in particular a dose of 45 mg HT/kg BW/day favors the recovery after ischemic stroke by improving stroke-associated learning and motor impairments. This effect is probably associated to an increase cerebral blood flow (CBF). Also, this study showed that HT induces an increase of functional and structural connectivity by enhancing transcription of the postsynaptic density protein 95 (Psd-95) marker, an anti-inflammatory activity by reducing ionized calcium-binding adapter molecule 1 (IBA-1) level as well as a neurogenic potential by raising the expression of brain derived neurotrophic factor (BDNF) in mice (Calahorra et al. 2019). In addition, increasing evidence reported the brain health benefits of HT supplementation that at a dose of (50 mg/kg of diet) improved cognitive functions and reduced Aβ42 and pE3-Aβ plaque in the cortex of TgCRND8 mice

(Nardiello et al. 2018). Moreover, it has been demonstrated that aged mice display a downregulation of Sirt1, CREB, Gap43, and GPx-1 gene expression in the brain tissue. In this way, supplementation of diet rich in olive polyphenols at a dose range of 1 to 10 mg/kg has positive long-term effects on cognition and energy metabolism in the aged brain (Reutzel et al. 2018). Finally, a recent interesting evidence has demonstrated the virucidal activity of Hidrox® at a dose of 20 mg/kg by inactivating severe acute respiratory syndrome coronavirus 2 (SARS-CoV-2) infection in time and concentration dependently, proving once again that Hidrox® is more potent than of pure HT (Takeda et al. 2021).

13.2.2 Tracking the Polyphenols Metabolic Fate in Human Tissues

The presence of polyphenols and related chemicals (compounds) in plants shows great variability in both quality and quantity. These plant secondary metabolites constitute one of the most diverse and widely distributed groups of natural compounds. Their metabolism is also quite variable and not always fully understood. Many of them are also present in plants as glycosides that, after human ingestion, are hydrolyzed by intestinal enzymes and/or microflora in the colon to the corresponding aglycones (Ozdal et al. 2016; Giovinazzo et al. 2015). The phenolic compounds that are taken up into the bloodstream undergo rapid o-glucuronidation, o-sulphonation and o-methylation in the liver (Cassidy et al. 2017). In these conjugated forms, polyphenols are found as the first metabolites in the plasma and then, by the action of the intestinal microbiome, they are fragmented into smaller phenolic acids which can be re-directed into the bloodstream and also excreted in the urine (Crozier et al. 2009). Not surprisingly, the degree of absorption, metabolism, conjugation, and excretion of polyphenols has been found to be highly variable both within and between individuals (Bondia-Pons et al. 2014). If we add to the above that the bioactivities of polyphenols are widely varied, it follows that it is essential, both in the therapeutic as well as toxicological studies, to be able to perform detailed investigations on the bioavailability, metabolism, and fate of these compounds possibly in those that are the target tissues in humans or at least in related organoid systems.

Thus, regardless of the purpose of the study, it is necessary to be provided with sensitive methods that can monitor, in blood (de Villiers et al. 2012), as well as in other tissues or biofluids, both specific phenolic compounds or their metabolites and, more in general, a phenolic profile. Unlike the determination of the total polyphenol content in a generic natural source extract, which is mainly carried out by spectrophotometric assay, the analysis of specific polyphenols that may be found in human tissues (or even organoids) samples requires the separation of the individual polyphenol species, which is commonly performed by high-performance liquid chromatography (HPLC) or capillary electrophoresis (CE), and their subsequent detection by different detectors (UV–vis, mass spectrometry). Coupling HPLC, or rather UPLC (ultra-high performance liquid chromatography), with mass spectrometry is nowadays a fairly sensitive analytical procedure that allows to obtain also structural information on single analytes and, in some instances, the elucidation of their structure.Prior to the introduction of UPLC, capillary electrophoresis (CE) received substantial attention as an alternative to the HPLC separation method; this was because the resulting separations were often faster and more efficient than the corresponding HPLC separations and because CE allows in one run the analysis of polyphenol compounds with different nature.However, compared to current standard UPLC methods, CE methods generally suffer from lower sensitivity and robustness and are therefore mainly used in that analytical fields where it still provides actual advantage over liquid chromatography, such as in chiral separation (Sun et al. 2012). When mass spectrometry is used for the identification of analytes after their separation, the choice of the analyte ionization mode turns out to be crucial to unambiguously identify molecular ions and their fragments. In particular, for the characterization of anthocyanin glycosides in the presence of the corresponding flavonol glycosides, it was proposed to apply the negative ionization mode as an alternative to the usual positive ionization mode; indeed, the MS spectra acquired in the negative ionization mode have been shown a valuable tool for differentiating anthocyanins from non-anthocyanin polyphenols (Olech et al. 2020a). An important issue in the UPLC-MS analysis of polyphenols is their quantification. The content of antioxidant metabolites in a given extract can be determined by UPLC-ESI–MS/MS analysis. In this case the triple quadrupole MS detector is operated in MRM (Multiple Reaction Monitoring) scan mode, which ensures high sensitivity and selectivity of the analysis (Elessawy et al. 2020). This technique allows you to quickly distinguish compounds that have the same parent ions (same molecular weight) but give different fragment ions (Olech et al. 2020b; Adebooye et al. 2018). Combining a specific mass of a precursor ion with the equally specific mass of one of its product ions is generally an unambiguous and sensitive method to selectively track and quantitate a compound of interest. Since two stages of mass selection are utilized, MRM assays are particularly useful for the specific analysis of target compounds in complex mixtures and matrices, such as tissue or organoid extracts. This is why MRM mode has become the preferred method for the quantitative analysis of known or target compounds. An additional aspect to account for in wanting to track polyphenols in human tissues or organoids concerns the step of extracting them from these tissues before proceeding to their qualitative and/or quantitative analysis. Differently from the analysis of polyphenols in plant extracts or food products (Ajila et al. 2011; van Duynhovena et al. 2011), standardized procedures for the preparation, extraction, and analysis of tissue (or organoid) samples from human sources are not yet available. Of the methods offered in the current literature (Yang et al. 2019; Shafaei et al. 2019), each may have specific advantages toward the extraction and detection of certain polyphenolic compounds and this must be accounted for prior to analysis. Since it is not uncommon that the choice of a given analytical procedure does not meet all the requirements of a specific experimental investigation, the best approach to characterize the polyphenolic composition of a sample of human tissues (or biofluids) may at present be to use an appropriate combination of different analytical assays.

13.3 The Hormesis Paradigm and Nrf-2 Dependent Vitagene System in Aging and Longevity

13.3.1 Cellular Stress Response, HSF Biology and the Vitagene Network

Cellular stress response is recognized as the ability of a cell to counteract stressful challenges and restore redox homeostasis during aging process (Fig. 13.1). This intracellular mechanism requires the activation of stress adaptive response pathways, in particular the Nrf2 signaling pathway that regulates the transcription of cytoprotective genes termed *vitagenes* (Calabrese et al. 2010b; Calabrese et al. 2012f) including pro-survival molecules such as heat shock protein 70 (Hsp70), heme oxygenase-1 (HO-1), glutathione, sirtuins and the thioredoxin/thioredoxin reductase system (Calabrese et al. 2008g; Bellia et al. 2011; Di Paola et al. 2011; Pennisi et al. 2011; Scapagnini et al. 2011) endowed with anti-oxidant and anti-apoptotic activities. In this context, Heat shock response (HSR), represents an evolutionarily conserved cell defense mechanism suitably orchestrate to ensure health of the proteome and successful aging (Calabrese et al. 2008h, 2008i 2009a, 2009b, 2009c, 2009d, 2010a, 2010b; Bellia et al. 2009). The production of heat shock proteins, including molecular chaperones, is essential for the folding and repair of damaged proteins, serving thus to maintain protein homeostasis and are critical to cell survival during proteotoxic stress (De Lorenzo et al. 2010; Calabrese et al. 2010c, 2010d, 2010e; Di Domenico et al. 2010; Perluigi et al. 2010; Calabrese et al. 2010a). Several lines of evidence have shown that HSR contributes to establishing a cytoprotective state in a wide range of human diseases, including inflammation, cancer, aging and neurodegenerative disorders (Siciliano et al. 2011). Therefore, there is now strong interest in discovering and developing novel pharmacological agents capable of inducing the HSR to prevent misfolding, aggregation, and proteome mismanagement during aging and related cognitive disorders (Zhang et al. 2011; Westerheide et al. 2012). Heat shock transcription factors (HSFs), vitally important for all organisms to survive to acute or chronic proteotoxic stress, exert also a crucial role for normal development and lifespan-enhancing pathways by expanding the HSF target range well beyond the heat shock genes. Consistent with this observation, HSFs targeting cellular stress response genes and regulate their ubiquitous expression at the transcriptional, translational and post-translational levels, while they are maintained in an inactive state under physiological conditions (Calabrese et al. 2012a, 2012b). Mammalian HSF family consists of four members: HSF1, HSF2, HSF3 and HSF4, which recognize by binding the heat shock element (HSE), composed of inverted repeats of a pentameric consensus nGAAn sequence. HSF1 is a master transcriptional regulator of HSR in humans and invertebrates (i.e., Drosophila, Caenorhabditis elegans, and yeast), whereas HSF-2, -3, and -4 are expressed in vertebrates and exhibit several functional properties in the regulation of stress and lifespan. These distinct HSFs are regulated by heat stress, protein–protein interactions, extensive post-translational modifications, proteasomal degradation and small-molecule activators and inhibitors. In addition to thermal stress, the inducible expression of heat shock proteins (i.e., Hsp27, Hsp40, Hsp70 and Hsp90) is also triggered by environmental redox alterations or exposure to ROS electrophiles which cause trimerization and phosphorylation of DNA binding of HSF1 resulting in accumulation into the nucleus where it binds HSE for the transcription of stress responsive genes, thus pointing to the importance of the cysteine redox state for the maintenance of this transcription factor in its active control state (Calabrese et al. 2012a). Likewise of other transcription factors, HSFs are composed of functional domains which have been well characterized for HSF1. The DNA-binding domain (DBD) is a signature domain highly conserved of HSFs for recognition of target-genes that contains a helix-turn-helix motif and intermolecular interactions of the hydrophobic heptad repeats (HR-A/B) forming coiled coil, characteristic for many leucine zippers (Fujimoto et al. 2010). The reversible conversion of HSF-1 from monomer to trimer involves respectively the transition from the inert state in which the intramolecular coiled-coil formed by the amino-terminal HR-A/B interacts with the carboxy-terminal heptad (HR-C) located between the regulatory and trans-activation domains (TADs), to the functional state comprising only intermolecular HR-A/B interactions. The trans-activation domain in all HSFs is positioned at the amino terminal region, and is composed of two bipartite modules—AD1 and AD2, which are rich in hydrophobic and acidic residues allowing rapid and prolonged response to a wide range of stressors (Fujimoto et al. 2010). Therefore, the N-terminal domains of HSF1 mediate trimerization and acquisition of DNA-binding ability and are concentration dependent, whereas the C-terminal domains of the protein facilitate transcriptional activation of stress target genes and also upregulate HSF1 (Calabrese et al. 2012a; Fujimoto et al. 2010). HSF1 possesses an intrinsic thermosensing role modulated via posttranslational modifications (Calabrese et al. 2011b; Fujimoto et al. 2010). Interestingly, pharmacological inhibition of Hsp90 activity has emerged as a potential target in treating neurodegenerative diseases especially AD and PD. In this context, recent evidence showed that the inhibition of Hsp90 by geldanamycin and its derivatives converts HSF1 into a DNA-binding trimeric state in cells suppressing atypical neuronal activity and improving protein aggregation and its related toxicity via the induction of HSP70 in AD. On the other hand, HSF1 is maintained in an inactive monomeric state by the interaction with Hsp90 in non-stressed cells (Alam et al. 2017). In the activation cycle, HSF1 undergoes extensive post translational modifications, such as phosphorylation, sumoylation and acetylation. Specifically, under unstressed conditions HSF1 is a monomeric phosphoprotein on several serine residues. Upon stress, phosphorylation on serine 303 which is located within the regulatory domain (RD) of HSF1 is required for sumoylation. SUMO proteins are transiently and covalently bound to a specific site at lysine 298 residues of multiple cellular proteins. The sumoylation mechanism described for the first time in HSF1 occurs on a consensus sequence consisting of the tetrapeptide ΨKxE, where Ψ is represented by a branched hydrophobic amino acid and K the SUMO acceptor that presents the phosphorylation-dependent sumoylation motif (PDSM), mediating sumoylation. The acetylation coincides with the attenuation phase of the HSF1 activation cycle. Accordingly, acetylation of HSF1 is modulated by the functional balance of acetylation by histone acetyl-transferase E1A binding protein (p300) and deacetylation by the NAD^+-dependent sirtuin, SIRT1. The upregulation of the deacetylase and longevity factor SIRT1 enhances and prolongs the HSF1 binding to the human HSP70 promoter maintaining it in a competent state. Conversely, downregulation of SIRT1 enhances the acetylation of HSF1 and the attenuation of DNA-binding without affecting the formation of HSF1 trimers (Calabrese et al. 2011b). It is noteworthy that SIRT1 is also implicated in caloric restriction and aging as well as in lifespan extension. Indeed, the age-dependent loss of SIRT1 and deregulated HSF1 activity leads to impairment of the heat shock response and proteostasis in senescent cells, thus connecting the heat shock response to nutrition and aging (Calabrese et al. 2011b). Emerging research is focusing growing interest on pharmacological or nutritional antioxidant therapy based on the current knowledge of the protective potential of small-molecule activators and inhibitors modulated by HSF1 including *vitagenes* (i.e., Hsp90, Hsp70 and SIRT1) (Calabrese et al. 2010a; Akerfelt et al. 2012; Fujimoto et al. 2010). Recent evidence demonstrated that dietary supplementation with vitamin C (1 g) significantly ameliorated the detrimental effects of heat stress on liver, heart and kidney tissues of chickens at cellular and molecular levels by upregulation of heat shock proteins (HSP60, HSP70 and HSP90), heat shock factors (HSF-1 and HSF-3) and enzymatic antioxidants such as malondialdehyde (MDA), superoxide dismutase (SOD), glutathione S-transferase (GST), glutathione peroxidase (GPX), catalase (CAT) (Albokhadaim et al. 2019). Moreover, it has been reported that resveratrol exerts potent anti-inflammatory activity and viral eradication by increasing the expression of HSF1, SIRT1 and p53 in vitro and in vivo (Wang et al. 2017; Zeng et al. 2017). Importantly, subsequent preclinical studies revealed that treatment with celastrol (a quinone methide triterpenoid) induces hyperphosphorylation of HSF1, enhanced binding of HSF1 to the HSE and transcriptional activation of endogenous heat shock genes in a similar manner to that induced by heat shock stress (Westerheide et al. 2004). Further studies have suggested that functional inhibition of Hsp90 by celastrol may be the initial event that triggers dissociation of HSF1 from the heat shock complex (Salminen et al. 2010; Hieronymus et al. 2006). It is interesting to note that antioxidant sulforaphane contained in broccoli (Brassica oleracea), induces nuclear accumulation of HSF1 and activation of heat shock response through the upregulation of Hsp27 in several cell culture models, including human cell lines, as well as in animals after a single oral dose of 10 μm of the isothiocyanate (Gan et al. 2010). Inflammation caused by microglial activation induces adaptive immune responses by producing pro-inflammatory mediators, including nuclear factor (NF)-κB, caspase-3, HSF1 and consequently increased release of HSP60. Consistent with this, curcumin at a dose of 5 μg/ml markedly inhibits microglial activation by downregulation of HSF1 and reduced HSP60 expression in cells (Ding et al. 2016). Lastly, recent evidence reported that chlorogenic acid, a polyphenol distributed in plants and plant-derived food, exerts antioxidant and protective activities against cell stress in hormetic heat conditions. Particularly, chlorogenic acid confers thermal stress tolerance and autophagy by the activation of hypoxia-inducible factor-1 (HIF-1) mediated by HSF1 and HSP-70/HSP-16 in C. elegans (Carranza et al. 2020).

In addition to HSFs some of the vitagenes are also upregulated within the "phase 2 response", known as "the electrophile counterattack response", a cytoprotective response that protects against various electrophiles and oxidants (Athanasiou et al. 2007; Mancuso et al. 2007; Calabrese et al. 2008a; Mancuso et al. 2008) (Fig. 13.1). Examples include heme oxygenase 1, thioredoxin and thioredoxin reductase, all of which can be upregulated by the transcription factor Nrf2 (Nuclear erythroid 2-related factor 2) co-ordinately with a battery of cytoprotective proteins, such as glutathione transferases (GST), UDP-glucuronosyltransferase, NAD(P)H:quinone oxidoreductase 1 (NQO1), epoxide hydrolase, ferritin, γ-glutamylcysteine synthetase, glutathione reductase, aldo–keto reductases, and glutathione conjugate efflux pumps (Athanasiou et al. 2007; Mancuso et al. 2007; Calabrese et al. 2008a; Mancuso et al. 2008). This elaborate network of protective mechanisms allows eukaryotic organisms to counteract the damaging effects of oxidants and electrophiles, major agents involved in the pathogenesis of cancer, atherosclerosis, neurodegeneration, and aging. The gene expression of these cytoprotective proteins is coordinately regulated by a common molecular mechanism that involves the Keap1/Nrf2/ARE pathway. The upstream regulatory regions of these genes contain single or multiple copies of the antioxidant/electrophile response elements (ARE, EpRE) with the consensus sequence 5'-A/CTGAC/GNNNGCA/G-3').

The major transcription factor that binds to the ARE is nuclear factor erythroid 2-related factor 2 (Nrf2), a basic leucine zipper transcription factor. Activation of gene expression requires that Nrf2 binds to the ARE in heterodimeric combinations with members of the small Maf family of transcription factors. Under basal conditions the pathway operates at low levels due to the repressor function of the cytosolic protein Kelch-like ECH-associated protein 1 (Keap1), which binds to the E3 ubiquitin ligase Cullin3-RING box1 (Cul3-Rbx1) and presents Nrf2 for ubiquitination and subsequent proteosomal degradation.

Vitagenes, containing in its promoter region the antioxidant response element (ARE), there has been a growing interest over the last years in the heme oxygenase (HO) system, the family of enzymes that control the initial and rate-limiting steps in heme catabolism (Athanasiou et al. 2007; Mancuso et al. 2007; Calabrese et al. 2008a; Mancuso et al. 2008). The heme oxygenases have been recognized as dynamic sensors of cellular oxidative stress and modulators of redox homeostasis throughout the phylogenetic spectrum. Heme oxygenases are located within the endoplasmic reticulum where they act in association with NADPH cytochrome P450 reductase to oxidize heme to biliverdin, free ferrous iron and carbon monoxide (CO). Biliverdin reductase further catabolizes biliverdin to the bile pigment, bilirubin, a linear tetrapyrrole which has been shown to effectively counteract nitrosative stress due to its ability to interact with nitric oxide (NO) and reactive nitrogen species (RNS) (Athanasiou et al. 2007; Mancuso et al. 2006b; Calabrese et al. 2008a; Mancuso et al. 2008). Bilirubin is then conjugated with glucuronic acid and excreted (Athanasiou et al. 2007; Mancuso et al. 2006b; Calabrese et al. 2008a; Mancuso et al. 2008). Bilirubin has been shown to serve as an endogenous scavenger for both NO and RNS, which may alter the redox status of the cell and originate the nitrosative stress (Athanasiou et al. 2007; Mancuso et al. 2006b; Calabrese et al. 2008a; Mancuso et al. 2008).

Despite this important antioxidant properties, if produced in excess, as in the case of haemolytic anaemia or sepsis, unconjugated bilirubin becomes neurotoxic through multiple mechanisms involving the disruption of cell membrane structure, the reduction of mitochondrial transmembrane potential and the activation of the apoptotic cascade (Athanasiou et al. 2007; Mancuso et al. 2006b; Calabrese et al. 2008a; Mancuso et al. 2008). Mammalian cells express at least two isoforms of heme oxygenase: HO-1 and HO-2. A third protein, HO-3, is determined to be a retrotransposition of the HO-2 gene (pseudogene) and has been found unique to rats. Although HO-1 and HO-2 catalyze the same reaction, they play different roles in protecting tissues against injury.

A convincing hypothesis suggests that HO-1 induction is one of the earlier cellular response to tissue damage and is responsible for the rapid clearance of the intracellular pro-oxidant heme and its transformation into CO and biliverdin, the latter being the precursor of the antioxidant bilirubin (Fig. 13.1). On the contrary, constitutively expressed HO-2 is primarily involved in maintaining cellular heme homeostasis as well as in the sensing of intracellular levels of gaseous compounds including NO and CO. Due to the antioxidant response element (ARE) contained in its promoter region, redox regulation of HO-1 gene is now well defined (Athanasiou et al. 2007; Mancuso et al. 2006b; Calabrese et al. 2008a; Mancuso et al. 2008). HO-1, in fact, can be induced by several stimuli associated with oxidative and/or nitrosative stress, such as heme, Aβ, dopamine analogues, H_2O_2, hyperoxia, UV light, heavy metals, prostaglandins, NO, peroxynitrite, Th1 cytokines, oxidized lipid products and lipopolysaccharide, as well as certain growth factors.

In this light, very promising ROS scavangers are the polyphenolic compounds contained in some herbs and spices, e.g., curcumin. Curcumin is the active antioxidant principle in *Curcuma longa*, a colouring agent and food additive commonly used in Indian culinary preparations. This polyphenolic substance has the potential to inhibit lipid peroxidation and to effectively intercept and neutralize ROS and RNS. In addition, curcumin has been shown to significantly increase HO-1 in astrocytes and vascular endothelial cells. This latter effect on HO-1 can explain, at least in part, the anti-oxidant properties of curcumin, in particular keeping in mind that HO-1-derived bilirubin has the ability to scavenge both ROS and RNS. Epidemiological studies suggested that curcumin, as one of the most prevalent nutritional and medicinal compounds used by the Indian population, is responsible for the significantly reduced (4.4- fold) prevalence of AD in India compared to United States (Mancuso et al. 2008; Scapagnini et al. 2011; Westerheide et al. 2012).

Based on these findings, Lim and colleagues have provided convincing evidence that dietary curcumin given to an AD transgenic APPSw mouse model (Tg2576) for 6 months resulted in a suppression of indices of inflammation and oxidative damage in the brain of these mice (Song et al. 2012). Furthermore, increasing evidence indicates that curcumin inhibits NFkB activation, efficiently preventing cell death (Begum et al. 2008; Gupta et al. 2011; Di Renzo et al. 2012). Whereas the acute induction of this enzyme in neural and other tissues is predominantly cytoprotective in nature, protracted or repeated up-regulation of the Hmox1 gene in astrocytes, oligodendroglia and possibly neurons may perpetuate cellular dysfunction and demise in many chronic degenerative and neuroinflammatory conditions long after provocative stimuli initiation. Within this context, heme-derived free ferrous iron, CO, and biliverdin/bilirubin are all biologically active substances that can either ameliorate or exacerbate neural injury contingently to the specific disease conditions, such as intensity and duration of HO-1 expression and/or the nature of the resulting redox milieu. In 'stressed' astroglia, HO-1 hyperactivity promotes mitochondrial sequestration of non-transferrin iron and macroautophagy and may thereby contribute to the pathological iron deposition and bioenergetic failure found in most age-related oxidant neurodegenerative disorders. Glial HO-1 expression may impact also cell survival and neuroplasticity by modulating brain sterol metabolism and proteosomal degradation of neurotoxic protein aggregates (Calabrese et al. 2012a).

13.3.2 Hormetics Compounds and Redoxomics

Nutraceutical compounds that trigger the hormetic response are called "hormetins" *(see the chapter on nutritional hormetins in this book).* Their hormetic mechanisms in ageing is to help the endogenous redox equilibrium to avoid imbalances. Hormetins are found in many fruits and vegetables as form of product of their secondary (or specialized) metabolism. Examples of hormetins are for example, group C and E vitamins, flavonoids, anthocyanins and other antioxidant polyphenols (resveratrol, gallic acid, etc.) (Calabrese et al.2012e), microminerals such as selenium and zinc which have been shown to induce a typical hormetic dose–response (Hayes 2007). The benefits of hormetins are obtained through mitigation of the oxidative stress pathways as maintenance and repair of DNA damage pathways and antioxidant protection pathways (Rattan 2008). Powerful life–prolonging properties were achieved with feeding various phytochemical at low dosage in the nematode worm Caenorhabditis elegans (Cypser et al. 2002).

Cysteine nitrosylation is rapidly reverted to reduced cysteine in vitro and in vivo by spontaneous non-chemical reduction operated by vitamin C (ascorbic acid). Anyway, since cysteine nitrosylation is part of the unfolded protein response, UPR, (Uehara et al. 2006), vitamin C as hermetic compound might delay or jeopardize the UPR. Nevertheless, in redox proteomics which uses SDS-PAGE to visualize the nitrosylated proteins at cysteine level a powerful technique is the biotin switch method (Forrester et al. 2009). In brief, in vitro, cellular protein lysates containing S-nitrosylated cysteines and not are first blocked in their reduced cysteines with S-methylthiolation agent, S-methylmethanethiosulfonate (MMTS). Secondarily, ascorbate is used to remove the nitro group from cysteines (SNO) to obtain a free thiol (trans-nitrosation reaction) and the byproduct O-nitroso-ascorbate. Free cysteines are then biotinylated with biotin-HPDP via disulfide exchange mechanism. Biotinylated proteins are affinity captured by streptavidin-agarose, eluted and analyzed by SDS-PAGE/immunoblotting. Nowadays, omics techniques can use label free single or multiplex cysteine isobaric labelling compounds to monitor and compare different samples at the same time (Pan et al. 2014; McDonagh et al. 2014).

In facts, using iodoacetyl-based Cys-reactive isobaric tags as high throughput methods (iodoTMT 6-plex,), quantitative cysteine proteome was detected in H9c2, rat cardiomyocytes, showing differences in multiplexed redox samples treated with normoxia or hypoxia in the absence and presence of S-nitrosoglutathione (GSNO) (Pan et al. 2014).Adult and old mice skeletal muscle were analyzed by comparative redox proteomics showing a decrease in redox-responsive proteins in aged skeletal muscle (McDonagh et al. 2014). Oxidation of total proteins increased in old mice skeletal muscle as compared with adult mice ones. Besides the hormetins that are compounds that ameliorate the oxidative stress, the hermetic compounds instead are activating, at an experiment level only, called "the hormetic concentration zone"), the Keap1/Nrf2/ARE pathway: i.e., sulforaphane, curcumin, carnosic acid, dimethyl fumarate, etc. (Calabrese et al. 2014).

13.3.3 Hormetic Compounds as Metal Chelatorsand ROS Scavengers

Notably is the fact that some of the hormetic compounds acts as direct iron/heavy metal chelators, e.g., curcumin (Rainey et al. 2019; Balasubramanyam et al. 2003). Free Fe(II/III) and Cu(II) in the brain originated a ROS cascade by metal catalyzed oxidation occurring in neurogenerative conditions (i.e., PD and AD). ROS can oxidize DNA bases and the repair mechanism of DNA is regulated by the so-called base excision/SSB repair (BER/SSBR) pathway which uses base excision activity of NEIL family DNA glycosylases (Hegde et al. 2010). Transition metals, hence, generates both ROS mediate DNA damages and direct inhibition of NEIL type DNA glycosydases (Hegde et al. 2011). In this case, curcumin have been proved to act directly as ROS scavenger and metal chelation preventing DNA glycosidase direct inhibition (Hegde et al. 2011).

Dietary heme intake coming from excessively consumed red meat can start oxidative damage during oxymyoglobin and deoxymyoglobin heme and cytochrome heme derived breakdown products (Tappel 2007). Free and coordinated heme in blood or at intestine level catalyze oxidative reactions, involving also heme catalyzed lipid peroxidation (Carlin et al. 1988). The role of heme–iron in aging, and neurodegeneration is the reflection of systemic and local malfunctioning of Fe homeostasis due to iron overload, considered the main contributor of neurodegenerative diseases (Gozzelino 2016). In vivo experiments have been demonstrated that mice fed with an enriched Fe-diet senesce faster than those with a restricted Fe diet (Arruda et al. 2013). In the brain iron is accumulated in the substantia nigra and basal ganglia. Imbalance in iron homeostasis involving these regions of the brain is present in rare human genetic mutation leading to "neurodegeneration with brain iron accumulation" (NBIA) disorders (Wong et al. 2014; Gregory et al. 2013).

As stated above, the catabolism of cellular heme in the brain is mediated by the isogene 1 of heme oxygenase (HO-1). The isozyme HO-1 contains a C-terminus

PEST (proline-glutamic acid-serine-threonine) domain that rapidly target the protein to degradation upon the sequence of event that starts with its oxidation, followed by phosphorylation of PEST domain and ubiquitination and proteasome degradation (Lin et al. 2008). However, HO-1elevated mRNA and protein level are stimulated by ROS, and its overexpression in astrocytes increase mitochondrial damage and mitophagy by enhancing mitochondrial permeability by opening the mitochondrial permeability transition pore and deviating the flow of oxidative redox species from the mitochondria to the rest of the cell (Schipper et al. 2015; Zukor et al. 2009).

Even diet containing elevated amount of starch/glucose (high-glucose, HG) can originate an oxidative stress-dependent Heme Oxygenase-1 de novo mRNA expression determining ultimately neuronal apoptosis (Yang et al. 2017) because HG stimulates extracellular ROS generation via NADPH oxidase activation and mitochondrial ROS release via opening of the mitochondrial permeability transition pore. NF-κB and AP-1 transcription factors play a role in HO-1 expression in various cell types (Hsieh et al. 2010).

Plant derived NF-κB inhibitors and AP-1 inhibitors (Golan-Goldhirsh et al. 2014, Choy et al. 2019) are known also as ROS scavengers. However, their molecular involvement in the exact cascade of event upstream or downstream the ROS generating/transduction event need to be established per each phytochemical compound also generating a nutraceutical effect by testing the concentration range that elicit an hormetic response. Mouse mesangial cells treated with 25 mM glucose for 24 h showed increased ROS generation, by upregulation of NADPH oxidase (NOX) 1 and mitochondrial superoxide (Song et al. 2016). Pretreatment with 50 μM delphinidin or 1 mM N-acetylcysteine (NAC) suppressed the ROS generation (Song et al. 2016). Delphinidin is a polyphenol, in particular a flavonoid. Among the flavonoids, the anthocyanins include besides delphinidin, also cyanidin and pelargonidin. Their mechanism of action includes both a direct chelation effects of redox generating metals and direct action of ascorbic acid. In particular, the ascorbic acid (Vit. C) is a ubiquitous antioxidant of soluble intra e extracellular spaces. It detoxifies activated oxygen species: acts as an antioxidant either by reducing superoxide, H_2O_2 and hydroxyl radicals, or by quenching singlet oxygen. Under exposure to heavy metals, the levels of the antioxidant power of Vit. C decreases as it is oxidized to dehydroascorbic acid. Metal catalyzed oxidation by Copper (II) or Fe (II) are first chelated by ascorbic acid dicarboxylic group prior its oxidation. Adding of flavonoids, i.e., cyanidin or delphinidin (flavonoids that containing a vicinal ortho-dihydroxy group) results in a mixed chelation of such metal ions to form a stable metal-anhydro chelate (flavonoid-metal-ascorbate) (Annamraju et al. 1997). This mechanism is known, e.g., inside plant vacuoles, as co-pigmentation, since different classes of flavonoids are complexing metals to detoxify those, preventing ROS formation (Takeda 2006). In particular, the blue color of the sepals of many plants are given by the intermolecular co-pigmentation mechanism in order to detoxify Al^{3+} (Takeda 2006). In Hydrangea (Hydrangea macrophylla), the sepal color can be regulated by the pH of the vacuole of the sepals, since to get blue color Al3 + is administrated to the roots. Upon translocation into sepals, it is sequestered in the vacuoles as 3-O-glucosyl-delphinidin based anthocyanins by co-pigmentation with other phenolic acids, e.g., 5-O-caffeoylquinic acid, 3-O-caffeoylquinic acid and/or 5-O-p-coumaroylquinic acid (Ito et al. 2018).

In mice cardiomyocites, delphinidin administration repressed angiotensin II induced cardiac hypertrophy by modulating oxidative stress and in particular by inhibiting the AMPK/NADPH oxidase (NOX)/mitogen-activated protein kinase (MAPK) signaling pathway (Chen et al. 2020). Another anthocyanin, the cyanidin-3-O-glucoside (C3G) is known also to have beneficial antioxidant and neuroprotective effects due to pleiotropic mechanism of actions (Zhang et al. 2019). Anthocyanin extract improved oxidative damage in HepG2 cells and prolonged the lifespan of *Caenorhabditis elegans* through MAPK and Nrf2 pathways (Yan et al. 2017).

13.3.4 Hormetic Compounds: Between Beneficial Nutraceutical and Pro-oxidative Effects

However, many polyphenols, including anthocyanins, show the hormetic behavior of the dose–response phenomenon by the inversion of the response between low and high doses (Halliwell et al. 2008). Nevertheless, anthocyanins are recognized as antioxidants and free radical scavengers, they are also able toprovoke*in vivo* pro-oxidant effects (Laughton et al. 1991, 1989; Sakanoet al. 2005). Among polyphenols, hydroxycinnamic acids (e.g., caffeic acid, catechol, gallic acid, ferulic acid, o/m/p-coumaric acid and chlorogenic acid) show prooxidant activity and cause DNA damage (Zheng et al. 2008). In general, the more hydroxyl substitutions in flavonoids, the stronger will be their antioxidant and prooxidant activities (Cao et al. 1997). To test the prooxidant activities of several polyphenols the redox system used was NADPH/peroxidase/H_2O_2 and DNA cleavage systems (Eghbaliferiz et al. 2016). Hence, the paradigm for the mechanism of action of polyphenols is from one hand through direct antioxidant activity, by their interaction with ROS, including both radical and nonradical oxygen species, such as $O_2{\cdot}^-$, H_2O_2, HOCl, $NONOO^-$(Noda et al. 1998; Nakagawa et al. 2002; Tsuda et al. 2000) and from another hand through their excessive dose abuse that can cause anti-hormetic effects leading to pro-oxidative reactions, e.g., quercetin, myricetin, kaempferol and caffeic, chlorogenic and ferulic acids (Azam et al. 2004; Watjen et al. 2005; Bouayed et al. 2010). Lately, the occidental world is re-discovering and validating scientifically the anti-ageing active ingredients from the Traditional Chinese Medicine, TCM (Shen et al. 2017). Anti-ageing ingredients from herbs or roots are related to plant secondary or better specialized metabolites: flavonoids, saponins, polysaccharides, alkaloids and others. Among them, astragaloside, *Cistanche tubulosa* acteoside, Ginsenoside Rg1, curcumin, resveratrol, carnosic acid and echinacoside possess in our judgement the best anti-ageing activity. Astragaloside (AST), Cistanche tubulosa acteoside for example, inhibit the telomerase activity (Guo et al. 2010). Besides, *Cistanche tubulosa*, a desert plant that paraxitize other's plant roots, produces also a wide range of metabolites containing phenylethanoid glycosides (cistantubuloside A, cistantubuloside B, echinacoside, cistanoside A, acteoside, isoacteoside, 2-acetylacteoside, cistanoside C and tubuloside), and it has been used in TCM for its many therapeutic effects including neuroprotection and anti-parkinsonian effects (Al-Snafi 2020). Ginsenoside Rg1, markedly decreases telomere shortening, reinforces telomerase activity inhibitor (Zhou et al. 2011). Curcumin, activates signalling pathways downstream of the anti-ageing modulators AMPK and NRF2, suppress inflammatory processes mediated by NF-kB signalling (Lim et al. 2001; Salvioli et al. 2007), while and its derivate, tetrahydrocurcumin,regulates sirtuins and FOXO1-responsive pathways (Argyropoulouet al. 2013). Phenylethanoids are a type of phenolic compounds characterized by a phenethyl alcohol structure. Tyrosol and hydroxytyrosol are examples of such compounds. Resveratrol, found in red wine, in yeast increases cell survival by stimulating SIRT1-dependent deacetylation of p53 (Howitz et al. 2003). Resveratrol, in this case, mimicked calorie restriction by stimulating also Sir2, increasing DNA stability and extending lifespan by 70% (Howitz et al. 2003).

Carnosic acid, and other phenolic diterpenes, has been identified in many aromatic plants because protects the chloroplasts from oxidative damage (i.e., drought stress induced), alongwise carotenoids, tocopherols, ascorbate, and glutathione/thioredoxin system (Munnè-Bosch et al. 2003). Carnosic acid, which displays high antioxidant properties in vitro and in vivo. In facts, carnosic acid could be considered a protecting agent of brain mitochondria since it activates the PI3K/Akt/Nrf2 signaling pathway, reduce the impact of pro-oxidant agents reduce the ROS by upregulating the Mn-SOD and increases the GSH levels (de Oliveira 2018). Another hydroxytyrol containing antioxidant compound is the "oleuropein" (elenolic acid linked to the orthodiphenol hydroxytyrosol by an ester bond, and to a molecule of glucose by a glycosidic bond) (Panizzi et al. 1960). Oleuropein is the bitter compound found in green olive skin, flesh, seeds, and leaves which seems inhibiting α-synuclein (αSN) aggregation and reducing its tendency to form toxic oligomers, e.g., in neuronal degeneration of Parkinson's disease (Mohammad-Beigi et al. 2019). However, in vivo, the fate of the oleuropein is given by the intestine derived hydrolytic products: 1) esterases can liberate the hydroxytyrosol and oleoside11-methyl ester glycoside and 2) a glycosidase can also further liberate the glucose. Therefore, it could be that the neuroprotective effect could be explicated by the hydroxytyrosol itself as it is the case of hypoxia derived injuries in rat brains (Gonzàlez-Correa et al. 2008).

Finally, the anti-aging, neuroprotective, immune system booster, echinacoside (ECH) a natural phenylethanoid glycoside containing caffeic acid and hydroxytyrosol glycosides, first isolated from Echinacea angustifolia DC, shows a plethora of ROS scavenging capabilities for the treatment aging (Nematal et al. 2011) and neurological disorders (Liu et al. 2018). ECH is partially hydrolyzed to aglycone in the rat body and it is found in the rat feces asacteoside, decaffeoylacteoside, lugrandosie, 3,4-dihydrophenyl ethanol (Liu et al. 2018). Several Parkinson's or Alzheimer's animal models induced by 6-OHDA, 1-methyl-4-phenyl-1,2,3,6-tetrahydropyridine (MPTP), d-galactose and β-amyloid Aβ-(25–35) as well as cerebral ischemia rats were used to estimate the neuroprotective effects of ECH (Zhao et al. 2016). In these cases, the ECH effects ranged from downregulating intracellular ROS, the expression of p38MAPK, caspase 3/8, NFkappaB, p52/p53 and upregulating the level of pERK MAPK, and SIRT1 (Wang et al. 2015; Zhang et al. 2017a, 2017b; Zhang et al. 2014; Zhu et al. 2011).

13.3.5 Nrf2/vitagene Pathway: A Crucial Hub in the Hormesis Paradigm

In recent decades, a novel cytoprotective panel has emerged suggesting a common mechanistic framework of hormetic dose responses following the redox activation of the Nrf2/vitagene signaling pathway to maintain or restore healthy aging (Fig. 13.1). The hormetic dose-responses are typically defined as overcompensation reactions to the direct alterations in cellular redox homeostasis by subtoxic or subthreshold doses of various ubiquitous environmental stressors. Indeed, hormesis-induced overcompensations often enhance cellular resiliency to prevent the occurrence of various toxic phenotypes, such as aging and aging-related disorders (Calabrese et al. 2018). According to the hormesis principle, exposure to low concentrations of stressors such as drugs, ionizing radiation, xenobiotics and natural polyphenols may protect against neuronal damage from a subsequent stressor agent (ROS), while at higher dose the toxic effect prevails (Calabrese EJ et al. 2012a, 2012b, 2012c, 2012d, 2021a). Cellular stressors shift the redox homeostasis of cells toward the oxidative state and in the process activate Nrf2 redox sensor to mediate a host of cytoprotective and resiliency responses, including antioxidation, detoxification, anti-inflammation, autophagy and mitochondrial biogenesis (Amara et al. 2020a; Ahmed et al. 2017). It is well established that under normal redox state, the repressor Keap1 is bound to Nrf2 sustaining its ubiquitination and degradation into the cytosol, maintaining Nrf2 at low levels. However, increased exposure to subtoxic hormetic levels of pro-oxidative stressors cause the oxidization of labile cysteine residues in Keap1 and consequently the loss of its capacity to mediate Nrf2 for degradation by proteosome. Therefore, Nrf2 accumulates and translocates into the nucleus where it binds to a region ofantioxidant response elements (ARE) upregulating numerous antioxidant pathways to restore redox homeostasis (e.g., HO-1, GCLM and GLLC), inhibits inflammation (TGF-B and NF-kB), and improves xenobiotic metabolism and excretion (NQO1, AKRIC1, and MRPI), apoptosis (BCL2 and BclxL), and autophagy (p62). Interestingly, transcription factor Nrf2 regulates a wide range of genes termed *vitagenes* (Calabrese et al. 2020a, 2020b; Scuto et al. 2020; Calabrese et al. 2014) (Fig. 13.1). The latter encodes for members of HSP family, such as heme-oxygenase-1 (HO-1) and HSP 70, thioredoxin/thioredoxin reductase, glutathione and sirtuins (Calabrese et al. 2012b; Cornelius et al. 2013, 2014; Trovato et al. 2014). Hence, moderate cellular stressors or inducers (i.e., polyphenols) exert a physiological modulation of Nrf2 and the activation of longevity *vitagenes* ensuring health benefits and a prolonged lifespan as well as cellular stress resilience in animal and humans (Pall et al. 2015; Amara et al. 2020b; Crespo et al. 2015).Accordingly, emerging scientific literature has investigated the dose–response and mechanistic features of some dietary polyphenols for which potential health benefits in a broad spectrum of neurological diseases have been reported (Leri et al. 2020; Cordaro et al. 2020a, 2020b; Scuto et al. 2019a). Compelling in vitro and in vivo evidence shown that natural polyphenols such as curcumin (Scuto et al. 2019b), ginsenoside (Chu et al. 2019), 2,7,2′-trihydroxy-4,4′7′-trimethoxy-1,1′-bephenanthrene (TTB) (Liu et al. 2019), berberine (Zhang et al. 2017a, 2017b), resveratrol (Plauth et al. 2016), green tea (Calabrese EJ et al. 2020a, 2020b) and sulforaphane display hormetic dose–response and have been receiving increasing attention in the field of neuroscience, including the areas of neuroprotection, as supplementation to drugs for Parkinson's disease (PD), Alzheimer's disease (AD), Huntington's disease (HD) and stroke by the activation of Nrf2 pathway (Alfieri et al. 2013; Zhao et al. 2017; Dwivedi et al. 2016). In line with the extensive evidence above mentioned, pharmacological studies demonstrated that redox environment induced by polyphenols could potentially protect the cell from oxidative stress during aging-related diseases. Intriguingly, recent preclinical and clinical studies have observed a paradox view of polyphenols that at very low concentration in the blood system act as pro-drugs generating H_2O_2 by inducing redox signaling through Nrf2 axis of cell adaptation and protection of oxidative stress (Eustress). On the other hand, high concentration of circulating polyphenols, can generate relatively increased levels of H_2O_2 and possible other derivatives, acting as cytotoxic agents affecting inflammation, apoptosis, necrosis and death (Distress) (Kanner 2020). Consistent with this concept, hormetic low concentrations (<50 μM) of cellular pro-oxidant resveratrol enhanced pool of endogenous reduced glutathione (GSH) and Sirt1 expression following the addiction of toxic 4-hydroxy-2-nonenal (HNE) via activation of the Nrf2 pathway in a concentration dependent manner (Plauth et al. 2016). Likewise, Calabrese and coworkers also revealed that a subtoxic dose of < 50 μM of sulforaphane induces chemoprotective effects mediated by the Nrf2/ARE pathway through a functional crosstalk with p53 and NF-kB pathways modulating the redox and inflammatory states of cells for potential implications on public health and clinical medicine (Calabrese EJ et al. 2021b). Furthermore, recent biomedical literature concerning the hormetic profile reported that sulforaphane activating Nrf2/ARE pathway induces an upregulation of phase II detoxification enzymes and antioxidant genes (e.g., *vitagenes*) in a wide range of chronic diseases including cancer and neurological disorders (Wang et al. 2018; Zhao et al. 2018; Calabrese EJ et al. 2021a, 2021b). Increasing evidence observed that low dose of berberine (1 μM) exhibitshormetic and neuroprotective effects against 6-hydroxydopamine (6-OHDA) and 1-methyl-4-phenyl-1,2,3,6-tetrahydropyridine-inducedneurotoxicity by up-regulating PI3K/AKT/Bcl-2 cell survival and Nrf2/HO-1 antioxidative signaling pathways in PC12 cells, rodent and zebrafish models of neurodegeneration (Kwon et al. 2010; Kim et al. 2014; Zhang et al. 2017a, 2017b) (Fig. 13.1). Moreover, alsoother protective agents can activate Nrf2 pathway, in particular, it has been observed that methylene blue (MB) prevented rotenone-induced neurotoxicity as well as lipid peroxidation in the ventral midbrain displaying a J-shaped hormetic dose–response profile troughan upregulation of Nrf2-dependent antioxidant *vitagenes* (i.e.,

HO-1, Trx, NQO1, Gclc, and Gclm) and a downregulation of a pro-inflammatory gene (*iNOS*) in rodent models of neurodegeneration and ulcerative colitis (Gureev et al. 2019; El-Sayed et al. 2019). Several experimental studies focused their interest on the hormetic role of Nrf2 in mediating radiation toxicity. The early study aboutionizing radiation was documented by Tsukimoto and coworkers showing that a low dose of ionizing radiation in the range of 0.1–0.5 Gy increased the cytosolic accumulation of Nrf2/vitagenes and the upregulation of GSH and HO-1in mouse macrophages (Tsukimoto et al. 2010). Recent evidence also reported the hormetic role of selenoproteins as potential radiation mitigators in redox pathways regulating the activity of selenoenzymes such as thioredoxin reductase-1 and glutathione peroxidases (Bartolini et al. 2020). Mechanistically, Nrf2 activation contribute to 2,2′-diselenyl dibenzoic acid (DSBA)-induced radioprotection in normal tissue against ionizing radiation-induced toxicity in vitro and in vivo (Bartolini et al. 2015, 2019). Finally, equally interesting studies demonstrated the radioprotective effects of repeated sulforaphane dose-dependent treatment (10 μM) resulting in a stronger Nrf2 response with a particular upregulation of NQO1 and HO-1 against toxic effects of free radicals induced by ionizing radiation in human fibroblasts (Mathew et al. 2014). Taken together, the emerging data above discussed shed light on the hormetic dose–response effects of many antioxidant agents including natural polyphenols that at low concentrations modulate Nrf2/vitagene pathway exerting preventive and therapeutic strategies to preserve health and extend lifespan during aging and age-related illnesses in humans.

13.4 Conclusions and Future Directions

The role of mitochondria in normal aging has been the focus of extensive research in the last decades and being complementary to and maturing from the free radical theory of aging, to the oxidative stress theory of aging, to the mitochondrial oxidative stress theory of aging, and today addressing the tight co-regulation of mitochondrial energy and redox signaling. In this scenario, it appears conceivably the function of the carnitine system as a prototypical vitagene operating at the functional interface of energy distribution between ancestral mechanisms of cell proliferation and differentiation and homeostatic mitochondrial dependent processes of cell survival which require energy for cellular stress response and redox homeostasis (Calabrese et al. 2011b). Very importantly, a new neuroprotective role has been elucidated for mitochondrial targeted compounds, such as nutritional antioxidants, carnitines or carnosine, which by intersecting convergent mechanisms that rely on cellular energy distribution and availability, such as cellular stress response pathways, DNA repair and molecular fidelity mechanisms improve health brain.In conclusion, the maintenance of optimum antioxidant potential appears be a promising novel therapeutic approach for those pathophysiological conditions, such as neurodegeneration or cancer, where hormetic stimulation of the vitagene pathway is strongly warranted. Maintenance of ROS is mostly depending on caloric metabolism, which is influenced by the diet. Apart metabolic restriction as self-mechanism of reduction in endogenous ROS production, hormetic compounds can help to induce the antioxidant system by functioning as ROS scavenging compound at optimal dose–response level (hormetic curve). At higher level the natural antioxidant hormetins can acts as pro-oxidants, unbalancing the ROS pathways and triggering aging at many levels: mitochondrial damage leads to apoptosis; telomerase activation leads to shortening of telomers; DNA damages lead to gene and epigenetic changes that can accelerate aging or cellular death.

Low dose response to antioxidant compounds cited here lead to the elicitation of survivor pathway (adaptative responses) helping their direct ROS scavenging action. In case of sirtuins, the adaptive response involve epigenetics. DNA methylation is involved in epigenetic changes that trigger gene activation/deactivation and chromatin remodeling. Sirtuin-1 (SIRT1) is a class-III histone deacetylase and hence involved also in chromatin remodeling. Sirtuin-1 activation and inhibition can be one of the potential treatment targets for diseases such as cancer and neurodegeneration (Calabrese et al. 2018). The hormetic compounds mentioned here have, in their low dosage administration, beneficial distinct nutraceutical effects depending on their chemical structure. Anyhow, depending on their hormetic curve dosage response, their negative pro-oxidative effects are affecting their nutraceutical performances. Hence future direction could foresee studies investigating their dose–response in relationship also to their natural or synthetic sources that might alter their biological function. The more chemical pure nutraceuticals are produced, by organic synthesis or by system biology, the more unique effects on the activated pathways could be drawn in a near future research.

Acknowledgements This study was supported by grants from "Piano di incentivi per la Ricerca, Linea Intervento 2 PIACERI, 2020–2022", University of Catania, Italy.

Compliance with Ethical Standards

Conflict of Interest The authors declare that they have no conflicts of interest.

References

Abdul HM, Calabrese V, Calvani M, Butterfield DA (2006) Acetyl-L-carnitine-induced up-regulation of heat shock proteins protects cortical neurons against amyloid-beta peptide 1–42-mediated oxidative stress and neurotoxicity: Implications for Alzheimer's disease. J Neurosci Res 84:398–408

Adebooye OC, Alashi AM, Aluko REJ (2018) Food Biochem:e12519

Ahmed SM, Luo L, Namani A, Wang XJ, Tang X (2017) Nrf2 signaling pathway: pivotal roles in inflammation. Biochim Biophys Acta Mol Basis Dis. 1863:585–597

Ajila CM, Brar SK, Verma M, Tyagi RD, Godbout S, Valéro JR (2011) Extraction and analysis of polyphenols: recent trends. Crit Rev Biotechnol 31:227–249

Al-Snafi AE (2020) Bioactive metabolites and pharmacology of cistanchetubulosa-a review. IOSR J Pharm 10:37–46

Alam Q, Alam MZ, Sait KHW, Anfinan N, Noorwali AW, Kamal MA, Khan MSA, Haque A (2017) Translational shift of HSP90 as a novel therapeutic target from cancer to neurodegenerative disorders: an emerging trend in the cure of alzheimer's and parkinson's diseases. Curr Drug Metab 18:868–876

Albokhadaim IF, Althnaian TA, El-Bahr SM (2019) Gene expression of heat shock proteins/factors (HSP60, HSP70, HSP90, HSF-1, HSF-3) and antioxidant enzyme activities in heat stressed broilers treated with vitamin C. Pol J Vet Sci 22:565–572

Alfieri A, Srivastava S, Siow RCM, Cash D, Modo M, Duchen MR, Fraser PA, Williams SCR, Mann GE (2013) Sulforaphane preconditioning of the Nrf2/HO-1 defense pathway protects the cerebral vasculature against blood-brain barrier disruption and neurological deficits in stroke. Free Radic Biol Med. 65:1012–1022

Amara I, Salah A, Timoumi R, Annabi E, Scuto M, Trovato A, Neffati F, Calabrese V, Abid-Essefi S (2020) Effect of di(2-ethylhexyl) phthalate on Nrf2-regulated glutathione homeostasis in mouse kidney. Cell Stress Chaperones 25:919–928

Amara I, Scuto M, Zappalà A, Ontario ML, Petralia A, Abid-Essefi S, Maiolino L, Signorile A, TrovatoSalinaro A, Calabrese V (2020) Hericium erinaceus prevents DEHP-induced mitochondrial dysfunction and apoptosis in PC12 cells. Int J Mol Sci 21:2138

Argyropoulou A, Aligiannis N, Trougakos IP, Skaltsounis A-L (2013) Natural compounds with anti-ageing activity. Nat Prod Rep 30:1412–1437

Arruda LF, Arruda SF, Campos NA, de Valencia FF, Siqueira EM (2013) Dietary iron concentration may influence aging process by altering oxidative stress in tissues of adult rats. PLoS One. 8:e61058

Athanasiou A, Clarke AB, Turner AE, Kumaran NM, Vakilpour S, Smith PA, Bagiokou D, Bradshaw TD, Westwell AD, Fang L, Lobo DN, Constantinescu CS, Calabrese V, Loesch A, Alexander SP, Clothier RH, Kendall DA, Bates TE (2007) Cannabinoid receptor agonists are mitochondrial inhibitors: a unified hypothesis of how cannabinoids modulate mitochondrial function and induce cell death. Biochem Biophys Res Commun 364:131–137

Azam S et al (2004) Prooxidant property of green tea polyphenols epicatechin and epigallocatechin-3-gallate: implications for anticancer properties. Toxicol in Vitro 18:555–561

Balasubramanyam M, Koteswari AA, Kumar RS, Monickaraj SF, Maheswari JU, Mohan V (2003) Curcumin-induced inhibition of cellular reactive oxygen species generation: novel therapeutic implications. J Biosci 28:715–721

Bartolini D, Commodi J, Piroddi M et al (2015) Glutathione S-transferase pi expression regulates the Nrf2-dependent response to hormeticdiselenides. Free Radic Biol Med 88:466–480

Bartolini D, Wang Y, Zhang J, et al (2019) A seleno-hormetine protects bone marrow hematopoietic cells against ionizing radiation-induced toxicities. PLoS ONE 14:e0205626

Bartolini D, Tew KD, Marinelli R, Galli F, Wang GY (2020) Nrf2-modulation by seleno-hormetic agents and its potential for radiation protection. BioFactors 46:239–245

Begum AN, Jones MR, Lim GP, Morihara T, Kim P, Heath DD et al (2008) Curcumin structure-function, bioavailability, and efficacy in models of neuroinflammation and Alzheimer's disease. J Pharmacol Exp Ther 326:196–208

Bellia F, Calabrese V, Guarino F, Cavallaro M, Cornelius C, De Pinto V, Rizzarelli E (2009) Carnosinase levels in aging brain: redox state induction and cellular stress response. Antioxid Redox Signal 11:2759–2775

Bellia F, Vecchio G, Cuzzocrea S, Calabrese V, Rizzarelli E (2011) Neuroprotection in oxidative driven diseases by carnosine. Mol Aspects of Med. 32:258–266

Bellumori M, Cecchi L, Innocenti M, Clodoveo ML, Corbo F, Mulinacci N (2019) The EFSA health claim on olive oil polyphenols: acid hydrolysis validation and total hydroxytyrosol and tyrosol determination in italian virgin olive oils. Molecules 24:2179

Berendsen AAM, van de Rest O, Feskens EJM, Santoro A, Ostan R, Pietruszka B, Brzozowska A, Stelmaszczyk-Kusz A, Jennings A, Gillings R, Cassidy A, Caille A, Caumon E, Malpuech-Brugere C, Franceschi C, de Groot LCPGM (2018) Changes in dietary intake and adherence to the NU-AGE diet following a one-year dietary intervention among European older adults-results of the NU-AGE randomized trial. Nutrients 10:1905

Bitler CM, Matt K, Irving M, Hook G, Crea R et al (2007) Olive extract supplement decreases pain and improves daily activities in adults with osteoarthritis and decrease plasma homocysteine in those with rheumatoid arthritis. Nutr Res 27:470–477

Bitler CM, Viale TM, Damaj B, Crea R (2005) Hydrolyzed olive vegetation water in mice has anti-inflammatory activity. J Nutr 135:1475–1479

Bondia-Pons I, Pöhö P, Bozzetto L, Vetrani C, Patti L, Aura AM, Annuzzi G, Hyötyläinen T, Rivellese AA, Orešič M (2014) Isoenergetic diets differing in their n-3 fatty acid and polyphenol content reflect different plasma and HDL-fraction lipidomic profiles in subjects at high cardiovascular risk. Mol Nutr Food Res 58:1873–1882

Bouayed J, Bohn T (2010) Exogenous antioxidants—Double-edged swords in cellular redox state. Health beneficial effects at physiologic doses versus deleterious effects at high doses. Oxid Med Cell Longev 3:228–237

Brunetti G, Di Rosa G, Scuto M, Leri M, Stefani M, Schmitz-Linneweber C, Calabrese V, Saul N. (2020) Healthspan maintenance and prevention of parkinson's-like phenotypes with hydroxytyrosol and oleuropein aglycone in C. elegans. Int J Mol Sci 21:2588.

CalabreseEJ II, Calabrese V (2012a) Hormesis: why it is important to biogerontologists. Biogerontology 13:215–235

Calabrese EJ, Calabrese V (2012b) Low dose radiation therapy (LD-RT) is effective in the treatment of arthritis: Animal model findings. Int J Radiat Biol 89:287–294

Calabrese EJ, Calabrese V (2012c) Reduction of arthritic symptoms by low dose radiation therapy (LD-RT) is associated with an anti-inflammatory phenotype. Int J Radiat Biol 89:278–286

Calabrese EJ, Iavicoli I, Calabrese V (2012d) Hormesis: Its impact on medicine and health. Hum Exp Toxicol 32:120–152

Calabrese EJ, Kozumbo WJ (2021a) The hormetic dose-response mechanism: Nrf2 activation. Pharmacol Res 167:105526

Calabrese EJ, Kozumbo WJ (2021b) The phytoprotective agent sulforaphane prevents inflammatory degenerative diseases and age-related pathologies via Nrf2-mediated hormesis. Pharmacol Res 163:105283

Calabrese EJ, Tsatsakis A, Agathokleous E, Giordano J, Calabrese V (2020) Does green tea induce hormesis? Dose Resp 18:1559325820936170

Calabrese V, Calabrese EJ, Blain R, Blain RB (2011a) The hormesis database: the occurrence of hormetic dose responses in the toxicological literature. Regul Toxicol Pharmacol 61:73–81

Calabrese V, Cornelius C, Cuzzocrea S, Iavicoli I, Rizzarelli E, Calabrese EJ (2011b) Hormesis, cellular stress response and vitagenes as critical determinants in aging and longevity. Mol Aspects Med 32:279–304

Calabrese V, Cornelius C, Dinkova-Kostova AT, Iavicoli I, Di Paola R, Koverech A, Cuzzocrea S, Rizzarelli E, Calabrese EJ (2012e) Cellular stress responses, hormetic phytochemicals and vitagenes in aging and longevity. Biochim Biophys Acta 1822:753–783

Calabrese V, Cornelius C, Leso V, Trovato-Salinaro A, Ventimiglia B, Cavallaro M, Scuto M, Rizza S, Zanoli L, Neri S, Castellino P (2012f) Oxidative stress, glutathione status, sirtuin and cellular stress response in type 2 diabetes. Biochim Biophys Acta. 1822:729–736

Calabrese V, Cornelius C, Mancuso C, Lentile R, Stella AM, Butterfield DA (2010a) Redox homeostasis and cellular stress response in aging and neurodegeneration. Methods Mol Biol 610:285–308

Calabrese V (2007a) Highlight Commentary on Redox proteomics analysis of oxidatively 3 modified proteins in G93A–SOD1 transgenic mice—A model of 4 familial amyotrophic lateral sclerosi. Free Radical Biol Med 43:160–162

Calabrese V, Bates TE, Mancuso C, Cornelius C, Ventimiglia B, Cambria MT, Di Renzo L, De Lorenzo A, Dinkova-Kostova AT (2008a) Curcumin and the cellular stress response in free radical-related diseases. Mol Nutr Food Res 52:1062–1073

Calabrese V, Butterfield DA, Scapagnini G, Stella AM, Maines MD (2006a) Redox regulation of heat shock protein expression by signaling involving nitric oxide and carbon monoxide: relevance to brain aging, neurodegenerative disorders, and longevity. Antioxid Redox Signal 8:444–477

Calabrese V, Butterfield DA, Stella AM (2008b) Aging and oxidative stress response in the CNS. In: Abel L, Perez-Polo JR, Steffen R (eds) Development and Aging Changes in the Nervous System. Handbook of Neurochemistry and Molecular Neurobiology, 3rd ed, pp 128–234. ISBN: 978–0–387–32670–2

Calabrese V, Calafato S, Cornelius C, Mancuso C, Dinkova-Kostova A (2008c) Heme oxygenase: a master vitagene involved in cellular stress response in: enzymes and the cellular fight against oxidation. In: Eleuteri AM (ed) Research Signpost, vol 37/661, no 2, Fort P.O., Trivandrum-695 023, Kerala, India. ISBN: 978–81–308–0239–8

Calabrese V, Calafato S, Puleo E, Cornelius C, Sapienza M, Morganti P, Mancuso C (2008d) Redox regulation of cellular stress response by ferulic acid ethyl ester in human dermal fibroblasts: role of vitagenes. Clin Dermatol 26:358–363

Calabrese V, Colombrita C, Sultana R, Scapagnini G, Calvani M, Butterfield DA, Stella AM (2006b) Redox modulation of heat shock protein expression by acetylcarnitine in aging brain: relationship to antioxidant status and mitochondrial function. Antioxid Redox Signal 8:404–416

Calabrese V, Sultana R, Scapagnini G, Guagliano E, Sapienza M, Bella R, Kanski J, Pennisi G, Mancuso C, Stella AM, Butterfield DA (2006c) Nitrosative stress, cellular stress response, and thiol homeostasis in patients with Alzheimer's disease. Antioxid Redox Signal 8:1975–1986

Calabrese V, Maines MD (2006d) Antiaging medicine: antioxidants and aging. Antioxid Redox Signal 8:362–364

Calabrese V, Cornelius C, Dinkova-Kostova AT, Calabrese EJ (2009a) Vitagenes, cellular stress response and acetylcarnitine: relevance to hormesis. BioFactors 35:146–160

Calabrese V, Cornelius C, Dinkova-Kostova AT, Calabrese EJ, Mattson MP (2010b) Cellular stress responses, the hormesis paradigm and vitagenes: novel targets for therapeutic intervention in neurodegenerative disorders. Antioxid Redox Signal 13:1763–1811

Calabrese V, Cornelius C, Giuffrida AM, Calabrese EJ (2010c) Cellular stress responses, mitostress and carnitine insufficiencies as critical determinants in aging and neurodegenerative disorders: role of hormesis and vitagenes. Neurochem Res 35:1880–1915

Calabrese V, Cornelius C, Maiolino L, Luca M, Chiaramonte R, Toscano MA, Serra A (2010d) Oxidative stress, redox homeostasis and cellular stress response in ménière's disease: role of vitagenes. Neurochem Res 35:2208–2217

Calabrese V, Cornelius C, Mancuso C, Barone E, Calafato S, Bates T, Rizzarelli E, Kostova AT (2009a) Vitagenes, dietary antioxidants and neuroprotection in neurodegenerative diseases. Front Biosci 14:376–397

Calabrese V, Cornelius C, Mancuso C, Ientile R, Giuffrida Stella AM, Butterfield DA (2008e) Redox homeostasis and cellular stress response in aging and neurodegeneration. In: Uppu RM, Murthy SN, Pryor WA, Parinandi NL (eds) Free Radical and Antioxidant Protocols, 2nd edn. Humana Press, LA

Calabrese V, Cornelius C, Rizzarelli E, Owen JB, Dinkova-Kostova AT, Butterfield DA (2009c) Nitric oxide in cell survival: a Janus molecule. Antioxid Redox Signal 11:2717–2739

Calabrese V, Cornelius C, Trovato A, Cambria MT, Lo Cascio MS, Di Rienzo L, Condorelli D, De Lorenzo A, Calabrese EJ (2010e) The hormetic role of dietary antioxidants in free radical-related diseases. Curr Pharm Des 16:8778–8783

Calabrese V, Crea R (2016) Potential prevention and treatment of Neurodegenerative diseases: olive polyphenols and hydroxytyrosol. Eur. J. Neurodegener. Dis. 5:81–108

Calabrese V, Guagliano E, Sapienza M, Panebianco M, Calafato S, Puleo E, Pennisi G, Mancuso C, Butterfield AD, Giuffrida Stella AM (2007b) Redox regulation of cellular stress response in aging and neurodegenerative disorders: role of vitagenes. Neurochem Research 32:757–773

Calabrese V, Cornelius C, Mancuso C, Pennisi G, Calafato S, Bellia F, Bates TE, Giuffrida Stella AM, Schapira T, DinkovaKostova AT, Rizzarelli E (2008f) Cellular stress response: a novel target for chemoprevention and nutritional neuroprotection in aging, neurodegenerative disorders and longevity. Neurochem Res 33:2444–2471

Calabrese V, Ientile R, Cornelius C, Scalia M, Cambria MT, Ventimiglia B, Pennini G, Mancuso C, Butterfield DA (2008g) Nutritional redox homeostasis and cellular stress response: differential role of homocysteine and acetylcarnitine. In: Surh YJ, Dong Z, Cadenas E, Packer L (eds) Dietary modulation of cell sygnaling pathways, CRC Press, New York, N.Y. (USA).

Calabrese V, Mancuso C, Calvani M, Rizzarelli E, Butterfield DA, Giuffrida Stella AM (2007c) Nitric oxide in the CNS: neuroprotection versus Neurotoxicity. Nat Neurosci 8:766–775

Calabrese V, Mancuso C, Cornelius C, Calafato M, Ventimiglia B, Butterfield DA, Dinkova-Kostova AT, Rizzarelli E (2008h) Reactive nitrogen species and cellular stress tolerance in aging and neurodegeneration: role of vitagenes. In: Alvarez S, Evelson P (eds) Free Radical Pathophysiology, vol 37/661, no 2, pp 345–367. Transworld Research Network, Kerala, India. ISBN 978–81–7895–311–3

Calabrese V, Mancuso C, Ravagna A, Perluigi M, Cini C, De Marco C, Butterfield DA, Giuffrida Stella AM (2007d) In vivo induction of heat shock proteins in the substantia nigra following L-DOPA administration is associated with increased activity of mitochondrial complex I and nitrosative stress in rats: regulation by glutathione redox state. J Neurochem 101:709–717

Calabrese V, Mancuso C, Sapienza M, Puleo E, Calafato S, Cornelius C, Finocchiaro M, Mangiameli A, Di Mauro M, Stella AM, Castellino P (2007e) Oxidative stress and cellular stress response in diabetic nephropathy. Cell Stress Chaperones 12:299–306

Calabrese V, Perluigi M, Cornelius C, Coccia R, Di Domenico F, Mancuso C, Pennisi G, Dinkova-Kostova AT (2009d) Phenolics in aging and neurodegenerative disorders. In: Fraga CG (ed) Phenolic Compounds of Plant Origin and Health: The Biochemistry behind their Nutritional and Pharmacological Value. Wiley & Sons, NY

Calabrese V, Signorile A, Cornelius C, Mancuso C, Scapagnini G, Ventimiglia B, Ragusa N, Dinkova-Kostova A (2008i) Practical approaches to investigate redox regulation of heat shock protein expression and intracellular glutathione redox state. Methods Enzymol 441:83–110

Calabrese V, Scuto M, Salinaro AT, Dionisio G, Modafferi S, Ontario ML, Greco V, Sciuto S, Schmitt CP, Calabrese EJ, Peters V (2020b) Hydrogen sulfide and carnosine: modulation of oxidative stress and inflammation in kidney and brain axis. Antioxidants (basel). 9:1303

Calabrese V, Santoro A, TrovatoSalinaro A, Modafferi S, Scuto M, Albouchi F, Monti D, Giordano J, Zappia M, Franceschi C, Calabrese EJ (2018) Hormetic approaches to the treatment of Parkinson's disease: perspectives and possibilities. J Neurosci Res 96:1641–1662

Calabrese V, Scapagnini G, Davinelli SG, Koverech A, Koverech CD, Pasquale AT, Salinaro SM, Calabrese EJ, Genazzani AR (2014) Sex hormonal regulation and hormesis in aging and longevity: role of vitagenes. J. Cell Commun. Signal. 8:369–384

Calahorra J, Shenk J, Wielenga VH, Verweij V, Geenen B, Dederen PJ, Peinado MÁ, Siles E, Wiesmann M, Kiliaan AJ (2019) Hydroxytyrosol, the major phenolic compound of olive oil, as an acute therapeutic strategy after ischemic stroke. Nutrients 11:2430

Cao G, Sofic E, Prior RL (1997) Antioxidant and prooxidant behavior of flavonoids: structure–activity relationships. Free RadicBiol Med. 22:749–760

Carlin G, Djursäter R, Arfors KE (1988) Inhibition of heme-promoted enzymatic lipid peroxidation by desferrioxamine and EDTA. Ups J Med Sci. 93:215–223

Carranza ADV, Saragusti A, Chiabrando GA, Carrari F, Asis R (2020) Effects of chlorogenic acid on thermal stress tolerance in C. elegans via HIF-1, HSF-1 and autophagy. Phytomedicine 66:153132

Cassidy A, Minihane AM (2017) The role of metabolism (and the microbiome) in defining the clinical efficacy of dietary flavonoids. Am J Clin Nutr 105:10–22

Chen Y, Ge Z, Huang S, Zhou L, Zhai C, Chen Y, Hu Q, Cao W, Weng Y, Li Y (2020) Delphinidin attenuates pathological cardiac hypertrophy via the AMPK/NOX/MAPK signaling pathway. Aging (albany NY). 12:5362–5383

Choy KW, Murugan D, Leong XF, Abas R, Alias A, Mustafa MR (2019) Flavonoids as natural anti-inflammatory agents targeting nuclear factor-kappa b (nfκb) signaling in cardiovascular diseases: a mini review. Front Pharmacol 10:1295
Christian M, Sharper V, Hoberman A, Seng J, Fu L, Covell D, Diener R, Bitler C, Crea R (2004) The toxicity profile of hydrolyzed aqueous olive pulp extract. Drug Chem Toxicol 27:309–330
Chu SF, Zhang Z, Zhou X, He WB, Chen C, Luo P, Liu DD, Ai QD, Gong HF, Wang ZZ, Sun HS, Feng ZP, Chen NH (2019) Ginsenoside Rg1 protects against ischemic/reperfusion-induced neuronal injury through miR-144/Nrf2/ARE pathway. Acta Pharmacol Sin 40:13–25
Claude S, Boby C, Rodriguez-Mateos A, Spencer JPE, Gérard N, Morand C, Milenkovic D (2014) Flavanol metabolites reduce monocyte adhesion to endothelial cells through modulation of expression of genes via p38-MAPK and p65-Nf-kB pathways. Mol Nutr Food Res 58:1016–1027
Cordaro M, Scuto M, Siracusa R, D'amico R, Filippo Peritore A, Gugliandolo E, Fusco R, Crupi R, Impellizzeri D, Pozzebon M, Alfonsi D, Mattei N, Marcolongo G, Evangelista M, Cuzzocrea S, Di Paola R (2020) Effect of N-palmitoylethanolamine-oxazoline on comorbid neuropsychiatric disturbance associated with inflammatory bowel disease. FASEB J 34:4085–4106
Cordaro M, Siracusa R, Fusco R, D'Amico R, Peritore AF, Gugliandolo E, Genovese T, Scuto M, Crupi R, Mandalari G, Cuzzocrea S, Di Paola R, Impellizzeri D (2020) Cashew (anacardium occidentale L.) nuts counteract oxidative stress and inflammation in an acute experimental model of carrageenan-induced paw edema. Antioxidants 9(8):660. https://doi.org/10.3390/antiox9080660
Cornelius C, Trovato Salinaro A, Scuto M, Fronte V, Cambria MT, Pennisi M, Bella R, Milone P, Graziano A, Crupi R, Cuzzocrea S, Pennisi G, Calabrese V (2013) Cellular stress response, sirtuins and UCP proteins in Alzheimer disease: role of vitagenes. Immun Ageing. 10:41
Cornelius C, Koverech G, Crupi R, Di Paola R, Koverech A, Lodato F, Scuto M, Salinaro AT, Cuzzocrea S, Calabrese EJ, Calabrese V (2014) Osteoporosis and alzheimer pathology: Role of cellular stress response and hormetic redox signaling in aging and bone remodeling. Front Pharmacol 5:120
Crespo MC, Tomé-Carneiro J, Burgos-Ramos E, Loria Kohen V, Espinosa MI, Herranz J, Visioli F (2015) One-week administration of hydroxytyrosol to humans does not activate Phase II enzymes. Pharmacol Res 95–96:132–137
Crozier A, Jaganath IB, Clifford MN (2009) Dietary phenolics: chemistry, bioavailability and effects on health. Nat Prod Rep 26:1001–1043
Cypser J, Johnson T (2002) Multiple stressors in Caenorhabditis elegans induce stress hormesis and extended longevity. Gerontol A Biol Sci Med Sci 57:B109–B114
De Lorenzo A, Noce A, Bigioni M, Calabrese V, Della Rocca DG, Di Daniele N, Tozzo C, Di Renzo L (2010) The effects of Italian Mediterranean organic diet (IMOD) on health status. Curr Pharm Des 16:814–824
de Oliveira MR (2018) Carnosic acid as a promising agent in protecting mitochondria of brain cells. Mol Neurobiol 55:6687–6699
de Villiers A, Alberts P, Tredoux AG, Nieuwoudt HH (2012) Analytical techniques for wine analysis: an African perspective; a review. Anal Chim Acta 730:2–23
Di Domenico F, Perluigi M, Butterfield DA, Cornelius C, Calabrese V (2010) Oxidative Damage in rat brain during aging: interplay between energy and metabolic key target proteins. Neurochem Res 35:2184–2192
Di Paola R, Impellizzeri D, Trovato SA, Mazzon E, Bellia F, Cavallaro M, Cornelius C, Vecchio G, Calabrese V, Rizzarelli E, Cuzzocrea S (2011) Administration of carnosine in the treatment of acute spinal cord injury. Biochem Pharm 82:1478–1489
Di Renzo L, Bianchi A, Saraceno R, Calabrese V, Cornelius C, Iacopino L, Chimenti S, De Lorenzo A (2012) 174G/C IL-6 gene promoter polymorphism predicts therapeutic response to TNF-α blockers. Pharmacogenet Genomics 22:134–142

Di Renzo L, Bertoli A, Bigioni M, Del Gobbo V, Premrov MG, Calabrese V, Di Daniele N, De Lorenzo A (2008) Body composition and -174G/C interleukin-6 promoter gene polymorphism: association with progression of insulin resistance in normal weight obese syndrome. Curr Pharm Des 14:2699–2706

Di Rosa G, Brunett, G, Scuto M, TrovatoSalinaro A, Calabrese EJ, Crea R, Schmitz-Linneweber C, Calabrese V, Saul N (2020) Healthspan Enhancement by Olive Polyphenols in C. elegans Wild Type and Parkinson's Models. Int J Mol Sci 21:3893.

Ding F, Li F, Li Y, Hou X, Ma Y, Zhang N, Ma J, Zhang R, Lang B, Wang H, Wang Y (2016) HSP60 mediates the neuroprotective effects of curcumin by suppressing microglial activation. Exp Ther Med 12:823–828

Dwivedi S, Rajasekar N, Hanif K, Nath C, Shukla R (2016) sulforaphane ameliorates okadaic acid-induced memory impairment in rats by activating the Nrf2/HO-1 antioxidant pathway. Mol Neurobiol 53:5310–5323

Eghbaliferiz S, Iranshahi M (2016) (2016) Prooxidant activity of polyphenols, flavonoids, anthocyanins and carotenoids: updated review of mechanisms and catalyzing metals. Phytother Res 30:1379–1391

Elessawy FM, Bazghaleh N, Vandenberg A, Purves RW (2020) Polyphenol profile comparisons of seed coats of five pulse crops using a semi-quantitative liquid chromatography-mass spectrometric method. Phytochem Anal 31:458–471

El Sayed N, Sayed AS (2019) Protective effect of methylene blue on TNBAS-induced colitis in rat mediated through the modulation of inflammatory and apoptotic signaling pathways. Arch Toxicol 93:2927–2942

Estruch R, Ros E, Salas-Salvado J, Covas MI, Corella D, Aros F, Gomez-Gracia E, Ruiz-Gutierrez V, Fiol M, Lapetra J, Lamuela-Raventos RM, Serra-Majem L, Pinto X, Basora J, Munoz MA, Sorli JV, Martinez JA, Fito M, Gea A, Hernan MA, Martinez-Gonzalez MA (2018) Primary prevention of cardiovascular disease with a mediterranean diet supplemented with extra-virgin olive oil or nuts. N. Engl. J. Med. 378:e34

Flanagan E, Lamport D, Brennan L, Burnet P, Calabrese V, Cunnane SC, de Wilde MC, Dye L, Farrimond JA, Emerson Lombardo N, Hartmann T, Hartung T, Kalliomäki M, Kuhnle GG, La Fata G, Sala-Vila A, Samieri C, Smith AD, Spencer JPE, Thuret S, Tuohy K, Turroni S, Vanden Berghe W, Verkuijl M, Verzijden K, Yannakoulia M, Geurts L, Vauzour D (2020) Nutrition and the ageing brain: Moving towards clinical applications. Ageing Res Rev. 62:101079

Forrester MT, Foster MW, Benhar M, Stamler JS (2009) Detection of protein S-nitrosylation with the biotin-switch technique. Free RadicBiol Med. 46:119–126

Fujimoto M, Hayashida N, Katoh T, Oshima K, Shinkawa T, Prakasam R, Tan K, Inouye S, Takii R, Nakai A (2010) A novel mouse HSF3 has the potential to activate nonclassical heat-shock genes during heat shock. Mol Biol Cell 21:106–116

Gan N, Wu YC, Brunet M, Garrido C, Chung FL, Dai C, Mi L (2010) Sulforaphane activates heat shock response and enhances proteasome activity through up-regulation of Hsp27. J Biol Chem 285:35528–35536

Giovinazzo G, Grieco F (2015) Functional properties of grape and wine polyphenols. Plant Foods Hum Nutr 70:454–462

Golan-Goldhirsh A, Gopas J (2014) Plant derived inhibitors of NF-κB. Phytochem Rev 13:107–121

González-Correa JA, Navas MD, Lopez-Villodres JA, Trujillo M, Espartero JL, De La Cruz JP (2008) Neuroprotective effect of hydroxytyrosol and hydroxytyrosol acetate in rat brain slices subjected to hypoxia-reoxygenation. Neurosci Lett 446:143–146

Gozzelino R (2016) the pathophysiology of heme in the brain. Curr Alzheimer Res 13:174–184

Gregory A, Hayflick S, Adam MP, Ardinger HH, Pagon RA, et al (2013) Neurodegeneration with Brain Iron Accumulation Disorders Overview. In GeneReviews® [Internet]

Guo L, Wei XD, Ou Q, Wang S, Zhu GM (2010) Effect of astragaloside on the expression of telomerase activity and klotho gene in aged HELF cells. Chin J Gerontol 13:1819–1822

Gupta SC, Kim JH, Kannappan R, Reuter S, Dougherty PM, Aggarwal BB Role of nuclear factor kappa B-mediated inflammatory pathways in cancer-related symptoms and their regulation by nutritional agents. Exp Biol Med (Maywood) 236:658–667.

Gureev AP, Shaforostova EA, Laver DA, Khorolskaya VG, Syromyanikov MY, Popvo VN (2019) Methylene blue elicits non-genotoxic 202 production and protects brain mitochondria from rotenone toxicity. J Appl Biomed 17:107–114

Halliwell B (2008) Are polyphenols antioxidants or pro-oxidants? what do we learn from cell culture and in vivo studies? Arch Biochem Biophys 476:107–112

Hayes DP (2007) Nutritionalhormesis. Eur J Clin Nutr 61(2):147–159

Hegde ML, Hegde PM, Holthauzen LM, Hazra TK, Rao KS, Mitra S (2010) Specific Inhibition of NEIL-initiated repair of oxidized base damage in human genome by copper and iron: potential etiological linkage to neurodegenerative diseases. J Biol Chem 285(37):28812–28825

Hegde ML, Hegde PM, Rao KS, Mitra S (2011) Oxidative genome damage and its repair in neurodegenerative diseases: function of transition metals as a double-edged sword. J Alzheimers Dis 24:183–198

Hieronymus H, Lamb J, Ross KN, Peng XP, Clement C, Rodina A, Nieto M, Du J, Stegmaier K, Raj SM et al (2006) Gene expression signature-based chemical genomic prediction identifies a novel class of HSP90 pathway modulators. Cancer Cell 10:321–330

Howitz KT, Bitterman KJ, Cohen HY, Lamming DW, Lavu S, Wood JG, Zipkin RE, Chung P, Kisielewski A, Zhang LL, Scherer B, Sinclair DA (2003) Small molecule activators of sirtuins extend Saccharomyces cerevisiae lifespan. Nature 425:191–196

Hsieh HL, Wang HH, Wu CY, Yang CM (2010) Reactive oxygen species-dependent c-Fos/activator protein 1 induction upregulates heme oxygenase-1 expression by bradykinin in brain astrocytes. Antioxid Redox Signal 13:1829–1844

Imran M, Nadeem M, Gilani SA, Khan S, Sajid MW, Amir RM (2018) Antitumor perspectives of oleuropein and its metabolite hydroxytyrosol: recent updates. J Food Sci 83:1781–1791

Ito T, Oyama KI, Yoshida K (2018) Direct observation of hydrangea blue-complex composed of 3-o-glucosyldelphinidin, $Al3^{+}$ and 5-O-Acylquinic Acid by ESI-Mass spectrometry. Molecules 23:1424

Kanner J (2020) Polyphenols by generating H2O2, affect cell redox signaling, inhibit PTPs and activate Nrf2 axis for adaptation and cell surviving: in vitro, in vivo and human health. Antioxidants (Basel). 9:797

Killeen MJ, Linder M, Pontoniere P, Crea R (2014) NF-κβ signaling and chronic inflammatory diseases: exploring the potential of natural products to drive new therapeutic opportunities. Drug Discov Today 19:373–378

Kim M, Cho KH, Shin MS, Lee JM, Cho HS, Kim CJ, Shin DH, Yang HJ (2014) Berberine prevents nigrostriatal dopaminergic neuronal loss and suppresses hippocampal apoptosis in mice with Parkinson's disease. Int J Mol Med 3:870–878

Kwon IH, Choi HS, Shin KS, Lee BK, Lee CK, Hwang BY, Lim SC, Lee MK (2010) Effects of berberine on 6-hydroxydopamine-induced neurotoxicity in PC12 cells and a rat model of Parkinson's disease. Neurosci Lett 486:29–33

Laughton MJ, Halliwell B, Evans PJ, Hoult JRS (1989) Antioxidant and pro-oxidant actions of the plant phenolics quercetin, gossypol and myricetin. Effects on lipid peroxidation, hydroxyl radical generation and bleomycin-dependent damage to DNA. Biochem Pharmacol 38:2859–2865

Laughton MJ, Evans PJ, Moroney MA, Hoult JR, Halliwell B (1991) Inhibition of mammalian 5-lipoxygenase and cyclo-oxygenase by flavonoids and phenolic dietary additives. Relationship to antioxidant activity and to iron ion-reducing ability. Biochem Pharmacol 42(9):1673–1681.

Leri M, Scuto M, Ontario ML, Calabrese V, Calabrese EJ, Bucciantini M, Stefani M (2020) Healthy effects of plant polyphenols: molecular mechanisms. Int J Mol Sci 21:1250

Lim GP, Chu T, Yang FS, Beech W, Frautschy SA, Cole GM (2001) The curry spice curcumin reduces oxidative damage and amyloid pathology in an Alzheimer transgenic mouse. J Neurosci 21:8370–8377

Lin PH, Chiang MT, Chau LY (2008) Ubiquitin-proteasome system mediates heme oxygenase-1 degradation through endoplasmic reticulum-associated degradation pathway. Biochim Biophys Acta. 1783(10):1826–1834

Liu J, Yang L, Dong Y, Zhang B, Ma X (2018) Echinacoside, an inestimable natural product in treatment of neurological and other disorders. Molecules (basel, Switzerland). 23:1213

Liu L, Zhao Z, Yin Q, Zhang X (2019) TTB protects astrocytes against oxygen-glucose deprivation/reoxygenation-induced injury via activation of Nrf2/HO-1 signaling pathway. Front Pharmacol 10:792

Lodi R, Tonon C, Calabrese V, Schapira AH (2006) Friedreich's ataxia: from disease mechanisms to therapeutic interventions. Antioxid Redox Signal 8:438–443

Luo S, Jiang X, Jia L, Tan C, Li M, Yang Q, Du Y, Ding C (2019) In vivo and in vitro antioxidant activities of methanol extracts from olive leaves on caenorhabditis elegans. Molecules 24:704

Luo C, Li Y, Wang H, Cui Y, Feng Z, Li H, Li Y, Wang Y, Wurtz K, Weber P, Long J, Liu J (2013) Hydroxytyrosol promotes superoxide production and defects in autophagy leading to anti-proliferation and apoptosis on human prostate cancer cells. Curr Cancer Drug Targets 13:625–639

Mancuso C, Capone C, Ranieri SC, Fusco S, Calabrese V, Eboli ML, Preziosi P, Galeotti T, Pani G (2008) Bilirubin as an endogenous modulator of neurotrophin redox signaling. J Neurosci Res 86:1212–1230

Mancuso C, Pani G, Calabrese V (2006) Bilirubin: an endogenous scavenger of nitric oxide and reactive nitrogen species. Redox Rep 11:207–213

Mancuso C, Perluigi M, Cini C, De Marco C, Giuffrida Stella AM, Calabrese V (2006) Heme oxygenase and cyclooxygenase in the central nervous system: a functional interplay. J Neurosci Res 84:1385–1391

Mancuso C, Scapagnini G, Curro D, Giuffrida Stella AM, De Marco C, Butterfield DA, Calabrese V (2007) Mitochondrial dysfunction, free radical generation and cellular stress response in neurodegenerative disorders. Front Biosci 12:1107–1123

Marseglia A, Xu W, Fratiglioni L, Fabbri C, Berendsen AAM, Bialecka-Debek A, Jennings A, Gillings R, Meunier N, Caumon E, Fairweather-Tait S, Pietruszka B, De Groot LCPGM, Santoro A, Franceschi C (2018) Effect of the NU-AGE diet on cognitive functioning in older adults: a randomized controlled trial. Front Physiol 9:349–349

Martinez-Lapiscina EH, Clavero P, Toledo E, Estruch R, Salas-Salvado J, San Julian B, Sanchez-Tainta A, Ros E, Valls-Pedret C, Martinez-Gonzalez MA (2013) Mediterranean diet improves cognition: the PREDIMED-NAVARRA randomised trial. J Neurol Neurosurg Psychiatr 84:1318–1325

Mathew ST, Bergström P, Hammarsten O (2014) Repeated Nrf2 stimulation using sulforaphane protects fibroblasts from ionizing radiation. ToxicolAppl Pharmacol. 276:188–194

McDonagh B, Sakellariou GK, Smith NT, Brownridge P, Jackson MJ (2014) Differential cysteine labeling and global label-free proteomics reveals an altered metabolic state in skeletal muscle aging. J Proteome Res 13:5008–5021

Miquel S, Champ C, Day J, Aarts E, Bahr BA, Bakker M, Bánáti D, Calabrese V, Cederholm T, Cryan J, Dye L, Farrimond JA, Korosi A, Layé S, Maudsley S, Milenkovic D, Mohajeri MH, Sijben J, Solomon A, Spencer JPE, Thuret S, Vanden Berghe W, Vauzour D, Vellas B, Wesnes K, Willatts P, Wittenberg R, Geurts L (2018) Poor cognitive ageing: Vulnerabilities, mechanisms and the impact of nutritional interventions. Ageing Res Rev 42:40–55

Mohammad-Beigi H, Aliakbari F, Sahin C, Lomax C, Tawfike A, Schafer NP, Amiri-Nowdijeh A, Eskandari H, Møller IM, Hosseini-Mazinani M, Christiansen G, Ward JL, Morshedi D, Otzen DE (2019) Oleuropein derivatives from olive fruit extracts reduce α-synuclein fibrillation and oligomer toxicity. J Biol Chem 294:4215–4232

Munné-Bosch S, Alegre L (2003) Drought-induced changes in the redox state of alpha-tocopherol, ascorbate, and the diterpene carnosic acid in chloroplasts of Labiatae species differing in carnosic acid contents. Plant Physiol 131:1816–1825

Nakagawa T, Yokozawa T (2002) Direct scavenging of nitric oxide and superoxide by green tea. Food Chem Toxicol 40(12):1745–1750

Nardiello P, Pantano D, Lapucci A, Stefani M, Casamenti F (2018) Diet supplementation with hydroxytyrosol ameliorates brain pathology and restores cognitive functions in a mouse model of amyloid-β deposition. J Alzheimers Dis 63:1161–1172

Nematalla KM, Ghada SAA, Yousef M, Shabib ZA (2011) Effect of Echinacea as antioxidant on markers of aging. Aust J Basic Appl Sci 5:18–26

Noda Y, Kaneyuki T, Igarashi K, Mori A, Packer L (1998) Antioxidant activity of nasunin, an anthocyanin in eggplant. Res Commun Mol Pathol Pharmacol Res Commun Mol Pathol Pharmacol 102:175–187

Numano F, Crea R, Wiedemann T, Bitler C (2007) Vegetation water composition for treatment of inflammatory skin conditions. US Patent. Appl. 20070020350

Olech M, Nowak R, Ivanova D, Tashev A, Boyadzhieva S, Kalotova G, Angelov G, Gawlik-Dziki U (2020a) Appl Sci 10:8921

Olech Marta, Kasprzak Kamila, Wójtowicz Agnieszka, Oniszczuk Tomasz, Nowak Renata, Waksmundzka-Hajnos Monika, Combrzyński Maciej, Gancarz Marek, Kowalska Iwona, Krajewska Anna, Oniszczuk Anna (2020) Polyphenol composition and antioxidant Potential of Instant Gruels Enriched with Lycium barbarum L. Fruit. Molecules 25(19):4538

Ozdal T, Sela DA, Xiao J, Boyacioglu D, Chen F, Capanoglu E (2016) The reciprocal interactions between polyphenols and gut microbiota and effects on bioaccessibility. Nutrients 8:78

Pall ML, Levine S (2015) Nrf2, a master regulator of detoxification and also antioxidant, anti-inflammatory and other cytoprotective mechanisms, is raised by health promoting factors. Sheng Li Xue Bao. 67:1–18

Pan KT, Chen YY, Pu TH et al (2014) Mass spectrometry-based quantitative proteomics for dissecting multiplexed redox cysteine modifications in nitric oxide-protected cardiomyocyte under hypoxia. Antioxid Redox Signal 20:1365–1381

Panizzi L, Scarpati ML, Oriente EG (1960) Structure of the bitter glucoside oleuropein. Note II. Gazzetta Chimica Italiana 90:1449–1485

Pennisi G, Cornelius C, Cavallaro MM, Trovato SA, Cambria MT, Pennisi M, Bella R, Milone P, Ventimiglia B, Migliore MR, Di Renzo L, De Lorenzo A, Calabrese V (2011) Redox regulation of cellular stress response in multiple sclerosis. Biochem Pharm 82:1490–1499

Perluigi M, Di Domenico F, Giorgi A, Schininà ME, Coccia R, Cini C, Bellia F, Cambria MT, Cornelius C, Butterfield DA, Calabrese V (2010) Redox proteomics in aging rat brain: Involvement of mitochondrial reduced glutathione status and mitochondrial protein oxidation in the aging process. J Neurosci Res 88:3498–3507

Perluigi M, Joshi G, Sultana R, Calabrese V, De Marco C, Coccia R, Cini C, Butterfield DA (2006) In vivo protective effects of ferulic acid ethyl ester against amyloid-beta peptide 1–42-induced oxidative stress. J Neurosci Res 84:418–426

Piroddi M, Depunzio I, Calabrese V, Mancuso C, Aisa CM, Binaglia L, Minelli A, Butterfield AD, Galli F (2007) Oxidatively-modified and glycated proteins as candidate pro-inflammatory toxins in uremia and dialysis patients. Amino Acids 32:573–592

Plauth A, Geikowski A, Cichon S, Wowro SJ, Liedgens L, Rousseau M, Weidner C, Fuhr L, Kliem M, Jenkins G, Lotito S, Wainwright LJ, Sauer S (2016) Hormeticshiftingof redox environment by pro-oxidative resveratrol protects cells against stress. Free RadicBiol Med. 99:608–622

Poon HF, Calabrese V, Calvani M, Butterfield DA (2006) Proteomics analyses of specific protein oxidation and protein expression in aged rat brain and its modulation by L-acetylcarnitine: insights into the mechanisms of action of this proposed therapeutic agent for CNS disorders associated with oxidative stress. Antioxid Redox Signal 8:381–394

Quach A, Tanaka T, Lu AT, Chen BH, Ferrucci L, Ritz B, Bandinelli S, Neuhouser ML, Beasley JM, Snetselaar L, Wallace RB, Tsao PS, Absher D, Assimes TL, Stewart JD, Li Y, Hou L, Baccarelli AA, Whitsel EA, Horvath S (2017) Epigenetic clock analysis of diet, exercise, education, and lifestyle factors. Aging (albany NY) 9:27

Rainey NE, Moustapha A, Saric A, Nicolas G, Sureau F, Petit PX (2019) Iron chelation by curcumin suppresses both curcumin-induced autophagy and cell death together with iron overload neoplastic transformation. Cell Death Discov 5:150

Rattan SI (2008) Hormesis in aging. Ageing Res Rev 7:63–78
Reutzel M, Grewal R, Silaidos C, Zotzel J, Marx S, Tretzel J, Eckert GP (2018) Effects of long-term treatment with a blend of highly purified olive secoiridoids on cognition and brain atp levels in aged nmri mice. Oxidative Med Cellular Longevity 2018:1–10. https://doi.org/10.1155/2018/4070935
Richard N, Arnold S, Hoeller U, Kilpert C, Wertz K, Schwager J (2011) Hydroxytyrosol is the major anti-inflammatory compound in aqueous olive extracts and impairs cytokine and chemokine production in macrophages. Planta Med 77:1890–1897
Rietjens SJ, Bast A, Haenen GR (2007) New insights into controversies on the antioxidant potential of the olive oil antioxidant hydroxytyrosol. J Agric Food Chem 555:7609–7614
Sakano M, Mizutani M, Murata M, Oikawa S, Hiraku Y, Kawanishi S (2005) Procyanidin B2 has anti- and pro-oxidant effects on metal-mediated DNA damage. Free Radic Biol Med 39:1041–1049
Salminen A, Lehtonen M, Paimela T, Kaarniranta K (2010) Celastrol: molecular targets of Thunder God Vine. Biochem Biophys Res Commun 394:439–442
Salvioli S, Sikora E, Cooper EL, Franceschi C (2007) Curcumin in cell death processes: a challenge for CAM of age-related pathologies. Evid Based Complementary Altern Med. 4:181–190
Samieri C (2018) Epidemiology and risk factors of Alzheimer's disease: a focus on diet. In: Perneczky R (ed) Biomarkers for Preclinical Alzheimer's Disease. Springer Nature
Santoro A, Pini E, Scurti M, Palmas G, Berendsen A, Brzozowska A, Pietruszka B, Szczecinska A, Cano N, Meunier N, de Groot CPGM, Feskens E, Fairweather-Tait S, Salvioli S, Capri M, Brigidi P, Franceschi C, Consortium N-A (2014) Combating inflammaging through a Mediterranean whole diet approach: the NU-AGE project's conceptual framework and design. Mech Ageing Dev 136–137:3–13
Scapagnini G, Caruso C, Calabrese V (2011) Therapeutic potential of dietary polyphenols against brain ageing and neurodegenerative disorders. Adv Exp Med Biol 698:27–35
Scapagnini G, Colombrita C, Amadio M, D'Agata V, Arcelli E, Sapienza M, Quattrone A, Calabrese V (2006) Curcumin activates defensive genes and protects neurons against oxidative stress. Antioxid Redox Signal 8:395–403
Scarmeas N, Anastasiou CA, Yannakoulia M (2018) Nutrition and prevention of cognitive impairment. Lancet Neurol 17:1006–1015
Schipper SHM, Song SW (2015) A heme oxygenase-1 transducer model of degenerative and developmental brain disorders. Int J Mol Sci 16:5400–5419
Scoditti E, Capurso C, Capurso A, Massaro M (2014) Vascular effects of the Mediterranean diet-part II: role of omega-3 fatty acids and olive oil polyphenols. Vascul Pharmacol 63:127–134
Scuto M, Trovato Salinaro A, Modafferi S, Polimeni A, Pfeffer T, Weigand T, Calabrese V, Schmitt CP, Peters V (2020) Carnosine activates cellular stress response in podocytes and reduces glycative and lipoperoxidative stress. Biomedicines 8:177
Scuto M, Di Mauro P, Ontario ML, Amato C, Modafferi S, Ciavardelli D, Salinaro AT, Maiolino L, Calabrese V (2019a) Nutritional mushroom treatment in meniere's disease with coriolus versicolor: a rationale for therapeutic intervention in neuroinflammation and antineurodegeneration. Int J Mol Sci 21(1):284
Scuto MC, Mancuso C, Tomasello B, Laura Ontario ML, Cavallaro A, Frasca F, Maiolino L, Trovato Salinaro A, Calabrese EJ, Calabrese V (2019b) Curcumin, hormesis and the nervous system. Nutrients 11(10):2417
Shafaei A, Croft K, Hodgson J, Boyce MC (2019) Simultaneous quantitative analysis of polyphenolic compounds in human plasma by liquid chromatography tandem mass spectrometry. J Sep Sci 42:2909–2921
Shannon OM, Lee V, Bundy R, Gillings R, Jennings A, Stephan B, Hornberger M, Balanos G, Paddick SM, Hanson S, Hardeman W, Holmes R, Garner N, Aldred S, Siervo M, Mathers JC, Minihane AM (2021) Feasibility and acceptability of a multi-domain intervention to increase Mediterranean diet adherence and physical activity in older UK adults at risk of dementia: protocol for the MedEx-UK randomised controlled trial. BMJ Open. 11:e042823

Shen CY, Jiang JG, Yang L, Wang DW, Zhu W (2017) Anti-ageing active ingredients from herbs and nutraceuticals used in traditional Chinese medicine: pharmacological mechanisms and implications for drug discovery. Br J Pharmacol 174(11):1395–1425

Siciliano R, Barone E, Calabrese V, Rispoli V, Butterfield DA, Mancuso C (2011) Experimental research on nitric oxide and the therapy of alzheimer disease: a challenging bridge. CNS Neurol Disord Drug Targets. 10:766–776

Siracusa R, Scuto M, Fusco R, Trovato A, Ontario ML, Crea R, Di Paola R, Cuzzocrea S, Calabrese V (2020) Anti-inflammatory and Anti-oxidant Activity of Hidrox® in Rotenone-Induced Parkinson's Disease in Mice. Antioxidants (basel). 9:824

Song W, Zukor H, Lin SH, Hascalovici J, Liberman A, Tavitian A, Mui J, Vali H, Tong XK, Bhardwaj SK, Srivastava LK, Hamel E, Schipper HM (2012) Schizophrenia-like features in transgenic mice overexpressing human HO-1 in the astrocytic compartment. J Neurosci 32:10841–10853

Song SE, Jo HJ, Kim YW, Cho YJ, Kim JR, Park SY (2016) Delphinidin prevents high glucose-induced cell proliferation and collagen synthesis by inhibition of NOX-1 and mitochondrial superoxide in mesangial cells. J Pharmacol Sci 130:235–243

Soni MG, Burdock GA, Christian MS, Bitler CM, Crea R (2006) Safety assessment of aqueous olive pulp extract as an antioxidant or antimicrobial agent in foods. Food Chem Toxicol 44:903–915

Sun J, Lin LZ, Chen P (2012) Study of the mass spectrometric behaviors of anthocyanins in negative ionization mode and its applications for characterization of anthocyanins and non-anthocyanin polyphenols. Rapid Commun Mass Spectrom 26:1123–1133

Takeda K (2006) Blue metal complex pigments involved in blue flower color. Proc JpnAcad Ser B Phys Biol Sci. 82:142–154

Takeda Y, Jamsransuren D, Matsuda S, Crea R, Ogawa H (2021) The SARS-CoV-2-Inactivating Activity of Hydroxytyrosol-Rich Aqueous Olive Pulp Extract (HIDROX®) and Its Use as a Virucidal Cream for Topical Application. Viruses 13:232

Tappel A (2007) Heme of consumed red meat can act as a catalyst of oxidative damage and could initiate colon, breast and prostate cancers, heart disease and other diseases. Med Hypotheses 68:562–564

Trovato Salinaro A, Cornelius C, Koverech G, Koverech A, Scuto M, Lodato F, Fronte V, Muccilli V, Reibaldi M, Longo A, Uva MG, Calabrese V (2014) Cellular stress response, redox status, and vitagenes in glaucoma: a systemic oxidant disorder linked to Alzheimer's disease. Front Pharmacol 5:129

Tsuda T, Kato Y, Osawa T. (2000) Mechanism for the peroxynitrite scavenging activity by anthocyanins. FEBS Lett 484:207–210

Tsukimoto M, Tamaishi N, Homma T, Kojima S (2010) Low-dose gamma-ray irradiation induces translocation of Nrf2 into nuclear in mouse macrophage RAW264.7 cells. J Radiat Res 51:349–353

Uehara, T, Nakamura, T, Yao, D, Shi, ZQ, Gu, Z, Ma, Y, Masliah, E, Nomura, Y, Lipton, S.A. (2006)S-nitrosylated protein-disulphide isomerase links protein misfolding to neurodegeneration. *Nature*. 441, 513–7.

van Duynhovena J, Vaughana EE, Jacobsa DM, Kempermana RA, van Velzena EJJ, Grossa G, Rogera LC, Possemiersd S, Smildec AK, Doréb J, Westerhuisc JA, Van de Wieled T (2011) PNAS 108:4531–4538

Vauzour D, Camprubi-Robles M, Miquel-Kergoat S, Andres-Lacueva C, Banati D, Barberger-Gateau P, Bowman GL, Caberlotto L, Clarke R, Hogervorst E, Kiliaan AJ, Lucca U, Manach C, Minihane A-M, Mitchell ES, Perneczky R, Perry H, Roussel A-M, Schuermans J, Sijben J, Spencer JPE, Thuret S, van de Rest O, Vandewoude M, Wesnes K, Williams RJ, Williams RSB, Ramirez M (2017) Nutrition for the ageing brain: towards evidence for an optimal diet. Ageing Res Rev 35:222–240

Visioli F, Bellomo G, Galli C (1998) Free radical-scavenging properties of olive oil polyphenols. BiochemBiophys Res Commun. 247:60–64

Visioli F, Wolfram R, Richard D, Abdullah MI, Crea R (2009) Olive phenolics increase glutathione levels in healthy volunteers. J Agric Food Chem 57:17939–17946

Wang N, Zhang F, Yang L, Zou J, Wang H, Liu K, Liu M, Zhang H, Xiao X, Wang K (2017) Resveratrol protects against L-arginine-induced acute necrotizing pancreatitis in mice by enhancing SIRT1-mediated deacetylation of p53 and heat shock factor 1. Int J Mol Med 40:427–437

Wang Y, Mandal AK, Son YO, Pratheeshkumar P, Wise JTF, Wang L, Zhang Z, Shi X, Chen Z (2018) Roles of ROS, Nrf2, and autophagy in cadmium-carcinogenesis and its prevention by sulforaphane. Toxicol Appl Pharmacol 353:23–30

Wang YH, Xuan ZH, Tian S, Du GH (2015) Echinacoside Protects against 6-Hydroxydopamine-Induced Mitochondrial Dysfunction and Inflammatory Responses in PC12 Cells via Reducing ROS Production. *Evid. Based Complement. Alternat. Med.*189239.

Watjen W et al (2005) Low concentrations of flavonoids are protective in rat H4IIE cells whereas high concentrations cause DNA damage and apoptosis. J Nutr 135:525–531

Westerheide SD, Bosman JD, Mbadugha BN, Kawahara TL, Matsumoto G, Kim S, Gu W, Devlin JP, Silverman RB, Morimoto RI (2004) Celastrols as inducers of the heat shock response and cytoprotection. J Biol Chem 279:56053–56060

Westerheide SD, Raynes R, Powell C, Xue B, Uversky VN (2012) HSF transcription factor family, heat shock response, and protein intrinsic disorder. Curr Protein Pept Sci 13:86–103

Wong BX, Duce JA (2014) The iron regulatory capability of the major protein participants in prevalent neurodegenerative disorders. Front Pharmacol 5:81

Yan F, Chen Y, Azat R, Zheng X (2017) Mulberry Anthocyanin Extract Ameliorates Oxidative Damage in HepG2 Cells and Prolongs the Lifespan of Caenorhabditis elegans through MAPK and Nrf2 Pathways. Oxid Med Cell Longev 7956158:12

Yang CM, Lin CC, Hsieh HL (2017) High-glucose-derived oxidative stress-dependent heme oxygenase-1 expression from astrocytes contributes to the neuronal apoptosis. Mol Neurobiol 54:470–483

Yang R, Xiu H, Zhou Q, Sun L, Mu H, Li H, Wang S, Li X, Chen W, Dong J (2019) Oxid Med Cell Longev 4809836:1–11

Zeng X, Pan X, Xu X, Lin J, Que F, Tian Y, Li L, Liu S (2017) Resveratrol Reactivates Latent HIV through Increasing Histone Acetylation and Activating Heat Shock Factor 1. J Agric Food Chem 65:4384–4394

Zhang Y, Ahn YH, Benjamin IJ, Honda T, Hicks RJ, Calabrese V, Cole PA, Dinkova-Kostova AT (2011) HSF1-dependent upregulation of Hsp70 by sulfhydryl-reactive inducers of the KEAP1/NRF2/ARE pathway. Chem Biol 18:1355–1361

Zhang C, Li C, Chen S, Li Z, Jia X, Wang K, Bao J, Liang Y, Wang X, Chen M, Li P, Su H, Wan JB, Lee SMY, Liu K, He C (2017a) Berberine protects against 6-OHDA-induced neurotoxicity in PC12 cells and zebrafish through hormetic mechanisms involving PI3K/AKT/Bcl-2 and Nrf2/HO-1 pathways. Redox Biol 11:1–11

Zhang D, Li H, Wang JB (2014) Echinacoside inhibits amyloid fibrillization of HEWL and protects against Abeta-induced neurotoxicity. Int J Biol Macromol 72:243–253

Zhang J, Wu J, Liu F, Tong L, Chen Z, Chen J, He H, Xu R, Ma Y, Huang C (2019) Neuroprotective effects of anthocyanins and its major component cyanidin-3-O-glucoside (C3G) in the central nervous system: an outlined review. Eur J Pharmacol 858:172500

Zhang J, Zhang Z, Xiang J, Cai M, Yu Z, Li X, Wu T, Cai D (2017b) Neuroprotective effects of echinacoside on regulating the stress-active p38mapk and nf-kappab p52 signals in the mice model of parkinson's disease. Neurochem Res 42:975–985

Zhang X, Cao J, Zhong L (2009) Hydroxytyrosol inhibits pro-inflammatory cytokines, iNOS, and COX-2 expression in human monocytic cells. Naunyn Schmiedebergs Arch Pharmacol 379:581–586

Zhao F, Zhang J, Chang N (2018) Epigenetic modification of Nrf2 by sulforaphane increases the antioxidative and anti-inflammatory capacity in a cellular model of Alzheimer's disease. Eur J Pharmacol 824:1–10

Zhao Q, Yang X, Cai D, Ye L, Hou Y, Zhang L, Cheng J, Shen Y, Wang K, Bai Y (2016) Echinacoside protects against MPP (+)-induced neuronal apoptosis via ROS/ATF3/CHOP pathway regulation. Neurosci Bull 32:349–362

Zhao X, Wang R, Xiong J, Yan D, Li A, Wang S, Xu J, Zhou J (2017) JWA antagonizes paraquat-induced neurotoxicity via activation of Nrf2. Toxicol Lett 277:32–40
Zheng L-F, Dai F, Zhou B, Yang L, Liu Z-L (2008) Prooxidant activity of hydroxycinnamic acids on DNA damage in the presence of Cu (II) ions: mechanism and structure–activity relationship. Food Chem Toxicol 46:149–156
Zhou Y, Jiang R, Yang B, Yao X, Wang P, Liu DF et al (2011) Changes of telomere and telomerase in effect of ginsenoside Rg1 to delay hematopoietic stem cell senescence. China J Chin Mater Med. 22:3172–3175
Zhu H, Cheng C, Zhang C, Wang Z (2011) Echinacoside suppresses cellular senescence of human fibroblastic cells by down-regulation of p53. J Chin Pharm Sci 20:523–528
Zukor ZH, Song SW, Liberman LA, Mui MJ, Vali VH, Fillebeen FC, Pantopoulos PK, Wu WuT, Guerquin-Kern G-K, Schipper SHM (2009) HO-1-mediated macroautophagy: a mechanism for unregulated iron deposition in aging and degenerating neural tissues. J Neurochem 109:776–791

Chapter 14
Nutrition and Exercise to Maintain Physical Functioning During Ageing

Pol Grootswagers, Marco Mensink, and Lisette de Groot

Abstract During ageing, losses in muscle mass and muscle strength are common. Unfavourable changes in muscle architecture and neurological activation of muscle tissue decrease the unit-specific muscular output, also known as muscle quality. Together, the losses in the muscular domains lead to lower physical functioning and thereby jeopardise one's independence and quality of life.

This chapter explains the age-related declines in muscle mass and quality, and how nutrition and exercise can be strategically used to prevent and counter these declines. Due attention is given to the role of nutrition and exercise in light of age-related syndromes and critical life events where muscular losses are accelerated.

Keywords Physical functioning · Exercise · Muscle quality · Nutrition · Sarcopenia · Frailty

Together with exercise, good nutrition is key for optimal muscle health and physical functioning. When people get older, their level of physical functioning becomes an important factor in their quality of life. The body's physical condition determines the level of independence and the activities one can join and is therefore of vital importance for healthy ageing. Thereby, physical functioning is the foundation of the extent to which an older person can still be physically, mentally, and socially active. The aim of this article is to explain how physical functioning normally declines during ageing and how this decline can be prevented or addressed with nutrition and exercise strategies.

P. Grootswagers (✉) · M. Mensink · L. de Groot
Division of Human Nutrition, Wageningen University, P.O. Box 17, 6700 AA Wageningen, The Netherlands
e-mail: pol.grootswagers@wur.nl

S. I. S. Rattan and G. Kaur (eds.), *Nutrition, Food and Diet in Ageing and Longevity*, Healthy Ageing and Longevity 14, https://doi.org/10.1007/978-3-030-83017-5_14

14.1 Age-Related Physical Decline

A high level of physical functioning is hard to maintain throughout ageing. During ageing, several biological processes have to be fought to preserve as much of the functional capacities a person has around the age of thirty. Estimations show that after the age of thirty, people lose around three to eight per cent of muscle mass every decade (Holloszy 2000, Melton et al. 2000). Such loss implies that if a person has 35 kg of muscle mass at the age of 30, this person might end up with between 23 and 30 kg muscle mass remaining at the age of 80. One of the reasons for this age-related loss in muscle mass is a decreased exposure to anabolic stimuli later in life, such as physical activity (McPhee et al. 2016), protein intake (Tieland et al. 2012a) and anabolic hormones (Morley 2017). Additionally, the anabolic response to these stimuli is blunted in older people, a phenomenon that is termed anabolic resistance (Burd et al. 2013).

Underneath this visible muscle mass decline, a change that is less obvious, but arguably as important, is happening: a declining muscle strength. During ageing, muscle strength is decreasing at a much faster rate than muscle mass does (Fig. 14.1). The estimated loss in muscle strength is between 10 and 20% per decade up to the age of 70 (Hughes et al. 2001) and between 20 and 30% in the decades thereafter (Goodpaster et al. 2006). The loss in muscle strength is considered a more important risk factor for functional decline than the loss in muscle mass (Schaap et al. 2013). Muscle strength might, therefore, be more relevant than muscle mass for quality of life of older adults. This notion is confirmed by the work of Balogun et al. (Balogun et al. 2017). They compared community-dwelling older adults in the lowest 20% of handgrip strength or lower-limb muscle strength with those with normal strength

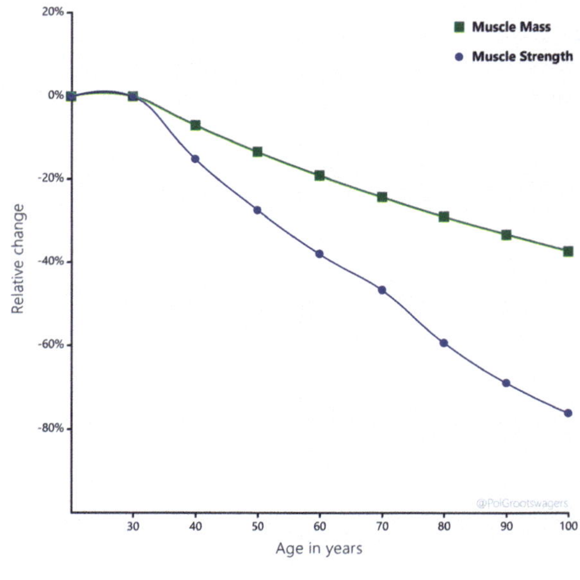

Fig. 14.1 Trajectories of ageing-induced losses in muscle mass (Holloszy 2000, Melton et al. 2000) and muscle strength (Hughes et al. 2001; Goodpaster et al. 2006). Note that the gap between muscle mass and strengths widens over the course of ageing, indicating loss of muscle quality

levels, and found a significant and clinically meaningful lower health-related quality of life over ten years of follow-up for those with low strength. The associations between low appendicular mean mass and health-related quality of life were weaker, indicating that muscle strength is a stronger determinant of quality of life than muscle mass (Balogun et al. 2017).

The observation that muscle strength is lost at a higher rate than mass reveals that muscle function is not solely determined by muscle mass. The functioning of a muscle beyond its mass can be called muscle quality (Barbat-Artigas et al. 2012). A pragmatic definition of muscle quality is function per mass and can be calculated as muscle strength or power per unit of mass, volume or cross-sectional area (Barbat-Artigas et al. 2012). A more sophisticated way of evaluating muscle quality is to examine the determinants of relative muscle strength themselves (Correa-de-Araujo et al. 2017). Four important determinants of muscle quality are type II muscle fiber atrophy, mitochondrial functioning, fat infiltration, and neuromuscular activation (Fig. 14.2). The bottom layer of Fig. 14.2 shows the lifestyle factors that can improve determinants of muscle quality (explained in part 4).

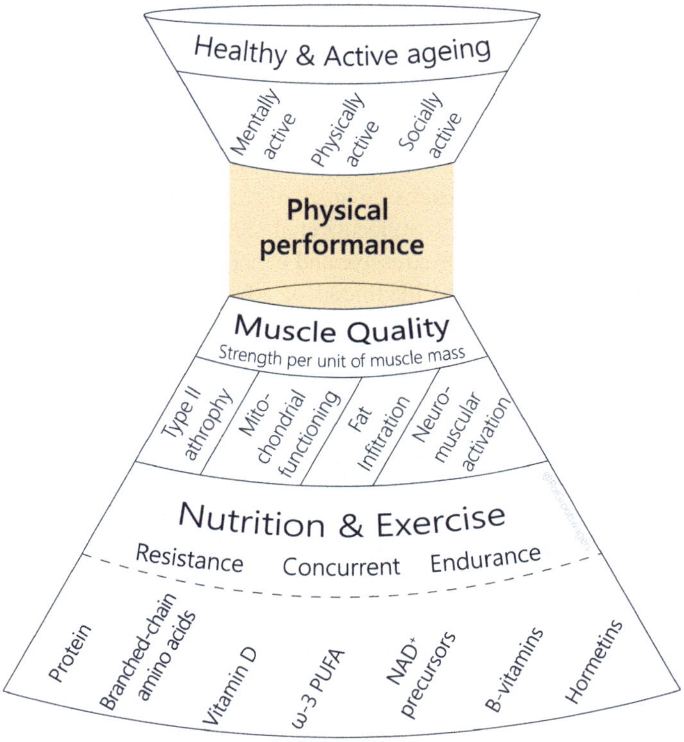

Fig. 14.2 Schematic representation of the possible influence of nutrition and exercise on healthy ageing via muscle quality-related mediators of physical performance

14.2 Ageing Syndromes

Over the last decades, several forms of physical dysfunction have been classified into ageing phenotypes. The most important syndromes that are not always related to diseases are malnutrition, sarcopenia and frailty. Here, we shortly touch upon these three phenotypes and describe how they can be addressed via nutritional therapies.

Malnutrition is defined as 'a state resulting from lack of intake or uptake of nutrition that leads to altered body composition (decreased fat free mass) and body cell mass leading to diminished physical and mental function and impaired clinical outcome from disease' (Cederholm et al. 2017) It can be assessed via multiple screening tools, such as MNA, MST, SNAQ^{RC} and NUFFE, which are tailored to various settings (Power et al. 2018). The estimated prevalence of malnutrition is 11% in community-dwelling older adults (Schilp et al. 2012) and even higher in more vulnerable populations, with 35% in those receiving home-care (Schilp et al. 2012) and 38% in patients admitted to geriatric wards (Kruizenga et al. 2016). Malnutrition is strongly related to morbidity (Abizanda et al. 2016), mortality (McMinn et al. 2011) and functional impairments (Kiesswetter et al. 2014).

A clinical nutritional treatment against malnutrition is oral nutritional supplementation (ONS). These supplements contain all macro- and micronutrients and have shown to be effective in increasing body weight (Milne et al. 2009). Novel formulations of ONS aim to additionally target the physical condition of patients via potent nutrients such as whey protein, leucine, HMB, ursolic acid, and vitamin D (Grootswagers et al. 2018; Deutz et al. 2016; Bauer et al. 2015). These novel formulations have shown promising results on hospital readmission, chair stand test, and gait speed (Grootswagers et al. 2018; Deutz et al. 2016; Bauer et al. 2015).

Sarcopenia is defined as 'a progressive and generalised skeletal muscle disorder that is associated with increased likelihood of adverse outcomes including falls, fractures, physical disability and mortality' (Cruz-Jentoft et al. 2019). It is prevalent in an estimated 10% of the older adult population (Shafiee et al. 2017), and has clear detrimental effects on quality of life and independent living of older adults (Dos Santos et al. 2017; Franzon et al. 2019; Tsekoura et al. 2017) New guidelines have put muscle function central in the sarcopenia diagnosis (Cruz-Jentoft et al. 2019), which can be measured as handgrip strength or chair rise test time. Sarcopenia can also be screened for via the SARC-F questionnaire (Malmstrom et al. 2016). The recommended treatment of sarcopenia consists of resistance exercise, ideally in combination with protein or leucine and vitamin D when deficient (Morley 2018, Morley et al. 2010). Recently, sarcopenia was assigned an ICD-10 code (Anker et al. 2016), which greatly advances the recognition of sarcopenia by the international medical community.

Frailty is defined as 'A medical syndrome with multiple causes and contributors that is characterised by diminished strength, endurance, and reduced physiologic function that increases an individual's vulnerability for developing increased dependency and/or death (Morley et al. 2013)' and is prevalent in around 10% in women and 5% in men (Collard et al. 2012). Frailty is usually assessed by Fried's criteria

(Fried et al. 2001). The distinction between frailty and sarcopenia lies in the multidimensional nature of frailty, covering not only physical declines but also cognitive and social deteriorations. Treatment options for frailty include physical exercise, protein and vitamin D (Artaza-Artabe et al. 2016).

Malnutrition, sarcopenia and frailty share common determinants, such as weight loss and muscle weakness. The overlap in the phenotypes leads to the co-existence of multiple phenotypes within the same patient (Cederholm 2015). Researchers, therefore, advised striving towards interventions that treat all three conditions, for instance, by targeting common risk factors (Verstraeten et al. 2021).

14.3 Critical Life Events

The gradual course of physical decline, as described in paragraph 1, reflect population means. However, on the individual level, such a gradual decline is most often no reality. During critical life events, muscle mass, muscle quality and physical functioning can decline rapidly. For instance, during ten days of bed rest, older adults lose around 1 kg of leg lean mass (Kortebein et al. 2007) and 11% of muscle strength (Kortebein et al. 2008). Importantly, the acute declines in muscle-related measures differ between individuals and are strongly related to the patient's age. For instance, for the same amount of bed rest, a young individual loses only 25% of what an older person loses (LeBlanc et al. 1992). The exact reasons for the large differences between young and older adults during similar periods of bed rest remain to be elucidated, but higher cortisol levels and reduced response to anabolic stimuli in older adults seem to play a role (English and Paddon-Jones 2010).

Critical life events are important time-frames for nutritional interventions. With nutrition and exercise, losses in muscle mass and muscle quality can be minimised, possibly preserving independence in daily activities. During and after a hospital stay and other forms of bed rest, careful monitoring of a patient's dietary intake, nutritional intake and physical activity is important. Lifestyle interventions within the capabilities of the patient, and the possibilities of the setting, are recommended to be administered when needed (English and Paddon-Jones, 2010). These interventions should try to include resistance exercise (or, if impossible, aerobic exercise or exercise mimetics) (Kouw et al. 2019), combined with supplementation with anabolic nutrients such as protein (English and Paddon-Jones 2010), leucine (English et al. 2015), or β-hydroxy β-methyl butyric acid (Deutz et al. 2013).

Often, these critical life events occur unannounced and are the result of acute injuries or diseases. In the case of elective surgeries, such as planned joint replacements, preparation for the critical life event is possible. This idea led to the concept of prehabilitation, or the *better in, better out concept* (Hoogeboom et al. 2014). In the weeks leading to surgery, patients are prepared via exercise training programs, often in combination with nutrition. Prehabilitation programs aim to increase muscle mass and function, to create extra capacity to better deal with the inevitable losses during hospital stays. The time window to train patients is short, on average around four

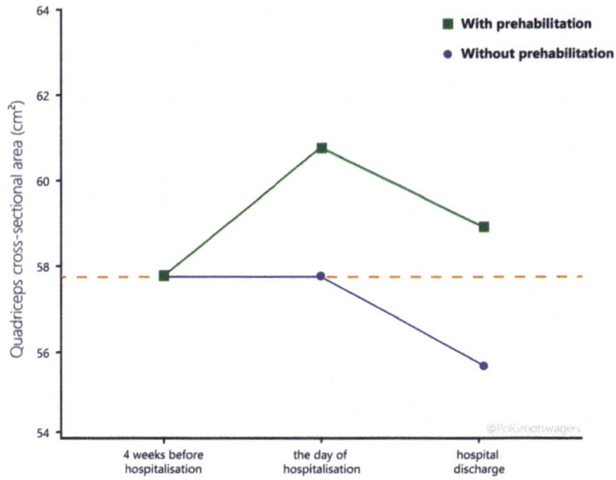

Fig. 14.3 The projected trajectory of the potential of prehabilitation to prevent net losses in muscle mass in older adults during hospitalisation, after plotting the increase found in Grootswagers et al. (Grootswagers et al. 2020)(a, + 5.4%) and the decrease adapted from Kouw et al.[12] (b, −3.4%). The orange dotted line represents the starting level of the quadriceps cross-sectional area

weeks (Moran et al. 2016; Hughes et al. 2019). A high-intensity aerobic interval and resistance exercise training, in combination with protein supplementation, improves muscle mass by 4% and muscle strength by 10%, just enough to prevent losses during hospital stay (Grootswagers et al. 2020) (Fig. 14.3).

14.4 Nutrition and Exercise

Physical functioning can be directly improved via exercise training (Chou et al. 2012). Therefore, exercise is the central aspect of the treatment of sarcopenia (Phu et al. 2015). The role of nutrition is often more indirect and supportive of exercise training, but the potential of nutrition deserves not to be overlooked. Some nutrients have clear anabolic properties and play important roles in muscle protein synthesis, muscle mass maintenance, and muscle repair. Other nutrients are involved in mechanisms that can improve the muscle output, such as force, power and endurance, by improving nutrient availability in the muscle, by improving the quality of the muscle tissue architecture or by improving the neurological activation mechanisms of the muscle.

The main anabolic nutrient is protein. The consumption of proteins elevates concentrations of amino acids and insulin in the bloodstream. The increased concentrations of these anabolic compounds are signaled by receptors located in the sarcolemma and increase the activity of the mammalian target of rapamycin (mTOR) pathway, which is the central pathway in muscle protein synthesis stimulation (McCarthy and Esser 2010) (Fig. 14.4). Leucine, an essential amino acid, has a direct stimulating effect on mTOR (Son et al. 2019) and is proposed as a key amino acid in muscle anabolism (Drummond and Rasmussen 2008). Interestingly, some of the pathways via which protein leads to muscle anabolism are proposed to underly an effect of protein on bone mass, too (**Text box 1**). Importantly, over the course of ageing, the response to anabolic stimuli is blunted. This anabolic resistance can

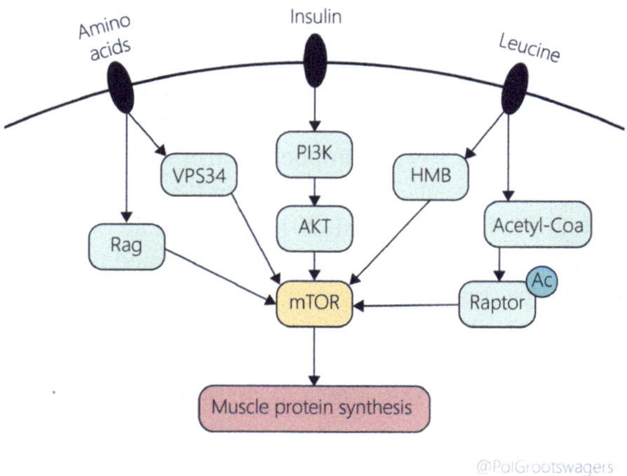

Fig. 14.4 Pathways via which nutrients can induce muscle protein synthesis

be overcome by increasing the nutrient dosage and by increasing physical activity (Burd et al. 2013).

Increasing muscle mass is not the only mode of action via which nutrition and exercise can improve physical performance during ageing. Nutrients are involved in many more pathways that determine muscle output, such as muscle architecture, muscle energetics and neurological activation of the skeletal muscle tissue. The most important determinants of muscle quality where nutrients can play a role are (1) type-II atrophy, (2) mitochondrial dysfunction, (3) fat infiltration and (4) decreased neurological activation. The way in which nutrition and exercise can improve each of these four aspects of muscle quality will be explained below.

14.4.1 Type-II Atrophy

Specific type II muscle fibre atrophy is an important determinant of age-related decreases in muscle mass and quality. In muscle tissue, we distinguish two fibre types: type I and type II (Schiaffino and Reggiani, 2011). Type I fibres have a higher abundance of mitochondria and oxidative enzymes (Schiaffino 2010) and are more abundant in the skeletal muscle tissue of endurance athletes (Wilson et al. 2012). The type II fibres (including type IIa, IIb and IIx) have a higher glycolytic enzyme activity than type I fibres have (Essén et al. 1975), but they are more prone to fatigue (Talbot and Maves, 2016). Type II fibres are more abundant in the skeletal muscle tissue of sprinters and weight lifters (Wilson et al. 2012). During ageing, specific atrophy of type II muscle fibre size occurs. Research shows that the size of type II muscle fibres is 10 to 40% smaller in older compared to young adults (Nilwik et al. 2013; Dreyer et al. 2006; Kosek et al. 2006; Verdijk et al. 2014). Interestingly, one study showed that the difference in type II fibre size fully explained between-group differences in quadriceps size (Nilwik et al. 2013).

The cross-sectional area of type II muscle fibres does not only correlate with muscle mass but also with leg strength (Verdijk et al. 2010). Targeting age-related type II fibre atrophy is, therefore, a potential strategy to improve muscle quality. Studies show promising results on improving type II fibre size via resistance exercise (Charette et al. 1991; Campbell et al. 1999; Hagerman et al. 2000; McGuigan et al. 2001; Claflin et al. 2011; Kryger and Andersen 2007; Suetta et al. 2008; Karavirta et al. 2011; Leenders et al. 2013; Frank et al. 2016; Holwerda et al. 2018; Pyka et al. 1994; Trappe et al. 2000). It should be noted that there is an equal amount of RCTs that fail to show improvements in type II fibre size (Taaffe et al. 1996; Sipila et al. 1997; Godard et al. 2002; Campbell et al. 2002; Brose et al. 2003; Slivka et al. 2008; Cooke et al. 2014; Snijders et al. 2019; Strandberg et al. 2019; Trappe et al. 2001; Mero et al. 2013; Mitchell et al. 2015; Bechshoft et al. 2017), which could be due to different training regimens or duration, sample characteristics, or low sample size. Other types of exercise, such as endurance exercise and electrostimulation, do not seem to improve type II fibre size.

Protein could theoretically augment resistance exercise-induced improvements in type II fibre size by allowing muscle protein synthesis after resistance exercise (Tipton et al. 1999). However, only one trial (Dirks et al. 2017) of a total of nine showed positive effects of protein supplementation on type II fibre size in older adults. Moreover, trials with leucine (Leenders et al. 2011), vitamin D (Vaes et al. 2018; Ceglia et al. 2013), antioxidants (Gouzi et al. 2019), creatine (Brose et al. 2003; Cooke et al. 2014) and chromium picolinate (Campbell et al. 1999, 2002) failed to improve type II fibre size in older adults, suggesting that this determinant of muscle quality should be targeted predominantly via resistance exercise. Resistance exercise could be combined with supplementation of anabolic nutrients to obtain postulated additional improvements in type II fibre size (Little and Phillips, 2009), but this strategy needs further investigation in older adults.

14.4.2 Mitochondrial Dysfunction

Apart from focussing on improving muscle strength via increasing the size of type II fibres, another promising strategy to improve muscle quality is by improving mitochondrial functioning in skeletal muscle tissue. Due to the specific atrophy of type II fibers, muscles of older adults have a higher proportion of type I fibres, which have a greater abundance of mitochondria (Schiaffino 2010). Yet, the number of mitochondria and the mitochondrial oxidative capacity are reduced in older individuals (Sun et al. 2016). Sarcopenic individuals show an even further reduced mitochondrial oxidative capacity (Migliavacca et al. 2019). In these individuals, the biosynthesis of nicotinamide adenine dinucleotide (NAD^+) is also repressed (Migliavacca et al. 2019). Studies show that lower mitochondrial functioning relates to lower physical performance in older adults (Santanasto et al. 2016; Coen et al. 2013).

Nutrition might stimulate mitochondrial functioning via multiple targets (Wesselink et al. 2019). For instance, dietary intake of NAD^+-precursors, such as vitamin B3 and tryptophan, can lead to increased levels of NAD^+ (Rajman et al. 2018; Liu et al. 2018). Moreover, certain bio-actives, such as ursolic acid, seem to directly improve mitochondrial biogenesis in mice models (Chen et al. 2017a). In older adults, the few trials that have tested the effects of different nutrients, or bio-actives, on mitochondrial functioning all found positive results (Pollack et al. 2017; Taub et al. 2016; Alway et al. 2017; de Oliveira et al. 2017). Regarding muscle quality, targeting the mitochondria might be the approach that is most adaptive to subtle changes in the environments, such as obtained by nutritional interventions. Apart from interventions with nutrients or bio-actives, endurance training and high-intensity interval training seem effective in improving mitochondrial functioning too.

14.4.3 Fat Infiltration

The third determinant of muscle quality is the infiltration of fat tissue in the muscle. The cross-sectional images of quadriceps muscle tissue presented in Fig. 14.5 show a clear presence of intramuscular adipose tissue (IMAT) in the 66-year-old person (B) compared to the 24-year-old person (A) (McLeod et al. 2016). While IMAT does contribute to quadriceps volume and to apparent muscle mass, it negatively affects muscle strength (Akazawa et al. 2017, Akazawa et al. 2018) or muscle endurance (Akazawa et al. 2017). Fat infiltration is, therefore, an illustrative cause of a loss in muscle quality, as it directly decreases the muscle strength per unit of muscle size. Fat

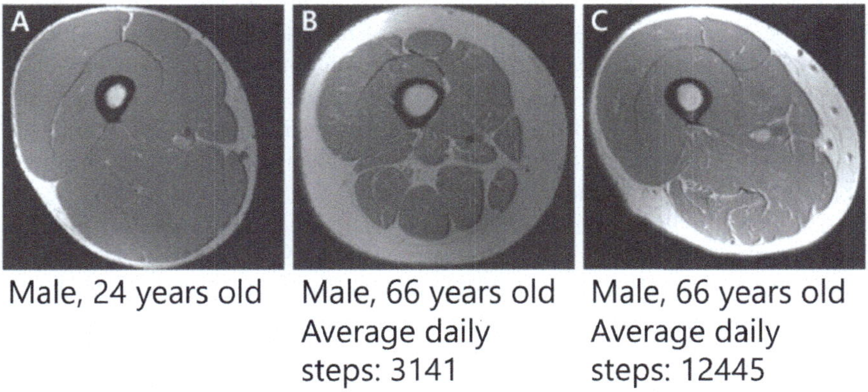

Fig. 14.5 Cross-sectional image of a quadriceps of (A) a 24-year old male, (B) a 66-year old male and (C) a 66-year old male with a high level of physical activity. Adapted from McLeod et al. 2016a (McLeod et al. 2016b), under CC BY 4.0 license (http://creativecommons.org/licenses/by/4.0/), and with changes in the number of panels, panel identification letters, and textual description within figures

infiltration may be an essential risk factor for functional decline (Visser et al. 2005). The first studies in older adults show that exercise programs can decrease IMAT (Ikenaga et al. 2017) or prevent IMAT accumulation (Goodpaster et al. 2008). That IMAT accumulation in the muscle is preventable by lifestyle is also visible in Fig. 5c. This quadriceps belongs to a 66-year-old male with a high level of physical activity. The quadriceps of this person shows amounts of IMAT that are more comparable to those of the 24-year-old male (A) than to those of the 66-year-old sedentary male (B). This suggests that physical activity might diminish the effects of ageing on IMAT accumulation. Yet, exercise interventions in older adults have yielded conflicting results.

Apart from physical activity, nutrition might play a role in IMAT accumulation too. Increased IMAT is observed in older adults at risk of malnutrition (Akazawa et al. 2019), with diabetes (Schafer et al. 2010), and it is related to increased insulin resistance (Albu et al. 2005; Goodpaster et al. 2000). These findings highlight the metabolic aspects of IMAT and suggest a role for nutritional interventions to prevent or decrease IMAT build up. The first dietary interventions aimed at lowering IMAT in older adults have been finished recently, with positive results for supplementing whey protein and vitamin D (Yamada et al. 2019; Englund et al. 2019).

14.4.4 Neurological Activation

In some cases, the loss in strength output of a muscle is not caused by morphological changes within the muscle tissue but by decreased neuromuscular activation and the loss of motor units (Clark and Fielding, 2012). Typically during ageing, we observe a loss in voluntary activation (Rozand et al. 2020). That is, the maximum force output generated by a patient's nervous system is lower than the maximum output that can be generated with electrostimulation of the muscle. Decreased voluntary activation is related to decreased physical functioning in older adults (Clark et al. 2011, 2010; Reid et al. 2014). Vitamin D (Dhesi et al. 2004) and homocysteine-lowering nutrients vitamin B6, B12 and folate (Longstreth et al. 1996a, Soumaré et al. 2006) are postulated to improve neuromuscular activation. However, the direct effect of these nutrients on neuromuscular outcomes are rarely measured (Kougias et al. 2018). Nutrients tested in randomised controlled trials that do show to improve neuromuscular activation in older adults are creatine (Stout et al. 2007), milk fat globule membrane (Minegishi et al. 2016) and N-3 fatty acids (Rodacki et al. 2012). The exercise regimens that are most effective in improving neuromuscular activation are resistance training and power training.

14.5 Future

The future of nutrition and exercise for the maintenance of physical functioning during ageing is exciting. Many nutrients have been identified as possible factors in physical functioning. In this final section of the chapter, an overview of the most promising nutrients is given, as well as special attention to the foreseen consequences of a transition to more plant based protein sources.

14.5.1 Vitamin B6, B12 and Folate

The three vitamins B6, B12 and folate act together in the one-carbon pathway, via which they suppress concentrations of homocysteine. Lowering homocysteine concentrations could be a successful strategy to improve physical functioning. Suppressed homocysteine relates to better performance in different domains of physical functioning (Vidoni et al. 2017; Kuo et al. 2007; van Schoor et al. 2012). Elevated homocysteine concentrations could hamper the functioning of mitochondria (Chen et al. 2017b, Ganapathy et al. 2011; Veeranki and Tyagi, 2013) and endothelia (Loscalzo 1996) and are related to increased white matter intensities, which affect the functioning of lower extremities (Longstreth et al. 1996b, Soumare et al. 2006). Therefore, these three B-vitamins could play a role in preserving functional capacities during ageing. Indeed, in the last years, a couple of studies have identified a possible role for B-vitamins in the maintenance of physical functioning during ageing (Behrouzi et al. 2019; Grootswagers et al. 2021; Balboa-Castillo et al. 2018; Struijk et al. 2018).

14.5.2 Omega-3 Fatty Acids and Vitamin D

Fatty fish species, such as salmon, herring and mackerel, are rich in omega-3 fatty acids and vitamin D, and intake of these types of fish are related to improved physical functioning in older adults (Struijk et al. 2018, Rondanelli et al. 2020). Omega-3 fatty acids have anti-inflammatory properties and could thereby target inflammageing (Dupont et al. 2019), one of the hallmarks of ageing. Additionally, omega-3 fatty acids can attenuate anabolic resistance by improving insulin sensitivity and endothelial functioning (Dupont et al. 2019, Smith et al. 2011). When combined with anabolic nutrients such as protein and amino acids, omega-3 fatty acids can improve muscle mass and functioning in older adults (Di Girolamo et al. 2014).

Vitamin D could influence muscle functioning by regulating protein synthesis and mitochondrial functioning (Domingues-Faria et al. 2017). Vitamin D status is related to improved physical functioning in older adults (Bischoff-Ferrari et al. 2004) and seems to be protective against falls (Bischoff-Ferrari et al. 2009). Where we already saw interactions between vitamin B6, B12 and folate, and between omega-3 fatty acids and protein intake, vitamin D has an interaction with calcium (Lips 2012). These interactions indicate the importance of food matrices: the natural form of foods in which nutrients present themselves together, such as protein, omega-3 fatty acids and vitamin D in fish, or vitamin D and calcium in dairy products. On a higher level, there seems to be an interaction between food items, clustered as food patterns, such as the Mediterranean diet, which is related to improved functionality at higher ages (Shahar et al. 2012; Critselis and Panagiotakos 2020; Milaneschi et al. 2011).

14.5.3 Hormetins

Hormetins are a special group of nutrients, as these components of food often do not have clear biological functions but do seem to play important roles in preventing age-related declines in functioning (Rattan 2008). These compounds cause mild stress in cells. Mild stress is postulated to yield beneficial effects via supercompensation, a phenomenon most known from exercise physiology. Exercise is a stressor for many body systems, but the recovery and supercompensation of the stress improve the functionality of these systems. The overall mechanism is called *hormesis* and works in a similar fashion, where mild stressors to the body, cells, or systems improve the functionality (Gems and Partridge 2008, Rattan 2008). Nutritional hormetins include flavonoids, resveratrol, antioxidants, and ursolic acid. Ursolic acid has shown potential in inhibiting muscle atrophy (Kunkel et al. 2011, 2012) and stimulating mitochondrial functioning (Chen et al. 2017a, Grootswagers et al. 2018). Resveratrol was found to improve mitochondrial functioning (Alway et al. 2017; Pollack et al. 2017) and type-II fiber size (Alway et al. 2017). Flavonoids are part of polyphenols and are found in dark chocolate, tea, coffee, fruits and vegetables, and have been related to improved mitochondrial functioning (Taub et al. 2016) and muscle mass maintenance (Salucci and Falcieri, 2020). There are many more compounds in natural foods that classify as hormetins, and the research outlook into these compounds in relation to ageing is exciting.

14.5.4 Protein Transition

Societal and governmental pressures lead to more and more people transiting from animal-based protein sources to plant-based protein sources. Animal protein has always been considered superior to plant protein in muscle anabolism (Gorissen and Witard, 2018). The quality of these animal proteins, which is based on amino acid composition and bio-availability, is overall higher than that of plant protein. It is therefore not surprising that studies have found greater acute muscle protein synthesis responses to animal products in comparison to soy, wheat or rice protein (Gorissen et al. 2016, Hartman et al. 2007, Phillips 2012, Tang et al. 2009, Wilkinson et al.

2007, Yang et al. 2012). Also, in longitudinal studies, indications for a muscle mass, preserving superiority of animal protein have been found (McLean et al. 2016). On the other hand, there have been some studies that found results suggesting an advantageous role for plant protein in the preservation of physical functioning (Dawson-Hughes et al. 2010; Behrouzi et al. 2019), possibly via alkaline properties (Dawson-Hughes et al. 2008). However, it is possible that in these studies, that were cross-sectional in nature, a confounding effect of dietary quality was present. After all, diets containing many plant proteins also contain many vitamins, minerals, fibers, hormetins and other bioactive compounds found in plants. There is a clear need for controlled, long-term, diet-based studies scrutinising the true consequences of shifting towards plant-based diets in older adults. These diet-based studies should follow a more holistic approach to balance the reductionistic dominance in current evidence – in other words, investigating meals or diets instead of single foods or ingredients (Burd et al. 2019).

14.5.5 Implementation of Effective Interventions

Over the years, many effective lifestyle treatments against malnutrition, sarcopenia and frailty have been identified. Unfortunately, many of these interventions do not find their way to practice (van Dongen 2019). Only by proper implementation of successful interventions in clinical practice, older adults can benefit from the wealth of knowledge that is being acquired by science. Proper implementation is thus very important but also a time-consuming activity. As an example, we showcase the implementation process of an intervention called ProMuscle. ProMuscle started as a clinical trial in 2010 and proved the efficacy of resistance training in combination with protein supplementation on lean body mass and physical performance in frail older adults (Tieland et al. 2012b, Tieland et al. 2012c). The clinical trial was translated to practice and piloted in 2014 (van Dongen et al. 2017) and after adaptations tested for effectiveness in a large practical study between 2016 and 2018 (van Dongen et al. 2020), where it proved to be effective. These positive effects in practice led to the awarding of an official entry in the Dutch Centre of Healthy Living database of governmental approved effective interventions. In future steps, the ProMuscle intervention will be further implemented into regional and national care patterns. With a path of 10 years between clinical study and the official recognition, this example clearly shows that a great effort is needed to successfully implement effective lifestyle treatments in practice. However, the societal impact and the benefits for older adults of proper implementation do evidently outweigh this effort.

14.6 Conclusion

To conclude, declining physical functioning during ageing is a serious problem. Lifestyle interventions have great potential in preventing and treating losses in muscle mass, muscle quality and physical functioning. These interventions should be well-timed and should employ the nutrition-exercise synergism as well as the combined effects of different nutrients in holistic approaches.

Textboxes:

1: Bone

Textbox 1

Aside from the declining muscle mass and muscle quality during ageing, decreases in bone health play important roles in age-related declines in physical abilities. Bone mass declines gradually after the age of 30, and in women, steeply after menopause. Reduced bone mass severely increases the risk of fractures and can result in osteoporosis, a geriatric phenotype prevalent in 9–38% of women and 1–8% of men (Wade et al. 2014). Nutrition plays an important role in bone mass retention during ageing. Notoriously, vitamin D and calcium are related to improved bone health, but supplementation studies are inconclusive (Zhao et al. 2017). Protein intake might play a more important role by stimulation IGF-1 and by increasing dietary calcium uptake (Dolan and Sale 2019), and higher protein intake is indeed related to reduced fracture risk (Groenendijk et al. 2019). Interestingly, vitamin D, calcium and protein are typically present in animal-based foods. A purely plant-based diet might thus jeopardise bone health, a notion that indeed has been observed in longitudinal observational studies, where vegans showed to have 15 times higher fracture risk compared to omnivores (Tong et al. 2020).

Compliance with Ethical Standards

Conflict of Interest All authors declare they have no conflict of interest.

References

Abizanda P, Sinclair A, Barcons N, Lizán L, Rodríguez-Mañas L (2016) Costs of malnutrition in institutionalized and community-dwelling older adults: a systematic review. J Am Med Dir Assoc 17:17–23

Akazawa N, Harada K, Okawa N, Tamura K, Moriyama H (2018) Muscle mass and intramuscular fat of the quadriceps are related to muscle strength in non-ambulatory chronic stroke survivors: a cross-sectional study. PLoS One 13:e0201789

Akazawa N, Okawa N, Hino T, Tsuji R, Tamura K, Moriyama H (2019) Higher malnutrition risk is related to increased intramuscular adipose tissue of the quadriceps in older inpatients: a cross-sectional study. Clin Nutr 39:2586–2592

Akazawa N, Okawa N, Tamura K, Moriyama H (2017) Relationships between intramuscular fat, muscle strength and gait independence in older women: a cross-sectional study. Geriatr Gerontol Int 17:1683–1688

Albu JB, Kovera AJ, Allen L, Wainwright M, Berk E, Raja-Khan N, Janumala I, Burkey B, Heshka S, Gallagher D (2005) Independent association of insulin resistance with larger amounts of intermuscular adipose tissue and a greater acute insulin response to glucose in African American than in white nondiabetic women. Am J Clin Nutr 82:1210–1217

Alway SE, McCrory JL, Kearcher K, Vickers A, Frear B, Gilleland DL, Bonner DE, Thomas JM, Donley DA, Lively MW, Mohamed JS (2017) Resveratrol enhances exercise-induced cellular and functional adaptations of skeletal muscle in older men and women. J Gerontol A Biol Sci Med Sci 72:1595–1606

Anker SD, Morley JE, von Haehling S (2016) Welcome to the ICD-10 code for sarcopenia. J Cachexia Sarcopenia Muscle 7:512–514

Artaza-Artabe I, Sáez-López P, Sánchez-Hernández N, Fernández-Gutierrez N, Malafarina V (2016) The relationship between nutrition and frailty: Effects of protein intake, nutritional supplementation, vitamin D and exercise on muscle metabolism in the elderly. A systematic review. Maturitas 93:89–99

Balboa-Castillo T, Struijk EA, Lopez-Garcia E, Banegas JR, Rodriguez-Artalejo F, Guallar-Castillon P (2018) Low vitamin intake is associated with risk of frailty in older adults. Age Ageing 47:872–879

Balogun S, Winzenberg T, Wills K, Scott D, Jones G, Aitken D, Callisaya ML (2017) Prospective associations of low muscle mass and function with 10-year falls risk, incident fracture and mortality in community-dwelling older adults. J Nutr Health Aging 21:843–848

Barbat-Artigas S, Rolland Y, Zamboni M, Aubertin-Leheudre M (2012) How to assess functional status: a new muscle quality index. J Nutr Health Aging 16:67–77

Bauer JM, Verlaan S, Bautmans I, Brandt K, Donini LM, Maggio M, McMurdo ME, Mets T, Seal C, Wijers SL, Ceda GP, de Vito G, Donders G, Drey M, Greig C, Holmback U, Narici M, McPhee J, Poggiogalle E, Power D, Scafoglieri A, Schultz R, Sieber CC, Cederholm T (2015) Effects of a vitamin D and leucine-enriched whey protein nutritional supplement on measures of sarcopenia in older adults, the PROVIDE study: a randomised, double-blind, placebo-controlled trial. J Am Med Dir Assoc 16:740–747

Bechshoft RL, Malmgaard-Clausen NM, Gliese B, Beyer N, Mackey AL, Andersen JL, Kjaer M, Holm L (2017) Improved skeletal muscle mass and strength after heavy strength training in very old individuals. Exp Gerontol 92:96–105

Behrouzi Pariya, Grootswagers Pol, Keizer Paul L C, Smeets Ellen T H C, Feskens Edith J M, Lisette CPGM, de Groot, Fred A van Eeuwijk, (2019) Dietary intakes of vegetable protein, folate, and vitamins B-6 and B-12 are partially correlated with physical functioning of dutch older adults using copula graphical models. J Nutr 150(3):634–643

Bischoff-Ferrari HA, Dawson-Hughes B, Staehelin HB, Orav JE, Stuck AE, Theiler R, Wong JB, Egli A, Kiel DP, Henschkowski J (2009) Fall prevention with supplemental and active forms of vitamin D: a meta-analysis of randomised controlled trials. Bmj 339:b3692

Bischoff-Ferrari HA, Dietrich T, Orav EJ, Hu FB, Zhang Y, Karlson EW, Dawson-Hughes B (2004) Higher 25-hydroxyvitamin D concentrations are associated with better lower-extremity function in both active and inactive persons aged > or =60 y. Am J Clin Nutr 80:752–758

Brose A, Parise G, Tarnopolsky MA (2003) Creatine supplementation enhances isometric strength and body composition improvements following strength exercise training in older adults. J Gerontol A Biol Sci Med Sci 58:11–19

Burd NA, Gorissen SH, van Loon LJ (2013) Anabolic resistance of muscle protein synthesis with aging. Exerc Sport Sci Rev 41:169–173

Burd NA, Mckenna CF, Salvador AF, Paulussen KJM, Moore DR (2019) Dietary protein quantity, quality, and exercise are key to healthy living: a muscle-centric perspective across the lifespan. Front Nutr 6

Campbell WW, Joseph LJ, Anderson RA, Davey SL, Hinton J, Evans WJ (2002) Effects of resistive training and chromium picolinate on body composition and skeletal muscle size in older women. Int J Sport Nutr Exerc Metab 12:125–135

Campbell WW, Joseph LJ, Davey SL, Cyr-Campbell D, Anderson RA, Evans WJ (1999) Effects of resistance training and chromium picolinate on body composition and skeletal muscle in older men. J Appl Physiol 1985(86):29–39

Cederholm T (2015) Overlaps between frailty and sarcopenia definitions. Nestle Nutr Inst Workshop Ser 83:65–69

Cederholm T, Barazzoni R, Austin P, Ballmer P, Biolo G, Bischoff SC, Compher C, Correia I, Higashiguchi T, Holst M (2017) ESPEN guidelines on definitions and terminology of clinical nutrition. Clin Nutr 36:49–64

Ceglia L, Niramitmahapanya S, da Silva Morais M, Rivas DA, Harris SS, Bischoff-Ferrari H, Fielding RA, Dawson-Hughes B (2013) A randomised study on the effect of vitamin D(3) supplementation on skeletal muscle morphology and vitamin D receptor concentration in older women. J Clin Endocrinol Metab 98:E1927–E1935

Charette SL, McEvoy L, Pyka G, Snow-Harter C, Guido D, Wiswell RA, Marcus R (1991) Muscle hypertrophy response to resistance training in older women. J Appl Physiol 1985(70):1912–1916

Chen J, Wong HS, Leong PK, Leung HY, Chan WM, Ko KM (2017a) Ursolic acid induces mitochondrial biogenesis through the activation of AMPK and PGC-1 in C2C12 myotubes: a possible mechanism underlying its beneficial effect on exercise endurance. Food Funct 8:2425–2436

Chen S, Dong Z, Zhao Y, Sai N, Wang X, Liu H, Huang G, Zhang X (2017b) Homocysteine induces mitochondrial dysfunction involving the crosstalk between oxidative stress and mitochondrial pSTAT3 in rat ischemic brain. Sci Rep 7:6932

Chou C-H, Hwang C-L, Wu Y-T (2012) Effect of exercise on physical function, daily living activities, and quality of life in the frail older adults: a meta-analysis. Arch Phys Med Rehabil 93:237–244

Claflin DR, Larkin LM, Cederna PS, Horowitz JF, Alexander NB, Cole NM, Galecki AT, Chen S, Nyquist LV, Carlson BM, Faulkner JA, Ashton-Miller JA (2011) Effects of high- and low-velocity resistance training on the contractile properties of skeletal muscle fibers from young and older humans. J Appl Physiol 1985(111):1021–1030

Clark DJ, Fielding RA (2012) Neuromuscular contributions to age-related weakness. J Gerontol A Biol Sci Med Sci 67:41–47

Clark DJ, Patten C, Reid KF, Carabello RJ, Phillips EM, Fielding RA (2010) Impaired voluntary neuromuscular activation limits muscle power in mobility-limited older adults. J Gerontol A Biol Sci Med Sci 65:495–502

Clark DJ, Patten C, Reid KF, Carabello RJ, Phillips EM, Fielding RA (2011) Muscle performance and physical function are associated with voluntary rate of neuromuscular activation in older adults. J Gerontol A Biol Sci Med Sci 66:115–121

Coen PM, Jubrias SA, Distefano G, Amati F, Mackey DC, Glynn NW, Manini TM, Wohlgemuth SE, Leeuwenburgh C, Cummings SR, Newman AB, Ferrucci L, Toledo FG, Shankland E, Conley KE, Goodpaster BH (2013) Skeletal muscle mitochondrial energetics are associated with maximal aerobic capacity and walking speed in older adults. J Gerontol A Biol Sci Med Sci 68:447–455

Collard RM, Boter H, Schoevers RA, OUDE VOSHAAR, R. C. (2012) Prevalence of frailty in community-dwelling older persons: a systematic review. J Am Geriatr Soc 60:1487–1492

Cooke MB, Brabham B, Buford TW, Shelmadine BD, McPheeters M, Hudson GM, Stathis C, Greenwood M, Kreider R, Willoughby DS (2014) Creatine supplementation post-exercise does not enhance training-induced adaptations in middle to older aged males. Eur J Appl Physiol 114:1321–1332

Correa-De-araujo R, Harris-Love MO, Miljkovic I, Fragala MS, Anthony BW, Manini TM (2017) The need for standardised assessment of muscle quality in skeletal muscle function deficit and other aging-related muscle dysfunctions: a symposium report. Front Physiol 8:87–87

Critselis E, Panagiotakos D (2020) Adherence to the Mediterranean diet and healthy ageing: current evidence, biological pathways, and future directions. Crit Rev Food Sci Nutr 60:2148–2157

Cruz-Jentoft AJ, Bahat G, Bauer J, Boirie Y, Bruyère O, Cederholm T, Cooper C, Landi F, Rolland Y, Sayer AA, Schneider SM, Sieber CC, Topinkova E, Vandewoude M, Visser M, Zamboni M, writing group for the european working group on sarcopenia in older P, The extended group for E (2019) Sarcopenia: revised European consensus on definition and diagnosis. Age Ageing 48:16–31

Dawson-Hughes B, Castaneda-Sceppa C, Harris SS, Palermo NJ, Cloutier G, Ceglia L, Dallal GE (2010) Impact of supplementation with bicarbonate on lower-extremity muscle performance in older men and women. Osteoporos Int 21:1171–1179

Dawson-Hughes B, Harris SS, Ceglia L (2008) Alkaline diets favor lean tissue mass in older adults. Am J Clin Nutr 87:662–665

de Oliveira GV, Morgado M, Conte-Junior CA, Alvares TS (2017) Acute effect of dietary nitrate on forearm muscle oxygenation, blood volume and strength in older adults: a randomised clinical trial. PLoS One 12:e0188893

Deutz NE, Matheson EM, Matarese LE, Luo M, Baggs GE, Nelson JL, Hegazi RA, Tappenden KA, Ziegler TR (2016) Readmission and mortality in malnourished, older, hospitalized adults treated with a specialized oral nutritional supplement: a randomized clinical trial. Clin Nutr 35(1):18–26. https://doi.org/10.1016/j.clnu.2015.12.010

Deutz NEP, Pereira SL, Hays NP, Oliver JS, Edens NK, Evans CM, Wolfe RR (2013) Effect of β-hydroxy-β-methylbutyrate (HMB) on lean body mass during 10 days of bed rest in older adults. Clin Nutr 32:704–712

Dhesi JK, Jackson SH, Bearne LM, Moniz C, Hurley MV, Swift CG, Allain TJ (2004) Vitamin D supplementation improves neuromuscular function in older people who fall. Age Ageing 33:589–595

di Girolamo FG, Situlin R, Mazzucco S, Valentini R, Toigo G, Biolo G (2014) Omega-3 fatty acids and protein metabolism: enhancement of anabolic interventions for sarcopenia. Curr Opin Clin Nutr Metab Care 17:145–150

Dirks ML, Tieland M, Verdijk LB, Losen M, Nilwik R, Mensink M, de Groot L, van Loon LJC (2017) Protein supplementation augments muscle fiber hypertrophy but does not modulate satellite cell content during prolonged resistance-type exercise training in frail elderly. J Am Med Dir Assoc 18:608–615

Dolan E, Sale C (2019) Protein and bone health across the lifespan. Proc Nutr Soc 78:45–55

Domingues-Faria C, Boirie Y, Walrand S (2017) Vitamin D and muscle trophicity. Curr Opin Clin Nutr Metab Care 20:169–174

dos Santos L, Cyrino ES, Antunes M, Santos DA, Sardinha LB (2017) Sarcopenia and physical independence in older adults: the independent and synergic role of muscle mass and muscle function. J Cachexia Sarcopenia Muscle 8:245–250

Dreyer HC, Blanco CE, Sattler FR, Schroeder ET, Wiswell RA (2006) Satellite cell numbers in young and older men 24 hours after eccentric exercise. Muscle Nerve 33:242–253

Drummond MJ, Rasmussen BB (2008) Leucine-enriched nutrients and the regulation of mammalian target of rapamycin signalling and human skeletal muscle protein synthesis. Curr Opin Clin Nutr Metab Care 11:222–226

Dupont J, Dedeyne L, Dalle S, Koppo K, Gielen E (2019) The role of omega-3 in the prevention and treatment of sarcopenia. Aging Clin Exp Res 31:825–836

English KL, Mettler JA, Ellison JB, Mamerow MM, Arentson-Lantz E, Pattarini JM, Ploutz-Snyder R, Sheffield-Moore M, Paddon-Jones D (2015) Leucine partially protects muscle mass and function during bed rest in middle-aged adults1,2. Am J Clin Nutr 103:465–473

English KL, Paddon-Jones D (2010) Protecting muscle mass and function in older adults during bed rest. Curr Opin Clin Nutr Metab Care 13:34–39

Englund DA, Kirn DR, Koochek A, Zhu H, Travison TG, Reid KF, von Berens A, Melin M, Cederholm T, Gustafsson T, Fielding RA (2019) Corrigendum: nutritional supplementation with physical activity improves muscle composition in mobility-limited older adults, the VIVE2 study: a randomised, double-blind, placebo-controlled trial. J Gerontol A Biol Sci Med Sci 74:1993

Essén B, Jansson E, Henriksson J, Taylor AW, Saltin B (1975) Metabolic characteristics of fibre types in human skeletal muscle. Acta Physiol Scand 95:153–165

Frank P, Andersson E, Ponten M, Ekblom B, Ekblom M, Sahlin K (2016) Strength training improves muscle aerobic capacity and glucose tolerance in elderly. Scand J Med Sci Sports 26:764–773

Franzon K, Zethelius B, Cederholm T, Kilander L (2019) The impact of muscle function, muscle mass and sarcopenia on independent ageing in very old Swedish men. BMC Geriatr 19:153

Fried LP, Tangen CM, Walston J, Newman AB, Hirsch C, Gottdiener J, Seeman T, Tracy R, Kop WJ, Burke G, McBurnie MA, cardiovascular health study collaborative research G (2001) Frailty in older adults: evidence for a phenotype. J Gerontol A Biol Sci Med Sci 56:M146–M156

Ganapathy PS, Perry RL, Tawfik A, Smith RM, Perry E, Roon P, Bozard BR, Ha YJ, Smith SB (2011) Homocysteine-mediated modulation of mitochondrial dynamics in retinal ganglion cells. Invest Ophthalmol vis Sci 52:5551–5558

Gems D, Partridge L (2008) Stress-response hormesis and aging: "that which does not kill us makes us stronger." Cell Metab 7:200–203

Godard MP, Gallagher PM, Raue U, Trappe SW (2002) Alterations in single muscle fiber calcium sensitivity with resistance training in older women. Pflugers Arch 444:419–425

Goodpaster BH, Chomentowski P, Ward BK, Rossi A, Glynn NW, Delmonico MJ, Kritchevsky SB, Pahor M, Newman AB (2008) Effects of physical activity on strength and skeletal muscle fat infiltration in older adults: a randomised controlled trial. J Appl Physiol 1985(105):1498–1503

Goodpaster BH, Park SW, Harris TB, Kritchevsky SB, Nevitt M, Schwartz AV, Simonsick EM, Tylavsky FA, Visser M, Newman AB (2006) The loss of skeletal muscle strength, mass, and quality in older adults: the health, aging and body composition study. J Gerontol A Biol Sci Med Sci 61:1059–1064

Goodpaster BH, Thaete FL, Kelley DE (2000) Thigh adipose tissue distribution is associated with insulin resistance in obesity and in type 2 diabetes mellitus. Am J Clin Nutr 71:885–892

Gorissen SHM, Horstman AMH, Franssen R, Crombag JJR, Langer H, Bierau J, Respondek F, van Loon LJC (2016) Ingestion of wheat protein increases in vivo muscle protein synthesis rates in healthy older men in a randomised trial. J Nutr 146:1651–1659

Gorissen SHM, Witard OC (2018) Characterising the muscle anabolic potential of dairy, meat and plant-based protein sources in older adults. Proc Nutr Soc 77:20–31

Gouzi F, Maury J, Heraud N, Molinari N, Bertet H, Ayoub B, Blaquiere M, Bughin F, de Rigal P, Poulain M, Pincemail J, Cristol JP, Laoudj-Chenivesse D, Mercier J, Prefaut C, Pomies P, Hayot M (2019) Additional effects of nutritional antioxidant supplementation on peripheral muscle during pulmonary rehabilitation in COPD patients: a randomized controlled trial. Oxid Med Cell Longev 2019:5496346

Groenendijk I, den Boeft L, van Loon LJC, de Groot LCPGM (2019) High versus low dietary protein intake and bone health in older adults: a systematic review and meta-analysis. Comput Struct Biotechnol J 17:1101–1112

Grootswagers P, et al (2020) A 4-week exercise and protein program improves muscle mass and physical functioning in older adults - a pilot study. Exp Gerontol 141:111094

Grootswagers P, Mensink M, Berendsen AAM, Deen CPJ, Kema IP, Bakker SJL, Santoro A, Franceschi C, Meunier N, Malpuech-Brugère C, Bialecka-Debek A, Rolf K, Fairweather-Tait S, Jennings A, Feskens EJM, Lisette CPGM, de Groot (2021) Vitamin B-6 intake is related to physical performance in European older adults: results of the New Dietary Strategies Addressing the Specific Needs of the Elderly Population for Healthy Aging in Europe (NU-AGE) study. Am J Clin Nutr 113(4):781–789

Grootswagers P, Smeets E, de Groot L (2018) A novel oral nutritional supplement improves gait speed in Dutch older adults with (risk of) undernutrition. Clin Nutr 37:S6

Hagerman FC, Walsh SJ, Staron RS, Hikida RS, Gilders RM, Murray TF, Toma K, Ragg KE (2000) Effects of high-intensity resistance training on untrained older men. I. Strength, cardiovascular, and metabolic responses. J Gerontol A Biol Sci Med Sci 55:B336–B346

Hartman JW, Tang JE, Wilkinson SB, Tarnopolsky MA, Lawrence RL, Fullerton AV, Phillips SM (2007) Consumption of fat-free fluid milk after resistance exercise promotes greater lean mass accretion than does consumption of soy or carbohydrate in young, novice, male weightlifters. Am J Clin Nutr 86:373–381

Holloszy JO (2000) The biology of aging. Mayo Clin Proc 75(Suppl, S3–8):discussion S8–9

Holwerda AM, Overkamp M, Paulussen KJM, Smeets JSJ, van Kranenburg J, Backx EMP, Gijsen AP, Goessens JPB, Verdijk LB, van Loon LJC (2018) Protein supplementation after exercise and before sleep does not further augment muscle mass and strength gains during resistance exercise training in active older men. J Nutr 148:1723–1732

Hoogeboom TJ, Dronkers JJ, Hulzebos EHJ, van Meeteren NLU (2014) Merits of exercise therapy before and after major surgery. Curr Opin Anaesthesiol 27:161–166

Hughes MJ, Hackney RJ, Lamb PJ, Wigmore SJ, Christopher Deans DA, Skipworth RJE (2019) Prehabilitation before major abdominal surgery: a systematic review and meta-analysis. World J Surg 43:1661–1668

Hughes VA, Frontera WR, Wood M, Evans WJ, Dallal GE, Roubenoff R, Fiatarone Singh MA (2001) Longitudinal muscle strength changes in older adults: influence of muscle mass, physical activity, and health. J Gerontol A Biol Sci Med Sci 56:B209–B217

Ikenaga M, Yamada Y, Kose Y, Morimura K, Higaki Y, Kiyonaga A, Tanaka H (2017) Effects of a 12-week, short-interval, intermittent, low-intensity, slow-jogging program on skeletal muscle, fat infiltration, and fitness in older adults: randomised controlled trial. Eur J Appl Physiol 117:7–15

Karavirta L, Hakkinen A, Sillanpaa E, Garcia-Lopez D, Kauhanen A, Haapasaari A, Alen M, Pakarinen A, Kraemer WJ, Izquierdo M, Gorostiaga E, Hakkinen K (2011) Effects of combined endurance and strength training on muscle strength, power and hypertrophy in 40–67-year-old men. Scand J Med Sci Sports 21:402–411

Kiesswetter E, Pohlhausen S, Uhlig K, Diekmann R, Lesser S, Uter W, Heseker H, Stehle P, Sieber CC, Volkert D (2014) Prognostic differences of the Mini Nutritional Assessment short form and long form in relation to 1-year functional decline and mortality in community-dwelling older adults receiving home care. J Am Geriatr Soc 62:512–517

Kortebein P, Ferrando A, Lombeida J, Wolfe R, Evans WJ (2007) Effect of 10 days of bed rest on skeletal muscle in healthy older adults. JAMA 297:1772–1774

Kortebein P, Symons TB, Ferrando A, Paddon-Jones D, Ronsen O, Protas E, Conger S, Lombeida J, Wolfe R, Evans WJ (2008) Functional impact of 10 days of bed rest in healthy older adults. J Gerontol A Biol Sci Med Sci 63:1076–1081

Kosek DJ, Kim JS, Petrella JK, Cross JM, Bamman MM (2006) Efficacy of 3 days/wk resistance training on myofiber hypertrophy and myogenic mechanisms in young vs. older adults. J Appl Physiol (1985) 101:531–544

Kougias DG, Das T, Perez AB, Pereira SL (2018) A role for nutritional intervention in addressing the aging neuromuscular junction. Nutr Res 53:1–14

Kouw IWK, Groen BBL, Smeets JSJ, Kramer IF, van Kranenburg JMX, Nilwik R, Geurts JAP, ten Broeke RHM, Poeze M, van Loon LJC, Verdijk LB (2019) One week of hospitalization following elective hip surgery induces substantial muscle atrophy in older patients. J Am Med Dir Assoc 20:35–42

Kruizenga H, van Keeken S, Weijs P, Bastiaanse L, Beijer S, Huisman-de Waal G, Jager-Wittenaar H, Jonkers-Schuitema C, Klos M, Remijnse-Meester W, Witteman B, Thijs A (2016) Undernutrition screening survey in 564,063 patients: patients with a positive undernutrition screening score stay in hospital 1.4 d longer. Am J Clin Nutr 103:1026–1032

Kryger AI, Andersen JL (2007) Resistance training in the oldest old: consequences for muscle strength, fiber types, fiber size, and MHC isoforms. Scand J Med Sci Sports 17:422–430

Kunkel SD, Elmore CJ, Bongers KS, Ebert SM, Fox DK, Dyle MC, Bullard SA, Adams CM (2012) Ursolic acid increases skeletal muscle and brown fat and decreases diet-induced obesity, glucose intolerance and fatty liver disease. PLoS One 7:e39332

Kunkel SD, Suneja M, Ebert SM, Bongers KS, Fox DK, Malmberg SE, Alipour F, Shields RK, Adams CM (2011) mRNA expression signatures of human skeletal muscle atrophy identify a natural compound that increases muscle mass. Cell Metab 13:627–638

Kuo HK, Liao KC, Leveille SG, Bean JF, Yen CJ, Chen JH, Yu YH, Tai TY (2007) Relationship of homocysteine levels to quadriceps strength, gait speed, and late-life disability in older adults. J Gerontol A Biol Sci Med Sci 62:434–439

Leblanc AD, Schneider VS, Evans HJ, Pientok C, Rowe R, Spector E (1992) Regional changes in muscle mass following 17 weeks of bed rest. J Appl Physiol 1985(73):2172–2178
Leenders M, Verdijk LB, van der Hoeven L, van Kranenburg J, Hartgens F, Wodzig WK, Saris WH, van Loon LJ (2011) Prolonged leucine supplementation does not augment muscle mass or affect glycemic control in elderly type 2 diabetic men. J Nutr 141:1070–1076
Leenders M, Verdijk LB, van der Hoeven L, van Kranenburg J, Nilwik R, Wodzig WK, Senden JM, Keizer HA, van Loon LJ (2013) Protein supplementation during resistance-type exercise training in the elderly. Med Sci Sports Exerc 45:542–552
Lips P (2012) Interaction between vitamin D and calcium. Scand J Clin Lab Invest Suppl 243:60–64
Little JP, Phillips SM (2009) Resistance exercise and nutrition to counteract muscle wasting. Appl Physiol Nutr Metab 34:817–828
Liu L, Su X, William JQ, Sheng H, Kristin K, David WF, Philip R, Le Z, Karthikeyani C, Eileen W, Marie M, Timothy JM, Joseph AB, Joshua DR (2018) Quantitative analysis of nad synthesis-breakdown fluxes. Cell Metab 27(5):1067-1080.e5
Jr Longstreth WT, Manolio TA, Arnold A, Burke GL, Bryan N, Jungreis CA, Enright PL, O'leary D, Fried L (1996a) Clinical correlates of white matter findings on cranial magnetic resonance imaging of 3301 elderly people. The Cardiovascular Health Study. Stroke 27:1274–1282
Longstreth WT, Manolio TA, Arnold A, Burke GL, Bryan N, Jungreis CA, Enright PL, Oleary D, Fried L (1996b) Clinical correlates of white matter findings on cranial magnetic resonance imaging of 3301 elderly people - The cardiovascular health study. Stroke 27:1274–1282
Loscalzo J (1996) The oxidant stress of hyperhomocyst(e)inemia. J Clin Investig 98:5–7
Malmstrom TK, Miller DK, Simonsick EM, Ferrucci L, Morley JE (2016) SARC-F: a symptom score to predict persons with sarcopenia at risk for poor functional outcomes. J Cachexia Sarcopenia Muscle 7:28–36
McCarthy JJ, Esser KA (2010) Anabolic and catabolic pathways regulating skeletal muscle mass. Curr Opin Clin Nutr Metab Care 13:230–235
McGuigan MR, Bronks R, Newton RU, Sharman MJ, Graham JC, Cody DV, Kraemer WJ (2001) Resistance training in patients with peripheral arterial disease: effects on myosin isoforms, fiber type distribution, and capillary supply to skeletal muscle. J Gerontol A Biol Sci Med Sci 56:B302–B310
McLean RR, Mangano KM, Hannan MT, Kiel DP, Sahni S (2016) Dietary protein intake is protective against loss of grip strength among older adults in the framingham offspring cohort. J Gerontol Ser-Biol Sci Med Sci 71:356–361
McLeod M, Breen L, Hamilton DL, Philp A (2016) Live strong and prosper: the importance of skeletal muscle strength for healthy ageing. Biogerontology 17:497–510
Mcminn J, Steel C, Bowman A (2011) Investigation and management of unintentional weight loss in older adults. BMJ 342:d1732
McPhee JS, French DP, Jackson D, Nazroo J, Pendleton N, Degens H (2016) Physical activity in older age: perspectives for healthy ageing and frailty. Biogerontology 17:567–580
Joseph Melton III K, Khosla S, Crowson CS, O'Connor MK, O'Fallon WM, Riggs BL (2000) Epidemiology of sarcopenia. J Am Geriat Soc 48(6):625–630
Mero AA, Hulmi JJ, Salmijarvi H, Katajavuori M, Haverinen M, Holviala J, Ridanpaa T, Hakkinen K, Kovanen V, Ahtiainen JP, Selanne H (2013) Resistance training induced increase in muscle fiber size in young and older men. Eur J Appl Physiol 113:641–650
Migliavacca E, Tay SKH, Patel HP, Sonntag T, Civiletto G, Mcfarlane C, Forrester T, Barton SJ, Leow MK, Antoun E, Charpagne A, Seng Chong Y, Descombes P, Feng L, Francis-Emmanuel P, Garratt ES, Giner MP, Green CO, Karaz S, Kothandaraman N, Marquis J, Metairon S, Moco S, Nelson G, Ngo S, Pleasants T, Raymond F, Sayer AA, Ming Sim C, Slater-Jefferies J, Syddall HE, Fang Tan P, Titcombe P, Vaz C, Westbury LD, Wong G, Yonghui W, Cooper C, Sheppard A, Godfrey KM, Lillycrop KA, Karnani N, Feige JN (2019) Mitochondrial oxidative capacity and NAD(+) biosynthesis are reduced in human sarcopenia across ethnicities. Nat Commun 10:5808

Milaneschi Y, Bandinelli S, Corsi AM, Lauretani F, Paolisso G, Dominguez LJ, Semba RD, Tanaka T, Abbatecola AM, Talegawkar SA, Guralnik JM, Ferrucci L (2011) Mediterranean diet and mobility decline in older persons. Exp Gerontol 46:303–308
Milne AC, Potter J, Vivanti A, Avenell A (2009) Protein and energy supplementation in elderly people at risk from malnutrition. Cochrane Database Syst Rev CD003288
Minegishi Y, Ota N, Soga S, Shimotoyodome A (2016) Effects of nutritional supplementation with milk fat globule membrane on physical and muscle function in healthy adults aged 60 and over with semiweekly light exercise: a randomised double-blind, placebo-controlled pilot trial. J Nutr Sci Vitaminol (tokyo) 62:409–415
Mitchell CJ, Oikawa SY, Ogborn DI, Nates NJ, Macneil LG, Tarnopolsky M, Phillips SM (2015) Daily chocolate milk consumption does not enhance the effect of resistance training in young and old men: a randomised controlled trial. Appl Physiol Nutr Metab 40:199–202
Moran J, Guinan E, McCormick P, Larkin J, Mockler D, Hussey J, Moriarty J, Wilson F (2016) The ability of prehabilitation to influence postoperative outcome after intra-abdominal operation: a systematic review and meta-analysis. Surgery 160:1189–1201
Morley JE (2017) Hormones and Sarcopenia. Curr Pharm Des 23:4484–4492
Morley JE (2018) Treatment of sarcopenia: the road to the future. J Cachexia Sarcopenia Muscle 9:1196–1199
Morley JE, Argiles JM, Evans WJ, Bhasin S, Cella D, Deutz NE, Doehner W, Fearon KC, Ferrucci L, Hellerstein MK, Kalantar-Zadeh K, Lochs H, Macdonald N, Mulligan K, Muscaritoli M, Ponikowski P, Posthauer ME, Rossi Fanelli F, Schambelan M, Schols AM, Schuster MW, Anker SD (2010) Nutritional recommendations for the management of sarcopenia. J Am Med Dir Assoc 11:391–396
Morley JE, Vellas B, van Kan GA, Anker SD, Bauer JM, Bernabei R, Cesari M, Chumlea WC, Doehner W, Evans J, Fried LP, Guralnik JM, Katz PR, Malmstrom TK, McCarter RJ, Gutierrez Robledo LM, Rockwood K, Von Haehling S, Vandewoude MF, Walston J (2013) Frailty consensus: a call to action. J Am Med Dir Assoc 14:392–397
Nilwik R, Snijders T, Leenders M, Groen BB, van Kranenburg J, Verdijk LB, van Loon LJ (2013) The decline in skeletal muscle mass with aging is mainly attributed to a reduction in type II muscle fiber size. Exp Gerontol 48:492–498
Phillips SM (2012) Nutrient-rich meat proteins in offsetting age-related muscle loss. Meat Sci 92:174–178
Phu S, Boersma D, Duque G (2015) Exercise and sarcopenia. J Clin Densitom 18:488–492
Pollack RM, Barzilai N, Anghel V, Kulkarni AS, Golden A, O'Broin P, Sinclair DA, Bonkowski MS, Coleville AJ, Powell D, Kim S, Moaddel R, Stein D, Zhang K, Hawkins M, Crandall JP (2017) Resveratrol improves vascular function and mitochondrial number but not glucose metabolism in older adults. J Gerontol A Biol Sci Med Sci 72:1703–1709
Power L, Mullally D, Gibney ER, Clarke M, Visser M, Volkert D, Bardon, L, de Van der Schueren MAE, Corish CA (2018) A review of the validity of malnutrition screening tools used in older adults in community and healthcare settings – A MaNuEL study. Clinical Nutrition ESPEN 24:1–13
Pyka G, Lindenberger E, Charette S, Marcus R (1994) Muscle strength and fiber adaptations to a year-long resistance training program in elderly men and women. J Gerontol 49:M22–M27
Rajman L, Chwalek K, Sinclair DA (2018) Therapeutic potential of NAD-boosting molecules: the in vivo evidence. Cell Metab 27:529–547
Rattan SI (2008) Hormesis in aging. Ageing Res Rev 7:63–78
Reid KF, Pasha E, Doros G, Clark DJ, Patten C, Phillips EM, Frontera WR, Fielding RA (2014) Longitudinal decline of lower extremity muscle power in healthy and mobility-limited older adults: influence of muscle mass, strength, composition, neuromuscular activation and single fiber contractile properties. Eur J Appl Physiol 114:29–39
Rodacki CL, Rodacki AL, Pereira G, Naliwaiko K, Coelho I, Pequito D, Fernandes LC (2012) Fish-oil supplementation enhances the effects of strength training in elderly women. Am J Clin Nutr 95:428–436

Rondanelli M, Rigon C, Perna S, Gasparri C, Iannello G, Akber R, Alalwan T , Freije Afnan M (2020) Novel insights on intake of fish and prevention of sarcopenia: all reasons for an adequate consumption. Nutrients 12(2):307

Rozand V, Sundberg CW, Hunter SK, Smith AE (2020) Age-related deficits in voluntary activation: a systematic review and meta-analysis. Med Sci Sports Exerc 52:549–560

Salucci S, Falcieri E (2020) Polyphenols and their potential role in preventing skeletal muscle atrophy. Nutr Res 74:10–22

Santanasto AJ, Coen PM, Glynn NW, Conley KE, Jubrias SA, Amati F, Strotmeyer ES, Boudreau RM, Goodpaster BH, Newman AB (2016) The relationship between mitochondrial function and walking performance in older adults with a wide range of physical function. Exp Gerontol 81:1–7

Schaap LA, Koster A, Visser M (2013) Adiposity, muscle mass, and muscle strength in relation to functional decline in older persons. Epidemiol Rev 35:51–65

Schafer AL, Vittinghoff E, Lang TF, Sellmeyer DE, Harris TB, Kanaya AM, Strotmeyer ES, Cawthon PM, Cummings SR, Tylavsky FA, Scherzinger AL, Schwartz AV (2010) Fat infiltration of muscle, diabetes, and clinical fracture risk in older adults. J Clin Endocrinol Metab 95:E368–E372

Schiaffino S (2010) Fibre types in skeletal muscle: a personal account. Acta Physiol (oxf) 199:451–463

Schiaffino S, Reggiani C (2011) Fiber types in mammalian skeletal muscles. Physiol Rev 91:1447–1531

Schilp J, Kruizenga HM, Wijnhoven HA, Leistra E, Evers AM, van Binsbergen JJ, Deeg DJ, Visser M (2012) High prevalence of undernutrition in Dutch community-dwelling older individuals. Nutrition 28:1151–1156

Shafiee G, Keshtkar A, Soltani A, Ahadi Z, Larijani B, Heshmat R (2017) Prevalence of sarcopenia in the world: a systematic review and meta- analysis of general population studies. J Diabetes Metab Disord 16:21

Shahar DR, Houston DK, Hue TF, Lee J-S, Sahyoun NR, Tylavsky FA, Geva D, Vardi H, Harris TB (2012) Adherence to mediterranean diet and decline in walking speed over 8 years in community-dwelling older adults. J Am Geriatr Soc 60:1881–1888

Sipila S, Elorinne M, Alen M, Suominen H, Kovanen V (1997) Effects of strength and endurance training on muscle fibre characteristics in elderly women. Clin Physiol 17:459–474

Slivka D, Raue U, Hollon C, Minchev K, Trappe S (2008) Single muscle fiber adaptations to resistance training in old (>80 yr) men: evidence for limited skeletal muscle plasticity. Am J Physiol Regul Integr Comp Physiol 295:R273–R280

Smith GI, Atherton P, Reeds DN, Mohammed BS, Rankin D, Rennie MJ, Mittendorfer B (2011) Dietary omega-3 fatty acid supplementation increases the rate of muscle protein synthesis in older adults: a randomised controlled trial. Am J Clin Nutr 93:402–412

Snijders T, Nederveen JP, Bell KE, Lau SW, Mazara N, Kumbhare DA, Phillips SM, Parise G (2019) Prolonged exercise training improves the acute type II muscle fibre satellite cell response in healthy older men. J Physiol 597:105–119

Son SM, Park SJ, Lee H, Siddiqi F, Lee JE, Menzies FM, Rubinsztein DC (2019) Leucine signals to mTORC1 via its metabolite acetyl-coenzyme A. Cell Metab 29:192-201.e7

Soumare A, Elbaz A, Ducros V, Tavernier B, Alperovitch A, Tzourio C (2006) Cross-sectional association between homocysteine and motor function in the elderly. Neurology 67:985–990

Soumaré A, Elbaz A, Ducros V, Tavernier B, Alpérovitch A, Tzourio C (2006) Cross-sectional association between homocysteine and motor function in the elderly. Neurology 67:985–990

Stout JR, Sue Graves B, Cramer JT, Goldstein ER, Costa PB, Smith AE, Walter AA (2007) Effects of creatine supplementation on the onset of neuromuscular fatigue threshold and muscle strength in elderly men and women (64–86 years). J Nutr Health Aging 11:459–464

Strandberg E, Ponsot E, Piehl-Aulin K, Falk G, Kadi F (2019) Resistance training alone or combined with N-3 PUFA-rich diet in older women: effects on muscle fiber hypertrophy. J Gerontol A Biol Sci Med Sci 74:489–494

Struijk EA, Lana A, Guallar-Castillon P, Rodriguez-Artalejo F, Lopez-Garcia E (2018) Intake of B vitamins and impairment in physical function in older adults. Clin Nutr 37:1271–1278
Suetta C, Andersen JL, Dalgas U, Berget J, Koskinen S, Aagaard P, Magnusson SP, Kjaer M (2008) Resistance training induces qualitative changes in muscle morphology, muscle architecture, and muscle function in elderly postoperative patients. J Appl Physiol 1985(105):180–186
Sun N, Youle RJ, Finkel T (2016) The mitochondrial basis of aging. Mol Cell 61:654–666
Taaffe DR, Pruitt L, Pyka G, Guido D, Marcus R (1996) Comparative effects of high- and low-intensity resistance training on thigh muscle strength, fiber area, and tissue composition in elderly women. Clin Physiol 16:381–392
Talbot J, Maves L (2016) Skeletal muscle fiber type: using insights from muscle developmental biology to dissect targets for susceptibility and resistance to muscle disease. Wiley Interdiscip Rev Dev Biol 5:518–534
Tang JE, Moore DR, Kujbida GW, Tarnopolsky MA, Phillips SM (2009) Ingestion of whey hydrolysate, casein, or soy protein isolate: effects on mixed muscle protein synthesis at rest and following resistance exercise in young men. J Appl Physiol 107:987–992
Taub PR, Ramirez-Sanchez I, Patel M, Higginbotham E, Moreno-Ulloa A, Roman-Pintos LM, Phillips P, Perkins G, Ceballos G, Villarreal F (2016) Beneficial effects of dark chocolate on exercise capacity in sedentary subjects: underlying mechanisms. A double blind, randomised, placebo controlled trial. Food Funct 7:3686–3693
Tieland M, Karin J, den Berg BV, van Loon Luc JC, de Groot Lisette CPGM (2012a) Dietary protein intake in community-dwelling, frail, and institutionalized elderly people: scope for improvement. Eur J Nutr 51(2):173–179
Tieland M, Dirks ML, van der Zwaluw N, Verdijk LB, van de Rest O, de Groot LC, van Loon LJ (2012b) Protein supplementation increases muscle mass gain during prolonged resistance-type exercise training in frail elderly people: a randomised, double-blind, placebo-controlled trial. J Am Med Dir Assoc 13:713–719
Tieland M, van de Rest O, Dirks ML, van der Zwaluw N, Mensink M, van Loon LJ, de Groot LC (2012c) Protein supplementation improves physical performance in frail elderly people: a randomised, double-blind, placebo-controlled trial. J Am Med Dir Assoc 13:720–726
Tipton KD, Ferrando AA, Phillips SM, Doyle D, JR (1999) Postexercise net protein synthesis in human muscle from orally administered amino acids. Am J Physiol 276:E628–E634
Tong TYN, Appleby PN, Armstrong MEG, Fensom GK, Knuppel A, Papier K, Perez-Cornago A, Travis RC, Key TJ (2020) Vegetarian and vegan diets and risks of total and site-specific fractures: results from the prospective EPIC-Oxford study. BMC Med 18:353
Trappe S, Godard M, Gallagher P, Carroll C, Rowden G, Porter D (2001) Resistance training improves single muscle fiber contractile function in older women. Am J Physiol Cell Physiol 281:C398-406
Trappe S, Williamson D, Godard M, Porter D, Rowden G, Costill D (2000) Effect of resistance training on single muscle fiber contractile function in older men. J Appl Physiol 1985(89):143–152
Tsekoura M, Kastrinis A, Katsoulaki M, Billis E, Gliatis J (2017) Sarcopenia and its impact on quality of life. Adv Exp Med Biol 987:213–218
Vaes AMM, Tieland M, Toussaint N, Nilwik R, Verdijk LB, van Loon LJC, de Groot L (2018) Cholecalciferol or 25-hydroxycholecalciferol supplementation does not affect muscle strength and physical performance in prefrail and frail older adults. J Nutr 148:712–720
van Dongen EJ, Leerlooijer JN, Steijns JM, Tieland M, de Groot LC, Haveman-Nies A (2017) Translation of a tailored nutrition and resistance exercise intervention for elderly people to a real-life setting: adaptation process and pilot study. BMC Geriatr 17:25
Van Dongen EJI (2019) Dietary protein and resistance exercise training for community-dwelling older adults : intervention adaptation, implementation and effectiveness. Wageningen University
van Dongen EJI, Haveman-Nies A, Doets EL, Dorhout BG, de Groot LCPGM (2020) Effectiveness of a diet and resistance exercise intervention on muscle health in older adults: promuscle in practice. J Am Med Dir Assoc 21:1065-1072.e3

van Schoor NM, Swart KM, Pluijm SM, Visser M, Simsek S, Smulders Y, Lips P (2012) Cross-sectional and longitudinal association between homocysteine, vitamin B12 and physical performance in older persons. Eur J Clin Nutr 66:174–181

Veeranki S, Tyagi SC (2013) Defective homocysteine metabolism: potential implications for skeletal muscle malfunction. Int J Mol Sci 14:15074–15091

Verdijk LB, Snijders T, Beelen M, Savelberg HH, Meijer K, Kuipers H, van Loon LJ (2010) Characteristics of muscle fiber type are predictive of skeletal muscle mass and strength in elderly men. J Am Geriatr Soc 58:2069–2075

Verdijk LB, Snijders T, Drost M, Delhaas T, Kadi F, van Loon LJ (2014) Satellite cells in human skeletal muscle; from birth to old age. Age (dordr) 36:545–547

Verstraeten LMG, van Wijngaarden JP, Pacifico J, Reijnierse EM, Meskers CGM, Maier AB (2021) Association between malnutrition and stages of sarcopenia in geriatric rehabilitation inpatients: Resort. Clin Nutr

Vidoni ML, Pettee Gabriel K, Luo ST, Simonsick EM, Day RS (2017) Vitamin B12 and homocysteine associations with gait speed in older adults: the baltimore longitudinal study of aging. J Nutr Health Aging 21:1321–1328

Visser M, Goodpaster BH, Kritchevsky SB, Newman AB, Nevitt M, Rubin SM, Simonsick EM, Harris TB (2005) Muscle mass, muscle strength, and muscle fat infiltration as predictors of incident mobility limitations in well-functioning older persons. J Gerontol A Biol Sci Med Sci 60:324–333

Wade S, Strader C, Fitzpatrick L, Anthony M, O'Malley C (2014) Estimating prevalence of osteoporosis: examples from industrialised countries. Arch Osteoporos 9:1–10

Wesselink E, Koekkoek WAC, Grefte S, Witkamp RF, van Zanten ARH (2019) Feeding mitochondria: potential role of nutritional components to improve critical illness convalescence. Clin Nutr 38:982–995

Wilkinson SB, Tarnopolsky MA, Macdonald MJ, Macdonald JR, Armstrong D, Philips SM (2007) Consumption of fluid skim milk promotes greater muscle protein accretion after resistance exercise than does consumption of an isonitrogenous and isoenergetic soy-protein beverage. Am J Clin Nutr 85:1031–1040

Wilson JM, Loenneke JP, Jo E, Wilson GJ, Zourdos MC, Kim JS (2012) The effects of endurance, strength, and power training on muscle fiber type shifting. J Strength Cond Res 26:1724–1729

Yamada M, Kimura Y, Ishiyama D, Nishio N, Otobe Y, Tanaka T, Ohji S, Koyama S, Sato A, Suzuki M, Ogawa H, Ichikawa T, Ito D, Arai H (2019) Synergistic effect of bodyweight resistance exercise and protein supplementation on skeletal muscle in sarcopenic or dynapenic older adults. Geriatr Gerontol Int 19:429–437

Yang Y, Churchward-Venne TA, Burd NA, Breen L, Tarnopolsky MA, Phillips SM (2012) Myofibrillar protein synthesis following ingestion of soy protein isolate at rest and after resistance exercise in elderly men. Nutr Metab 9

Zhao JG, Zeng XT, Wang J, Liu L (2017) Association between calcium or vitamin D supplementation and fracture incidence in community-dwelling older adults: a systematic review and meta-analysis. JAMA 318:2466–2482

Part III
Diet and Culture

Chapter 15
Dietary Patterns and Healthy Ageing

Ligia J. Dominguez, Eleonora Baiamonte, Martina Guarrera, Angela Parisi, Federica Tagliaferri, and Mario Barbagallo

Abstract Ageing is a complex process in which multiple factors are involved that can contribute to determine whether a person will or not be affected by diseases that are more frequently observed in advanced age. The factors involved comprise genetic, environmental, behavioural, and dietary factors, which influence pathways that regulate the ageing process and the life expectancy, rendering longevity a multi-faceted phenomenon. Even if a miraculous elixir or pill is not yet available, there is general agreement that nutrition has a major impact on the overall mortality and on the development of age-related chronic non-communicable diseases. Nutrition research has focused for decades on single nutrients in relation to health outcomes, although people eat food and combinations of foods rather than nutrients in isolation. Even if research on specific nutrients is scientifically valid and may provide key information on the mechanisms of effects, recent attention to the complex synergistic interactions among nutrients, other food constituents, and whole foods, has led to a growing interest in the total dietary patterns. This chapter describes some specific dietary patterns that have been associated with an increased life expectancy and with reductions of incident chronic diseases, thus consenting people to live a longer and healthier life. We describe the main characteristics and available evidence for Mediterranean, vegetarian, Japanese, and Okinawa dietary patterns, confirming the powerful role that nutrition plays in healthy ageing.

Keywords Ageing · Diet · Lifestyle · Longevity · Cardiovascular · Cancer · Chronic · Dementia · Alzheimer · Dietary pattern · Morbidity · Mortality · Food · Health · Longevity · Nutrition · Blue zone · Culture

15.1 Introduction

Having made life longer in the past hundred and fifty years is a great collective achievement of humanity. This success reflects the advances in social and economic

---

L. J. Dominguez (✉) · E. Baiamonte · M. Guarrera · A. Parisi · F. Tagliaferri · M. Barbagallo
Geriatric Unit, Department of Internal Medicine and Geriatrics, University of Palermo, Palermo, Italy
e-mail: ligia.dominguez@unipa.it

S. I. S. Rattan and G. Kaur (eds.), *Nutrition, Food and Diet in Ageing and Longevity*, Healthy Ageing and Longevity 14, https://doi.org/10.1007/978-3-030-83017-5_15

development as well as in medicine, in particular resulting in marked reduction of fatal childhood illness, maternal mortality, as well as mortality at older ages (Beard et al. 2016). Life expectancy had not changed from ancient times in Europe at the time of the Roman Empire up to the end of the nineteenth century (Harman 2001). Contrariwise, currently most European citizens and persons living in Japan may expect to live over eighty years and worldwide now most people can expect to live beyond sixty years (Beard et al. 2016). Furthermore, a person aged sixty year today could expect to live, on average, additional twenty-two years. Worldwide, from 1990 to 2013, life expectancy at birth rose by 6.2 years (95% UI 5.6–6.6), from 65.3 years in 1990 to 71.5 years in 2013 (Murray et al. 2015).

The reduction in infectious diseases with the advances in diagnosis strategies, antibiotics, vaccines, and hygienic conditions among others, has been accompanied by a rise in the development of non-communicable diseases (NCDs) more frequently observed in old age, such as cardiovascular disease (CVD), cancer, and neurodegenerative diseases and responsible for about 70% of global deaths annually, including premature deaths (between the ages of 30 and 70 years) (Murray et al. 2015). This apparent dilemma of lengthening the average life duration and increasing degenerative illness is foreseeable, because a long-lived organism is exposed for a longer time to risk factors that provoke such diseases. This occurs not only in industrialized nations but also in developing and underdeveloped areas where the populations are more numerous (Dominguez et al. 2006; Eyowas et al. 2019).

Thus, mortality attributable to NCDs has incessantly increased and continues to grow throughout the world. Not only deaths but also morbidity, disability and deterioration of the quality of life are growing linked to the development of these frequent diseases (Murray et al. 2015). Because NCDs are more frequently seen in old age, persistent ageing of the world population is one of the prime factors to explain the explosion of NCDs. The consequences for health, health systems, their workforce, and budgets are profound (Beard et al. 2016). Thus, the goals of medicine might not only focus at the extension of life but also at ensuring an old age as free as possible of morbidity and disability.

A fundamental determinant of this alarming scenario has been recognized in the modifications in lifestyle of entire populations, which entail a large incidence of risk factors. Unhealthy diet, overeating, sedentary lifestyle, and cigarette smoking will be the true architects of the increased incidence of obesity, hypertension, diabetes, and dyslipidaemia, all strong risk factors for CVD and some types of cancer.

In 2019, 7.9 million deaths and 187.7 million disability-adjusted life years (DALYs) were attributable to dietary risk factors. High intake of sodium and low intake of whole grains and fruits were leading dietary risks for deaths and DALYs worldwide. NCDs dietary risk factors increased significantly during 1990–2019 (Qiao et al. 2021). Therefore, greater efforts are needed to raise public awareness of interventions and improve dietary practices to reduce the disease burden caused by suboptimal dietary intake.

For decades nutrition research was focused on single nutrients in relation to health outcomes. Albeit people eat food and not isolated nutrients, the practice of referring to nutrition as a composite of nutrients and other biochemicals rather than as food is still widespread among scientists, food industry, governmental guidelines, and the lay public. That approach used to limit dietary guidance to recommendations on some nutrients, which had inadvertent significant implications for science, industry, and the public. Even if research on specific nutrients is scientifically valid and may provide key information on the mechanisms of effects, consumers and the food industry may misinterpret information that may not be correct or relevant. Conversely, recent attention to the complex synergistic interactions among nutrients, other food constituents, and whole foods, has led to a growing interest in total dietary patterns (Jacobs and Orlich 2014). Several a priori and a posteriori methods have been developed to appraise the adherence to a dietary pattern instead of only taking into account single nutrients or foods, including the summation of food or nutrient recommendations met; data-driven approaches (i.e., principal components and cluster analyses) which describe actual intake patterns in the population; and reduced rank regression, which defines linear combinations of food intakes that maximally explain intermediate markers of disease. Overall, the focus on dietary patterns has been rewarding in demonstrating the powerful protective associations of healthy or prudent dietary patterns, and the higher health risk associations of Western or meat and refined grains patterns. Although, there is no single optimal method for describing a dietary pattern, due mainly to the complexity of the multiple combinations and interactions, this approach has been very useful and appreciated helping to decipher which dietary patterns are likely to be associated with healthy ageing, such as those included in this chapter. Thus, the research on dietary pattern has a great potential for nutrition policy, especially demonstrating the key role of total diet in health promotion. While the advances in understanding what foods combinations are best for health is certainly a step forward in nutrition research, other major challenges persist, such as how to make available high-quality food to multitudes, that is at the same time sustainable (Willett et al. 2019).

This chapter describes some specific dietary patterns which have been shown, by experimental, epidemiological and clinical evidences, to be associated with an increased life expectancy, and reductions of incident NCDs, thus consenting people to live a longer and healthier life. We describe the main characteristics and evidence of Mediterranean, vegetarian, Japanese, and Okinawa dietary patterns. In fact, geographical areas where the world's longest-lived persons have been identified, the so called "Blue Zones" (Poulain et al. 2004), follow healthy diets with the principles of Mediterranean, Japanese, or vegetarian dietary patterns, confirming the powerful role that nutrition plays in healthy ageing.

15.2 Mediterranean Dietary Pattern

A large body of research, initiated with the Seven Countries Studies (Keys 1995) and continued uninterrupted, has consistently shown that the traditional dietary habits and lifestyle followed by population living in the Mediterranean basin (Mediterranean diet, MeDiet), may play a key role in the prevention of NCDs and premature mortality (Trichopoulou et al. 2014; Carlos and Fuente-Arrillaga et al. 2018; Dinu et al. 2018; Martinez-Gonzalez et al. 2019). MeDiet has been included in the 2015–2020 Dietary Guidelines for Americans among the most healthy dietary pattern recommended (Tagtow et al. 2016) and has been associated with greater nutrient adequacy addressing deficit of fibre, calcium, magnesium, potassium, all nutrients identified as of public health concern (Serra-Majem et al. 2003). This dietary pattern, mostly based on plant-derived foods (but also admitting low amounts of animal food), favouring seasonal and local food consumption and production, emerges as an eating model that also takes into consideration environmental concerns. The traditional MeDiet is not merely a cluster of foods, but also a cultural model that comprises the way food is selected, processed, and distributed, together with other foundations of lifestyle. These characteristics have led UNESCO in 2010 to include MeDiet on the list of the intangible cultural heritage of humanity (Dernini and Berry 2015). It comprises an abundant daily consumption of seasonal and colourful vegetables with none or minimal culinary interventions, fresh fruit of the season consumed at the end of every meal, nuts and seeds (as snacks and as part of recipes), legumes several times per week, unprocessed cereals every day, high consumption of olive oil as the main source of fat (for cooking and added raw for seasoning), herbs and spices to season recipes, moderate amount of fish (2–3 times per week), dairy products (milk, yogurt, cheese) allowed daily in limited quantities, eggs (3–4 per week), sweets only a few times a week, red meat and meat products with extreme moderation (1–2 times per month), drinking plenty of water, as opposed to wine, which is consumed in moderation with meals always respecting beliefs of each community (Trichopoulou et al. 2014). MeDiet is composed mainly of unprocessed nutrient-dense foods, in contrast to "empty calories" of westernized diets rich in processed foods that are full of calories but poor in nutrients, undeniably associated with an increased risk of NCDs and obesity. In addition to the nutritional components, MeDiet comprises historical knowledge, practices, skills, and traditions, transmitted from generation to generation, as well as physical activity and consuming meals in the company of others.

Regrettably, a large portion of the current populations from Mediterranean countries, where the benefits of MeDiet were first described, are moving away from this dietary pattern, mainly for portion sizes, proportions of food groups, and used of industrial food (Kyriacou et al. 2015). This is particularly true for children and adolescents (Archero et al. 2018), possibly due to the ubiquitous diffusion of Westernization and globalization of food production and consumption, which has homogenized eating behaviour worldwide in present times.

Extensive medical literature has shown that a higher adherence to the MeDiet was associated with reduction of total mortality (Bonaccio and Castelnuovo et al. 2018; Eleftheriou et al. 2018); reduced incidence of CVD and related mortality (Martinez-Gonzalez et al. 2019); lower rates of cognitive decline, dementia, and depression (Lai et al. 2014; Singh et al. 2014; Dominguez et al. 2019); reduced cancer-related mortality and incidence of breast, colorectal, head and neck, respiratory, gastric, bladder, and liver cancer (Morze et al. 2021); weight loss and reduction in BMI (Esposito et al. 2011); reduced incidence of type 2 diabetes (T2D) (Schwingshackl et al. 2015); reduced risk of chronic obstructive respiratory disease (Fischer et al. 2019); as well as reduced incident fragility fractures (Benetou et al. 2013; Byberg et al. 2016; Haring et al. 2016).

Besides its multiple health beneficial effects, MeDiet emphasizes the respect of seasonal crop cycles, territory and biodiversity, warranting a correct balance between nature, mankind and renewal of resources (Dernini and Berry 2015). For all these reasons, the MeDiet is a healthy and sustainable dietary model, proving to be one of the most sustainable nutritional patterns for both, the environment and our health.

15.3 Vegetarian Dietary Patterns

In general, vegetarian diets are adopted following ethical, health, environmental, religious, political, cultural, aesthetic, economic, and culinary influences. They do not follow specific ethnic or geographical locations, although some countries have more persons following this type of diet, but identify eating practices that totally or partially exclude foods of animal origin. Vegetarian diets are characterized by a high and varied consumption of plant-based foods including fruits, vegetables, cereal grains, nuts, and seeds, mushrooms, vegetable oil, herbs, and spices. There are several variations of the vegetarian dietary pattern, including vegan (excludes all animal products including honey), vegan raw (as vegan with food cooked at temperature below 48 °C), lacto-vegetarian (includes dairy products but not eggs), ovo-vegetarian (includes eggs but not dairy products), and lacto-ovo vegetarian (includes dairy products and eggs), pesco-vegetarian (includes fish and seafood) but all avoid red meat consumption and may abstain from by-products of animal slaughter (i.e. rennet and gelatine) (Melina et al. 2016). The avoidance of foods with animal origin at a population level may have positive environmental consequences. It is now accepted that a plant-based diet produces significant less greenhouse gas emissions compared to animal-based diets, thus reducing the environmental damage (McMichael et al. 2007; Rosi et al. 2017; Willett et al. 2019).

One of the "Blue Zones" described by Poulain et al. is in Loma Linda in California, where a group of near 9,000 people from the Seventh Day Adventist community seem to live longer than other Californians showing a lower incidence of cancer and CVD (Poulain et al. 2004). Results from the large Adventist Health Study 2 (AHS-2) showed that participants with higher adherence to a vegetarian dietary pattern had a significant lower hazard ratio for all-cause mortality when considering all kinds of vegetarians after a follow-up of 5.79 years. However, considering different types of vegetarianism, that is, vegans, lacto-ovo-vegetarians, and semi-vegetarians compared to non-vegetarians, only pesco-vegetarians had lower mortality, while the rest of the groups had no benefit. The significant associations were larger and more often significant for men than for women (Orlich et al. 2013). There has been some concern about the increased risk of vitamin B12 and omega-3 polyunsaturated fatty acids (PUFA) deficit among vegetarians (Li 2014). Recent analyses including prospective studies other than the Adventist Health Study, have demonstrated significant although smaller risk reductions for CVD mortality, cerebrovascular disease, T2D, and chronic kidney disease. In addition, there are yet concerns about different dietary deficiencies with strict vegan diets for protein, omega-3 PUFA, vitamin D, vitamin B12, iron, calcium, and zinc (Rocha et al. 2019). A systematic review including 86 cross-sectional and 10 cohort prospective studies found that among cross-sectional studies there was an overall reduction in BMI, total cholesterol, LDL-cholesterol, and blood glucose concentrations in vegetarians and vegans compared to omnivores. In prospective studies, the analyses found a significant reduced risk of incidence and/or mortality from ischemic heart disease (IHD) (−25%) and of incidence of total cancer (−8%), but not of total CV and cerebrovascular diseases, all-cause mortality and cancer-related mortality. Considering only vegans, there was a decreased risk of incident total cancer (−15%), but the number of studies was limited (Dinu et al. 2017). A meta-analysis of eight studies among 183,321 participants, found a significant heterogeneity, particularly evident for studies of Seventh Day Adventists (SDA) cohorts. In fact, SDA studies reported greater effect size as compared to non-SDA studies that were significant only for reduction of IHD. The authors concluded that there is modest CV benefit, but no clear reduction in overall mortality associated with a vegetarian diet. The benefit was driven mainly by studies in SDA, whereas the effect of vegetarian diet in other cohorts remains unproven (Kwok et al. 2014). Noteworthy, Adventists also follow lifestyle components, such as exercise and avoiding smoking that can contribute to protection against NCDs. A recent analysis comparing AHS-2 participants and non-smokers control US populations found a significant lower all-cause and cancer-related mortality, as well as breast, colorectal, and lung cancer incidents in AHS-2 participants vs. controls (Fraser et al. 2020). Other analyses from the AHS-2 including 125,000 participants showed that vegetarianism has a potential protection against obesity, T2D, and metabolic syndrome risk (Tonstad et al. 2009; Rizzo et al. 2011). However, it is not yet completly clear whether the reduction in T2D risk in vegetarians can be attributed to the absence of meat or to the higher intake of plant foods. As regards dementia, an earlier analysis of two cohorts from the AHS compared participants consuming vegetarian diets vs. those consuming meat. In a small cohort meat-eaters were twice more likely to become demented as their vegetarian controls. A second analysis in a larger cohort showed no significant differences between vegetarians and non-vegetarians. An important limitation was the lack of clear indication of any standardized cognitive assessment during the studies (Giem et al. 1993). A more recent study aimed to evaluate the influence of lifestyle risk factors and religious living on chronic neurological diseases, because SDA do not consume tobacco, alcohol, or pork, and many adhere to lacto-ovo-vegetarian diet, and Baptists discourage excessive use of alcohol and tobacco. Comparing the hospital admission rates for some major neurological diseases among 6,532 SDA and 3,720 Baptists, the incidence for dementia or Alzheimer's disease was significantly lower for members of both communities, while the incidence for Parkinson's disease and epilepsy were not significantly different vs. the general population (Thygesen et al. 2017). More studies are still needed to understand the interaction between such lifestyle and other components of the religious belief systems.

As the consumption of meat, especially red and processed meat, has been associated with an increase in mortality and other negative health outcomes (Wang et al. 2016), it is difficult to define whether the benefit of vegetarian diets derives from the greater consumption of plant-food, lack of consumption of meat, or substitution by other foods. Nevertheless, the health benefits of adhering to vegetarian dietary patterns continue to be reported. Mounting evidence favours a whole food plant-based diet that minimizes processed foods and emerges as a potential universal healthy diet. A consensus agreement on the specific definitions of vegetarian diets is currently lacking, which can help explain the heterogeneity of some results. It is also imperative to indicate the integration of vitamin B12, vitamin D and other nutrients if a deficiency is suspected.

15.4 Japanese Dietary Pattern

Japan is the country with the highest life expectancy, the most accelerated ageing, and the largest number of centenarians in the world (Ikeda et al. 2011). In 2002, Japanese female life expectancy rose at a consistent rate of about 3 months per year for the preceding 160 years (Oeppen and Vaupel 2002).The Seven Countries Studies already had shown that, together with the MeDiet, the Japanese dietary pattern was associated with low rates of IHD and all-cause mortality compared to the US and Finish diets (Keys et al. 1984; Willett 1994). More recent cohort observational studies have confirmed these findings, demonstrating that adherence to a Japanese diet results in a lower risk of cardiovascular mortality (Shimazu et al. 2007; Okada et al. 2018). A study by Okada et al. developed a Japanese diet score derived from seven food groups (beans and bean products, fresh fishes, vegetables, Japanese pickles, fungi, seaweeds, and fruits). There was a significant association between a higher adherence to the Japanese diet score and lower all-cause and CVD mortality, especially in women (Okada et al. 2018). Possibly the association of adherence to the traditional Japanese diet and low rates of incident IHD may be due to the high consumption of omega-3 PUFA (Okuyama et al. 1996). Yet, the risk of stroke and cerebrovascular disease is quite high in Japanese metropolitan areas, where excessive salt consumption, stress, smoking, and alcohol abuse may account for the higher prevalence of hypertension (Shimazu et al. 2007; Iso 2011).

Due to the exceptional longevity of Japan, interest has increased in the scientific community attempting to clarify what this Japanese super-longevity is due to. The Japanese dietary pattern mainly comprises small portions of fresh seasonal traditional foods (fish, rice, seaweeds, soy and derivatives, vegetables and green tea), practicing the traditional habit of "hara hachi bu", which means eating until you are 80% full (Pignolo 2019, Galioto et al. 2008). Carbohydrates are largely represented, making up to 60 to 65% of total calories, owing to the high consumption of rice; fat intake represents 20 to 25% of total energy, while proteins, especially of plant origin, provides the remaining 5 to 10%. Compared to Mediterranean people, Japanese consume lower amount of fats and the omega-6/omega-3 ratio is about 2 to 3 (Tokudome et al. 2004). An important disadvantage of diets in Japan and in general in eastern Asia is their high sodium content. It is accepted that a dietary with an excess of salt is associated with higher risk for the development of hypertension, but also possibly for the increased risk of stomach cancer (Rust and Ekmekcioglu 2017).

Besides the healthy components of the Japanese diet, there are other lifestyle features and social conditions that go together with it that help explain Japan's exceptional longevity. The Japanese population achieved longevity in a fairly short time through a rapid reduction in mortality rates for NCDs from the 1950s to the 1960s, followed by a large reduction in stroke mortality rates, which was a high cause of mortality in the 1950s. Health improvements in the Japanese population continued after the mid-1960s through the execution of preventive community public health measures against NCDs mortality, both primary and secondary, as well as the use of advanced medical technologies by means of a total coverage insurance system. This included inequalities reduction by similar educational opportunities and financial access at a population level. Currently, the accelerated ageing holds new major challenges to sustain the improvement in ageing population health (Ikeda et al. 2011).

The archipelago of Japan composed of over 3,000 islands encompasses historical, cultural, and religious influences and contacts with many other nations; all these attributes have contributed to the genesis of the Japanese unique and peculiar dietary style, conformed of an extensive variety of foods and characteristic gastronomy. In 1980 The Japanese Ministry of Agriculture, Forestry and Fisheries proposed the concept of "Japanese Diet" for the first time, recognizing the general superiority in terms of health compared to Western diets, a concept that, as mentioned, had already been raised by the Seven Countries Study (Keys et al. 1984).

Afterwards, in 2005, the Japanese government started a series of programs on dietary recommendations proposing to represent the traditional Japanese healthy diet in a sort of inverted food pyramid, reminiscent of a spinning top. Running people on the summit of the spinning top emphasized the importance of regular physical activity and drinking water. This inverted cone was divided in four layers: the highest comprised grain-based dishes (i.e., rice, noodles); the following down layer made of plant-based dishes (i.e., raw and cooked vegetable and soups); the third made of fish, eggs, meat and soy-bean dishes; and the fourth comprised dairy products and fruit (Yoshiike et al. 2007).

A study examined the association between adherence to the Japanese Food Guide Spinning Top and total and cause specific mortality; 36,624 participants aged 45 to 75 years, healthy at baseline were recruited from 11 public health centres across Japan. Higher scores on the food guide (better adherence) were associated with significant lower total mortality. The score was also inversely associated with CVD-related mortality and with cerebrovascular disease. There was some evidence, though not significant, of an inverse association for cancer mortality (Kurotani et al. 2016). Supporting these results, the onset of obesity-related tumours was lower in the Japanese population, compared with the US population (Tokudome et al. 2000).

Regrettably, Japanese modern lifestyle has become markedly westernized. Using a 12-item self-administered Japanese diet score (JDS), Kanauchi et al. evaluated a sample of 1458 Japanese citizens aged 18 to 84 years. They found that as much as 47.7% had a low adherence to JDS and only 11.1% showed high adherence. After multivariate adjustments, younger and physically inactive persons, as well as heavy drinkers were those most significantly associated with low adherence to JDS. They proposed a Japanese diet pyramid similar to the MeDiet pyramid to be used as a tool for nutrition education and dietary guidance (Kanauchi and Kanauchi 2019). In the past, there was a greater prevalence of stomach cancer in Japan (excluding the Okinawa prefecture) compared to North American and European countries, which has been linked to the high consumption of salt and salt preserved foods (Hirohata and Kono 1997). In recent years, this issue has progressively improved linked to the recommendations for healthy eating and control of excess consumption of salt (Inoue and Tsugane 2005). Another study aimed to evaluate the association between the Japanese dietary pattern and disability-free survival (DFS) time in a sample of 9456 Japanese older adults, followed for a 10-year period. Using a previously validated Japanese Diet Index (JDI) to assess adherence, the authors reported that a higher JDI score was significantly associated with longer DFS time compared to the lowest JDI. Each 1-SD increase of the JDI score was associated with 3.7 additional months of life without disability regardless of sex and chronic conditions at baseline (Zhang et al. 2019). Regarding cognitive decline, a clinical trial, involving 1006 community-dwelling Japanese participants, aged 60 to 79 years followed up for a median of 15 years, found that a high intake of soybeans and derivatives, vegetables, seaweeds, fish and fruit and a low intake of rice was associated with a reduced risk of developing Alzheimer's disease (Ozawa et al. 2013).

15.5 Okinawan Dietary Pattern

This dietary pattern is followed by populations with the longest life expectancies in the world, most of them maintaining active lives. In fact, Okinawa is part of the five "Blue Zones", being home to some of the world's longest-lived people with the lowest risk of aging-associated diseases (Poulain et al. 2004). The Okinawa diet is a nutrient-dense, low-calorie diet original from the Japanese Ryukyu Islands, of which Okinawa is the largest. The Okinawa diet has drawn particular attention in recent years because native people from the island were found to have exceptional longevity in relatively good health and the highest concentration of centenarians in the world (Miyagi et al. 2003). Therefore, they have been intensively investigated in an attempt to explain their long lifespan and in particular their lengthy health span. Nevertheless, their life expectancy has fallen in recent years, possibly owing to a decreasing use of the traditional local diet, as well as other genetic, lifestyle, and environmental factors. Most of the calories (80%) of the Okinawan diet come from vegetable sources, specifically, whole grains, fruits, legumes (mainly soy), relatively small amounts of fish, and limited amounts of lean meats (Willcox et al. 2009).

Compared to the average Japanese diet, the traditional Okinawa diet has twenty percent lower amount of calories and contained a large proportion of coloured vegetables (particularly sweet potatoes – *Ipomoea batatas*). Sweet potato, the main carbohydrate source of Okinawa diet, is rich in antioxidants and has a low-glycaemic index, while the Japanese diet takes carbohydrates from rice. Okinawa diet has low amounts of saturated fats linked to a limited consumption of meat, eggs, and dairy products. This dietary pattern may be considered a natural example of caloric restriction with optimal nutrition, which has been proposed as a potential modifier of longevity (Testa et al. 2014; Willcox and Willcox 2014). Centenarians from Okinawa follow the Confucian teaching of "hara hachi bu" (Pignolo 2019, Galioto et al. 2008), which is indeed a sort of caloric restriction. Among Okinawans, the onset of colon, prostate and breast cancers is near 50% lower than in the rest of Japan (Willcox, Willcox et al. 2009); this effect could be largely due to the limited intake of saturated fats, the very low omega-6/omega-3 ratio, and the large amounts of antioxidant and anti-inflammatory compounds consumed with the diet (Okuyama et al. 1996; Willcox et al. 2009).

Although the risk of stroke and cerebrovascular disease is high among residents of Japanese metropolitan areas with lower adherence to the traditional diet and lifestyle (Shimazu et al. 2007; Iso 2011), the traditional Okinawa diet, further reduces the risk of IHD and protects from the onset of strokes and cerebrovascular disease (Galioto et al. 2008; Willcox et al. 2009; Willcox and Willcox 2014). The prevalence of obesity in Japan is only 3 to 4%, which is much lower than the prevalence in European countries and in the US, where obesity is a major determinant of the high prevalence of NCDs. Prevalence of obesity and T2D are also very low in centenarians from Okinawa (Willcox and Willcox 2014).

15.6 Conclusions

The field of nutrition has rapidly changed over the past century. Currently, it is shifting the focus from managing diseases of overt nutrient deficits to a new paradigm of diseases related to lifestyle factors, mainly excess caloric intake, sedentary lives and anxiety. Advances in the nutrition field, technology and food industry have contributed to decreasing the frequency of diseases of deficit in most regions, while simultaneously witnessing the growing challenges of rising rates of obesity, NCDs, and ageing. The initial reductionist approach, driven by the ambition to understand the mechanisms responsible for the effects of individual nutrients at the cellular and molecular levels has evolved into a more holistic aim of understanding the role of combinations of nutrients and foods in dietary patterns as key determinants of health and disease. These archetypes of eating are also linked to social, behavioural, cultural, geographical, and environmental factors that impact health maintenance and prevention of disease risk.

We have reviewed the characteristics of the dietary patterns, which hitherto have accumulated more evidence on their associations with prolonged survival and decreased incidence of NCDs. World areas where exceptional longevity has been observed, the so called "Blue Zones", follow this type of dietary patterns, namely Mediterranean, Japanese/Okinawan, and vegetarian. These dietary patterns share common features that have been linked to healthy ageing mechanisms, i.e., use of plant-based foods and whole grain products, which are plenty of fibre, vitamins, minerals, and phytonutrients with antioxidant and anti-inflammatory actions; consumption of high mono- and polyunsaturated fats with cardio-protective potential; moderate intake of protein, most of plant origin, with the exception of moderate consumption of fish; use of herbs and spices, rich in antioxidant and anti-inflammatory effects; moderate o no alcohol consumption; low or no consumption of red and processed meat, processed and ultra-processed foods and sugar, which have been associated with an increased risk of obesity, T2D, CVD, and some types of cancer. The beneficial dietary components, as well as the other lifestyle determinants (i.e., physical activity) together with moderation in food consumption observed in populations adhering to these dietary patterns, may subject the body to a positive stress (hormesis) able to counteract the ageing process (Rattan 2014).

Thus, accumulating scientific evidence point to the crucial role of diet in the modification of ageing mechanisms and in the genesis of age-associated diseases ranging from CVD to cancer. Further undertakings are needed to integrate these healthy dietary and lifestyle choices into daily living in communities worldwide to make healthy eating more accessible and sustainable.

Compliance with Ethical Standards

Conflict of Interest All authors declare they have no conflict of interest.

References

Archero F, Ricotti R et al (2018) Adherence to the mediterranean diet among school children and adolescents living in Northern Italy and unhealthy food behaviors associated to overweight. Nutrients 10(9):1322

Beard JR, Officer A et al (2016) The World report on ageing and health: a policy framework for healthy ageing. Lancet 387(10033):2145–2154

Benetou V, Orfanos P et al (2013) Mediterranean diet and incidence of hip fractures in a European cohort. Osteoporos Int 24(5):1587–1598

Bonaccio M, Di Castelnuovo A, et al (2018) Mediterranean diet and mortality in the elderly: a prospective cohort study and a meta-analysis. Br J Nutr 120(8):841–854

Byberg L, Bellavia A et al (2016) Mediterranean diet and hip fracture in Swedish men and women. J Bone Miner Res 31(12):2098–2105

Carlos S, De La Fuente-Arrillaga C et al (2018) Mediterranean diet and health outcomes in the SUN Cohort. Nutrients 10(4):439

Dernini S, Berry EM (2015) Mediterranean diet: from a healthy diet to a sustainable dietary pattern. Front Nutr 2:15

Dinu M, Abbate R et al (2017) Vegetarian, vegan diets and multiple health outcomes: a systematic review with meta-analysis of observational studies. Crit Rev Food Sci Nutr 57(17):3640–3649

Dinu M, Pagliai G et al (2018) Mediterranean diet and multiple health outcomes: an umbrella review of meta-analyses of observational studies and randomised trials. Eur J Clin Nutr 72(1):30–43
Dominguez LJ, Barbagallo M et al (2019) Dietary patterns and cognitive decline: key features for prevention. Curr Pharm Des 25(22):2428–2442
Dominguez LJ, Galioto A et al (2006) Ageing, lifestyle modifications, and cardiovascular disease in developing countries. J Nutr Health Aging 10(2):143–149
Eleftheriou D, Benetou V et al (2018) Mediterranean diet and its components in relation to all-cause mortality: meta-analysis. Br J Nutr 120(10):1081–1097
Esposito K, Kastorini CM et al (2011) Mediterranean diet and weight loss: meta-analysis of randomized controlled trials. Metab Syndr Relat Disord 9(1):1–12
Eyowas FA, Schneider M et al (2019) Multimorbidity of chronic non-communicable diseases and its models of care in low- and middle-income countries: a scoping review protocol. BMJ Open 9(10):e033320
Fischer A, Johansson I et al (2019) Adherence to a mediterranean-like diet as a protective factor against COPD: a nested case-control study. COPD 16(3–4):272–277
Fraser GE, Cosgrove CM et al (2020) Lower rates of cancer and all-cause mortality in an adventist cohort compared with a US Census population. Cancer 126(5):1102–1111
Galioto A, Dominguez LJ et al (2008) Cardiovascular risk factors in centenarians. Exp Gerontol 43(2):106–113
Giem P, Beeson WL et al (1993) The incidence of dementia and intake of animal products: preliminary findings from the Adventist Health Study. Neuroepidemiology 12(1):28–36
Haring B, Crandall CJ et al (2016) Dietary patterns and fractures in postmenopausal women: results from the women's health initiative. JAMA Intern Med 176(5):645–652
Harman D (2001) Aging: overview. Ann N Y Acad Sci 928:1–21
Hirohata T, Kono S (1997) Diet/nutrition and stomach cancer in Japan. Int J Cancer Suppl 10:34–36
Ikeda N, Saito E et al (2011) What has made the population of Japan healthy? Lancet 378(9796):1094–1105
Inoue M, Tsugane S (2005) Epidemiology of gastric cancer in Japan. Postgrad Med J 81(957):419–424
Iso H (2011) Lifestyle and cardiovascular disease in Japan. J Atheroscler Thromb 18(2):83–88
Jacobs DR Jr, Orlich MJ (2014) Diet pattern and longevity: do simple rules suffice? a commentary. Am J Clin Nutr 100(Suppl 1):313S–319S
Kanauchi M, Kanauchi K (2019) Proposal for an empirical Japanese diet score and the Japanese diet pyramid. Nutrients 11(11):274
Keys A (1995) Mediterranean diet and public health: personal reflections. Am J Clin Nutr 61(6 Suppl):1321S–1323S
Keys A, Menotti A et al (1984) The seven countries study: 2,289 deaths in 15 years. Prev Med 13(2):141–154
Kurotani K, Akter S et al (2016) Quality of diet and mortality among Japanese men and women: Japan Public Health Center based prospective study. BMJ 352:i1209
Kwok CS, Umar S et al (2014) Vegetarian diet, seventh day adventists and risk of cardiovascular mortality: a systematic review and meta-analysis. Int J Cardiol 176(3):680–686
Kyriacou A, Evans JM et al (2015) Adherence to the mediterranean diet by the greek and cypriot population: a systematic review. Eur J Public Health 25(6):1012–1018
Lai JS, Hiles S et al (2014) A systematic review and meta-analysis of dietary patterns and depression in community-dwelling adults. Am J Clin Nutr 99(1):181–197
Li D (2014) Effect of the vegetarian diet on non-communicable diseases. J Sci Food Agric 94(2):169–173
Martinez-Gonzalez MA, Gea A et al (2019) The mediterranean diet and cardiovascular health. Circ Res 124(5):779–798
McMichael AJ, Powles JW et al (2007) Food, livestock production, energy, climate change, and health. Lancet 370(9594):1253–1263

Melina V, Craig W et al (2016) Position of the academy of nutrition and dietetics: vegetarian diets. J Acad Nutr Diet 116(12):1970–1980

Miyagi S, Iwama N et al (2003) Longevity and diet in Okinawa, Japan: the past, present and future. Asia Pac J Public Health 15(Suppl):S3–S9

Morze J, Danielewicz A et al (2021) An updated systematic review and meta-analysis on adherence to mediterranean diet and risk of cancer. Eur J Nutr 60(3):1561–1586

Murray CJ, Barber RM et al (2015) Global, regional, and national disability-adjusted life years (DALYs) for 306 diseases and injuries and healthy life expectancy (HALE) for 188 countries, 1990–2013: quantifying the epidemiological transition. Lancet 386(10009):2145–2191

Oeppen J, Vaupel JW (2002) Demography. Broken limits to life expectancy. Science 296(5570):1029–1031

Okada E, Nakamura K et al (2018) The Japanese food score and risk of all-cause, CVD and cancer mortality: the Japan Collaborative Cohort Study. Br J Nutr 120(4):464–471

Okuyama H, Kobayashi T et al (1996) Dietary fatty acids–the N-6/N-3 balance and chronic elderly diseases. Excess linoleic acid and relative N-3 deficiency syndrome seen in Japan. Prog Lipid Res 35(4):409–457

Orlich MJ, Singh PN et al (2013) Vegetarian dietary patterns and mortality in Adventist Health Study 2. JAMA Intern Med 173(13):1230–1238

Ozawa M, Ninomiya T et al (2013) Dietary patterns and risk of dementia in an elderly Japanese population: the Hisayama Study. Am J Clin Nutr 97(5):1076–1082

Pignolo RJ (2019) Exceptional human longevity. Mayo Clin Proc 94(1):110–124

Poulain M, Pes GM et al (2004) Identification of a geographic area characterized by extreme longevity in the Sardinia island: the AKEA study. Exp Gerontol 39(9):1423–1429

Jie Q, Lin X, Wu Y, Xin H, Xiaowen P, Xu J, Wu JY, Ren Y, Shan PF (2021) Global burden of non-communicable diseases attributable to dietary risks in 1990–2019. J Hum Nutr Diet. https://doi.org/10.1111/jhn.12904 (online ahead of printing)

Rattan SI (2014) Molecular gerontology: from homeodynamics to hormesis. Curr Pharm Des 20(18):3036–3039

Rizzo NS, Sabate J et al (2011) Vegetarian dietary patterns are associated with a lower risk of metabolic syndrome: the adventist health study 2. Diabetes Care 34(5):1225–1227

Rocha JP, Laster J et al (2019) Multiple health benefits and minimal risks associated with vegetarian diets. Curr Nutr Rep 8(4):374–381

Rosi A, Mena P et al (2017) Environmental impact of omnivorous, ovo-lacto-vegetarian, and vegan diet. Sci Rep 7(1):6105

Rust P, Ekmekcioglu C (2017) Impact of salt intake on the pathogenesis and treatment of hypertension. Adv Exp Med Biol 956:61–84

Schwingshackl L, Missbach B et al (2015) Adherence to a Mediterranean diet and risk of diabetes: a systematic review and meta-analysis. Public Health Nutr 18(7):1292–1299

Serra-Majem L, Ribas L et al (2003) Nutrient adequacy and mediterranean diet in Spanish school children and adolescents. Eur J Clin Nutr 57(Suppl 1):S35–S39

Shimazu T, Kuriyama S et al (2007) Dietary patterns and cardiovascular disease mortality in Japan: a prospective cohort study. Int J Epidemiol 36(3):600–609

Singh B, Parsaik AK et al (2014) Association of mediterranean diet with mild cognitive impairment and Alzheimer's disease: a systematic review and meta-analysis. J Alzheimers Dis 39(2):271–282

Tagtow A, Rahavi E et al (2016) Coming together to communicate the 2015–2020 dietary guidelines for Americans. J Acad Nutr Diet 116(2):209–212

Testa G, Biasi F et al (2014) Calorie restriction and dietary restriction mimetics: a strategy for improving healthy aging and longevity. Curr Pharm Des 20(18):2950–2977

Thygesen LC, Gimsing LN et al (2017) Chronic neurodegenerative illnesses and epilepsy in Danish adventists and baptists: a nationwide cohort study. J Alzheimers Dis 56(4):1429–1435

Tokudome SY, Ichikawa et al (2004) The Mediterranean vs the Japanese diet. Eur J Clin Nutr 58(9):1323; author reply 1324–1325

Tokudome S, Nagaya T et al (2000) Japanese versus mediterranean diets and cancer. Asian Pac J Cancer Prev 1(1):61–66
Tonstad S, Butler T et al (2009) Type of vegetarian diet, body weight, and prevalence of type 2 diabetes. Diab Care 32(5):791–796
Trichopoulou A, Martinez-Gonzalez MA et al (2014) Definitions and potential health benefits of the Mediterranean diet: views from experts around the world. BMC Med 12:112
Wang X, Lin X et al (2016) Red and processed meat consumption and mortality: dose-response meta-analysis of prospective cohort studies. Public Health Nutr 19(5):893–905
Willcox BJ, Willcox DC (2014) Caloric restriction, caloric restriction mimetics, and healthy aging in Okinawa: controversies and clinical implications. Curr Opin Clin Nutr Metab Care 17(1):51–58
Willcox DC, Willcox BJ et al (2009) The Okinawan diet: health implications of a low-calorie, nutrient-dense, antioxidant-rich dietary pattern low in glycemic load. J Am Coll Nutr 28(Suppl):500S–516S
Willett W, Rockstrom J et al (2019) Food in the anthropocene: the EAT-Lancet commission on healthy diets from sustainable food systems. Lancet 393(10170):447–492
Willett WC (1994) Diet and health: what should we eat? Science 264(5158):532–537
Yoshiike N, Hayashi F et al (2007) A new food guide in Japan: the Japanese food guide Spinning Top. Nutr Rev 65(4):149–154
Zhang S, Tomata Y et al (2019) The Japanese dietary pattern is associated with longer disability-free survival time in the general elderly population in the ohsaki cohort 2006 Study. J Nutr 149(7):1245–1251

Chapter 16
Ketogenic Diet, Circadian Rhythm and Aging

Anita Jagota and Sushree Abhidhatri Sharma

Abstract Circadian rhythms are manifested as a series of autonomous and endogenous oscillators and are generated by the molecular circadian clock system in order to coordinate the internal time with the exogenous environmental changes in a 24 hour (h) daily cycle. In mammals, Suprachiasmatic nucleus (SCN) is regarded as the master clock, as it is entrained by the environmental cues and in turn synchronizes the peripheral clocks. Desynchronization between master clock and peripheral clocks give rise to many cellular, physiological and pathological alterations. In recent years, the focus is on establishing dietary regimes as therapeutic interventions. Ketogenic diet (KD) is one such dietary regimen characterized by relatively higher fat and protein proportions in diet than the carbohydrates, which recapitulates certain metabolic aspects of fatty acid metabolism and production of ketone bodies (KBs). KD which is comprised of saturated triglycerides polyunsaturated fatty acids (PUFAs) appears to be beneficial in improving age associated perturbations. This chapter focuses on the therapeutic effects of KD on healthy aging as well as unravelling the connection between KD and circadian rhythm, hence further establishing rationale for the KD regime as a therapeutic intervention.

Keywords Ketogenic diet · Circadian rhythm · Aging · Longevity · Ketone bodies · Lifespan

16.1 Introduction

A self-entrained endogenous biological clock (circadian clock) is present in every living organism from unicellular to multicellular in order to maintain the synchrony with the day and night cycle. Biological clock has its own rhythm and periodicity called biological rhythm. When this biological rhythm persists under constant conditions and having periodicity of approximately one day or 24 h, it is called as circadian

---

A. Jagota (✉) · S. A. Sharma
Neurobiology and Molecular Chronobiology Laboratory, Department of Animal Biology, School of Life Sciences, University of Hyderabad, Hyderabad 500046, India
e-mail: ajsl@uohyd.ernet.in

S. I. S. Rattan and G. Kaur (eds.), *Nutrition, Food and Diet in Ageing and Longevity*, Healthy Ageing and Longevity 14, https://doi.org/10.1007/978-3-030-83017-5_16

rhythm (Jagota et al. 2000). In mammals the circadian system is comprised of a hierarchy of oscillators, in which SCN of the hypothalamus is regarded as the master clock regulating downstream oscillators in peripheral tissues called as peripheral clocks such as liver, kidney, intestine etc. (Ko and Takahashi 2006). The master pacemaker coupled population of neuronal circadian oscillators perceives the external photic cues (light) and in turn synchronizes the peripheral clocks. Thus SCN plays a major role in orchestrating cellular and metabolic processes by relaying temporal information to the entire body via humoral and neural communication (Walker et al. 2020). In SCN, the circadian molecular machinery consists of two tightly interlocked feedback loops that comprise of core clock genes such as *Clock, Bmal1, Per1, Per2, Per3, Cry1, Cry2* etc.

Aging is an inevitable phenomena characterized by progressive decline in physiological functions and cognitive impairments. Several evidences have suggested that aging is associated with dysfunction in 24-h circadian rhythms due to loss of synchronization between the master clock and peripheral clocks leading to a myriad of complications in cellular, hormonal and metabolic processes. Reports from our laboratory suggest that there are alterations in daily rhythms of serotonin (Jagota and Kalyani 2008; 2010), antioxidant enzymes (Manikonda and Jagota 2012), leptin (Reddy and Jagota 2014), clock genes (Mattam and Jagota 2014), serotonin metabolism (Reddy and Jagota 2015), NO and Socs expression rhythms (Vinod and Jagota 2016; 2017) with aging. Further we have also reported various herbal therapeutic interventions such as curcumin and hydro-alcoholic leaf extract of *Withania somnifera* towards the restoration of various clock genes, immune genes, *Sirt1* and nuclear factor erythroid 2-related factor 2(*Nrf2*) upon aging (Kukkemane and Jagota 2019; Thummadi and Jagota 2019; Kukkemane and Jagota 2020).

Studies on the dietary regime and its possible therapeutic intervention has become pivotal in the field of research over the last few decades. Diet has been very vital as it is regarded as the non-photic cue, therefore the nutrient composition and dietary pattern play major role in regulating metabolic, cellular processes by entraining as well as reinstating peripheral clocks. Many dietary interventions have been identified by various scientists such as ketogenic diet (KD), low-fat diet (LFD), low-carbohydrate diet, high-protein diet (HPD) and high-fat diet (HFD) etc. KD has been used to treat pathological conditions like polycystic ovary syndrome, acne, neurological diseases, obesity, metabolic syndrome, diabetes mellitus type 2, cancer, respiratory and cardiovascular disease etc. KD can be an effective therapy in amyotrophic lateral sclerosis (ALS), AD, PD and some mitochondriopathies (Paoli et al. 2014). Therefore in this chapter, we focus on the role of KD in restoration of circadian function, and acting as an anti-aging regime, thus promoting longevity (Fig. 16.1).

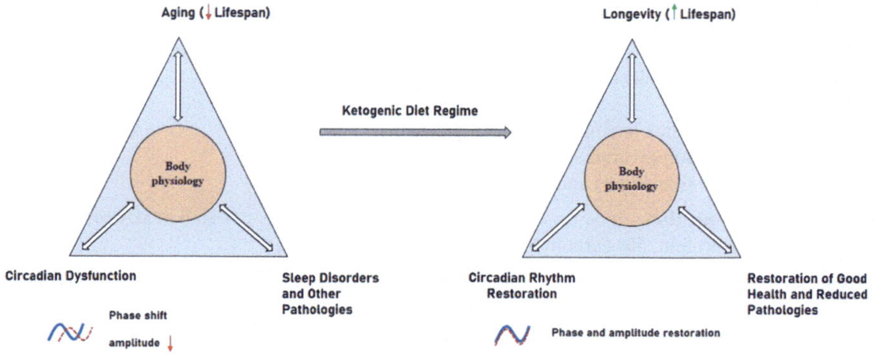

Fig. 16.1 Effects of ketogenic diet (KD) regime on age induced circadian dysfunction, longevity and lifespan

16.2 Ketogenic Diet

KD, modified dietary regime has been well established and very popular as a therapy for epilepsy since 1920s (Paoli et al. 2014). KD got the clinical recognition as a dietary intervention over the last few decades (Veyrat-Durebex et al. 2018) (Fig. 16.2). There

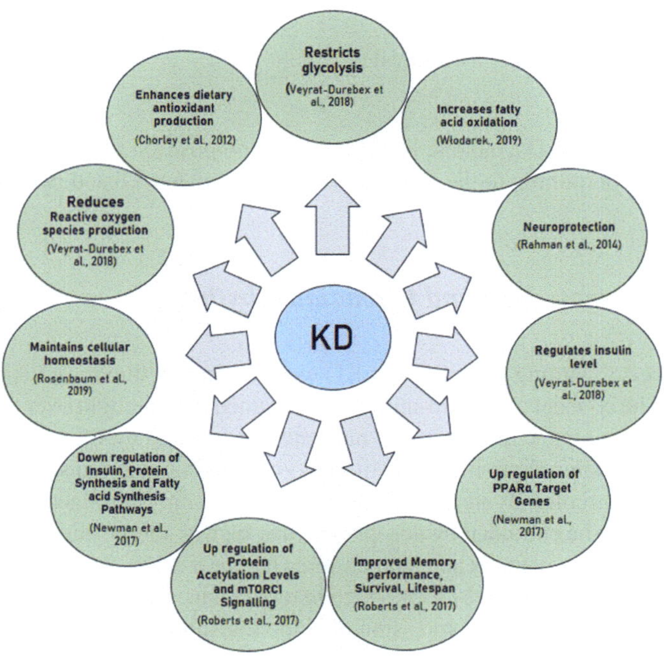

Fig. 16.2 Pleotropic effects of ketogenic diet (KD) on body physiology: KD has beneficial effects on the mechanisms involved in maintaining a proper body physiology

Table 16.1 Various types of ketogenic diet

Modified Atkins diet Specificity: 65% of the calories from fat More palatable More suitable for patients having behavioural problems as well as children and can be administrated at home Side Effects: Fewer side effects and tolerability	Auvin (2016)
Isocaloric ketogenic diet Specificity: 15% protein, 5% carbohydrate, 80% fat	Hall et al. (2016)
Classic ketogenic diet Specificity: 4:1 ratio of fat to carbohydrate and protein Long-chain triglycerides (LCT) derived from standard foods which will give 80–90% of the calories Side Effects: Hypocalcaemia, hyperuicemia, metabolic acidosis, formation of kidney stones, dyslipidaemia, gastrointestinal complications such as diarrhoea, constipation, and vomiting	de Lima et al. (2017)
Medium-chain triglyceride diet Specificity: Less fat compared to carbohydrate and protein More palatable Comprises of octanoic (C8) and decanoic (C10) fatty acids Produces more ketones per kilocalorie of energy Absorbed and delivered easily to the liver through portal blood Side Effects: Abdominal discomfort and bloating complications	Khabbush et al. (2017)
Low glycemic index ketogenic diet Specificity: Low glycemic index foods Brings out slight changes in blood glucose level More favourable and palatable Side Effects: Fewer side effects and tolerability	Barzegar et al. (2019)

are various types of KD available based on their composition (Table 16.1). Further various reports pertaining to the side effects of KD have been listed in Table 16.2.

16.2.1 Ketogenic Diet and Circadian Rhythm

Various reports regarding the effects of KD on circadian rhythm have been listed in Table 16.3. The concept of chrono-nutrition has gained the limelight over the last few years. The digestion and absorption of nutrients is regulated by CTS as the process relies on the circadian rhythm of digestive enzymes, membrane transporters and the migrating motility complex etc. Peripheral clocks are not only entrained by the master clock but can also be entrained by non-photic cues such as sound, humidity, different kinds of diets, dietary patterns and temperature (Jagota et al. 2019). KD might lead to shifting of the peripheral/slave clock which in turn can increase intestinal absorption of nutrients through the over-expression of some transporters. Itokawa et al. (2013) demonstrated timing of nutrient intake and its composition caused a phase shift in the rhythm of expression of *Per2* in the liver. Further, *Per2* expression in liver is essential during ketogenesis as it is regarded as the direct regulator of expression of a

Table 16.2 Ketogenic diet intervention: side effects

Fatty acid oxidation defects, diabetes mellitus, carnitine deficiencies, certain mitochondrial diseases, organic acidurias, pyruvate carboxylase deficiency, concomitant steroid use, critical liver disorders and acute intermittent porphyria	Neal (2012)
Hyperketonemia, acidosis	Paoli et al. (2014)
Prolonged deficiency in energy supply as well as protein intake	Taylor et al. (2018)
Gastrointestinal discomfort in patients suffering from epileptic syndrome	McDonald and Cervenka (2018)
Alteration in lipid profile leading to dyslipidemia including an array of clinical complications such as hyperlipidemia, hypertriglyceridemia and hypercholesterolemia	Barzegar et al. (2019)
Hyperuricemia, hypomagnesemia, hepatitis, hyponatremia and pancreatitis Atherosclerosis, cardiomyopathy, impaired hepatic functions, neuropathy of the optic nerve, anemia, nephrolithiasis, reduction of mineral bone density, and deficiencies of vitamins and mineral components in case of sustained administration	Włodarek (2019)

Table 16.3 Effects of ketogenic diet intervention on circadian timing system (CTS)

Time-of-day-dependent hypothermia and expression of *Cirbp* (Induced) Circadian genes' expression in mouse liver (affected)	Oishi et al. (2013)
The phase of circadian rhythms (delayed) AMP-activated protein kinase (AMPK) in mouse liver (not affected)	Genzer et al. (2015)
Expression of the muscle circadian gene Slc25a25 involved in muscle thermogenesis (induced)	Nakao et al. (2017)
Susceptibility to SUDEP in Scn1aR1407X/+ mice (influenced)	Teran et al. (2019)
Diurnal rhythm of associated biomarkers (Induced) Liver injury in Klf10 deficient mice (worsens)	Leclère et al. (2020)

rate-limiting enzyme carnitine palmitoyltransferase1A (*Cpt1a*), which transfers long-chain fatty acids to the inner mitochondrial membrane for ß-oxidation. *Per2* is also an indirect regulator of rate-limiting enzyme mitochondrial hydroxymethyl glutaryl-CoA synthase (*Hmgcs2*) for ketogenesis from the resulting acetyl-CoA (Chavan et al. 2016).

Tognini et al. (2017) demonstrated that KD affected the peripheral clocks such as liver and intestine differently. KD regime led to increase in the amplitude of various clock control genes (CCGs) expression in liver as well as tissue-specific oscillation of Peroxisome proliferator-activated receptor alpha (PPARα) and its target genes. However, the amplitude of CCGs was not drastically altered in the intestine. β-hydroxybutyrate (β-OHB) levels in the serum and intestine were found to oscillate robustly in a circadian manner. Oishi et al. (2009) reported that KD resulted in not only phase advance but also in shortening of free running period in mice. KD led to an increase in lipolysis, loss of body weight and advanced the onset of both daily locomotor activity and behavioural rhythms in mice and was also responsible in regulation of gene expression of core clock in peripheral tissues.

Upon KD regime, in Kcna1 knockout epileptic mice, there was phase shift of Sirtuin1 (SIRT1) mediated diurnal rest activity rhythms. There was reduction of number and periodicity of seizures, restoring normal behavioural rhythms (Fenoglio-Simeone et al. 2009). The circulating levels of ketone bodies (KBs) have been reported to increase with KD regime. Fasting induces ketogenic genes under regulation of mTOR–PPARα axis regulated by CTS. KD has also been reported to elevate transcriptional activation of CCG's by targeting CLOCK:BMAL1 complex and upregulates *DBP*, a clock output gene in the liver, kidney, heart, and adipose tissue (Oishi et al. 2009).

16.2.2 Ketogenic Diet and Aging

Various studies related to the effects of KD on aging, lifespan and longevity have been listed in Table 16.4. The key pathways and the molecular mechanisms to improve lifespan and longevity are controlled by behavioural, pharmacological and dietary factors. KD when administered alternatively with the control diet on weekly basis (Cyclic KD), helped to reduce obesity, midlife mortality and improved memory performance though the maximum lifespan remained unaffected. Further KD intervention led to down regulation of insulin, protein synthesis, and fatty acid synthesis pathways, whereas PPARα target genes were up regulated which were consistent across tissues, and preserved in old age, thus KD was found to improve survival, memory, and lifespan in aging mice (Newman et al. 2017). Another study reported that when 12 months old C57BL/6 mice were given KD, it significantly maintained physiological functions, improved longevity and lifespan compared to controls. Further KD was found to elevate protein acetylation levels and regulate tissue specific mTORC1 signalling (Roberts et al. 2017). KD has been reported to result in reduction in the oxidative and endoplasmic stress and turnover of proteins. It also improved antioxidant cellular status and preserved oxidative muscle fibers in mice, thus beneficial in maintaining muscle mass, function and preventing sarcopenia with the progression of age (Wallace et al. 2021)

16.2.3 Ketogenic Diet Modulation of Epigenetic Mechanisms

Acetylation and deacetylation of histones are two key mechanisms of histone modifications. Recent research in the field of aging has shown that inhibition of histone deactylases (HDACs) and increased histone acetylation plays protective role in various age associated diseased models. β-OHB led to inhibition of class I and class II HDACs in turn influencing age-associated gene expression (Han et al. 2020). Recent

Table 16.4 Effects of ketogenic diet Intervention on Aging, Longevity and Lifespan

Ketogenic diets: an historical antiepileptic therapy with promising potentialities for the aging brain	Balietti et al. (2010a, b) (Review)
Succinic dehydrogenase (SDH) activity (increased) SDH-positive mitochondria density in cerebellar Purkinje cells of late-adult rats (recovered)	Balietti et al. (2010a, b)
Brain glucose and ketone uptake in aged rats (increased)	Roy et al. (2012)
Cognitive function in elderly adults (Improved)	Ota et al. (2016)
Can ketones compensate for deteriorating brain glucose uptake during aging? Implications for the risk and treatment of Alzheimer's disease	Cunnane et al. (2016) (Review)
Markers of mitochondrial content in a tissue specific manner in adult mice (increased)	Zhou et al. (2017)
Improvement in age-related cognitive functions and life expectancy by ketogenic diets	Astrup and Hjorth (2017) (Review)
Hippocampal transporter levels (altered) adiposity in aged rats (reduced)	Hernandez et al. (2018)
The therapeutic potential of ketogenic diet throughout life: focus on metabolic, neurodevelopmental and neurodegenerative disorders	Kraeuter et al. (2019) (Review)
Aging-associated myocardial remodeling and dysfunction in mice (Reduced)	Yu et al. (2020)
Risk factors for Alzheimer's disease (Reduced)	Neth et al. (2020)
The ketogenic diet all grown up—Ketogenic diet therapies for adults	Husari and Cervenka (2020) (Review)
Neurocognitive impairment in adults aging with HIV (Improved)	Morrison et al. (2020)
Aging-Related Sarcopenic Obesity (Beneficial)	Suchkov et al. (2021) (Review)
Markers of mitochondrial mass in a tissue specific manner in aged mice (Induced)	Zhou et al. (2021)

reports have indicated an important aspect of epigenetic modifications by ketone bodies i.e. β-hydroxybutyrylation which involves processes like DNA methylation and histone covalent post-translational modifications (PTMs) such as histone lysine acetylation (β-hydroxybutyrylation at lysins of histones (Kbhb)), histone methylation, phosphorylation and ubiquitination. β-hydroxybutyrylation of histones is also associated with active gene promoters (Dabek et al. 2020). KD therapy modulates epigenetic modifications by histone phosphorylation, acetylation and DNA methylation (Ruan and Crawford 2018). KD treatment was beneficial for the restoration of DNA methylation status in rodent models. KD intervention led to increased levels of $NAD+$ modulating $NAD+$-dependent enzymes, including Sirtuins which are involved in deacetylation processes (Elamin et al. 2018). Further DNA methylation maybe blocked by elevated levels of adenosines during intake of KD. Hence KD intervention recapitulates a novel epigenetic mechanism which may be helpful in slowing down the aging process and preventing age associated dysfunctions (Paoli et al. 2014).

16.2.4 Effects of Ketogenic Diet on Longevity

Douris et al. (2015) demonstrated that the group of mice fed lifelong with KD had reduced catabolic processes as well as shift in amino acid metabolism, and conserving amino acid levels. They also showed increased levels of energy expenditure, elevated levels of circulating β-OHB, triglycerides, improved glucose homeostasis, up regulated fibroblast growth factor 21 expression in the liver but down-regulated lipogenic enzymes such as stearoyl-CoA desaturase-1 and reduced fat mass compared to chow-fed controls. Thus, long-term KD intervention led to persistent health promoting metabolic changes without affecting survival in case of mice. Further KD might affect quality as well as quantity of mitochondria, hence beneficial in ameliorating oxidative stress by increased levels of hepatic superoxide dismutase 1, catalase antioxidant protein, elevated volume of mitochondria in liver and gastrocnemius as well as increased median lifespan (762 days) compared to control group (624 days) (Parry et al. 2018). Switch between glycolysis and ketosis is helpful in promoting survival by accelerating metabolism through fat oxidation during fasting. KD along with time restricted feeding (TRF) have proven to be more beneficial (Hernandez et al. 2020).

16.2.5 Clinical Studies

Rosenbaum et al. (2019) demonstrated that in humans after switching from baseline diet (BD) (15% protein, 50% carbohydrate, 35% fat) to isocaloric KD (15% protein, 5% carbohydrate, 80% fat) there was increase in the gastric inhibitory peptide, fasting ketones, glucagon, low-density lipoprotein cholesterol, free fatty acids, adiponectin, C-reactive protein and inflammatory markers. On the other hand, fasting insulin, triglycerides, C-peptides, fibroblast growth factor 21 and insulin-mediated anti-lipolysis were significantly decreased. However, parameters associated with glucose homeostasis were diet and test meal dependent. In another study, Phillips et al. (2018) reported that physiological ketosis was found to be maintained in group of Parkinson's Disease (PD) patients following KD regime with improvements in both motor and nonmotor symptoms. Various clinical studies have suggested that KD has been beneficial in alleviating many pathophysiological conditions in case of age associated neurodegenerative diseases such as PD, Alzheimer's Disease (AD), Dementia etc. thus improving survival and lifespan (Davis et al. 2021).

16.3 Mechanism of Action of KD

The mechanism of KD action is mainly through the process of ketogenesis involving oxidation of fatty acids and ketogenic amino acids which leads to the production of Ketone bodies (KBs) such as β-OHB, acetoacetate (ACA) and acetone (Longo et al. 2019). The latter is the least abundant. The site for the production of these KBs are mainly mitochondria of hepatocytes as well as kidney epithelia, astrocytes and enterocytes (Grabackaet al. 2016).

KD is often considered as a biochemical model of fasting. Since it has low carbohydrate content, administration of KD forces the body physiology to enter into a fasting state by switching the energy production from glucose metabolism to energy production through metabolism of fatty acids (FAs) and KBs formation. KD has proved to be efficient in improving metabolism, energy production and various physiological parameters without the need of starvation. KD administration leads to reduced blood glucose concentration and insulin-to-glucagon ratio. It also causes increased concentration of glucagon, reduced glucose concentration and inhibition of glycogenesis (Włodarek 2019).

The mechanism of action of KD based on literature review has been schematically represented in Fig. 16.3.

16.3.1 *Ketone Body: Energy Metabolism*

Brain is comparatively less capable in utilizing FAs, therefore it has to utilize KBs synthesized in the liver as an energy source during intake of KD (McDonald and Cervenka 2018). It is reported that β-OHB and ACA can enter into the brain with the help of proton-linked, monocarboxylic acid transporters crossing the blood brain barrier (BBB). The expression of these particular transporters depends on the level of ketosis. Reports have suggested that KBs are efficient source of energy rather than glucose as they can undergo metabolism faster compared to glucose and because of their ability to enter directly into the Krebs cycle bypassing the glycolytic pathway (Elamin et al. 2018; McDonald and Cervenka 2018). KD also inhibits glycolysis and FA utilization, therefore glycolytic ATP production is drastically reduced and ATP generation by mitochondrial oxidation is dramatically elevated (Veyrat-Durebex et al. 2018). Study has indicated that KD up regulates the expression of hippocampal genes for mitochondrial enzymes of energy metabolism (Bough et al. 2006).

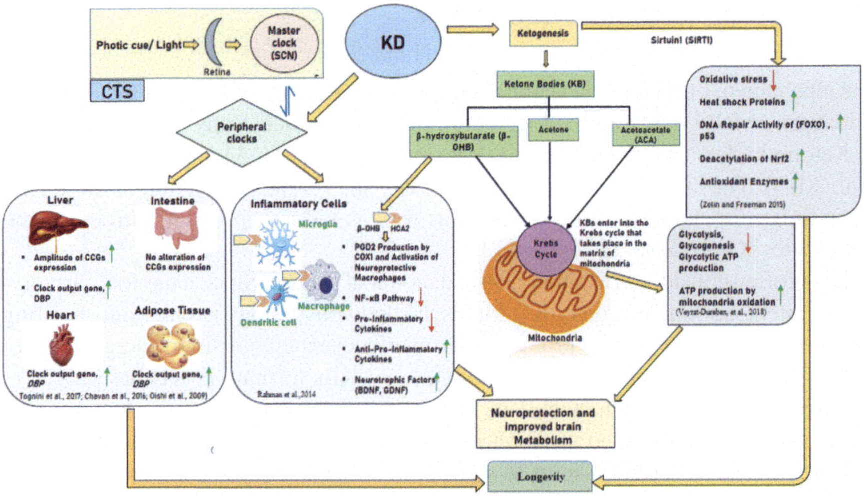

Fig. 16.3 Schematic representation of the effects of ketogenic diet (KD) on circadian rhythm dysfunction, longevity and lifespan: peripheral clocks which are under the regulation of master clock (SCN) are also entrained by KD which in turn give signals to SCN. KD intervention led to increase in the amplitude of various clock control genes (CCGs) expression in the liver, whereas expression of CCGs remained unaltered in intestine. Up regulation of *DBP*, a clock output gene in the liver, kidney, heart, and adipose tissue was observed after KD therapy contributing to increased longevity. β-hydroxybutarate (β-OHB), one of the ketone bodies (KBs) binds with the hydroxy-carboxylic acid receptor 2 (HCA2) receptor present on the inflammatory cells (Peripheral clocks) such as microglia, macrophages and dendritic cells resulting into production of Prostaglandin D2 (PGD2) by COX1 which activates neuroprotective macrophages. Further KD helps in reducing pro-inflammatory cytokines levels, down regulating NF-κB pathway, increasing anti-inflammatory cytokines levels as well as neurotrophic factors leading to increased longevity, lifespan. KD acts through Sirtuin1 (SIRT1) to reduce oxidative stress, elevate antioxidant status, increase levels of heat shock proteins, enhance DNA repair activity of (FOXO), p53, increase deacetylation of nuclear factor erythroid 2-related factor 2 (Nrf2) which are beneficial for improving longevity and lifespan

16.4 Mechanisms Involved in Therapeutic Effects of KD Enhancing Lifespan and Longevity

16.4.1 Reducing Inflammation

Inflammation is one of the hallmarks of aging and the link between aging and inflammation is depicted as "Inflammaging". Many age-associated diseases are associated with elevated level of inflammatory cells such as macrophages, increased circulatory levels of pro-inflammatory molecules. KD inhibits nuclear factor kappa-light-chain-enhancer of activated B cells (NF-kB) activation, which is very crucial in neuroinflammation condition (Rahman et al. 2014). KD also down regulates expression of nitric oxide synthase and COX2 as well as of pro-inflammatory cytokines like IL-1b, IL-6, CCL2/MCP-1,TNF-α (Dupuis et al. 2015). Reports have also suggested that, β-OHB may alleviate upsurge of inflammatory cytokines (Hao et al. 2012). Moreover, an innate immune sensor NOD-like receptor 3 (NLRP3) inflammasome which regulates caspase-1 activation and the release of pro-inflammatory cytokines, such as IL-1β and IL-18 by reducing the K+ efflux from cells is prohibited by ketones (Veyrat-Durebex et al. 2018; McDonald and Cervenka 2018).

16.4.2 Reducing Oxidative Stress

ROS are responsible for the damage to DNA, lipids, and proteins leading to apoptosis of cells which further play a vital role in pathophysiology of many age associated disorders. KD intervention increases ketone body formation leading to ketosis thus resulting in ROS. Such a change could involve by passing complex 1 in electron transport chain in mitochondria, improving mitochondrial respiration as well as increasing NAD^+/NADH ratio which may further improve cellular respiration, redox reactions and mitochondrial biogenesis (Yang and Sauve 2016).

Elamin et al. (2018) reported that rats administrated with KD had increased ratio of NAD^+/NADH in brain cortex and hippocampus after two days of intervention. KD induces SIRT1 which has pleotropic action such as limiting oxidative stress, increasing heat shock proteins production, enhancing DNA repairing activity of fork head transcription factor (FOXO) and protein p53 as well as deacetylating Nrf2 (Zelin and Freeman 2015). Nrf2 induces the enzymes such as glutathione reductase, peroxiredoxin, thioredoxin which help in synthesis of active form of endogenous antioxidant agents and alleviates oxidative stress by enhancing the expression of an antioxidant protein heme oxygenase-1 (HO-1) (Chorley et al. 2012). It is also reported that KD improves the level of glutathione and glutathione peroxidase (GSH-Px) in the hippocampus (Bough et al. 2006).

16.5 Neuroprotective Activity

Neuroinflammation is evident with the progression of age giving rise to many pathological conditions such as PD and AD. Oxidative stress and mitochondrial dysfunctions play essential role in creating neuroinflammatory environment, which ultimately leads to neurodegeneration. KD exerts its neuroprotective effect by increased production of KBs, which lead to more ATP production and less ROS production, hence improving the efficiency of mitochondrial respiration. Reports have also suggested that KD is rich in PUFAs, which are believed to protect the neurons by reducing free radicals, incident of inflammatory environment and controlling nerve-cell excitability. Further it modulates neuronal membrane receptors leading to increased concentration of brain ATP, phosphocreatine and induces mitochondrial biogenesis (Bough et al. 2006). The immune cells present in the brain express GPR109A, which is a G protein-coupled receptor called hydroxy-carboxylic acid receptor 2 (HCA2) (Yang et al. 2019). KD targets the neuroinflammatory process, as β-OHB bind to HCA2 receptor present on microglia, dendritic cells, and macrophages resulting into Prostaglandin D2 (PGD2) production by COX1 and activation of neuroprotective macrophages, eventually ameliorating neuroinflammation (Taggart et al. 2005; Yang et al. 2019). In vitro studies have shown that β-OHB inhibited histone deacetylases (HDACs) such as 1, 3 and 4 (class I and II a), restoring memory function and improving synaptic plasticity (Bough et al. 2006).The neuroprotective effect of KD has been also depicted by its ability to reduce the glycemic index i.e. reduced glycolysis, which act as caloric restriction contributing to increase in lifespan of various species (Włodarek 2019). It has also been suggested that KD modifies gut microbiota which help in protecting against various seizures (Olson et al. 2018). Moreover KD improves resistance and adaptability of neuronal cells against stress and metabolic challenges (Veyrat-Durebex et al. 2018). It also stabilizes synaptic plasticity, activates ATP-sensitive potassium channels, enhances neurotrophic factors activity and fights against neurotoxic environment (McDonald and Cervenka 2018; Włodarek 2019).

16.6 Neurotrophic Factors

KD regulates the level of neuroprotective mediators such as brain-derived neurotrophic factor (BDNF), neurotrophin-3 (NT-3) and glial cell line-derived neurotrophic factor (GDNF). Few have reported that KD regime may result in the circadian shift in the expression of BDNF in the liver though BDNF level is actually reduced in striatum (Włodarek 2019) (Fig. 16.3).

16.7 Conclusion

KD is a modified dietary regime, which advances the onset of locomotor activity rhythms and daily behavioural rhythms in mice. KD affects circadian rhythm of peripheral clocks differentially by regulating core clock mechanism in the peripheral tissues. It also contributes towards improving memory, cognitive abilities and lifespan during aging process. It has been proven to be beneficial in mitigating various factors contributing to aging, age-associated neurodegenerative disorders as well as other pathologies like cancer, respiratory, cardiovascular disease, obesity, metabolic syndrome, and diabetes mellitus type 2. Hence KD may be considered as the potential therapeutic intervention as it improves physiological functions, maintains brain health thus leading to a healthy survival strategy and longevity.

Acknowledgements AJ is thankful to Prof. Suresh Rattan and Prof. Gurcharan Kaur for giving this opportunity and sincere patience during preparation of manuscript. The work is supported by DBT (102/IFD/SAN/5407/2011-2012), ICMR (Ref. No. 55/7/2012-/BMS), and UPE II and Grants to AJ. SAS is thankful to ICMR for SRF.

Compliance with Ethical Standards

Conflict of Interest All authors declare they have no conflict of interest.

References

Astrup A, Hjorth MF (2017) Improvement in age-related cognitive functions and life expectancy by ketogenic diets. Nat Rev Endocrinol 13:695–696

Auvin S (2016) Non-pharmacological medical treatment in pediatric epilepsies. Revue Neurologique 172(3):182–185

Balietti M, Casoli T, Di Stefano G, Giorgetti B, Aicardi G, Fattoretti P (2010) Ketogenic diets: an historical antiepileptic therapy with promising potentialities for the aging brain. Ageing Res Rev 9:273–279

Balietti M, Giorgetti B, Di Stefano G, Casoli T, Platano D, Solazzi M, Fattoretti P (2010) A ketogenic diet increases succinic dehydrogenase (SDH) activity and recovers age-related decrease in numeric density of SDH-positive mitochondria in cerebellar Purkinje cells of late-adult rats. Micron 41:143–148

Barzegar M, Afghan M, Tarmahi V, Behtari M, Rahimi Khamaneh S, Raeisi S (2019) Ketogenic diet: overview, types, and possible anti-seizure mechanisms. Nutr Neurosci 24:307–316

Bough KJ, Wetherington J, Hassel B, Pare JF, Gawryluk JW, Greene JG, Shaw R, Smith Y, Geiger JD, Dingledine RJ (2006) Mitochondrial biogenesis in the anticonvulsant mechanism of the ketogenic diet. Ann Neurol Official J Am Neurol Assoc Child Neurol Soc 60:223–235

Chavan R, Feillet C, Costa SS, Delorme JE, Okabe T, Ripperger JA, Albrecht U (2016) Liver-derived ketone bodies are necessary for food anticipation. Nat Commun 7:10580

Chorley BN, Campbell MR, Wang X, Karaca M, Sambandan D, Bangura F, Xue P, Pi J, Kleeberger SR, Bell DA (2012) Identification of novel NRF2-regulated genes by ChIP-Seq: influence on retinoid X receptor alpha. Nucleic Acids Res 40:7416–7429

Cunnane SC, Courchesne-Loyer A, St-Pierre V, Vandenberghe C, Pierotti T, Fortier M, Croteau E, Castellano CA (2016) Can ketones compensate for deteriorating brain glucose uptake during aging? Implications for the risk and treatment of Alzheimer's disease. Ann N Y Acad Sci 1367:12–20

Dąbek A, Wojtala M, Pirola L, Balcerczyk A (2020) Modulation of cellular biochemistry, epigenetics and metabolomics by ketone bodies. Implications of the ketogenic diet in the physiology of the organism and pathological states. Nutrients 12:788

Davis JJ, Fournakis N, Ellison J (2021) Ketogenic diet for the treatment and prevention of dementia: a review. J Geriatr Psychiatry Neurol 34:3–10

de Lima PA, Prudêncio MB, Murakami DK, de Brito Sampaio LP, Neto AM, Damasceno NR (2017) Effect of classic ketogenic diet treatment on lipoprotein subfractions in children and adolescents with refractory epilepsy. Nutrition 33:271–277

Douris N, Melman T, Pecherer JM, Pissios P, Flier JS, Cantley LC, Locasale JW, Maratos-Flier E (2015) Adaptive changes in amino acid metabolism permit normal longevity in mice consuming a low-carbohydrate ketogenic diet. Biochimica et Biophysica Acta 1852:2056–2065

Dupuis N, Curatolo N, Benoist JF, Auvin S (2015) Ketogenic diet exhibits anti-inflammatory properties. Epilepsia 56:e95–e98

Elamin M, Ruskin DN, Masino SA, Sacchetti P (2018) Ketogenic diet modulates NAD^+-dependent enzymes and reduces DNA damage in hippocampus. Front Cell Neurosci 12:263

Fenoglio-Simeone KA, Wilke JC, Milligan HL, Allen CN, Rho JM, Maganti RK (2009) Ketogenic diet treatment abolishes seizure periodicity and improves diurnal rhythmicity in epileptic Kcna1-null mice. Epilepsia 50:2027–2034

Genzer Y, Dadon M, Burg C, Chapnik N, Froy O (2015) Ketogenic diet delays the phase of circadian rhythms and does not affect AMP-activated protein kinase (AMPK) in mouse liver. Mol Cell Endocrinol 417:124–130

Grabacka M, Pierzchalska M, Dean M, Reiss K (2016) Regulation of ketone body metabolism and the role of PPARα. Int J Mol Sci 17:2093

Hall KD, Chen KY, Guo J, Lam YY, Leibel RL, Mayer LE, Reitman ML, Rosenbaum M, Smith SR, Walsh BT, Ravussin E (2016) Energy expenditure and body composition changes after an isocaloric ketogenic diet in overweight and obese men. Am J Clin Nutr 104(2):324–333

Han YM, Ramprasath T, Zou MH (2020) β-hydroxybutyrate and its metabolic effects on age-associated pathology. Exp Mol Med 52:548–555

Hao J, Liu R, Turner G, Shi FD, Rho JM (2012) Inflammation-mediated memory dysfunction and effects of a ketogenic diet in a murine model of multiple sclerosis. PloS One 7:e35476

Hernandez AR, Hernandez CM, Campos KT, Truckenbrod LM, Sakarya Y, McQuail JA, Carter CS, Bizon JL, Maurer AP, Burke SN (2018) The antiepileptic ketogenic diet alters hippocampal transporter levels and reduces adiposity in aged rats. J Gerontol Ser A 73:450–458

Hernandez A, Truckenbrod L, Federico Q, Campos K, Moon B, Ferekides N, Hoppe M, D'Agostino D, Burke S (2020) Metabolic switching is impaired by aging and facilitated by ketosis independent of glycogen. Aging (Albany NY) 12:7963

Husari KS, Cervenka MC (2020) The ketogenic diet all grown up—Ketogenic diet therapies for adults. Epilepsy Res 162:106319

Itokawa M, Hirao A, Nagahama H, Otsuka M, Ohtsu T, Furutani N, Hirao K, Hatta T, Shibata S (2013) Time-restricted feeding of rapidly digested starches causes stronger entrainment of the liver clock in PER2::LUCIFERASE knock-in mice. Nutr Res 33:109–119

Jagota A, Kalyani D (2008) Daily serotonin rhythms in rat brain during postnatal development and aging. Biogerontology 9:229–234

Jagota A, Kalyani D (2010) Effect of melatonin on age induced changes in daily serotonin rhythms in suprachiasmatic nucleus of male Wistar rat. Biogerontology 11:299–308

Jagota A, Horacio O, Schwartz WJ (2000) Morning and evening circadian oscillations in the suprachiasmatic nucleus in vitro. Nat Neurosci 3:372–376

Jagota A, Thummadi NB, Kukkemane K (2019) Circadian regulation of hormesis for health and longevity. In: The science of hormesis in health and longevity. Academic Press, pp 223–233

Khabbush A, Orford M, Tsai YC, Rutherford T, O'Donnell M, Eaton S, Heales SJ (2017) Neuronal decanoic acid oxidation is markedly lower than that of octanoic acid: a mechanistic insight into the medium-chain triglyceride ketogenic diet. Epilepsia 58:1423–1429

Ko CH, Takahashi JS (2006) Molecular components of the mammalian circadian clock. Hum Mol Genet 15(suppl_2):R271–R277

Kraeuter AK, Guest PC, Sarnyai Z (2019) The therapeutic potential of ketogenic diet throughout life: focus on metabolic, neurodevelopmental and neurodegenerative disorders. Rev Biomarker Stud Aging Anti-Aging Res 1178:77–101

Kukkemane K, Jagota A (2019) Therapeutic effects of curcumin on age-induced alterations in daily rhythms of clock genes and Sirt1 expression in the SCN of male Wistar rats. Biogerontology 20:405–419

Kukkemane K, Jagota A (2020) Therapeutic effects of hydro-alcoholic leaf extract of Withania-somnifera on age-induced changes in daily rhythms of Sirt1, Nrf2 and Rev-erbα in the SCN of male Wistar rats. Biogerontology 21:593–607

Leclère PS, Rousseau D, Patouraux S, Guérin S, Bonnafous S, Gréchez-Cassiau A, Ruberto AA, Luci C, Subramaniam M, Tran A, Delaunay F (2020) MCD diet-induced steatohepatitis generates a diurnal rhythm of associated biomarkers and worsens liver injury in Klf10 deficient mice. Sci Rep 10:12139

Longo R, Peri C, Cricrì D, Coppi L, Caruso D, Mitro N, De Fabiani E, Crestani M (2019) Ketogenic diet: a new light shining on old but gold biochemistry. Nutrients 11:2497

Manikonda PK, Jagota A (2012) Melatonin administration differentially affects age-induced alterations in daily rhythms of lipid peroxidation and antioxidant enzymes in male rat liver. Biogerontology 13:511–524

Mattam U, Jagota A (2014) Differential role of melatonin in restoration of age-induced alterations in daily rhythms of expression of various clock genes in suprachiasmatic nucleus of male Wistar rats. Biogerontology 15(3):257–268

McDonald TJ, Cervenka MC (2018) Ketogenic diets for adult neurological disorders. Neurotherapeutics 15(4):1018–1031

Morrison SA, Fazeli PL, Gower B, Willig AL, Younger J, Sneed NM, Vance DE (2020) Cognitive effects of a ketogenic diet on neurocognitive impairment in adults aging with HIV: a pilot study. J Assoc Nurses AIDS Care JANAC 31:312–324

Nakao R, Shimba S, Oishi K (2017) Ketogenic diet induces expression of the muscle circadian gene Slc25a25 via neural pathway that might be involved in muscle thermogenesis. Sci Rep 7:2885

Neal E (2012) Dietary treatment of epilepsy: practical implementation of ketogenic therapy. Wiley, Hoboken

Neth BJ, Mintz A, Whitlow C, Jung Y, Sai KS, Register TC, Kellar D, Lockhart SN, Hoscheidt S, Maldjian J, Heslegrave AJ (2020) Modified ketogenic diet is associated with improved cerebrospinal fluid biomarker profile, cerebral perfusion, and cerebral ketone body uptake in older adults at risk for Alzheimer's disease: a pilot study. Neurobiol Aging 86:54–63

Newman JC, Covarrubias AJ, Zhao M, Yu X, Gut P, Ng CP, Huang Y, Haldar S, Verdin E (2017) Ketogenic diet reduces midlife mortality and improves memory in aging mice. Cell Metab 26:547–557

Oishi K, Uchida D, Ohkura N, Doi R, Ishida N, Kadota K, Horie S (2009) Ketogenic diet disrupts the circadian clock and increases hypofibrinolytic risk by inducing expression of plasminogen activator inhibitor-1. Arterioscler Thromb Vasc Biol 29(10):1571–1577

Oishi K, Yamamoto S, Uchida D, Doi R (2013) Ketogenic diet and fasting induce the expression of cold-inducible RNA-binding protein with time-dependent hypothermia in the mouse liver. FEBS Open Bio 3:192–195

Olson CA, Vuong HE, Yano JM, Liang QY, Nusbaum DJ, Hsiao EY (2018) The gut microbiota mediates the anti-seizure effects of the ketogenic diet. Cell 173:1728–1741

Ota M, Matsuo J, Ishida I, Hattori K, Teraishi T, Tonouchi H, Ashida K, Takahashi T, Kunugi H (2016) Effect of a ketogenic meal on cognitive function in elderly adults: potential for cognitive enhancement. Psychopharmacology 233:3797–3802

Paoli A, Bianco A, Damiani E, Bosco G (2014) Ketogenic diet in neuromuscular and neurodegenerative diseases. BioMed Res Int 2014:474296

Parry HA, Kephart WC, Mumford PW, Romero MA, Mobley CB, Zhang Y, Roberts MD, Kavazis AN (2018) Ketogenic diet increases mitochondria volume in the liver and skeletal muscle without altering oxidative stress markers in rats. Heliyon 4:e00975

Phillips MC, Murtagh DK, Gilbertson LJ, Asztely FJ, Lynch CD (2018) Low-fat versus ketogenic diet in Parkinson's disease: a pilot randomized controlled trial. Mov Disord 33:1306–1314

Rahman M, Muhammad S, Khan MA, Chen H, Ridder DA, Müller-Fielitz H, Pokorná B, Vollbrandt T, Stölting I, Nadrowitz R, Okun JG (2014) The β-hydroxybutyrate receptor HCA 2 activates a neuroprotective subset of macrophages. Nat Commun 5:1–11

Reddy MY, Jagota A (2015) Melatonin has differential effects on age-induced stoichiometric changes in daily chronomics of serotonin metabolism in SCN of male Wistar rats. Biogerontology 16:285–302

Reddy VDK, Jagota A (2014) Effect of restricted feeding on nocturnality and daily leptin rhythms in OVLT in aged male Wistar rats. Biogerontology 15:245–256

Roberts MN, Wallace MA, Tomilov AA, Zhou Z, Marcotte GR, Tran D, Perez G, Gutierrez-Casado E, Koike S, Knotts TA, Imai DM (2017) A ketogenic diet extends longevity and healthspan in adult mice. Cell Metab 26:539–546

Rosenbaum M, Hall KD, Guo J, Ravussin E, Mayer LS, Reitman ML, Smith SR, Walsh BT, Leibel RL (2019) Glucose and lipid homeostasis and inflammation in humans following an isocaloric ketogenic diet. Obesity (Silver Spring Md) 27:971–98

Roy M, Nugent S, Tremblay-Mercier J, Tremblay S, Courchesne-Loyer A, Beaudoin JF, Tremblay L, Descoteaux M, Lecomte R, Cunnane SC (2012) The ketogenic diet increases brain glucose and ketone uptake in aged rats: a dual tracer PET and volumetric MRI study. Brain Res 1488:14–23

Ruan HB, Crawford PA (2018) Ketone bodies as epigenetic modifiers. Curr Opin Clin Nutr Metab Care 21:260–266

Suchkov S, Salmi TS, Bai CH, Alizargar J, Wu JP (2021) Ketogenic diet is good for aging-related sarcopenic obesity. In: Ketogenic diet. IntechOpen

Taggart AK, Kero J, Gan X, Cai TQ, Cheng K, Ippolito M, Ren N, Kaplan R, Wu K, Wu TJ, Jin L (2005) (D)-β-hydroxybutyrate inhibits adipocyte lipolysis via the nicotinic acid receptor PUMA-G. J Biol Chem 280:26649–26652

Taylor MK, Sullivan DK, Mahnken JD, Burns JM, Swerdlow RH (2018) Feasibility and efficacy data from a ketogenic diet intervention in Alzheimer's disease. Alzheimer's Dementia Transl Res Clin Interv 4:28–36

Teran FA, Kim Y, Crotts MS, Bravo E, Emaus KJ, Richerson GB (2019) Time of day and a ketogenic diet influence susceptibility to SUDEP in $Scn1a^{R1407X/+}$ mice. Front Neurol 10:278

Thummadi NB, Jagota A (2019) Aging renders desynchronization between clock and immune genes in male Wistar rat kidney: chronobiotic role of curcumin. Biogerontology 20:515–532

Tognini P, Murakami M, Liu Y, Eckel-Mahan KL, Newman JC, Verdin E, Baldi P, Sassone-Corsi P (2017) Distinct circadian signatures in liver and gut clocks revealed by ketogenic diet. Cell Metab 26:523–538

Veyrat-Durebex C, Reynier P, Procaccio V, Hergesheimer R, Corcia P, Andres CR, Blasco H (2018) How can a ketogenic diet improve motor function? Front Mol Neurosci 11:15

Vinod C, Jagota A (2016) Daily NO rhythms in peripheral clocks in aging male Wistar rats: protective effects of exogenous melatonin. Biogerontology 17:859–871

Vinod C, Jagota A (2017) Daily Socs1 rhythms alter with aging differentially in peripheral clocks in male Wistar rats: therapeutic effects of melatonin. Biogerontology 18:333–345

Walker WH, Walton JC, DeVries AC, Nelson RJ (2020) Circadian rhythm disruption and mental health. Transl Psychiatry 10:1–13

Wallace MA, Aguirre NW, Marcotte GR, Marshall AG, Baehr LM, Hughes DC, Hamilton KL, Roberts MN, Lopez-Dominguez JA, Miller BF, Ramsey JJ (2021) The ketogenic diet preserves skeletal muscle with aging in mice. Aging Cell 20:e13322

Włodarek D (2019) Role of ketogenic diets in neurodegenerative diseases (Alzheimer's disease and Parkinson's disease). Nutrients 11:169

Yang H, Shan W, Zhu F, Wu J, Wang Q (2019) Ketone bodies in neurological diseases: focus on neuroprotection and underlying mechanisms. Front Neurol 10:585

Yang Y, Sauve AA (2016) NAD+ metabolism: bioenergetics, signaling and manipulation for therapy. Biochimica et Biophysica Acta (BBA)-Proteins Proteomics 1864:1787–1800

Yu Y, Wang F, Wang J, Zhang D, Zhao X (2020) Ketogenic diet attenuates aging-associated myocardial remodeling and dysfunction in mice. Exp Gerontol 140:111058

Zelin E, Freeman BC (2015) Lysine deacetylases regulate the heat shock response including the age-associated impairment of HSF1. J Mol Biol 427:1644–1654

Zhou Z, Hagopian K, López-Domínguez JA, Kim K, Jasoliya M, Roberts MN, Cortopassi GA, Showalter MR, Roberts BS, González-Reyes JA, Baar K (2021) A ketogenic diet impacts marker of mitochondrial mass in a tissue specific manner in aged mice. Aging 13:7914–7930

Zhou Z, Hagopian K, Roberts M, Tran D, Perez G, Kim K, Ramsey JA (2017) A ketogenic diet increases markers of mitochondrial content in a tissue specific manner in adult mice. FASEB J 31:lb482

Chapter 17
Diet According to Traditional Chinese Medicine for Health and Longevity

Yifang Zhang

Abstract Diet nutritional therapy is a primary mode of healing used in Traditional Chinese medicine (TCM) to prevent and treat disease. While modern medicine has only recently recognized that proper nutrition is foundational to human health, they often fail to understand its therapeutic importance. For centuries, TCM physicians have integrated diet therapies into their healing practices providing a detailed knowledge of the healing power of common foods, including the integration of herbal medicines into dietetics. This chapter addresses the key philosophies underlying Chinese medical dietetics, underscores the importance of a healthy digestive system as the basis of human health, and provides guidance on how foods and botanicals can maintain, restore, and build health and achieve healthy longevity.

Keywords Aging · Lifespan · Food · Diet · Health · Yin-yang

17.1 Introduction

The chapter lays out how to think about and pragmatically adapt one's diet according to TCM to optimize for health and longevity. It first goes through the basic theories of the TCM related to diet, such as understanding food nature and flavor from the theories of the Yin-Yang, the Five Elements and using the meridian and organ theory to explain the pertinence of foods. It then explains the TCM aspects of the digestive system and details of the different body constitution, and allows the reader to self-assess where they stand. Building on those first two parts, it then explains how to regulate one's diet based on one's body constitution, age, gender and according to seasons and geographic location. Finally yet importantly, this chapter shares with the readers some of TCM's best wisdom on diet for health and longevity. For example, providing vivid examples on how to maintain one's brain function while ageing, how to regulate and boost immunity, how to prevent illness, how to tackle common or serious illness and it arises and so on.

Y. Zhang (✉)
Shanghai East International Medical Center, 24F, Building B, 551 South Pudong Road, Shanghai, China

S. I. S. Rattan and G. Kaur (eds.), *Nutrition, Food and Diet in Ageing and Longevity*, Healthy Ageing and Longevity 14, https://doi.org/10.1007/978-3-030-83017-5_17

There is a Chinese proverb saying, diet is linked with both the body and soul. Food and medicine share the same roots, which means that foods and daily diet not only provide nutrition and delicious flavors, but also bring energy, happiness, and are important for optimal health. In China, functional nutrition draws on thousands of years of culture, based on TCM, which recognizes that dietary regimens and herbal tonics are a source of healing and defense, can strengthen the body, maintain fitness, prevent and cure illnesses, and contribute greatly to longevity and prosperity. The approach of TCM is, at its core, using nature, alternative-healing methods, tailored to the individual's body and needs. However, people may ask how can one maintain good health, prevent and treat illness, and at the same time enjoy delicious cuisine?

The Yellow Emperor's Canon of Medicine is the earliest existing Chinese medical classic, and the source of Chinese medical theory and herbs. TCM doctors have been applying its theory in hospitals and clinics. The Yellow Emperor's Canon of Medicine states that medicine inhibits illness. Five Chinese traditional grains (rice, red beans, wheat, soybeans, millet) can be used to nourish the body. There are five kinds of fruits (peach, plum, apricot, chestnut, Jujube) which can be used to assist the five kinds of grains to fulfil such functions. The five kinds of livestock (cow, goat, pig, and chicken) can be used to supplement the internal organs. Five kinds of vegetables (sunflowers, leaves of bean, onions, scallion, Chinese chive) can be used to enrich the viscera. When the energy and flavor of foods harmonize, people can strengthen their body essence and increase their Qi (energy) (Yang et al. translation Li and Liu 2005).

Based on the above classic literature, we have learned the following three lessons. First, the choice of daily food ingredients should include a rich variety of grains, vegetables, fruits, and proteins to balance nutrition. Second, foods have their own energy and flavors, suited for different health conditions. Third, balanced energy and flavors of foods have therapeutic effects on the human body, especially strengthening the essence and increasing Qi. TCM find that there is a close relationship between people's spirit and health. Food can nourish the spirit, which influences our health status. TCM widens the spectrum, and creates a link between essence, Qi and spirit with food. Moreover, essence, Qi and spirit have a particularly important impact on human health and longevity.

17.2 TCM Basic Theories Related to Diet

The TCM concept of a healthy diet includes establishing a broad, balanced food base, making food easy to digest and absorb, while also ensuring that the food quantities are suited to each individual, and taking into consideration the climate change and regional differences. TCM finds food affects our daily Yin (interior energy)-Yang (exterior energy) balance. Because foods are categorized as either Yin or Yang in their nature, they can be used as an effective healing method to maintain and re-build equilibrium of body energy.

Building up a healthy diet is a way to attain physical, emotional, and mental wellness. To achieve a healthy balance, one needs to understand TCM concepts of foods and herbs, and healthy diet first.

17.2.1 Understanding of Food Nature from TCM Theories of Yin-Yang

Choosing the foods and herbs that are right for us is essential to achieving and maintaining good health. Daily diet is closely related to the temperature change of the day and the alternation of the mild, cold and hot seasons. Therefore, TCM uses Yin and Yang to define the cold, hot, warmth, and coolness of the four energies of food. Yin and Yang are the two fundamental principles or forces in the universe, ever opposing and supplementing each other. This ancient philosophical concept has become an important component of the basic theory of TCM. In the beginning, Yin and Yang described a location in relation to the sun. A place exposed to the sun is Yang, and a place without exposure is Yin. The southern side of a mountain, for example, is Yang, while its northern side is Yin. Thus the ancient Chinese people, during their everyday life and work, came to understand that all aspects of the natural world could be seen as having a dual aspect. For example: day and night, brightness and dimness, movement and stillness, upward and downward directions, heat and cold, etc. (Zhang et al. 1989).

The terms Yin and Yang express these dual and opposite qualities of food and how to achieve balance. This is an inherent property and not necessarily dependent on the food temperature. Yang food means that its energy creates warmth or heat when taking them. For instance, eating chili warms the body. "Flavor" can also be defined as a Yin and Yang. Yang taste includes pungent, sweet or bland. Yang cooking or preparation methods include stir-frying, stewing, baking, deep-frying, roasting, grilling or barbecue (Leggett 1994). These foods and herbs make our energy rise to the body's surface.

Yin food means that the energy of foods is cool or cold, and the tastes sour, bitter or salty. Yin cooking or preparation methods include steam or eating raw (Leggett 1994). Yin foods restrain the body's energy and cause it to descend. Many foods have very mild Yin or Yang qualities therefore defined as "neutral". Examples of such foods include rice, corn, kale and carrots.

Even if we do not have detailed knowledge about the Yin and Yang of food, if we consume a broad range of foods (rainbow-colored food), we can quite naturally get a balance of Yin and Yang.

17.2.2 Understanding of Food Flavor from TCM Theories of Five Elements

Through taste and experience, the wise ancestors not only found the food for human survival, but also found the medical herbs to treat illness. Ancestors discovered the method of determining the action and properties of foods and herbs through tasting and the cold and heat reaction of the human body after eating, and gradually formed a relatively perfect theory of the properties of TCM herbs, based on "four Qi and five flavors". Because different tastes have different effects on the human body, through the theory of Five Elements, we can better understand the relationship between food flavor and the human body.

In Chinese, the Five Elements are Wu Xing—Wu means five and Xing means movement and change. The Five Elements are wood, fire, earth, metal and water. These elements each have their own special properties, which are at the root of the ancient philosophical concepts used in TCM.

Although the Chinese theory of Five Elements and the Greek theory of Four Elements are different in their history of formation, the rudiments of both belong to the earliest atomic theory. In order to explain the material world around them, ancient philosophers made a generalization and deduction about the respective properties of the substances and their interactive relationships. According to Chinese theory, wood, fire, earth, metal and water are the five basic substances that constitute the material world. They each have their own specific properties, but they also have interactive generation and restriction functions, and are in a constant state of motion and change (Zhang et al. 1989).

Five Elements theory believes that food has different colors and bring varied tastes. Tastes have certain effects on the body as described below.

Sour: Such as hawthorn, lemon and dried dark plum. This taste helps with digestive absorption, resisting fatty foods and preventing indigestion. It generates fluids and Yin, and stops discharge, perspiration, chronic cough and diarrhea. It also has an astringent effect on emissions, including sperm and frequent urination. It helps our body consolidate essential substances, preventing them from escaping. Sour foods can also bring disordered Qi back to normal. Modern research shows sour flavors to be generally cleansing and detoxifying. However, we have to limit intake when ulcer or stones are present.

Bitter: Such as aloe vera and green tea. This taste clears away heat and dries dampness. It can control abnormally ascending Qi and purge any pathogenic fire effect. In certain combinations, it can also improve the body's Yin. Bitter foods can be used to treat most cases of excess and acute damp-heat or heat-fire. These foods should be limited if a weakness of Qi and Blood is present.

Sweet: Such as jujube and honey. Serving to nourish, moisten, moderate and invigorate the body, sweet foods can also regulate Qi, Blood and function of the viscera. They strengthen deficiency syndrome and alleviate dryness. Sweet foods work in coordination with the spleen and stomach. They can help relieve pain and spasms, and reduce cough, ulcer and constipation. An excess of sweets should be avoided when suffering from damp, phlegm and water retention conditions.

Spicy (Pungent): Such as ginger, Sichuan pepper and clove. This taste disperses and promotes movement of Qi and Blood circulation. It stimulates digestion and helps break through blockage. It treats syndromes of the exterior, and expels stagnation of Qi, Blood and pathogens. Spicy foods must be used carefully as many people cannot tolerate them.

Salty: Such as marine algae and kelp. These foods can promote moisture and have a softening effect. In particular, these foods regulate the moisture balance flow downwards in the body. They also move Qi downward, increase urine and bowel movements, and are used to treat constipation and swelling. They promote the action of the kidney system, allowing beneficial foods to be fully absorbed and functional, and improving concentration. Salty foods soften nodes and masses, and disperse accumulations in hardening muscles and glands.

Bland: Such as dolichos seed and pearl barley. This taste promotes urination and treats edema.

17.2.3 Using Meridian and Organ Theory to Explain the Pertinence of Food

Over its long history, Chinese medicine has come to realize that different foods enter specific meridian pathways, directing their effect towards particular organs, that is to say knowledge of the metabolic trend of food in the human body. This information directs the therapeutic use of the food (Zhang et al. 1989). When we know which meridian or organ a food will target, this is useful in treating a disorder of that particular meridian or organ. For instance, onion enters the lung meridian and lung, while lychee targets the liver meridian and liver. Peppers affect the stomach meridian and stomach; sunflower seeds, the spleen meridian and spleen; kidney beans, naturally, the kidney meridian and kidney; and coffee, the heart and its meridian.

17.3 TCM Aspects of the Digestive System and Body Constitution

Over thousands of years TCM has accumulated knowledge about using foods and herbs therapeutically, providing natural and alternative way to boost energy and immunity, regulate one's mood, enhance one's beauty, reduce toxins and treat other health conditions, externally and internally. TCM believes that the human body is made up of five systems. These five systems are related to the digestion and absorption of food, the transportation and metabolism of nutrients. In this paper, we focus on the structure and function of the digestive system and body constitution.

17.3.1 Understanding the Digestive System

"We are what we eat" was a popular saying in the 1990's, although it is only partially true. "We are what we eat and absorb" would be a better expression of the facts. What we absorb is dependent upon how healthy our spleen system is. According to Chinese medical theory, the spleen is the "Agricultural Bureau" of the body, controlling where energy "seeds" are planted and how the health of the body is harvested. The spleen is the post-natal base of life, the source of acquired constitution. This means that once all the congenital factors of the body have been determined, one's health development relies on the spleen to do the rest of the work. The spleen directs digestion and absorption, governs Blood flow within the vessels. Therefore, the spleen system is considered the "source of Qi and Blood." In TCM, the spleen and stomach make up the digestive system, which includes the pancreas, small intestine, and part of the stomach functions. It is related to digestion, water metabolism, and hemopoiesis in modern medicine. Below we provide a self-assessment for people to understand the digestive system.

Questions Relating to the Spleen System (Zhang 2016).

A Do you have severe food allergies or do you get food poisoning more than twice a year?

☐yes ☐ no

B Do you have sensitivities to certain foods?

☐ yes ☐ no

C Do you often get bloated after eating?

☐yes ☐no

D Do you have heartburn?

☐yes ☐no

E Do you often have diarrhea?

☐yes ☐ no

F Do you frequently have bad breath?

☐yes ☐ no

G Do you often have an upset stomach or nausea?

☐yes ☐ no

H Do you bruise easily?

☐yes ☐no

I Do you dislike the wet season or damp weather?

☐ yes ☐no

J Do you have muscle problems (weakness, tightness, stiffness, knots, muscle tears)

☐ yes ☐no

If the number of yes responses is 4 or more, one's digestive system is relatively weak, and cannot tolerate excessively cold, hot, acid and heavy food. These people needs to adjust their digestive system. By contrast, if the number of yes responses is less than 4, these people merely need to maintain a healthy diet.

17.3.2 Understanding Body Constitution

Body constitution comprises our physical state, including the morphology, function of our internal systems, and metabolism, along with our mental and spiritual states. As we pass through life, everyone's physical constitution goes through periods of relative balance and imbalance, for example, passing from hot to cold or strong to weak. An imbalance of our physical constitution can mark a transitional stage, when we are shifting away from health towards disorder, but before disease develops. Therefore, maintaining balance in our physical constitution can prevent or lessen disease and promote recovery from illness. TCM strives to balance the physical constitution, mitigate shock from the outside environment, and dissolve toxic substances within the body.

The features of one's constitution can be detected in three areas: the physical build of the person, the body's internal functions, and the psychological state. It also depends on the stage of life the person is in, such as puberty or menopause.

The constitution has two origins: congenital natural disposition and post-natal lifestyle (i.e. nature and nurture). Many factors influence the formation of the constitution, such as the parents' health, physically and mentally, the time of conception, and the mother's condition during pregnancy. These are all part of the congenital natural disposition of one's constitution. However, most of the influence comes from our own actions and lifestyle. We care more about how each person's digestive system operates, in order to cooperate best with a good choice of foods.

Below we have provided a self-assessment for people to learn more about cold and hot, weakness and overly strong, neutral and mixes of body constitution.

Questions Relating to Temperature: Neutral, Cold, Hot or Mixes Constitution (Zhang and Yao 2012).

1) Are you sensitive to cold or heat?

□normal (1)

□sensitive to cold (2)

□sensitive to heat (3)

2) What do you prefer to drink?

□depends on season (1)

□warm/hot drinks (2)

□cold drinks (3)

3) Do you sweat a lot?

□normal (1)

□less than average (2)

□more than average (3)

4) How do you classify your thirst?
☐normal (1)
☐not often thirsty (2)
☐often thirsty (3)
5) How is your complexion?
☐shining and rosy (1)
☐pale and puffy (2)
☐flushed (3)
6) Which season do you prefer?
☐average (1)
☐prefer summer (2)
☐prefer winter (3)
7) What is your pulse rate (beats per minute)?
☐from 60 to 80 (1)
☐less than 60 (2)
☐over 80 (3)
8) Do you like tea or coffee?
☐up to two cups of coffee or tea everyday (1)
☐three or more cups of tea everyday (2)
☐three or more cups of coffee everyday (3)
9) What kind of food do you prefer?
☐depends on season (1)
☐a light taste or raw food (2)
☐spicy or strongly flavored (3)

Assessment:

Cold or hot

Neutral: 6 or more responses of (1)
Cold: 6 or more of (2)
Hot: 6 of more of (3)
Mixes: if fewer than 6 of any one response
Questions Relating to Humidity: Neutral, Damp, Dry or Mixes Constitution
1) Do you enjoy a rainy or dry environment?
☐normal (1)
☐dry (2)
☐raining (3)
2) What kind of taste do you usually have in your mouth?
☐normal (1)

☐sticky and sweet (2)
☐dry or bitter (3)
3) What is your skin condition?
☐normal or mixed (1)
☐oily (2)
☐dry or cracking (3)
4) How would you characterize your excretion? (discharge from eyes, ears and skin; for women, include monthly period).
☐comfortable amount (1)
☐quite a lot (2)
☐scant or absent (3)
5) Do you smoke or drink alcohol?
☐occasionally (1)
☐frequently (2)
☐refrain from both (3)
6) What is your tolerance for dairy products?
☐ average (1)
☐less than average (2)
☐more than average (3)
7) How do you feel in general?
☐happy and relaxed (1)
☐heavy, sleepy; fullness of chest and stomach (2)
☐irritable, anxious; dry lips and throat (3)
8) How would you characterize your bowel movements and urine output?
☐normal (1)
☐loose stool or turbid urine (2)
☐dry stool, constipation or scanty urine (3)
9) How would you describe your build?
☐average (1)
☐heavy build (2)

☐slim (3)

Assessment:

Damp or dry

Neutral: 6 or more responses of (1)
Damp: 6 or more of (2)
Dry: 6 or more of (3)
Mixes: if fewer than 6 of any one response.

Questions Regarding Your Response to Adversity: Neutral, Weak, Overly Strong or Mixes Constitution.

1) Do you feel energetic?
☐average (1)
☐more than average (2)
☐less than average (3)
2) What is your tongue like when you get up in the morning?
☐pink body and thin fur (1)
☐dark or purple body and thick fur (2)
☐pale or deeper red body and no fur (3)
3) What kind of food do you prefer?
☐mixes, with more vegetables and less meat (1)
☐mostly meat (2)
☐vegetarian (3)
4) How often is your elimination?
☐normal (1)
☐infrequent (2)
☐too frequent (3)
5) How often do you get a cold every year?
☐once or a few times (1)
☐never (2)
☐often (3)
6) How often do you get excited?
☐normal (1)
☐frequently (2)
☐seldom (3)
7) How do your muscles feel?
☐ normal (1)
☐tight and sore (2)
☐soft and weak (3)
8) How quickly do you feel shortness of breath when hiking?
☐15 min to half an hour (1)
☐more than half an hour (2)
☐after a few minutes (3)
9) How does your head often feel?
☐normal (1)
☐pressure or sharp headache (2)
☐lightheaded or dizziness (3)

Assessment:

Weak or overly strong

Neutral: 6 or more responses of (1)
Weak: 6 or more of (3)
Overly Strong: 6 or more of (2)
Mixes: if fewer than 6 of any one response.

In completing the above self-assessment, we encounter some pairs of concepts: cold or hot, damp or dry, weak or overly strong. All of these values in a medial range are normal. It is normal to feel cold in the winter and hot in the summer. However, if one always feel cold even in warm seasons, then should seek the underlying reasons and try some food remedies. The approach toward damp and dry is similar to cold and hot. People feel wet in humid season and dehydrated in the dry season. These feelings are normal within boundaries. Dampness nourishes our inside and moistens the surface of skin while dryness limits the growth of mold. However, too much dampness makes skin oily and develop acne. In contrast, too much dryness causes skin to wrinkle and crack. Therefore, it is best to maintain a neutral status.

After an assessment of body constitution, one can choose the right foods and herbs to achieve and maintain a healthy balance. Let us take damp, dry and neutral constitutions as examples. Neutral type means that one is in healthy balance. In order to maintain this state, this person should eat a broad range of foods, and drink water according to the climate and level of perspiration. However, if one has a damp constitution, too much wet is carried inside. One has to add specific foods to the diet (such as pearl barley, azuki beans, corn) to reduce dampness. As for the dry type, one should take foods such as lily bulb, Chinese wolfberry, honey or lemon to nourish and moisten the body.

17.4 How to Match Diet with Body Variety and Avoid Blight

Chinese dietary therapy (shiliao) applies foods and herbs to achieve health promotion, prevention and treating diseases, rehabilitation, and healing from diseases.

When applying food and diet for those purposes we need to think of the points below.

17.4.1 Food According to Constitution, Age and Sex

Food and dietary recommendations by TCM theory are usually made according to the person's individual constitution and condition.

It is important to ensure that people's diet are aligned with their personal nature. People with a Yang constitution are usually by nature more active and forceful. They do not tolerate heat well, so in the warmer summer months, they must eat more Yin foods and reduce their intake of hot or spicy foods. The more gentle nature of these Yin foods will bring one's Yang nature into balance. By contrast, those with a Yin constitution are often quieter and less energetic. They have a low tolerance for cold, so should consume more Yang foods, which are warm or hot, and should add spice to boost their energy.

Two women with the same illness and same age group can have completely different food therapies because their constitutions are opposite. For treating migraine, a person feeling hot, sharp pains, a flushed face, and frequent perspiration, should use chamomile due to its cooling and bitter qualities and its ability to target the liver meridian. A person with a splitting headache on the top of head or the whole head, spreading over the eyes and temple areas, feeling cold from their body to their head, induced by cold and rain, and likes pressure and warmth, needs to utilize dry ginger, and angelica dahurica which are warm, hot and spicy. A student of mine applies dietary therapy to treat more than 60 women suffering from migraines with over 80% experiencing positive results.

Children, adults and seniors may not use the same ingredients or the same dosage to treat a symptom because their constitutions are different. From the age of 24 to 35, people should focus on preventative measures. They should do all things in moderation (for example, food consumption or working hours), and keep a balanced emotional state. For this group it is best to choose more ingredients that are neutral. From the age of 36 to 65, people should focus on preventing chronic and more serious illnesses, like high blood pressure, high cholesterol and high blood sugar. It is important to take notice of any hereditary diseases and take extra steps toward prevention. This age group should also be sure to consume foods that contain Qi tonics and nourish Blood and body fluids.

At the age of 66 and above, health care should focus on harmonizing the organ systems. Special attention should be given to the digestive, circulatory, cardiovascular and metabolic systems, regardless of whether or not they suffer from ailments. In the younger age group, people may tend to consume a great deal of fruits or vegetables as a main source of fiber. In China, people over the age of 80 do not consume much fruit during the winter months. Their source of fiber will be more from root vegetables, like potatoes, yam, sweet potatoes and taro, as this is better for their digestive system. This traces back to the famous Dr. Chen from the Song dynasty (960–1279 AD), who encouraged elderly people to eat foods that are warm, cooked and soft, and to avoid foods that are glutinous, hardened, raw or cold, because elderly people have weaker Blood and essence, and their digestive system function decreases.

Many healthy longevity people that I interviewed, believe that people over 90 should always eat a highly digestible, low fat, low calorie, low sugar, vitamin rich diet with plenty of dietary fiber, and eat more vegetable oil. They should avoid eating a lot of greasy or fatty food, and only occasionally eat deep-fried, preserved, and overly spicy, salty, or sweet foods. They should also eat 20 percent less and chew their food well before swallowing.

Although the above rules apply generally, because women and men have different body constitutions, their health care can also be different. Since women have changes such as menstruation, pregnancy and post-partum in their life (Zhang 1990), they often need special foods to provide balance. For instance, Chinese angelica is used much more for women. As for men, they have prostate glands and different reproductive structures than women, and they produce sperm and have regular ejaculation. Chinese chive seed is used mostly for men in regulating these functions.

17.4.2 Considering Season and Locality

It is important to match one's diet to the season and personal nature. In the broadest terms, this involves eating Yang foods to warm and energize the body during the winter, then eating calming Yin foods to cool the body and reduce heat in the summer.

Be sure to select seasonal produce to get the most nutrients and health benefits. For instance, some fruits that are seasonal in the summer can assist with cooling the body. Due to globalization and technological advancements in agriculture, some food items formerly available only during the summer may now be purchased year-round. However, this disrupts the natural way and is not conducive to promoting health. Moreover, people should keep their local environment in mind. If one lives in an area that is hot and humid, they need to focus on consuming foods that can help reduce water retention and cool.

Human physical and emotional status is greatly influenced by climate and the environment, its rhythm and seasons. When applying foods to assist human health, one should eat less cold food in the winter, less hot food in the summer, less spicy and pungent food in the autumn dry season and less heavy and oily food in humid environments.

Geographical features of a place have a strong impact on both body constitution and food choices. People from the north can take more hot food than people who have grown up in the south. People from humid regions can tolerate wet climates better than people from dry regions.

In order to get the best quality of ingredients one should select foods that are still produced in the area from which they traditionally originate. For instance, the best quality wolfberries come from Ningxia and Gansu provinces in China. The best jujubes (red dates) are from Hebei and Shandong province.

However, this theory is just a general guidance, and when people become ill (or unbalanced), certain routine foods must not be eaten or eaten less quantity while eating alternative sorts of food until they recover from imbalance. Generally speaking, warm and hot therapeutic foods, such as ginger, fennel, jujube and walnut have the function of warming the interior, dispelling cold and assisting Yang, and can be used to harmonize cold constitutions, treat cold and Yin syndromes (weakness and dampness). The cool and cold nature of therapeutic foods, such as mung beans, watermelon, pear and chrysanthemum, have the function of clearing heat, purging fire, cooling blood and detoxifying, and can be used to harmonize hot constitutions, and treat heat and Yang syndromes (excess and hot-dryness).

17.4.3 Avoiding Certain Foods

Some foods can be harmful or cause illness. In some people or under certain conditions, foods can cause acute or immediate reactions and problems. Examples include allergies, food poisoning, symptoms related to lactose intolerance, etc. Timing and amount of food consumption can also negatively impact the digestive system, such as prolonged hunger or long periods between meals, or eating and drinking too much at one meal (Zhang et al. 1990). Over-eating one type of food also has a negative effect. An example is a person who exercises regularly and eats healthy in every way, including a heavily vegetable-based diet, except eats a large quantity of dark chocolate every day and then later diagnosed with high cholesterol. Over the long-term, food consumption can even help in negating inherited disorders. The Chinese saying "illness comes from mouth" (bingcongkouru 病从口入) is still true for many diseases. Heart diseases, a fatty liver, high blood pressure, high cholesterol and weight gain very much have to do with our food intake. If we can reduce certain foods in the early stages of those illnesses, we can really take control of the condition and eventually return to a healthy state.

Another way that food can harm is when people eat foods that don't agree with their body. As we know, TCM believes that individual bodies may be more inclined to "hot" or "cold" constitutions. Those who are on the hot side may experience more constipation, heartburn or mouth ulcers. If this is the case, they should avoid foods that raise the heat in the body, such as spicy foods, coffee or hot soups. By contrast, people who are in the cold spectrum may have an upset stomach from drinking too many cold drinks, eating cold natured food like persimmon, eating ice cream, and so on, and therefore should avoid raw foods or other cold foods that make them feel ill, even if they enjoy the flavor.

To avoid excess in eating certain types of foods, one should be mindful of flavor preference or aversion. Continually eating from only one flavor group can negatively impact various organs. For example, eating only spicy foods can make you sweat too much and reduce water content in the body, making the lung body fluid dry out, it also scatters Qi, people should avoid spicy foods when suffering Qi weakness. Likewise, too much salt can negatively impact the kidney as they struggle to filter properly. People who eat too many sweet things, such as cookies or anything with added sugar, will often face problems with their pancreas from over stimulation, too much insulin production and may cause muscle weakness and put on weight.

Undigested food becoming stagnant in the digestive tract can lead to chronic ailments. This may happen if food is not digested thoroughly the first time, if it is not passed in a timely manner, or if particles become stuck in the intestines. Signs that food has accumulated in the digestive tract may include poor appetite, belching or gas with a distinct smell, bloating or diarrhea, and in severe cases, painful heartburn with a bitter taste in the throat and mouth. When not resolved, gastritis, irritable bowel syndrome (IBS), pancreatitis and gallstones are likely to occur; polyps can also develop in the colon, leading to colon cancer.

17.5 Diet for Boosting Healthy Life and Longevity

Following the discussion about the TCM concepts and theory in Sects. 17.1, 17.2 and 17.3, this section provides some recommendations for memories, immunity and longevity by introducing some diet and recipes.

Foods and herbs have specific therapeutic actions beyond their temperature, taste, and the meridians traveled. Food may either strengthen a particular substance or function (Qi, Blood, Yin and Yang), or reduce and regulate the influence of a pathological condition (Qi or Blood stagnation, dampness, heat or cold). Lychee, for example, reduces cold and regulates Blood circulation and therefore can treat pigmentation on the face; kidney beans can enhance Yang.

Chinese herbal diet (yaoshan) refers to a diet remedy made up of foods and herbs with therapeutic effects. Specifically, the herbal diet recipes can be either in the form of original food, or processed and mixed from food and herbal materials (Zhang et al. 1993). According to the form and processing method of recipes, a herbal diet can be divided into the following two categories. Type1 is liquid form, such as juice or smoothies obtained from freshly squeezed food. It can also be porridge, tea, soup and paste formed by boiling, and steam. Type 2 is solid form, such as pancake and steamed bread.

There are two principles that dictate how to pair foods and herbs. One focuses on matching two foods that have a similar function of Yin or Yang. These form natural and regular pairings, just like the way you might pair red meat with red wine. In TCM, an example would be ginger with brown sugar, or silver fungus with honey or crystal sugar. The second principle is to pair foods that have differing but complimentary effects on the body. This may involve tonifying/strengthening Qi or Blood through Yin and Yang properties for the most effective balance. For example, taking ginger (which strengthens Yang) and dates (to tonify Blood) works to strengthen the body overall.

17.5.1 Maintaining Brain Function During Aging

Aging is an unavoidable phenomenon. It is characterized by both internal and external changes. On the physical side, there is a decline in physical ability, weakening of the function of the sense organs, and onset of disorders like cardiac problems, diabetes, and blood pressure changes. There are also changes mentally, there may be memory loss or a decrease in mental and logical sharpness.

It is impossible to stay young forever. However, by regulating eating and lifestyle, one can maintain optimal health. Changing diet cannot only extend life, but it also will make one feel and look younger than their actual age. Both physical and mental body could improve through foods and diet. Strengthening Qi, Yang and nourishing Blood can keep one young mentally, maintaining concentration and a sharp mind. Toward this end, one could often take ginseng, astragalus root, walnuts, pistachios, chestnuts, raspberries, cassia fruits, black sesame seeds, cherries, angelica root, coconuts, cinnamon powder, gingko nuts, rosemary, saffron crocus, pumpkin seeds, wolfberries, longan fruit, schisandra berries and pu'er tea.

17.5.2 Regulating and Boosting Immunity

If the immune system is depressed or overly sensitive, it becomes prone to numerous diseases, ranging from the common cold to cancer. The onset of respiratory infection tends to be higher when combined with a sensitive or weakened immune system; in the long-term this can lead to auto-immune and allergy-originating diseases. TCM believes that the lung, the spleen and the kidney Qi is our body's defense, preventing illness from invading deep into body. To strengthen the Qi, and ultimately overall immunity, one should eat healthy foods, and make simple dietary changes to balance all the functions. In doing so, we can prevent some respiratory illnesses and the onset of allergies. Toward this end, one could often take ginseng, astragalus root, fig, ginger, licorice, garlic, jujubes, lotus seeds, marine algae, sea buckthorn, silver ear, onion, lingzhi (reishi) and shiitake mushroom, and five leaf gynostemma herb.

17.5.3 Achieving Healthy Longevity

Throughout China's history, the combination of health foods and diet therapy has been used to achieve the goal of healthy longevity. According to my interviews with more than 80 elderly people over the age of 70, and reviewing more than 2000 of my own clinical documents, in addition to some literature. I have a concluded that food ingredients and diet patterns are closely related to health and longevity, which is reflected in two aspects of health preservation and outcome of my clinical treatments.

17.5.3.1 Diet Therapy for Illness Prevention

Through the proper choice of foods and herbs one can live a balanced, healthy life, and avoid illness. This highlights the preventative side of TCM, which should be appealing to all: An ounce of prevention is worth a pound of cure!

In southern China, most people start to pay attention to health preservation and choose special diet in their 40 to 50s. People in this age group clearly feel a certain level of body and mental functions deterioration, and often have disorders in their nutrient absorption and organ functions. This health transition will affect people's work performance, and ultimately their quality of life. In the past one can eat whichever food he prefers, of any quantity at any time of day, or likewise consume only one meal per day, or even fast for three to five days. Once one has reached middle age, one would feel dizzy, sweaty and unable to work if one skips a meal; drinking alcohol would cause a headache and heart palpitations; and eating spicy food may lead to a stomachache and diarrhea. Some people seek the advice of TCM doctor, to identify their body constitutions, and the functional status of the digestive system and to get recommendations for suitable food and diets. Their bodily functions may restore to normal in a few months after taking professional advices and a food remedies. Some of them get even more benefits after taking systematic measures, for instance, eating small portion of meals up to four or five times each day (applicable to people with weak spleen and stomach), eating cold food only in warm weather (for people with cold constitution, choose cooked food, warm food in winter), and following the seasonal changes to adjust diets. Undertaking such methods for several years, people could reduce the incidence of respiratory and digestive system illness. Other common symptoms are hot flashes, sweating, insomnia and losing temper easily in the age group between 50 and 60s, which is a result of declining human hormone levels. However, if people maintain a healthy diet these symptoms can be reduced or disappear, therefore benefit the quality of life for advance age.

For those who would like to apply food therapy before reaching the middle age, they could think of the two aspects presented below.

(1) Methods of healthcare and prevention based on previous medical experience and weaknesses.

From people's history of illness, doctors can detect the varied weakness in each patient. For example, when contracting a cold or flu, the disease may affect the nose and sinuses in one patient, but affect the stomach and appetite in another. If cold always affects one's nose and sinuses, their lung Qi needs to be strengthened. If a patient does not suffer from colds frequently, their condition is normal; but if a patient catches cold 4 to 5 times a year or chronically, they must pay attention to it. To strengthen lung and prevent illness, people can eat almonds, pears, lotus roots, or lotus seeds.

If cold or stress always affects one's stomach, causing symptoms such as stomachache, bloating, and diarrhea, the stomach and spleen function needs to be strengthened. Patients can eat ginger, whole grains such as glutinous rice, buckwheat, congee, or oats, root vegetables such as yams, or beans like lentils.

(2) Methods of healthcare and prevention based on family history.

If patient has a family history of a certain condition, one can take measures to prevent the same condition occurring in his or her lifetime. Family history is a good indicator of potential health issues, so people should look to their parents and grandparents (or even immediate relatives) to identify potential weak areas. If a patient has a family history of joint problems, such as the hip or toe joints, or the development of osteoporosis (during menopause), they must tonify the kidney system. Some preventative methods for joint problems are delaying menopause by tonifying the kidney and related organ systems with food (starting from the age of around 42 or 45). For example, one can eat more chestnuts, mulberry, hazel, sour jujube, and drink herbal wine from time to time. If patient has a family history of allergic rhinitis and asthma, they must tonify the lung system and related organ systems with food. For example, one can eat more fresh and dried ginger, almonds, juicy pear and gingko nut to maintain a good function of the lung.

Regulating the diet, especially using tonics in autumn and winter, is one of the dietary traditions in China, through by which people can cope with winter cold and emergencies as well as maintain health. Those who take tonics in autumn and winter can also prevent common diseases, such as by enhancing the lung and kidney respiratory functions, support heart function, and optimize health of the individual to prevent flu, bronchitis, asthma and arrhythmia.

Enhancing spleen (stomach) and kidney system are most effective for healthy longevity. The classic works of Yellow Emperor's Canon of Medicine holds that the key factor in the internal causes of human aging is the degeneration of vitality in the kidney system. If one can maintain the function and adjust the balance of Yin and Yang in the kidney system, will be effective in delaying aging.

The kidney system (a broad TCM definition including development, spine, marrow, bladder) is the administrative organ with the greatest power. The kidney system is the congenital base of life and stores essence. The kidney system acts as the strength and intelligence of the entire body, and is responsible for its overall body constitution. It is a storage facility for good essence, and it governs the growth and development of the body as well as the maturation of the reproductive systems. It is the congenital base of life, roots of Yin–Yang energy, and grows partly based on heritage. People's health management influence the kidney's overall status (Zhang 2010).

The kidney system of TCM is related to modern medicine functions including reproduction endocrine, bones, brain, and spinal nerves, etc. If someone regulates the function of the kidney system, can maintain concentration and a good memory, it would be beneficial to the quality of sleeping, healthy bones and hearing, and prevent early Alzheimer's disease and renal hypertension.

In TCM, the spleen system and stomach system are the roots of the acquired constitution, the source of Qi and Blood. The digestive system, represented by the spleen and stomach includes the function of the pancreas of modern medicine, which is related to endocrine, nutritional and metabolic diseases with high clinical morbidity and mortality, such as diabetes, fatty liver, and obesity.

Once people reach middle age, they start to feel that skipping breakfast or eating too fast affects their physical and mental condition throughout the day. Generally, breakfast should be rich in variety and nutritious, as the entire day's store of Yang energy and one's mental energy is dependent on the various nutrients absorbed in the morning. Lunch should also include sufficient nutritious food, since it maintains people's energy in the afternoon and evening. Dinner may be simple, since eating foods that are hard to digest or overeating can affect sleep.

Research on intestinal microbes in recent years has revealed that the balance status of intestinal flora will influence the central nervous system, reproductive system, endocrine and nutrient metabolism (people's tolerance of food, waste excretion). These studies have shown that human behaviors, emotions, personality characteristics and cognitive abilities are not only regulated by the brain, but also affected by the gut-brain axis; not only by their brain nerves, but also by intestinal microbes. Human pain perception and pain responses, emotions, and cognitive abilities are closely related to intestinal microbes (Collen 2015). TCM believes that the structure and function of the intestine are related to the major systems of the spleen, the liver and the kidney. In other words, whether the microbes are balanced depends on whether people can correctly choose healthy food to prevent food poisoning. The five organ systems defined by TCM also have effects on the digestive function of synergy.

Appetite is one of the key signs of human health. The quality of appetite is not only related to the activities of the digestive system, but also to people's overall physiological function. Many diseases can cause changes in appetite. Actually abnormal appetite is the first symptom of the disease in clinic. Doctors often say that those patients who have a good appetite should recover easily from illness. Another sign of status of normal digestion is bowel movement conditions. For middle-aged and elderly people, a slightly soft stool is better than constipation. People with constipation, such as excessive force during defecation, may induce heart disease. To treat constipation with diet therapies would work. Healthy people can eat vegetables that are more cruciferous. Those with conditions of less teeth, weak spleen energy and stomach dysfunction, can choose root vegetables with short and soft fibers, such as yam, sweet potato, potato, taro, etc. The methods of cooking could be steaming, making soup; cutting vegetables into small pieces to mix with rice, a pancake or a smoothie with turnips, greens, beans, celery vegetables. TCM doctors often recommend that patients with constipation should have hemp seeds, almonds, honey, sesame seeds, and pine nuts.

17.5.3.2 Diet Therapy for Treating Serious Diseases

This section introduces some methods and principles for diet therapy using the example of the top four mortality rates in Shanghai, namely cardiovascular and cerebrovascular diseases, tumors, respiratory diseases, endocrine nutrition and metabolic diseases. Many people suffer from hypertension, hyperlipidemia, heart and brain ischemia, tumors, allergic asthma, insulin resistance, diabetes before they pay attention to health diet. At this stage, TCM doctors prescribe foods and herbs according to the severity of diseases. In the acute stage, doctors diagnose the type of disease and work out the best treatments as soon as possible, supplemented by food therapy.

Patients in recovery stage can accelerate the pace if they can apply both medical treatments and food therapy. For some patients who have to take long-term or lifetime medication such as type 1 and type 2 diabetes, they can introduce diet and herbs as a complimentary remedy to slow the progression of the disease and prevent complications. In the early stage of hypertension without family history, patients can try to replace some medicines with foods and herbs, and postpone the time required for chemical intervention.

17.5.3.3 Diet Therapy for Treating Common Diseases

There are six groups of commonly used foods and herbs of which can protect one from sickness and prolong life. Doctors can apply their functions based on the TCM perspective. The following six categories of foods are classified according to the food property of neutral, cool and warm. Neutral foods are moderate and can be taken daily or weekly for everyone. Cool effect foods are applied to balance hot condition, and warm effect foods to cold condition. (Table 17.1).

(1) Nuts and Seeds

Nuts and seeds are the essence of plants, providing abundant antioxidants and anti-inflammatory elements to people because they are rich in nutrients and contain large amounts of protein, trace elements, and vitamins.

Peanuts are prominent longevity food. Chinese medicine proposes that most nuts and seeds are neutral or warm in temperature. They supply Qi and Blood, strengthen Yin and Yang, moisturize the skin, quench thirst, moisten dryness, remove annoyance, relieve asthma, and reverse Qi reflux. This food remedy is suitable for one who shows signs of has an early onset of aging, and night and frequent urination, fatigue, dizzy spells, dry throat and mouth, dry cough, dry or peeling skin, dry itching, constipation and other similar symptoms.

Those patients who have symptoms of excessive phlegm and congestions, who easily feel hot and sweaty or who are Yin-Yang balanced should eat less nuts and seeds or eat cautiously.

Since most nuts and seeds contain a higher portion of fat, high oil content, easily oxidable, people eat too much of nuts leading to weight gain. It is advisable to consume moderate amounts of nuts and seeds, preferably 20 to 30 g per day.

(2) Fruits

Fruits are rich in beta-carotene and vitamin c, the key antioxidants that help people absorb special enzymes, reduce blood cholesterol levels, and lower the risk of hypertension.

However, fruits are high in sugar and acid. People should not take them as staple foods. Fruits with higher medicinal value can be used both as food and herbs, likewise lemon, cranberries, apples, mulberries, raspberries, pomegranate, schisandra berries, longan fruit, hawthorn, wolfberries, and fresh jujube. Many fruits are often used in their dried form such as lemon, mulberries, raspberries, schisandra berries, longan fruit, hawthorn, wolfberries and jujube. Fruits can

Table 17.1 Property of six groups of foods (Dou 1981; Nanjing University of Traditional Chinese medicine 2006; Tang 2004)

	Neutral	Cool	Warm
Nuts and Seeds	Hazel Almond Sesame Flaxseed Pumpkin seed Sunflower seed Peanut		Walnut Aniseed Pine nut Chestnut Coconut milk
Fruits	Cranberry Wolfberry [a]Fresh jujube	Lemon Apple Mulberry	Raspberry Pomegranate Schisandra berry Longan fruit Hawthorn, [a]dry jujube
Allium			Garlic Garlic/green chive Spring onion /Onion Coriander (cilantro)
Bean products Bean sprouts Fermented food	Yellow, black soybean Green soybean Soymilk [a]Azuki bean [a]Black bean sauce	Tofu Pu'er tea Green tea Mung bean [a]Azuki bean, black bean sauce	Vinegar
Whole grains	Corn [a]Sweet potato Rye [a]Oats	Wheat Barley Millet Buckwheat	[a]Oats Glutinous rice [a]Sweet potato
Algae cruciferous vegetables	[a]Radish seed Chinese cabbage Cabbage	Nori Kelp Marine algae Wakame Bok choy Shepherd's purse [a]Radish	Mustard Mustard seed

[a]Due to differences in origin and production process, there are two temperature possibilities

increase organ functions, stabilize emotions, regulate Yin and Yang, tonify Qi, and nourish the Blood. Those patients who experience dizziness, infertility, early onset of aging, and night and frequent urination can choose the fruit family. They eat the raw and fresh fruits in harvest or in the warm season and the cooked or dried fruits in the other seasons or in cold weather. When their digestive functions are back to normal, patients can eat raw and fresh fruits; while the function weak, eat cooked or dried fruits.

The phrase "an apple a day keeps the doctor away" is a folk proverb that emphasizes the apple's comprehensive conditioning and capabilities of health care. Eating apples on a regular basis not only increases nutrition of the heart and the lung by cooling their heat, but also has a positive impact on aiding digestion. Because apples have both strengthening and regulating functions, they nourish body fluids and remove summer heat, and help relieve restlessness, thirst and irritability. As apples benefit the stomach, they help drinkers sober up from alcohol. Apples have been used to treat lack of appetite, dry mouth, poor digestion, diarrhea and bloating in the abdomen after overeating or overdrinking alcohol.

(3) Allium

Allium plants are also taken as vegetables and condiments, and are particularly effective for the balance of probiotics in the body. Garlic, chives, spring onion, onions, and coriander (cilantro) all contain dietary fibers and organosulfur compounds, can kill viruses and therefore limit fungal infection.

TCM thinks that allium plants are warm in nature and spicy in taste, and can stimulate the body's Qi, Blood, and Yang, increase functions of the kidney, the liver, the heart, the lung and the digestive system, help body Qi movement and Blood circulation, and open the pores and nasal orifices. Spring onion, garlic green chives are commonly eaten daily.

Spring onion has the function of improving immunity on the body surface and regulates the opening and closing of sweat pores. Spring onion is useful for treating cold and flu, and inhibiting the onset of water retention above the chest, as well as relieving cold abdominal pain. This plant also has the function of detoxifying. Spring onion removes irritants from mosquito bites, and prevents constipation and reduced urine. Doctors also apply spring onion to treat light symptom of dysentery and acne.

Garlic warms the digestive system and moves body Qi. It treats food retention, cold sensations pain in the abdomen, and diarrhea. Garlic detoxifies, kills parasites, and inhibits the flu virus, tuberculosis, dysentery and vaginal discharge. It also helps reduce swelling and edema. Garlic juice can be applied topically on the affected area to treat fungal infection of hands, and itchiness, due to athlete's foot. We can eat raw or cooked garlic. Garlic leaves and garlic oil are alternatives for those who dislike garlic.

Chinese medicine suggests that onions have the effects of strengthening the stomach and regulating the movement of Qi, detoxifying insect bites, and killing insects. It is commonly used for treating hyperlipidemia, poor appetite, abdominal bloating, trauma ulcers, and trichomonas vaginitis in China.

Green chives (garlic chives) warm and tonify the liver and the kidney, reinforce Yang energy, strengthen the stomach, enliven circulation, and dissipate Blood stasis. Doctors apply green chives to treat menstrual and discharge problems, sore back and knees, incontinence or polyuria caused by kidney Yang weakness.

(4) Bean products, bean sprouts and fermented food

Beans are rich in protein, fiber, and carbohydrates while being very low in fat and sugar, they are warehouses of nutrients needed by the human body (Mitchell 2009). This includes yellow, black, green soybeans, black beans, azuki, mung beans, sword beans, and peas.

Soybeans and soymilk have the function of strengthening the digestive system and moistening dryness. They are more useful for treating chronic and weak cough, chronic diarrhea, etc. Soymilk can promote weight gain in people who are too thin by improving their digestion. People who have arthritis and joint pain may have some relief by eating proper dosage of soy products. Since tofu contains gypsum, the energy of tofu (as defined by TCM) is cooler than soymilk. Therefore, tofu can be used to treat acute eye infection or the lung infection (asthma and cough with green yellow mucus). Tofu will ameliorate conditions of people who have stomach heat alongside bad breath. Black soybeans have a function of strengthening the digestive system and kidney Yin, helping Blood circulation, reducing toxins and water retention.

Those who do not like beans or cannot consume many beans because of health conditions can eat more fermented foods like sauerkraut, miso soup, kimchi, yogurt, natto, dried black bean sauce and vinegar. The beneficial bacteria inside fermented foods can improve immunity, dispel the onset of the flu, inhibit chronic inflammation, support the digestive system, help Blood circulation, ease discomfort in the chest, and relieve palpitations. Natto kinase and vinegar may also soften Blood vessels and dissolve thrombi.

Sprouts contain high levels of concentrated enzymes that make them easier to digest than that of mature beans. Bean sprouts (soybean and mung bean) are used to treat common warts, reduce heat to resolve early stage of urinary tract infection and relieve alcohol hangover.

The active ingredients in tea (green, black, or pu'er), tea polyphenols, are antioxidant substances. Those residents who consume tea often have lower risk of cancer, indicating that tea polyphenols can eliminate free radicals and therefore prevent cancer.

Some doctors recommend Pu'er tea for taking care of digestive health, especially for people with weight issues or difficulty with digesting mixed protein foods. If there is an infection in the colon, with diarrhea and abdominal pain, Pu'er tea can help by reducing the degree of pain and stopping diarrhea. It can help produce the lining inside of the digestive track to protect excess acidity, and such acidity causes ulcers and bacterial imbalances that can result in infection. When people drink excess alcohol and experience a hangover or liver discomfort, Pu'er tea can help reduce the symptoms and eliminate the liver toxins. It can also help improve the symptoms of indigestion, heavy bloating, and refresh the body and breath. People can use this tea as a mouthwash to eliminate bleeding and soreness of the teeth and gums. Pu'er tea can help the movement of Qi and body fluid, especially in the lung and spleen systems. Drinking it for long period can reduce blood lipids and blood cholesterol, help stabilize blood pressure and prevent artery hardening because it expel mucous and phlegm from inside of organs and Blood vessels. Pu'er cools heat and nourish body fluids, expels summer heat, quenches thirst, and treats constipation and facial pimples (Zhang 2021).

Green tea clears the eyes and head, and aids concentration and alertness. Green tea is a good alternative to coffee for caffeine to aid waking and generate early morning energy. Specifically, early morning tiredness may be caused by blockages in energy flow, which green tea helps to eliminate. It is known to reduce hot feelings in the body manifested as dizziness, thirst and irritability and refreshing the body. Some people, primarily those who have a hot constitution, can use green tea to increase urination, as it helps low output of urine and edema.

(5) Whole Grains

Grains, here mainly rice and cereals, include glutinous rice, wheat, barley, oats, rye, corn, millet, and buckwheat. Grains contain a large amount of soluble and insoluble fibers, B vitamins, and proteins. With mild taste, grains are digestible and absorbable, and an indispensable source of essential nutrients for children and the elderly. Grains can strengthen the spleen and stomach systems. Since properties of grains are multifold, including neutral, cool, or warm, they provide easy choices for individuals and eating in various seasons. For instance, oats and glutinous rice are warm in temperature, so people can have more in the winter.

Oats are rich in protein, calcium, riboflavin, and thiamine, and they are the best cereals of antioxidant food. Daily intake of oats can strengthen the spleen and stomach system, tonifying Qi and Blood.

Buckwheat contains more protein than oat and wheatgrains (Mitchell 2009). Its taste is sweet and its energy is cool. It is used to clear away heat, dampness and toxin, and help in the recovery of people's intestinal function. Both internal and external usage can stop yellow color sweating, treat dysentery, erysipelas and furuncle of heat. The leaves and branch of buckwheat have a function to strengthen the vascular system and prevent bleeding.

Millet is easy to digest, and hence plays a role in benefiting the spleen and stomach system for elderly people who suffer from stomach ailments. Its taste is sweet and its energy cool. If people make millet into form of porridge, there appears a layer of delicate sticky substance on top, commonly known as rice oil. Chinese medicine considers that rice oil is extremely rich in nutrients and has the strongest nourishing power. They even claim that "rice oil can substitute for ginseng soup".

Sweet potato's properties are neutral in energy and taste sweet, and its function is related to the kidney, the spleen systems. Sweet potatoes strengthen the spleen and stomach system by creating energy, smoothing bowel movement, and therefore doctors use it to treat burning sensations in the stomach and esophagus, constipation, bloody stool and diarrhea.

(6) Algae food and cruciferous vegetables.
Seaweed are spore plants, including Marine algae, kelp, nori, and wakame, etc. Seaweed contains high amounts of minerals such as calcium, iron, magnesium, iodine, phytonutrients lignans, B vitamin folate (Mitchell 2009) etc. Seaweed properties are cold, taste are bitter and salty, it enters the liver, stomach and kidney systems. Marine algae and kelp are noted for the ability to dispel phlegm and soften lumps. Patients who have an enlarged thyroid or have lumps, or muscular nodules can use marine algae as a remedy. Middle age or elderly patients whose immune system are weak, warts are more likely to grow, or hernias to develop. Doctors can use marine algae and kelp to treat. Topically use of marine algae can reduce pigmentation, wrinkles and under-eye bags.
Cruciferous vegetables are common vegetables, and including bok choy, Chinese cabbage, mustard, cabbage, shepherd's purse and radish.
Many herbs belong to this category, such as white mustard seeds and radish seeds. Its medical function is prevention of cancer. In Chinese herbal medicine, radish seeds are neutral in temperature, and sweet and spicy in taste. White mustard seeds are hot in temperature, and spicy in taste. Both of them can be used to regulate Qi, and Blood and expel Phlegm. They can adjust the metabolism of Qi, Blood, and the body fluids. Doctors use them to treat indigestion, fullness of the abdomen, dysuria, constipation, and edema.
Shepherd's purse has function of adjustment of digestive, reproductive system. Its medical usage is to stop hemorrhages, treat vomiting blood, blood in the stool, heavy menstruation, prolonged lochiorrhea, and so on.

17.6 Conclusion

Chinese medical philosophy of Yin-Yang and Five Elements provides unique guidance on understanding of food property and body constitutions. Human health is closely related to natural environment particularly through everyone's diet.

The theory of diet therapy in TCM believes that people should apply food property to balance Yin-Yang, and maintain health. Because people have different constitutions, such as age, gender, health status, and live in varied seasons and geographic location, they need to create a justified and tailored diet.

Enhancing the spleen (stomach) and kidney system are most effective way for health and longevity. Recent research in the literature on the gut-brain axis has revealed that human body and spiritual health influenced by the digestive system, particularly intestinal microbes. Therefore, food therapy is very effective and need to be further development from mixed disciplines.

Compliance with Ethical Standards

Conflict of Interest I have no conflict of interest.

References

Collen A (2015) 10% Human How your body's microbes hold the key to health and happiness. HarperCollins, New York

Dou G (1981) Dietary treatment guidelines. Jiangsu Science and Technology Press, Nanjing (Books available only in Chinese)

Leggett D (1994) Helping ourselves: a guide to traditional Chinese food energetics. Meridian Press, Basildon

Mitchell D (2009) The complete book of nutritional healing. St. Martin's Press, New York

Nanjing University of Traditional Chinese Medicine (2006) Chinese materia medica dictionary. Shanghai Science and Technology Press, Shanghai (Books available only in Chinese)

Tang Z (2004) Botanic functional food. Chemical Industry Press, Beijing (Books available only in Chinese)

Yang M, et al translation Li Z, Liu X (2005) Yellow emperor's canon of medicine plain conversation. Xi'an World Publishing Corporation, Shaanxi

Zhang Y (2010) Managing your emotional health using traditional Chinese medicine. Reader's Digest, New York

Zhang Y (2016) Your guide to holistic beauty using the wisdom of Traditional Chinese medicine. Better Link Press, New York

Zhang Y (2021) A comprehensive handbook of Traditional Chinese medicine. Better Link Press, New York

Zhang Y, Yao Y (2012) Your guide to health with foods and herbs using the wisdom of Traditional Chinese medicine. Better Link Press, New York

Zhang E et al (1989) Basic theory of Traditional Chinese medicine (I). Publishing House of Shanghai College of Traditional Chinese Medicine, Shanghai

Zhang E et al (1990) Basic theory of Traditional Chinese medicine (II). Publishing House of Shanghai College of Traditional Chinese Medicine, Shanghai

Zhang E et al (1993) Chinese medicated diet. Publishing House of Shanghai College of Traditional Chinese Medicine, Shanghai

Chapter 18
Indian Traditional Foods and Diets: Combining Traditional Wisdom with Modern Science of Nutraceuticals and Functional Foods

Jyoti S. Gokhale, S. S. Lele, and Laxmi Ananthanarayan

Abstract Indian traditional foods emphasize not only on providing the energy needs of the body but also aim to improve the overall health and prevent diseases in the Indian population. Indian traditional diets which are predominantly vegetarian majorly comprise of whole grains, pulses, nuts, vegetables, fruits, spices, herbs, dairy products, and fermented foods which themselves can be considered as 'functional' being high in a variety of phytochemicals with nutraceutical potential. India, being a diverse country with various cultures, languages, climates, religion, and communities, exemplifies great variety in cuisines and their preparation methods. Along with their intrinsic nutritive nature, the traditional foods also enrich the diet with different bioactive compounds. Many factors have resulted in transition in the dietary habits of the Indian population especially in metropolitan cities going away from traditional dietary practices with direct impact on the health of the people. Therefore, a revival in the eating of traditional diets based on appreciation of their health beneficial properties backed by scientific studies is necessary. Indian tradition also considers foods with functional ingredients to be consumed during various stages of life depending on the physiological needs. The major classes of bioactives/phytochemicals present in the Indian traditional foods along with their health benefits have been included and examples of traditional Indian foods with potential as functional foods have been specified. This awareness of traditional wisdom behind healthful properties of traditional foods has opened up many new avenues for scientific investigations which can lead to validation and proper documentation of the claims made.

Keywords Bioactive · Phytochemicals · Functional ingredient · Indian traditional foods · Food diversity · Health benefits

J. S. Gokhale · S. S. Lele · L. Ananthanarayan (✉)
Department of Food Engineering and Technology, Institute of Chemical Technology, Mumbai 400019, India
e-mail: l.ananthanarayan@ictmumbai.edu.in

S. I. S. Rattan and G. Kaur (eds.), *Nutrition, Food and Diet in Ageing and Longevity*, Healthy Ageing and Longevity 14, https://doi.org/10.1007/978-3-030-83017-5_18

18.1 Introduction

India is known for diversity in terms of language, food, clothing and fine arts like music and dance as well. At present there are 29 states, 8 union territories and a population over 1.39 billion eating nearly 5 billion meals a day. Traditional Indian food preparations are influenced by availability of raw materials and climate. People of the northern states of India are generally wheat eaters whereas, people belonging to the southern parts daily consume rice as staple food. Across India pulses (whole and split), lentils, millets and sprouts are consumed regularly, although some of the ingredients used and method of preparation are quite different in various regions. Since India ranks second in the world in fruit and vegetable production, Indian traditional diets predominantly include plant sources. Although a significant fraction of the population consume fish, meat, eggs, poultry, and other non-vegetarian foods, rarely does anyone follow an exclusively non-vegetarian diet. This is perhaps due to limitations in the availability of fish and meat and sometimes due to high cost of these items. Further there are restrictions on consumption of such products due to religious beliefs and traditions. Interestingly, milk although scientifically a commodity of animal origin, is considered as "vegetarian" not just by tradition but also by the Indian Food Law. Perhaps India is the only country in the world where a processed food of animal origin has a red dot and vegetarian food has a green dot used as a symbol on the package, but milk is a "green dot" product!

The combined effect of these factors is that Indians have been consuming lots of fruits and vegetables, drinking water as the main beverage and consuming milk and dairy products in their regular diet. Another interesting fact is regular consumption of yoghurt-like fermented milk product, called *dahi*. Even today, in most rural parts of India, *dahi* is made in every household and consumed daily during meals. Some prefer to consume it in the form of *chhas* or *lassi* which are beverage forms made from *dahi*. This probiotic food product is rich in *lactobacilli* and forms part of daily diet across India in all states typically consumed during meals. Survivability of the probiotic culture in acidic pH of the stomach and further in conditions prevailing in the gut to reach the colon in an active form is probably aided by consuming these cultures as part of a meal. Most of the Indian festivals and religious rituals are linked with the season and there are traditions which recommend a specific food and diet to mark the celebration. Just to cite one example, in January when the Sun enters Capricorn zodiac, it is celebrated as *Makar Sankranti* festival. This day marks the end of winter solstice during mid-winter period in India and is celebrated by different names across India, but the commonality is that all the states have a tradition of consuming lot of sesame seeds and jaggery in various formats. For example, '*tilgul*' a popular traditional confection of Maharashtra which is sentimentally exchanged by one another during *Makar Sankranti* festival is made from sesame seeds and jaggery which gives a good supply of energy from sugar and oil and are suitable for the winter season. Thus, Indian traditions nurture both values and good habits for healthy living.

Indian people also follow the dietary recommendations proposed by Ayurveda. The literal meaning of Ayurveda is "knowledge of life". There are three main books in Ayurveda written by the great authors *Charak*, *Sushrut*, and *Vagbhat* (Anonymous 2021). In addition, there are few more documents written over the centuries. All this literature is in Sanskrit language, mostly in poetic form for ease of memorization. The first and largest book is by *Charak* which focuses on "fire of digestion". It is a collection of knowledge accrued over thousands of years. Ayurveda describes diagnostic methods, botanical extracts in various forms to be used as medicine, body cleansing methods, surgery, and many other aspects. Ayurveda is not just alternate medicine, but it also talks of personalized food based on constitution of each individual, food based on climate, food for patients with specific ailment and food for different age groups. Main thrust of Ayurveda is on preventing diseases and maintaining health by balancing diet with lifestyle and personal constitution (Lad 2017). In fact, Ayurveda has also described correct way of cooking food so as to get maximum benefits (Lad and Lad 2010). This suggests that although the term nutraceutical is recently coined ancient Indian traditions had similar concept of food being used as medicine!

The term nutraceutical was coined by Stephen DeFelice, founder, and chairman of the Foundation for Innovation in Medicine (FIM), in 1989 from "nutrition" and "pharmaceuticals" (Kalra 2003). According to DeFelice, nutraceutical can be defined as, 'a food (or part of a food) that provides medical or health benefits, including the prevention and/or treatment of a disease'. The health benefits of functional foods are mainly because of the bioactive compounds such as phytochemicals, vitamins, peptides, found naturally in them, or extracted from other sources and added into them, whereas nutraceuticals are themselves concentrated forms of the bioactive substances which show disease preventive or curative effect. The terms 'bioactive' or 'functional ingredient' are used to represent a substance with significant biological activity having a positive impact on health. Other terms, such as phytochemicals or phytoconstituents, represent natural chemical entities of plant origin. Nutraceuticals are provided in pharmaceutical form whereas functional foods are given in food matrix. One of the important prerequisites, both in nutraceuticals and functional foods, is that they should provide the functional ingredient in sufficient quantity to prevent or cure a disease. In general, there is no need for doctor's recommendation for taking functional foods and they can be a part of our daily diet, whereas it is advisable to consult a doctor before taking any nutraceutical preparation. For example, if anyone wants to consume probiotic curd in their daily diet, they need not take it under medical supervision but if an individual wants to take Vitamin D supplement, it is preferable to be taken under doctor's advice.

In the last few decades, transformation in the food habits, reduced physical work, and more sedentary jobs have led to many lifestyle-associated diseases such as diabetes, obesity, hypercholesterolemia, hypertension to name some. Nowadays, malnutrition is not only seen in economically backward people but also in urban and affluent sections of society due to lack of timely and nutritious foods which has led to micronutrient deficiencies (Rishi 2020). Deficiencies of calcium, vitamin B12, vitamin D, iodine are very commonly witnessed in the population which has increased the importance of nutraceutical and functional food market. Moreover, the technological advances, sociodemographic profiles, changing agricultural practices, lifestyle changes especially in metro cities and availability of processed foods has led to a transition in the dietary habits of people away from traditional eating patterns to consumption of junk foods and refined foods loaded with empty calories. This has further strengthened the need for foods with increased health benefits such as fiber containing nutritious foods having fewer calories, with low glycemic response and a high satiety value. Thus, these are the major drivers in the growth, current development and production witnessed in the global functional food and nutraceuticals market. The age-old sayings “Let food be thy medicine and medicine be thy food” and “Prevention is better than cure” are gaining new perspectives once again in the minds of consumers nowadays. Thus, nutraceuticals and functional foods are now looked upon as important items on the monthly expenses list, indispensable from health viewpoint.

The global nutraceutical market is estimated to reach from $247 billion in 2019 to $336 billion in 2023 with compound annual growth rate (CAGR) of 8%. Currently, USA, Japan, and Europe share 90% part of the global market whereas India’s, global nutraceutical market share of 2% in 2019 is expected to reach 3.5% by 2023 with CAGR of 21% and a total estimated valuation of $11 billion (Rishi 2020). Thus, nutraceuticals and functional foods are set to become essentially accepted and a regular part in the Indian household in the near future.

The recent outlook on foods which has shifted from merely providing sustainable life and growth to providing preventive rather curative action against various forms of diseases has been a conventionally recognized belief in the Indian traditional food system. Thus, recent recognition of the role of bioactives in the form of nutraceuticals or the role of functional foods truly aligns with the design of Indian traditional foods. Indian traditional foods emphasize not only on providing the energy needs of the body but also aim to improve the overall health and prevent diseases in the Indian population. The Indian traditional wisdom of food choices can easily be correlated with the buzz word of modern science such as “nutraceutical” or “functional food”. Traditional wisdom can help to understand or claim the benefits of different foods, but this must be validated by supporting it with robust scientific evidence and then making these foods available in consumer acceptable and convenient formats. For example, *ashwagandha* root which has been prominently advised to be important in improving strength and vitality is now getting acceptance in sports nutrition. Similar is the case with *Haldi* milk which is part of a traditional Indian diet and consumed for its health benefits but is now getting globally accepted as ‘*turmeric latte*’. Thus, in both the examples, it is traditional knowledge which has led to a way of offering the health benefits in a new format. This chapter gives a bird’s eye view about the traditional wisdom of the Indian foods which has been in existence for more than 5000 years. Thus, it is an attempt to educate the reader about the nutritional preeminence of many Indian traditional foods backed by scientific evidence. It also covers recent literature about the bioactive substances found in Indian traditional foods which are predominantly plant based and which today are being acknowledged as nutraceuticals or functional foods.

18.2 Indian Traditional Foods

Traditional foods of any country define the culture of the nation and are closely linked to the health and wellbeing of the population. They also reflect the agricultural patterns, traditional beliefs with respect to festivals, seasonal variability, availability, and accessibility to foods along with processing/preparation techniques followed through ages. The Indian traditional food system includes thousands of Indian cuisines, region specific foods, based on understanding of what to eat when, and guided by knowledge of Ayurveda and use of herbs for medicinal benefits which has existed for centuries. Therefore, Indian traditional wisdom is deeply entrenched in promoting health. Along with this, Indian cooking also considers cuisines for different seasons, aptness of the cuisine according to time of the day, pre-existing health condition and even mood. Existing traditional wisdom behind Indian foods thus offers scope for undertaking proper scientific investigations in order to strengthen the health claims associated with them.

Indian traditional diets majorly comprise of whole grains, nuts, pulses, vegetables, fruits, spices, herbs, milk, dairy products, and fermented foods which themselves can be considered as "functional" being high in fibers, antioxidants, micronutrients, probiotics, prebiotics which are the popular choices for health campaigns nowadays globally. Many Indian traditional foods impart significant physiological health benefits ranging from improving immune system, and gut health, while reducing oxidative stress, cholesterol, risk of heart diseases, inflammatory diseases and demonstrating anti-obesity effect. The functional (bioactive) components abundantly available in Indian traditional foods which positively impacts human health can be in the form of vitamins, minerals, dietary fibers, lignin, polyphenols, flavonoids, fatty acids, essential amino acids, peptides, lactic acid bacteria to name some.

Indian diets are rich in plant-based ingredients and constituents with potential nutraceutical benefits as we understand today. They have existed for centuries even before knowledge of 'nutraceuticals' and 'functional foods' was recognized. Thus, it is necessary for today's global generation to have knowledge of the heritage of Indian foods. This chapter mainly focuses on the integration of traditional wisdom of Indian foods with modern sciences using existing scientific literature as a base. Hence, the traditional and ethnic foods with their nutritional and health beneficial aspects unleash a colossal opportunity for many researchers throughout the world and provide for the possible reach of traditional foods in the health sector through conduct of sound scientific investigations.

India, being a diverse country with various cultures, languages, climates, religion, and communities, exemplifies great variety in cuisines and their preparation methods. Predominantly Indian population follows a vegetarian diet though some regions especially the coastal regions follow a non-vegetarian diet where the availability of fish and other marine foods is abundant. But in India even those who follow a non-vegetarian diet eat such foods occasionally while a greater part of their diets includes vegetarian food sources. The Indian diet therefore comprises of a healthful fusion of many food ingredients of plant and animal origin. Indian traditional foods

Table 18.1 Some examples of region-specific Indian Traditional Foods. (Ananthanarayan et al. 2019)

Meal type	Food items
	North Region
Breakfast	*Chole Bhatura, Kulcha, Daal Puri, Daal phara, Aloo tikki*
Staples	*Chapati/ Roti, Naan roti, Paratha, Rumali roti,* Stuffed *paratha, Makke di roti,* Plain rice, *Vegetable pulav, Khichdi, Aloo curry, Baingan bharata,* Stuffed vegetable, *Saag, Murgh Masallam, Paneer tikka masala, Tandoori* fish *tikka, Daal fry* with *tadaka, Kadhi pakod, Rajma*
Savouries	*Kachori, Samosa, Mathri, Besansev, Golgappe*
Confections	*Sooji halwa, Petha, Shahi tukra, Phirni, Kheer, Jalebi, Gajar ka halwa, Panjeeri*
Beverages	*Shikanji,* Sugarcane juice, *Jal jeera, Lassi, Chai*
	East Region
Breakfast	*Cheera Doi, Pitha, Luchi cholar dal, Ghoogni, Alu pithka, Mangsher singhara, Momos*
Staples	*Luchi, Moghlai porotha, Petai porotha, Komol chawal, Ghee bhat, Khalo daal, Til ko alu, Bengena pitika, Begun bhaja, Macher paturi, Muri ghonto, Mochar ghonto, Macher jhal*
Savouries	*Chanachur, Chaal bhaja, Daalmuth, Chira bhaja, Kardoi, Nimki, Gur tokti*
Confections	*Balooshahi, Chhena gaja, Chhena poda, Goja, Gaja, Gokul peethe, Malpua, Mishti doi, Pantua, Patisapta, Puli pithe Payesh, Peda, Rosogulla, Nolen gurer shondesh, Thekua*
Beverages	*Gondhoraj ghol, Aam pora shorbat, Bael shorbat, Sattu sherbet*
	South Region
Breakfast	*Dosa, Iddiyappam, Idli, Nandu* omelette, *Puttu, Kuzi paniyaram, Upma, Uttapam, Medu wada, Mysore bonda*
Staples	*Akki roti, Jolada roti, Obbattu, Parottha, Bisibele bhat, Hyderabadi biryani, Pongal, Puliyodharai*
Savouries	*Murukku, Kara boondi,* banana chips, Tapioca chips, *masala* groundnut
Confections	*Payasam, Sakkara Pongal, Jangiri, Mysore pak, Unniyappam*
Beverages	*Nannari sherbet, Panakam,* Buttermilk, *Neera, Filter kaapi*
	West Region
Breakfast	*Pohe, Dhokla, Khandvi, Thalipeeth, Pav bhaji, Vada pav, Kombdi vade*
Staples	*Bajri no rotlo, Bhakari, Jawar no rotlo, Pooranpoli, Amti, Ambottik, Bombil* fry, *Crab xecxec, Daalithoy, Turiyapatra subji, Undhiyu, Veg kolhapuri, Vindaloo*
Savouries	*Chakali, chivda, chorafali, khakra, thepla, farsipuri, fafda*
Confections	*Laddu, Modak, Mohan thaal, Shankarpali, Shira, Shrikhand, Soanpapadi, Sukhdi*
Beverages	*Chhas, Panna, Feni, Kokam sherbat*

which have evolved over a long period of time include staple foods from cereal based items; accompanying dishes including vegetable and legume/*dal* preparations, pickles, chutneys, *papad*, *ghee* or oil as a medium of cooking, foods prepared from fermented batter, milk and nonmilk-based sweets and various snack items. Table 18.1 enlists some examples of different food items consumed under different meal types based on specific parts of the country.

In general, the meal patterns in India are comprised of fiber-rich foods such as *roti*, *chapatti*, fruits and vegetables, protein sources such as legumes/*dals*, energy sources such as rice with moderate quantity of vegetable oil/ *ghee* (clarified butter) used in the preparation, a variety of spices used for flavoring and curds as a source of probiotic. It can be observed that region wise there are differences in the types of food consumed and their preparation methods. Rice (*Oryza sativa*) consumption predominates in South India whereas North Indian staple foods are prepared from wheat (*Triticum aestivum*). Millets are preferred by few of the states where these crops are grown and harvested.

Though the Indian traditional food is a nutritious way of following a healthy lifestyle, the globalization of food chain, increase in disposable incomes, enormous increase in number of processed foods available, easy access to fast foods and other factors have resulted in transition in the dietary habits of the Indian population with direct impact on the health of the people. Nutritional quality of the diet is decreasing at an alarming rate which is adversely affecting the health of the population. Thus, there is an imperative need for educating the consumer about negative impact of changed dietary habits and the nutritional superiority of many traditional foods supported by scientific evidence.

18.3 Indian Traditional Foods as Functional Foods

Nutrition is a globally dynamic area of scientific investigation which commences with history, geography, culture, resources, and genetics and goes till the development of different food products with their nutritional benefits for the consumer and overall communities. Internationally, food tradition is playing important role in this type of research, but it has always been considered as a belief-based tradition without any scientific evidence about any of the nutritional or health aspects. Recently, the interest in traditional and ethnic foods has been increasing leading to increase inclaims about their nutritional and health promoting benefits. Nowadays, the preparation of traditional foods has deviated from the ways it used to be prepared in the earlier times. In principle, there is a body of knowledge about the traditional food, but it is strewn globally and some or most of it is in the minds of persons belonging to older generation or in the form of notes maintained by these people. This knowledge about traditional foods and its benefits may have been reported or published in different locally published magazines, books, or journals but it would be hard to salvage this if at all it is retrievable. Thus, much of this knowledge will dissipate if not captured in time. For a long time, the immense experience and knowledge gained by many inhabitants who often for good reasons were accustomed to certain diets has been overlooked. It is only in the past decade or so that scientists have seriously started their research to find out about the claims made for traditional foods based on the food composition and their preparation methods.

Indian traditional diets mainly comprise dishes prepared from cereals, millets, pulses/ legumes, nuts, fruits, vegetables, spices and herbs, food acidulants, milk and milk products. Cereals, millets, and pulses have been an integral part of Indian foods and are majorly used in the preparation of all meals of the day. Whole grain cereals and millets provide macronutrients viz. carbohydrate, fats, and proteins along with vitamins and minerals. Oilseeds/nuts provide fat soluble vitamins A, D, E, K and water-soluble B vitamins, minerals, and protein. Staple foods are mainly prepared from a number of different cereals and millets whereas a variety of legumes and vegetables are used to prepare the accompanying dishes. A variety of fruits are consumed as such as per seasonal availability without processing. Acidulants, spices and herbs are added in most food preparations to provide taste and aroma to the preparation. Cow milk is essentially traditional to Indian diets especially for infants and children, but adults too consume it universally. *Dahi* which is a fermented curd made from milk is consumed during meals.

18.3.1 Cereals and Millets

Whole cereals with their low glycemic index help in reduction of blood sugar levels, and they further aid in blood pressure regulation, thyroid, cardiovascular and celiac diseases. In India pearl millet (*Pennisetum glaucum*), sorghum (*Sorghum bicolor*), finger millet i.e., *ragi* (*Eleusine coracana*), little millet (*Panicum sumatrense*), foxtail millet (*Setaria italica*), kodo millet (*Paspalum scrobiculatum*), barnyard millet (*Echinochloa esculenta*) and borso millet (*Panicum millaceum*) are cultivated from ancient times. In India, these millets are consumed in the form of breads known as *bhakri* (Maharashtra), *rotlo* (Gujrat), and *roti*. Also, these can be served in the form of *khichdi* prepared with legumes, spices, and *ghee*. In South India, fermented form of millets is consumed as *dosa, idli, appam* and *uthappam*. In Karnataka, *ragi mudda, jolyade rotti* are quite popular. In Maharashtra, millets are also enjoyed during harvest festival wherein they are roasted and consumed with spices and lime juice (Samarth et al. 2018). Millets are also consumed in the form of *Thalipith* which is a multigrain spicy flat bread. In some regions of India, millets are consumed in the form of *laddus*, *chakli* and *chat* (Singh et al. 2016). All these dishes with specific preparation procedures are super healthy and can be served to any age group. Consumption of millet-based food preparations provide additional benefit by being gluten-free which can be tolerated by individuals suffering from celiac disease.

Millets are endowed with phytochemicals such as flavonones, tannins, and flavonoids. Finger millet is rich in polyphenols along with high content of calcium, and dietary fiber (Mathanghi and Sudha 2012). The phytochemical constituents reported in pearl millet are tannins, steroids, phenols, alkaloids, terpenoids, cardiac glycosides, and balsams. Kodo millet is found to have good phenolic and flavonoid content (Sharma and Saxena 2016). Foxtail millet is one example whose composition is not very different from quinoa (*Chenopodium quinoa*) and is one of the oldest surviving crops, highly adaptable to climatic changes as well as drought resistant. Thus, instead of looking at quinoa as a superfood, contemporary recipes with foxtail millet can be developed as a functional food.

Ambali, a finger-millet based fermented semi-liquid product, is a food product from southern India especially Karnataka and Tamilnadu. This is considered as a geriatric product due to high calcium and low resistant starch content of the finger millet. It is consumed as such or with buttermilk added for taste and its consumption is preferred in the summer season. The leucine to lysine ratio in the finger millet is an indicator of its pellagragenic character. Pellagra is a disease caused due to niacin deficiency with symptoms of diarrhea, dermatitis, and dementia. The fermentation carried out during the preparation of *Ambali* decreases the leucine to lysine ratio thereby reducing the pellagragenic character and increases the concentration of thiamine, riboflavin and tryptophan and bioavailability of minerals (Sarkar et al. 2015).

Poitabhat of Assam, *Panta Bhat* of Bengal and *Pokhalo* of Odisa are popular fermented rice-based foods consumed during lunch and breakfast. This is consumed with water along with cooked vegetables. The fermentation process of rice is reported to increase vitamin B complex, and vitamin K. Fermented sour rice is energy dense and body rehydrating food and it also controls the bowel movement and prevents constipation. About 100 g of sour rice was found to contain 73.91 mg iron, 303 mg sodium, 839 mg potassium, and 850 mg calcium (Ray et al. 2016). While diverse food cultures or habits exist in the Indian subcontinent, where people in different regions consume different kinds of staple foods based on crop availability locally, yet every food brings in its own bioactives and health beneficial properties.

Khichdi is a good example of an Indian one pot meal and is a highly sought comfort food. It is prepared by cooking rice and *dal* together and tempering with selected spices. On the other hand, the Indian *thali* is a multiple course meal served all together comprising of staples such as rice, *roti,* and *dal* with many accompaniments of vegetables, salads, legume-based preparations, and curd-based preparations. In Indian *thali*, legumes have importance as they are a main source of proteins mainly for the vegetarians (Ananthanarayan et al. 2019). Further complimentary food ingredients which are used in one meal to fulfill the needs of essential amino acids include combinations such as rice + *dal* which arethe staple foods of India.

18.3.2 Legumes/ Pulses

Legumes are good sources of proteins (20–45/100 g), dietary fibers (30.5/100 g FM) and minerals like calcium, magnesium (56 mg/100 g), potassium (955 mg/100 g), phosphorus (451 mg/100 g), magnesium (122 mg/100 g) and iron (7.5 mg/100 g)

(Faris et al. 2013). Recent literature shows that bioavailability of phytochemicals present in the legumes can be increased by soaking, fermenting, and sprouting, and Indian legume dishes are prepared by using one of these methods (Faris et al. 2013). Legumes like chickpea (*Cicer arietinum*), pigeon pea (*Cajanus cajan*), green gram (*Vigna radiata*), black eye beans (*Vigna unguiculata*), kidney bean (*Phaseolus vulgaris*), black gram (*Vigna mungo*), red lentil (*Lens culinaris*), split Bengal gram (*Cicer arietinum*), green peas (*Pisum sativum*), white peas (*Lathyrus sativus*), horse gram (*Macrotyloma uniflorum*) are majorly used in India.

In different parts of India, legumes are consumed in different forms such as in south region legumes are generally used in *sambar* and pulses in combination with cereals are used for the making of popular fermented foods such as *idli, dosa, uttapam*. The combination of cereal and pulses can be seen to be complimentary as rice lacks the essential amino acid lysine whereas pulses lack in methionine. The fermentation of *idli* batter is known to significantly reduce antinutrients (phytates and tannins) associated with the legume which in turn increases the bioaccessability of zinc and iron (Hemalatha et al. 2007). This is improved by virtue of the formation of organic acids during the fermentation, which form soluble ligands with minerals, thereby preventing the formation of insoluble ligands of minerals with phytates. Few reports suggest that the preparation of *idli* with parboiled rice unlike regular rice can help to preserve the vitamins. Thus, *idli* prepared with parboiled rice fermented with right proportion of black gram, and then steamed can deliver healthy and tasty meal option (Srinivasan 2010). The main process of leavening in *idli* batter formation is brought about by heterofermentative lactic acid bacteria while the acidity is achieved by homofermentative bacteria. The literature also suggests that fermentation increases the levels of amylase, protease, total acids, soluble solids, lysine, cysteine, methionine, nonprotein nitrogen, soluble vitamins such as folate, vitamin A, vitamin B1, vitamin B2, and vitamin B12 content, with reduction in antinutrient phytic acid (Ray et al. 2016).

Glycemic index (GI) is an important nutritional concept which plays significant role in the physiological effects of foods in relation to diseases like cardiovascular diseases and *Diabetes mellitus*. Glycemic load (GL) reflects the total glycemic impact of ingested carbohydrate-rich diet. Studies have shown that along with lesser physical activity, foods with high GI and GL are responsible for risk of Type II Diabetes, weight management and prevalence of childhood obesity. Many Indian traditional staple foods are high in GI and GL, but combinations of different dishes are shown to lower the GI of the overall meal and thus play important role in managing good health. For example, *idli*, a high GI food (68), when consumed in combination with *sambhar*, which is low in GI (24), lowers the overall GI and GL of the meal. Similarly, *dosa*, high in GI (62), in accompaniment with mint *chutney* (GI = 27) decreases the GI and GL of the meal (Ruchi et al. 2014). Further the ingredients used in the preparation of these accompaniments are also having bioactives with nutraceutical properties as described ahead in this chapter.

In the last decade, lot of research has been carried out to explore the health benefits of legumes. The research shows that legumes are power sources of nutrients which help to boost immunity, reduce the risk of cardiovascular diseases, few types of cancers such as colon, breast and prostate and help to manage obesity due to their satiety value (Faris et al. 2013). Few examples can be considered in this regard. Green pigeon pea seeds are considered superior to dry splits in nutrition hence in western part of India (Maharashtra and Gujrat), these legume pods are boiled, and seeds are eaten during celebration of harvest season. It contains phytochemicals like alkaloids, flavonoids, tannins, saponins, and terpenes (Oke 2014; Saxena et al. 2010). Besides a high nutritional value, pigeon pea is also used in traditional folk medicine in India, China, Philippines, and some other nations. Literature on this aspect shows that pigeon pea can prevent and curing a number of human ailments such as bronchitis, coughs, pneumonia, respiratory infections, dysentery, menstrual disorders, sores, wounds, abdominal tumors, tooth ache, and diabetes (Saxena et al. 2010).

Chickpea also known as *chana,* is another example of a legume which can be discussed as India is a leading producer of chickpea i.e., 66% of the world production is from India (Rachwa-Rosiak et al. 2015). It is a power source of proteins and is consumed in the form of *bhuna chana*, *chana chat*, *pakode/ bhajis* and variety of *farsans.* It contains 20.9–25.27% (d/w) of protein, 6.49–9.94% (d/w) of crude fiber, essential amino acids (39.89/100 g protein) and endogenous amino acids (58.64/100 g protein) (Rachwa-Rosiak et al. 2015). It also contains lectins and agglutinins which can be anticancer, immunomodulatory, anti-obesity in nature (Nida et al. 2017; Sharma et al. 2019). In North India, *sattu,* made from chickpea, is consumed as a healthy drink which is also known as '*Desi Horlicks*'. This *sattu* can be consumed as raw powder or can be used for making different recipes (Sharma et al. 2019). It has high protein percent, high antioxidant content and phenolic content (Nida et al. 2017). *Besan laddus*, *puranpoli* are few chickpea-based traditional confections of India.

Kulith pithala and *Kulith chilka,* traditional recipes from coastal Maharashtra, *Huralisaru* with rice from Tamilnadu and *Muthirathoran* from Kerala are few examples of foods made from horse gram. Horse gram is an excellent source of protein (15–30%), carbohydrates (50–60.9%), essential amino acids, carotene, phosphorus, iron, and vitamins such as thiamine, riboflavin, niacin, and vitamin C (Prasad and Singh 2015). It helps to treat kidney stones, urinary diseases, piles, common cold, throat infection, and fever (Oke 2014; Prasad and Singh 2015).Coconut based cow pea (*Vigna unguiculata*) curry, cow pea rice, *vanpayar thoran* from south India contains cowpea as a main ingredient whichhas 23–25% protein and 50–67% carbohydrate. Presence of significant amounts of protein, calories, and some water-soluble vitamins, makes cowpea a promising food ingredient (Devi et al. 2015). *Chakali* is one of the traditional foods of central region of India, which is made from *harbara dal*, rice, *mung dal*, and *uraddal* (Jagdale and Ghodke 2020). It can be considered as a very good example of multigrain traditional food product existing even before the advent of multigrain *atta* or multigrain bread which are recently making a place in many food markets.

Sprouting is one of the processing techniques traditionally used in India and it has been shown to make all the legumes more nutritious and easier to digest (Devi et al. 2015). Melatonin is a neurohormone with antioxidant properties and its secretion is influenced by nutritional factors, meal timing, diet type along with light–dark cycle. Sprouted lentils (*Lens culinaris*) have shown to increase concentration of bioavailable melatonin. Thus, intake of sprouted lentils can modulate oxidative stress and prevent aging and related diseases (Rebollo-Hernanz et al. 2020).

18.3.3 Nuts

Nuts have been part of Indian diets for ages. For example, the commonly consumed peanut (*Arachis hypogaea*) contains vitamin E, chlorogenic acid, caffeic acid, coumaric acid, ferulic acid, flavonoids, tocopherols (α, γ and δ), tocotrienols (γ and δ) and stilbene (Davis and Dean 2016; Asibuo et al. 2008). Less commonly consumed nuts are almonds (*Prunus dulcis*), cashew nuts (*Anacardium occidentale)*, hazelnuts (*Corylus*), walnuts *(Juglans regia)*, brazil nuts (*Bertholletia excelsa*), pistachios (*Pistacia vera*). Major bioactives present in the nuts are fatty acids, fibers, vitamins, minerals, phenolics. The important fat soluble bioactives such as fatty acids, tocols, phytosterols, sphingolipids, carotenoids, alkyl phenols are present in tree nuts. Tree nuts have shown to reduce the risk of cardiovascular diseases, hypertension and have cholesterol lowering effect (Alasalvar et al. 2020).

Coconut (*Cocos nucifera*) based cuisines are majorly consumed in the coastal region of India. Different dishes like coconut-based gravies, coconut *laddu*, coconut milk-based *solkadhi*, coconut *chutney* are few examples. The fat content of coconut elevates the high-density cholesterol, a good cholesterol, known to reduce the risk of cardiovascular diseases and fiber content exhibits anti-obesity and antidiabetic effects. Coconut also showed anticancer and antilipidemic properties with its protein, polyphenols, and flavonoid content (Kaur et al. 2019).

18.3.4 Fruits and Vegetables

Climatic and physio-geographical conditions in India are suitable for growing various kinds of horticultural crops like fruits, vegetables, nuts, spices, plantation crops. Over the last decade, production of fruits has increased from 50.9 to 97.35 MT since 2004–05 to 2017–18. Globally, India ranks second in the production of fruits and vegetables (Glance 2018). While a great variety of tropical and semi-tropical fruits are cultivated on a large scale, there are some less explored fruits with nutraceutical potential a few of which are discussed below.

The amla fruit (*Phyllanthus emblica*) is consumed in raw, cooked, and pickled forms wherein *amla* candy, *amla* squash, *amla* barfi and *amla* juice are consumed by Indian population as health foods and a natural source of ascorbic acid. The fruit extract has shown antilipidemic, antidiabetic, anti-inflammatory, anticancer, and antiulcer activities. The high content of vitamin C and polyphenols makes it a medicinal plant and it is also used in ayurvedic preparations like *Chavyanprash* and *Trifala* (Sarkar et al. 2015).

Bael fruit (*Aegle marmelos*) is another therapeutically important fruit which contains many bioactives like alkaloids, terpenoids, coumarins, tannins, phenyl-propanoids, flavonoids, and polysaccharides. It shows different therapeutic effects like antibacterial, anti-inflammatory, hepatoprotective, antifertility and antiarthritic. The extracts of the fruit can be used in the preparation of different functional foods and nutraceuticals (Shashikumar et al. 2018). *Jambul* (*Syzygium cumini*), also known as *Jamun, Jambol*, is a seasonal fruit of India. Barks, seeds, and leaves of *Jambol* have shown to possess antidiarrheal, antibacterial, and anti-inflammatory effects. The extract of *Jamun seed* demonstrates antidiabetic effect (Bitencourt 2020).

Raw banana (*Musa acuminata*) is consumed in the southern region of India as a vegetable preparation. Studies have shown that raw banana products improve gastrointestinal symptoms, glycemic metabolism, weight control and renal, liver complications related to diabetes. Thus, it can also be considered as a functional ingredient in different functional foods (Falcomer et al. 2019).

India has 12% share in the worldwide production of vegetables and 5–6% of total vegetable production is contributed by gourd (Cucurbitaceae) family. In India, most consumed vegetables from this family are bitter gourd (*Momordica charantia*), bottle gourd (*Lagenaria siceraria*), ash gourd (*Benincasa hispida*), snake gourd (*Trichosanthes cucumerina*), ridged gourd (*Luffa acutangula*) and pumpkin (*Cucurbita pepo*). Ash gourd and bottle gourd have shown to prevent cancer and cardiovascular diseases whereas bitter gourd plays an important role in the prevention of neurodegenerative diseases. Literature suggests that bottle gourd and bitter gourd can be used to cure Diabetes mellitus, obesity, hyperlipidemia, and viral and bacterial infections (Palamthodi and Lele 2014).

Bitter gourd is widely consumed vegetable preparation in Indian diet though its bitterness limits its use. It contains charantin, polypeptides and vicine as major bioactives which have been shown to have antidiabetic effect. Research investigations have been conducted to reduce the bitterness of the bitter gourd extract and improve the stability of the bioactives. Nanoliposomes have been studied as a nutraceutical delivery system of bitter gourd bioactives (Erami et al. 2019).

18.3.5 Spices and Herbs

Spices play a vital role in flavoring of Indian traditional foods. A perfect blend of aromatic spices in any cuisine is a critical component of its preparation. Mustard seeds (*Brassica nigra*), cumin seeds (*Cuminum cyminum*), asafoetida (*Ferulaassafoetida*), turmeric (*Curcuma longa*), chili pepper (*Capsicum frutescens*), black pepper (*Piper nigrum*), fenugreek seeds (*Trigonellafoenum-graecum*), curry leaves (*Murraya koenigii*), ginger (*Zingiber officinale*), garlic (*Allium sativum*) are majorly used spices in the Indian cuisine. *Garam masala* which is a combination of five different spices and *sambhar masala* powder are commonly used spice blends in

Indian foods. These spices create hot, sour, sweet, savory, and aromatic sensation in one meal. Some spices are used while cooking, whereas others are used as a seasoning. Along with providing flavor, taste and aroma to the diet, spices are shown to produce positive physiological effects on the human body. Since the last three decades, experimental documentation of the beneficial effects of spices is ongoing. The most common effect of spices is its stimulating action on digestion. Based on this, spices are used in traditional medicine for correcting digestive disorders. They are shown to stimulate bile acid production which helps in fat digestion and absorption. Few of them show stimulation of digestive enzyme such as lipase. Thus, there is abundant literature demonstrating the health beneficial effects of spices used in Indian traditional foods such as improving gut health, immune function, in weight management, in reduction of blood cholesterol and reducing the risk of heart diseases (Srinivasan 2010). Thus, with 5000 years long history of use, spices can be considered as first-ever recorded functional foods and they can be used for extraction of bioactives/ functional ingredient for use in nutraceutical preparation.

It has been reported that Asian population shows lesser incidences of neurodegenerative diseases as compared to the rest of the world who do not consume as many spices in their daily diet. Literature supports the fact that nutraceuticals derived from spices such as turmeric, black pepper, red pepper (*Capsicum annuum*), clove (*Syzygium aromaticum*), ginger, garlic, coriander (*Coriandrum sativum*), cinnamon (*Cinnamomum verum*) prevent neurodegenerative disorders by targeting inflammatory pathways. Out of these, majorly curcumin from turmeric is vastly explored and available in different commercial forms. There is scope for exploring other spices for the production of new generation nutraceuticals whose benefits can be established with scientific evidence (Peter and Babu 2012).

Turmeric is majorly used spice in Indian staple foods. Curcumin, the main bioactive component of turmeric, has shown a wide spectrum of pharmacological effects such as anti-inflammatory, antimicrobial, antirheumatic, immunomodulatory and anticancer effects (Afolayan et al. 2018). It also shows antioxidant properties. The major challenge of using curcumin as a nutraceutical is its poor oral bioavailability. Research showed that co-administration of curcumin with natural compounds such as piperine from black pepper, quercetin from different fruits and vegetables and silibinin from milk thistle (*Silybum marianum*) decreased metabolism of curcumin by enzyme inhibition and thereby increased the amount of curcumin available for absorption (Liu et al. 2020). *Molagurasam*, a popular thin soup-like preparation of South India which is traditionally prescribed for respiratory infections may have unknowingly demonstrated its efficacy through use of pepper (*molagu*) and turmeric in combination thereby enhancing the bioavailability of curcumin along with accompanying nutraceutical benefits.

Fresh curry leaves, having citrusy aromatic notes, have been used in Indian cooking especially in central and south Indian cuisines. They are used directly in the preparation of different curries and used in the form of the powder which can be served with other accompaniments. They contain good amount of polyphenols which show antioxidant properties. Alcoholic extract of curry leaves has shown antioxidant properties and can be used as nutraceutical preparation (Ningappa et al. 2008).

Ginger and garlic are commonly used in many Indian dishes to improve the taste, flavor and they also provide many health beneficial effects. Bioactives present in garlic have shown to decrease triglycerides, cholesterol and decrease hypertension whereas ginger helps to obtain relief from cough, cold and flu. Coriander, cumin, and fennel (*Foeniculum vulgare*) are also majorly used spices in Indian cuisines which contain quercetin as an active component which plays a major role in prevention of cancer. Fenugreek seeds show antidiabetic and hypocholesteremic properties (Peter and Babu 2012).

18.3.6 Food Acidulants

Lime (*Citrus aurantiifolia*), tamarind (*Tamarindus indica*), *kokum* (*Garcinia indica*), and *amchur* powder (dry raw mango powder) are known food acidulants which are used in many Indian culinary preparations to impart sour taste to the food products. It has been reported that the organic acids present in these ingredients help in the absorption of zinc and iron from food grains and beta-carotene from green leafy vegetables. Thus, these food acidulants which are used in the Indian cuisine to impart sour taste also help in improving the bioaccessability of the micronutrients from grains and vegetables (Srinivasan 2010).

Dried rinds of *kokum* are used as a food acidulant in the coastal regions of India. *Kokum sharbat* and *kokum aagal* are two products from Konkan region of Maharashtra. *Kokum* is also commonly used in curry preparations. The dark red fruit of *kokum* has shown anti-obesity properties due to the presence of hydroxycitric acid whereas antioxidant, antibacterial, gastroprotective and chemo-preventive effects are exhibited by garcinol. *Kokam solkadhi* is a delightful appetizer drink prepared with coconut milk and *kokam*. It is majorly consumed in coastal region of India. The phytoconstituents present in *kokam*, majorly hydroxy citric acid, phenolics and anthocyanins, have shown to exhibit antibacterial, antifungal, anti-ulcerogenic, cardioprotective, chemo-preventive, antioxidant, and anti-obesity effects (Baliga et al. 2011).

18.3.7 Milk and Milk Products

Milk and milk products play very important role in the Indian diet. '*Haldi* milk' which is consumed by most of the Indian population for centuries is now globally accepted as '*Turmeric latte*' and has therapeutic value because of curcumin from turmeric. Indian diets, especially in central and southern parts of India, usually have one serving of *Dahi* (yogurt) almost every day. *Dahi* is a naturally fermented milk product with lactic acid bacteria. Literature suggests that it is a good source of folic acid, vitamin B complex and riboflavin while it also provides probiotic effect thus maintaining the gut health. Diacetyl, hydrogen peroxide, and reuterin found in *dahi* inhibit *E. coli*, *Bacillus subtilis* and *Staphylococcus aureus*. *Dahi* helps to control diarrhea in children. *Chilika* curd is a famous dish of Odisa which is prepared from concentrated buffalo milk. *Chilika*'s shelf life is higher than regular *dahi* due to the antifungal properties of dendrocin, a constituent of the bamboo (*Bambusoideae*) leaves used during the preparation (Sarkar et al. 2015).

Ghee is commonly used ingredient in all Indian traditional foods being a major source of energy. Recent literature shows that milk fat contains conjugated linoleic acid, butyric acid and vitamins which have anticarcinogenic properties. Though higher consumption of *ghee* was presumed to be one of the causes of heart ailments however paradoxically *ghee* was found to show hypocholesteremic effects (Srinivasan 2010).

Colostrum is another unconventional traditional food source in India which is used to provide immunity. It is the breast fluid produced by mammals before breast milk is released and considered to be very nutritious and contains high levels of antibodies which helps fight infections. *Kharwas* is a colostrum-based sweet dish prepared by steaming a mixture of colostrum, milk, jaggery/sugar until it coagulates into a semisolid mass which is moulded. It has been reported that colostrum has a high amount of lactoferrin, lactoperoxidase, and lysozyme, which have antimicrobial and antiviral properties. Thus, colostrum contributes to the wellness status of healthy individuals and improves healing in individuals with immune deficiency condition. The three major components viz. lactoperoxidase, lactoferrin and lysozyme show major effects by regulating bacterial growth, acting on peptidoglycan layer of Gram-negative bacteria causing bacterial lysis. Literature suggests that colostrum is active against Herpes virus I and II and against gastric ulcer caused by *Helicobacter pylori*. Some colostrum antibodies are reported to be active against the virulence factors produced by enterohaemorrhagic strains produced by *Escherichia coli*, particularly intimin, hemolysin etc. (Conte and Scarantino 2013; Silva et al. 2019). In addition to the general food categories described above. Table 18.2 summarizes the bioactives present in specific Indian traditional foods along with their health benefits.

18.4 Indian Traditional Foods at Different Life Stages

Indian tradition also considers foods with functional constituents during various stages of life. This is because in different stages of life cycle nutritional requirements, physiological functions and challenges faced by health are different. Table 18.3 gives some details of Indian traditional foods with functional constituents which are recommended to specific age groups based on physiological status at different stages of life.

For instance, infant foods such as *Bharadi/ pej* which improve sleeping pattern, boost immunity, is rich in protein, and good for bone development are recommended. Pregnant women are advised to take *ragi* based products such as *ragi laddu*, and ragi *sattva* being rich in calcium, phosphorous, magnesium. Other phytochemicals present in these products show anti-inflammatory, and anticancer effects. Lactating women

Table 18.2 Traditional Indian foods with their bioactive constituents and health benefits

Product	Region	Preparation steps	Bioactives	Major findings	Reference
Axone	Nagaland	Soyabean fermentation	- Probiotics	- Antimicrobial, fibrinolytic, hypocholesteraemic and antioxidant	Chaudhary et. al. (2018)
Bamboo Shoot Curry	Tamil Nadu, North Eastern States	Soaking of bamboo pieces then boiling followed by adding spices and seasoned lentils	- High edible fibre content - Vitamin C and Vitamin E	– Low fat food – Recommended for patients having piles and burning sensation during urination	Sarkar et al. (2015)
Banana Stem Juice	South India	Extraction of juice pressing of innermost trunk of the stem	- Alpha amylase and alpha glucosidase inhibitors	- Hyperglycaemic effect	Sarkar et al. (2015)
Chilika Curd	Odisha	Boiling of buffalo milk and then pouring into pre-treated bamboo-containers covered with leaves and incubated	- Dendrocin - Thermotolerant *Lactobacilli* - Vitamin B complex, folic acid, and riboflavin - Diacetyl - Hydrogen peroxide - Reuterin	- Antifungal, antioxidant, and probiotic properties	Sarkar et al. (2015)
Dhokla	Gujarat	Fermentation of Bengal gram and rice	- Folic acid	- Antioxidant property - Important for age related diseases and oxidative stress-induced degenerative diseases	Sarkar et al. (2015)

(continued)

Table 18.2 (continued)

Product	Region	Preparation steps	Bioactives	Major findings	Reference
Dosa	South India	Fermentation of rice and de-husked black gram	- Free amino acids, enzymes, vitamins, folic acid - Antimicrobial and antioxidant ingredients	- Improves bioavailability of zinc and iron - Helps in managing pre- and post-diabetic condition - Helps in increasing fertility, foetus weight, and breast milk - Provides sufficient energy for extended physical stamina - Helps in the treatment of rheumatism and neural disorders	Hemalatha et al. (2007)
Enduri Pitha	Odisha	Fermentation of batter of parboiled rice and black gram in turmeric leaf and then steaming	- Probiotics - Phytochemicals	- Increases immunity - Fermentation seems to enhance the nutritional quality of the blend of black gram and rice	Sarkar et al. (2015)
Germinated H*urihuttu*	Karnataka and Tamilnadu	Flour of popped finger millet	- Reducing sugar - Amylase enzyme	- Dietetic foods for anaemia patients and geriatric food formulation - Improves the bioavailability of iron and zinc	Sarkar et al. (2015)

(continued)

Table 18.2 (continued)

Product	Region	Preparation steps	Bioactives	Major findings	Reference
Ginna or *Kharwas*	Pan India	Colostrum-based Indian sweet	- Immunoglobulins - Iron binding lactoferrin protein - Vitamin A - Minerals	- Antibacterial, antiviral - Anti-inflammatory - Useful in Gut disorders	Sarkar et al. (2015)
Gundruk	North East India	Fermentation of leaves of *rayosag*, mustard and cauliflower	- Ascorbic acid - Carotene - Lactic acid - Lactic acid bacteria	- Good appetizer - Anticarcinogenic	Sarkar et al. (2015)
Haria	West Bengal	Fermentation of rice	- Maltotetrose, Maltotriose, and Maltose, pyranose derivatives - Lactic acid bacteria	- Antioxidant - Symbiotic activity	Ghosh et al. (2015)
Hawaijar	Manipur	Alkaline fermentation of soybean	- Soluble proteins - Trypsin, glutamine, cystine, lysine and leucine	- High fibrinolytic activity - Anticancer, antiosteoporosis, antidiabetic and hypocholesterolaemic - Antioxidant	Sarkar et al. (2015);
Hentak	Himachal Pradesh	Non-salted dried fermentation of leafy vegetables such as cauliflower, mustard, radish, *rayosag*	- Ascorbic acid, lactic acid - Carotene	- Anticancer properties	Chaudhary et al. (2018)

(continued)

Table 18.2 (continued)

Product	Region	Preparation steps	Bioactives	Major findings	Reference
Idli	South India	Fermentation of rice and black gram *dal* and then steaming	- Essential amino acids - Folate, Vitamin A, B1, B2, and B12, - Acetoin and volatile fatty acids	- Antiobesity - Dietary supplement for children - Increases bioavailability of zinc and iron	Hemalatha et al. (2007)
Jackfruit Seed *Chutney*	Karnataka	Pasting boiled jackfruit seeds and mixing with chili, onion, garlic, and grated coconut	- Phenolic compounds - Prebiotics	- Help in intestinal microbial balance	Sarkar et al. (2015)
Kambu Koozh	South India	Non-alcoholic natural fermentation of pearl millet porridge	- Probiotics - Fibre and proteins	- Prevention of diarrhoea and constipation	Chaudhary et al. (2018)
Khatti Rabdi	Rajasthan	Lactic acid fermentation of milk with pearl millet	- Fibre, proteins - Probiotics	- Low glycaemic index - Helpful in managing noninsulin dependent *Diabetes mellitus*	Sarkar et al. (2015)
Koshimbir	Maharashtra	Mixing of sprouted legumes, coconut, *dahi*	- Vitamin C - Reduced tannin or phytic acid content	- Increases availability of iron and zinc	Sarkar et al. (2015)
Mustard Paste	West Bengal, Odisha	Usually eaten in curries or even consumed raw	- Dithiolthiones - Sinapic acid	- Inactivates the mutagenicity of food mutagens such as tryptophan pyrolysate - rotective effect against liver toxicity	Mayengbam et al. (2014)

(continued)

Table 18.2 (continued)

Product	Region	Preparation steps	Bioactives	Major findings	Reference
Neem Flower *Chutney*	South India	Grinding of roasted neem flower, red chili, urad dal, coconut, tamarind, and salt	- Unsaturated fatty acids such as linolenic, linoleic, and oleic acids	- Removes worm - Reduces the blood sugar level - High antioxidant properties	Sarkar et al. (2015)
Paan	Pan India	Betel leaves along with areca nut and lime	- Chavibetol - Codinene - Hydroxychavicol - Calcium	- Mouth-freshener - Digestive-stimulating effects - Aromatic and carminative - Aphrodisiac and antiseptic - Responsible for preventing osteoporosis among the economically weaker sections of the population in India	Sarkar et al. (2015)
Peanut *Chikki*	Pan India	Roasted peanuts mixed with *Jaggery* syrup and molded	- Insoluble polysaccharides (cellulose and hemicellulose) - soluble oligosaccharides (raffinose, stachyose and verbascose), - Protein and iron - Monounsaturated fatty acid	- Instant energy source - Cardioprotective	Bonku and Yu (2020)

(continued)

Table 18.2 (continued)

Product	Region	Preparation steps	Bioactives	Major findings	Reference
Selroti	Sikkim and Darjeeling	Rice-based fermentation	- Digestible proteins - *Lactobacilli, Pediococci, Enterococci*, and *Leuconostoc*	- Symbiotic food due to the presence of prebiotic and probiotic	Sarkar et al. (2015)
Tikhur Barfi or *Halwa*	North India	Cooking of *tikhur* starch with sugar, water/ milk with dry fruits to yield *barfis* on cooling	- Starch and volatile oils	- Consumed by individuals during fast as it is rich in energy - Used in peptic ulcers, dysentery, diarrhoea, and colitis and herbal tonic for patients suffering from tuberculosis - Used as a cooling food during summer	Chandel et al. (2018)
Til Laddu	Pan India	Mixing of sesame seeds with *jaggery* and shaping into balls	- Magnesium, copper, calcium, iron, zinc - Vitamin B6 - Polyunsaturated fatty acids - Sesamol and sesamin	- Antioxidant and antidepressant properties - Mitigate the effects of anxiety, nerve and bone disorders, poor circulation, lowered immunity, and bowel problems - Relieve lethargy, fatigue, and insomnia, while promoting strength and vitality, and enhancing blood circulation	Monteiro et al. (2014)

Table 18.3 Indian traditional Foods as functional foods for different life stages

S no	Food item	Preparation	Bioactive compounds	Targeted age group	Reference
1	*Rajgira Kheer*	Amaranth seeds made into a pudding with milk and sugar	- High iron and calcium content	Weaning food Food for Calcium deficient people	Raghuvanshi and Bhati (2019)
2	*Rajgira Khichdi*	Steamed Amaranth seeds, smashed potatoes tempered with clarified butter and cumin seeds	- High iron and calcium content	Weaning food Food for Calcium deficient people	Raghuvanshi and Bhati (2019)
3	*Bhardi/ Pej*	Cooked and smashed *moong dal* tempered with carrom, cumin seeds and salt	- Flavonoids, - Isoflavonoids - Phenolics	Infants	Ganesan and Xu (2018)
4	*Ragi Satva*	Mixing of *ragi*flour, with milk, jaggery and then cooking	- Calcium - Phosphorous - Magnesium - Potassium	Infant, Pregnant, and lactating mothers	Chandra et al. (2016)
5	*Ambali*	Mixing of finger millet flour with water to make a thick batter followed by fermentation and cooking	- Calcium, Thiamine and Riboflavin - Rich in protein, low resistant starch - Thiamine, riboflavin, tryptophan	Paediatric food and Geriatric food	Sarkar et al. (2015)
6	Spinach *Roti*	Boiling the spinach in water followed by kneading dough with boiled spinach	- Protein and iron rich	Teen age girls during menstruation and lactating women	Sarkar et al. (2015)
7	Horse gram soup	1 spoon horse gram flour + 1 tsp black pepper powder + 1/4tsp asafoetida, *haldi*, red chilli + add water to make soup like consistency + salt to taste Give it a boil for 5 min. serve hot	- Protein source - Antioxidant	Pregnant woman and elderly people	Prasad and Singh (2015)

(continued)

Table 18.3 (continued)

S no	Food item	Preparation	Bioactive compounds	Targeted age group	Reference
8	*Rajgira Laddu*	Mixing of amaranth seeds with jaggery and shaping into balls	- Iron - Protein - Calcium	Pregnant and lactating mothers	Raghuvanshi and Bhati (2019)
9	*Ragi Laddu*	Mixing of *ragi* flour, *dinka* with jaggery and shaping into balls	- Calcium - Phosphorous - Magnesium - Potassium	Pregnant and lactating mothers	Chandra et al. (2016)
10	*Haldi* milk	Mixing of *haldi* with warm milk	- Curcumin - Calcium	Pregnant women	Jain et al. (2011)
11	*Dinka Laddu*	Mixing of *kharak*, dried coconut, gum acacia, jaggery and garden cress seeds with dates and then shaping into balls	- Fibre, calcium, vitamins, iron, minerals	Lactating mothers	Jain et al. (2011); Rutu et al. (2017)
12	Fenugreek *Laddu*	Mixing of fenugreek seed powder with gum acacia, *kharak*, jaggery, dates and then shaping into balls	- Phytoestrogens - Steroidal sapogenins	Lactating mothers	Jain et al. (2011); Snehlata and Payal (2012)
13	Carrom seeds based *laddu*	Mixing of carrom seeds, grated coconut, dry roasted poppy seeds and heating in clarified butter. Add heated jaggery and then shaping into balls	- Camphene, pinene, myrcene, limonene, terpinene - terpinen-4-ol, thymol and carvacrol	Lactating mothers and Postpartum diet	Bairwa et al. (2012)
14	*Khaskhas kheer*	Mixing of soaked poppy seeds with milk, rice, sugar, cardamom, and cook. Garnish with almonds and pistachios	- Morphine, codeine, thebaine, Oripavine, noscapine and papaverine	Lactating mothers	Knutsen et al. (2018)
15	*Methi Kheer*/ Porridge	Soak *methi* seeds overnight, grind it coarsely, Add milk, grated coconut, nuts and *jaggery*. Add cardamom for flavour, boil it for 15–20 min	- Trimethylamine, Neurin, Trigonelline, Choline, Gentianine, Carpaine and Betain - Isoleucine, 4-Hydroxyisoleucine, Histidine, Leucine, lysine, L-tryptophan, Arginine	Women during prenatal phase and elderly people	Abdou and Fathey (2018)

are suggested to take foods which improve lactation, cure backache, and improve bone strength. *Dinka laddu*, fenugreek *laddu*, *khaskhas kheer* and *haldi* milk are few examples of traditional foods advised for pregnant and lactating mothers. Foods containing high amount of protein, calcium and iron are advised for elderly people.

Amaranth (*Amaranthus*) grain is a pseudo-cereal having properties of both cereal and leguminous seed and it is gluten-free. In India, it is known as "*Rajgira*" and "*Chaulai*". Amaranth *kheer*, Amaranth *laddu* and Amaranth *khichri* are some of the traditional cuisines of India especially given as weaning food and for preschoolers. Amaranth has lysine content twice that of wheat and thrice that of maize protein. It contains high levels of riboflavin, niacin, ascorbic acid, calcium and magnesium and low levels of antinutritional factors. It has almost 20% of the daily recommended amount of calcium, iron, magnesium, folate per 100 g. Thus, amaranth cuisines are considered as best suitable breakfast for preschoolers, celiac patients due to gluten free nature and osteoporotic patients being high in calcium content. *Amaranth laddu* is made with *jaggery* and thus high content of calcium, iron and protein makes it suitable for pregnant and lactating women. *Amaranth kheer* and *khichri* were found to be suitable for preschool children as amaranth is easily chewable and contains high amount of protein and calcium. It has been reported that one serving of 250 to 300 g of amaranth *kheer* and *upma* can provide two thirds requirement of calcium in adults and thus can be suitable to combat calcium deficiency. It can also be used as a weaning food for infants (Raghuvanshi and Bhati 2019).

18.5 Nutraceuticals in Indian Traditional Foods

India hasa rich tradition of herbal and ayurvedic medicines, which is part of an ancient healing science. Ancient ayurvedic practices have been the cornerstone of wellness in India. Thus, India has played a very significant role in showing evidence-based nutraceuticals to the world with its solid reference of Ayurveda which can be modernized through evidence-based methodologies. India is poised to be a front-runner in the nutraceutical ingredient market due to its vast biodiversity, robust agricultural research facilities and increasing interest in this field. Nutraceuticals are also playing substantial role in evidence-based medicine. Major classes of bioactives/ phytochemicals present in the Indian traditional foods which can be perceived as nutraceuticals along with their health benefits are given in Table 18.4, of which few classes have been briefly described in the following sections.

18.5.1 Dietary Fibers

Dietary fibers which are also known as roughage are indigestible part of food which body cannot digest or absorb. Depending on the water solubility they are grouped into two classes i.e., water soluble dietary fibers such as pectin, β-glucan, and water

Table 18.4 Bioactives with their food sources, health benefits and examples of traditional foods as potential functional foods

Bioactives	Food sources	Potential Health Benefits	Examples of potential Traditional Functional foods
Dietary fibres			
Insoluble fibre	Wheat bran, corn bran, fruit skin	- Gut health - May reduce risk of cancers	- *Roti* - Corn *subji*
Soluble fibre	Peas, beans, apples, citrus fruits	- May reduce risk of coronary heart disease (CHD) and some types of cancers	- Rajma *subji* - Chole *subji*
Fatty acids			
MUFAs	Tree nuts	- May reduce risk of CHD	- Peanut *chikki* - Dry fruit *sheera*
PUFA: ALA	Walnuts, flaxseeds	- Eye and heart health - Mental function	- Flaxseed *laddu* - Walnut *burfi*
PUFA: DHA/EPA	Fish oils	- Heart health - May reduce the risk of CHD	- Fish fry - Fish *curry*
Flavanoids			
Flavanols	Tea, coffee, apples, grapes, onions	- Heart health - Antioxidant	- *Chai* - *Onion chutney*
Flavanones	Citrus fruits	- Antioxidant	- Orange juice
Procyanidins and proanthocyanidins	Cinnamon, apples, strawberries, tea, grapes	- Urinary tract health and heart health	- Cinnamon tea
Isothiocyanates			
Sulforaphane	Cauliflower, cabbage, horseradish, spinach	- Detoxification - Antioxidant	- *Palak* soup - *Sarson ka saag*
Minerals			
Calcium	Spinach, *ragi*, milk, yogurt, amaranth	- May reduce the risk of osteoporosis	- *Paneer* - Amaranth *laddu*
Magnesium	Spinach, pumpkin, cereals, almonds, beans	- Immune health and bone health - Maintenance of normal muscle and nerve function	- *Palak paratha* - *Bhopala thalipeeth*
Potassium	Cereals, beans, banana, citrus fruits, potatoes, green leafy vegetables	- May reduce the risk of high blood pressure and stroke	- *Alu subji* - Banana *shikran*
Selenium	Whole grains, garlic, egg, fish	- Antioxidant - Immune health and prostate health	- Garlic *chutney* - *Thalipeeth*

(continued)

Table 18.4 (continued)

Bioactives	Food sources	Potential Health Benefits	Examples of potential Traditional Functional foods
Natural Pigments			
Anthocyanins	Red grapes, blackberries, blueberries, red cabbage	- Antioxidant - Brain function	- Berry Jam - Grape juice
Beta-carotene	Carrots, pumpkin, potatoes, tomatoes, spinach	- Antioxidant - Conversion to Vitamin A	- Carrot *Koshimbir* - Palak *subji*
Betalains	Beetroot, amaranth, *Malabar* spinach	- Antioxidant - Anti-inflammatory - Detoxification	- Beet root *koshimbir* - *Mayalu pakode*
Lutein, zeaxanthin	Spinach, egg, corn, citrus fruits, carrots, asparagus	- Eye health	- *Dal palak* - Corn *bhel*
Lycopene	Tomatoes, watermelon, strawberry, pink guava	- Prostate health	- Tomato *rasam* - Pink guava *koshimbir*
Phenolic acids			
Caffeic acid, ferulic acid	Apples, pears, nuts, whole grains, coffee	- Antioxidant - Eye and heart health	- Green tea - *Shengdana chutney*
Plant Sterols/ Stanols			
Free stanols/ sterols	Corn, wheat	- May reduce risk of CHD	- *Upma* - *Daliya*
Prebiotics			
Inulin, fructo-oligosaccharide, polydextrose	Whole grains, onions, garlic, sweet potatoes, apples	- Gut health- - Supports calcium absorption	- *Bhakari* - *Ratala kees*
Probiotics			
Yeast, *Lactic acid bacteria*, *Bifidobacteria*	Yogurt	- Gut health and immune health	- *Masala Chaas* - Curd rice
Phytoestrogens			
Lignans	Flaxseeds, nuts, lentils, cauliflower, carrot	- Heart and immune healths	- Carrot *halwa* - Lentil spinach *curry*
Sulphides/Thiols			
Diallyl sulphide	Garlic, onions	- Detoxification - Heart, immune and gut health	- *Kanda lasoon chutney* - Onion *pakode*

(continued)

Table 18.4 (continued)

Bioactives	Food sources	Potential Health Benefits	Examples of potential Traditional Functional foods
Dithiolthiones	Cruciferous vegetables	- Detoxification - Heart, immune and gut health	- Cauliflower *subji* - *Gobi paratha*
Vitamins			
Vitamin A	Carrot, spinach, sweet potato, milk	- Contributes to cell integrity - Eye, bone, and immune health	- *Palak khichdi* - *Basundi*
Thiamine	Lentils, peas, brown rice, pistachios	- Mental function - Regulate metabolism	- *Dal makhana* - Pea *pulav*
Riboflavin	Eggs, green leafy vegetables, dairy products	- Supports cell growth - Regulate metabolism	- *Methi paratha* - Ragi *kheer*
Niacin	Poultry, fish, nuts, eggs	- Supports cell growth - Regulate metabolism	- Dry fruit *chikki* - *Masala chicken*
Pantothenic acid	Sweet potato, lentils	- Regulates metabolism and hormone synthesis	- *Moong dal Khichdi* - *Massor dal curry*
Pyridoxine	Beans, legumes, whole grains, fish, meat	- Immune health - Regulate metabolism	- Whole wheat *roti* - *Mataki usal*
Folate	Beans, legumes, Citrus fruits, green leafy vegetables	- Prevents neurodegenerative disorder in the new-born - Immune health	- *Palak Chilla* - *Chana masala*
Cyanocobalamin	Egg, milk, poultry, meat	- Mental function - Blood cell formation - Regulates metabolism	- Rice *Kheer* - *Chicken masala*
Biotin	Dairy products, eggs, fish	- Regulates metabolism and hormone synthesis	- *Haldi* milk - Paneer
Vitamin C	Guava, red/green pepper, citrus fruit, strawberries	- Antioxidant - Maintenance of bone and immune health	- *Amla candy* - Lemon juice
Vitamin D	Sunlight, fish, milk	- Regulates calcium and phosphorous absorption - Cell growth	- Masala *chaas* - *Lassi*
Vitamin E	Sunflower seeds, almonds, hazelnuts	- Antioxidant - Immune and heart health	- Almond *burfi* - Mixed nut *chikki*

insoluble dietary fibers such as lignin, cellulose. They mainly help in lowering the glycemic response of foods and prevent constipation. Indian staple foods such as cereals, pulses, grains, vegetables, and fruits are good sources of dietary fibers (Dhingra et al. 2012). *Ragi*, lentils, barley (*Hordeum vulgare*) are sources of β-glucan which is water soluble (Kudake et al. 2018) whereas whole grains and grain flour, variety of fruits and vegetables contain water insoluble dietary fibers. In India, fruits and few vegetables are preferred to be eaten in the raw form. Different fruits and vegetables with edible peels such as apple (*Malus domestica*), cucumber (*Cucumis sativus*), carrot possess good content of dietary fibers which can be extracted from fruit pomace which is a waste generated in fruit processing.

18.5.2 Fatty Acids

Certain omega-3 fatty acids (α-linolenic acid) and omega-6-fatty acids (linoleic acid) cannot be synthesized by our body and are considered as essential fatty acids. The ideal ratio of omega-6 fatty acids to omega-3 fatty acids is 4:1 or lower. These fatty acids play important role in the synthesis of long chain polyunsaturated fatty acids (PUFAs) such as eicosapentaenoic acid and docosahexaenoic acid. But many diets including Indian diet have very high ratio of the same as most vegetarian foods contain high quantity of omega-6 than omega-3 fatty acids. Major sources of omega-3 fatty acids are flaxseed (*Linum usitatissimum*) and seafood (Patel et al. 2020).

18.5.3 Minerals

Minerals are the micronutrients which are required in small amounts for the development of body. They are classified in two categories namely macro-minerals and microminerals. Calcium, magnesium, potassium, sodium, chloride, phosphorus and sulfur are macro-minerals; while trace minerals are iodine, zinc, selenium, iron, manganese, copper, cobalt, molybdenum, fluoride, chromium and boron (Kudake et al. 2017; Gharibzahedi and Jafari 2017). Minerals play an important role in cell signaling, nerve signal transmission, cell communication, osmotic balance. They are also important in many metabolic processes, enzyme activation and inhibition. Minerals such as Ca, Mg, Mn, P, B and F help in bone development and strengthening, and Ca, P and F are required for teeth. Minerals like Co, I and Fe help during erythrogenesis (Gharibzahedi and Jafari 2017). These minerals can be incorporated in our diet through different vegetables, fruits, grains, cereals, and pulses. Nutrient profile of many processed foods is being improved through fortification done with minerals. Adequate consumption of minerals prevents body from several disorders. Few minerals such as calcium, iron are available as supplements in the pharmaceutical form such as capsule, tablet and syrup.

18.5.4 Natural Pigments

Carotenoids are lipid-soluble natural pigments, present in red, orange-colored fruits and vegetables such as oranges (*Citrus X sinensis*), tomatoes (*Solanum lycopersicum*), carrots (*Daucus carota* subsp. sativus). β-carotene is dietary source of provitamin A which plays important role in eye health (Mezzomo and Ferreira 2016). Lutein and zeaxanthin, present in green leafy vegetables, reduce the risk of age-related macular degeneration and cataracts (Becerra et al. 2020). Other natural pigments like betalains, and anthocyanins play role in scavenging free radicals, help in brain function and detoxification of body (Joaquín-Cruz et al. 2015; Sawicki, and Wiczkowski 2018).

18.5.5 Probiotics and Prebiotics

Probiotics are live microorganisms which when ingested in adequate amount confer gut health benefits whereas prebiotics are food for gut microbiota. Indian traditional foods are rich sources of probiotics and prebiotics. Curd and buttermilk which are consumed by majority of the population are probiotic sources and whole grains, onions (*Allium cepa*), garlic, banana are few examples of prebiotics consumed in traditional foods. Metabolic syndrome represents a set of metabolic abnormalities and clinical factors including insulin resistance, dyslipidaemia, hypertension, abdominal obesity all of which may increase the risk for cardiovascular disease and type 2 diabetes mellitus. The modulation of gut flora by supplementation with probiotics and prebiotics may have a potential beneficial influence on management of metabolic syndrome (Xavier-Santos et al. 2020). However extensive research studies are required to explore the therapeutic effect of probiotics and prebiotics from Indian traditional foods in the treatment of metabolic syndrome.

Further, para-probiotics and postbiotics are emerging options as functional foods conferring health benefits. Para-probiotics are "inactivated (non-viable) microbial cells, which, when consumed demonstrate health beneficial properties" while postbiotics are "metabolic by-products of LAB which are excreted by the cells during growth" (Aguilar-Toaláet al. 2018; Taverniti and Guglielmetti 2011). There are some commonly consumed traditional food preparations in India made by cooking fermented substrates. *Idli* is one such example wherein a fermented rice and black gram batter is steam cooked. While the cooking will inactivate the microbial cells (LAB is involved in *idli* batter fermentation) the steamed *idli* is still likely to provide some health beneficial properties based on concept of para-probiotics and postbiotics. A similar idea can be proposed with respect to *kadhi,* a traditional preparation made by heating sour curds and tempering it with spices. While curds will be a source of LAB cultures the *kadhi* prepared by heating curds could be a potential source of para-probiotics and postbiotics. However, this again calls for extensive research investigations to validate this hypothesis by firstly evaluating the probiotic potential of LAB cultures present in these fermented substrates and thereafter evaluating the effect of thermal treatment used in their preparation on such health beneficial properties.

18.5.6 Antioxidants

Antioxidants inhibit oxidation reaction caused by free radicals and prevent cell destruction. They give protection from oxidative stress which helps in prevention of heart diseases, cancer, arthritis, stroke, respiratory diseases and immune deficiency (Goufo and Trindade 2014). Different nuts commonly consumed contain variety of compounds with antioxidant potential. Black gram consists of *o*-dihydroxy phenols, flavonols, tannins (Kamboj and Nanda 2018) whereas chickpea consists of flavonols, flavone glycosides, oligomeric as well as polymeric pro-anthocyanidins, cinnamic acid, salicylic acid, hydroxycinnamic acid, *p*-coumaric acid, gallic acid, caffeic acid, vanillic acid, ferulic acid, anise acid, tannic acid, piperonyl, and chlorogenic acid (Gupta et al. 2017; Rachwa-Rosiak et al. 2015). Rice contains phenolic acids, flavonoids, anthocyanins, pro-anthocyanidins, tocopherols, tocotrienols, oryzanol, and phytic acid (Goufo and Trindade 2014). Carotene, lycopene, betanin are mainly present in vegetables having antioxidant properties. Vitamins also possess antioxidant properties. Thus, majority of the Indian foods contain bioactives with antioxidant potential *(also see the chapter on nutritional hormetins in this book.)*

18.6 Summary and Future Perspectives

Traditional foods have nourished mankind for centuries. India with its great diversity of food cultures and its people following a predominantly vegetarian diet has for several decades been following a diet rich in a variety of phytoconstituents long before the terms like “nutraceuticals” and “functional foods” were coined. With its warm tropical climate and agricultural activities India also enjoys a great biodiversity in the form of a variety of fruits, vegetables, grains, spices, and herbs each of them being a powerhouse of phytochemicals. Indian traditional wisdom practiced with respect to dietary habits believes in consuming a variety of food sources of seasonal and local availability. This ensures consumption of a great variety of phytochemicals which today we recognize as having nutraceutical value. Traditional wisdom has also acknowledged the important role of food in maintaining health. Complimentary to this is the ancient health science of Ayurveda which also gives dietary recommendations to manage health. However, this valuable knowledge of health promoting properties of Indian foods has not been extensively or systematically documented while most of it has just been transmitted through generations by word of mouth.

While the Indian people have been habitually following this healthy diet for all these years however of late due to changing lifestyles, increasing participation of women in professional domain, increasing affordability and easy availability of junk food, paucity of time for food preparation, the dietary habits of people especially in metro cities have changed and consumption of refined foods with high energy density but poor nutritional value has increased. Lifestyle associated diseases are concomitantly on the rise. Therefore, it is imperative that recognition be given to the importance of nutraceuticals and functional foods. There is a need to scientifically explore the nutraceutical potential of traditional food ingredients and to acknowledge some of the traditional foods as functional foods. This knowledge when based on sound scientific data will find greater acceptance with the present generation. Instead of purchasing expensive formats of nutraceuticals and functional food products the individuals can then rely on traditional foods to supply them with health beneficial constituents which will be both acceptable and affordable. What we know today in this regard is just the tip of the iceberg while a lot remains to be discovered. A systematic exploration of all types of Indian foods consumed in different parts of the Indian subcontinent for establishing their phytochemical profile needs to be undertaken. The effect of raw material quality and variety on the bioactive profile needs to be considered. Robust clinical studies need to be undertaken to establish their health beneficial properties. This will help to rediscover the value of these foods and promote them for present and future generations. The bioavailability of the phytoconstituents and factors affecting it needs to be understood. The ability of other food constituents in promoting/inhibiting the absorption of phytochemicals has to be studied. Certain combinations of ingredients used in the preparation of traditional foods or certain combinations of preparations eaten together as a meal can improve the bioavailability or have some complimentary effect which also needs to be evaluated. There is also possibility of employing appropriate processing technologies to transform traditional forms of products into convenient modern formats for today's consumer. The processing and storage stability of the bioactive constituents in these food products must be therefore established if one has to prepare such products on a large scale for wide distribution.

There are a number of edible plant sources such as fruits and vegetables which are relatively unexplored, not widely cultivated and only consumed locally and seasonally by a small community of people. Such natural food sources can also have a hitherto unexplored therapeutic potential, about which some knowledge may be present in restricted local communities. This also needs to be investigated. The wastes generated in food processing industries can themselves be a source of bioactives either directly or through suitable conversion techniques employed. This can help in the recovery of valuable by products in the form of bioactives. Thus, many avenues of research and development are opening up for authenticating the value of traditional wisdom and for exploring the health beneficial properties of traditional foods.

Compliance with Ethical Standards

Conflict of Interest All authors declare they have no conflict of interest.

References

Abdou RM, Fathey M (2018) Evaluation of early postpartum fenugreek supplementation on expressed breast milk volume and prolactin levels variation. Egyptian Pediatric Assoc Gazette 66(3):57–60

Afolayan FI, Erinwusi B, Oyeyemi OT (2018) Immunomodulatory activity of curcumin-entrapped poly d, l-lactic-co-glycolic acid nanoparticles in mice. Integrative Med Res 7(2):168–175

Aguilar-Toalá JE, Garcia-Varela R, Garcia HS, Mata-Haro V, González-Córdova AF, Vallejo-Cordoba B, Hernández-Mendoza A (2018) Postbiotics: an evolving term within the functional foods field. Trends Food Sci Technol 75:105–114

Alasalvar C, Salvadó JS, Ros E (2020) Bioactives and health benefits of nuts and dried fruits. Food Chemistry 314:126192

Ananthanarayan L, Dubey KK, Muley AB, Singhal RS, (2019) Indian traditional foods: preparation, processing and nutrition. In: Traditional Foods, pp 127–199. Springer, Cham

Anonymous (2021). The Ayurvedic Institute. https://www.ayurveda.com/

Asibuo JY, Akromah R, Safo-Kantanka O, Adu-Dapaah HK, Ohemeng-Dapaah S, Agyeman A (2008) Chemical composition of groundnut, Arachis hypogaea (L) landraces. African J Biotechnol 7(13)

Bairwa R, Sodha RS, Rajawat BS (2012) Trachyspermum ammi. Pharmacogn Rev 6(11):56

Baliga MS, Bhat HP, Pai RJ, Boloor R, Palatty PL (2011) The chemistry and medicinal uses of the underutilized Indian fruit tree Garcinia indica Choisy (kokum): a review. Food Res Int 44(7):1790–1799

Becerra MO, Contreras LM, Lo MH, Díaz JM, Herrera GC (2020) Lutein as a functional food ingredient: Stability and bioavailability. J Funct Foods 66:103771

Bitencourt PE (2020) Nanoparticle formulation of *Syzygium cumini*, antioxidants, and diabetes: Biological activities of *S. cumini* nanoparticles. In: Diabetes, pp 343–350. Academic Press

Bonku R, Yu J (2020) Health aspects of peanuts as an outcome of its chemical composition. Food Sci Human Wellness 9(1):21–30

Chandel N, Kumar V, Ram DS (2018) Physicochemical and functional properties of *Curcuma angustifolia* (*Tikhur*): an underutilized starch. J Pharmacognosy Phytochemistry 7:2438–2440

Chandra D, Chandra S, Sharma AK (2016) Review of Finger millet (*Eleusine coracana* (L.) Gaertn): a powerhouse of health benefiting nutrients. Food Science and Human Wellness 5(3):149–155

Chaudhary A, Sharma DK, Arora A (2018) Prospects of Indian traditional fermented food as functional foods. Indian J Agric Sci 88(10):1496–1501

Conte F, Scarantino S (2013) A study on the quality of bovine colostrum: physical, chemical and safety assessment. Int Food Res J 20(2):925–931

Davis JP, Dean LL (2016) Peanut composition, flavor and nutrition. In: Peanuts, pp 289–345. AOCS Press

Devi CB, Kushwaha A, Kumar A (2015) Sprouting characteristics and associated changes in nutritional composition of cowpea (*Vigna unguiculata*). J Food Sci Technol 52(10):6821–6827

Dhingra D, Michael M, Rajput H, Patil RT (2012) Dietary fiber in foods: a review. J Food Sci Technol 49(3):255–266

Erami SR, Amiri ZR, Jafari SM (2019) Nanoliposomal encapsulation of Bitter Gourd (*Momordica charantia*) fruit extract as a rich source of health-promoting bioactive compounds. LWT 116:108581

Falcomer AL, Riquette RFR, de Lima BR, Ginani VC, Zandonadi RP (2019) Health benefits of green banana consumption: a systematic review. Nutrients 11(6):1222

Faris MEAIE, Takruri HR, Issa AY (2013) Role of lentils (*Lens culinaris* L.) in human health and nutrition: a review. Mediterranean J Nutrition Metabolism 6(1):3–16

Ganesan K, Xu B (2018) A critical review on phytochemical profile and health promoting effects of mung bean (Vigna radiata). Food Sci Human Wellness 7(1):11–33

Gharibzahedi SMT, Jafari SM (2017) The importance of minerals in human nutrition: bioavailability, food fortification, processing effects and nanoencapsulation. Trends Food Sci Technol 62:119–132

Ghosh K, Ray M, Adak A, Dey P, Halder SK, Das A, Jana A, Parua S, Mohapatra PKD, Pati BR, Mondal KC (2015) Microbial, saccharifying and antioxidant properties of an Indian rice based fermented beverage. Food Chem 168:196–202

Glance HSAA (2018) Horticulture Statistics Division Department of Agriculture. Cooperation & Farmers 'Welfare Ministry of Agriculture and Farmers' Welfare Government of India

Goufo P, Trindade H (2014) Rice antioxidants: phenolic acids, flavonoids, anthocyanins, proanthocyanidins, tocopherols, tocotrienols, γ-oryzanol, and phytic acid. Food Sci Nutr 2(2):75–104

Gupta RK, Gupta K, Sharma A, Das M, Ansari IA, Dwivedi PD (2017) Health risks and benefits of chickpea (*Cicer arietinum*) consumption. J Agric Food Chem 65(1):6–22

Hemalatha S, Platel K, Srinivasan K (2007) Influence of germination and fermentation on bioaccessibility of zinc and iron from food grains. Eur J Clin Nutr 61(3):342–348

Jagdale YD, Ghodke SV (2020) Development of innovative flour based Indian traditional product: multigrain Chakli. Int Res J Eng Technol 7(5):4161–4168

Jain N, Sharma V, Ramawat KG (2011) Therapeutic potentials of medicinal plants traditionally used during postpartum period and their molecular targets. J Ecobiotechnol 3(10):30–39

Joaquín-Cruz E, Dueñas M, García-Cruz L, Salinas-Moreno Y, Santos-Buelga C, García-Salinas C (2015) Anthocyanin and phenolic characterization, chemical composition and antioxidant activity of chagalapoli (*Ardisia compressa* K.) fruit: a tropical source of natural pigments. Food Res Int 70:151–157

Kalra EK (2003) Nutraceutical-Definition and Introduction. AAPS Pharmsci 5(3):27–28

Kamboj R, Nanda V (2018) Proximate composition, nutritional profile and health benefits of legumes-a review. Legume Res Int J 41(3):1–8

Kaur K, Chhikara N, Sharma P, Garg MK, Panghal A (2019) Coconut meal: Nutraceutical importance and food industry application. Foods Raw Mater 7(2):419–427

Knutsen HK, Alexander J, Barregård L, Bignami M, Brüschweiler B, Ceccatelli S, Cottrill B, Dinovi M, Edler L, Grasl-Kraupp B (2018) Update of the Scientific Opinion on opium alkaloids in poppy seeds. EFSA J 16(5):e05243

Kudake DC, Bhalerao PP, Chaudhari NS, Muley AB, Talib MI, Parate VR (2018) Fortification of wheat flour with ragi flour: effect on physical, nutritional, antioxidant and sensory profile of noodles. Current Res Nutrition Food Sci 6(1):165–173

Kudake DC, Pawar AV, Muley AB, Parate VR, Talib MI (2017) Enrichment of wheat flour noodles with oat flour: effect on physical, nutritional, antioxidant and sensory properties. Int J Curr Microbiol App Sci 6(12):204–213

Lad U, Lad V (2010) Ayurvedic Cooking for Self-Healing. Motilal Banarsidass Publishe, 3rd (edn.)

Lad V (2017) Ayurveda: The Science of Self-Healing - A Practical Guide, 6th edn. Motilal Banarsidass Publisher

Liu Z, Smart JD, Pannala AS (2020) Recent developments in formulation design for improving oral bioavailability of curcumin: a review. J Drug Delivery Sci Technol: 102082

Mathanghi SK, Sudha K (2012) Functional and phytochemical properties of finger millet (*Eleusine coracana* L.) for health. Int J Pharmaceutical Chem Biolo Sci 2(4):431–438

Mayengbam S, Aachary A, Thiyam-Holländer U (2014) Endogenous phenolics in hulls and cotyledons of mustard and canola: a comparative study on its sinapates and antioxidant capacity. Antioxidants 3(3):544–558

Mezzomo N, Ferreira SR (2016) Carotenoid's functionality, sources, and processing by supercritical technology: a review. J Chem 3164312:1–16

Monteiro ÉMH, Chibli LA, Yamamoto CH, Pereira MCS, Vilela FMP, Rodarte MP, de Oliveira Pinto MA, Henriques DP, do Amaral M, Silvério MS, de Matos Araújo ALS, Da Luz André de Araújo, A (2014) Antinociceptive and anti-inflammatory activities of the sesame oil and sesamin. Nutrients 6(5):1931–1944

Nida F, Jaiswal M, Agrahari K, Singh A (2017) Enrichment of sattu products by using nutrient rich powders. Int J Home Sci 3(1):272–276
Ningappa MB, Dinesha R, Srinivas L (2008) Antioxidant and free radical scavenging activities of polyphenol-enriched curry leaf (*Murraya koenigii* L.) extracts. Food Chemistry 106(2):720–728
Oke DG (2014) Proximate and phytochemical analysis of *Cajanus cajan* (Pigeon pea) leaves. Chem Sci Trans 3(3):1172–1178
Palamthodi S, Lele SS (2014) Nutraceutical applications of gourd family vegetables: *Benincasa hispida*, *Lagenaria siceraria* and *Momordica charantia*. Biomed Prev Nutr 4(1):15–21
Patel A, Rova U, Christakopoulos P, Matsakas L (2020) Introduction to essential fatty acids. In: Nutraceutical Fatty Acids from Oleaginous Microalgae: A Human Health Perspective, Book Editor(s): Patel A. and Matsakas L.
Peter KV, Babu KN (2012) Introduction to herbs and spices: medicinal uses and sustainable production. In: Handbook of herbs and spices. Woodhead Publishing, pp 1–16
Prasad SK, Singh MK (2015) Horse gram-an underutilized nutraceutical pulse crop: a review. J Food Sci Technol 52(5):2489–2499
Rachwa-Rosiak D, Nebesny E, Budryn G (2015) Chickpeas-composition, nutritional value, health benefits, application to bread and snacks: a review. Crit Rev Food Sci Nutr 55(8):1137–1145
Raghuvanshi RS, Bhati D (2019) Development of breakfast recipes from amaranth grains for preschoolers, celiac and osteoporotic subjects. Pantnagar J Res 17(3):267–272
Ray M, Ghosh K, Singh S, Mondal KC (2016) Folk to functional: an explorative overview of rice-based fermented foods and beverages in India. J Ethnic Foods 3(1):5–18
Rebollo-Hernanz M, Aguilera Y, Herrera T, Cayuelas LT, Dueñas M, Rodríguez-Rodríguez P, Ramiro-Cortijo D, Arribas SM, Martín-Cabrejas MA (2020) Bioavailability of melatonin from lentil sprouts and its role in the plasmatic antioxidant status in rats. Foods 9(3):330
Rishi A (2020) Nutraceuticals in India: A game changing opportunity in the healthcare and food industry. ETHealthWorld. https://health.economictimes.indiatimes.com/news/pharma/nutraceuticals-in-india-a-game-changing-opportunity-in-the-healthcare-and-food-industry/74512687
Ruchi V, Mohan V, Ramya Bai M, Sudha V (2014) Glycemic index of Indian cereal staple foods and their relationship to diabetes and metabolic syndrome. Wheat and Rice in Disease Prevention and Health, pp 333–346
Rutu S, Pearl S, Shekhar A (2017) A shelf-life study of nutritious ladoo. Int J Food Nutr Sci 6(3):57–60
Samarth AG, More DR, Hashmi SI, Sontakke MD (2018) Studies on screening of sweet sorghum cultivars for tender sweet sorghum (Hurda). J Pharmacognosy Phytochemistry 7(2):2698–2700
Sarkar P, DH LK, Dhumal C, Panigrahi SS, Choudhary R (2015) Traditional and ayurvedic foods of Indian origin. J Ethnic Foods 2(3):97–109
Sawicki T, Wiczkowski W (2018) The effects of boiling and fermentation on betalain profiles and antioxidant capacities of red beetroot products. Food Chem 259:292–303
Saxena KB, Kumar RV, Sultana R (2010) Quality nutrition through pigeon pea- a review. Health 2(11):1335–1344
Sharma A, Mazumdar B, Keshav A (2019) Formulation, standardization and characterization of novel sattu beverage enriched with beetroot juice. J Food Sci Technol 57(3):1–8
Sharma S, Saxena J (2016) Phytochemical screening and quantitative estimation of total phenolic content and total flavonoid content of grains of *paspalum scrobiculatum*. Asian J Pharm Clin Res 9(6):73–76
Shashikumar JN, Champawat PS, Mudgal VD, Jain SK, Deepak S, Mahesh K (2018) A review: food, medicinal and nutraceutical properties of fenugreek (*Trigonella Foenum* - Graecum L.). Int J Chem Stud 6(2):1239–1245
Silva EGDSO, Rangel AHDN, Mürmam L, Bezerra MF, Oliveira JPFD (2019) Bovine colostrum: benefits of its use in human food. Food Sci Technol 39:355–362
Singh V, Sati V, Agarwal S (2016) Potential of germinated sorghum (Sorghum bicolor) for utilization as a health food. Int J Sci Environ Technol 5(6):4203–4212

Snehlata HS, Payal DR (2012) Fenugreek (*Trigonella foenum*-graecum L.): an overview. Int J Current Pharmaceutical Rev Res 2(4):169–187
Srinivasan K (2010) Traditional Indian functional foods. Functional Foods of the East. Nutraceutical Sci Technol Ser 10:51–76
Taverniti V, Guglielmetti S (2011) The immunomodulatory properties of probiotic microorganisms beyond their viability (ghost probiotics: proposal of para-probiotic concept). Genes Nutr 6(3):261–274
Xavier-Santos D, Bedani R, Lima ED, Saad SMI (2020) Impact of probiotics and prebiotics targeting metabolic syndrome. J Funct Foods 64:103666

Chapter 19
Diet and Circadian Rhythms: Implications for Aging and Longevity

Anita Jagota, Zeeshan Akhtar Khan, and M. Sultan Khan

Abstract Circadian rhythms are maintained by the interaction of external environmental cues with body's molecular clock machinery and help to optimize physiological functions by temporally coordinating them at the cellular, tissue, organ and behavioural level of an organism. Feeding-fasting pattern is one of the most important external cues that affect the robustness of the biological rhythms. Upon aging due to loss of indispensable neurons in the master clock- Suprachiasmatic nucleus (SCN), these rhythms get compromised and so does the temporal coordination thus leading to various age-related pathologies. Irregular eating-fasting patterns can also temporally disrupt the coordination between metabolism and physiology, leading to the onset of many chronic diseases and early aging. Hence, avoiding irregular feeding-fasting habits and maintaining strong rhythmic cycles following optimum amplitude and phase of rhythms can help in healthy aging and preventing diseases such as sleep disorders, cardiovascular, metabolic disorders, diabetes, obesity, breast cancer, inflammation, hypertension, neurodegeneration etc. Therefore, synchronizing the external cues and timings of signals from master clock along with time restricted eating patterns can help in sustaining a robust circadian clock. Such synchronization will help in preventing the diseases and improving their prognosis. In this chapter, we aim to discuss the role of diet in restoration of age-induced circadian dysfunction.

Keywords Circadian clock · Dietary intervention · Metabolism · Physiology · Sleep · Fasting

Z.A. Khan and M.S. Khan—Equal Contribution.

A. Jagota (✉) · Z. A. Khan · M. S. Khan
Neurobiology and Molecular Chronobiology Laboratory, Department of Animal Biology, School of Life Sciences, University of Hyderabad, Hyderabad 500 046, India
e-mail: ajsl@uohyd.ernet.in

S. I. S. Rattan and G. Kaur (eds.), *Nutrition, Food and Diet in Ageing and Longevity*, Healthy Ageing and Longevity 14, https://doi.org/10.1007/978-3-030-83017-5_19

19.1 Introduction

Circadian rhythms are evolutionarily conserved biological rhythms in almost all organisms ranging from archaea to mammals. Circadian (Latin: *circa* – approximately; *diēm*– day) rhythms are approximately 24 hour(h) rhythmic oscillations that regulate the physiology of an organism at molecular and behavioural level (Whitehead et al. 2009; Edgar et al. 2012). The adaptation to external environmental conditions using clock system allows an organism to predict changes and give an advantage to survival (Kondratov et al. 2006). Many diseases are associated with disruptions in the circadian clock like psychological disorders, sleep disorders, metabolic disorders, cardiovascular diseases, cancer, insomnia, fatigue, disorientation and hormonal profile shifts in night shift workers etc. (Hofman et al. 2006; Gibson et al. 2009). During aging, there is a shift in both amplitude and phases of circadian rhythm (Gibson et al. 2009; Jagota 2012).

Circadian rhythms have evolved as an adaptation to the oscillations found in the environment and to get entrained by them, notably to the day-night cycle (Jagota 2006). In the case of mammals, the main circadian rhythm regulator is SCN containing ~20,000 neurons located in the hypothalamic region. It acts as a timekeeper in controlling and synchronizing the circadian period of many physiological and behavioral functions and entraining them to 24 h light and dark cycles (Jagota et al. 2000; Jagota 2006; Takahashi et al. 2017). Three major pathways, Retino-hypothalamic tract (RHT), Geniculo-hypothalamic tract (GHT), and Retino-raphe pathway (RRP) receive the information from the environment through several cues such as photic (light) and non-photic cues (food, sound, humidity etc.). All these pathways diverge from the retinal ganglion cells before they enter into the SCN. Among these three, RHT is a photic pathway that has a direct synaptic contact of retinal ganglionic cells with the SCN; GHT is an indirect photic pathway where the retina conveys input signals to Intergeniculate Nucleus (IGL) via a separate branch of RHT that overlaps with the RHT terminals in the SCN; RRP is the third major input pathway which participates in the non-photic regulation of the SCN, where neuronal fibres from raphe nuclei end in the core region of the SCN (Jagota 2012).The auto-regulatory transcriptional and translational feedback loops drive the coordinated expression of genes such as (*Clock*)-circadian locomotor output cycles kaput, *(Bmal1)-Brain and muscle aryl hydrocarbon receptor nuclear translocator-like protein 1), (Per1)-Period1, (Per2)-Period2, (Per3)-Period3, (Cry1)-Cryptochrome1, (Cry2)-Cryptochrome2, (Ror)-retinoic acid-related orphan nuclear receptors,* and *(Rev-Erb)-reverse erythroblastosis virus* etc. at both levels (Takahashi 2017).

The SCN plays the role of central clock or the relay centre of information. It regulates the release of neurohormone melatonin, the messenger of time from pineal gland. Melatonin then synchronizes the peripheral clocks with the central clock (Vriend and Rieter 2014). Every mammalian cell is autonomous and has its own clock machinery constituting the peripheral clock system controlled by the SCN through both sympathetic as well as parasympathetic pathways (Kalsbeek et al.

2010). The endogenous rhythms of clock gene expression have been reported by researchers in cell culture of peripheral clocks like liver, adrenal glands, hormones like adrenocorticotropic hormone (ACTH) and glucocorticoids (Cailotto et al. 2005; Mahoney et al. 2010).

19.2 Food as an Important Clock Regulator

Food has been reported as an important non-photic *zeitgeber* (ZT) or time giver to synchronize an organism's biological rhythms and shifting the timing of food leads to the shift in the activity of an organism (Carneiro et al. 2012). In SCN lesioned mice, food restriction has demonstrated to induce locomotor behaviours as well as temperature rhythms (Froy et al. 2010). Food timings entrain circadian clocks in different brain regions and most peripheral organs, thereby synchronizing their daily rhythms. This entrainment of peripheral clocks to mealtime is accomplished by multiple feeding-related signals, including absorbed nutrients and metabolic hormones, acting in parallel or series in a tissue-specific manner. Signals that synchronize circadian clock in the brain with feeding time are presumed to generate the circadian rhythms of food-anticipatory activity (FAA) that emerge when food is restricted to a fixed daily mealtime. Such FAA is regulated and paced by the food-entrainable oscillator (FEO) located outside the SCN (Flores et al. 2016; Chaudhari et al. 2017). Several timekeeping mechanisms involved in the FAA occurring before food intake include neuronal activation, molecular clock entrainment, hormonal cues, and metabolic regulation (Tahara et al. 2013; Challet 2019). The synchronization between food entrainable clock and central clock system is driven by a complex mechanism including humoral and enzymatic regulatory pathway (Challet 2019). β-hydroxybutyrate (β-OHB) synthesized in the liver during fasting conditions has been proposed to be a key candidate molecule in those pathways. Apart from being an energy source, β-OHB has many cellular signalling actions and participates in FAA modulation and its production is clock regulated involving *Per2* (Newman et al. 2017).

19.3 Food and Melatonin Hormone

The synthesis and release of melatonin from the pineal gland is under regulation of the SCN. The photic cues are received by the SCN clock and relayed to pineal gland through a multi-synaptic pathway. Melatonin is a "night-time hormone" in all the animals (diurnal, nocturnal and crepuscular) irrespective of their activity niche and thus called messenger of darkness (Jagota et al. 2012; 2019). Melatonin is synthesised from serotonin through N-acetylation followed by methylation reaction in the presence of arylalkylamine N-acetyltransferase (AANAT; rate limiting enzyme) and hydroxyl indole-O-methyl-transferase (HIOMT) respectively (Jagota et al. 2012).

Endogenous melatonin rhythm modulates feeding and fasting behaviour and anticipation of meal time. Melatonin cues modulate the phase and amplitudes of various hormonal rhythms including rhythms of cortisol and insulin (Challet et al. 2019). Reduced levels of melatonin has been associated with aging (Reddy and Jagota 2015). Melatonin is an important hormetin (Jagota et al. 2019), and its manipulation with dietary interventions in elderly may lead to restoration of circadian clock functions and improvement of health and wellbeing.

19.4 Circadian Dysfunction with Aging

Aging is an inevitable unidirectional natural process in the lifespan of an organism. The disruption of the endogenous nature of the molecular clock, deteriorated temporal synchronisation of the oscillators and an overall change in the physiology dampens the circadian rhythm as the organism ages (Jagota et al. 2000; Manoogiana and Panda 2017; Jagota et al. 2019). A phase advance in body temperature and melatonin secretion rhythms has been associated with aging in humans. Alterations in the daily rhythms of clock genes (Mattam and Jagota 2014), serotonin metabolism (Kalyani and Jagota 2008; 2010; Reddy and Jagota 2015), antioxidant enzymes (Manikonda and Jagota 2012), leptin (Reddy and Jagota 2014), nitric oxide (NO), and Suppressor of cytokine signaling (*Socs*) (Vinod and Jagota 2016; 2017) expression has been reported from our laboratory. Additionally, alterations in daily rhythms of Sirtuin1 (Sirt1), Nuclear factor erythroid 2–related factor 2 (Nrf2), Rev-erbα and inflammatory markers including Nfkb1, Tumor necrosis factor alpha (Tnf-α), Interleukin-6 (Il-6), Toll-like receptor 4 (Tlr4) and Toll-like receptor 9 (Tlr9) affecting circadian timing system (CTS) were observed with aging (Thummadi and Jagota 2019; Kukkemane and Jagota 2020). As endogenous rhythms dampen and deteriorate with age, the contribution by the external cues and its timing starts playing an increasingly critical role in maintaining the amplitude and phase of an organism's circadian clock (Manoogian and Panda 2017; Jagota et al. 2020).

19.5 Circadian Rhythms, Metabolism and Homeostasis with Aging

The circadian clock controls energy metabolism and maintains homeostasis in peripheral tissues through the controlled expression of various metabolic hormones like leptin, ghrelin, secretin etc. involved in different metabolic pathways. This helps in maintaining the normal physiological functions and healthy aging (Green et al. 2008; Froy 2009).The circadian clock is itself under metabolic regulation and the disturbances induced by a nutrient imbalance results in circadian dysfunction (Chaix et al. 2014). Many hormones such as insulin, glucagon adiponectin, corticosterone, and ghrelin have been found to show circadian expression and oscillation (Ando et al. 2005; Yang et al. 2006). The temporal regulation of insulin which shows a peak in its production at 1700 h and a nadir at 0400 h is achieved by both the patterns of feeding-fasting and SCN signalling (Sadacca et al. 2011; Vieira et al. 2015). Leptin, a well-known appetite repressor also shows circadian rhythmicity in its expression. The removal of the SCN abolishes the rhythmicity of its secretion in rodents leading to uncontrolled feeding behaviour, obesity, pathologies, and early aging effects suggesting the role of the clock system in aging, metabolism and maintaining homeostasis (Kalra et al. 2003). The receptors for leptin and ghrelin have been demonstrated in SCN (Yi et al. 2006; Zigman, et al. 2006) thus, establishing a direct link between the main circadian clock (SCN) and metabolism (Prosser et al. 2003). Metabolism is also influenced by the microbiome which makes an important component of the gastrointestinal tract (GIT). The microbiome shows rhythmic oscillations in its composition as per the requirement in GIT for proper metabolism. Erratic feeding patterns can dampen the taxonomic diversity and disrupt the oscillating rhythm of the microbiome, contributing to metabolic disorders like intestinal dysbiosis, obesity, and early aging (Voigt et al. 2016). Forced feeding-fasting patterns achieved by different strategies like intermittent fasting (IF), periodic fasting (PF), and calorie restriction (CR) have been shown to restore some of these oscillations to normal and achieve healthy aging (Zarrinpar et al. 2014).

19.6 Chrononutrition: Timing of Food as a Therapeutic Intervention

Due to demanding work pressure and changing food habits, unhealthy and unscheduled meals including junk food have now become a part of our lifestyle. Such unscheduled meals with unhealthy amounts of sugar, salts, caffeine, processed meat, fats and an inadequate intake of fruits, green leafy vegetables, cereals etc. are the major risk factors for developing age-related pathologies and accelerate the aging process (Micha et al. 2017). Patterns of feeding and fasting can potentially contribute to the development of chronic pathologies and thus have an influential impact on human health and onset of diseases (Zarrinpar et al. 2014). Although clock oscillators in our body can recuperate from mild alteration in our daily feeding times, chronic imbalanced and unscheduled feeding behaviour results in untimed cues, circadian dysfunction and disease pathologies later in life with unhealthy aging (Asher and Sassone-Corsi 2015; Lopez-Minguez et al. 2019). Thus, the temporal attributes of food and its role in health and disease are as vital as the qualitative and quantitative nutritional aspects (Gupta et al. 2017; Kant et al. 2018).

19.6.1 Food, Energy Metabolism, Circadian Dysfunction, and Aging

Various mitochondrial rate-limiting enzymes are rhythmically expressed (Neufeld-Cohen et al. 2016). In circadian mutant mice, enforced feeding-fasting patterns can reinstate rhythmic expression of some of metabolites, such as Acylcarnitine carrier protein and Acyl CoA Dehydrogenase (Manoogian and Panda 2016; Neufeld-Cohen et al. 2016). Mice with a deleted exon-19 of the *Clock* gene shows an altered feeding rhythm, over-eating, obesity and other metabolic syndromes like high blood leptin, lipids and glucose levels (Turek et al. 2005). Such mutant ($Clock^{\Delta 19}$) mice also showed a decrease in the expression of hypothalamic peptides like ghrelin and orexin, which are important for energy balance (Turek et al. 2005). *Per2* mutations abolished rhythmicity for glucocorticoids, feeding patterns and caused obesity (Yang et al. 2009).

Aging leads to significant disruption in adipogenesis by affecting circadian components, *Bmal1, Rev-erbα* causing multiple pathologies like obesity, cardiovascular diseases and decline in longevity (Duez et al. 2008). Peroxisome proliferator-activated receptor α (PPARα) is important for transcriptional activation of *Bmal1* and genes responsible for the metabolism of lipids and glucose (Lefebvre et al. 2006; Lamia et al. 2008). Impairment of PPARα leads to defective fatty acid oxidative pathways, enhanced inflammatory response, oxidative stress and renal fibrosis and impaired energy metabolism (Adnan 2007). Adenosine 5′-monophosphate–activated protein kinase (AMPK), an energy sensor of cells integrates the circadian clock with metabolism by regulating the response to feeding and modulating NAD^+ levels and SIRT1 activity (Hardie et al. 2006; Canto et al. 2009). Peroxisome proliferator-activated receptor gamma coactivator 1-alpha (PGC-1α), a transcriptional co-activator of *Bmal1* and *Rev-erbα* also regulates energy metabolism. Disrupted expressions of AMPK and PGC-1α leads to abnormal daily diurnal rhythms, body temperature energy imbalance, metabolic disorders, multiple pathologies and early aging (Grimaldi et al. 2007). Another important protein found to display a link between metabolism and the circadian clock of mammals is SIRT1. The influence of nutrient state and the circadian clock on insulin sensitivity is via SIRT1 (Bass and Takahashi 2010). It is an NAD^+ dependent histone deacetylase that plays an important role in extending the life span in yeast, *Caenorhabditis elegans*, Drosophila and mice (Mair et al. 2008; Canto et al. 2009). SIRT1 can interact with CLOCK directly and deacetylate BMAL1 and PER2 in cultured fibroblasts (Asher et al. 2008; Nakahata et al. 2008). The information on the food, circadian clock, and metabolic factors discussed here has been compiled as a schematic diagram in Fig. 19.1.

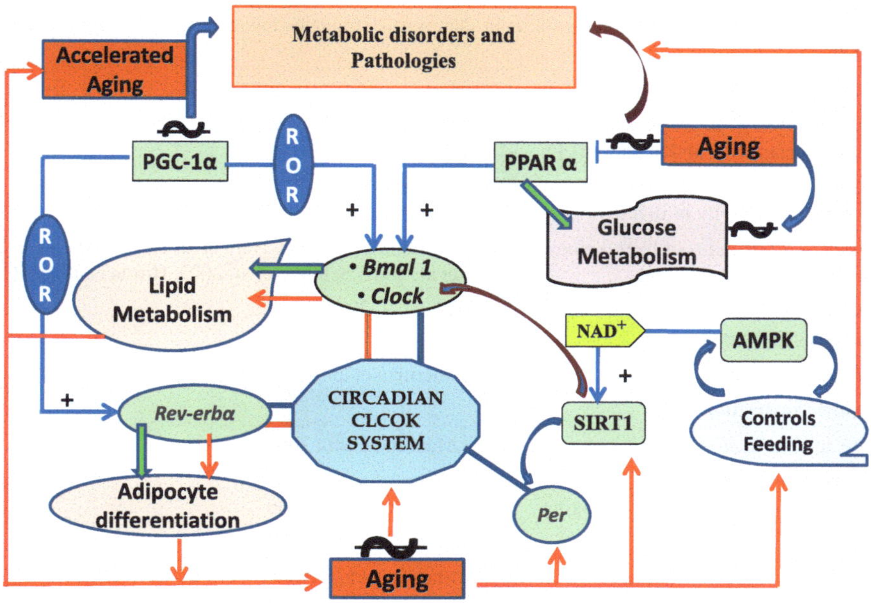

Fig 19.1 Simplified schematic representation of the link between circadian clock, metabolism and aging. Aging leads to significant disruption in glucose and lipid metabolism by affecting circadian components like *Bmal1, Clock, Rev-erbα* etc. and metabolic components like Peroxisome proliferator-activated receptor α (PPARα) and Peroxisome proliferator-activated receptor gamma coactivator 1-alpha (PGC-1α). Also, feeding and fasting cycle, controlled by Adenosine 5′-monophosphate–activated protein kinase (AMPK), with the involvement of NAD$^+$, and SIRT1 gets desynchronised upon aging. This age induced desynchronization between clock and metabolism leads energy imbalance, metabolic disorders and accelerated aging.  = Abolition of rhythms; (+) = Induces the expression

19.7 Various Dietary Interventions: Circadian Clock, Aging and Longevity

Dietary intervention is an ancient practice and various researchers using experimental models from invertebrates (*C. elegans*) to mammals (rat/mice) have demonstrated mean life-span extension by using different strategies of dietary interventions (Di Francesco et al. 2018). Such strategies (Table 19.1) achieved either by fasting for a certain period or reducing the calorie intake are discussed below.

19.7.1 Restricted Feeding (RF)

Restricting food to a particular time of the day while still ensuring nutritional adequacy is called restricted feeding (RF) (Froy et al. 2010). RF controls the FAA,

Table 19.1 Various dietary interventions in therapeutic strategies towards circadian dysfunction and aging

Restricted feeding (RF)	
Hepatic P450 levels, body temperature, locomotor activity, and heart rate (↑)	Hara et al. 2001, Hirao et al. 2006
Corticosterone secretion, gastrointestinal motility and activity of digestive enzymes	Stephan 2002
Core clock apparatus of liver, kidney, heart, and pancreas, life span (↑)	Damiola et al. 2000, Hara et al. 2001, Stokkan et al. 2001, Schibler et al. 2003, Hirota et al. 2004
Expression of c-myc & p53 (↓) and increases lifespan (↑)	Wu et al. 2004
Obesity (↓) & aging in rat (↓)	Sherman et al. 2012, Chaix et al. 2014
Caloric restriction (CR)	
Life span in *C. elegans*, *Drosophila*, rodents, and monkeys (↑)	Mair and Dillin 2008
Phosphorylation of CREB, and aging	Ripperger et al. 2006
Clock, memory & aging	Etchegaray et al. 2006
Life span in the clinical trials (↑)	Mattson et al. 2017
Oxidative stress, aging, mitochondrial function, and inflammation (↓)	Lopez-Lluch et al. 2019
Arterial hypertension (↓)	An et al. 2020
Intermittent fasting (IF)	
Life span as compared to food given ad libitum (↑)	Goodrick et al. 1990, Mattson et al. 2005
Glucose metabolism (↑), cardio-protection, Aging, neuro-protection (↑)	Anson et al. 2003, Contestabile et al. 2004; Mattson and Wan 2005
Resistance to aging & cancer (↑)	Descamps et al 2005, Mattson and Wan 2005
Cardiovascular diseases (↓)	Varady et al. 2007
Aging and Cognitive performance (↑)	Singh et al. 2012
Abdominal fat and blood pressure in humans (↓)	Harvie et al. 2011
Visceral fat, and insulin resistance (↓)	Barnosky et al. 2014
Inflammatory diseases (↓)	Johnson et al. 2007

which includes corticosterone secretion, gastrointestinal motility and activity of digestive enzymes before meals (Stephan et al. 2002). RF is dominant over the SCN and able to drive the rhythms in clock mutant rodents and animals with lesioned SCN regardless of photic cues (Hara et al. 2001; Stephan 2002). But in some cases, RF only affects the clock system in peripheral tissues such as the heart, kidney, pancreas and liver, causing their uncoupling from the SCN, followed by metabolic disorders, pathologies and early aging (Schibler et al. 2003; Hirota et al. 2004). This suggests the nutritional regulation of clock oscillators in peripheral tissues and its involvement in metabolic disorders and aging (Lin et al. 2008). Damiola et al. 2000 demonstrated when the food availability is made normal, the SCN, whose phase remains unaffected, resets the peripheral oscillators and synchronization of physiology. *Per2* mutant mice do not show wheel-running food anticipation (Mistlberger 2006). In one of the interesting studies, the survival time of mice inoculated with osteosarcoma was prolonged under RF (Wu et al. 2004). RF has been found to modify the expression of genes responsible for carcinogenesis and tumor progression such as c-myc and p53 (Filipski et al. 2005) but whether RF affects life span in a real sense or not is still not clear.

19.7.2 Intermittent Fasting (IF)

IF is a cyclic pattern of eating and fasting, one of the regimen of IF is alternate day fasting (ADF), where food is available *ad-libitum* every alternate day (Froy et al. 2010). Two more variations of IF include: Periodic Fasting (PF) which lasts for 2 or more days followed by the next cycle after 6–7 days (Longo et al. 2016; Vargas et al. 2020) and Time restricted feeding (TRF), where food is taken in a specific time window of 8 h or less (Mattson et al. 2017). Recently, IF has gained popularity due to its various beneficial effects on health (De Cabo et al. 2019). Animals on IF exhibit increased life span, improved glucose metabolism, cardio physiology, neuroprotection and resistance towards cancer (Descamps et al. 2005; Mattson et al. 2005; Varady et al. 2007). Rev-erbα, an important component of the circadian clock, also regulates the expression of genes involved in metabolism and inflammation. It was observed that there is an average 11 fold increase in its levels at the end of the 4th week during 30-days of IF (Mindikoglu et al. 2020). RF schedules elicit a phase shift in molecular and metabolic machinery components of peripheral clocks. Likewise, imposed periods of extended daily fasting, independent of dietary composition and calorie intake have significant metabolic and lifespan benefits (Mitchell et al. 2018).

19.7.3 Time-Restricted Feeding (TRF)

A feeding regime like Time-restricted feeding (TRF) has shown beneficial effects in animals and is believed to compensate and consolidate the circadian rhythms (Chaix et al. 2014; De Cabo and Mattson 2019). Flies on TRF showed a sustained nocturnal sleep which is about double the total sleep duration of the flies with food ad libitum (Gill et al. 2015). Nocturnal rodents given TRF with food ad libitum during the light phase (normally their rest period) show altered expression of clock genes and clock-controlled genes (CCGs) in the peripheral tissues without affecting SCN (Damiola et al. 2000; Stokkan et al. 2001). A differential adaption to the novel feeding regimens has been observed in peripheral clocks where Kidneys, lungs, or heart adapts to changes slower than the liver. TRF of normal diets improves energy metabolism in humans in short-term studies and contributes to a healthy life span even though calories are not restricted (Sutton et al. 2018; Jamshed et al. 2019). High-fat diets (HFD), representing the western diet model, has been reported to cause circadian dysfunction by disrupting the feeding-fasting rhythms and gene expression rhythms (Hatori and Panda 2015; Potter et al. 2016). A very recent report has shown that an extended HFD and night restricted feeding (NtRF) regimen on senescence-accelerated mouse strain, SAMP8 ameliorates age-related phenotypes. The positive impact of HFD has been manifested till metabolic perturbations kicked in. So, the HFD and chrono-nutritional feeding in combination can be an effective anti-aging strategy (Oike et al. 2020) ensuring a healthy life span.

19.7.4 Caloric Restriction (CR)

CR has almost similar beneficial effects on health as that mediated through IF and leads to circadian clock plasticity by chromatin remodelling (Nakahata et al. 2007). CR can induce expression of brain-derived neurotrophic factor (BDNF) in the hippocampus, phosphorylation of cAMP response element-binding protein (CREB), dendritic spine density and transcription of BDNF (Whitmore et al. 2000; Eide et al. 2001; Etchegaray et al. 2006). CR not only reduces the risk of age-associated circadian dysfunction and disorders but also significantly impedes aging and enhances longevity (Taormina et al. 2014). When *Drosophila* was subjected to caloric restriction, expression of several clock genes showed increase in amplitude (Katewa et al. 2015). CR affects significantly the expression of several genes in both central clock-SCN and the peripheral clocklike the liver (Patel et al. 2016). The importance of meal timing by studies on α-MUPA mice have shown that reduced calories alone were not able to sustain rhythms unless feeding was spontaneously timed at night, or the day through a RF protocol. Further, to investigate the involvement of the circadian clock in impacting the metabolic activity and life span via CR, core clock gene knockouts were used (*Bmal1* in mice and *Per* and *Timeless* (*Tim*) in Drosophila); circadian clock disruption along with multiple metabolic disorders and increased life span was noted in such experiments (Patel et al. 2016; Katewa et al. 2015). In mammals, CR mediates decline in blood IGF-1 level and the effect was compromised in mice deficient for BMAL1, an important circadian transcriptional factor. With CR, the diurnal activity and sleep pattern dampens in fruit flies as seen in humans with night sleep pattern. An overall diagrammatic representation of effects of different feeding regimens on aging and longevity has been summarised as Fig. 19.2.

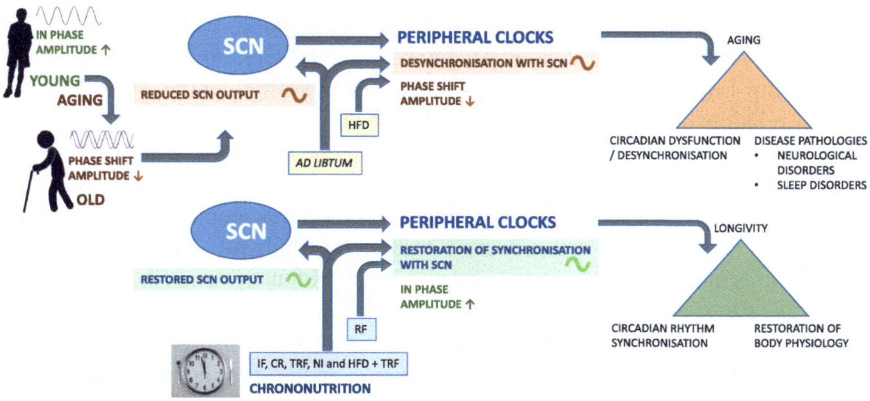

Fig 19.2 A diagrammatic representation of the effect of feeding regimens on aging and longevity through resetting of central and peripheral clock system. Intermittent fasting, caloric restriction, time restricted feeding, and high fat diet with TRF helps in resetting the circadian rhythms in both peripheral and the central clock systems but RF resets the rhythms only in the peripheral clocks

19.8 Nutritional Epigenetic: Aging and Clock

The nutritional modulation of the circadian clock has been reported to be linked with the epigenetic regulation of various clock genes. The modifications by sirtuins, a class III NAD$^+$ dependent diet-sensitive histone deacetylase enzymes, are of vital importance to genome stability. *Sirt1* is believed to be a key epigenetic regulator protecting the mammals from events and consequences that ultimately lead to aging (Hudec et al. 2020). Higher *Silent information regulator 2* (*Sir2;* mammalian *Sirt1* equivalent) expression has been reported to extend the lifespan in *Drosophila melanogaster* and *Sirt1* knock-outs die young owing to developmental defects. Single nucleotide polymorphism (SNPs) gene variants for core clock genes have been implicated in age related disorders and individual dietary response in humans. Among the epigenetic mechanisms that control circadian rhythms, microRNAs are the least studied ones compared to SNPs and such studies can be a key to novel chrono-therapeutic interventions (Mico et al. 2016).

19.9 Conclusion

Dietary interventions are promising and easy-to-adapt strategies for the modulation and prevention of circadian dysfunction and senescence in humans of different ages. Recently, adjustment of caloric intake using different strategies like IF, PF and TRF have emerged as potential strategies towards treatment of metabolic syndromes like cardiovascular diseases, cancer and possibly neurodegenerative diseases etc. and helping in the synchronization of the circadian clock system. RF entrains peripheral clocks suggesting their role via temporal food consumption, whereas, CR and IF appears to synchronize the central pacemaker in the SCN, suggesting their role by involving low calories intake in the entrainment of the central clock system. A direct relationship between feeding time and onset of chronic diseases such as obesity, breast cancer and inflammatory and metabolic disorders, neurological and sleep disorders explains that metabolic state is linked to sensitization in different parts of the brain, especially the hypothalamus and hippocampus, to maintain the coordination between the neuroendocrine system, metabolism, and energy balance. The CTS, thus influences and resets a wide variety of output systems like cellular and physiological systems to perform in a more synchronized manner hence maintaining the robust circadian rhythms using dietary interventions can ensure better tissue and body homeostasis and mediating aging attenuation and promoting longevity.

Acknowledgements AJ is thankful to Prof. Suresh Rattan and Prof. Gurcharan Kaur for giving this opportunity and sincere patience during preparation of manuscript. The work is supported by DBT (102/IFD/SAN/5407/2011-2012), ICMR (Ref. No. 55/7/2012-/BMS), and UPE II and Grants to AJ. ZAK and SMK are thankful to BBL Fellowships from University of Hyderabad.

Compliance with Ethical Standards

Conflict of Interest All authors declare they have no conflict of interest.

References

An HS, Lee JY, Choi EB, Jeong EAS, Kim HJ, Park KE, Jin KA, Lee Z, J.E. Koh J.S, (2020) Caloric restriction reverses left ventricular hypertrophy through the regulation of cardiac iron homeostasis in impaired leptin signalling in mice. Sci Rep 10:7176

Ando H, Yanagihara H, Hayashi Y, Obi Y, Tsuruoka S, Takamura T, Kaneko S, Fujimura A (2005) Rhythmic messenger ribonucleic acid expression of clock genes and adipocytokines in mouse visceral adipose tissue. Endocrinology 146:5631–5636

Anson RM, Guo Z, de Cabo R, Iyun T, Rios M, Hagepanos A, Ingram DK, Lane MA, Mattson MP (2003) Intermittent fasting dissociates beneficial effects of dietary restriction on glucose metabolism and neuronal resistance to injury from calorie intake. Proc Natl Acad Sci USA 100:6216–6220

Asher G, Sassone-Corsi P (2015) Time for food: the intimate interplay between nutrition, metabolism, and the circadian clock. Cell 161:84–92

Asher G, Gatfield D, Stratmann M, Reinke H, Dibner C, Kreppel F, Mostoslavsky R, Alt FW, Schibler U (2008) SIRT1 regulates circadian clock gene expression through PER2 deacetylation. Cell 134:317–328

Barnosky AR, Hoddy KK, Unterman TG, Varady KA (2014) Intermittent fasting vs daily calorie restriction for type 2 diabetes prevention: a review of human findings. Transl Res J Lab Clin Med 164:302–311

Bass J, Takahashi JS (2010) Circadian integration of metabolism and energetic. Science 330(6009):1349–1354

Cailotto C, La Fleur SE, Van Heijningen C, Wortel J, Kalsbeek A et al (2005) The suprachiasmatic nucleus controls the daily variation of plasma glucose via the autonomic output to the liver: are the clock genes involved? Eur J Neurosci 22:2531–2540

Canto C, Gerhart-Hines Z, Feige JN, Lagouge M, Noriega L, Milne JC, Elliott PJ, Puigserver P, Auwerx J (2009) AMPK regulates energy expenditure by modulating NAD^+ metabolism and SIRT1 activity. Nature 458:1056–1060

Carneiro BTS, Araujo JF (2012) Food entrainment: major and recent findings. Front Behav Neurosci 6:83

Chaix A, Zarrinpar A, Miu P, Panda S (2014) Time-restricted feeding is a preventative and therapeutic intervention against diverse nutritional challenges. Cell Metab 20:991–1005

Challet E (2019) The circadian regulation of food intake. Nat Rev Endocrinol 15:393–405

Chaudhari A, Gupta R, Makwana K, Kondratov R (2017) Circadian clocks, diets and aging [JB]. Nutr Healthy Aging 4(2):101–112

Contestabile A, Ciani E (2004) Dietary restriction differentially protects from neurodegeneration in animal models of excitotoxicity. Brain Res 1002:162–166

Damiola F, Le Minli N, Preitner N, Kornmann B, Fleury-Olela F, Schibler U (2000) Restricted feeding uncouples circadian oscillators in peripheral tissues from the central pacemaker in the suprachiasmatic nucleus. Genes Dev 14:2950–2961

de Cabo R, Mattson MP (2019) Effects of intermittent fasting on health, aging, and disease. N Engl J Med 381(26):2541–2551

Descamps O, Riondel J, Ducros V, Roussel AM (2005) Mitochondrial production of reactive oxygen species and incidence of age-associated lymphoma in OF1 mice: effect of alternate-day fasting. Mech Ageing Dev 126:1185–1191

Di Di Francesco A, Germanio C, de Bernier M, Cabo RA (2018) A time to fast. Science 362:770–775

Duez H, Staels B (2008) Rev-erb alpha gives a time cue to metabolism. FEBS Lett 582:19–25

Edgar RS, Green EW, Zhao Y, van Ooijen G, Olmedo M, Qin X, Xu Y, Pan M, Valekunja UK, Feeney KA, Maywood ES, Hastings MH, Baliga NS, Merrow M, Millar AJ, Johnson CH, Kyriacou CP, O'Neill JS, Reddy AB (2012) Peroxiredoxins are conserved markers of circadian rhythms. Nature 485:459–464

Eide EJ, Virshup DM (2001) Casein kinase I: another cog in the circadian clockworks. Chronobiol Int 18:389–398

Erol A (2007) The Functions of PPARs in aging and longevity. PPAR Res 2007:1–10

Etchegaray JP, Yang X, DeBruyne JP, Peters AH, Weaver DR, Jenuwein T, Reppert SM (2006) The polycomb group protein EZH2 is required for mammalian circadian clock function. J Biol Chem 281:21209–21215

Filipski E, Innominato PF, Wu M, Li XM, Iacobelli S, Xian LJ, Levi F (2005) Effects of light and food schedules on the liver and tumor molecular clocks in mice. J Natl Cancer Inst 97:507–517

Flores DEFL, Bettilyon CN, Jia L, Yamazaki S (2016) The running wheel enhances food anticipatory activity: an exploratory study. Front Behav Neurosci 10:143

Froy O (2009) Metabolism and circadian rhythms-implications for obesity. Endocr Rev 31:1–24

Froy O, Miskin R (2010) Effect of feeding regimens on circadian rhythms: implications for aging and longevity. Aging 2(1):7–27

Gibson EM, Williams WP 3rd, Kriegsfeld LJ (2009) Aging in the circadian system: considerations for health, disease prevention, and longevity. Exp Gerontol 44:51–56

Gill S, Panda S (2015) A smartphone app reveals erratic diurnal eating patterns in humans that can be modulated for health benefits. Cell Metab 22(5):789

Goodrick CL, Ingram DK, Reynolds MA, Freeman JR, Cider N (1990) Effects of intermittent feeding upon body weight and lifespan in inbred mice: interaction of genotype and age. Mech Ageing Dev 55:69–87

Green CB, Takahashi JS, Bass J (2008) The meter of metabolism. Cell 134:728–742

Grimaldi B, Sassone-Corsi P (2007) Circadian rhythms: metabolic clockwork. Nature 447:386–387

Guan XM, Hess JF, Yu H, Hey PJ, van der Ploeg LH (1997) Differential expression of mRNA for leptin receptor isoforms in the rat brain. Mol Cell Endocrinol 133:1–7

López-Lluch G, Del Pozo-Cruz J, Sánchez-Cuesta A, Cortés-Rodríguez AB, Navas P (2019) Bioavailability of coenzyme Q10 supplements depends on carrier lipids and solubilization. Nutrition 57:133–140

Gupta NJ, Kumar V, Panda S (2017) A camera-phone based study reveals erratic eating pattern and disrupted daily eating-fasting cycle among adults in India. PLoS ONE 12:e0172852

Hara R, Wan K, Wakamatsu H, Aida R, Moriya T, Akiyama M, Shibata S (2001) Restricted feeding entrains liver clock without the participation of the suprachiasmatic nucleus. Genes Cells 6:269–278

Hardie DG, Hawley SA, Scott JW (2006) AMP-activated protein kinase–the development of the energy sensor concept. J Physiol 574:7–15

Harvie MN, Pegington M, Mattson MP, Frystyk J, Dillon B, Evans G, Cuzick J, Jebb SA, Martin B, Cutler RG (2011) The effects of intermittent or continuous energy restriction on weight loss and metabolic disease risk markers: a randomized trial in young overweight women. Int J Obes (lond) 35:714–727

Hatori M, Panda S (2015) Response of peripheral rhythms to the timing of food intake. Meth Enzymol 2015(552):145–161

Hirao J, Arakawa S, Watanabe K, Ito K, Furukawa T (2006) Effects of restricted feeding on daily fluctuations of hepatic functions including p450 monooxygenase activities in rats. J Biol Chem 281:3165–3171

Hirota T, Fukada Y (2004) Resetting mechanism of central and peripheral circadian clocks in mammals. Zoolog Sci 21:359–368

Hofman MA, Swaab DF (2006) Living by the clock: the circadian pacemaker in older people. Ageing Res Rev 5:33–51

Hudec M, Dankova P, Solc R, Bettazova N, Cerna M (2020) Epigenetic regulation of circadian rhythm and its possible role in diabetes mellitus. Int J Mol Sci 21(8):3005

Jagota A (2006) Suprachiasmatic nucleus: the center for circadian timing system in mammals. Proc Indian Natl Sci Acad 71:275–288

Jagota A (2012) Age-induced alterations in biological clock: therapeutic effects of melatonin. In: Brain aging and therapeutic interventions. Springer, Netherlands, pp 111–129

Jagota A, Kalyani D (2008) Daily serotonin rhythms in rat brain during postnatal development and ageing. Biogerontology 9:229–234

Jagota A, Kalyani D (2010) Effect of melatonin on age induced changes in daily serotonin rhythms in suprachiasmatic nucleus of male Wistar rat. Biogerontology 11:299–308

Jagota A, Horacio O, Schwartz WJ (2000) Morning and evening circadian oscillations in the suprachiasmatic nucleus in vitro. Nat Neurosci 3(4):372–376

Jagota A, Thummadi NB, Kukkemane K (2019) Circadian regulation of hormesis for health and longevity. In: The science of hormesis in health and longevity. Elsevier, pp 223–233

Jamshed H, Beyl R, Della Manna D, Yang E, Ravussin E, Peterson C (2019) Early time-restricted feeding improves 24-hour glucose levels and affects markers of the circadian clock, aging, and autophagy in humans. Nutrients 11(6):1234

Johnson JB, Summer W, Cutler RG, Martin B, Hyun DH, Dixit VD, Pearson M, Nassar M, Tellejohan R, Maudsley S (2007) Alternate day calorie restriction improves clinical findings and reduces markers of oxidative stress and inflammation in overweight adults with moderate asthma. Free Radical Biol Med 42:665–674

Kalra SP, Bagnasco M, Otukonyong EE, Dube MG, Kalra PS (2003) Rhythmic, reciprocal ghrelin and leptin signalling: new insight in the development of obesity. RegulPept 111:1–11

Kalsbeek A, Bruinstroop E, Yi CX, Klieverik LP, La Fleur SE, Fliers E (2010) Hypothalamic control of energy metabolism via the autonomic nervous system. Ann NY Acad Sci 1212:114–129

Kant AK (2018) Eating patterns of US adults: meals, snacks, and time of eating. Physiol Behav 193:270–278

Katewa SD, Akagi K, Bose N, Rakshit K, Camarella T, Zheng X, Hall D, Davis S, Nelson CS, Brem RB, Ramanathan A, Sehgal A, Giebultowicz JM, Kapahi P (2015) Peripheral circadian clocks mediate dietary restriction-dependent changes in lifespan and fat metabolism in drosophila. Cell Metab 23(1):143

Kondratov RV, Kondratova AA, Gorbacheva VY, Vykhovanets OV, Antoch MP (2006) Early aging and age-related pathologies in mice deficient in BMAL1, the core component of the circadian clock. Genes Dev 20:1868–1873

Kukkemane K, Jagota A (2020) Therapeutic effects of hydro-alcoholic leaf extract of With ania-somnifera on age-induced changes in daily rhythms of Sirt1, Nrf2 and Rev-erbα in the SCN of male Wistar rats. Biogerontology 21(5):593–607

Lamia KA, Storch KF, Weitz CJ (2008) Physiological significance of a peripheral tissue circadian clock. Proc Natl Acad Sci USA 105:15172–15177

Lefebvre P, Chinetti G, Fruchart JC, Staels B (2006) Sorting out the roles of PPAR alpha in energy metabolism and vascular homeostasis. J Clin Invest 116:571–580

Lin JD, Liu C, Li S (2008) Integration of energy metabolism and the mammalian clock. Cell Cycle 7:453–457

Longo VD, Panda S (2016) Fasting, circadian rhythms, and time-restricted feeding in healthy lifespan. Cell Metab 23:1048–1059

Lopez-Minguez J, Gómez-Abellán P, Garaulet M (2019) Timing of breakfast, lunch, and dinner. Effects on obesity and metabolic risk. Nutrients 11(11):2624

Mahoney CE, Brewer D, Costello MK, Brewer JM, Bittman EL (2010) Lateralization of the central circadian pacemaker output: a test of neural control of peripheral oscillator phase. Am J Physiol 299:R751–R761

Mair W, Dillin A (2008) Aging and survival: the genetics of life span extension by dietary restriction. Annu Rev Biochem 77:727–754

Manikonda PK, Jagota A (2012) Melatonin administration differentially affects age-induced alterations in daily rhythms of lipid peroxidation and antioxidant enzymes in male rat. Biogerontology 13(5):511–524

Manoogian ENC, Panda S (2016) Circadian clock, nutrient quality, and eating pattern tune diurnal rhythms in the mitochondrial proteome. Proc Natl Acad Sci 13(12):3127–3129

Manoogian ENC, Panda S (2017) Circadian rhythms, time-restricted feeding, and healthy aging. Ageing Res Rev 39:59–67

Mattam U, Jagota A (2014) Differential role of melatonin in restoration of age-induced alterations in daily rhythms of expression of various clock genes in suprachiasmatic nucleus of male Wistar rats. Biogerentology 15:257–268

Mattson MP, Wan R (2005) Beneficial effects of intermittent fasting and caloric restriction on the cardiovascular and cerebrovascular systems. J NutrBiochem 16:129–137

Mattson MP, Longo VD, Harvie M (2017) Impact of intermittent fasting on health and disease processes. Ageing Res Rev 39:46–58

Micha R, Peñalvo JL, Cudhea F, Imamura F, Rehm CD, Mozaffarian D (2017) Association between dietary factors and mortality from heart disease, stroke, and type 2 diabetes in the United States. JAMA 317:912–924

Micó V, Díez-Ricote L, Daimiel L (2016) Nutrigenetics and nutrimiromics of the circadian system: the time for human health. Int J Mol Sci 17(3):299

Mindikoglu AL, Abdulsada MM, Jain A, Choi JM, Jalal PK, Devaraj S, Mezzari MP, Petrosino JF, Opekun AR, Jung SY (2020) Intermittent fasting from dawn to sunset for 30 consecutive days is associated with anticancer proteomic signature and upregulates key regulatory proteins of glucose and lipid metabolism, circadian clock, DNA repair, cytoskeleton remodelling, immune system and cognitive function in healthy subjects. J Proteomics 217:103645

Mistlberger RE (2006) Circadian rhythms: perturbing a food-entrained clock. Curr Biol 16:R968–R969

Mitchell SJ (2018) Daily fasting improves health and survival in male mice independent of diet composition and calories. Cell Metab 29(1):221–228

Nakahata Y, Grimaldi B, Sahar S, Hirayama J, Sassone-Corsi P (2007) Signalling to the circadian clock: plasticity by chromatin remodelling. CurrOpin Cell Bio 19:230–237

Nakahata Y, Kaluzova M, Grimaldi B, Sahar S, Hirayama J, Chen D, Guarente LP, Sassone-Corsi P (2008) The NAD^+ dependent deacetylase SIRT1 modulates CLOCK-mediated chromatin remodelling and circadian control. Cell 134:329–340

Neufeld-Cohen A, Robles MS, Aviram R, Manella G, Adamovich Y, Ladeuix B, Nir D, Rousso-Noori L, Kuperman Y, Golik M, Mann M, Asher G (2016) Circadian control of oscillations in mitochondrial rate-limiting enzymes and nutrient utilization by PERIOD proteins. Proc Natl Acad Sci USA 113:E1673–E1682

Newman JC, Verdin E (2017) Beta-Hydroxybutyrate: a signalling metabolite. Annu Rev Nutr 37:51–76

Oike H, Ogawa Y, Azami K (2020) Long-term feeding of a high-fat diet ameliorated age-related phenotypes in SAMP8 mice. Nutrients 12(5):1416

Patel SA, Chaudhari A, Gupta R, Velingkaar N, Kondratov RV (2016) Circadian clocks govern calorie restriction-mediated life span extension through BMAL1- and IGF-1-dependent mechanisms. FASEB J 30(4):1634

Potter GD, Cade JE, Grant PJ, Hardie LJ (2016) Nutrition and the circadian system. Br J Nutr 116(3):434–442

Prosser RA, Bergeron HE (2003) Leptin phase-advances the rat suprachiasmatic circadian clock in vitro. Neurosci Lett 336:139–142

Reddy VDK, Jagota A (2014) Effect of restricted feeding on nocturnality and daily leptin rhythms in OVLT in aged male Wistar rats. Biogerontology 15:245–256

Reddy MY, Jagota A (2015) Melatonin has differential effects on age-induced stoichiometric changes in daily chronomics of serotonin metabolism in SCN of male Wistar rats. Biogerontology 16:285–302

Ripperger JA, Schibler U (2006) Rhythmic CLOCK-BMAL1 binding to multiple E-box motifs drives circadian *Dbp* transcription and chromatin transitions. Nat Genet 38:369–374

Sadacca LA, Lamia KA, DeLemos AS, Blum B, Weitz CJ (2011) An intrinsic circadian clock of the pancreas is required for normal insulin release and glucose homeostasis in mice. Diabetologia 54:120–124

Schibler U, Schibler U, Ripperger J, Ripperger J, Sa B, Sa B (2003) Peripheral circadian oscillators in mammals: time and food. J Biol Rhythms 18:250–260

Sherman H, Genzer Y, Cohen R, Chapnik N, Madar Z, Froy O (2012) Timed high-fat diet resets circadian metabolism and prevents obesity. FASEB J 26(8):3493

Singh R, Lakhanpal D, Kumar S, Sharma S, Kataria H, Kaur M, Kaur G (2012) Late-onset intermittent fasting dietary restriction as a potential intervention to retard age-associated brain function impairments in male rats. Age (dordr) 34:917–933

Stephan FK (2002) The "other" circadian system: food as a *Zeitgeber*. J Biol Rhythms 17(284):292

Stokkan KA, Yamazaki S, Tei H, Sakaki Y, Menaker M (2001) Entrainment of the circadian clock in the liver by feeding. Science 291:490–493

Sutton EF, Beyl R, Early KS, Cefalu WT, Ravussin E, Peterson CM (2018) Early time-restricted feeding improves insulin sensitivity, blood pressure, and oxidative stress even without weight loss in men with prediabetes. Cell Metab 27:1212–1221.e1213

Tahara Y, Shibata S (2013) Chronobiology and nutrition. Neuroscience 253:78–88

Takahashi JS (2017) Transcriptional architecture of the mammalian circadian clock. Nat Rev Genet 18:164–179

Taormina G, Mirisola MG (2014) Calorie Restriction in mammals and simple model organisms. BioMed Res Int 10:1155

Thummadi NB, Jagota A (2019) Aging renders desynchronization between clock and immune genes in male Wistar rat kidney: chronobiotic role of curcumin. Biogerontology 20(4):515–532

Turek FW, Joshu C, Kohsaka A, Lin E, Ivanova G, McDearmon E, Laposky A, Losee-Olson S, Easton A, Jensen DR, Eckel RH, Takahashi JS, Bass J (2005) Obesity and metabolic syndrome in circadian *Clock* mutant mice. Science 308:1043–1045

Varady KA, Hellerstein MK (2007) Alternate-day fasting and chronic disease prevention: a review of human and animal trials. Am J ClinNutr. 86:7–13

Vieira E, Merino B, Quesada I (2015) Role of the clock gene Rev-erbα in metabolism and in the endocrine pancreas. Diabetes Obes Metab 17:106–114
Vinod C, Jagota A (2016) Daily NO rhythms in peripheral clocks in ageing male Wistar rats: protective effects of exogenous melatonin. Biogerontology 17:859–871
Vinod C, Jagota A (2017) Daily Socs1 rhythms alter with ageing differentially in peripheral clocks in male Wistar rats: therapeutic effects of melatonin. Biogerontology 18:333–345
Voigt RM, Summa KC, Forsyth CB, Green SJ, Engen P, Naqib A, Vitaterna MH, Turek FW, Keshavarzian A (2016) The circadian clock mutation promotes intestinal dysbiosis. Clin Exp Res 40:335–347
Vriend J, Reiter RJ (2014) Melatonin feedback on clock genes : a theory involving the proteosome. J Pineal Res 58:1–11
Whitehead K, Pan M, Masumura K-I, Bonneau R, Baliga NS (2009) Diurnally entrained anticipatory behavior in archaea. PLoS ONE 4(5):e5485
Whitmore D, Cermakian N, Crosio C, Foulkes NS, Pando MP, Travnickova Z, Sassone-Corsi P (2000) A clock work organ. Biol Chem 381:793–800
Wu MW, Li XM, Xian LJ, Levi F (2004) Effects of meal timing on tumor progression in mice. Life Sci 75:1181–1193
Yang X, Downes M, Yu RT, Bookout AL, He W, Straume M, Mangelsdorf DJ, Evans RM (2006) Nuclear receptor expression links the circadian clock to metabolism. Cell 126:801–810
Yang S, Liu A, Weidenhammer A, Cooksey RC, McClain D, Kim MK, Aguilera G, Abel ED, Chung JH (2009) The role of *mPer2* clock gene in glucocorticoid and feeding rhythms. Endocrinology 150:2153–2160
Yi CX, van der Vliet J, Dai J, Yin G, Ru L, Buijs RM (2006) Ventromedial arcuate nucleus communicates peripheral metabolic information to the suprachiasmatic nucleus. Endocrinology 147:283–294
Zarrinpar A, Chaix A, Yooseph S, Panda S (2014) Diet and feeding pattern affect the diurnal dynamics of the gut microbiome. Cell Metab 20:1006–1017
Zigman JM, Jones JE, Lee CE, Saper CB, Elmquist JK (2006) Expression of ghrelin receptor mRNA in the rat and the mouse brain. J Comp Neurol 494:528–548

Chapter 20
Longevity Foods in Myth, Legend and History

Ilia Stambler

Abstract Human beings have always had a tremendous interest in longevity foods, the foods that may be most beneficial for healthy longevity. This interest is natural and may be deeply rooted in human thought and experience. The present historical-cultural analysis will survey some of the historical traditions, myths, legends, cognitive, social, cultural and ethical norms related to the varied concepts of longevity foods, across the world, from antiquity to the early modern period. This work will provide a multi-disciplinary framework for the discussion of longevity nutrition, by presenting the search for longevity foods as intertwined in a rich historical, socio-cultural, ideological and cognitive milieu conducive to the pursuit of healthy longevity.

Keywords Longevity foods · Ayurveda Rasayana · Traditional Chinese medicine · Internal and external alchemy · Magic · Religion · Ethics · Moderation · Dietary restriction · Rejuvenation

20.1 Longevity Foods and Culture

Human beings have always had a tremendous interest in longevity foods, the foods that may be most conducive to healthy longevity. This interest is natural and may be inherent in human thought and experience. Foods are the most obvious means to sustain life and influence life's powers. And if any and all foods can sustain life, then it could be straightforwardly inferred that some foods may be especially potent for sustaining life, for enhancing health and longevity. Thus, the interest in longevity foods has been perennial and universal across cultures. Yet, often this interest has been expressed as a demand for 'the right recipe,' asking what it is exactly that we need to eat to prolong the days of our life in good health. Often it has amounted to an even simpler demand for the right 'nutritious pill'—conveniently prepackaged, preferably at a discounted price—that one could swallow, without even the need to bother with recipes and preparations.

I. Stambler (✉)
Vetek (Seniority) Association—The Movement for Longevity and Quality of Life, Tel-Aviv, Israel

S. I. S. Rattan and G. Kaur (eds.), *Nutrition, Food and Diet in Ageing and Longevity*, Healthy Ageing and Longevity 14, https://doi.org/10.1007/978-3-030-83017-5_20

However, a cultural study would reveal that the search for longevity foods is much more complex than simply listing the right nutritious ingredients. Rather, it involves the construction of elaborate social practices, normative behaviors and even rituals for the obtaining, preparing and consuming of the foods. It creates articulate belief systems, even entire historical and communal identities, philosophical worldviews, cosmologies, mythologies and religions, woven around the foods. It entails special patterns and modes of narration and cognition. In short, it involves the vast historical and cultural lore integral and often determinative for the conceptions of longevity foods. Often the cultural and ideological motifs were the primary sources for creating the longevity recipes.

The present historical-cultural analysis will survey some of these historical traditions, myths, legends, cognitive, social and cultural norms, related to the diverse concepts of longevity foods. The aim of this work is to advance the academic and public interest in longevity foods, in healthy and life-prolonging nutrition. Yet, it aims to raise this interest not by providing statistical correlations and normative recommendations for 'the right recipes,' but by presenting the search for longevity foods as intertwined in a rich historical, socio-cultural, ideological and cognitive milieu conducive to the pursuit of healthy longevity.

20.2 Longevity Foods and Magic: The Power of Imagination

The discovery and development of longevity foods is now firmly associated with scientific thought, with the standards of scientific methodology and evidence. Yet, historically, the establishment of certain foods as longevity foods was firmly rooted in magical thinking, sometimes coupled with and sometimes decoupled from sound empirical observations of the foods' benefits. Yet, the element of magical thinking has been fundamentally present and needs to be recognized.

A classical explanation of magical thinking can be found in Sir James George Frazer's seminal anthropological work *The Golden Bough: A Study in Magic and Religion*, first published in 1890 (Frazer 1925, pp. 11–12):

> If we analyse the principles of thought on which magic is based, they will probably be found to resolve themselves into two: first, that like produces like, or that an effect resembles its cause; and, second, that things which have once been in contact with each other continue to act on each other at a distance after the physical contact has been severed. The former principle may be called the Law of Similarity, the latter the Law of Contact or Contagion. From the first of these principles, namely the Law of Similarity, the magician infers that he can produce any effect he desires merely by imitating it: from the second he infers that whatever he does to a material object will affect equally the person with whom the object was once in contact, whether it formed part of his body or not. Charms based on the Law of Similarity may be called Homoeopathic or Imitative Magic. Charms based on the Law of Contact or Contagion may be called Contagious Magic. ... Homeopathic magic is founded on the association of ideas by similarity; contagious magic is founded on the association of ideas by contiguity. ... Both branches of magic, the homoeopathic and the contagious, may

conveniently be comprehended under the general name of Sympathetic Magic, since both assume that things act on each other at a distance through a secret sympathy.

The Golden Bough gives a striking example of employing sympathetic/homeopathic magic—based on the principle "like produces like"—for increasing longevity. As Frazer reports (p. 36):

> To ensure a long life the Chinese have recourse to certain complicated charms, which concentrate in themselves the magical essence emanating, on homoeopathic principles, from times and seasons, from persons and from things. ... Amongst the clothes there is one robe in particular on which special pains have been lavished to imbue it with this priceless quality. It is a long silken gown of the deepest blue colour, with the word "longevity" embroidered all over it in thread of gold. As the garment purports to prolong the life of its owner, he often wears it, especially on festive occasions, in order to allow the influence of longevity, created by the many golden letters with which it is bespangled, to work their full effect upon his person.

Not just garments, but also other objects, including foods, could carry such a symbolic-magical function, wishfully thought to work their imaginary effects to extend a person's longevity. In this mode of thought, food is not just a mere foodstuff, but an empowering sign and symbol. The actual empirical influence, if it is indeed observed, is secondary to the original symbolic, magical and ritualistic function of the food.

The magical "association of ideas"—either based on "similarity" or "contiguity"—may be considered a ramification of basic human imagination, the metaphorical and metonymical thought processes that seek analogies and connections in the world of experience. The commonality of physical forms and spatial relations may be deeply engraved in the human mind (Johnson 1987). Thus it should not be surprising that human imagination made various symbolic associations between human life prolongation and various observable properties among the common foods. Various purported healing and life-prolonging powers of foods have been derived from their observable properties, such as "thorniness and odor," "size," "age," "movement," "quaking and rustling of leaves," presence of "seeds" and "exudations," "shape," "color," "place and method of growth," etc. (Simoons 1998). Foods with attractive attributes may have been believed to attract beneficial influences, while unattractive foods may have been hoped to repel threats (or vice versa). Associative imagination has been the basis of the ancient medical doctrine of "signatures" according to which the intrinsic powers of entities, either healing or impairing, are signified by their ostensible similarities to the object of their influence (Vannier 1945). By seeking "symbolic analogies" ("sympathies" or "affinities") for poisons and remedies, the proponents of this doctrine hoped to discover the means that would provide nourishment, while absorbing or expelling the poisons or excess matters, and thus purifying the body and inducing it into a state of stable equilibrium. Beside the theorizing, analogizing and imagining, actual effects from partaking of certain foods, either beneficial or harmful, could also have been observed, and formed the basis for dietary recommendations. And, of course, for many foods, the imagination may have contributed to the observed effects, due to what is now commonly referred to as "psychosomatic" or "placebo" influences. But for many foods, such empirical considerations appear to be only additional, and often secondary or even negligible, compared to the magical associations.

Based on such magical associative thought principles, or the sympathetic "association of ideas," not necessarily empirically valid, but rather intuitive and ostensible, there have been promulgated a vast assortment of foods presumably extending healthy longevity.

The magical justifications for selecting longevity foods are salient in Traditional Chinese Medicine, in particular in its branches stemming from the Taoist tradition of external alchemy or external elixir production (Waidan) and internal alchemy or body-internal elixir production (Neidan) (Kohn 2001). In traditional Chinese culture, longevity, even extreme longevity, has always been a defining, all pervasive pursuit, especially in Taoism (allegedly emerging c. 600 BCE with the teaching of Lao Tse), but apparently also earlier. In the traditional Chinese household pantheon, "Longevity" (Shou, 寿) is one of the three most venerated deities, alongside "Happiness" (Fu, 福) and "Prosperity" (Lu, 禄), altogether referred to as the "three lucky star gods—Fu, Lu, Shou." According to Chinese legend, the Yellow Emperor, Huangdi (Huang-ti), fabled to have originated many fields of Chinese culture around 2600–2700 BCE, also possessed the secret of immortality. He is not to be confused with China's first historical emperor and seeker of immortality recipes, Qin Shi Huangdi, 259–210 BCE. Taoism built on the ancient legend and magic, yet developed an elaborate prolongevity system, with many practitioners and many books discussing the achievement of extreme longevity on Taoist principles (Gruman 1966; Kohn 2001). In this tradition, the main means for attaining extreme longevity was through consuming special foods, often with salient imaginative magical characteristics.

Gerald Gruman, in his classical work *A History of Ideas About the Prolongation of Life: The Evolution of Prolongevity Hypotheses to 1800* (1966) lists some of the attributes and properties of foods commonly believed to extend longevity (the "prolongevity foods") in Taoist practice (Gruman 1966, p. 44):

> These prolongevity foods or "hsien [immortal] medicines" were chosen for a variety of reasons. Many of the organic products were singled out in accord with the phoenix theme, that there are plants and animals enjoying a much greater life span than that of man: included here would be items like tortoise broth, crane eggs, and pine resin. Eggs of all sorts were valued in line with the Taoist regard for the perfect vitality of the embryo. Peaches were associated with the fruit in the Western Paradise of Hsi Wang-mu. Great numbers of herbs and minerals were venerated for such properties as a red color (like cinnabar), a resemblance to man or an animal (e.g., various roots), a slippery fluidlike texture or a translucent, glowing appearance. At the top of the list were such valuable minerals as pearls, mica, jade, silver, gold, and cinnabar.

Most of these properties (e.g. redness, fluidity, glow) were imaginatively associated with vitality, with a greater content of the vital energy "chi," hence foods possessing such properties were singled out as longevity foods. The ancient text entitled "Ts'an T'ung Ch'i" ("The akinness of the three," c. 142 CE) attributed to the famous Taoist alchemist and inventor Wei Boyang (Wei Po-Yang), notes the principles for selecting such foods implicitly: "One should make inferences from clues and signs just as he would from the appearance of strangers he meets. The thing to do is to compare things by classes and to trace their beginnings and ends" (Wei Po-Yang 1932, p. 253).

The imaginative inferences are exemplified in the ancient Taoist text *Baopuzi* (Pao-p'u-tzu) or "The Book of the Master Who Embraces Simplicity" attributed to the alchemist Ge Hong (Ko Hung, 283–343 CE). In the chapter "The Genie's Pharmacopoeia," the book provides an extensive list of longevity foods and medicines, along with ways of their procurement, including cinnabar, gold, silver, excrescences, jades, mica, pearls, realgar, brown hematite, quartz, rock crystal, geodes, sulfur, wild honey, laminar malachite, pine and cypress resins, truffles, yellow dock, liriope, tree sesame, salomonia, goldthread, fern, mulberries, lycium (box-thorns), etc. etc. (Ware 1966, p. 178). Most of these substances are characterized by the natural properties of durability that could presumably be transferred to the persons who consume them. Of special interest is the valorization of mushrooms ("Zhi," sometimes translated as "excrescences") as longevity foods. *Baopuzi* lists several types of the longevity mushrooms or "excrescences": "rock, wood, herb, flesh, and the tiny, each of them has almost a hundred species" (Ware 1966, p. 179). Their potency is said to range from extending human longevity by hundreds of years to granting potential immortality. The metaphoric association of substances of such diverse types, commonly designated as "longevity mushrooms," seems not obvious. Yet presumably their common signs of durability, etymological or visual similarities, or perhaps some psychoactive properties, apparent to the imaginative magicians, justified their classification as a single general category of longevity foods. The magical mental association as a basis for the longevity food conception was particularly prominent in the Chinese tradition. But similar instances of association of ideas can be found throughout the world—in India, Europe, the Americas and elsewhere (Andrews 2000; Simoons 1998; Schultes et al. 2001).

The element of magical thinking has been persistent and pervasive in the history of "longevity foods." This element needs to be recognized, and should not be underestimated. We may rather ask ourselves, whether in our presumably more enlightened and evidence-based age, the same magical patterns of thought are still not strongly present. Apparently, large segments of the public still exercise the same type of magical justifications in their nutritional choices as our distant ancestors. Though there may be also more modern developments of magical thinking. Thus, marketers often advertise 'anti-aging brands' showcasing robust, successful and happy elderly consumers of the brands, yet without much actual evidence of the products efficacy. And yet some parts of the public believe that by partaking of these brands, they too will become as robust, successful and happy as people in the advertisements. Is it not yet another case of belief in "contact magic" based on the magical "Law of Contagion"? And when some parts of the popular-scientific community believe that life-prolonging effects of certain nutrients as observed in cells or simple model organisms, have immediate relevance and instructive value for humans—is it not a case of belief in "imitative magic" based on the magical "Law of Similarity"? And when some longevity advocates are convinced that by speaking so much more about longevity research, actual longevity therapies will necessarily arise much faster and the actual human longevity will become much greater—is it not yet another potential case of "sympathetic" magical thinking, when deeds are believed to be directly engendered by words? Even such commonplace phrases as "you are what you eat" may be yet another example of "sympathetic" magical perception, when certain ostensible attributes are projected on the entire object. Such imaginative magical thinking may be an inherent part of our human nature, but its presence needs to be recognized and distinguished from empirical scientific evidence, in order to better navigate our rational search for truly effective longevity foods.

20.3 Longevity Foods and Religion: The Power of Regulation

Magical thinking is not the only cognitive construct that has been strongly associated with the conceptions of longevity foods. Frazer distinguished magical thinking from religious thinking, and both from scientific thinking. Magical thinking sought connections and influences between phenomena in the natural world, even though those were erroneously deduced by mere imagination. Science also sought connections and influences in the natural world, though based on observation, evidence and logic. On the other hand, religious ideology has aimed to produce beneficial effects in our world by propitiating the supernatural realm, by following certain ideations and practices to attract the favors of deities and the assistance of divine powers, while avoiding their punishments and repelling their curses. Though, of course, combinations of different modes of thought have often been present, between magic and religion, but also between magic and science, as well as between science and religion. The connection to religious devotional practices and philosophical ideations is strongly felt in the religious traditions connected to longevity foods. The life-prolonging morality and self-control, but also altered states of mind, often feature as parts of such religious ideations and practices.

There have been many examples of narratives on great longevity, even radical life extension and immortality, and the dietary practices necessary for their attainment, in the religious traditions of Sumer, Egypt, Persia, China, Japan, Korea, ancient Greece and the Roman Empire, in Judaism, Christianity, Islam, Hinduism, as well as more modern religious movements (Stambler 2014). Rejuvenation, resurrection and extreme longevity thanks to special foods are also recurrent themes in the native myths of North, Central and South America, and in Oceania and Africa. There are stories, in the vast majority of these native mythological traditions, that humans were originally immortal, and succumbed to death only because of loss of self-control, negligence, ill-will or accident (Grimal 1989). Loss of self-control was an especially fatal human fault responsible for the loss of longevity foods and the resulting loss of the powers of longevity. Several characteristic examples will be provided further on.

Notably, in these traditions, prolongevity substances, especially plants, are almost invariably perceived as foods. Insofar as foods were generally vital for the survival, logically, also life-prolonging properties were presumed to be present in special kinds of foods. It may just have been difficult to imagine other means of consumption of putative longevity remedies. Hence longevity foods are often indistinguishable from longevity drugs.

Of further note is the fact that many substances associated with longevity were "entheogenic"—a term commonly denoting psychoactive substances used in religious practices, originating from the Greek words "entheos" (inspired by god) and "genesthai" (coming into being). Consuming such substances, in the religious practitioners' mind, enabled their connection to the gods or the godly realm, drew to them the beneficial divine powers or repelled harmful demonic influences. The connection with the high supernatural realm was often perceived as altered states of mind, including hallucinations. Or rather, the altered states of mind, such as hallucinations, were taken as signs of a connection with the higher otherworldly spiritual realm. The connection to the spiritual realm would hopefully enable the adept to derive from it special powers, including longevity. Yet, in religious practices, self-control and the dominion of the mind over the body were even more ubiquitous attributes associated with the consumption of longevity foods, as will be exemplified shortly. Often the powers of self-control or mind-over-body control were linked with altered states of mind, as indeed strong self-control can be considered a rather unusual or "altered" state of mind.

A clear and rather simple connection between longevity foods, altered states of mind and religion, can be seen in the ancient native American traditions, from North America through South America. In these traditions, some of the longevity-granting, rejuvenating and healing plants included cocoa, cactus, aloe (octli), ayahuasca (caapi jungle vine), and manioc (Grimal 1989; Andrews 2000; Schultes et al. 2001). Many of these supposedly life-prolonging plants were also psychoactive (Andrews 2000; Schultes et al. 2001). That may provide an additional insight into the connection of physical and mental spheres in the pursuit of longevity. Some of the explanations for the connection may be that psychoactive plants appeared to provide a conduit with the spiritual world and with the gods, and thus to help the practitioners draw in divine powers of vitality from the other world and partake in the vitalities of the gods they contacted. Though there could be a more mundane explanation, such as that certain physical sensations (weightlessness, sense of inclusiveness, etc.) induced in the psychoactive substance users, became associated in their mind with vitality and longevity. In any case, the connection of the pursuit of longevity with the mental state is again apparent.

Yet, the mental-physical connection may be even more encompassing, comprising issues of morality, self-control and social regulation. Thus, an American Indian legend tells that people would have been immortal had they followed a wise man's instructions, and welcomed an old man carrying a basket of rotting flesh, rather than a pleasant looking young man carrying sweets (Grimal 1989, p. 489). Thus, foolish indulgence and lack of foresight and discipline, in particular in relation to food, became the original sins that brought about human mortality. The instructions on the ethical behavior in relation to food, longevity and mortality, have been elaborated in virtually every religious tradition.

Thus, one of the earliest representations of rejuvenation and life extension in relation to foods, as well as one of the earliest known works of literature and religious instruction, is the Sumero-Babylonian *Epic of Gilgamesh*, a story about the hero's struggle with death (the most complete version has been dated from c. 1300 BCE to 650 BCE, but the story possibly originated as early as about 3000 BCE). According to the *Epic of Gilgamesh*: "There is a plant like a thorn with its root [deep down in the ocean], Like unto those of the briar (in sooth) its prickles will scratch [thee], (Yet) if thy hand reach this plant, [thou'lt surely find life (everlasting)]" (Campbell 1928, pp. 55–56, Tablet 11 "The Flood"). The plant has been likened to box-thorn and dog-rose (Veenker 1981). Gilgamesh initially procures the plant, but then loses it and with it loses the hope of immortality. Yet, the loss of this particular plant, and of the immortality it may have conferred, is only an outcome of a more profound human failing. The primary cause for the loss of immortality was the lack of self-control. When first tested for his worthiness by the immortal sage Utnapishtim, Gilgamesh loses the chance of immortality by not being able to exercise enough self-control to fight his urge to sleep. And then, after procuring the plant of immortality from the bottom of the ocean, Gilgamesh is not watchful enough and loses it to a serpent. Thus, the intrinsic human inability of self-restraint prevents further human development toward a more powerful, robust and long-lived being. The plant (foodstuff), regardless of any empirical qualities, serves to convey the allegory.

There are striking parallels between the description of the immortalizing plant, the flood and the story of the extremely long-lived Utnapishtim in the epic of *Gilgamesh* and the biblical stories (with the composition sometimes dated c. 1300 BCE to 450 BCE) about the "tree of life" and about the extreme longevity of antediluvian patriarchs (Genesis 2:9, 3:22–24, 5:1–32). Notably, the access to the tree of life was lost due to the lack of discipline of Adam and Eve. Also the great longevity of antediluvian patriarchs was annulled due to human depravity that brought about the deluge.

In the ancient cultural traditions, there were many "recipe" type referrals to longevity foods that did not seem to bear a direct relation to questions of morality, self-control, mental state or religious precepts. But some indirect relations could nonetheless be inferred, even in those less obvious cases. Thus, in one of the earliest known Egyptian medical papyruses, "The Edwin Smith Surgical Papyrus" (commonly dated to the period of the New Kingdom of Egypt, c. 1500 BCE), there is a "Recipe for Transforming an Old Man into a Youth." The recipe involved the use of bruised and dried "hemayet-fruit." Some recent identifications of this remedy vary from fenugreek to almond (Ghalioungui 1973; Jacobs and Hart 2019). The remedy would not only have a cosmetic anti-aging effect—remove wrinkles, beautify the skin, remove blemishes, disfigurements, and "all signs of age"—but it would also have a true rejuvenating effect, as it would remove "all weaknesses which are in the flesh" (Breasted 1930, pp. 506–507). Alongside the purely material recipes, the book abounds in religious incantations designed to ward off evil influences. In yet another ancient Egyptian medical papyrus, "The Ebers Papyrus" (c. 1500–1600 BCE), there are described cosmetic anti-aging treatments to prevent the graying of hair (for example by the use of honey, onion water, donkey liver and crocodile fat), and to stimulate hair growth (for example by the use of flaxseed oil, gazelle excrements and snake fat). Yet, actual treatment of aging as a medical condition was also mentioned: when a physician examines a person whose "heart is feeble as if the frailty of old age has overcome him … then say thou: 'It is an accumulation of morbid juices.' He shall not wilfully minimize the danger, nor put his trust in feeble remedies" (Bryan 1930, p. 142). Large portions of the "Ebers Papyrus" are dedicated to combating diseases due to unhealthy and immoderate eating, such as constipation, that could be indirectly related to moral instruction (Bryan 1930; Joachim 1890). In this medical text, invocations of healing gods—Isis, Osiris and Ra—are also a part of preparing medicines. Another famous instance of the interest in healing and rejuvenation in ancient Egypt was the lore of Imhotep, the legendary high priest and chief minister to the pharaoh Djoser, and the reputed builder of the first step pyramid (c. 2650–2600 BCE). Among other accomplishments, Imhotep was said to be skilled in the art of rejuvenation and later deified as a god of medicine (Hurry 1926). Thus, the connection of medical practice with religious instruction can be perceived.

Notably, in ancient Egypt, medicine was largely under the control of the powerful priesthood who appropriated the responsibility for the physical preservation of the king and his subjects, as well as for the magical and religious rituals aimed to ensure the appropriate order of personal, public and environmental affairs. Thus, the religious priestly component of ancient Egyptian medicine was prominent, even though the functions of a physician, a magician and a priest were not always uniformly co-exercised by the same practitioners (Ghalioungui 1973). The preservation from ravishes of old age in this life and the preservation of the physical body as a necessary condition for the preservation of the soul in the afterlife, may have been parts of the more general concern with preservation and perpetuation in that culture. In ancient Egypt, the religion, including its dietary regimens, may have served as a means to discipline the citizens, to control and regulate the social order. The desire for social order and preservation may have coalesced with the goals of health and life prolongation. Arguably, the preeminent preoccupation of the Egyptians with immortality was inseparably linked to their static cosmology, to their obsession with preservation, balance and constancy in all spheres including the society. Moreover, a case may yet be made that many of their pioneering technologies, from pyramid construction to embalming and surgery to prolongevity food recipes and regimens, may have been at least partly driven by their general desire for individual and social preservation. Thus, in the Egyptian tradition too, longevity foods could be seen as a part and parcel of much broader social and ideological frameworks.

20.4 Longevity Foods and Myth: The Power of Self-control

Also in Hinduism (or rather in the variety of religions of India designated by this term), the pursuit of great longevity and the foods that may help achieve this goal, have been persistent themes since a very early time, and have been related to deep social, ideological and moral foundations. In India, the immortal Rishis, Arhats, and the

Ciranjivas (the "extremely long-lived persons") have been revered. Their extreme longevity is often attributed to "Amrit"— अमृत—or the "nectar of immortality," a revered and desired substance. In the legend ("purana") of "Churning the Ocean of Milk," the different supernatural races—Asuras ("demons") and Devas ("gods")—work together to create this "nectar of immortality" (Lidke and Dirnberger 2009). In the Rigveda, one of the earliest known Vedic collections of India (c. 1700–1100 BCE), the entire Book 9 is composed of hymns praising the immortality-giving "Soma" plant (Griffith 1891). Recent identifications of "Soma" range from fly-agaric, ephedra and cannabis, to sacred lotus, heather and honey (Flattery and Schwartz 1989). In the ancient Indian epic of the Ramayana (often dated c. 400 BCE, and sometimes purported to relate to events occurring 4000 and even 5000 BCE), the monkey king Hanuman uses the Sanjeevani plant (translated as "One that infuses life") to revive Rama's younger brother Lakshman who was severely wounded by Ravan (Griffith 1895). The life-giving Sanjeevani plant (sometimes spelled "sanjivani") has been commonly identified as the lycophyte *Selaginella bryopteris*, growing at the Dunagiri (also called Dronagiri or Mahodaya) mountain in the Himalayas (Ganeshaiah et al. 2009). As narrated in the "Shiva Purana" legend (c. 800–1000 CE), at the war between asuras and devas, during the rule of the asura emperor Jalandhara, the asuras were resuscitated using "mritasanjivani vidya"—the sacred knowledge of revival, which might involve the use of the Sanjeevani plant, and other components unattainable to us, such as "drops of water infused with mantras." In the opposing camp, the devas too were revived by medicinal herbs brought from the Dronagiri mountain (Shastri 1970, pp. 870–871). Note that the mythological traditional narratives about the Amrit and Sanjeevani and the ritualistic formulas of the Soma may have served social functions: to maintain the social order, to bind the society together through narrative and ritual, to create and preserve the common historical identity. Thus, once again, the longevity foods and medicines function as elements of a vast cultural lore.

The traditional Indian medicine of Ayurveda, or "the science of (long) life," also has a strong sanction from religion. According to the Ayurveda tradition, the principles of Ayurveda were created by Brahma, and underwent a chain of transmission among deities, from Brahma to Prajapati to the Ashvins, to Indra, who then taught this art to Dhanvantari, an incarnation of Lord Vishnu, the protector of life and the giver of Ayurveda on Earth (Bhishagratna 1907, p. 8). Ayurveda includes a special field of Rasayana (meaning, in Sanskrit, the path "āyana" of essence or juice "rasa"), which is mainly dedicated to rejuvenation, and mainly utilizes edible remedies or foodstuffs (e.g. juices, soups, mixed decoctions, pastes and cakes). According to one of the earliest Ayurvedic texts, *The Sushruta Samhita* (Sushruta's Compilation of Knowledge, c. 800–300 BCE, attributed to Sushruta, an alleged disciple of Dhanvantari), life can be normally prolonged to 100 years. Yet, with the use of certain Rasayana remedies, prepared from certain plants, such as Vidanga-Kalpa and Brahmi Rasayana, life can be prolonged to 500 or 800 years (Bhishagratna 1911, pp. 518, 525). And the use of the "Soma plant, the lord of all medicinal herbs [24 candidate plants are named, such as Chandramah, Amsuman and Munjavan species], is followed by rejuvenation of the system of its user and enables him to witness ten thousand summers on earth in the full enjoyment of a new (youthful) body" (Bhishagratna 1911, p. 536).

Of special interest in the text are the instructions for the preparation and application of the Soma elixir and other Rasayana treatments, integrating the physical, intellectual and moral domains. Mental and spiritual aptitude is considered vital for the successful treatment. As it is said of the Soma: "The Soma plants are invisible to the impious or to the ungrateful as well to the unbeliever in the curative virtues of medicine and to those spiteful to the Brahmanas" (Bhishagratna 1911, p. 538). A person preparing for the treatment must perform a cleansing, and undergo the treatment with mental clarity and spiritual awareness, including the recital of incantations (mantras), exercising "a quiet control over the mind" and renouncing "all passions and anger" (p. 535). Without those components a remedy would not be effective, "as the application of a dye to a piece of dirty cloth will prove non-effective" (p. 515). The application of Rasayana foodstuffs must be adjusted with the rest of the diet, such as "meals without any salts" (p. 520), as well as with the environment, the season and the general regimen of life. One cannot just consume Rasayana remedies mindlessly and expect favorable results, insofar as the treatment is "a combination of medicine and mantra" (p. 527). The physical effects of rejuvenation are necessarily accompanied by moral and intellectual effects. For example, the use of Svetavalguja Rasayana "would make the life of its user sinless, and extend it in the full glow of health and vigour and in the sound enjoyment of a vigorous memory and of all his intellectual faculties to a hundred green summers" (p. 522). The remedies are to be applied "in a spirit of self-control, whereby he would be able to acquire longevity" (p. 527). Thus, moral virtues, especially the virtue of self-control are perceived in the Ayurveda tradition as integral for the effective longevity nourishment.

The religious devotion and the pursuit of rejuvenation and radical life extension are also present in another foundational text of Ayurveda, *The Charaka Samhita* (Charaka's Compilation of Knowledge, c. 300–100 BCE). Like Sushruta, Charaka attributes the origins of Ayurveda to the gods. According to the *Charaka Samhita*, the normal human life-span is 100 years. Yet, the users of an Amalaka Rasayana could live many hundreds of years and the users of the Amalakayasa Brahma Rasayana could reach the life-span of 1000 years (Van Loon 2003, p. 446, 455). The great sages, who grasped perfectly the knowledge of Ayurveda, "attained the highest well-being and nonperishable life-span" (Van Loon 2003, p. 107). This Ayurvedic text too emphasizes the vital importance of the synergy of the body and mind, spirituality and morality, social order and environmental soundness, for the successful longevity nourishment, beside the simple listing of nutritional ingredients. As the principle of synergy is succinctly described in the *Charaka Samhita*: "Thus the person subdued with malaise, depression, sleep, drowsiness, lassitude, lack of enthusiasm, dyspnea, incapability in physical and mental activities, loss of memory, intellect and lustre, becomes resort of illness and thus does not enjoy the normal life-span. Hence, looking to these defects, one should abstain from all the said unwholesome diet and activities so that he becomes fit for using the rasayana treatment" (p. 449). And furthermore,

"The formulations meant for providing longevity and alleviating senility and diseases succeed (only) in patients having purified mind and body and controlled self" (p. 467).

Of course, there may have been a considerable empirical basis for at least some of the longevity claims of at least some of the Rasayana remedies. Without at least some minimal shown efficacy, these remedies may have had difficulty surviving in medical study and practice for hundreds of years, to the present time. This empirical basis now continues to be evaluated, as many traditional Ayurvedic medicines are currently practiced and undergo testing and validation processes (Chaudhary et al. 2010). Yet, it is important to realize that empirical evidence for the efficacy of particular food ingredients, and of particular methods of their preparation, is not the only component in the construction of the longevity food traditions, often even not the main one. Religious and moral motivations for human development and for the maintenance of social order, the traditional narratives for the construction of personal as well as ethnic or national identities, the characteristic mental and psychological perceptions, are also indispensable components for any longevity food recipe.

Buddhism, having originated in India around the sixth century BCE, too had a strong connection to the pursuit of longevity (Maher 2009), with many common elements with other traditional Indian sources. Material means, in particular foodstuffs, for rejuvenation and life extension, have been developed and recommended by Buddhist physicians. But in this tradition too, the simple concoction of the remedy is only a part of the longevity food conception, other parts are from the social and mental domains: the ideological motivations, narrative traditions, communal rituals to maintain social coherence, and the self discipline to create and fulfill personal development.

In Buddhist tradition, the Great Buddha who grants Longevity is Amitābha, the Buddha of Infinite Light, also known as Amitāyus, the Buddha of Infinite Life. Those who invoke him will reach longevity in this realm, and will be reborn in Amitabha's Pure Land or the Land of Bliss (Sukhāvatī or Dewachen in Tibetan Buddhism) where they will enjoy virtually unlimited longevity. This pure and egalitarian land of longevity was created by Amitabha's avowed devotion and perseverance, through the adherence to the Buddhist merits, such as faith, energy, mindfulness, concentration, and wisdom.

In the Pure Land, divine plants grow that have life-sustaining and healing properties. There are mentioned palm trees, lotus trees and "the tree of awakening." However, these are not simple foods: they provide sustenance and healing, but not necessarily through ingestion (Gomez 1996). The mental nourishment, or food as mental nourishment, is also integral for the overall environment of the Land of Bliss that permits deathlessness. This is clearly seen as an ideal, a vision toward which a devotee should strive. The food in the Land of Bliss corresponds to the spiritual and beautifying nature of that world. Thus, in the treatise Sukhāvatī-vyūha Sūtra ("The Sutra Displaying the World of Bliss"), it is said that "living beings in the Land of Bliss do not consume coarse food, or food consumed in lumps or morsels; rather whatever fare they wish to eat, that exact same dish they visualize in their minds as already consumed. And thus their body is satisfied, their limbs are satisfied. Furthermore, they do not need to ingest any of this nourishment in order to feel satisfied" (Gomez 1996, p. 88).

Thus the Buddhist virtue of detachment may be implied, or satisfaction without addiction. The life-prolonging nutrition is not just moderate (subject to a conscious act of control of some degree), but the degree of conscious control is complete. No longer does the food dictate the nourishment to the person, but the person dictates the nourishment by one's own wish. With such a high degree of mental control, the lifespan also becomes unlimited, as the person frees oneself from deadly material limitations. This interpretation also seems to be supported, in the Sutra, by the parable of a prisoner who, even though offered "much food and drink of various kinds, pure and excellent," would not be able to "relish this food, consume it, or feel any satisfaction from it" (p. 105). The real benefit and enjoyment from food would accrue only if the prisoner is released from "the prison," by implication set free from the dependence on and addiction to the limiting and damaging environment. Thus a moral element is strongly implied.

Related to the Indian tradition, in the ancient Iranian (Aryan) religious tradition, there are remarkable references to great longevity and longevity foods. There are strong commonalities between these traditions, apparently due to the common Indo-European origins. This commonality is exemplified in the similar terms and names of the gods. Though, curiously, some terms assume rather opposite meanings. Thus, in the Indian tradition, the "devas" are perceived as benevolent gods, and the "asuras" are the demons. In contrast, in the Iranian tradition, "ahura" are the gods (cf. the benevolent gods Aesir in Norse religions), while the "daeva" are evil spirits (Parpola 2015). However, many terms are very similar. Thus, what is called in the Indian tradition the life-prolonging "Soma" is referred to as "Haoma" in ancient Iranian (Aryan) religious sources, such as *Avesta,* the sacred text of the Iranian Zoroastrian religion (c. 1200–200 BCE) (Darmesteter 1883; Flattery and Schwartz 1989). The Iranian religious tradition also has a strong instructive moral sentiment regarding longevity. According to the *Avesta*, during the rule of the mythical king Jamshid (Yima), people knew no disease, aging and death (Darmesteter 1880). The legendary "cup of Jamshid" that this king possessed was said to be a container for the elixir of immortality and at the same time a means to reveal the world knowledge and gain wisdom and truth (Wilson 1999; Bennett 2018; Nematollahi 2018). The cup of Jamshid can be seen as a precious item in a great treasury of folklore telling of "magic vessels"—from the Cauldron of Rebirth in Celtic myths to the Holy Chalice (or the Holy Grail) in Christian legends (Forlong 1883; Matthews 1997). Those vessels were believed to impart special powers to the entities contained or prepared in them, especially the powers of revival and life extension, but also required moral worth to obtain and keep them. Possessing the magic cup was insufficient to keep prolonging Jamshid's life indefinitely. According to "The Book of Kings" (*Shah Nameh*) by the Persian poet Ferdowsi (c. 940–1020 CE), Jamshid became proud and his reign of prosperity and longevity was terminated, and he himself was killed by the demonic king Zahhak (Zimmern 1883). This story once again emphasized that longevity could not be just provided by the dishes that are eaten or vessels that contain the foods, but by the moral attitude of the people.

20.5 Longevity Foods and Ethics: The Power of Balance

Longevity foods have been incorporated as parts of the codes of moral conduct in almost every religion. Thus, in the Jewish religious rules of conduct (Halakhah)—"tumah" (the unholiness, evil or impurity), generally and in connection to foods, means simply "the negation of life," hence the prohibition of murder and of bloodshed, and the laws of "tumah ve'taharah" or ritual purity, also in relation to foods (Berlin 1997). A variety of foods have been recommended for their presumed healing and life-prolonging properties, as a part of religious teaching in the Bible and the Talmud (Rosner 1995).

There are, in the Jewish religions tradition, references to longevity foods that are of a more mythical and mystical nature. For example, in the Jewish oral tradition, "Luz" (almond) is a very fraught mythical and mystical concept, associated with life prolongation, even immortality. It denotes the source of resurrection and regeneration, as well as an endocrine gland and a sprout. "Luz" is also the name of the blessed land of the immortals, sometimes said to be located in the northern area of Israel (Kohler 1906). The traditional narrative about "Luz" may even foreshadow some notions of regenerative biotechnology. Thus, Jacob used "Luz" (almond) rods for "bioengineering," to change the color of his sheep (Genesis 30:37–39). Moreover, there is an extensive Jewish oral tradition about the "Etzem Luz"—עצם לוז—the bone of resurrection, the indestructible part of the human body from which the resurrection will proceed. The properties of the luz bone could have been tested in experiments, as for example related in the story of Rabbi Joshua and Emperor Hadrian in the Talmud (Kohler 1906). The description of the miraculous properties of Almond, as linked with longevity and regeneration in those stories, is mainly mythical and mystical.

Yet, in other instances of the Jewish religious tradition, there is a strong rationalist attitude toward longevity foods, associated with a clear moral imperative. The Jewish religious tradition is generally quite supportive of the efforts for life preservation and longevity (Stambler 2017). The principle "ve-chai bahem"—וחי בהם—viz. the obligation to live by the commandments and not to die by them (Leviticus 18:5) has been strongly emphasized. In line with this obligation, the great Jewish rationalist religious philosopher and physician Maimonides (1135–1204, Rabbi Mosheh ben Maimon) posited a rational moral imperative for the prolongation of life. In his "Responsum on Longevity" Maimonides stated: "It is written: 'When you build a new house, you should make a parapet for your roof so that you bring not bloodshed upon your house should any man fall therefrom" [Deut. 22:8]. This phrase proves that preparing oneself, and adopting precautionary measures—in that one is careful before undertaking dangerous enterprises—can prevent their occurrence. … This demonstrates, however, that there is no firmly determined time for death. Moreover, the elimination of harmful things is efficacious in prolonging life, whereas the undertaking of dangerous things is the basis for shortening life" (Rosner 1998). This statement expresses the rational moral imperative for life prolongation. The actual means for life prolongation were envisioned by Maimonides as maintaining an ideal nutritional balance in the body, fully replenishing the vital substances expended in life processes with new necessary nutritional substances. Such a maintenance of the nutritional balance, according to Maimonides, could in principle prolong life indefinitely (without a definite endpoint of life). In Maimonides' own words: "For us Jews, there is no predetermined end point of life. The living being exists as long as replenishment is provided [for that amount of] its substantive moisture [i.e. bodily humors] that dissolves" (Rosner 1998). Thus the concept of nutritional balance becomes linked with the principles of ethical conduct, derived from religious teachings. Balance and moderation are related in Jewish tradition. Thus a dictum attributed to Maimonides spells out "Health" ("BRIUT" in Hebrew) as an abbreviation of "Restrain anger, reduce eating, increase movement" ("bolem rogzo, yafhit okhlo, veyagbir tnuato").

A similar link between religion, ethical instruction and prolongevity nutrition, can be also observed in the works of medieval Islamic and Christian scholars who contemplated the pursuit of longevity. Thus according to the seminal Islamic alchemist Abu Mūsā Jābir ibn Hayyān (also known as Jabir in Arabic and Geber in Latin, c. 721–815), the prolongation of healthy life may be achieved by a balancing or equilibration of "elements" ("natures") in the human body. "This equilibrium once obtained," he wrote "they will no longer be subject to change, alteration or modification and neither they nor their children ever will perish" (Gruman 1966). The preservation of balance of particular elements (natures or humors) has become the foundational principle for the scholastic theory of longevity also in the works of medieval Christian philosophers, alchemists and physicians, such as the Italian theologian and alchemist Thomas Aquinas (1225–1274) whose *Summa Theologica* (1265–1274) includes discussions of alchemy, health and longevity; the English philosopher and alchemist Roger Bacon (c. 1219–1292) who treated on longevity in his *Opus Majus* (1266); the German friar and alchemist Albertus Magnus (1193–1280) who wrote about the prolongation of life in *On Youth and Old Age and On Life and Death*; the Italian physician and alchemist Arnaldus de Villa Nova (c. 1240–1311) whose original and translated medical books such as *Opera Medica Omnia* include the subjects of old age and long life, and many others (Stambler 2019a). In these and many other religious thinkers, the virtues of balance, moderation and self-control generally, and the same virtues exercised in relation to nutrition in particular, were inseparable elements of the moral religious code of behavior.

The notions of balance, self-control and moderation, almost always went hand in hand in the religious ethical rules related to longevity foods and life-prolonging nutrition. Yet, the element of moderation, or self-limitation, appears to be by far the strongest valorized premise in these rules. According to many of those rules, in order to extend health and longevity, it does not matter so much what the person eats, but rather what the person does not eat, or limits oneself in eating. A vast array of taboo foods have been proposed in various religions, that were seen not just as spiritually harming, but also health-impairing and life-shortening. For example, in some cultures—in ancient Egypt, India, Greece and Rome—the eating of beans was prohibited or discouraged, especially black beans, as they were associated with the world of the dead or believed to contain the souls of the dead, and thus could be harmful or polluting for the living. In ancient Egypt, the eating of beans was prohibited to the priests rather due to their sacredness as symbols of the inseparably related death and life through resurrection. Yet in other cultures, especially among native Americans, beans were permitted for consumption and even revered for their life-giving germinative power (Simoons 1998; Andrews 2000; Ghalioungui 1973). Paradoxically, often the same observed properties of certain foods were reasons for their designation either as taboo or medicine. Thus, the strong pungent taste and smell of garlic and onions, in oral traditions across the Old World, associated them with evil influences, as being desirable for the demons and offensive for the gods, discouraging their consumption. Yet in other instances, often in the same parts of the world, those very same properties of strong pungent smell and taste were believed to repel evil influences and strengthen the body of the consumer. Accordingly, in those traditions, garlic and onions were seen as beneficial or medicinal and recommended as foods or protective amulets. Thus, food recommendations based on magical "association of ideas" could be quite ambiguous. In China, in the Taoist prolongevity tradition, any person wishing to achieve great longevity was discouraged from eating "the five grains" (rice, millet, wheat, oats, and beans) as those cereals were associated with corruptible earth and believed to nourish "the three worms" or "the three corpses"—the destructive demonic entities residing in the human body and bringing about early death (Gruman 1966; Ware 1966). Yet, in other instances, in cultures across the world, including China, those same cereals were revered as essential staples of life, originating from benevolent gods (Andrews 2000). In the Americas, some "pseudo-cereals" like quinoa and amaranth were considered divine and life-prolonging.

Another interesting case of dietary restriction concerns eating plants of the nightshade family (*Solanaceae*, commonly containing the alkaloid solanine). These include plants with obvious intoxicating properties, such as mandrake, henbane, thorn apple, belladonna (deadly nightshade) and tobacco, but also common foods, such as eggplants, peppers, tomatoes and potatoes (Schultes et al. 2001; Andrews 2000). Some of the clearly toxic, stupefying and hallucinogenic plants were associated with witchcraft and demonic possession and avoided as foods. Yet also some of the common foods of this family, especially those brought late to the European diet, such as potatoes and tomatoes, were at least initially viewed with suspicion, perhaps as any newcomers might be and perhaps also due to the family association with the obviously poisonous plants. Curiously, some other intoxicating and psychoactive plant foods and drinks of other types, such as grape wine, grain beer and honey mead in Eurasia and ayahuasca brew in America, with their exhilarating, mind-altering properties, were often sanctioned as beneficial and even sacred substances, and often associated with health and longevity (Schultes et al. 2001; Andrews 2000). Once again, magical imagination combined with sensory experience may have led to ambiguous and sometimes contradictory dietary recommendations. Yet, as it can be seen in almost any case of specific taboo foods, whose consumption was prohibited or limited—the magical imaginary inferences, religious moral prescriptions and real-life cautionary experiences have been intertwined.

Beside the restrictions and limitations on specific foods, just moderation as such, or the limitation in consuming any foods, is probably the most persistent element in longevity nutrition guides throughout the centuries. Fasting, so common in virtually any religion, is one of the most prevalent ramifications of dietary moderation. In various religious traditions, fasting, apart from spiritual purification and connection with the gods, was also directly associated with physical effects, including longevity, even extreme longevity. Here again, the power of fast was not seen as due to some special "physiological mechanism" as we might seek today, but due to special powers of the devoted mind and communion with the gods that led to the life-prolonging effects. It may have been assumed that if the person were able to rule over the constraints of one's passions, even the passion for food, those passions would no longer hold sway over the person, and this would enable the person to overcome the present constraints, including the constraints of a short lifespan.

This line of thought is clearly perceived in Indian mythology. For example, several asuras were allowed the gift of extreme vitality thanks to their powers of self-control, austerity and perseverance, including prolonged fasting. Among the asuras who achieved great vitality by means of austerity were Hiranyakashipu and Arunasura. The story of Arunasura, as related in the great purana *Srimad Devi Bhagavatam* (c. 800–1000 CE), is instructive considering the underlying mindset (Swami Vijñanananda 1922, pp. 1046–1053):

> With a view to conquer the Devas, he [Arunasura] went to the banks of the Ganges in the Himâlayâs, practised a very hard Tapasyâ [penance], to Brahmâ, taking Him to be the Protector of the Daityas [an asura race]. First influenced by Tamo Guṉa [the evil quality of ignorance], he withheld in his body the five Vâyus [primary breaths] and partook only the dry leaves and repeated the Gâyatrî Mantra [the universal protective prayer] and practised austerities. Thus he practised for full ten thousand years. Then for another ten thousand years the Daitya lived drinking some drops of water only; then for another ten thousand years he remained by inhaling air only; and then for another ten thousand years he did not take any thing and thus practised he his wonderful Tapasyâ.

Though Arunasura's initial wish that he "shall not die" is refused by Brahma as "an impossibility"—his austerities are nevertheless rewarded by Brahma by the boon of incredible invulnerability to many causes of death. Arunasura's death "shall not be caused by any war, nor by any arms or weapons, nor by any man or any woman, by any biped or quadruped or any combination of two." Thus, through the power of self-control, particularly fasting, Arunasura advances greatly in his quest to conquer death. But alas, deathlessness is "an impossibility." Hence he is eventually killed by the contrivance of the devas via the pestilence of insects. The allegory is deep, suggesting the ultimate vulnerability and fallibility of sentient living beings, but also implying that the struggle for great vitality and longevity is not futile and the gift of vitality may be awarded for perseverance, self-control and self-denial. This is the power of 'extreme moderation' believed to be necessary for extreme longevity, as paradoxical as the notion may seem.

20.6 Longevity Foods and History: The Power of Moderation

In various forms and expressions, the need for moderation, particularly moderation in diet, has remained the absolute consensus among the seekers of extended longevity for centuries (Shapin and Martyn 2000). Almost as invariably, the recommendation of moderation was woven into a fabric of moral instruction for the most beneficial behavior, for the individual as well as for the society.

One of the earliest emphatic instances of this stance can be found in the words of Lao-Tse, the great legendary teacher of Taoism (c. sixth century BCE), in his treatise *Tao Te Ching* (Lao-Tse 1891, Chap. 59 "Guarding the Tao," pp. 102–103):

> For regulating the human in our constitution and rendering the proper service to the heavenly, there is nothing like moderation. It is only by this moderation that there is effected an early return (to man's normal state). That early return is what I call the repeated accumulation of the attributes (of the Tao). With that repeated accumulation of those attributes, there comes the subjugation (of every obstacle to such return). Of this subjugation we know not what shall be the limit; and when one knows not what the limit shall be, he may be the ruler of a state. He who possesses the mother of the state may continue long. His case is like that (of the plant) of which we say that its roots are deep and its flower stalks firm:—this is the way to secure that its enduring life shall long be seen.

The consensus about the importance of moderation has existed in the East and in the West. Thus, one of the founding figures of Western medicine, Hippocrates (c. 460–370 BCE) gave prescriptions for healthy longevity, such as "exertion, food, drink, sleep, sexual activity, in moderation" (Smith 1994, p. 263). Aristotle (384–322 BCE) advised on the need for moderation to preserve health in old age, for example in his treatises *On Length and Shortness of Life* and *On Youth, Old Age, Life and Death, and Respiration* (Barnes 1984). Also Cicero (106–43 BCE), in *On Old Age* (*De Senectute*), advised that "we must look after our health, use moderate exercise, take just enough food and drink to recruit, but not to overload, our strength" (Cicero 1900, p. 59). Generally, moderation in food had been a central tenet of the art of "gerocomia" ("gerocomica" or "gerontocomia" from the Greek "care for the aged") since the writing of the Greco-Roman physician Galen (Aelius/Claudius Galenus, c. 129–217 CE). These principles were stated in Galen's book *De tuenda Sanitate. Gerontocomia* (5th book *On the Preservation of Health. Gerontocomia*) (Galen 1725). This tradition continued for centuries to come.

In the European works on gerocomia, throughout the Middle Ages to the early modern period, the recommendation on dietary moderation was commonly a part of the more general moral instruction in Christian humility and moderation. Indeed, in the Christian tradition, "the seven deadly sins," i.e. qualities that literally cause death—pride, greed, lust, envy, gluttony, anger, and sloth—have excess as their common basis. The avoidance of the sins corresponds to the moral and spiritual pursuit of longevity. With moderation, the corresponding "seven heavenly virtues" or "goods" can be attained—humility, liberality, chastity, kindness, abstinence, patience, diligence—that can be life-prolonging. As discussed by Thomas Aquinas in his fundamental work on Christian theology, *Summa Theologica* (1265–1274), the vices (e.g. pride, gluttony, lust, greed) are simply goods in excess (honor, appetite, sexual intercourse, riches) (Aquinas 1920, Summa Theologica, 2:1, Question 84, Article 4). Thomas Aquinas generally was keenly interested in longevity, as a concomitant of moral living. As he stated: "in the state of innocence man would have been immortal" (Summa Theologica, 1, Question 97, Article 1) and "Death and other bodily defects are the result of sin" (2:1, Question 85, Articles 5–6). As moderation in food was seen as necessary for longevity, the complete avoidance of food, in the state of absolute purity, would be tantamount to immortality. In Aquinas' words: "It would seem that in the state of innocence man did not require food. For food is necessary for man to restore what he has lost. But Adam's body suffered no loss, as being incorruptible. Therefore he had no need of food" (2:1, Question 97, Article 3). Aquinas also proposed practical spiritual prescriptions for longevity, such as honoring one's parents (2:2, Question 122, Article 5). Thus, the instruction on moderation, especially as regards the avoidance of the deadly sin of gluttony, has been a part of the general Christian ethics and theology.

The connection between moderation in food and religious ethics appears as a dominant motif in the available European works on longevity through the early modern period and later. A prominent example is the book by the Flemish priest Leonardus Lessius (1554–1623) entitled *Treatise of Health and Long Life* (*Hygiasticon*, first published in 1613) positing the rules of moderate living (Lessius 1743). Notably, however, it has never been agreed what exactly a "moderate" measure is and how it is to be determined. Akin to the general religious doctrines that admitted of multiple interpretations, also the idea of moderation allowed a vast space for interpretation and controversy. Some rules of thumb for a correct "moderate" diet were sometimes proposed. Thus, for example, Lessius' "Seven Rules for the better Discovery of this right Measure" posited that the meals should be less than would render the person "incapable or unfit for his mental employments." The meals taken should not produce "dullness or heaviness of disposition." If changing the diet, "it is to be done cautiously and by Degrees." "For those, who are much advanced in Years, and for those also, that are of weak complexions, twelve or fourteen Ounces of Food a-Day are judged sufficient." Any type of food may be good in small measure: "Any Sort of Food that is common to one suits agreeably enough with hale Constitutions, if so be not too much of it be taken at one Time." The diet must be simple and uniform: "Every man should above all Things forbear Variety of Dishes, and the luxurious Artfulness of Cookery." Finally, appetites should not be excessively aroused by the "Fancy of Imagination" (Lessius 1743, pp. 14–32). These rules are instructive, as even now, as 400 years ago, our concepts of moderation still largely rely on intuitions and subjective sensations, rather than precise measurements. They are also instructive in that they again emphasize that the mental state and perception are integral for any notion of longevity nutrition.

Despite the uncertainties regarding the exact "moderate" measure, the importance of moderation in diet, of consuming less than people usually do, has been emphasized by many Western authors writing about longevity throughout the Middle Ages and the early modern period. These include the Italian professor Gabriele Zerbi (1445–1505) who wrote a definitive work on Gerocomia, entitled *Gerontocomia, scilicet de senium cura atque victu* (1489, "Gerontocomia, or, care and nutrition for old age," written in Rome upon the request of Pope Innocent VIII, 1432–1492). Another stalwart of this tradition was the Italian long-lived writer Luigi Cornaro (1467–1566) who wrote the massively popular *Discorso sulla vita sobria* (Discourse on a sober life, first published in 1566) (Cornaro 1743). Also the German physician Johann Heinrich Cohausen (1665–1750), in his famous prolongevity work *Hermippus Redivivus or the Sage's Triumph over Old Age and the Grave* (1742) argued that the "title to long life" is given by "temperance" (Cohausen 1744, p. 54). Many more examples of this tradition can be cited (Grmek 1958; Stambler 2019b).

The emphasis on moderation strongly influenced the ideology and practice of early modern hygienists who strove to extend longevity and preserve health in old age. Perhaps the most notable in this hygienic tradition was Christoph Wilhelm Hufeland (1762–1836), the renowned German hygienist, physician to the King of Prussia Friedrich Wilhelm III, and to Goethe and Schiller. In his book *Macrobiotics or the Art of Prolonging Human Life* (1796), Hufeland coined a particular term for the pursuit of healthy longevity—"macrobiotics" which has survived to the present (Wilson 1854). Hufeland specifically distinguished the art of longevity extension from the general medical art that commonly aims to treat individual diseases and symptoms and mainly considers short term effects. Hufeland asserted that the aim of macrobiotics is more long-term and holistic. As he wrote: "The object of medical art is health; that of the macrobiotic, long life. The means employed in the medical art are regulated according to the present state of the body and its variations; those of the macrobiotic, by general principles" (Wilson 1854, pp. vii–viii). Some of the general principles that determine human longevity, according to Hufeland, include: "the innate quantity of vital power," "firmness of organization of the vital organs," and the rates of "consumption" vs. "renovation" ("restoration" or "regeneration") of the vital force and of the organs (Wilson 1854, p. 40). Moderation, in Hufeland, is an absolutely crucial means for the conservation of vital power. "Strengthening, carried too far," he wrote "may tend to accelerate life, and consequently, to shorten its duration" (Wilson 1854, p. viii). Moreover, "the more intensively a being lives, the more will its life lose its extension" (Wilson 1854, p. 42).

Similar principles were professed by several European hygienists of the eighteenth century, such as the German-Latvian proponent of healthy longevity, Johann Bernhard Fischer (1685–1772), who served as "Archiatrus" (head of the ministry of medicine) of the Russian Empire and authored the book *On Old Age, its Degrees and Diseases* (*De Senior Eiusque Gradibus et Morbis,* first published in 1754, republished in 1760). In turn, the traditions of longevity hygiene of the eighteenth century formed the basis for the emergence of "Medicine for the aged" (*médecine de vieillards)* in France in the nineteenth century, and "Geriatrics" in the early twentieth century in the US (Grmek 1958; Stambler 2019b). In virtually all the works related to aging discussed so far, there has been expressed a strong urge for dietary moderation as a part of a prudent and ethical living.

Until the nineteenth century, perhaps one of the very few dissenters from this consensus about the need for moderation was the renowned French lawyer, physician and gastronome Jean Anthelme Brillat-Savarin (1755–1826). In his *Physiologie du goût* (*The Physiology of Taste*, first published in 1825), Brillat-Savarin spoke of the "Longevity of Gourmands" and praised nutritional abundance (Brillat-Savarin 1854, pp. 194–196):

> I am happy, I cannot be more so, to inform my readers that good cheer is far from being injurious, and that all things being equal, gourmands live longer than other people. This was proved by a scientific dissertation recently read at the academy, by Doctor Villermet [the hygienist Louis René Villermé, 1782–1863]. … Those who indulge in good cheer, are rarely, or never sick. … as all portions of their organization are better sustained, nature has more resources, and the body incomparably resists destruction.

In the longevity food culture up to the beginning of the twentieth century, that would be a rare exception that would celebrate hedonistic enjoyment, abundance and diversity of foods, rather than limitation and self-denial. Historically, in the older period, well until the beginning of the twentieth century, the vast majority of the authors rather valorized moderation and austerity in food consumption. It may be interesting to observe whether and to which extent the valorization of moderation was gradually superseded or supplemented by valorization of enjoyment, abundance and diversity of consumption as a means toward healthy longevity, in the more recent period. The evolution of ideas about longevity foods in the later modern period would require a more thorough investigation, beyond the scope of this work that focuses on the older magical, mythological, religious and historical traditions, and should rather not encroach into the territory of more up-to-date reviews. The question of evidence regarding the actual longevity benefits of limited vs. unlimited consumption is also beyond the scope of this work. The main argument of this work remains that, historically, the conception of longevity foods has not been a purely scientific one, but also a psychological, social and meta-scientific construct, involving questions of social good and ethical personal conduct.

20.7 Conclusion: The Need to Promote Longevity Nourishment in Social Context

As I have argued, beside purely scientific and evidential elements, the discourse on longevity foods has historically comprised a vast range of accompanying elements, from the social and mental domains, including cultural, educational, ethical, philosophical and religious aspects. The inclusion of such aspects into the discussion of longevity foods has created rich and colorful contexts, often involving quite fateful behaviors, movements, passions, special interests and conflicts. A more thorough elaboration of those contexts and conflicts would also go beyond the presently feasible scope.

As humorously noted by a mid-twentieth century German physician and longevity popularizer Gerhard Venzmer (Venzmer 1937, p. 127, my translation):

> Unfortunately, one cannot write about nutrition without making some part of humanity his bitter enemies. A fanatic of a 'correct nutrition' makes no concessions, for him 'his' way of

> nutrition is the only right one, and whoever thinks otherwise is an idiot, and possibly also a 'traitor to national health' or a 'paid agent of certain industries'. ... It is a shame how the fanatics of particular forms of diet, with every bite they take, constantly worry whether the food they consume is the 'right one.' They eventually feed themselves so 'healthy' that, through sheer anxiety, they cannot enjoy the 'health' that they buy so dearly.

Indeed, far from being a set of straightforward recipes detailing the lists of the correct foodstuffs that need to be consumed, their amounts and preparation procedures, longevity foods have often functioned as social actors, rallying points, slogans and advertisements, signs and symbols of individual and communal identity, directives for the personal and social rules of conduct. They involved not just "psychological-somatic" aspects, but much more encompassing "psychological-ideological-cognitive-social-economic-environmental-somatic" and many more integrally interrelated aspects. The concepts of longevity foods were adjusted to diverse social and ideological outlooks, depending on particular contexts and environments. It is important, for a thorough presentation of the subject, to consider longevity foods in such a rich multi-disciplinary context and promote a multi-disciplinary discussion. The present work, even though only initiatory and incomplete, has aimed to stimulate such a multi-disciplinary discourse.

Such a multi-faceted consideration may have practical implications, also in terms of health research and public health. A multi-disciplinary discourse may eventually contribute to a stronger scientific evidence for the effects of specific longevity foods, to their more efficient research, development and utilization. Thus, it may be often beneficial to distinguish actual scientific evidence for a particular longevity food, from the magical thinking that may be involved in its promulgation. Yet, in order to enable this distinction, it is important to recognize that there could be an element of magical thinking. Also to discern scientific evidence among special social and ideological interests, it is important to realize that there may be such social and ideological interests, not necessarily related to science. On the positive side, it may be important to empower individuals and the society, to engage social, psychological and cognitive resources that may be determinative for the effective development and beneficial and universal utilization of longevity foods. For that purpose, a better understanding of those multi-domain human resources and determinants is required. It is hoped that the present work will contribute to such a multi-disciplinary understanding, and thus to a more thorough engagement of the human capacity for the benefit of providing health and life-prolonging nourishment for all.

Compliance with Ethical Standards

Conflict of Interest The author declares that he has no conflict of interest.

References

Andrews T (2000) Nectar and ambrosia: an encyclopedia of food in world mythology. ABC-CLIO, Santa Barbara, California

Aquinas T (1920) The Summa Theologiæ of St. Thomas Aquinas. Second and revised edition, literally translated by fathers of the English Dominican Province (first published in 1265–1274). Reprinted in The New Advent, USA. http://www.newadvent.org/summa/1097.htm. Accessed 1 Jan 2021

Barnes J (ed) (1984) The complete works of Aristotle. The revised Oxford translation. Princeton University Press, Princeton

Bennett C (2018) Liber 420: Cannabis, magickal herbs and the occult. Trine Day, Walterville, Oregon

Berlin M (ed) (1997) Tameh met (unholiness of death), Tumah (unholiness). In: Talmudic encyclopedia. A digest of Halachic literature and Jewish law from the Tannaitic period to the present time, vol 19 (Hebrew). Talmudic Encyclopedia Institute, Jerusalem, pp 450–507

Bhishagratna KKL (1907) An English translation of the Sushruta Samhita, based on original Sanskrit text (Transl), vol 1. Calcutta. Reprinted in Internet Archive. https://archive.org/details/englishtranslati01susruoft. Accessed 1 Jan 2021

Bhishagratna KKL (1911) An English translation of the Sushruta Samhita, based on original Sanskrit text (Transl), vol 2. Calcutta. Reprinted in Internet Archive. https://archive.org/details/englishtranslati00susruoft. Accessed 1 Jan 2021

Breasted JH (ed) (1930) The Edwin Smith surgical papyrus (Transl). The University of Chicago Press, Chicago

Brillat-Savarin JA (1854) The physiology of taste; or, transcendental gastronomy. Translated by Fayette Robinson. Lindsay & Blakiston, Philadelphia (first published in 1825). Reprinted in Project Gutenberg. http://www.gutenberg.org/cache/epub/5434/pg5434.html. Accessed 1 Jan 2021

Bryan CP (1930) The Papyrus Ebers. Translated from the German version. Geoffrey Bles, London

Campbell R (1928) The epic of Gilgamesh (Transl edn). Luzac & Co., London. Reprinted in Sacred Texts. http://www.sacred-texts.com/ane/eog/index.htm. Accessed 1 Jan 2021

Chaudhary A, Singh N, Kumar N (2010) Pharmacovigilance: boon for the safety and efficacy of Ayuvedic formulations. J Ayurveda Integr Med 1(4):251–256

Cicero MT (1900) Two essays on old age and friendship. Translated from the Latin of Cicero by E. S. Shuckburgh. McMillan and Co., New York

Cohausen JH (1744) Hermippus Redivivus or the sage's triumph over old age and the grave, wherein a method is laid down for prolonging the life and vigour of man, including a commentary upon an ancient inscription, in which this great secret is revealed, supported by numerous authorities. The whole is interspersed with a great variety of remarkable and well attested relations. J. Nourse, London (first published in Latin in 1742)

Cornaro L (1743) On the benefits of a sober life. In: A treatise of health and long life, with the sure means of attaining it. In 2 books. The first by Leonard Lessius. The Second by Lewis Cornaro, a noble Venetian. Translated into English by Timothy Smith, Apothecary. C. Hitch, London

Darmesteter J (1880) The Zend Avesta, part 1, The Vendidad, Fargard 2, Yima (Gamshêd) (Transl). In: Müller FM (ed) The sacred books of the East, vol 4. Clarendon Press, Oxford. Reprinted in Sacred Texts. https://www.sacred-texts.com/zor/sbe04/sbe0408.htm. Accessed 1 Jan 2021

Darmesteter J (1883) The Zend Avesta, part 2, 9 Gos Yast (Transl). In: Müller FM (ed) The sacred books of the East, vol 23. Clarendon Press, Oxford. Reprinted in Sacred Texts. https://www.sacred-texts.com/zor/sbe23/sbe2314.htm. Accessed 1 Jan 2021

Flattery DS, Schwartz M (1989) Haoma and harmaline: the botanical identity of the Indo-Iranian sacred hallucinogen "soma" and its legacy in religion, language, and Middle-Eastern folklore. University of California Press, Berkeley

Forlong JGR (1883) Rivers of life, or, Sources and streams of the faiths of man in all lands; showing the evolution of faiths from the rudest symbolisms to the latest spiritual developments. Turnbull and Spears, London

Frazer JG (1925) The golden bough. A study in magic and religion, vol 1. The Macmillan Company, New York (Abridged edition, first published in 1890)

Galen (1725) De tuenda Sanitate Gerontocomia. In: Floyer J (ed) Medicina gerocomica, or, the Galenic art of preserving old men's healths. J. Isted, London

Ganeshaiah KN, Vasudeva R, Shaanker RU (2009) In search of Sanjeevani. Curr Sci 97(4):484–489

Ghalioungui P (1973) The house of life per Ankh. Magic and medical science in ancient Egypt. B.M. Israel, Amsterdam

Gomez LO (1996) The land of the bliss: the paradise of the Buddha of measureless light (Transl). University of Hawaii Press, Honolulu

Griffith RTH (1891) The hymns of the Rigveda (Transl). E.J. Lazarus and Co., Benares. Reprinted in Sacred Texts. http://www.sacred-texts.com/hin/rigveda/. Accessed 1 Jan 2021

Griffith RTH (1895) Ramayan of Valmiki (Transl). E.J. Lazarus and Co., Benares. Reprinted in Sacred Texts. http://www.sacred-texts.com/hin/rama/index.htm. Accessed 1 Jan 2021

Grimal P (ed) (1989) Larousse world mythology. Gallery Books, New York, pp 452, 468, 487, 489, 493, 502, 522, 525, 545

Grmek MD (1958) On ageing and old age. Basic problems and historic aspects of gerontology and geriatrics. Monographiae Biologicae, Den Haag 5(2):1–106

Gruman GJ (1966) A history of ideas about the prolongation of life. The evolution of prolongevity hypotheses to 1800. Trans Am Philos Soc 56:1–102

Hurry JB (1926) Imhotep: the vizier and physician of king Zoser and afterwards the Egyptian god of medicine. Oxford University Press, Oxford

Jacobs S, Hart K (2019) Nefertiti's Secret. A 3600-year-old Egyptian papyrus reveals a modern wrinkle cure. Lulu Publishing, USA

Joachim H (1890) Papyrus Ebers. Das Älteste Buch Über Heilkunde (The Ebers papyrus, the oldest book on medicine) (Transl edn). Georg Reimer, Berlin

Johnson M (1987) The body in the mind. The University of Chicago Press, Chicago

Kohler K (1906) Luz. In: Jewish Encyclopedia, in 12 volumes. Funk and Wagnalls Company, New York. Reprinted in Jewish Encyclopedia online. https://jewishencyclopedia.com/articles/10200-luz. Accessed 1 Jan 2021

Kohn L (2001) Daoism and Chinese culture. Three Pines Press, Cambridge, Massachusetts

Lao-Tse (1891) The Tao Teh King. The Tao and its characteristics. Translated by James Legge. In: Müller FM (ed) The sacred books of the East, vol 39. Clarendon Press, Oxford. Reprinted in Sacred Texts. https://www.sacred-texts.com/tao/sbe39/sbe39066.htm. Accessed 1 Jan 2021

Lessius L (1743) Hygiasticon, or a treatise of the means of health and long life. In: A treatise of health and long life, with the sure means of attaining it. In 2 books. The first by Leonard Lessius. The Second by Lewis Cornaro, a noble Venetian. Translated into English by Timothy Smith, Apothecary. C. Hitch, London

Lidke JS, Dirnberger JW (2009) Churning the Ocean of Milk: Hindu Tantrism and radical life technologies. In: Maher DF, Mercer C (eds) Religion and the implications of radical life extension. Macmillan Palgrave, New York, pp 97–110

Maher DF (2009) Two wings of a bird: radical life extension from a Buddhist perspective. In: Maher DF, Mercer C (eds) Religion and the implications of radical life extension. Macmillan Palgrave, New York, pp 111–121

Matthews J (ed) (1997) Sources of the Grail. An anthology. Lindisfarne Press, New York

Nematollahi N (2018) The true meaning of the Cup of Jamsheed. In: Conference proceedings of the Middle East studies association, San Antonio, Texas

Parpola A (2015) The roots of Hinduism: the early Aryans and the Indus civilization. Oxford University Press, Oxford

Po-Yang W (1932) An ancient Chinese treatise on alchemy entitled Ts'an T'ung C'hi, written by Wei Po-Yang about 142 A.D., now translated from the Chinese into English by Lu-Ch'iang

Wu, with an introduction and notes by Tenney L. Davis. Massachusetts Institute of Technology, Cambridge Massachussets, Isis, vol 18, no 2, pp 210–289
Rosner F (1995) Medicine in the Bible and the Talmud. Ktav, Hoboken, New Jersey
Rosner F (1998) Moses Maimonides' responsum on longevity. In: Rosner F (ed) The medical legacy of Moses Maimonides. Ktav, Hoboken, New Jersey, pp 246–258
Schultes RE, Hofmann A, Rätsch C (2001) Plants of the gods: their sacred healing and hallucinogenic powers. Healing Arts Press, Rochester, Vermont
Shapin S, Martyn C (2000) How to live forever: lessons of history. Br Med J 321:1580–1582
Shastri JL (1970) Siva Purana - English Translation, vol 2. Motilal Banarsidass Publishers, Delhi
Simoons FJ (1998) Plants of life, plants of death. The University of Wisconsin Press, Madison
Smith WD (ed) (1994) Hippocrates, vol 7, Epidemics. Loeb classical library. Harvard University Press, Cambridge, Massachusetts
Stambler I (2014) A history of life-extensionism in the twentieth century. Longevity History, Rishon Lezion. http://www.longevityhistory.com/. Accessed 1 Jan 2021
Stambler I (2017) Longevity and the Jewish tradition. In: Longevity promotion: multidisciplinary perspectives. Longevity History, Rishon Lezion. http://www.longevityhistory.com/. Accessed 1 Jan 2021
Stambler I (2019a) Medical alchemy in history. In: Dupre M, Gu D (eds) The encyclopedia of gerontology and population aging. Springer, New York
Stambler I (2019b) Gerocomia in the history of aging care. In: Dupre M, Gu D (eds) The encyclopedia of gerontology and population aging. Springer, New York
Vannier L (1945) La Tradition Scientifique de l'Homœopathie (The scientific tradition of homeopathy). In: Médecine Officielle et Médecines Hérétiques (Official medicine and heretical medicines). Plon, Paris, pp 63–116
Van Loon G (ed) (2003) Charaka Samhita. Handbook on Ayurveda, vol 1. P.V. Sharma and Chaukhambha Orientalia Publishers, Durham, North Carolina
Veenker RA (1981) Gilgamesh and the magic plant. Biblical Archeologist 44(4):199–205
Venzmer G (1937) Lang leben und jung bleiben! (Live long and stay young) Franckhsche Verlag, Stuttgart
Swami Vijñanananda (1922) Srimad Devi Bhagavatam. The tenth book, Chap. XIII. On the account of Bhramari Devi. In: Basu BD (ed) The sacred books of the Hindus, vol 26. Panini office, Allahabad. Reprinted in Sacred Texts. (Transl). https://sacred-texts.com/hin/db/bk10ch13.htm. Accessed 1 Jan 2021
Ware JR (1966) Alchemy, medicine, religion in the China of A.D. 320: The Nei P'ien of Ko Hung (Pao-p'u tzu). Massachusetts Institute of Technology Press, Cambridge, Massachussets
Wilson PL (1999) Ploughing the clouds: the search for Irish soma. City Lights, San Francisco
Wilson E (ed) (1854) Hufeland's art of prolonging life. Ticknor, Reed, and Fields, Boston (originally C.W. Hufeland, Makrobiotik; oder, Die Kunst das menschliche Leben zu verlängern, Jena, 1796)
Zimmern (1883) Ferdowsi. The Epic of Kings. Stories retold from Firdusi. T Fisher Unwin, London. Reprinted in Sacred Texts. (Transl edn). http://www.sacred-texts.com/neu/shahnama.txt. Accessed 1 Jan 2021

Part IV
Nutritional and Dietary Interventions

Chapter 21
Nutritional Regulation of Aging and Longevity

Alexey Moskalev

Abstract Nutrition (diet composition and eating regimen) is one of the key factors of longevity. A balanced diet allows you to avoid essential nutrient deficiencies and hazardous substances to metabolic health. A healthy diet, containing enough dietary fibers, polyphenols, and metabiotics, maintains an optimal quantitative and qualitative balance of the microbiota. Aging is a complex of chronic inflammation and stress processes, so anti-inflammatory nutrients, antimutagens, and hormetins are an integral part of a longevity diet. One of the factors of inflammation is a hyperglycemic diet. Controlling the levels of simple carbohydrates (glucose, fructose, galactose) in the diet can reduce the level of glycation in tissues, which is a fundamental mechanism of aging. Several biologically active substances in food are capable of targeted suppression of aging-associated signaling pathways and activating pro-longevity stress-resistance. Diet personalization is the future of preventive medicine, taking into account markers of metabolic health, the nutritional density of the diet, and markers of biological age.

Keywords Health · Aging · Longevity · Hyperglycemia · Carbohydrates · Hormetin

21.1 Introduction

Nutrition is an important determinant of longevity and the risks of age-related diseases. In the 2018 Global Nutritional Quality Report noted that malnutrition is a major risk factor for illness and death, more severe than air pollution or tobacco smoking (see: https://globalnutritionreport.org/reports/global-nutrition-report-2018). Studies carried out in the United States, the population of which, on the one hand, is the most studied, and on the other hand, for the most part does not adhere to the principles of healthy eating, show that a healthy lifestyle adds 12–14 years to lifespan (Li et al. 2018). A balanced diet offers more promise in

A. Moskalev (✉)
Institute of Biology of FRC "Komi Science Center", Ural Division of RAS, Syktyvkar, Russia
e-mail: amoskalev@list.ru

S. I. S. Rattan and G. Kaur (eds.), *Nutrition, Food and Diet in Ageing and Longevity*, Healthy Ageing and Longevity 14, https://doi.org/10.1007/978-3-030-83017-5_21

terms of prolonging life than dietary supplements or pharmaceuticals. For example, unlike complex polyphenols, individual substances such as resveratrol, tocopherol, polyphenol-containing grape skin extract, antioxidant SkQ1, and metformin do not lead to any differences between the lifespan of control and experimental SHR mice (Panchenko et al. 2019).

Analysis of data on the composition of 1000 raw foods, made it possible to identify foodstuff with the most beneficial combination of useful nutrients (Kim et al. 2015). Based on this paper, a Nutritional score, and a list of the 100 "healthiest" foods were created (see: http://www.bbc.com/future/story/20180126-the-100-most-nutritious-foods). The top "superfoods" for example include almonds, cherimoya fruit, fish (sea bass, flounder), chia and pumpkin seeds, Swiss chard, celery and parsley, lard. Potentially unhealthy foods include those with an excess of simple carbohydrates (Seidelmann et al. 2018), animal proteins (Song et al. 2016; Virtanen et al. 2019), highly processed (Rico-Campa et al. 2019), fried (Sun 2019), salty (Khan et al. 2019) food, as well as fast food and sugary drinks (Barrington and White 2016), that increase the mortality.

In addition to the mortality risks associated with nutrition, it is of great interest to study the effect of diet on biological age. According to the most accurate method to date for measuring biological age and mortality risks by DNA methylation, the calculated parameter AgeAccelGrim is closely related to nutritional quality. Healthy diet contributes to slower aging on this parameter (Lu et al. 2019). It is important to note that a healthy diet determines not only the duration, but also the quality of life. People who follow a healthy diet are not only much healthier but much less likely to show signs of depressed mood and chronic depression than people who eat junk food (Firth et al. 2019).

21.2 Principles of a Diet That Promote Healthy Longevity

There are 7 basic principles of a diet that could promote healthy longevity.

21.2.1 Balanced Nutrition

It is important to have a balanced diet. Essential amino acids and fatty acids, "complex" carbohydrates, soluble and insoluble dietary fiber, minerals and vitamins, biologically active substances (terpenes, carotenoids, polyphenols) are unevenly distributed among different food products. Therefore, food should be as varied as possible, according to the principle of "all just a little bit". Refined foods (white flour, white rice, sugar, refined vegetable oil) that are low in nutritional value and have an excess of calories should be avoided.

21.2.1.1 Proteins

Proteins, especially those containing 8 essential proteinogenic amino acids (tryptophan, valine, threonine, isoleucine, leucine, lysine, phenylalanine, methionine), are the most important component of the diet. However, excess protein food can create additional risks at certain age periods of life. It was found a relationship between protein intake and mortality in different age groups (Levine et al. 2014). According to the same study, elevated protein intake (20% of the average intake of 1,823 cal per day) at the age of 50–65 leads to a 75% increase in overall mortality and a fourfold increase in cancer deaths over the next 18 years. However, the same protein intake (20%) over the age of 65 leads to fewer cancer deaths. It is worth noting a fivefold increase in mortality from diabetes mellitus at any age with increased protein intake.

In a meta-analysis of 31 prospective cohort studies, normal total protein intake was associated with a lower risk of all-cause mortality, especially if the most of this was from plant sources (Naghshi 2020). An additional 3% of energy from plant proteins per day was associated with a 5% reduction in the risk of all-cause death. At the same time, some authors question such conclusions, pointing out the concomitant consumption of other useful biologically active substances, such as dietary fibers and phytonutrients (Chalvon-Demersay et al. 2017). Animal protein may not be overly harmful, but simply be accompanied by additional amounts of potentially detrimental substances: for example, heme iron as an oxidant in red meat, carnitine and choline from meat and eggs (converted by intestinal bacteria into TMA and then to atherogenic TMAO the liver), pro-inflammatory Omega-6 arachidonic acid. Processed meat contains exessive salt and fats, as well as carcinogenic compounds: heterocyclic amines, polycyclic aromatic hydrocarbons, and N-nitroso compounds (Song et al. 2016).

Despite the need for essential amino acids, it excessive consumption carries certain health risks. On the one hand, essential amino acids such as branched-chain amino acids BCAA (valine, isoleucine, valine) and methionine play a protective role against liver fibrosis and sarcopenia. At the same time, their excess in the diet contributes to the risk of metabolic syndrome: hypertension, diabetes mellitus, obesity, atherosclerosis, thrombosis, autoimmune diseases. Elevated BCAAs are associated with obesity, insulin resistance and decreased lifespan in mice (Solon-Biet et al. 2019). Decreased consumption of BCAA restored of metabolic health in obese mice (Cummings et al. 2018). In a human subject with ≥7% weight gain at follow-up, dietary BCAA intake was associated with a higher risk of metabolic syndrome (MetS) (Hosseinpour-Niazi 2020). It is worth noting as one of the promising interventions to prevent excess BCAA that cold hardening and the brown adipose tissue that develops thereby promote the removal of BCAA from the bloodstream through their utilization as an energy substrate in mitochondria (Yoneshiro et al. 2019).

As for methionine, its long-term dietary restriction (MR) to the required minimum increased the lifespan of mice (Miller et al. 2005) and rats (Orentreich et al. 1993). Short-term MR improves metabolic health in the mice (Yu et al. 2018). In mice MR produced more substantial effects than BCAA leucine restriction (LR) on the body mass and glucose homeostasis and reduced hepatic lipogenic gene expression, which was absent with the LR diet (Lees et al. 2017). Dietary restriction of the essential amino acid methionine, the reduction of which has anti-aging and anti-obesity properties, affects cancer outcome through reproducible changes in one-carbon metabolism (Gao et al. 2019).

In a mouse model, a decrease in protein and amino acid intake, including BCAA and methionine, contributed to an increase in the levels of the metabolic hormone FGF21, a sensor of amino acids GCN2 kinase, an improvement in metabolic parameters, and a decrease in mTOR activity (Green and Lamming 2019). Furthermore, diets with lower levels of sulfur-containing amino acids (SAA) (close to the minimum recommended levels) are associated with a reduced risk of cardiometabolic disease (Dang 2020). A low SAA diet uses plant-based protein sources rather than meat-based foods.

21.2.1.2 Lipids

Saturated fat should not be the main source of the diet. For example, a high fat diet alters the activity of intestinal IgA+ immune cells in mice (Luck et al. 2019). IgA is an essential link that controls intestinal and adipose tissue inflammation, intestinal permeability, microbial invasion, and the composition of the gut microbiome. In addition, specialized enteroendocrine (EEC) cells in the gut are known to recognize nutrients and send endocrine signals to the brain. If you eat exclusively fatty foods, the intestinal microflora changes and EEC cells are turned off due to the stress of the endoplasmic reticulum that has arisen in them, which blocks the normal exchange of signals between the intestine and the brain (Ye et al. 2019). A fatty diet promoted the proliferation of Acinetobacter bacteria, which caused the EEC to be silenced (Ye et al. 2019). In this regard, it is skeptical to use a high-fat keto diet without any therapeutic indications.

Cholesterol is contained in the membranes of all animal cells, participates in the synthesis of steroid hormones, is necessary for the synthesis of bile acids and vitamins of group D (under the influence of sunlight in the skin), protects erythrocytes from destruction. However, elevated total blood cholesterol levels are related to morbidity and mortality from coronary heart disease in middle-aged men (Tomas Abadal et al. 2001). Regular consumption of red meat, fatty dairy products, and eggs increases the risk of colon cancer (Wang et al. 2018). Cholesterol itself is a tumorigenic factor that stimulates the proliferation of intestinal stem cells. Regular red meat consumption also elevates the risk of ischemic heart disease and diabetes (Papier et al. 2021).

However, moderate consumption of animal fats did not increase the risk of coronary heart disease (Chowdhury et al. 2014). In seniors over 85, high cholesterol in the blood was not associated with an increased risk of cardiovascular disease or death (Weverling-Rijnsburger et al. 1997). Probably, without systemic inflammatory processes, cholesterol in the bloodstream is not dangerous for blood vessels, however, it's very excessive consumption can have negative consequences.

Monounsaturated fatty acids have sufficient fluidity (due to the presence of one double bond in the molecule), but at the same time less oxidizability compared to polyunsaturated ones with two bonds. Also, monounsaturated fatty acids predominate in the membranes of long-lived lines of model animals (Han et al. 2017). In addition, extra virgin olive oil, first cold pressed, contains a whole range of substances with potential geroprotective effects - oleic acid, hydroxytyrosol, oleocanthal, oleanolic and olive acids, erythrodiol, beta-sitosterol, vitamins E and K1. Extra virgin olive oil can be considered as an antidepressant and tranquilizer, affecting the monoaminergic and endocanabioid systems (Han et al. 2017; Perveen 2013). Olive oil in mice improved cognitive function and inhibited neurodegeneration (Lauretti et al. 2017). It has an anti-mutagenic effect (Aly et al. 2018). Olive oil extended lifespan in a rat model (Baati et al. 2012). Finally, plant-based monounsaturated fats are associated with lower mortality in humans (Guasch-Ferre et al. 2019). Twelve weeks of a plant-based MUFA diet reduces liver fat and improves liver and overall insulin sensitivity (Errazuriz et al. 2017).

In contrast, polyunsaturated omega-6 are highly oxidizable molecules, and their oxidation products are cytotoxic (Iuchi et al. 2019) and cancerogenic (Bartsch et al. 1999). Omega-6 source rapeseed oil in large quantities may increase the risk of neurodegenerative disorders (Lauretti et al. 2017). However, it's important to note that essential omega-6 linoleic acid has long-term benefits for the prevention of type 2 diabetes (Fan et al. 2017) and risks of overall and cardiovascular mortality in men (Virtanen et al. 2018).

A study of the effect of the lifetime consumption of various sources of fats rich in monounsaturated (olive oil), omega-6 polyunsaturated (sunflower oil) or omega-3 polyunsaturated fatty acids (fish oil) on liver aging in male Wistar rats showed that extra virgin olive oil led to the lowest oxidative and ultrastructural changes in the liver. Sunflower oil caused fibrosis, ultrastructural changes, and oxidative processes. Fish oil increased age-related oxidation decreased the activity of the mitochondrial electron transport chain but increased the relative telomere length. Age-related gene expression changes in animals fed extra virgin olive oil and fish oil were mainly associated with mitochondrial function and oxidative stress pathways, followed by control of the cell cycle and telomere length. Thus, extra virgin olive oil can be considered a source of dietary fat that provide better protection to the liver during the aging process (Varela-Lopez et al. 2018). While olive oil (without heat treatment) should be consumed in a tablespoon every day, it is better to get the omega-3 available to the body from 2 servings of fatty fish per week, and omega-6 - in the form of a handful of fresh nuts and a pinch of seeds (sesame, linseed) per day.

Undoubtedly, some omega-3 and omega-6 are essential, in fact, vitamins that our body needs in small quantities. Even though omega-6 are generally considered to be pro-inflammatory, they include the essential fatty acid linoleic acid, which is essential for regular consumption. It is found mainly in vegetable oils. Among the omega-3s, eicosapentaenoic and docosahexaenoic acids, found in seafood, are essential. Thus, it is important to consume both omega-3 and omega-6, the only question is to achieve their optimal ratio in the diet 1:3 or even better 1:1.

Modern epidemiological studies essentially clarify the existing paradigm. On the one hand, consumption of trans fatty acids, animal MUFAs, some PUFAs (α-linolenic acid and arachidonic acid) is associated with higher mortality; on the other hand, dietary intake of marine omega-3 PUFAs and replacement of PUFAs with plant MUFAs or linoleic acid were associated with lower overall and cardiovascular mortality (Zhuang et al. 2019).

Regarding the fat composition of the diet, it is worth mentioning that it should not be used ready-made commercial food, since the margarine contained in it includes trans fats that can cause cytotoxic stress of the endoplasmic reticulum (Oteng and Kersten 2020). Deodorized (hot steam treated) oils contain toxic contaminants glycidyl ethers (Shimamura 2020). In addition, products of animal origin must be purchased fresh, since during long-term storage, even in a frozen state, oxysterols accumulate in them, which accelerate atherogenic changes in the vessels (Vejux 2020).

21.2.1.3 Carbohydrates

The Glycemic Index (GI) is a scale that measures how quickly specific foods release glucose into the bloodstream. Depending on the GI, carbohydrates are usually divided into undesirable "refined" or "fast" and beneficial "slow" or "complex". Fast foods include glucose, sucrose, amylopectin, fructose, lactose, and galactose. They are also found in added sugars, starchy vegetables, pasta, breads, sweets, whole milk, honey, and fruits. "Slow-release carbohydrates" - in whole grains, legumes, non-starchy vegetables. The product's GI should preferably not exceed an average threshold of 70. Low-GI diets in patients with impaired glucose tolerance or diabetes were effective at reducing glycated hemoglobin (HbA1c), fasting glucose, BMI, total cholesterol, and LDL (Zafar et al. 2019). Compared to diets providing a high glycemic index, low glycemic index protocols resulted in significantly more pronounced decreases in serum triglycerides and HOMA-index (Schwingshackl et al. 2015). On the contrary, high GI was associated with increased risks of colorectal, bladder and kidney cancers (Turati 2019).

Dietary simple sugars alter microbial ecology in the gut and promote colitis in mice (Khan 2020). In addition to GI, it is worth considering the glycemic load - the amount of available carbohydrates in the product. This value should not exceed 20 (see: https://www.diabetes.co.uk/diet/glycemic-load.html). Thus, GI is correlated with the best indicators of metabolic health markers. However, the evidence for the association of GI with endpoints such as disease is rather limited. It cannot be concluded that a diet high in carbohydrates or an increase in the percentage of total energy intake in the form of carbohydrates increases the likelihood of obesity (Sartorius 2018). In this connection other measures of dietary quality, such as fiber may be more likely to predict health outcomes (Vega-Lopez et al. 2018).

However, fiber should be consumed in whole foods rather than supplements. Even relatively short periods of inulin supplementation in mice with intact gut microbiome adversely affect liver metabolism and function (Pauly 2020). Raw vegetables and fruits not only carry prebiotics (fibers, fructans, pectins) to feed our microflora, they are the main source of this microflora (live bacteria, viruses, fungi) in our intestines, that is, a probiotic (Wassermann et al. 2019).

In terms of macronutrients, a balanced diet might look like this. 10–12% of the average daily calorie intake can come from proteins, mostly plant (legumes twice a week; whole grain cereals from sorghum, amaranth, canihua, quinoa, oats, barley, buckwheat; seeds in vegetable salads; nuts for a snack), but every day and something from sources of animal proteins - for example, once a week, shrimp, scallops, poultry meat, a piece of goat or sheep cheese, a couple of eggs, alternating each time. Approximately 55% of calories are from carbohydrates. Carbohydrates are whole grain cereals, boiled sweet potatoes, as well as vegetables (beets, carrots, eggplants, zucchini, various types of cabbage) and fruits. Bread, desserts, sugary drinks, pasta and potatoes should be avoided, as these are quickly digestible carbohydrates that sharply and for a long time increase blood sugar levels (they have a high glycemic index and glycemic load). The remaining 35% of calories are fats, mostly monounsaturated (from olive oil, almonds and avocado), but also polyunsaturated (from fish, seafood, flaxseed oil, seeds, and nuts), and saturated - about a tablespoon of butter in a side dish.

21.2.2 Prebiotics and Metabiotics

Human is a "superorganism", a consortium of numerous living things from different kingdoms - fungi, protozoa, bacteria, archaea, and viruses. In the first years of life, a complex community of microorganisms is formed in our colon, to which immune tolerance is formed. Its basic composition does not change very much during the life. That is why taking antibiotics can provoke risks, for example, of diabetes (Mikkelsen et al. 2015). Imbalance in the composition and quantity of intestinal microflora also increases the risk of cardiovascular diseases, inflammation (in the intestine and systemic), neurodegeneration and carcinogenesis.

The invasion of foreign strains of even beneficial bacterial species can lead to inflammatory reactions or their penetration beyond the intestinal barrier into the tissues of our body. For this reason, it should not be necessary recommend taking probiotics (cultures of living bacteria) in the form of dietary supplements or overly enriched products. The usefulness of such an intervention in the delicate ecological balance of our microbial communities has not been adequately confirmed - moreover, after prolonged periods of using probiotics, harmful consequences may appear (Shenderov 2013). However, there are indications for the use of probiotics, for example asthma, cystic fibrosis, respiratory infections (Stavropoulou and Bezirtzoglou 2020).

The role of microbiota homeostasis in aging and age-related diseases is becoming increasingly evident. Older mice have fewer *Akkermansia muciniphila* bacteria than younger mice. This reduction in number causes inflammation, which ultimately induces insulin resistance and could leads to type 2 diabetes (Naito et al. 2018). *A. muciniphila* ameliorates the age-related decline in colonic mucus thickness (van der Lugt et al. 2019). Depression is linked with a depletion of the butyrate-producing bacteria *Faecalibacterium* and *Coprococcus* (Valles-Colomer et al. 2019).

Decrease in the proportion of the bacterial family *Prevotellaceae* and bacterial genera producing butyrate is a marker of Parkinson's disease (Haikal et al. 2019).

The *Lactobacillus rhamnosus GG* strain reduces appetite, alcohol-induced microflora disorders, the accumulation of fat, markers of metabolic syndrome, and stops inflammation in the liver. This strain has also been shown to be effective in reducing the symptoms of anxiety and depression. The effect is vagus-mediated and executed through GABA receptors (Capurso 2019).

Food should be prebiotic, that is supporting of beneficial intestinal microflora (fibrous vegetables, greens, legumes, berries, fruits, nuts) and metabiotic (containing products of fermentation by bacteria, such as cheese, tofu, miso, kombucha, Pu-erh, sauerkraut, kimchi, natto, wine, beer).

Metabiotics are fermentation products containing metabolites of beneficial bacteria. The effect of metabiotic is associated with the fact that microorganisms live in whole communities, where the waste product of one species is a necessary substrate for the existence of other species. Metabiotic products do not contain living microorganisms, but only traces of their activity, which can be useful for our own microflora, as well as intestinal cells and other tissues.

Bacteria associate with the metabolism of dietary carbohydrates (into short-chain fatty acids and gases), proteins, plant polyphenols, bile acids, and vitamins (Rowland et al. 2018). The microbiota can alter the activity of cannabinoid and opioid receptors in the gut, taste buds in the mouth. Through toxins released by the microbiota, for example, in response to fasting, it negatively affects mood and increases the desire to consume food (Alcock et al. 2014). Most of the peripheral pool of neurotransmitters and biogenic amines, such as serotonin, dopamine, norepinephrine, GABA, is produced by the intestinal microbiota (Aliper et al. 2016). Fortunately, most of these neuroactive compounds do not penetrate the blood-brain barrier, however, in the case of systemic inflammation, its permeability can increase.

The reciprocal relationship between gut microbial metabolism and mental health is one of the most intriguing and controversial topics in microbiome research. A catalog of the neuroactive potential of intestinal prokaryotes has been created (Valles-Colomer et al. 2019). Microbial synthesis of the dopamine metabolite 3,4-dihydroxyphenylacetic acid was positively correlated with mental quality of life. In contrast, microbial synthesis of gamma-aminobutyric acid has been implicated in depression (Valles-Colomer et al. 2019).

The gut microbiome regulates host glucose homeostasis via peripheral serotonin (Elhassan et al. 2019). Peripheral dopamine controlled by gut microbes inhibits invariant natural killer T cell-mediated hepatitis (Dang 2020). Gamma-aminobutyric acid-producing lactobacilli positively affect metabolism and depressive-like behavior in a mouse model of metabolic syndrome (Patterson et al. 2019). Beneficial bacteria provide our body with potential geroprotectors such as vitamin K2 (MK-7), enterolactone, butyrate and spermidine.

The benefits of a diet high in fiber in the colon have been well documented in epidemiological studies. They provide resistance to influenza through the activation of immune cells (Trompette et al. 2018). Human food enzymes are unable to digest some complex carbohydrates and plant polysaccharides, in particular fructose-oligosaccharides (fructans and inulin), xyloglucans, galacto-oligosaccharides (sialylated oligosaccharides), soluble and insoluble dietary fiber (cellulose), hemicellulose. Instead, these polysaccharides are metabolized by microbes that form short-chain fatty acids (SCFAs), including acetate, propionate, and butyrate (Holscher 2017).

Typically, the benefits of the colon microbiota are attributed to short-chain fatty acids, in particular butyrate. By digesting dietary fiber, microbes create butyrate, which is a source of energy for mitochondria (Donohoe et al. 2011), enhances their biogenesis (Clark and Mach 2017) and is an epigenetic regulator. Butyrate is an inhibitor of the histone deacetylase HDAC (Davie 2003), which plays a key role in aging at the cellular level (Vizioli et al. 2020). It is also a ligand for G protein coupled receptors (Bourassa et al. 2016). Receptors GPR41 and GPR43 are expressed in adipocytes, colon epithelial cells and peripheral blood mononuclear cells. Knockout mouse studies have shown the involvement of GPR41 and GPR43 in chronic inflammatory diseases such as obesity, colitis, asthma, and arthritis (Ang and Ding 2016). Lys14 of histone H3 (H3K14) was found to be a propionylation and butyrylation site *in vivo*. H3K14pr and H3K14bu labels are set by histone acetyltransferases. The promoters of active genes are preferably enriched with these labels and are recognized by acylation-specific reading proteins. Also, propionyl-CoA appears to be able to stimulate transcription (Kebede et al. 2017).

Butyrate, a major metabolite of intestinal bacteria and crucial energy source for gut epithelial cells, also possesses anti-inflammatory properties (Saemann et al. 2000). Butyrate inhibits the accumulation and effects of neurotoxic ceramide, formed during intestinal permeability disorders in dysbiosis (Amano et al. 2019). Not only dietary fiber is a source of butyrate. Human intestine bacteria *Intestinimonas* strain AF211 synthetize butyrate from lysine and the Amadori product fructoselysine (Bui et al. 2015). Colonocytes of the large intestine, which are in constant contact with butyrate, need it as an energy resource, but it is destructive for interstitial stem cells, and they hide in crypts (Snippert 2016). Dietary indigestible fibers alter the composition of the gut microbiota and its metabolic profile in such a way that the abundance of long-chain fatty acids also increases. They promote the activity of autoimmune suppressive TH2 cells (Berer et al. 2018).

Commensal bacteria, including *Escherichia coli*, stimulate HDAC activity by metabolizing phytate and producing inositol 1,4,5-triphosphate (InsP3). Intestinal effects of InsP3 and ingestion of phytate promote intestinal regeneration after injury (Ma et al. 2020). However, SCFAs are not the only beneficial gut microflora products. Brial et al. identify the inverse correlation between diabetes and serum 4-cresol, a metabolite produced by intestinal bacteria (Brial et al. 2020). Microbial tryptophan catabolites from proteolysis affect the health of the host (Roager and Licht 2018). It is assumed that these metabolites activate the immune system by binding to the cyclic hydrocarbon receptor (AHR), strengthen the intestinal epithelial barrier, stimulate gastrointestinal motility, as well as the secretion of intestinal hormones, have anti-inflammatory, antioxidant effects in the systemic circulation, and presumably modulate the intestinal microbial composition.

During aging, goblet cells are loosing, which secrete mucus to protect the epithelium. Indoles, which are secreted by the gut microbiota, act through the aryl carbohydrate receptor and the anti-inflammatory cytokine IL-10, restoring depleted goblet cells in older animals (Powell et al. 2020). Indole prevents steatohepatitis (Altunina et al. 2020). Indolepropionic acid, a gut microbiota-produced metabolite, preserves β-cell function and prevents T2D (de Mello et al. 2017). *Bifidobacterium*, *Allobaculum*, and *Parabacteroides* in the intestine produce hydrogen (Nishimura 2018), a potent antioxidant, attenuating risk factors of metabolic syndrome (LeBaron et al. 2020).

Another useful microbiota product is polyamine spermidine. It is also found among foods: it is most of all in sprouts of cereals and legumes. Spermidine intake is associated with a reduction in overall mortality and mortality from cardiovascular disease and cancer in humans (Kiechl et al. 2018). Experiments in mice have shown that spermidine can increase lifespan, prevent the development of liver fibrosis and malignant hepatoma, one of the most common forms of cancer, through the induction of chaperone-mediated autophagy (Shen et al. 2017). Spermidine delays tumor onset in a mouse model of induced lung cancer (Eisenberg et al. 2016). Supplementation with autophagy-activating spermidine has been shown to protect against neurodegeneration and cognitive decline in animal models (Buttner et al. 2014; Wang et al. 2015). Spermidine has a cardioprotective effect (Eisenberg et al. 2016). It lowers blood pressure and has a vasoprotective effect (Eisenberg et al. 2017). Spermidine promotes hair growth (Ramot 2011). The molecular mechanism of the geroprotective action of spermidine is that it inhibits the activity of EP300 acetyltransferase, which suppresses autophagy (Madeo et al. 2018). Spermidine reverses D-galactose-induced skeletal muscle atrophy in aging rats by enhancing autophagy and decreasing apoptosis through the AMPK-FOXO3a signaling pathway (Fan et al. 2017). By its mechanism of action, it is also a mimetic of low-calorie nutrition (Minois 2014).

Vitamin K2 is produced by the gut microbiota. Vitamin K2 contributes to the retention of calcium in the bone mass, preventing its redistribution into blood vessels and the risk of calcification (Schwalfenberg 2017; Vossen et al. 2015). Microbiota can transform polyphenols into bioavailable phenolic acids with geroprotector and anti-inflammatory effects, reducing cardio-metabolic risks (Correa et al. 2019).

The gut microbiota can also be negative. A high-fat diet encourages the proliferation of *Acinetobacter* bacteria, which suppresses EEC cells in the intestines, which blocks the normal exchange of signals between the intestine and the brain (Ye et al. 2019).

Microbial-derived acetate contributes to the development of metabolic syndrome (Perry et al. 2016). In male rats resistant to insulin and prone to obesity, it was found that the consumption of fatty foods in the blood plasma and faeces of animals increased 1.5–3 times the rate of formation and concentration of acetate, but not propionate or butyrate. This acetate is of microbial origin. In animals that received fatty foods, the secretion of insulin is several times higher than in the control. This insulin spike is triggered by acetate through its effects on the centers in the brain that activate the parasympathetic nervous system. At the same time, there is a surge in gastrin, which makes you feel hungry (Perry et al. 2016). By the way, acetate formed during alcohol metabolism in the liver appears to be the cause of additive behavior, modulating gene expression in the brain via acetyl CoA synthetase 2 (Ghrayeb et al. 2019).

After glycolysis, with an insufficient amount of oxygen in the cell, L-lactate is formed, which acidifies the blood, but in the liver it can be converted into glucose or used by the brain as a metabolism regulator and a source of energy (Riske et al. 2017). However, lactobacilli and some other lactic acid bacteria are able to secrete another type of lactate, D-lactate (Vitetta et al. 2017). This substance exhibits cardio and neurotoxicity, contributing to mitochondrial dysfunction (Ling et al. 2012). Pickles, yoghurt, sour milk, tomatoes, apples, beer and wine lead to some increase in D-lactate (Kowlgi and Chhabra 2015).

Carnitine and choline are major precursors for the gut microbiota-dependent generation of a TMA, which in the liver converts into an atherogenic metabolite, TMAO. Trimethylamine-N-oxide (TMAO) promotes age-related vascular oxidative stress and endothelial dysfunction in mice and healthy humans (Brunt et al. 2020). Studying the intestinal microbiome in subjects characterized by phenotypes of TMAO producers, identified 39 functional taxonomic units that correlated with TMAO productivity, including *Emergencia timonensis* and *Ihubacter massiliensis* (Wu et al. 2020). In randomized controlled trials, chronic consumption of red meat in the diet increases systemic TMAO levels by rising microbial production of TMA/TMAO from carnitine, but not choline. Stopping red meat consumption reduces plasma TMAO within 4 weeks (W. Wang et al. 2019).

The severity of the effect of pathogenic microbiota on health is due to the degree of permeability of the gastrointestinal and blood–brain barrier. Factors that damage the barrier function of the intestinal epithelium contribute to the development of inflammatory reactions to food antigens. For example, reovirus infection triggers inflammatory responses to dietary antigens (Bouziat et al. 2017). Enteropathogenic *E. coli* disrupts tight junction barrier function (Shifflett et al. 2005). Fructose stimulateing lipogenesis in liver, disrupting intestinal barrier and provoking inflammation (Lambertz et al. 2017; Todoric et al. 2020). Hyperglycemia drives intestinal barrier dysfunction (Thaiss et al. 2018). Proinflammatory cytokines disrupt epithelial barrier function (Bruewer et al. 2003). Ethanol impairs intestinal barrier function (Elamin 2014). Theoretically, it is possible to develop approaches to correct this condition. Normalization of intestinal microflora - a way to eliminate increased intestinal permeability and prevent autoimmune disorders (Mu et al. 2017). Some isoflavones, such as puerarin, can improve barrier function (Che et al. 2020).

21.2.3 Anti-inflammatory Foods

Aging is a chronic inflammatory process. Overeating, intestinal dysbiosis, and vitamin D deficiency as dietary factors contribute to chronic inflammation (Moskalev et al. 2020). Chronic inflammation, in turn, underlies age-related diseases such as cardiovascular, cognitive (e.g., depression), sleep disorders, and neoplastic diseases.

At the same time, it is known that regular consumption of some foods promotes inflammation, while others reduce it. It is well known that Mediterranean diet has anti-inflammatory effect, which is characterized by the use of olive oil as the main source of fat, a high consumption of fruits, nuts, vegetables, legumes, whole grains, spices, and herbs, as well as a moderate consumption of dairy products (mainly cheese and yogurt), fish, poultry, and a reduced consumption of red meat and processed foods (Fernandes 2020).

Antioxidants reduce inflammation. Diet antioxidant status can be analyzed using a database (see: https://www.superfoodly.com/orac-values). Direct anti-inflammatory effect is exerted by such biologically active food substances as alpha-lipoic acid, glycine, genistein, kaempferol, apigenin, oridonin, gamma-tocotrienol, phenethyl caffeic acid ester, hydroxytyrosol, procyanidins, resveratrol, quercetin, theaflavin, epigalquinate, rutin, isothiocyanates, capsaicin, zinc and magnesium (Altunina et al. 2020; Iddir 2020; Joseph et al. 2016; Liu and Dudley 2020; Proshkina 2020b). The specially developed DDI index can be used to evaluate the diet in terms of its inflammatory status (Cavicchia et al. 2009).

21.2.4 Antimutagenic Foods

The environmental insults and internal errors in cell metabolism can lead to DNA damage and mutations that accelerate the aging of the body (Moskalev et al. 2013).

Mutagenic factors are:

- Ionizing radiation (air travel, X-ray exposure, space radiation, background radiation level);
- UVB light;
- Reactive forms of oxygen, nitrogen and chlorine (free radicals), including inflammation;
- Viruses;
- Mutagenic chemicals (benzpyrene, acrolein, nitroso compounds, polycyclic hydrocarbons);
- Cell division (replication errors, telomere shortening).

Antimutagenic foods reducs the risk of DNA damage and the accumulation of mutations. According to their mechanism of action, they either prevent DNA damage (ROS-scavengers), are cofactors of repair enzymes, or, as hormetins, induce the expression of their own antioxidant and repair enzymes in the cell. Antimutagenic substances in food include a wide range of compounds (Proshkina et al. 2020). In particular, in food products, cysteine (red peppers, onions, garlic, Brussels sprouts, broccoli, poultry, fermented milk products), gallic acid (chicory, grapefruit, cloudberries, tea, oregano, apples, pomegranates), lipoic acid (spinach, broccoli, tomatoes, peas), polyphenols (legumes, tomatoes, various berries, red wine, cocoa, chocolate, tea, curry), organic forms of selenium (Brazil nuts, cashews, fish, turkey, chicken, brown rice, legumes, mushrooms, oatmeal, spinach) (Sloczynska et al. 2014).

21.2.5 Hormetic Foods

Stress response proteins play a key role in the maintenance of vital functions at the cellular level. Lack of some essential nutrients, DNA damage, hypoxia, exposure to toxins, high or low temperatures are perceived by cells as stress factors. They can cause significant damage, and if the cell did not appropriately respond to them and repair itself, it would die. Activating the stress response and recovery system can not only reverse damage, but also put the system at a higher level of protection against new spontaneous errors and damage, including those associated with aging. Stress resistance mechanisms can be induced by factors that promote moderate stress, which is not accompanied by significant damage, but is able to activate a protective response. This phenomenon is called "hormesis", and its agents are called "hormetins" (Rattan and Demirovic 2009). *(also see the chapter on nutritional hormetins in this book).*

Hormetins moderately activate the cell's own stress response mechanisms: mild DNA damage inducers quercetin (capers, onions, cranberries, plums, blueberries, currants, cherries, apples) (Yamashita and Kawanishi 2000), sulforaphane (Sestili et al. 2010) (chopped broccoli) or curcumin (curry, turmeric) (Ting et al. 2015), mitohormetins resveratrol (red grapes, peanuts) (Gueguen 2015), carnosine (turkey, chicken) (Calabrese et al. 2012) and N -acetylglucosamine (mushrooms, cartilage, shrimp) (Matsumura 2020).

21.2.6 Low Glycemic Load

The glycemic index is an estimate of the rate at which various foods can raise blood sugar levels. The higher the glycemic index, the more harmful the use of the product is considered in terms of the appearance of excess weight and insulin resistance. When determining the glycemic load, not only the rate of absorption of glucose after consumption of a particular product is considered, but also the amount of carbohydrates in it. That is, the rate of increase in blood sugar level is determined, and how much this level will rise and, accordingly, how long it will hold at a high level before the body can reduce it to normal levels.

The diet should be low in glycation end products, low insulin- and glycemic index and glycemic load. These are products containing slowly digestible carbohydrates (durum wheat pasta, whole grain cereals, legumes, vegetables, herbs). On the contrary, it is better to avoid sweet, starchy vegitables, white rice. When preparing food, it is worth avoiding frying or baking, in which glycation end products are formed in large quantities.

Glucose is proinflammatory (Dandona et al. 2007), and a low-glycemic diet can help reduce inflammation (Steckhan et al. 2016). There are databases of food glycemic indices and glycemic loads (Foster-Powell et al. 2002). Insulinemia contributes to obesity and thus can provoke inflammation. High insulin also causes high blood pressure. The insulin index of the diet can be also assessed (Neuhouser et al. 2006).

Glycation end products cause extracellular matrix stiffness (Fedintsev and Moskalev 2020) and inflammation via RAGEs (Alexey Moskalev et al. 2021). There is a database that allows the analysis of glycation end products in the diet (Uribarri et al. 2010). In addition to glucose, glycation end products are provoked by methylglyoxal (Baig et al. 2017), fructose, galactose, ribose, or glyceraldehyde (Syrovy 1994). Fructose should be more controlled in food from this point of view because it is 10 times more reactive than glucose (Gugliucci 2017). In turn, galactose is generally used in experiments as a factor causing accelerated aging (Azman and Zakaria 2019).

21.2.7 Foods Containing Potential Geroprotectors

Many biologically active substances of natural origin in experiments on laboratory organisms showed geroprotective properties, increasing lifespan (Barardo et al. 2017; A. Moskalev et al. 2015; Proshkina et al. 2020b). Food biologically active substances can inhibit major aging-associated signaling pathways such as PI3K/mTor (Huang 2013), p38 (Guerra and Issinger 2019), NF-kB (Nam 2006).

There are preclinical studies showing NAD+ precursors as a powerful and safe geroprotector (Braidy and Liu 2020). If we trace where our body can get it or what depletes it, it turns out that it is possible to regulate its level through nutrition (Fig. 21.1).

Another notable geroprotector in food is spermidine (Madeo et al. 2018), which has been shown to slow down the aging of the nervous system, boosting autophagy (Bhukel et al. 2017).

Table 21.1 Potential geroprotectors in food

Geroprotector (food)	Target genes
Apigenin (oranges, apples, cherries, grapes, onions, parsley, broccoli, sweet green peppers, celery, barley, tomatoes, tea), curcumin (curry), fisetin (strawberries, apples, persimmons and onions), indoles indole-3- carbinol and 3,3'-diindolylmethane (broccoli, cauliflower and white cabbage, Brussels sprouts), isoflavones genistein and deguelin (soy), quercetin (tea, onions, red grapes and apples), resveratrol (red grapes), tocotrienol (brown rice), caffeine (coffee), epigallocatechin gallate (green tea), capsaicin (chili pepper)	mTOR, PI3K
Proanthocyanidins (grape seeds, blueberries, blackberries, apples, peaches, pears, nectarines, kiwi, mangoes, dates, bananas, sorghum, barley, walnuts, cashews), ginsenoside (ginseng), apigenin (see above), astosanthin (lozenges), shrimp), L-theanine (tea)	p38
Isothiocyanates (mustard), capsaicin (red pepper), glycyrrhizic acid (licorice), alpha lipoic acid (spinach), kaempferol (cabbage), apigenin (parsley), rutin (tea, berries), theaflavins (tea), hydroxytyrosol (olive butter)	NF-kB

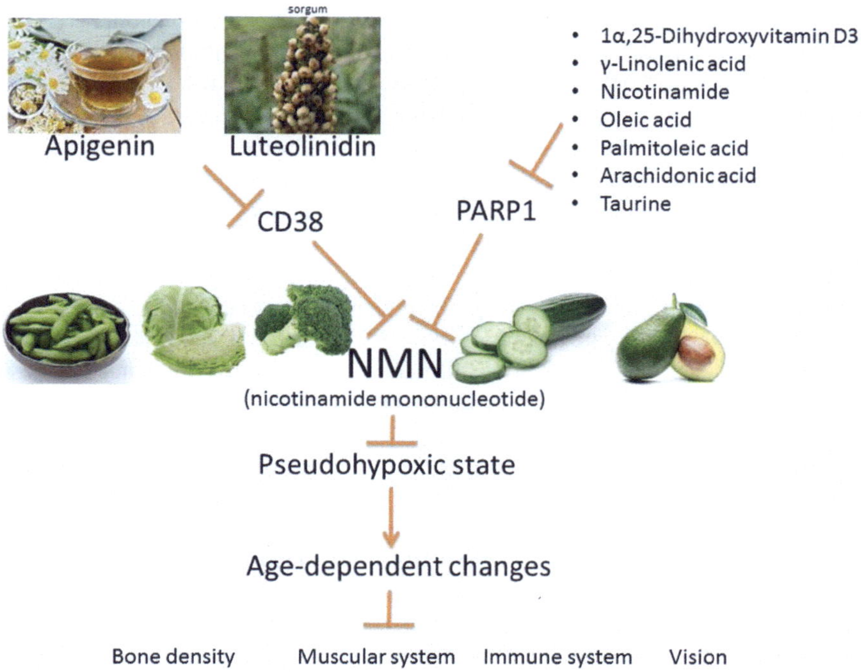

Fig. 21.1 NAD + regulation by food

21.3 Aging Epigenetics and Nutrition

Epigenetic changes are inheritable but reversible modifications of DNA and chromatin that do not change the primary nucleotide sequence but cause a change in gene activity. For each organism in any of its cells is one genome, but many epigenomes. Epigenetic regulation is carried out at the level of DNA (methylation of CpG islands), chromatin (modification of histone proteins), and RNA (methylation of RNA, microRNA, and long non-coding RNAs). Epigenetic regulation plays a decisive role in development. In the early stages of embryonic development, cells actively demethylate their DNA. Multicellular organisms differentiate cell types, tissues, and organs by regulating gene expression through epigenetic mechanisms such as DNA methylation, histone methylation and acetylation.

With age, there is a global decrease in the level of DNA methylation, gene-specific hypermethylation and chromatin remodeling (Tollefsbol 2018). As a result, age-dependent tissue-specific overexpression/repression of about 10% of genes of the entire genome occurs, which plays a decisive role in reducing the regenerative and reparative abilities of cells and tissues and causes a functional decline in age-related diseases. Epigenetic age, or methylation clocks, is becoming the industry standard for assessing the aging rate and biological age of an individual (Moskalev 2020).

The rate of aging-associated epigenetic changes is influenced by both hereditary factors (associated, for example, with the efficiency of assimilation of B vitamins: B6, B9, B12, betaine) and nutrition. For instance, vitamins B6, B9, B12, C, A play an important role in regulating the degree of DNA methylation (Ferreira et al. 2020). The degree of histone acetylation is also regulated by functional nutrition. In particular, the high degree of histone acetylation under the influence of histone acetyltransferases (HATs) accelerates aging, and the donor of acetate groups is acetyl-CoA, which occurs in large quantities in the cell during a high-calorie fatty diet. Spermidine (found in fermented foods, mushrooms, sprouts) and a low-calorie diet reduce HATs activity (Pietrocola et al. 2015). The high activity of HDAC 1–3 deacetylases also increases the aging rate (McIntyre et al. 2019), therefore their inhibitors (3,3'-di-indolylmethane, isothiocyanates (sulforaphane), Se-methyl-1-selenocysteine in broccoli; equol, genistein in legumes; apigenin in citrus fruits and parsley; chrysin in olive oil; quercetin and ursolic acid in apples and green tea; allicin, diallyl disulfide in garlic; caffeic acid and catechins green tea; coumarin/hydroxycinnamic acid in cinnamon; curcumin in curry) may be considered as the potential geroprotectors. On the other hand, the increased activity of deacetylases of histones of the SIRT family slows down aging. SIRT1 activators are pterostilbene, piceatannol, resveratrol (blueberries, blackberries, black grapes, red wine), fisetin (fresh cucumbers, strawberries), quercetin (cranberries, onions, apples, blueberries).

The SIRT-activating diet called the Sirtfood Diet is gaining popularity (Pallauf 2013), which for example includes regular consumption of polyphenol-rich foods: legumes, capers, blueberries, green tea, olive oil, parsley, arugula, strawberries, citrus fruits, apples, kale, black coffee and curry.

21.4 Antinutrients

Many nutritionists recommend avoiding phytic acid, which is stored as a source of phosphorus in the seeds and nuts. Since it interferes with the assimilation of some macro- and microelements - magnesium, calcium, iron, zinc, copper. However, it is a chelator of excess transition metals, which cause glycation through the Maillard reaction. Clinical studies have shown its benefit in reducing glycation levels in diabetic patients (Sanchis et al. 2018). The medicinal value of phytic acid outweighs its negative effects (Abdulwaliyu et al. 2019). The most rational thing is not to pre-soak whole grains, legumes or nuts containing phytic acid, but to use them as a separate meal. In this case, they will not interfere with the absorption of trace elements from other foods.

Lectins are proteins with a high degree of stereospecificity in recognition of various sugar structures and the formation of reversible bonds when interacting with glycoconjugate complexes. They are found in abundance in plants (legumes, nightshades, whole grains, peanut), animals and many other species, and are known to agglutinate various erythrocyte blood groups. Lectins are easily destroyed even after 10 minutes of heat treatment. So if you don't eat raw legumes, grains, and potato skins, you won't get harmful dosages. There are really toxic lectins - ricin, phytohemagglutinin, but you are unlikely to encounter them in toxic amounts. In contrast, some lectins are beneficial with anti-cancer, immunomodulatory, and antimicrobial properties (Mishra 2019).

Moreover, as it becomes clear now, plant lectins inactivate pathogenic microflora, stimulate the phagocytic activity of cellular immunity, and act as "natural vaccines".

21.5 Diet Regimen and Longevity

It should be noted that it is not only the qualitative and quantitative composition of the diet that matters, but also the diet regimen. For example, a 10-h Time-limited Meal (TRE) in metabolic syndrome (MetS) patients promotes weight loss, reduces waist circumference, body fat percentage, and visceral fat. TRE in MetS lowers blood pressure, atherogenic lipids and glycated hemoglobin (Wilkinson et al. 2020). On the other hand, the use of TRE in old age is less beneficial for health than regular exercise (Schafer et al. 2019).

In contrast, fractional meals often recommended by nutritionists have mixed health effects. A meta-analysis comparing 15 studies did not support the conclusion about the benefits of split meals for maintaining optimal body weight (Schoenfeld et al. 2015). In one of the latest clinical studies in patients with type 2 diabetes mellitus, the state of metabolic health markers was compared in two groups: the first ate thoroughly twice a day, the second - six times a day in small portions. Patients in the first group had a decrease in body weight, the amount of fat in the liver, and fasting plasma glucose levels, as well as an increase in the efficiency of glucose absorption, compared with those who divided the same number of calories into six meals (Kahleova et al. 2014). With frequent snacks, the level of saturation hormones (insulin and leptin) is constantly high. As a result, the regulatory parts of the brain get used to this evenly high background and become insensitive to it. That is, tissue resistance to insulin and leptin signals is formed. And this leads to a deterioration in the regulation of both eating behavior and metabolism in general. Being hungry all the time increases your risk of obesity, type 2 diabetes, and cardiovascular problems. Returning to the traditional diet (less often, but full), it is possible to normalize the level of these hormones in the blood, the sensitivity of tissues to them and the energy balance in the body. Some foods are addictive and overeating. A rating of such food has been compiled (Schulte et al. 2015).

21.6 Diet of Long Living Persons

The traditional diet in Okinawa differs significantly from the general Japanese one and is based on the use of sweet potatoes (completely replaces rice), leafy vegetables, yellow root vegetables, soy products (miso and tofu) and medicinal plants. Seafood and algae, lean pork, fruits, spices (primarily turmeric), tea, and weak alcohol are consumed in moderation. There are practically no dairy products in the diet. Salt replaced with spices. Thus, the traditional Okinawan diet is low in calories, contains a low ratio of omega 6/omega 3 fatty acids (close to 1: 1), has a low glycemic index, includes a little meat, but a lot of plant fiber. The Okinawan diet slows down the development of cardiovascular disease and inflammation (Willcox et al. 2009).

A diet close to the Paleolithic or Mediterranean reduces the risk of all-cause mortality. The Mediterranean diet has proven to be the most effective. The Paleolithic diet was common among the ancient hunter-gatherers before the advent of agriculture. The structure of the diet is characterized by a predominantly plant-based diet, with a wide variety of fruits, nuts, and vegetables. It also includes lean meats and is low in dairy, grains, sugar, and salt (Whalen et al. 2017). The Mediterranean is like the Paleolithic, but allows a moderate consumption of grains, dairy products, and alcohol.

The Mediterranean diet is the traditional diet of the Mediterranean countries - Italy and Greece. It is characterized by high consumption of olive oil (as the main dietary fat), vegetables, legumes, whole grains, fruits and nuts, moderate consumption of poultry and fish, low consumption of fatty dairy products and red meat, and moderate consumption of wine as the main source of alcohol, consumed exclusively at the same time as food. As it turned out, it is in Italy and Greece that people who adhere to traditional food habits are less likely to suffer from neoplastic and cardiovascular diseases. The Mediterranean diet does not reduce body fat in general, but it does reduce the proportion of fat in the liver, which leads to metabolic syndrome and chronic disease (Gepner et al. 2019). Unlike the Western diet, it protects against steatohepatitis (Shively et al. 2019). In the EPIC-Potsdam cohort, the Mediterranean Diet associated with lower risk of type 2 diabetes in the overall population and of myocardial infarction in women (Galbete et al. 2018).

The Nordic or Scandinavian diet is based on staples such as cabbage, rye bread, root vegetables, apples, pears, northern berries, oatmeal, shellfish, and fish. According to research, adherence to such a diet reduces the risk of death. In the EPIC-Potsdam cohort, the Nordic diet showed a possible beneficial effect on myocardial infarction in the overall population and for stroke in men (Galbete et al. 2018). Traditional healthy Nordic foods was found to be related to lower mortality among middle-aged Danes, in particular among men (Olsen et al. 2011). It should be noted that a common feature of traditional diets is the absence of processed food, the freshness of the products (which is achieved due to the locality and seasonality of the diet).

21.7 Conclusions

Nowadays, the principles of healthy eating are common to all. However, each person is unique and carries own risks of genetic predisposition, has a individual metabolic rate, physical, and mental activity. A personalized approach to the diet for longevity can be expected in the future. It will consider genetic predisposition, gender, biological age, physical and intellectual activity, concomitant diseases, clinical analyzes (excess or lack of essential fatty and amino acids, vitamins, minerals in the blood, markers of metabolism, microbiota, biological age). Regular monitoring of all these data, together with the technologies of synthetic food products and 3D-printing, will allow to personally approach the needs of the body, eliminate harmful influences, introduce potential geroprotectors into the diet, and make healthy food the most aesthetic and tasty. In principle, it is technologically possible already now.

Compliance with Ethical Standards

Conflict of Interest The author has no conflict of interest.

References

Abdulwaliyu I et al (2019) Investigation of the medicinal significance of phytic acid as an indispensable anti-nutrient in diseases. Clin Nutrition Exp 28:42–61

Alcock J, Maley CC, Aktipis CA (2014) Is eating behavior manipulated by the gastrointestinal microbiota? Evolutionary pressures and potential mechanisms. BioEssays 36(10):940–949

Aliper A et al (2016) In search for geroprotectors: in silico screening and in vitro validation of signalome-level mimetics of young healthy state. Aging (albany NY) 8(9):2127–2152

Altunina NV, Lizogub VG, Bondarchuk OM (2020) Alpha-lipoic acid as a means of influence on systemic inflammation in type 2 diabetes mellitus patients with prior myocardial infarction. J Med Life 13(1):32–36

Aly FM, Othman A, Haridy MAM (2018) Protective effects of fullerene C60 nanoparticles and virgin olive oil against genotoxicity induced by cyclophosphamide in rats. Oxid Med Cell Longev 2018:1261356

Amano H et al (2019) Telomere dysfunction induces sirtuin repression that drives telomere-dependent disease. Cell Metab 29(6):1274–1290, e9

Ang Z, Ding JL (2016) GPR41 and GPR43 in Obesity and Inflammation - Protective or Causative? Front Immunol 7:28

Anonymous '2018 Global Nutrition Report. https://globalnutritionreport.org/reports/global-nutrition-report-2018. Accessed.

Anonymous 'Glycemic Load', <https://www.diabetes.co.uk/diet/glycemic-load.html>, accessed.

Anonymous 'The world's most nutritious foods. http://www.bbc.com/future/story/20180126-the-100-most-nutritious-foods. Accessed

Azman KF, Zakaria R (2019) D-Galactose-induced accelerated aging model: an overview. Biogerontology 20(6):763–782

Baati T et al (2012) The prolongation of the lifespan of rats by repeated oral administration of [60]fullerene. Biomaterials 33(19):4936–4946

Baig MH et al (2017) Methylglyoxal and Advanced Glycation End products: Insight of the regulatory machinery affecting the myogenic program and of its modulation by natural compounds. Sci Rep 7(1):5916

Barardo D et al (2017) The DrugAge database of aging-related drugs. Aging Cell 16(3):594–597

Barrington WE, White E (2016) Mortality outcomes associated with intake of fast-food items and sugar-sweetened drinks among older adults in the Vitamins and Lifestyle (VITAL) study. Public Health Nutr 19(18):3319–3326

Bartsch H, Nair J, Owen RW (1999) Dietary polyunsaturated fatty acids and cancers of the breast and colorectum: emerging evidence for their role as risk modifiers. Carcinogenesis 20(12):2209–2218

Berer K et al (2018) Dietary non-fermentable fiber prevents autoimmune neurological disease by changing gut metabolic and immune status. Sci Rep 8(1):10431

Bhukel A, Madeo F, Sigrist SJ (2017) Spermidine boosts autophagy to protect from synapse aging. Autophagy 13(2):444–445
Bourassa MW et al (2016) Butyrate, neuroepigenetics and the gut microbiome: can a high fiber diet improve brain health? Neurosci Lett 625:56–63
Bouziat R et al (2017) Reovirus infection triggers inflammatory responses to dietary antigens and development of celiac disease. Science 356(6333):44–50
Braidy N, Liu Y (2020) NAD+ therapy in age-related degenerative disorders: A benefit/risk analysis. Exp Gerontol 132:110831
Brial F et al (2020) The natural metabolite 4-cresol improves glucose homeostasis and enhances beta-cell function. Cell Rep 30(7):2306–2320, e5
Bruewer M et al (2003) Proinflammatory cytokines disrupt epithelial barrier function by apoptosis-independent mechanisms. J Immunol 171(11):6164–6172
Brunt VE et al (2020) Trimethylamine-N-oxide promotes age-related vascular oxidative stress and endothelial dysfunction in mice and healthy humans. Hypertension 76(1):101–112
Bui TP et al (2015) Production of butyrate from lysine and the Amadori product fructoselysine by a human gut commensal. Nat Commun 6:10062
Buttner S et al (2014) Spermidine protects against alpha-synuclein neurotoxicity. Cell Cycle 13(24):3903–3908
Calabrese V et al (1822) (2012) Cellular stress responses, hormetic phytochemicals and vitagenes in aging and longevity . Biochim Biophys Acta 5:753–783
Capurso L (2019) Thirty years of lactobacillus rhamnosus GG: a review. J Clin Gastroenterol 53(Suppl 1):S1–S41
Cavicchia PP et al (2009) A new dietary inflammatory index predicts interval changes in serum high-sensitivity C-reactive protein. J Nutr 139(12):2365–2372
Chalvon-Demersay T et al (2017) A systematic review of the effects of plant compared with animal protein sources on features of metabolic syndrome. J Nutr 147(3):281–292
Che S et al (2020) Puerarin prevents epithelial tight junction dysfunction induced by ethanol in Caco-2 cell model. J Functional Foods 73:104079
Chowdhury R et al (2014) Association of dietary, circulating, and supplement fatty acids with coronary risk: a systematic review and meta-analysis. Ann Intern Med 160(6):398–406
Clark A, Mach N (2017) The Crosstalk between the Gut Microbiota and Mitochondria during Exercise. Front Physiol 8:319
Correa TAF et al (2019) The Two-Way Polyphenols-Microbiota Interactions and Their Effects on Obesity and Related Metabolic Diseases. Front Nutr 6:188
Cummings NE et al (2018) Restoration of metabolic health by decreased consumption of branched-chain amino acids. J Physiol 596(4):623–645
Dandona P et al (2007) Proinflammatory effects of glucose and anti-inflammatory effect of insulin: relevance to cardiovascular disease. Am J Cardiol 99(4A):15B-26B
Dang Y et al (2020) Berberine ameliorates cellular senescence and extends the lifespan of mice via regulating p16 and cyclin protein expression. Aging Cell 19(1):e13060
Davie JR (2003) Inhibition of histone deacetylase activity by butyrate. J Nutr 133(7 Suppl):2485S-S2493
de Mello VD et al (2017) Indolepropionic acid and novel lipid metabolites are associated with a lower risk of type 2 diabetes in the Finnish Diabetes Prevention Study. Sci Rep 7:46337
Dong Z et al (2020) Association of sulfur amino acid consumption with cardiometabolic risk factors: Cross-sectional findings from NHANES III. EClinicalMedicine 19:100248
Donohoe DR et al (2011) The microbiome and butyrate regulate energy metabolism and autophagy in the mammalian colon. Cell Metab 13(5):517–526
Eisenberg T et al (2017) Dietary spermidine for lowering high blood pressure. Autophagy 13(4):767–769
Eisenberg T et al (2016) Cardioprotection and lifespan extension by the natural polyamine spermidine. Nat Med 22(12):1428–1438

Elamin E et al (2014) Ethanol impairs intestinal barrier function in humans through mitogen activated protein kinase signaling: a combined in vivo and in vitro approach. PLoS One 9(9):e107421

Elhassan YS et al (2019) Nicotinamide riboside augments the human skeletal muscle NAD⁺ metabolome and induces transcriptomic and anti-inflammatory signatures in aged subjects: a placebo-controlled, randomized trial. bioRxiv:680462

Errazuriz I et al (2017) Randomized controlled trial of a MUFA or fiber-rich diet on hepatic fat in prediabetes. J Clin Endocrinol Metab 102(5):1765–1774

Fan J et al (2017) Spermidine coupled with exercise rescues skeletal muscle atrophy from D-gal-induced aging rats through enhanced autophagy and reduced apoptosis via AMPK-FOXO3a signal pathway. Oncotarget 8(11):17475–17490

Fedintsev A, Moskalev A (2020) Stochastic non-enzymatic modification of long-lived macromolecules - a missing hallmark of aging. Ageing Res Rev 62:101097

Fernandes J et al (2020) Is olive oil good for you? a systematic review and meta-analysis on anti-inflammatory benefits from regular dietary intake. Nutrition 69:110559

Ferreira LMR et al. (2020) Intermediary metabolism: an intricate network at the crossroads of cell fate and function. Biochim Biophys Acta Mol Basis Dis 1866(10):165887

Firth J et al (2019) The effects of dietary improvement on symptoms of depression and anxiety: a meta-analysis of randomized controlled trials. Psychosom Med 81(3):265–280

Foster-Powell K, Holt SH, Brand-Miller JC (2002) International table of glycemic index and glycemic load values: 2002. Am J Clin Nutr 76(1):5–56

Galbete C et al (2018) Nordic diet, Mediterranean diet, and the risk of chronic diseases: the EPIC-Potsdam study. BMC Med 16(1):99

Gao X et al (2019) Dietary methionine influences therapy in mouse cancer models and alters human metabolism. Nature 572(7769):397–401

Gepner Y et al (2019) The beneficial effects of Mediterranean diet over low-fat diet may be mediated by decreasing hepatic fat content. J Hepatol 71(2):379–388

Ghrayeb A, Gottlieb E, Mor I (2019) Alcohol-derived acetate modulates brain function. Nat Metab 1(11):1036–1037

Green CL, Lamming DW (2019) Regulation of metabolic health by essential dietary amino acids. Mech Ageing Dev 177:186–200

Guasch-Ferre M et al (2019) Associations of monounsaturated fatty acids from plant and animal sources with total and cause-specific mortality in two US prospective cohort studies. Circ Res 124(8):1266–1275

Gueguen N et al (2015) Resveratrol directly binds to mitochondrial complex i and increases oxidative stress in brain mitochondria of aged mice. PLoS One 10(12):e014429

Guerra B, Issinger OG (2019) Natural compounds and derivatives as ser/thr protein kinase modulators and inhibitors. Pharmaceuticals (Basel) 12(1):4

Gugliucci A (2017) Formation of fructose-mediated advanced glycation end products and their roles in metabolic and inflammatory diseases. Adv Nutr 8(1):54–62

Haikal C, Chen QQ, Li JY (2019) Microbiome changes: an indicator of Parkinson's disease? Transl Neurodegener 8:38

Han S et al (2017) Mono-unsaturated fatty acids link H3K4me3 modifiers to C. elegans lifespan. Nature 544(7649):185–190

Holscher HD (2017) Dietary fiber and prebiotics and the gastrointestinal microbiota. Gut Microbes 8(2):172–184

Hosseinpour-Niazi S et al (2020) Weight gain, but not macronutrient intake, modifies the effect of dietary branch chain amino acids on the risk of metabolic syndrome. Diabetes Res Clin Pract 161:108039

Huang S (2013) Inhibition of PI3K/Akt/mTOR signaling by natural products. Anticancer Agents Med Chem 13(7):967–970

Iddir M et al (2020) Strengthening the immune system and reducing inflammation and oxidative stress through diet and nutrition: considerations during the COVID-19 Crisis. Nutrients 12(6)

Iuchi K et al (2019) Oxidized unsaturated fatty acids induce apoptotic cell death in cultured cells. Mol Med Rep 19(4):2767–2773

Joseph SV, Edirisinghe I, Burton-Freeman BM (2016) Fruit polyphenols: a review of anti-inflammatory effects in humans. Crit Rev Food Sci Nutr 56(3):419–444
Kahleova H et al (2014). https://www.ncbi.nlm.nih.gov/pmc/articles/PMC4079942/
Kebede AF et al (2017) Histone propionylation is a mark of active chromatin. Nat Struct Mol Biol 24(12):1048–1056
Khan S et al (2020) Dietary simple sugars alter microbial ecology in the gut and promote colitis in mice. Sci Transl Med 12(567)
Khan SU et al (2019) Effects of nutritional supplements and dietary interventions on cardiovascular outcomes: an umbrella review and evidence map. Ann Intern Med 171(3):190–198
Kiechl S et al (2018) Higher spermidine intake is linked to lower mortality: a prospective population-based study. Am J Clin Nutr 108(2):371–380
Kim S et al (2015) Uncovering the nutritional landscape of food. PLoS One 10(3):e0118697
Kowlgi NG, Chhabra L (2015) D-lactic acidosis: an underrecognized complication of short bowel syndrome. Gastroenterol Res Pract 2015:476215
Lambertz J et al (2017) Fructose: a dietary sugar in crosstalk with microbiota contributing to the development and progression of non-alcoholic liver disease. Front Immunol 8:1159
Lauretti E, Iuliano L, Pratico D (2017) Extra-virgin olive oil ameliorates cognition and neuropathology of the 3xTg mice: role of autophagy. Ann Clin Transl Neurol 4(8):564–574
LeBaron TW et al (2020) The Effects of 24-Week. High-Concentration Hydrogen-Rich Water on Body Composition, Blood Lipid Profiles and Inflammation Biomarkers in Men and Women with Metabolic Syndrome: A Randomized Controlled Trial', Diabetes Metab Syndr Obes 13:889–896
Lees EK et al (2017) Direct comparison of methionine restriction with leucine restriction on the metabolic health of C57BL/6J mice. Sci Rep 7(1):9977
Levine ME et al (2014) Low protein intake is associated with a major reduction in IGF-1, cancer, and overall mortality in the 65 and younger but not older population. Cell Metab 19(3):407–417
Li Y et al (2018) Impact of healthy lifestyle factors on life expectancies in the US population. Circulation 138(4):345–355
Ling B et al (2012) D-Lactate altered mitochondrial energy production in rat brain and heart but not liver. Nutr Metab (lond) 9(1):6
Liu M, Dudley SC, Jr. (2020) Magnesium, oxidative stress, inflammation, and cardiovascular disease. Antioxidants (Basel) 9(10)
Lu AT et al (2019) DNA methylation GrimAge strongly predicts lifespan and healthspan. Aging (albany NY) 11(2):303–327
Luck H et al (2019) Gut-associated IgA(+) immune cells regulate obesity-related insulin resistance. Nat Commun 10(1):3650
Ma L et al (2020) Indole alleviates diet-induced hepatic steatosis and inflammation in a manner involving myeloid cell 6-Phosphofructo-2-Kinase/Fructose-2,6-Biphosphatase 3. Hepatology 72(4):1191–1203
Madeo F et al (2018) Spermidine delays aging in humans. Aging (albany NY) 10(8):2209–2211
Matsumura T et al (2020) N-acetyl-l-tyrosine is an intrinsic triggering factor of mitohormesis in stressed animals. EMBO Rep 21(5):e49211
McIntyre RL et al (2019) From molecular promise to preclinical results: HDAC inhibitors in the race for healthy aging drugs. EMBO Mol Med 11:e9854
Mikkelsen KH et al (2015) Use of antibiotics and risk of type 2 diabetes: a population-based case-control study. J Clin Endocrinol Metab 100(10):3633–3640
Miller RA et al (2005) Methionine-deficient diet extends mouse lifespan, slows immune and lens aging, alters glucose, T4, IGF-I and insulin levels, and increases hepatocyte MIF levels and stress resistance. Aging Cell 4(3):119–125
Minois N (2014) Molecular basis of the 'anti-aging' effect of spermidine and other natural polyamines - a mini-review. Gerontology 60(4):319–326
Mishra A et al (2019) Structure-function and application of plant lectins in disease biology and immunity. Food Chem Toxicol 134:110827
Moskalev A (2020) The challenges of estimating biological age Elife 9

Moskalev A, Stambler I, Caruso C (2020) Innate and adaptive immunity in aging and longevity: the foundation of resilience. Aging Dis 11(6):1363–1373

Moskalev A et al (2015) Geroprotectors.org: a new, structured and curated database of current therapeutic interventions in aging and age-related disease. Aging (albany NY) 7(9):616–628

Moskalev AA et al (2013) The role of DNA damage and repair in aging through the prism of Koch-like criteria. Ageing Res Rev 12(2):661–684

Moskalev A, Stambler I, Caruso C (2021) Innate and adaptive immunity in aging and longevity: the foundation of resilience. Aging Disease 12(1),0

Mu Q et al (2017) Leaky Gut As a Danger Signal for Autoimmune Diseases. Front Immunol 8:598

Naghshi S et al (2020) Dietary intake of total, animal, and plant proteins and risk of all cause, cardiovascular, and cancer mortality: systematic review and dose-response meta-analysis of prospective cohort studies BMJ 370:m2412

Naito Y, Uchiyama K, Takagi T (2018) A next-generation beneficial microbe: Akkermansia muciniphila. J Clin Biochem Nutr 63(1):33–35

Nam NH (2006) Naturally occurring NF-kappaB inhibitors. Mini Rev Med Chem 6(8):945–951

Neuhouser ML et al (2006) Development of a glycemic index database for food frequency questionnaires used in epidemiologic studies. J Nutr 136(6):1604–1609

Nishimura N et al. (2018) Transplantation of high hydrogen-producing microbiota leads to generation of large amounts of colonic hydrogen in recipient rats fed high amylose maize starch. Nutrients 10(2)

Olsen A et al (2011) Healthy aspects of the Nordic diet are related to lower total mortality. J Nutr 141(4):639–644

Orentreich N et al (1993) Low methionine ingestion by rats extends life span. J Nutr 123(2):269–274

Oteng AB, Kersten S (2020) Mechanisms of action of trans fatty acids. Adv Nutr 11(3):697–708

Pallauf K et al (2013) Nutrition and healthy ageing: calorie restriction or polyphenol-rich "MediterrAsian" diet? Oxid Med Cell Longev 2013:707421

Panchenko AV et al (2019) Comparative analysis of experimental data about the effects of various polyphenols on lifespan and aging. Adv Gerontol 32(3):325–330

Papier K et al (2021) Meat consumption and risk of 25 common conditions: outcome-wide analyses in 475,000 men and women in the UK Biobank study. BMC Med 19(1):53

Patterson E et al (2019) Gamma-aminobutyric acid-producing lactobacilli positively affect metabolism and depressive-like behaviour in a mouse model of metabolic syndrome. Sci Rep 9(1):16323

Pauly MJ et al (2020) Inulin supplementation disturbs hepatic cholesterol and bile acid metabolism independent from housing temperature. Nutrients 12(10)

Perry RJ et al (2016) Acetate mediates a microbiome-brain-beta-cell axis to promote metabolic syndrome. Nature 534(7606):213–217

Perveen T et al (2013) Role of monoaminergic system in the etiology of olive oil induced antidepressant and anxiolytic effects in rats. ISRN Pharmacol 2013:615685

Pietrocola F et al (2015) Spermidine induces autophagy by inhibiting the acetyltransferase EP300. Cell Death Differ 22(3):509–516

Powell DN et al (2020) Indoles from the commensal microbiota act via the AHR and IL-10 to tune the cellular composition of the colonic epithelium during aging. Proc Natl Acad Sci U S A 117(35):21519–21526

Proshkina E, Shaposhnikov M, Moskalev A (2020a) Genome-protecting compounds as potential geroprotectors. Int J Mol Sci 21(12)

Proshkina E et al (2020b) Terpenoids as potential geroprotectors. Antioxidants (Basel) 9(6)

Ramot Y et al (2011) Spermidine promotes human hair growth and is a novel modulator of human epithelial stem cell functions. PLoS One 6(7):e22564

Rattan SI, Demirovic D (2009) Hormesis can and does work in humans. Dose Response 8(1):58–63

Rico-Campa A et al (2019) Association between consumption of ultra-processed foods and all cause mortality: SUN prospective cohort study. BMJ 365:1949

Riske L et al (2017) Lactate in the brain: an update on its relevance to brain energy, neurons, glia and panic disorder. Ther Adv Psychopharmacol 7(2):85–89
Roager HM, Licht TR (2018) Microbial tryptophan catabolites in health and disease. Nat Commun 9(1):3294
Rowland I et al (2018) Gut microbiota functions: metabolism of nutrients and other food components. Eur J Nutr 57(1):1–24
Saemann MD et al (2000) Anti-inflammatory effects of sodium butyrate on human monocytes: potent inhibition of IL-12 and up-regulation of IL-10 production. FASEB J 14(15):2380–2382
Sanchis P et al (2018) Phytate decreases formation of advanced glycation end-products in patients with type II diabetes: randomized crossover trial. Sci Rep 8(1):9619–9719
Sartorius K et al (2018) Does high-carbohydrate intake lead to increased risk of obesity? a systematic review and meta-analysis. BMJ Open 8(2):e018449
Schafer MJ et al (2019) Late-life time-restricted feeding and exercise differentially alter healthspan in obesity. Aging Cell 18(4):e12966
Schoenfeld BJ, Aragon AA, Krieger JW (2015) Effects of meal frequency on weight loss and body composition: a meta-analysis. Nutr Rev 73(2):69–82
Schulte EM, Avena NM, Gearhardt AN (2015) Which foods may be addictive? the roles of processing, fat content, and glycemic load. PLoS One 10(2):e0117959
Schwalfenberg GK (2017) Vitamins K1 and K2: the emerging group of vitamins required for human health. J Nutr Metab 2017:6254836
Schwingshackl L, Hobl LP, Hoffmann G (2015) Effects of low glycaemic index/low glycaemic load vs. high glycaemic index/ high glycaemic load diets on overweight/obesity and associated risk factors in children and adolescents: a systematic review and meta-analysis. Nutr J 14:87
Seidelmann SB et al (2018) Dietary carbohydrate intake and mortality: a prospective cohort study and meta-analysis. Lancet Public Health 3(9):e419–e428
Sestili P et al (2010) Sulforaphane induces DNA single strand breaks in cultured human cells. Mutat Res 689(1–2):65–73
Shen P et al (2017) Piceatannol extends the lifespan of Caenorhabditis elegans via DAF-16. BioFactors 43(3):379–387
Shenderov BA (2013) Metabiotics: novel idea or natural development of probiotic conception. Microb Ecol Health Dis 24
Shifflett DE et al (2005) Enteropathogenic E. coli disrupts tight junction barrier function and structure in vivo. Lab Invest 85(10):1308–1324
Shimamura Y et al (2020) Does external exposure of glycidol-related chemicals influence the forming of the hemoglobin adduct, N-(2,3-dihydroxypropyl)valine, as a Biomarker of Internal Exposure to Glycidol? Toxics 8(4)
Shively CA et al (2019) Mediterranean versus western diet effects on caloric intake, obesity, metabolism, and hepatosteatosis in nonhuman primates. Obesity (silver Spring) 27(5):777–784
Sloczynska K et al (2014) Antimutagenic compounds and their possible mechanisms of action. J Appl Genet 55(2):273–285
Snippert HJ (2016) Colonic crypts: safe haven from microbial products. Cell 165(7):1564–1566
Solon-Biet SM et al (2019) Branched chain amino acids impact health and lifespan indirectly via amino acid balance and appetite control. Nat Metab 1(5):532–545
Song M et al (2016) Association of animal and plant protein intake with all-cause and cause-specific mortality. JAMA Intern Med 176(10):1453–1463
Stavropoulou E, Bezirtzoglou E (2020) Probiotics in medicine: a long debate. Front Immunol 11:2192
Steckhan N et al (2016) Effects of different dietary approaches on inflammatory markers in patients with metabolic syndrome: a systematic review and meta-analysis. Nutrition 32(3):338–348
Sun Y et al (2019) Association of fried food consumption with all cause, cardiovascular, and cancer mortality: prospective cohort study. BMJ 364:k5420
Syrovy I (1994) Glycation of albumin: reaction with glucose, fructose, galactose, ribose or glyceraldehyde measured using four methods. J Biochem Biophys Methods 28(2):115–121

Thaiss CA et al (2018) Hyperglycemia drives intestinal barrier dysfunction and risk for enteric infection. Science 359(6382):1376–1383
Ting CY et al (2015) Curcumin triggers DNA damage and inhibits expression of DNA repair proteins in human lung cancer cells. Anticancer Res 35(7):3867–3873
Todoric J et al (2020) Fructose stimulated de novo lipogenesis is promoted by inflammation. Nat Metab 2(10):1034–1045
Tollefsbol TO (2018) Translational Epigenetics Series. In: Moskalev A, Vaiserman AM (eds) Epigenetics of Aging and Longevity (4; Boston: Academic Press), ii
Tomas Abadal L et al (2001) Risk factors and coronary morbimortality in a Mediterranean industrial cohort over 28 years of follow-up. The Manresa Study. Rev Esp Cardiol 54(10):1146–1154
Trompette A et al (2018) Dietary fiber confers protection against flu by shaping Ly6c(-) patrolling monocyte hematopoiesis and CD8(+) T Cell metabolism. Immunity 48(5):992–1005 e8
Turati F et al (2019) Glycemic index, glycemic load and cancer risk: an updated meta-analysis Nutrients 11(10)
Uribarri J et al (2010) Advanced glycation end products in foods and a practical guide to their reduction in the diet. J Acad Nutr Diet 110(6):911–916, e12
Valles-Colomer M et al (2019) The neuroactive potential of the human gut microbiota in quality of life and depression. Nat Microbiol 4(4):623–632
van der Lugt B et al (2019) Akkermansia muciniphila ameliorates the age-related decline in colonic mucus thickness and attenuates immune activation in accelerated aging Ercc1 (-/Delta7) mice. Immun Ageing 16:6
Varela-Lopez A et al (2018) Gene pathways associated with mitochondrial function, oxidative stress and telomere length are differentially expressed in the liver of rats fed lifelong on virgin olive, sunflower or fish oils. J Nutr Biochem 52:36–44
Vega-Lopez S, Venn BJ, Slavin JL (2018) Relevance of the glycemic index and glycemic load for body weight, diabetes, and cardiovascular disease. Nutrients 10(10)
Vejux A et al (2020) 7-Ketocholesterol and 7beta-hydroxycholesterol: In vitro and animal models used to characterize their activities and to identify molecules preventing their toxicity. Biochem Pharmacol 173:113648
Virtanen HEK et al (2019) Dietary proteins and protein sources and risk of death: the Kuopio Ischaemic Heart Disease Risk Factor Study. Am J Clin Nutr 109(5):1462–1471
Virtanen JK et al (2018) Serum n-6 polyunsaturated fatty acids and risk of death: the Kuopio Ischaemic Heart Disease Risk Factor Study. Am J Clin Nutr 107(3):427–435
Vitetta L et al (2017) Probiotics, D-Lactic acidosis, oxidative stress and strain specificity. Gut Microbes 8(4):311–322
Vizioli MG et al (2020) Mitochondria-to-nucleus retrograde signaling drives formation of cytoplasmic chromatin and inflammation in senescence. Genes Dev 34(5–6):428–445
Vossen LM et al (2015) Menaquinone-7 supplementation to reduce vascular calcification in patients with coronary artery disease: rationale and study protocol (VitaK-CAC Trial). Nutrients 7(11):8905–8915
Wang B et al (2018) Phospholipid remodeling and cholesterol availability regulate intestinal stemness and tumorigenesis. Cell Stem Cell 22(2):206–220, e4
Wang IF, Tsai KJ, Shen CK (2015) Spermidine on neurodegenerative diseases. Cell Cycle 14(5):697–698
Wang W et al (2019) Inhibiting the formation of advanced glycation end-products by three stilbenes and the identification of their adducts. Food Chem 295:10–15
Wassermann B, Muller H, Berg G (2019) An apple a day: which bacteria do we eat with organic and conventional apples? Front Microbiol 10:1629
Weverling-Rijnsburger AW et al (1997) Total cholesterol and risk of mortality in the oldest old. Lancet 350(9085):1119–1123
Whalen KA et al (2017) Paleolithic and mediterranean diet pattern scores are inversely associated with all-cause and cause-specific mortality in adults. J Nutr 147(4):612–620

Wilkinson MJ et al (2020) Ten-hour time-restricted eating reduces weight, blood pressure, and atherogenic lipids in patients with metabolic syndrome. Cell Metab 31(1):92–104 e5

Willcox DC et al (2009) The Okinawan diet: health implications of a low-calorie, nutrient-dense, antioxidant-rich dietary pattern low in glycemic load. J Am Coll Nutr 28(Suppl):500S-S516

Wu SE et al (2020) Microbiota-derived metabolite promotes HDAC3 activity in the gut. Nature 586(7827):108–112

Yamashita N, Kawanishi S (2000) Distinct mechanisms of DNA damage in apoptosis induced by quercetin and luteolin. Free Radic Res 33(5):623–633

Ye L et al. (2019) High fat diet induces microbiota-dependent silencing of enteroendocrine cells Elife 8

Yoneshiro T et al (2019) BCAA catabolism in brown fat controls energy homeostasis through SLC25A44. Nature 572(7771):614–619

Yu D et al (2018) Short-term methionine deprivation improves metabolic health via sexually dimorphic, mTORC1-independent mechanisms. FASEB J 32(6):3471–3482

Zafar MI et al (2019) Low-glycemic index diets as an intervention for diabetes: a systematic review and meta-analysis. Am J Clin Nutr 110(4):891–902

Zhuang P et al (2019) Dietary fats in relation to total and cause-specific mortality in a prospective cohort of 521 120 individuals with 16 years of follow-up. Circ Res 124(5):757–768

Chapter 22
Gerosuppressive and Senolytic Nutrients

Jan O. Nehlin

Abstract The aging process is intimately regulated by the contribution of interactions between genetic, epigenetic and environmental factors. Every individual is distinguished by a specific personal lifestyle, since conception, that influences the state of health, either leading to the preservation of a good state of health or accelerated aging trajectories that could hasten the occurrence of disease. The intake of specific nutritional factors, investigated in large epidemiological studies, have long been considered associated with specific health outcomes and prevention of health problems. As a result of the aging process, the occurrence of senescent cells within the body can cause damage to organs and tissues leading to many age-associated pathologies. The advent of the nutritional geroscience field is contributing to reveal how a diverse number of phytochemicals can help to reduce the rate of cell proliferation, enhance cellular function, eradicate senescent cells, or suppress, at varying degrees, their abnormal secretomes, mitigating the extent of damage posed by them in the body. A summary of senotherapeutic biomolecules acting as senolytics and gerosuppressors will be here presented, to strengthen our knowledge that natural compounds present in the diet, when given at the right time, at the right dose and in combination, could extend the duration of health span and the overall lifespan.

Keywords Health · Disease · Aging · Senescence · Gerosuppressor · Senolytic

22.1 Introduction

Since conception, humans are exposed to a nutritional environment that will shape their future health trajectories, during all their lifetimes. The interplay between nutritional lifestyle, parental epigenetic factors, and inherited gene variants will determine how well the body will function and cope with all the stresses that will be confronted with time.

J. O. Nehlin (✉)
Department of Clinical Research, Copenhagen University Hospital Amager
and Hvidovre-Copenhagen, Kettegård Alle 30, 2650 Hvidovre, Denmark
e-mail: jan.nehlin@regionh.dk

S. I. S. Rattan and G. Kaur (eds.), *Nutrition, Food and Diet in Ageing and Longevity*, Healthy Ageing and Longevity 14, https://doi.org/10.1007/978-3-030-83017-5_22

Traditional medicine, through thousands of years of experience, has contributed to unravel the identities of natural products that could have a role in preserving health, mitigating symptoms of disease, accelerated healing, and general well-being (Mukherjee et al. 2017; Wang et al. 2018). The chemical identification of the active ingredients with both beneficial and detrimental roles has been ongoing for decades.

The health benefits of Mediterranean and Asian diets have been confirmed in many clinical trials and epidemiological surveys. These diets are characterized by several features, including low meat consumption, fish consumption, the intake of oils instead of fats as lipid sources, moderate amounts of red wine, and significant amounts of fresh fruit and vegetables. Polyphenols have been shown to have multiple health effects as demonstrated by several relevant population studies and clinical trials (Leri et al. 2020; Cannataro et al. 2021; Shen et al. 2017; Gensous et al. 2020; Quach et al. 2017). The cellular mechanisms by which polyphenols exert their function has been recently reviewed (Russo et al. 2020; Mansoori et al. 2021; Jantan et al. 2021) as well as the role of polyphenols against aging and cancer (Bian et al. 2020; Russo et al. 2020).

The role of the microbiota needs to be highlighted in relation to healthy aging, because it is intimately associated with the specific intake of nutrients, and long-term health consequences (Fragkou et al. 2021, Tavella et al. 2021), as well as parameters such as eating time (Taetzsch et al. 2021).

22.2 General Considerations

Senescent cells arise in the body during development, stress, disease and aging (Childs et al. 2017). Several mechanisms can lead to senescence including replicative senescence (RS), stress-induced senescence (SIPS), oncogene-induced senescence (OIS), mitochondrial dysfunction-associated senescence (MiDAS), and many more (Di Micco et al. 2021; Calcinotto et al. 2019).

The immune system targets senescent cells that harbor damage that could pose a potential risk if allowed to bypass the senescent state (Prata et al. 2018, Kale et al. 2020). However, with time, immune function gradually decays giving rise to immunosenescence. The immune cells responsible for targeting and eliminating senescent cells become reduced in number and function, thereby allowing senescent cells to accumulate in the body (Rodrigues et al. 2021, Prata et al. 2018, Kale et al. 2020). The accumulation of senescent cells results in the characteristic low-grade chronic inflammation that is associated with age, which causes organ dysfunction (Childs et al. 2017; Furman et al. 2019).

Aging is a universal living condition that is accompanied by an increased risk of age-associated diseases, all of which are gathered under the concept of geriatric syndrome. The World Health Organization (WHO) has categorized senescence as a significant risk factor behind the origin of older age diseases (Code MG2A: Old age; International Classification of Diseases (ICD) 11th (June 2018). Senolytics (see

Sects. 22.3 and 22.4) are considered an aid to prolong the state of disease-free lifetime in people (Khaltourina et al. 2020).

22.3 Definitions and Classification of Senotherapeutics

Senotherapeutics or gerotherapeutics are compounds, medicines or protocols that exert senotherapy, an intervention targeting senescent cells, being considered an emerging strategy for the extension of health span, and prevention or treatment of age-associated diseases. At present, senotherapeutics can be classified in:

(1) Senolytics; (2) Gerosuppressors (SASP modulators or inhibitors); (3) Geroprotectors; (4) Epigenetic modifiers; (5) Gene therapeutics

The first four classes of senotherapeutics include nutritional components, whereas gene therapy strategies are mostly aimed at editing genes that might predispose to accelerated aging, less successful aging and even, enhance polymorphic variants to increase the length of health span, postpone the appearance of aging-associated diseases and extend lifespan.

The present work will consider the nutritional aspects involved in aging and focus on those compounds that exhibit senolytic and gerosuppressive functions. The nutritional geroscience field aims for the identification and characterization of nutritional compounds that have an effect on the function and/or viability of senescent cells, as well as health promoting activities that reduce or postpone the genesis of senescence. Some recent works have presented a list of senotherapeutics that are either in the discovery phase or are at different stages of clinical trials (Prasanna et al. 2021; Kirkland and Tchkonia 2020; Paez-Ribes et al. 2019; Morsli and Bellantuono 2021; Short et al. 2019).

22.4 Senolytics

Senolytics are defined as a new class of drugs that selectively kill senescent cells. They comprise small molecules that specifically induce cell death in senescent cells, targeting a number of targets including survival pathways and anti-apoptotic mechanisms (Zhu et al. 2015; Yosef et al. 2016). Several recent reviews have dealt with the topic of senolytic compounds (Kirkland and Tchkonia 2020; Martel et al. 2020; Robbins et al. 2021; Short et al. 2019).

Most studies on senolytics so far have focused on the effects of senescent cell eradicators in cell culture and mice studies. Such studies have rendered a great deal of knowledge as regards the specificity and extent of senolytic compounds on the alleviation of the senescent cell burden on specific tissues and their corresponding pathologies, including idiopathic pulmonary fibrosis (Justice et al. 2019), renal fibrosis post-acute kidney injury (Li et al. 2021), diabetic kidney disease (Hickson et al. 2019), eye age-macular degeneration (Lee et al. 2021), bone loss (Farr et al. 2017; Chandra et al. 2020), cardiac ischemia–reperfusion injury (Dookun et al. 2020b), cardiovascular disease (Dookun et al. 2020a, Song et al. 2020), osteoarthritis (Dai et al. 2020; Jeon et al. 2017), diabetes (Palmer et al. 2021, Thompson et al. 2019), cognitive decline (Bussian et al. 2018), uterine fibrosis (Cavalcante et al. 2020), cancer (Picallos-Rabina et al. 2020), frailty (Xu et al. 2018), etc. The first documented study to report the elimination of senescent cells in vivo in humans by senolytics was the report by Justice et al., in idiopathic pulmonary disease (Justice et al. 2019).

Many ongoing pre-clinical and clinical trials (Kirkland and Tchkonia 2020; Prasanna et al. 2021) will provide unequivocal evidence of the role of senolytics in a variety of human aging-associated diseases (Childs et al. 2017).

Several phytochemicals belonging to the flavonoid family have been documented to exhibit potent senolytic activities (Yousefzadeh et al. 2018). A list of foods describing their flavonoid contents and amounts is available (Haytowitz et al. 2018). Early epidemiological studies already showed that the intake of foods containing flavonoids such as quercetin, kaempferol, myricetin, apigenin, and luteolin, was significantly inversely associated with mortality from coronary heart disease (Hertog et al. 1993).

Many senolytics can upregulate SIRT1 activity, a sirtuin that belongs to the family of class III histone deacetylases, whose enzymatic activity is dependent on NAD^+. SIRT1 is implicated in the cellular response to caloric restriction and lifespan extension (Iside et al. 2020). Several natural compounds that activate the Nrf2 (nuclear factor erythroid-derived 2-related factor 2) pathway, controlling the cellular response to stress caused by reactive oxygen species (ROS), have been found to be senolytic (Malavolta et al. 2018).

While senolytics seem to be very promising compounds, there are concerns as regards their possible side effects or disadvantages, if their administration exceeds specific concentrations, and duration, and whether the recipient suffers or not from an age-associated disease (Martin et al. 2021, Raffaele et al. 2021). For such reasons, the National Institute on Aging, USA, has called for standardization in the investigations and measures regarding senotherapeutic use, including reports on the balance between senolytic and cytotoxic effects, markers of specific senescent cell types linked to the specific interventions, interactions with coexisting diseases and their effects at various ages, especially in older people (Romashkan et al. 2021).

Senolytics can be classified as being of nutritional origin or pharmaceutical origin. Both classes will be addressed in this summary, with focus on senolytics from natural sources.

22.4.1 Nutritional Senolytics

Nutritional senolytics are compounds that originate from natural sources and exhibit senolytic activity. The most studied nutritional senolytics are presented, but the list is likely to grow further as more research will reveal whether known or novel polyphenols possess senolytic activities.

Below, a list of prominent, well-documented, nutritional senolytic compounds (in alphabetical order) aims to describe their origin, targets and possible uses in senotherapy.

22.4.1.1 Allicin

Allicin (diallyl thiosulfinate) is a volatile sulfur compound present in garlic (*Allium sativum*), a vegetable with multiple health effects (Borlinghaus et al. 2021). The health-promoting and disease-preventing effects of garlic on many human common diseases, such as cancer, cardiovascular and metabolic disorders, blood pressure, and diabetes, through its antioxidant, anti-inflammatory, and lipid-lowering properties, have been documented in vitro, in vivo, and clinical studies (Ansary et al. 2020). Allicin has antiproliferative, anticlonogenic, and senolytic effects. In addition, allicin decreased cell viability and induced apoptosis by loss of $\Delta\Psi m$, caspase-3, caspase-8, and caspase-9 activation, upregulation of NOXA, P21, and BAK, as well as down-regulation of BCL-XL expression in breast cancer cell lines (Rosas-Gonzalez et al. 2020).

22.4.1.2 Curcumin and Curcumin Analogues

Curcumin is a yellow polyphenolic pigment from the turmeric (*Curcuma longa L.*) rhizome that has traditionally been used in culinary and food coloring, and as an ingredient in Ayurveda and Chinese medicine health promotion and disease prevention (Sharifi-Rad et al. 2020). Numerous studies suggest that curcumin has some health benefits in delaying aging and may be useful in preventing and treating age-related diseases (Cherif et al. 2019, Zia et al. 2021; Sharifi-Rad et al. 2020; Bielak-Zmijewska et al. 2019). Curcumin has been shown to have senolytic activity (Yousefzadeh et al. 2018). Among four commonly used curcumin analogs, EF24, HO-3867, 2-HBA and dimethoxycurcumin (DIMC), EF24 is the most potent and broad-spectrum senolytic agent. EF24 exerts selective lysis of senescent cells through the induction of apoptosis in a reactive oxygen species (ROS) production independent manner but associated with an increase in the proteasome degradation of the Bcl-2 anti-apoptotic protein family proteins known to play an important role in protecting senescent cells from apoptosis (Li et al. 2019).

22.4.1.3 Fisetin

Fisetin, with the chemical name 2-(3,4-Dihydroxyphenyl)-3,7-dihydroxy-4*H*-1-benzopyran-4-one, a flavonol, is a secondary metabolite of many plants, occurring in their green parts, fruits, as well as in barks and hardwood, that has anti-inflammatory, chemopreventive, chemotherapeutic and recently also senotherapeutic roles. Fisetin is present in strawberries, apples, persimmons, grapes, onions, kiwi, kale, nuts, smoke tree (*Cotinus coggygria*), etc. (Grynkiewicz and Demchuk 2019; Mehta et al. 2018) and influences many cellular pathways (Kashyap et al. 2019). Fisetin is considered a senolytic because it was found to selectively induce apoptosis in senescent but not proliferating human umbilical vein endothelial cells. However, it was not shown to be a senolytic in a senescent human lung fibroblast strain, or primary human preadipocytes (Zhu et al. 2017). Fisetin was found to be the most potent senolytic in a panel of ten flavonoid polyphenols using senescent mouse and human cells. Moreover, when given to wild-type mice late in life, fisetin restored tissue homestasis, reduced age-related pathologies, and extended median and maximum lifespan (Yousefzadeh et al. 2018). Fisetin is one of the senolytics that has been shown to have toxic effects when administered in high dosis (200 μM) in rat, inducing mitochondrial pro-oxidant activity (Constantin et al. 2011). Fisetin has other side effects such as the promotion of hair growth by activation of telomerase reverse transcriptase (TERT) expression in skin hair follicle bulge stem cells (Kubo et al. 2020). A summary of the in vivo evidence for fisetin targeting multiple age-associated dysfunctions and pathologies in mammals has been presented (Morsli and Bellantuono 2021).

22.4.1.4 Luteolin

Luteolin, with the chemical name 2-(3,4-Dihydroxyphenyl)-5,7-dihydroxy-4-chromenone, is a flavone, subclass of flavonoids, which is found in many vegetables, spices, fruits and medicinal herbs. It has many health benefits, and appears especially suited to alleviate the symptoms of glycolipid metabolism disorders such as insulin resistance, diabetes, and obesity (Wang et al. 2021). It was shown to have weak senolytic activity, in a comparative study with other flavonoids (see 4.1.3) (Yousefzadeh et al. 2018) but it has powerful anti-inflammatory actions in concert with IL-37 and IL-38 interleukins (Conti et al. 2021). Luteolin diminishes human neutrophil inflammatory responses by inhibiting Raf1-MEK-1-Erk, significantly inhibited superoxide anion generation, ROS production, and neutrophil extracellular trap (NET) formation in human neutrophils. The increase in elastase release, CD11b expression, and chemotaxis was also inhibited by luteolin (Yang et al. 2018). Apart from being anti-oxidative and anti-inflammatory, luteolin plays an important role in defending plants, for example against UV radiation, suggesting it could be useful in skin protection against UV-induced photoaging (Gendrisch et al. 2020).

22.4.1.5 Ortho-Vanillin

ortho-Vanillin (o-Vanillin; 2-hydroxy-3-methoxybenzaldehyde) is an organic solid present in the extracts and essential oils of many plants. o-Vanillin is an isomer of meta-Vanillin, a phenolic aldehyde.

Vanillin (4-hydroxy-3-methoxybenzaldehyde) that is a major component of the bean and pod of some plant species of the Vanilla genus, and is also synthesized on a large scale for use as a flavoring agent in food, fragrance and pharmaceutical industries. Vanillin exhibits antioxidant, antimicrobial, analgesic, anti-carcinogenic, anti-mutagenic and anti-sickling biological effects (Marton et al. 2016). o-Vanillin cleared senescent intervertebral disc (IVD) cells and reduced the senescence-associated secretory phenotype (SASP) (see Sect. 22.5) associated with inflammation and back pain. It also increased metabolic activity, caspase 3/7 activity, and apoptosis in cells from degenerate IVDs, but not in cells from non-mildly-degenerate IVDs (Cherif et al. 2019). o-vanillin exerts its actions through increased pro-apoptotic pathways and reduction of expression of senescence-associated genes. (Cherif et al. 2020). The vanillin analogues o-vanillin and 2,4,6- trihydroxybenzaldehyde were also shown to be cytotoxic against cultured human melanoma A375 cells, and are therefore being evaluated as potential anticancer drugs (Marton et al. 2016).

22.4.1.6 Piperlongumine

Long pepper (*Piper longum L.*), commonly known as "Pippali", is a pepper plant found in India and southeast Asia, that contains a vast number of health promoting compounds. Piperine and piperlongumine are its two major piperidone alkaloids (Yadav et al. 2020). Piperlongumine (also called piplartine or piperlongumin) has many pharmacological activities and is commonly used in Ayurvedic medicine (Bezerra et al. 2013). Piperlongumine has been shown to exert senolytic activity (Wang et al. 2016) with at least 172 different senolytic targets. One of them is oxidation resistance 1 (OXR1), an important antioxidant protein that regulates the expression of a variety of antioxidant enzymes and provides senescent cells with high resistance to oxidative stress. Piperlongumine induced OXR1 degradation through the ubiquitin–proteasome system in a senescent cell-specific manner (Zhang et al. 2018). Among other relevant targets, piperlongumine can also bind to annexin A1 (ANXA1), an endogenous anti-inflammatory mediator with therapeutic potential in cancer (Henrique et al. 2020).

22.4.1.7 Quercetin

Quercetin is a polyphenolic flavonoid, chemically known as 2-(3,4-dihydroxyphenyl)-3,5,7-trihydroxychromen-4-one, that is enriched in apples, berries, broccoli, cabbage, capers, cauliflower, cherries, citrus fruits, coriander, honey, nuts, red onions, tea, etc. (Batiha et al. 2020).

Quercetin was the first nutritional compound shown to have senolytic activity, and it was most effective against senescent human endothelial cells and mouse bone marrow mesenchymal stem cells. Quercetin was found to lack senolytic efficacy in senescent preadipocytes and mouse embryonic fibroblasts, but when combined with the anticancer drug Dasatinib, it was shown to be effective as senolytic in several types of senescent cells (Zhu et al. 2015).

In animal models, a senolytic cocktail of Dasatinib and quercetin (see also Sect. 22.4.2), which caused the elimination of senescent cells from adipose tissue, resulted in improved physical function and lifespan extension (Xu et al. 2018).

A clinical trial of Quercetin (Q) and Dasatinib (D) (a pharmaceutical senolytic; see Sect. 22.4.2) reduced the senescence burden in individuals with diabetic kidney disease (Hickson et al. 2019). Quercetin together with fisetin are considered as potentially useful adjuvants chemotherapeutic agents in the treatment of cancers, in view of their roles in modulating many cancer signalling pathways (Kashyap et al. 2019). The senolytic drug combination, D+Q, is known to reduce senescent cell abundance in aged mice, extending lifespan (Xu et al. 2018). D+Q senolytic treatment also has been shown to reduce intestinal senescence and inflammation while altering specific microbiota signatures (Saccon et al. 2021). A summary of the in vivo evidence of D+Q targeting multiple age-associated dysfunctions and pathologies in mammals, has been presented (Morsli and Bellantuono 2021).

22.4.1.8 Sulforaphane

The isothiocyanate sulforaphane (SF) is one of the most potent naturally occurring Phase 2 enzymes inducers derived from *Brassica* vegetables like broccoli, cabbage, brussel sprouts, etc. (Yuanfeng et al. 2021). Sulforaphane-induced cell cycle arrest, senescence by upregulation of cell cycle inhibitors p21 and p27, DNA hypomethylation and changes in microRNA profile in breast cancer cells (Lewinska et al. 2017). Another compound from cruciferous vegetables, Phenethyl isothiocyanate (PEITC) modulates the senescence effectors p16, p53, and p21, and induces increased staining of senescence-associated SA-β-Gal senescence biomarker (Malavolta et al. 2018).

22.4.1.9 Other Senolytic Compounds Based on Natural Products

Novel senolytic compounds were searched from natural products using chemoinformatic tools. Hinokitiol found in the roots of the Hinoki tree, preussomerin C from the *Lasiodiplodia theobromae* fungus, and tanshinone from *Salvia miltiorrhiza* Bunge roots, could be considered senolytic compound candidates since they share similarities in structure with senolytic leads such as tunicamycin, ginsenoside Rb1, ABT-737, rapamycin, navitoclax, timosaponin A-III, digoxin, roxithromycin, and azithromycin, and targets involved in senescence pathways with potential use in the treatment of age-related diseases (Barrera-Vazquez et al. 2021). An extract from the plant *Solidago virgaurea subsp. alpestris*, also known as goldenrod, is traditionally used as an anti-inflammatory herbal medicine, and was shown to exhibit weak senolytic activity (Lammermann et al. 2018).

Triptolide (TPL) is a diterpenoid extracted from the plant, *Tripterygium wilfordii Hook F*, which is a traditional Chinese medicinal herb. Bioactive TPL showed immunosuppressive, anti-fertility and anti-cystogenesis activities. TPL accelerated liver cancer cell line HepG2 cell senescence by regulating the AKT pathway. TPL could also enhance cellular senescence and inhibit tumor growth by negatively regulating human telomerase reverse transcriptase (hTERT) signaling pathway (Li et al. 2017).

Senotherapeutics have also been identified in marine dietary algae such as (2R*, 3S*, 6R*, 7S*, 10R*, 13R*)-7,13-Dihydroxy-2,6-cyclo-1(9),14-xenicadiene-18,19-dial derived from *Dilophus fasciola*, Laurendecumenyne A from *Laurencia decumbens* and 4-Bromo-3-ethyl-9-[(2E)- 2-penten-4-yn-1-yl]-2,8-dioxabicyclo[5.2.1]decan-6-ol from *Laurencia sp.* to be potent inhibitors of multiple target senescent-cell anti-apoptotic pathway proteins (Salekeen et al. 2021).

22.4.2 Pharmaceutical Senolytics

Pharmaceutical senolytics are compounds that have been chemically synthesized and that, originally, were used as experimental cancer drugs. Cancer cells avoid apoptosis and can undergo tumorigenesis, in part by upregulation of pro-survival proteins of the Bcl-2 protein family (Bcl-2, Bcl-XL, Bcl-w, Mcl-1, and A1). Upon further investigation, some of the cancer drugs were found to target senescent cells in tumors, such as navitoclax (previously called ABT-263) that targets Bcl-2, Bcl-xl, and Bcl-w (Zhu et al. 2016; Prasanna et al. 2021). Combinations of pharmaceutical senolytics (e.g. Dasatinib) and nutritional senolytics (e.g. quercetin) have proven very effective in anti-senescent cell elimination (Saccon et al. 2021; Morsli and Bellantuono 2021). These findings have triggered further work to try to repurpose other existing pharmaceuticals to target aging pathways (Vaiserman et al. 2021).

Since pharmaceutical senolytics are drugs, that are not present in nutrients or foods, they will not be addressed specifically in this chapter.

22.4.3 Other Senolytic Strategies

The present work is focused exclusively on nutritional factors affecting senescent cells, however, many non-nutritional approaches are being considered as well, to influence and/or erradicate senescent cells. Examples include exercise (Chen et al. 2021b), fibrates (Nogueira-Recalde et al. 2019), hyperbaric oxygen therapy (Hachmo et al. 2020), chemically-based prodrugs such as a non-toxic derivative of the compound 5-Fluorouridine, a senescence-specific killing compound 1 (SSK1) derived from the drug gemcitabine targeting lysosomal beta-galactosidase (Cai et al. 2020), galactose-modified cytotoxic prodrugs such as Duocarmycin (Guerrero et al. 2020), cardiac glycosides targeting the Na^+/K^+ ATPase pump such as digoxin, in combination with senogenic compounds Gemcitabine or Doxorubicin (Triana-Martinez et al. 2019; Martin et al. 2020), inhibitors of the HSP90 chaperone family such as 17-Dimethylaminoethylamino-17-demethoxygeldanamycin (17-DMAG) (Fuhrmann-Stroissnigg et al. 2017), a FOXO4 peptide that perturbs the FOXO4 interaction with p53 (Baar et al. 2017), CD153 vaccination to remove senescent T cells (Yoshida et al. 2020), etc.

Finally, the role of senescent cell eradication by the immune system (Prata et al. 2018; Kale et al. 2020) and nanomaterial-based delivery systems capable of preferentially killing senescent cells (nano-senolytics) and/or modulating their proinflammatory secretome (nano-senomorphics/nano-senostatics) are areas of active research (Adamczyk-Grochala and Lewinska 2020).

22.5 Gerosuppressors and Senomorphics

Gerosuppressors, also called senomorphics, are compounds or biomedical strategies that suppress, ameliorate, prevent or reverse the senescent state, inhibiting or reducing senescence-inducing triggers such as telomere damage, stress-induced senescence and gene-induced senescence (Martel et al. 2020). Senescent cells exhibit a characteristic secretory profile. This specific secretome is being currently referred to as the "Senescence-associated secretory phenotype" or SASP (Coppe et al. 2008; Kuilman and Peeper 2009). Many different types of compounds and strategies can inhibit or modulate the SASP, including medicines such as statins (Liu et al. 2015), glucocorticoids (Laberge et al. 2012), JAK1/2 inhibitors such as Ruxolitinib (Xu et al. 2015), NF-kB, p38 MAP kinase and MK2 kinase inhibitors (Alimbetov et al. 2016), splicing factor PTBP1 depletion (Georgilis et al. 2018), transgenic cells seeking IL-6 producing senescent cells (Qudrat et al. 2017), etc. Several senomorphic molecules, at an early characterization stage, have been recently presented (Mongelli et al. 2020).

The SASP is important to study because it is the origin of the heightened pro-inflammatory environment normally found in aged people. Thus, senotherapeutics targeting or modulating the SASP are emerging as alternative therapies to the senolytics (Birch and Gil 2020).

The SASP has been found to be senescence-process specific and cell-type specific, which means that no individual senescent cell would show the exact same secretome profile and the secretome changes over time (Basisty et al. 2020; Schafer et al. 2020).

The senolytic compounds described in Sect. 22.4 can also exert modulation of SASP, to various degrees, and will not be addressed in this section, as their main function is to have senolytic activity.

Below, a list of prominent nutritional gerosuppressors (in alphabetical order) aims to describe their origin, targets and possible uses in senotherapy.

22.5.1 Apigenin

Apigenin (4′,5,7,-trihydroxyflavone) is a flavonoid found in certain herbs, fruits, and vegetables (Cannataro et al. 2021; Jantan et al. 2021). Apigenin has multiple health-promoting effects and therapeutic functions (Salehi et al. 2019; Jantan et al. 2021). It can inhibit UV-induced cytotoxicity and prevent signs of skin aging in vivo (Choi et al. 2016), Apigenin can attenuate inflammation, which is associated with many chronic diseases of aging. It was shown to strongly inhibit the secretion of IL-6, a prominent cytokine expressed by senescent cells (Laberge et al. 2012). Apigenin suppressed the SASP in three human fibroblast strains induced to senescence by ionizing radiation, constitutive MAPK (mitogen-activated protein kinase) signaling, oncogenic RAS, or replicative exhaustion. The mechanism involves suppression of IL-1α signaling through IRAK1 and IRAK4, p38-MAPK, and NF-κB. Expression and secretion of one SASP factor, CXCL10 (IP10), was strongly inhibited by apigenin. Apigenin is a promising natural product for reducing the impact of senescent cells on age-related diseases such as cancer (Perrott et al. 2017).

Apigenin is an inhibitor of NAD+ase CD38 enzyme (Escande et al. 2013). Declining tissue nicotinamide adenine dinucleotide (NAD) levels are linked to aging and its associated diseases. Pro-inflammatory M1-like macrophages express high levels of the NAD-consuming enzyme CD38 (Covarrubias et al. 2020) and targeting NAD^+ metabolism has emerged as a potential therapeutic approach to ameliorate aging-related disease (Covarrubias et al. 2021).

22.5.2 *Avenanthramide*

Oats are whole grains that contain several nutrients that modulate directly the innate and adaptive immunity, and indirectly, elicit changes in the gut microbiota and related metabolites (Chen et al. 2021a). One of the oat's nutrients, avenanthramide C (Avn C), was validated as a new senomorphic compound, acting as an inhibitor of SASP and causing a reduction in the levels of markers of senescence. Avn C inhibited the activities of nuclear factor κB (NF-κB) and p38 mitogen-activated protein kinase, and the secretion of inflammatory cytokines. Avn C-induced inhibition of the SASP is triggered by senescence-related stress (Lim et al. 2020).

22.5.3 *Epigallocatechin-3-Gallate*

Green tea (*Camellia sinesis L.*) is widely known for its anti-cancer and anti-inflammatory properties. Its main antioxidant agents are catechins and derivatives including epicatechin, epigallocatechin, epicatechin gallate and epigallocatechin-3-gallate (EGCG). EGCG has potent health properties, among which is the prevention of several types of cancer (Musial et al. 2020). EGCG is also present in smaller amounts in fruits like apple and plums, onions, hazelnuts, pecans and carob (Haytowitz et al. 2018). EGCG was able to extend the lifespan in obese rats by improving free fatty acids metabolism and reducing the levels of inflammatory molecules and oxidative stress (Yuan et al. 2020). In 3T3-L1 pre-adipocytes, induced to senescence, EGCG could diminish IL-6 protein expression and CDKN1a (p21) mRNA expression, and increase mitochondrial SIRT3 and NRF2 mRNA expression. SIRT3 activating compounds, such as EGCG, may delay senescence of cells and senescence-induced inflammatory processes (Lilja et al. 2020). EGCG downregulated the PI3K/Akt/mTOR and AMPK signaling pathway and suppressed ROS, iNOS, Cox-2, NF-kB, SASP and p53-mediated cell cycle inhibition in pre-adipocytes, and suppressed the accumulation of anti-apoptotic protein Bcl-2 in senescent cells thereby promoting apoptosis-mediated cell death. EGCG acts as an mTOR inhibitor, SASP modulator as well as a potential senolytic agent thereby indicating its multi-faceted attributes that could be useful for developing anti-aging or age-delaying therapies (Kumar et al. 2019).

22.5.4 Ginsenoside F1

Ginseng has been used as a traditional herb in Asian countries for thousands of years, and contains a large number of active ingredients including steroidal saponins, protopanaxadiols, and protopanaxatriols, collectively known as ginsenosides (Ratan et al. 2021). Ginsenoside F1 is a ginsenoside found in Panax species that grow in the mountainous regions of East Asia. Ginsenoside F1 suppresses the SASP from astrocytes induced by D-galactose via suppressing p38MAPK-dependent NF-κB activity (Hou et al. 2018).

22.5.5 Metformin

Metformin (dimethylbiguanide) has become the preferred first-line oral blood glucose-lowering agent to manage type 2 diabetes. *Galega officinalis* (also known as goat's rue), a traditional herbal medicine in Europe, was found to be rich in guanidine, and shown to lower blood glucose. Guanidine derivatives, including metformin, were synthesized and some (not metformin) were used to treat diabetes in the 1920s and 1930s but were discontinued due to toxicity and the increased availability of insulin (Bailey 2017). Subsequent research led to the discoveries that metformin can improve nutrient sensing, enhance autophagy and intercellular communication, protect against macromolecular damage, delay stem cell aging, modulate mitochondrial function, regulate transcription, and reduce telomere attrition and senescence (Kulkarni et al. 2020). Specifically for aging, metformin leads to decreased insulin levels, decreased IGF-1 signalling, inhibition of mTOR, inhibition of mitochondrial complex 1 in the electron transport chain and reduction of endogenous production of reactive oxygen species (ROS), activation of AMP-activated kinase (AMPK), and reduction in DNA damage (Barzilai et al. 2016) and many more (Morsli and Bellantuono 2021). A large study known as "Targeting aging by metformin" (TAME) is a trial including more than 3000 individuals, ages 65–79, where the effects of metformin on the progression of heart disease, cancer and dementia will be monitored (Kulkarni et al. 2020).

22.5.6 *Puerarin*

Plants from the genus Pueraria are well known for the health and cosmetic benefits. The medicinally important plants of this genus are commonly known as kudzu, and the predominant phytochemical constituents are isoflavones, also known as phytoestrogens (Wang et al. 2020).

Puerarin (daidzein-8-C-glucoside) is a major isoflavone found in *Pueraria lobata* (Kudzu, Kuzu, Gegen), an edible legume. It possesses a variety of pharmacological actions (Ahmad et al. 2020, Bharti et al. 2020). Tuber and leaf extracts of Pueraria tuberosa contain several bioactive constituents such as daidzein, genistein, quercetin, irisolidone, biochanin A, biochanin B, isoorientin, and mangiferin, which possess an extensive range of pharmacological activities (Bharti et al. 2020). It can prevent the aging-phenotype of human dermal fibroblasts Puerarin blocks aging phenotype in cultured human dermal fibroblasts (Kamiya et al. 2021).

22.5.7 *Rapamycin (Sirolimus) and Rapalogues*

Rapamycin is not a nutritional gerosuppressor but originates from a natural source. It is one of the most studied senotherapeutics to date. The compound was isolated from *Streptomyces hygroscopicus*, a bacteria found in a soil sample at Rapa Nui, Easter island. Initially, it was known as having antifungal activities (called sirolimus), but later on it was found to have immunosuppressive and antiproliferative properties, so it was used successfully to reduce organ rejection with kidney transplantation. Its roles in cancer suppression led to the discovery that rapamycin is an inhibitor of the mammalian/mechanistic target of rapamycin (mTOR) and is one of the few drugs that is able to extend lifespan in several organisms (Seto 2012, Selvarani et al. 2020). A summary of the in vivo evidence for Rapamycin targeting multiple age-associated dysfunctions and pathologies in mammals has been reviewed (Morsli and Bellantuono 2021).

mTOR is a key component of cellular metabolism that integrates nutrient sensing with cellular processes that fuel cell growth and proliferation, playing a central role in the regulation of the aging process and lifespan (Papadopoli et al. 2019). The effects of rapamycin on various metabolic pathways have been summarized (Morsli and Bellantuono 2021). The low bioavailability of rapamycin led to the design of rapamycin analogues, called "rapalogues", such as mTOR inhibitors temsirolimus (CCI-779), everolimus (RAD001), and ridaforolimus/deforolimus (MK-8669/AP23573), by chemical design. Rapalogues have a safer toxicity profile but still exhibit a number of drawbacks (Viana et al. 2018). Second generation rapalogues, dual mTOR and PI3K inhibitors, mTORC1 and mTORC2 dual inhibitors, have been designed and await further characterization (Leontieva et al. 2015; Zhang et al. 2011). Several polyphenols have been found to inhibit or influence mTOR, such as honokiol, curcumin (see Sect. 22.4.1.2), epigallocatechin-3-gallate (see Sect. 22.5.3), theaflavin digallate, quercetin (see Sect. 22.4.1.7), punicalagin, fisetin (see Sect. 22.4.1.3), oleoeuropein (see Sect. 22.5.9) and resveratrol (see Sect. 22.5.8) (Pazoki-Toroudi et al. 2016; Kumar et al. 2019).

22.5.8 Resveratrol and Resveralogues

Resveratrol is a non-flavonoid polyphenol stilbenoid (3,5,4′-trihydroxy-*trans*-stilbene) and phytoalexin produced by plants in response to injury e.g. UV or pathogen attack, and is often found in grapes, raisins, berries and peanuts. Resveratrol has the ability to remove ROS, inhibit cyclooxygenase (COX), and trigger anti-inflammatory pathways via SIRT1 activation. Multiple studies show that resveratrol can suppress SASP. However, lifespan extension has yet not been confirmed (Grinan-Ferre et al. 2021).

Epigenetic modifications are associated with genome stability, gene transcription, and metabolic regulation. Acetylation is one of the most characterized histone modifications. Histone acetyltransferase (HAT) and histone deacetylase (HDAC) enzymes control the levels of histone acetylation, modulating gene expression (Cavalli and Heard 2019). Sirtuins (SIRT) 1–7 are enzymes classified as class III HDACs.

A recent work presented ongoing clinical trials using resveratrol as well as other phytochemicals stimulating SIRT1 activity (Iside et al. 2020). The parent resveratrol molecule has been modified to give rise to at least 24 different "resveralogues", structurally related to resveratrol. Replacement of the 3,5-dihydroxy substituents with 3,5-dimethoxy groups significantly enhances SIRT1 activity, and reduces toxicity. At 100 μM many of the compounds, including resveratrol, induce senescence in primary human fibroblast MRC5 cells in culture. However, at lower concentrations (10 μM), most compounds rescued subpopulations of cells within the culture from senescence (Birar et al. 2020).

22.5.9 Secoiridoids

Iridoids are cyclopentane [c] pyran monoterpenoids present in plants and insects, considered as defensive compounds. Cleavage of a bond in the cyclopentane ring leads to the formation of a subclass known as secoiridoids, bioactive compounds with a large variety of pharmacological properties including anti-diabetic, anti-inflammatory, immunosuppressive, neuroprotective, anti-cancer and anti-obesity. The olive tree *Olea europaea L.* is particularly rich in oleuropein (OL), dimethyl-OL, and ligstroside secoiridoids, and their hydrolysis derivatives are mostly OL-aglycone, oleocanthal (OLE), oleacein (OLA), elenolate, oleoside-11-methyl ester, elenoic acid, hydroxytyrosol (HTy), and tyrosol (Ty) (Castejon et al. 2020). The presence of OLE and OLA in extra-virgin olive oil has been investigated in relation to healthy aging (Nikou et al. 2019). OLE and ligstroside have been found to protect against mitochondrial dysfunction in models of Alzheimer's disease and brain aging (Grewal et al. 2020).

22.6 Other Nutritional and Dietary Interventions

22.6.1 Geroprotectors

Many natural products do not necessarily fall into the categories of senolytics and gerosuppressors per se, but exert their health-promoting and anti-aging activities as contributors to protect cellular integrity. A geroprotector is a compound that protects the integrity of cellular structures and macromolecules. Several gerosuppressors that modulate the SASP can also act as geroprotectors, such as rapamycin and metformin (Morsli and Bellantuono 2021).

Nutraceutical compounds encompassing phytochemicals, probiotic bacteria and omega-3-fatty acids have shown promising anti-immunosenescence and anti-cellular senescence potential in immune cell (Sharma and Padwad 2020).

There are many nutritional anti-aging compounds. Anti-carcinogenic, anti-inflammatory, anti-viral, anti-microbial, anti-proliferative and antioxidant activities can be found in many phytochemicals, that either directly or indirectly can be considered geroprotective. Examples are found in the Citrus family (oranges, lemons, grapefruits, etc.) (Barreca et al. 2020), foods in the Mediterranean diet (Cannataro et al. 2021), edible flowers (Zheng et al. 2021), honey (Jaganathan and Mandal 2009), olive oils (Nikou et al. 2019), mushrooms (Abdelshafy et al. 2021), cocoa (Martin and Ramos 2021), etc.

Alpha-ketoglutarate (AKG), an intermediate of the tricarboxylic acid cycle that leads to energy production, biosynthesis of certain amino acids, collagen biosynthesis, epigenetic regulation of gene expression, regulation of redox homeostasis, and detoxification of hazardous substances, can extend lifespan and delay the onset of age-associated decline in experimental animal models, and is being considered a geroprotective drug (Bayliak and Lushchak 2021).

Nicotinamide (NAM) is a form of vitamin B_3 found mainly in meat, fish, cereals, nuts, and mushrooms, as well as to a lesser extent in some vegetables and used as a dietary supplement and medication. Nicotinamide mononucleotide (NMN) is a nucleotide derivative of niacin (one of three forms of vitamin B_3). NMN is a precursor of NAD+, an important cofactor in redox reactions in energy metabolism. Aging is accompanied by a gradual decline in tissue and cellular NAD^+ levels in multiple model organisms, including rodents and humans. Therefore, supplementation with NMN is geroprotective (Covarrubias et al. 2021).

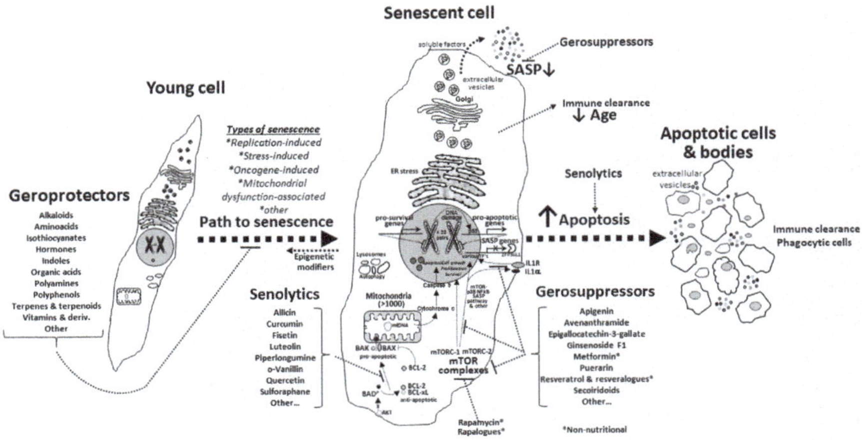

Fig. 22.1 Schematic representation of the main modes of action of several nutritional senolytic compounds and gerosuppressors on senescent cells. Nutritional compounds may have multiple targets and multiple functions. The specific senescent state can vary from one cell to another, and can also vary between cell/tissue types. A nutritional senotherapeutic may have different effects depending on the specific type of senescent cell. Abbreviations: mTOR (mammalian/mechanistic target of rapamycin), SASP (Senescence-associated secretory phenotype). The full name and description of genes and proteins can be found in https://www.genecards.org/

22.6.2 *Epigenetic Modifiers*

Very few studies have yet explored the role of specific nutrients in epigenetic modification to achieve rejuvenation as a means of anti-aging strategies. One recent study has shown that a 1-year Mediterranean-like diet, in a pilot study of elderly healthy subjects from the NU-AGE study (60 Italians, 60 Poles), resulted in epigenetic rejuvenation (Gensous et al. 2020).

In the trial "Thymus Regeneration, Immunorestoration, and Insulin Mitigation" (TRIIM), nine healthy men were given a cocktail of human growth hormone (hGH), metformin, dehydroepidandrosterone (DHEA), vitamin D3, and zinc for 1-year, that shed about 2.5 years off their biological ages, according to an analysis of their epigenome. (Fahy et al. 2019; Bartke et al. 2021) (Fig. 22.1).

22.7 Concluding Remarks and Future Perspectives

The nutritional geroscience field is relatively young. The impact that pharmacologically-synthesized senomorphics and dietary components, alone or in combination, might have on the short- and long-term survival and secretory function of senescent cells awaits further characterization. The cellular and systemic divergence of phytochemical targets suggests that the best way to assess the contribution of individual nutrients or compounds, alone or in combination, is to design long-term longitudinal clinical trials, whereby individuals of different genetic backgrounds, lifestyles and environments, are followed to assess health outcomes, disease emergence, duration of health span and lifespan.

Senescent cells arise by means of different processes, and their resulting SASP profiles will undoubtedly be heterogenous. It is therefore expected that their response to senolytics and senomorphic compounds will vary. Clinical intervention studies will be needed to understand the precise contribution of single or combined dietary phytochemicals, to the rate of aging in each one of the aging trajectories that lead to specific age-associated diseases and their accompanying disabilities.

It is becoming more evident that most of the compounds tested to date, that have anti-senescence phenotype properties, have multiple targets, and their use might influence cells and body processes other than only senescent cells. This implies that safety issues need to always be considered when administering senotherapeutics, including estimation of the senescence burden of an individual before any clinical intervention.

The significance of hormesis to the use of senotherapeutics will require understanding the effects of different doses, frequency and time of administration (Santoro et al. 2020; Rattan 2008).

The list of natural compounds that can exert senolytic and senomorphic activities is likely to grow in the years to come, some of which could have novel targets on pre-senescent and senescent cells. It is important to assess the chemical structure of the natural compound including its modifications such as glycosylation, etc. that could have implications for their activities in vivo, implying that chemical synthesis of the basic structure of the compounds alone may not be sufficient to result in fully active senotherapeutics (Huang et al. 2016).

Finally, it would be ideal to investigate the role of natural and synthetic compounds on non-senescent cells, at all stages of their lives, that could help to delay senescence and improve resilience, without significantly compromising basic body function. The pursuit of an extended health span by means of dietary interventions would facilitate health care and reduce or postpone disease burden among the elderly.

Acknowledgements I wish to thank Juliette Tavenier and Ove Andersen for stimulating discussions. The author wishes to acknowledge the contributions of many nutrition researchers, about the role of nutritional factors on health and aging, that could not be included here due to space limitations.

Compliance with Ethical Standards

Funding The support from Amager and Hvidovre- Copenhagen University hospital is gratefully acknowledged.

Conflict of Interest I have no conflict of interests to report.

References

Abdelshafy AM, Belwal T, Liang Z, Wang L, Li D, Luo Z, Li L (2021) A comprehensive review on phenolic compounds from edible mushrooms: Occurrence, biological activity, application and future prospective. Crit Rev Food Sci Nutr. https://doi.org/10.1080/10408398.2021.1898335. Online ahead of print

Adamczyk-Grochala J, Lewinska A (2020) Nano-based theranostic tools for the detection and elimination of senescent cells. Cells 9:2659

Ahmad B, Khan S, Liu Y, Xue M, Nabi G, Kumar S, Alshwmi M, Qarluq AW (2020) Molecular mechanisms of anticancer activities of puerarin. Cancer Manag Res 12:79–90

Alimbetov D, Davis T, Brook AJ, Cox LS, Faragher RG, Nurgozhin T, Zhumadilov Z, Kipling D (2016) Suppression of the senescence-associated secretory phenotype (SASP) in human fibroblasts using small molecule inhibitors of p38 MAP kinase and MK2. Biogerontology 17:305–315

Ansary J, Forbes-Hernandez TY, Gil E, Cianciosi D, Zhang J, Elexpuru-Zabaleta M, Simal-Gandara J, Giampieri F, Battino M (2020) Potential health benefit of garlic based on human intervention studies: a brief overview. Antioxidants (Basel) 9:619

Baar MP, Brandt RMC, Putavet DA, Klein JDD, Derks KWJ, Bourgeois BRM, Stryeck S, Rijksen Y, Van Willigenburg H, Feijtel DA, Van Der Pluijm I, Essers J, Van Cappellen WA, Van IWF, Houtsmuller AB, Pothof J, De Bruin RWF, Madl T, Hoeijmakers JHJ, Campisi J, De Keizer PLJ (2017) Targeted apoptosis of senescent cells restores tissue homeostasis in response to chemotoxicity and aging. Cell 169:132–147.e16

Bailey CJ (2017) Metformin: historical overview. Diabetologia 60:1566–1576

Barreca D, Mandalari G, Calderaro A, Smeriglio A, Trombetta D, Felice MR, Gattuso G (2020) Citrus flavones: an update on sources, biological functions, and health promoting properties. Plants (Basel) 9:288

Barrera-Vazquez OS, Gomez-Verjan JC, Magos-Guerrero GA (2021) Chemoinformatic screening for the selection of potential senolytic compounds from natural products. Biomolecules 11:467

Bartke A, Hascup E, Hascup K, Masternak MM (2021) Growth hormone and aging: new findings. World J Mens Health 39:454–465

Barzilai N, Crandall JP, Kritchevsky SB, Espeland MA (2016) Metformin as a tool to target aging. Cell Metab 23:1060–1065

Basisty N, Kale A, Jeon OH, Kuehnemann C, Payne T, Rao C, Holtz A, Shah S, Sharma V, Ferrucci L, Campisi J, Schilling B (2020) A proteomic atlas of senescence-associated secretomes for aging biomarker development. PLoS Biol 18:e3000599

Batiha GE, Beshbishy AM, Ikram M, Mulla ZS, El-Hack MEA, Taha AE, Algammal AM, Elewa YHA (2020) The pharmacological activity, biochemical properties, and pharmacokinetics of the major natural polyphenolic flavonoid: quercetin. Foods 9:374

Bayliak MM, Lushchak VI (2021) Pleiotropic effects of alpha-ketoglutarate as a potential anti-ageing agent. Ageing Res Rev 66:101237

Bezerra DP, Pessoa C, de Moraes MO, Saker-Neto N, Silveira ER, Costa-Lotufo LV (2013) Overview of the therapeutic potential of piplartine (piperlongumine). Eur J Pharm Sci 48:453–463

Bharti R, Chopra BS, Raut S, Khatri N (2020) Pueraria tuberosa: a review on traditional uses, pharmacology, and phytochemistry. Front Pharmacol 11:582506

Bian Y, Wei J, Zhao C, Li G (2020) Natural polyphenols targeting senescence: a novel prevention and therapy strategy for cancer. Int J Mol Sci 21:684

Bielak-Zmijewska A, Grabowska W, Ciolko A, Bojko A, Mosieniak G, Bijoch L, Sikora E (2019) The role of curcumin in the modulation of ageing. Int J Mol Sci 20:1239

Birar VC, Sheerin AN, Ostler EL, Faragher RGA (2020) Novel resveratrol derivatives have diverse effects on the survival, proliferation and senescence of primary human fibroblasts. Biogerontology 21:817–826

Birch J, Gil J (2020) Senescence and the SASP: many therapeutic avenues. Genes Dev 34:1565–1576

Borlinghaus J, Foerster Nee Reiter J, Kappler U, Antelmann H, Noll U, Gruhlke MCH, Slusarenko AJ (2021) Allicin, the odor of freshly crushed garlic: a review of recent progress in understanding Allicin's effects on cells. Molecules 26(6):1505

Bussian TJ, Aziz A, Meyer CF, Swenson BL, van Deursen JM, Baker DJ (2018) Clearance of senescent glial cells prevents tau-dependent pathology and cognitive decline. Nature 562:578–582

Cai Y, Zhou H, Zhu Y, Sun Q, Ji Y, Xue A, Wang Y, Chen W, Yu X, Wang L, Chen H, Li C, Luo T, Deng H (2020) Elimination of senescent cells by beta-galactosidase-targeted prodrug attenuates inflammation and restores physical function in aged mice. Cell Res 30:574–589

Calcinotto A, Kohli J, Zagato E, Pellegrini L, Demaria M, Alimonti A (2019) Cellular senescence: aging, cancer, and injury. Physiol Rev 99:1047–1078

Cannataro R, Fazio A, la Torre C, Caroleo MC, Cione E (2021) Polyphenols in the Mediterranean diet: from dietary sources to microRNA modulation. Antioxidants (Basel) 10:328

Castejon ML, Montoya T, Alarcon-de-la-Lastra C, Sanchez-Hidalgo M (2020) Potential protective role exerted by secoiridoids from Olea europaea L. in cancer, cardiovascular, neurodegenerative, aging-related, and immunoinflammatory diseases. Antioxidants (Basel) 9(2):149

Cavalcante MB, Saccon TD, Nunes ADC, Kirkland JL, Tchkonia T, Schneider A, Masternak MM (2020) Dasatinib plus quercetin prevents uterine age-related dysfunction and fibrosis in mice. Aging (Albany NY) 12:2711–2722

Cavalli G, Heard E (2019) Advances in epigenetics link genetics to the environment and disease. Nature 571(7766):489–499

Chandra A, Lagnado AB, Farr JN, Monroe DG, Park S, Hachfeld C, Tchkonia T, Kirkland JL, Khosla S, Passos JF, Pignolo RJ (2020) Targeted reduction of senescent cell burden alleviates focal radiotherapy-related bone loss. J Bone Miner Res 35:1119–1131

Chen O, Mah E, Dioum E, Marwaha A, Shanmugam S, Malleshi N, Sudha V, Gayathri R, Unnikrishnan R, Anjana RM, Krishnaswamy K, Mohan V, Chu Y (2021a) The role of oat nutrients in the immune system: a narrative review. Nutrients 13(4):1048

Chen XK, Yi ZN, Wong GT, Hasan KMM, Kwan JS, Ma AC, Chang RC (2021b) Is exercise a senolytic medicine? A systematic review. Aging Cell 20(1):e13294

Cherif H, Bisson DG, Jarzem P, Weber M, Ouellet JA, Haglund L (2019) Curcumin and o-Vanillin exhibit evidence of senolytic activity in human IVD cells in vitro. J Clin Med 8:433

Cherif H, Bisson DG, Mannarino M, Rabau O, Ouellet JA, Haglund L (2020) Senotherapeutic drugs for human intervertebral disc degeneration and low back pain. Elife 9:e54693

Childs BG, Gluscevic M, Baker DJ, Laberge RM, Marquess D, Dananberg J, van Deursen JM (2017) Senescent cells: an emerging target for diseases of ageing. Nat Rev Drug Discov 16:718–735

Choi S, Youn J, Kim K, Da Joo H, Shin S, Lee J, Lee HK, An IS, Kwon S, Youn HJ, Ahn KJ, An S, Cha HJ (2016) Apigenin inhibits UVA-induced cytotoxicity in vitro and prevents signs of skin aging in vivo. Int J Mol Med 38:627–34

Constantin RP, Constantin J, Pagadigorria CL, Ishii-Iwamoto EL, Bracht A, de Castro CV, Yamamoto NS (2011) Prooxidant activity of fisetin: effects on energy metabolism in the rat liver. J Biochem Mol Toxicol 25:117–126

Conti P, Caraffa A, Gallenga CE, Ross R, Kritas SK, Frydas I, Younes A, di Emidio P, Ronconi G, Pandolfi F (2021) Powerful anti-inflammatory action of luteolin: potential increase with IL-38. Biofactors 47:165–169

Coppe JP, Patil CK, Rodier F, Sun Y, Munoz DP, Goldstein J, Nelson PS, Desprez PY, Campisi J (2008) Senescence-associated secretory phenotypes reveal cell-nonautonomous functions of oncogenic RAS and the p53 tumor suppressor. PLoS Biol 6:2853–2868

Covarrubias AJ, Kale A, Perrone R, Lopez-Dominguez JA, Pisco AO, Kasler HG, Schmidt MS, Heckenbach I, Kwok R, Wiley CD, Wong HS, Gibbs E, Iyer SS, Basisty N, Wu Q, Kim IJ, Silva E, Vitangcol K, Shin KO, Lee YM, Riley R, Ben-Sahra I, Ott M, Schilling B, Scheibye-Knudsen M, Ishihara K, Quake SR, Newman J, Brenner C, Campisi J, Verdin E (2020) Senescent cells promote tissue NAD(+) decline during ageing via the activation of CD38(+) macrophages. Nat Metab 2:1265–1283

Covarrubias AJ, Perrone R, Grozio A, Verdin E (2021) NAD(+) metabolism and its roles in cellular processes during ageing. Nat Rev Mol Cell Biol 22:119–141

Dai H, Chen R, Gui C, Tao T, Ge Y, Zhao X, Qin R, Yao W, Gu S, Jiang Y, Gui J (2020) Eliminating senescent chondrogenic progenitor cells enhances chondrogenesis under intermittent hydrostatic pressure for the treatment of OA. Stem Cell Res Ther 11:199

di Micco R, Krizhanovsky V, Baker D, D'adda di Fagagna F (2021) Cellular senescence in ageing: from mechanisms to therapeutic opportunities. Nat Rev Mol Cell Biol 22(2):75–95

Dookun E, Passos JF, Arthur HM, Richardson GD (2020a) Therapeutic potential of senolytics in cardiovascular disease. Cardiovasc Drugs Ther. https://doi.org/10.1007/s10557-020-07075-w. Online ahead of print

Dookun E, Walaszczyk A, Redgrave R, Palmowski P, Tual-Chalot S, Suwana A, Chapman J, Jirkovsky E, Donastorg Sosa L, Gill E, Yausep OE, Santin Y, Mialet-Perez J, Andrew Owens W, Grieve D, Spyridopoulos I, Taggart M, Arthur HM, Passos JF, Richardson GD (2020b) Clearance of senescent cells during cardiac ischemia-reperfusion injury improves recovery. Aging Cell 19(10):e13249

Escande C, Nin V, Price NL, Capellini V, Gomes AP, Barbosa MT, O'Neil L, White TA, Sinclair DA, Chini EN (2013) Flavonoid apigenin is an inhibitor of the NAD+ ase CD38: implications for cellular NAD+ metabolism, protein acetylation, and treatment of metabolic syndrome. Diabetes 62:1084–1093

Fahy GM, Brooke RT, Watson JP, Good Z, Vasanawala SS, Maecker H, Leipold MD, Lin DTS, Kobor MS, Horvath S (2019) Reversal of epigenetic aging and immunosenescent trends in humans. Aging Cell 18:e13028

Farr JN, Xu M, Weivoda MM, Monroe DG, Fraser DG, Onken JL, Negley BA, Sfeir JG, Ogrodnik MB, Hachfeld CM, Lebrasseur NK, Drake MT, Pignolo RJ, Pirtskhalava T, Tchkonia T, Oursler MJ, Kirkland JL, Khosla S (2017) Targeting cellular senescence prevents age-related bone loss in mice. Nat Med 23:1072–1079

Fragkou PC, Karaviti D, Zemlin M, Skevaki C (2021) Impact of early life nutrition on children's immune system and noncommunicable diseases through its effects on the bacterial microbiome, virome and mycobiome. Front Immunol 12:644269

Fuhrmann-Stroissnigg H, Ling YY, Zhao J, McGowan SJ, Zhu Y, Brooks RW, Grassi D, Gregg SQ, Stripay JL, Dorronsoro A, Corbo L, Tang P, Bukata C, Ring N, Giacca M, Li X, Tchkonia T, Kirkland JL, Niedernhofer LJ, Robbins PD (2017) Identification of HSP90 inhibitors as a novel class of senolytics. Nat Commun 8:422

Furman D, Campisi J, Verdin E, Carrera-Bastos P, Targ S, Franceschi C, Ferrucci L, Gilroy DW, Fasano A, Miller GW, Miller AH, Mantovani A, Weyand CM, Barzilai N, Goronzy JJ, Rando TA, Effros RB, Lucia A, Kleinstreuer N, Slavich GM (2019) Chronic inflammation in the etiology of disease across the life span. Nat Med 25:1822–1832

Gendrisch F, Esser PR, Schempp CM, Wolfle U (2020) Luteolin as a modulator of skin aging and inflammation. Biofactors 47:170–180

Gensous N, Garagnani P, Santoro A, Giuliani C, Ostan R, Fabbri C, Milazzo M, Gentilini D, di Blasio AM, Pietruszka B, Madej D, Bialecka-Debek A, Brzozowska A, Franceschi C, Bacalini MG (2020) One-year Mediterranean diet promotes epigenetic rejuvenation with country- and sex-specific effects: a pilot study from the NU-AGE project. Geroscience 42:687–701

Georgilis A, Klotz S, Hanley CJ, Herranz N, Weirich B, Morancho B, Leote AC, D'Artista L, Gallage S, Seehawer M, Carroll T, Dharmalingam G, Wee KB, Mellone M, Pombo J, Heide D, Guccione E, Arribas J, Barbosa-Morais NL, Heikenwalder M, Thomas GJ, Zender L, Gil J (2018) PTBP1-mediated alternative splicing regulates the inflammatory secretome and the pro-tumorigenic effects of senescent cells. Cancer Cell 34:85–1029

Grewal R, Reutzel M, Dilberger B, Hein H, Zotzel J, Marx S, Tretzel J, Sarafeddinov A, Fuchs C, Eckert GP (2020) Purified oleocanthal and ligstroside protect against mitochondrial dysfunction in models of early Alzheimer's disease and brain ageing. Exp Neurol 328:113248

Grinan-Ferre C, Bellver-Sanchis A, Izquierdo V, Corpas R, Roig-Soriano J, Chillon M, Andres-Lacueva C, Somogyvari M, Soti C, Sanfeliu C, Pallas M (2021) The pleiotropic neuroprotective effects of resveratrol in cognitive decline and Alzheimer's disease pathology: from antioxidant to epigenetic therapy. Ageing Res Rev 67:101271

Grynkiewicz G, Demchuk OM (2019) New perspectives for fisetin. Front Chem 7:697

Guerrero A, Guiho R, Herranz N, Uren A, Withers DJ, Martinez-Barbera JP, Tietze LF, Gil J (2020) Galactose-modified duocarmycin prodrugs as senolytics. Aging Cell 19:e13133

Hachmo Y, Hadanny A, Abu Hamed R, Daniel-Kotovsky M, Catalogna M, Fishlev G, Lang E, Polak N, Doenyas K, Friedman M, Zemel Y, Bechor Y, Efrati S (2020) Hyperbaric oxygen therapy increases telomere length and decreases immunosenescence in isolated blood cells: a prospective trial. Aging (Albany NY) 12:22445–22456

Haytowitz DB, Wu X, Bhagwat S (2018) USDA Database for the Flavonoid Content of Selected Foods. Agricultural Research Service, U.S. Department of Agriculture, Nutrient Data Laboratory, Release 3.3, March 2018. https://www.ars.usda.gov/ARSUserFiles/80400535/Data/Flav/Flav3.3.pdf

Henrique T, Zanon CF, Girol AP, Stefanini ACB, Contessoto NSA, da Silveira NJF, Bezerra DP, Silveira ER, Barbosa-Filho JM, Cornelio ML, Oliani SM, Tajara EH (2020) Biological and physical approaches on the role of piplartine (piperlongumine) in cancer. Sci Rep 10:22283

Hertog MG, Feskens EJ, Hollman PC, Katan MB, Kromhout D (1993) Dietary antioxidant flavonoids and risk of coronary heart disease: the Zutphen elderly study. Lancet 342:1007–1011

Hickson LJ, Langhi Prata LGP, Bobart SA, Evans TK, Giorgadze N, Hashmi SK, Herrmann SM, Jensen MD, Jia Q, Jordan KL, Kellogg TA, Khosla S, Koerber DM, Lagnado AB, Lawson DK, Lebrasseur NK, Lerman LO, Mcdonald KM, Mckenzie TJ, Passos JF, Pignolo RJ, Pirtskhalava T, Saadiq IM, Schaefer KK, Textor SC, Victorelli SG, Volkman TL, Xue A, Wentworth MA, Wissler Gerdes EO, Zhu Y, Tchkonia T, Kirkland JL (2019) Senolytics decrease senescent cells in humans: preliminary report from a clinical trial of Dasatinib plus Quercetin in individuals with diabetic kidney disease. EBioMedicine 47:446–456

Hou J, Cui C, Kim S, Sung C, Choi C (2018) Ginsenoside F1 suppresses astrocytic senescence-associated secretory phenotype. Chem Biol Interact 283:75–83

Huang G, Lv M, Hu J, Huang K, Xu H (2016) Glycosylation and activities of natural products. Mini Rev Med Chem 16:1013–1016

Iside C, Scafuro M, Nebbioso A, Altucci L (2020) SIRT1 activation by natural phytochemicals: an overview. Front Pharmacol 11:1225

Jaganathan SK, Mandal M (2009) Antiproliferative effects of honey and of its polyphenols: a review. J Biomed Biotechnol 2009:830616

Jantan I, Haque MA, Arshad L, Harikrishnan H, Septama AW, Mohamed-Hussein ZA (2021) Dietary polyphenols suppress chronic inflammation by modulation of multiple inflammation-associated cell signaling pathways. J Nutr Biochem 93:108634

Jeon OH, Kim C, Laberge RM, Demaria M, Rathod S, Vasserot AP, Chung JW, Kim DH, Poon Y, David N, Baker DJ, van Deursen JM, Campisi J, Elisseeff JH (2017) Local clearance of senescent cells attenuates the development of post-traumatic osteoarthritis and creates a pro-regenerative environment. Nat Med 23:775–781

Justice JN, Nambiar AM, Tchkonia T, Lebrasseur NK, Pascual R, Hashmi SK, Prata L, Masternak MM, Kritchevsky SB, Musi N, Kirkland JL (2019) Senolytics in idiopathic pulmonary fibrosis: results from a first-in-human, open-label, pilot study. EBioMedicine 40:554–563

Kale A, Sharma A, Stolzing A, Desprez PY, Campisi J (2020) Role of immune cells in the removal of deleterious senescent cells. Immun Ageing 17:16

Kamiya Y, Odama M, Mizuguti A, Murakami S, Ito T (2021) Puerarin blocks aging phenotype in cultured human dermal fibroblasts. PLoS One 16(4):e0249367

Kashyap D, Garg VK, Tuli HS, Yerer MB, Sak K, Sharma AK, Kumar M, Aggarwal V, Sandhu SS (2019) Fisetin and quercetin: promising flavonoids with chemopreventive potential. Biomolecules 9:174

Khaltourina D, Matveyev Y, Alekseev A, Cortese F, Iovita A (2020) Aging fits the disease criteria of the international classification of diseases. Mech Ageing Dev 189:111230

Kirkland JL, Tchkonia T (2020) Senolytic drugs: from discovery to translation. J Intern Med 288:518–536

Kubo C, Ogawa M, Uehara N, Katakura Y (2020) Fisetin promotes hair growth by augmenting TERT expression. Front Cell Dev Biol 8:566617

Kuilman T, Peeper DS (2009) Senescence-messaging secretome: SMS-ing cellular stress. Nat Rev Cancer 9:81–94

Kulkarni AS, Gubbi S, Barzilai N (2020) Benefits of metformin in attenuating the hallmarks of aging. Cell Metab 32:15–30

Kumar R, Sharma A, Kumari A, Gulati A, Padwad Y, Sharma R (2019) Epigallocatechin gallate suppresses premature senescence of preadipocytes by inhibition of PI3K/Akt/mTOR pathway and induces senescent cell death by regulation of Bax/Bcl-2 pathway. Biogerontology 20:171–189

Laberge RM, Zhou L, Sarantos MR, Rodier F, Freund A, de Keizer PL, Liu S, Demaria M, Cong YS, Kapahi P, Desprez PY, Hughes RE, Campisi J (2012) Glucocorticoids suppress selected components of the senescence-associated secretory phenotype. Aging Cell 11:569–578

Lammermann I, Terlecki-Zaniewicz L, Weinmullner R, Schosserer M, Dellago H, De Matos Branco AD, Autheried D, Sevcnikar B, Kleissl L, Berlin I, Morizot F, Lejeune F, Fuzzati N, Forestier S, Toribio A, Tromeur A, Weinberg L, Higareda Almaraz JC, Scheideler M, Rietveld M, El Ghalbzouri A, Tschachler E, Gruber F, Grillari J (2018) Blocking negative effects of senescence in human skin fibroblasts with a plant extract. NPJ Aging Mech Dis 4:4

Lee KS, Lin S, Copland DA, Dick AD, Liu J (2021) Cellular senescence in the aging retina and developments of senotherapies for age-related macular degeneration. J Neuroinflammation 18:32

Leontieva OV, Demidenko ZN, Blagosklonny MV (2015) Dual mTORC1/C2 inhibitors suppress cellular geroconversion (a senescence program). Oncotarget 6:23238–23248

Leri M, Scuto M, Ontario ML, Calabrese V, Calabrese EJ, Bucciantini M, Stefani M (2020) Healthy effects of plant polyphenols: molecular mechanisms. Int J Mol Sci 21:1250

Lewinska A, Adamczyk-Grochala J, Deregowska A, Wnuk M (2017) Sulforaphane-induced cell cycle arrest and senescence are accompanied by DNA hypomethylation and changes in microRNA profile in breast cancer cells. Theranostics 7:3461–3477

Li C, Shen Y, Huang L, Liu C, Wang J (2021) Senolytic therapy ameliorates renal fibrosis postacute kidney injury by alleviating renal senescence. FASEB J 35:e21229

Li R, Zhang X, Tian X, Shen C, Zhang Q, Zhang Y, Wang Z, Wang F, Tao Y (2017) Triptolide inhibits tumor growth by induction of cellular senescence. Oncol Rep 37:442–448

Li W, He Y, Zhang R, Zheng G, Zhou D (2019) The curcumin analog EF24 is a novel senolytic agent. Aging (Albany NY) 11:771–782

Lilja S, Oldenburg J, Pointner A, Dewald L, Lerch M, Hippe B, Switzeny O, Haslberger A (2020) Epigallocatechin gallate effectively affects senescence and anti-SASP via SIRT3 in 3T3-L1 preadipocytes in comparison with other bioactive substances. Oxid Med Cell Longev 2020:4793125

Lim JS, Lee DY, Kim HS, Park SC, Park JT, Kim HS, Oh WK, Cho KA (2020) Identification of a novel senomorphic agent, avenanthramide C, via the suppression of the senescence-associated secretory phenotype. Mech Ageing Dev 192:111355

Liu S, Uppal H, Demaria M, Desprez PY, Campisi J, Kapahi P (2015) Simvastatin suppresses breast cancer cell proliferation induced by senescent cells. Sci Rep 5:17895

Malavolta M, Bracci M, Santarelli L, Sayeed MA, Pierpaoli E, Giacconi R, Costarelli L, Piacenza F, Basso A, Cardelli M, Provinciali M (2018) Inducers of Senescence, toxic compounds, and senolytics: the multiple faces of Nrf2-activating phytochemicals in cancer adjuvant therapy. Mediators Inflamm 2018:4159013

Mansoori S, Dini A, Chai SC (2021) Effects of tart cherry and its metabolites on aging and inflammatory conditions: efficacy and possible mechanisms. Ageing Res Rev 66:101254

Martel J, Ojcius DM, Wu CY, Peng HH, Voisin L, Perfettini JL, Ko YF, Young JD (2020) Emerging use of senolytics and senomorphics against aging and chronic diseases. Med Res Rev 40:2114–2131

Martin MA, Ramos S (2021) Impact of cocoa flavanols on human health. Food Chem Toxicol 151:112121

Martin N, Huna A, Bernard D (2021) Elimination of senescent endothelial cells: good or bad idea? Trends Cell Biol 5:327–330

Martin N, Soriani O, Bernard D (2020) Cardiac glycosides as senolytic compounds. Trends Mol Med 26:243–245

Marton A, Kusz E, Kolozsi C, Tubak V, Zagotto G, Buzas K, Quintieri L, Vizler C (2016) Vanillin analogues o-Vanillin and 2,4,6-trihydroxybenzaldehyde inhibit NFkB activation and suppress growth of A375 human melanoma. Anticancer Res 36:5743–5750

Mehta P, Pawar A, Mahadik K, Bothiraja C (2018) Emerging novel drug delivery strategies for bioactive flavonol fisetin in biomedicine. Biomed Pharmacother 106:1282–1291

Mongelli A, Atlante S, Barbi V, Bachetti T, Martelli F, Farsetti A, Gaetano C (2020) Treating senescence like cancer: novel perspectives in senotherapy of chronic diseases. Int J Mol Sci 21:7984

Morsli S, Bellantuono I (2021) The use of geroprotectors to prevent multimorbidity: opportunities and challenges. Mech Ageing Dev 193:111391

Mukherjee PK, Harwansh RK, Bahadur S, Banerjee S, Kar A, Chanda J, Biswas S, Ahmmed SM, Katiyar CK (2017) Development of Ayurveda - tradition to trend. J Ethnopharmacol 197:10–24

Musial C, Kuban-Jankowska A, Gorska-Ponikowska M (2020) Beneficial properties of green tea catechins. Int J Mol Sci 21:1744

Nikou T, Liaki V, Stathopoulos P, Sklirou AD, Tsakiri EN, Jakschitz T, Bonn G, Trougakos IP, Halabalaki M, Skaltsounis LA (2019) Comparison survey of EVOO polyphenols and exploration of healthy aging-promoting properties of oleocanthal and oleacein. Food Chem Toxicol 125:403–412

Nogueira-Recalde U, Lorenzo-Gomez I, Blanco FJ, Loza MI, Grassi D, Shirinsky V, Shirinsky I, Lotz M, Robbins PD, Dominguez E, Carames B (2019) Fibrates as drugs with senolytic and autophagic activity for osteoarthritis therapy. EBioMedicine 45:588–605

Paez-Ribes M, Gonzalez-Gualda E, Doherty GJ, Munoz-Espin D (2019) Targeting senescent cells in translational medicine. EMBO Mol Med 11:e10234

Palmer AK, Tchkonia T, Kirkland JL (2021) Senolytics: potential for alleviating diabetes and its complications. Endocrinology 162:bqab058

Papadopoli D, Boulay K, Kazak L, Pollak M, Mallette F, Topisirovic I, Hulea L (2019) mTOR as a central regulator of lifespan and aging. F1000Res 8 (F1000 Fac Rev):998:1–21

Pazoki-Toroudi H, Amani H, Ajami M, Nabavi SF, Braidy N, Kasi PD, Nabavi SM (2016) Targeting mTOR signaling by polyphenols: a new therapeutic target for ageing. Ageing Res Rev 31:55–66

Perrott KM, Wiley CD, Desprez PY, Campisi J (2017) Apigenin suppresses the senescence-associated secretory phenotype and paracrine effects on breast cancer cells. Geroscience 39:161–173

Picallos-Rabina P, Da Silva-Álvarez S, Antelo-Iglesias L, Triana-Martínez F, Collado M (2020) Senotherapy of cancer. In: Muñoz-Espin D, Demaria M (eds) Senolytics in disease, ageing and longevity. Springer, pp. 85-99. ISBN 978-3-030-44903-2

Prasanna PG, Citrin DE, Hildesheim J, Ahmed MM, Venkatachalam S, Riscuta G, Xi D, Zheng G, Van Deursen J, Goronzy J, Kron SJ, Anscher MS, Sharpless NE, Campisi J, Brown SL, Niedernhofer LJ, O'loghlen A, Georgakilas AG, Paris F, Gius D, Gewirtz DA, Schmitt CA, Abazeed ME, Kirkland JL, Richmond A, Romesser PB, Lowe SW, Gil J, Mendonca MS, Burma S, Zhou D, Coleman CN (2021) Therapy-induced senescence: opportunities to improve anti-cancer therapy. J Natl Cancer Inst, djab064. https://doi.org/10.1093/jnci/djab064. Online ahead of print

Prata L, Ovsyannikova IG, Tchkonia T, Kirkland JL (2018) Senescent cell clearance by the immune system: emerging therapeutic opportunities. Semin Immunol 40:101275

Quach A, Levine ME, Tanaka T, Lu AT, Chen BH, Ferrucci L, Ritz B, Bandinelli S, Neuhouser ML, Beasley JM, Snetselaar L, Wallace RB, Tsao PS, Absher D, Assimes TL, Stewart JD, Li Y,

Hou L, Baccarelli AA, Whitsel EA, Horvath S (2017) Epigenetic clock analysis of diet, exercise, education, and lifestyle factors. Aging (Albany NY) 9:419–446
Qudrat A, Wong J, Truong K (2017) Engineering mammalian cells to seek senescence-associated secretory phenotypes. J Cell Sci 130:3116–3123
Raffaele M, Kovacovicova K, Frohlich J, Lo Re O, Giallongo S, Oben JA, Faldyna M, Leva L, Giannone AG, Cabibi D, Vinciguerra M (2021) Mild exacerbation of obesity- and age-dependent liver disease progression by senolytic cocktail dasatinib + quercetin. Cell Commun Signal 19:44
Ratan ZA, Haidere MF, Hong YH, Park SH, Lee JO, Lee J, Cho JY (2021) Pharmacological potential of ginseng and its major component ginsenosides. J Ginseng Res 45:199–210
Rattan SI (2008) Hormesis in aging. Ageing Res Rev 7:63–78
Robbins PD, Jurk D, Khosla S, Kirkland JL, Lebrasseur NK, Miller JD, Passos JF, Pignolo RJ, Tchkonia T, Niedernhofer LJ (2021) Senolytic drugs: reducing senescent cell viability to extend health span. Annu Rev Pharmacol Toxicol 61:779–803
Rodrigues LP, Teixeira VR, Alencar-Silva T, Simonassi-Paiva B, Pereira RW, Pogue R, Carvalho JL (2021) Hallmarks of aging and immunosenescence: connecting the dots. Cytokine Growth Factor Rev 59:9–21
Romashkan S, Chang H, Hadley EC (2021) National institute on aging workshop repurposing drugs or dietary supplements for their senolytic or senomorphic effects: considerations for clinical trials. J Gerontol A Biol Sci Med Sci. 76:1144–1152
Rosas-Gonzalez VC, Tellez-Banuelos MC, Hernandez-Flores G, Bravo-Cuellar A, Aguilar-Lemarroy A, Jave-Suarez LF, Haramati J, Solorzano-Ibarra F, Ortiz-Lazareno PC (2020) Differential effects of alliin and allicin on apoptosis and senescence in luminal A and triple-negative breast cancer: Caspase, DeltaPsim, and pro-apoptotic gene involvement. Fundam Clin Pharmacol 34:671–686
Russo GL, Spagnuolo C, Russo M, Tedesco I, Moccia S, Cervellera C (2020) Mechanisms of aging and potential role of selected polyphenols in extending healthspan. Biochem Pharmacol 173:113719
Saccon TD, Nagpal R, Yadav H, Cavalcante MB, Nunes ADC, Schneider A, Gesing A, Hughes B, Yousefzadeh M, Tchkonia T, Kirkland JL, Niedernhofer LJ, Robbins PD, Masternak MM (2021) Senolytic combination of Dasatinib and Quercetin alleviates intestinal senescence and inflammation and modulates the gut microbiome in aged mice. J Gerontol A Biol Sci Med Sci, glab002. https://doi.org/10.1093/gerona/glab002. Online ahead of print
Salehi B, Venditti A, Sharifi-Rad M, Kregiel D, Sharifi-Rad J, Durazzo A, Lucarini M, Santini A, Souto EB, Novellino E, Antolak H, Azzini E, Setzer WN, Martins N (2019) The therapeutic potential of apigenin. Int J Mol Sci 20:1305
Salekeen R, Barua J, Shaha PR, Islam KMD, Islam ME, Billah MM, Rahman SMM (2021) Marine phycocompound screening reveals a potential source of novel senotherapeutics. J Biomol Struct Dyn 1–15. https://doi.org/10.1080/07391102.2021.1877822. Online ahead of print
Santoro A, Martucci M, Conte M, Capri M, Franceschi C, Salvioli S (2020) Inflammaging, hormesis and the rationale for anti-aging strategies. Ageing Res Rev 64:101142
Schafer MJ, Zhang X, Kumar A, Atkinson EJ, Zhu Y, Jachim S, Mazula DL, Brown AK, Berning M, Aversa Z, Kotajarvi B, Bruce CJ, Greason KL, Suri RM, Tracy RP, Cummings SR, White TA, Lebrasseur NK (2020) The senescence-associated secretome as an indicator of age and medical risk. JCI Insight 5:e133668
Selvarani R, Mohammed S, Richardson A (2020) Effect of rapamycin on aging and age-related diseases-past and future. Geroscience 43:1135–1158
Seto B (2012) Rapamycin and mTOR: a serendipitous discovery and implications for breast cancer. Clin Transl Med 1:29
Sharifi-Rad J, Rayess YE, Rizk AA, Sadaka C, Zgheib R, Zam W, Sestito S, Rapposelli S, Neffe-Skocinska K, Zielinska D, Salehi B, Setzer WN, Dosoky NS, Taheri Y, el Beyrouthy M, Martorell M, Ostrander EA, Suleria HAR, Cho WC, Maroyi A, Martins N (2020) Turmeric and its major compound curcumin on health: bioactive effects and safety profiles for food, pharmaceutical, biotechnological and medicinal applications . Front Pharmacol 11:01021

Sharma R, Padwad Y (2020) Nutraceuticals-based immunotherapeutic concepts and opportunities for the mitigation of cellular senescence and aging: a narrative review. Ageing Res Rev 63:101141

Shen CY, Jiang JG, Yang L, Wang DW, Zhu W (2017) Anti-ageing active ingredients from herbs and nutraceuticals used in traditional Chinese medicine: pharmacological mechanisms and implications for drug discovery. Br J Pharmacol 174:1395–1425

Short S, Fielder E, Miwa S, von Zglinicki T (2019) Senolytics and senostatics as adjuvant tumour therapy. EBioMedicine 41:683–692

Song P, Zhao Q, Zou MH (2020) Targeting senescent cells to attenuate cardiovascular disease progression. Ageing Res Rev 60:101072

Taetzsch A, Roberts SB, Bukhari A, Lichtenstein AH, Gilhooly CH, Martin E, Krauss AJ, Hatch-Mcchesney A, Das SK (2021) Eating timing: associations with dietary intake and metabolic health. J Acad Nutr Diet 121:738–748

Tavella T, Rampelli S, Guidarelli G, Bazzocchi A, Gasperini C, Pujos-Guillot E, Comte B, Barone M, Biagi E, Candela M, Nicoletti C, Kadi F, Battista G, Salvioli S, O'Toole PW, Franceschi C, Brigidi P, Turroni S, Santoro A (2021) Elevated gut microbiome abundance of Christensenellaceae, Porphyromonadaceae and Rikenellaceae is associated with reduced visceral adipose tissue and healthier metabolic profile in Italian elderly. Gut Microbes 13:1–19

Thompson PJ, Shah A, Ntranos V, Van Gool F, Atkinson M, Bhushan A (2019) Targeted elimination of senescent beta cells prevents type 1 diabetes. Cell Metab 29(5):1045–1060.e10

Triana-Martinez F, Picallos-Rabina P, Da Silva-Alvarez S, Pietrocola F, Llanos S, Rodilla V, Soprano E, Pedrosa P, Ferreiros A, Barradas M, Hernandez-Gonzalez F, Lalinde M, Prats N, Bernado C, Gonzalez P, Gomez M, Ikonomopoulou MP, Fernandez-Marcos PJ, Garcia-Caballero T, del Pino P, Arribas J, Vidal A, Gonzalez-Barcia M, Serrano M, Loza MI, Dominguez E, Collado M (2019) Identification and characterization of Cardiac Glycosides as senolytic compounds. Nat Commun 10:4731

Vaiserman A, Koliada A, Lushchak O, Castillo MJ (2021) Repurposing drugs to fight aging: the difficult path from bench to bedside. Med Res Rev 41:1676–1700

Viana SD, Reis F, Alves R (2018) Therapeutic use of mTOR inhibitors in renal diseases: advances, drawbacks, and challenges. Oxid Med Cell Longev 2018:3693625

Wang J, Wong YK, Liao F (2018) What has traditional Chinese medicine delivered for modern medicine? Expert Rev Mol Med 20:e4

Wang S, Zhang S, Wang S, Gao P, Dai L (2020) A comprehensive review on Pueraria: Insights on its chemistry and medicinal value. Biomed Pharmacother 131:110734

Wang Y, Chang J, Liu X, Zhang X, Zhang S, Zhang X, Zhou D, Zheng G (2016) Discovery of piperlongumine as a potential novel lead for the development of senolytic agents. Aging (Albany NY) 8:2915–2926

Wang Z, Zeng M, Wang Z, Qin F, Chen J, He Z (2021) Dietary Luteolin: a narrative review focusing on its pharmacokinetic properties and effects on glycolipid metabolism. J Agric Food Chem 69:1441–1454

Xu M, Pirtskhalava T, Farr JN, Weigand BM, Palmer AK, Weivoda MM, Inman CL, Ogrodnik MB, Hachfeld CM, Fraser DG, Onken JL, Johnson KO, Verzosa GC, Langhi LGP, Weigl M, Giorgadze N, Lebrasseur NK, Miller JD, Jurk D, Singh RJ, Allison DB, Ejima K, Hubbard GB, Ikeno Y, Cubro H, Garovic VD, Hou X, Weroha SJ, Robbins PD, Niedernhofer LJ, Khosla S, Tchkonia T, Kirkland JL (2018) Senolytics improve physical function and increase lifespan in old age. Nat Med 24:1246–1256

Xu M, Tchkonia T, Ding H, Ogrodnik M, Lubbers ER, Pirtskhalava T, White TA, Johnson KO, Stout MB, Mezera V, Giorgadze N, Jensen MD, Lebrasseur NK, Kirkland JL (2015) JAK inhibition alleviates the cellular senescence-associated secretory phenotype and frailty in old age. Proc Natl Acad Sci U S A 112:E6301–E6310

Yadav V, Krishnan A, Vohora D (2020) A systematic review on Piper longum L.: bridging traditional knowledge and pharmacological evidence for future translational research. J Ethnopharmacol 247:112255

Yang SC, Chen PJ, Chang SH, Weng YT, Chang FR, Chang KY, Chen CY, Kao TI, Hwang TL (2018) Luteolin attenuates neutrophilic oxidative stress and inflammatory arthritis by inhibiting Raf1 activity. Biochem Pharmacol 154:384–396

Yosef R, Pilpel N, Tokarsky-Amiel R, Biran A, Ovadya Y, Cohen S, Vadai E, Dassa L, Shahar E, Condiotti R, Ben-Porath I, Krizhanovsky V (2016) Directed elimination of senescent cells by inhibition of BCL-W and BCL-XL. Nat Commun 7:11190

Yoshida S, Nakagami H, Hayashi H, Ikeda Y, Sun J, Tenma A, Tomioka H, Kawano T, Shimamura M, Morishita R, Rakugi H (2020) The CD153 vaccine is a senotherapeutic option for preventing the accumulation of senescent T cells in mice. Nat Commun 11:2482

Yousefzadeh MJ, Zhu Y, McGowan SJ, Angelini L, Fuhrmann-Stroissnigg H, Xu M, Ling YY, Melos KI, Pirtskhalava T, Inman CL, McGuckian C, Wade EA, Kato JI, Grassi D, Wentworth M, Burd CE, Arriaga EA, Ladiges WL, Tchkonia T, Kirkland JL, Robbins PD, Niedernhofer LJ (2018) Fisetin is a senotherapeutic that extends health and lifespan. EBioMedicine 36:18–28

Yuan H, Li Y, Ling F, Guan Y, Zhang D, Zhu Q, Liu J, Wu Y, Niu Y (2020) The phytochemical epigallocatechin gallate prolongs the lifespan by improving lipid metabolism, reducing inflammation and oxidative stress in high-fat diet-fed obese rats. Aging Cell 19:e13199

Yuanfeng W, Chengzhi L, Ligen Z, Juan S, Xinjie S, Yao Z, Jianwei M (2021) Approaches for enhancing the stability and formation of sulforaphane. Food Chem 345:128771

Zhang X, Zhang S, Liu X, Wang Y, Chang J, Zhang X, Mackintosh SG, Tackett AJ, He Y, Lv D, Laberge RM, Campisi J, Wang J, Zheng G, Zhou D (2018) Oxidation resistance 1 is a novel senolytic target. Aging Cell 17:e12780

Zhang YJ, Duan Y, Zheng XF (2011) Targeting the mTOR kinase domain: the second generation of mTOR inhibitors. Drug Discov Today 16:325–331

Zheng J, Lu B, Xu B (2021) An update on the health benefits promoted by edible flowers and involved mechanisms. Food Chem 340:127940

Zhu Y, Doornebal EJ, Pirtskhalava T, Giorgadze N, Wentworth M, Fuhrmann-Stroissnigg H, Niedernhofer LJ, Robbins PD, Tchkonia T, Kirkland JL (2017) New agents that target senescent cells: the flavone, fisetin, and the BCL-XL inhibitors, A1331852 and A1155463. Aging (Albany NY) 9:955–963

Zhu Y, Tchkonia T, Fuhrmann-Stroissnigg H, Dai HM, Ling YY, Stout MB, Pirtskhalava T, Giorgadze N, Johnson KO, Giles CB, Wren JD, Niedernhofer LJ, Robbins PD, Kirkland JL (2016) Identification of a novel senolytic agent, navitoclax, targeting the Bcl-2 family of anti-apoptotic factors. Aging Cell 15:428–435

Zhu Y, Tchkonia T, Pirtskhalava T, Gower AC, Ding H, Giorgadze N, Palmer AK, Ikeno Y, Hubbard GB, Lenburg M, O'Hara SP, Larusso NF, Miller JD, Roos CM, Verzosa GC, Lebrasseur NK, Wren JD, Farr JN, Khosla S, Stout MB, McGowan SJ, Fuhrmann-Stroissnigg H, Gurkar AU, Zhao J, Colangelo D, Dorronsoro A, Ling YY, Barghouthy AS, Navarro DC, Sano T, Robbins PD, Niedernhofer LJ, Kirkland JL (2015) The Achilles' heel of senescent cells: from transcriptome to senolytic drugs. Aging Cell 14:644–658

Zia A, Farkhondeh T, Pourbagher-Shahri AM, Samarghandian S (2021) The role of curcumin in aging and senescence: molecular mechanisms. Biomed Pharmacother 134:111119

Chapter 23
Role of Short Peptides as an Important Nutritional Element in Maintenance of Body Homeostasis

Svetlana V. Trofimova and Vladimir Kh. Khavinson

Abstract This book chapter outlines the physiological and molecular mechanisms of metabolism of short peptides as well as its role in maintenance of homeostasis. Many of short peptides for oral use do not decompose in the gastrointestinal tract and absorb unchanged. Such peptides, which do not hydrolyze in a gastrointestinal tract, get to a blood plasma and then to various tissues and organs. Short peptides permeate through cytoplasmic membrane into the nucleus and nucleolus of a cell, where they able to bind to DNA an to regulate gene expression epigenetically. Interaction DNA-peptide probably is the earliest evolution form. It gives explanation of high biological activity and absolute safety of short peptides, and its' successful use in form of substance endowed with significant physiological effect.

Keywords Short peptides · Nutrition · Homeostasis · Gastrointestinal tract · Gene expression

23.1 Introduction

Numerous studies on regulatory peptides in recent decades have led to a re-examination of ideas about regulation mechanisms of physiological functions, maintenance of homeostasis processes and adaptation of the body's functional systems to the environment. Nutrition is one of the most important environmental factors that affect the human body throughout its life. Nutrients, being converted through metabolism, ensure the body's vital activity, affecting the quality and duration of life. Eating disorders always lead to some kind of negative consequences. Therefore, various pathological processes in the human body, as well as aging processes, can be considered from the point of view of disorders in the metabolism of nutrients at the organ, tissue and cellular level of the organization of living matter.

S. V. Trofimova (✉) · V. Kh. Khavinson (✉)
Saint Petersburg Institute of Bioregulation and Gerontology, St. Petersburg, Russia
e-mail: vladimir@khavinson.ru

V. Kh. Khavinson
Pavlov Institute of Physiology of the RAS, Saint-Petersburg, Russia

S. I. S. Rattan and G. Kaur (eds.), *Nutrition, Food and Diet in Ageing and Longevity*, Healthy Ageing and Longevity 14, https://doi.org/10.1007/978-3-030-83017-5_23

In this regard, a rational diet adequate to age, professional activity, and health status is considered as the most important factor in the prevention of most age-associated human diseases.

Modern nutritional science considers two aspects: the epidemiology of nutrition (correlations between the characteristics of national nutrition and the presence of metabolic pathology in people from different countries) and the biochemical transformation of nutrients in the body (intake of biologically active substances with food, i.e. signaling molecules). From the point of view of molecular biology, information on the metabolic status of an organism is contained in the oldest (intracellular) and evolutionarily younger systems of regulation (nervous, immune, and endocrine), the functional activity of which is carried out through signaling molecules. These molecules carry out epigenetic regulation of gene expression to solve the general problems of the body's adaptation to changes in the supply of nutrients from the environment.

A large number of scientific works are devoted to the study of the pharmacokinetics of various biological substances when they are administered orally (Bai et al. 1995; Diao and Meibohm 2013; Vargas-Bello-Pérez et al. 2019). Among them, the study of their epigenetic activity has taken the leading place in recent years. Thus, according to a number of authors, it has been established that fatty acids and amino acids entering the blood and various tissues can interact with cellular targets (membrane and nuclear structures) and epigenetically regulate gene expression and protein synthesis (Ryan and Seeley 2013; Boyko et al. 2007). Among the signaling molecules necessary to maintain homeostasis in the body, a special group is made up of regulatory peptides of exogenous and endogenous origin (Anisimov and Khavinson 2010; Liu et al. 2016).

23.2 Biology of Endogenous Polypeptides

The functions of endogenous polypeptide molecules (growth factors, pancreatic polypeptide, calcitonin, angiotensin, peptide hormones of the hypothalamus, etc.) in the regulation of metabolism and neuro-immune-endocrine interactions are well studied. In recent years, more and more attention of scientists has been attracted to short peptides formed by the degradation of proteins that enter the body with food. A growing body of evidence indicates that regulatory oligopeptides are involved in the processes of growth, development, and regeneration (Khavinson et al. 2020; Sinjari et al. 2020; Khavinson et al. 2011a). Many of them are well-studied structures that regulate various physiological functions of the body. It is assumed that at the level of oligopeptides there is a unified system of regulation of both embryonic, growth, regeneration types and the functioning of the formed organism (Khavinson et al. 2019, 2021; Caputi et al. 2019). Thus, intensive studies of regulatory peptides have led to a radical re-examination of ideas about the regulation mechanisms of physiological functions and the principles of homeostasis process coordination in functional systems of the body.

It is known that the intestinal regulatory systems carry out the transport of peptides formed from proteins during digestion much faster than the absorption of a free amino acid mixture, to which the body did not adapt during evolution. The small intestine has di- and tripeptide transporters providing absorption of short peptides (Bai et al. 1995; Shen et al. 2001). The rate of transport of some dipeptides exceeds the rate of transport of those amino acids of which they are composed. Peptidases of the enterocyte brush border cleave a significant part (about 40–60%) of short peptides only to di- and tripeptides.

However, with age, in various parts of the digestive tract, there is a steady increase in involutional processes that contribute to the disruption of formation and absorption of short peptides, which ultimately leads to a violation of the adaptation processes and homeostasis of the functional systems of the entire organism.

All these data served as the basis for the development of biologically active food supplements based on short peptides. The most famous and "oldest" short peptide on the pharmaceutical market in the world is the dipeptide carnosine (β-Ala-His) (Mendelson 2008; Derave et al. 2019; Boldyrev et al. 2013). So, according to research, oral administration of carnosine has an antioxidant effect, affecting lipid peroxidation products, oxygen anions and other free radicals. It reduces their number to the level necessary for the full functioning of the signaling systems, which has a positive effect on the body's functions (Liu et al. 2016). An example of another mild peptide is the Asn-Leu-Pro-Arg (NLPR) peptide, which has neurotrophic effects. It was established that oral administration of the tetrapeptide Asn-Leu-Pro-Arg (NLPR) in rats with weakened memory is accompanied by an increase in susceptibility and presence of a behavioral response, and also enhances expression of nerve growth factor (NGF) in the brain. It is assumed that NLPR can improve memory by initiating the expression of NGF, i.e. is a potential drug candidate for treatment of memory impairment (Zhou et al. 1994).

A significant contribution to the study of the action mechanism of short peptides was made by studies carried out at the St. Petersburg Institute of Bioregulation and Gerontology, where various short peptides involved in maintaining the structural and functional homeostasis of cell population were synthesized and studied (Khavinson 2014). Among these peptides are short peptides Regevil (Vilon, Lys-Glu (KE)) and Epimental (Epitalon Ala-Glu-Asp-Gly (AEDG)) which have a normalizing effect on the function of the body's immune and endocrine systems (Anisimov and Khavinson 2010; Khavinson and Malinin 2005). Peptides KE and AEDG have a big spectrum of biological activity: they increase average and maximum life expectancy, reduce the incidence of malignant tumors, promote an increase in telomere length, overcome the Hayflick cell division limit, restore the functional activity of cells of the immune system and the endocrine system (Sevostianova et al. 2013; Khavinson et al. 2012a; Anisimov and Khavinson 2010; Khavinson and Malinin 2005). Peptide Lys-Glu-Asp-Ala (KEDA), which restores the intensity and rhythm of protein synthesis in the culture of hepatocytes of old rats, is also of interest (Timofeeva et al. 2005).

It should be noted that metabolism of short peptides obtained by enzymatic hydrolysis in the gastrointestinal tract of proteins that enter the body with food and their synthetic analogs seems to be uniform, therefore the general biological role of short peptides in the regulation of homeostasis, regardless of their origin, is further examined. The metabolism of short peptides can be divided into 3 main stages: absorption into the blood in the gastrointestinal tract, interaction with targets of various organs and tissues, and elimination/resorption in the kidneys.

23.3 Metabolism of Short Peptides in the Gastrointestinal Tract and Their Absorption into the Bloodstream

Protein molecules that enter the body with food, under the action of enzymes of the gastric (pepsin, rennin, gastrixin) and intestinal (aminopeptidase, enteropeptidase) tract, are split mainly into di- and tripeptides. These short peptides are divided into two groups: resistant to hydrolysis and able to break down into amino acids. Hydrolysis of labile short peptides occurs on the surface of the brush border membranes of enterocytes (membrane digestion) or inside enterocytes using cytosolic peptidases (intracellular digestion) (Ugolev et al. 1975; Timofeeva et al. 2005, 2000; Shen et al. 2001).

Peptides resistant to hydrolysis (glycyl-glycine, proline-containing peptides, glycyl-sarcosine, carnosine, etc.) penetrate into the circulatory system in an unchanged form, which is confirmed by the data of various researchers (Matthews and Payne 1975). According to Addison et al. (1975), all dipeptides have the same transport mechanism from the gastrointestinal tract into the blood. It was established that the Ala-Gly-Gly (AGG) tripeptide and some dipeptides are not hydrolyzed in the small intestine and can enter the bloodstream unchanged. In addition, it has been shown that dipeptides, in particular the Gly-Pro (GP) peptide, can penetrate from the gastrointestinal tract not only into the bloodstream, but can also be transported even through the blood–brain barrier in an unchanged form (Boyko et al. 2000). In another study, it was shown that in addition to passive transfer in the alkaline border of enterocytes, there is a system of active transport of di- and tripeptides (Dyer et al. 1990).

When studying the morpho-physiological characteristics of the transport of the cyclic nonapeptide Cys-Tyr-Phe-Gln-Asn-Cys-Pro-Arg-Gly (CYFQNCPRG) through the intestinal epithelium, its partial absorption also occurred while maintaining physiological activity. The localization of the label to this nonapeptide in the cytoplasm of enterocytes and the intercellular space in the basal region of the intestinal epithelium of rats and frogs has been shown using the methods of electron and fluorescence confocal microscopy (Natochin et al. 2004; Prutskova and Seliverstova 2012).

These data are consistent with the results of studies of short peptides developed under the guidance of prof. V.Kh. Khavinson. Numerous studies have shown that short peptides KE, AEDG, KEDA are not hydrolyzed in the stomach, duodenum, jejunum and ileum, and are only slightly hydrolyzed in the colon and liver. The degree of hydrolysis of these peptides was determined by cleavage by L-amino acid oxidase. In addition, it was shown that the KEDA peptide is not subject to hydrolysis in blood plasma, which confirms the assumption about the possibility of transport of peptides in the bloodstream to specific target organs for them (Tutelyan et al. 2003; Timofeeva et al. 2005).

The results showing that the peptides KE, AEDG, KEDA are able to regulate the activity of digestive enzymes in old animals are important. It was found that oral administration of the KEDA peptide in old rats for 2 weeks increased the activity of sucrase and maltase in the duodenum by 60%, maltase in the jejunum and colon by 3 times, and aminopeptidase M and glycyl-L-leucinepeptidase by 2 times (Timofeeva et al. 2005). The KE peptide in a similar experiment increased the activity of maltase (by 1.2 times in the duodenum, by 1.5 times in the jejunum, by 1.3 times in the ileum) and alkaline phosphatase (by 2.2 times in the jejunum). and the ileum and 1, 8 times—in the duodenum) (Khavinson et al. 2001). It is important to note that the peptides KEDA and KE promoted an increase in the activity of gastrointestinal enzymes in old animals to values typical for young animals. In addition, oral administration of the KE and AEDG peptides for 1 month in aged rats improved the absorption of glucose and glycine in the medial region under the action of the KE peptide, and in the proximal and distal parts of the small intestine under the influence of the AEDG peptide (Khavinson et al. 2002).

23.4 Tissue-Specific and Gene-Specific Interaction of Short Peptides with Target Cells in Various Organs

Through the bloodstream, short peptides penetrate into various organs and tissues. Probably, short peptides, in accordance with their structure, have an affinity for certain organs and tissues in which they accumulate and exhibit the greatest biological activity. It was found that short peptides KE and AEDG have a pronounced tissue-specific effect on the thymus and pineal gland (Anisimov and Khavinson 2010, Khavinson et al. 2011a). It is assumed that the interaction of short peptides with target tissues is based on their ability to penetrate the cytoplasmic and nuclear membrane into the nucleus and nucleolus and interact with DNA, epigenetically regulating the expression of genes encoding a number of signaling molecules and protein markers of cell functional activity (differentiation factors, proliferation, apoptosis, transcription) (Khavinson et al. 2012a; Fedoreeva et al. 2011; Tünnemann et al. 2006; Khavinson et al. 2011b).

It was found that short peptides synthesized on the basis of the Tat-protein (activator of transcription of the viral genome of human immunodeficiency HIV-1) are able to penetrate into the cell. These peptides have been combined into the cell-penetrating peptides (CPP) group (Tünnemann et al. 2006). Penetration into the cell through the membrane is most often characteristic of alkaline peptides containing an excess of positively charged amino acid residues in the structure. The advantage of these peptides is that they easily overcome the acidic glycocalyx layer that is adjacent to the cell membrane (Futaki et al. 2003; Duchardt et al. 2007). For synthetic alkaline and amphiphilic peptides containing several lysine residues in the structure, the ability not only to penetrate into the cell, but also to form complexes with DNA and RNA was noted. It was found that the binding of these peptides to DNA leads to strengthening of its double helix (Kubo et al. 2012).

The direct interaction of the peptide with the membrane is determined by the electrostatic interaction of positively charged side groups of amino acid residues of arginine and lysine with negative carboxyl groups of phosphatidylserine, exposed on the outer side of the cytoplasmic membrane (Denisov et al. 1998). For negatively charged (carboxyl) side groups of peptides, the binding sites are positively charged groups of phosphatidylcholine and phosphatidylethanolamine. Thus, pinocytosis may be the main mechanism for the penetration of short peptides through the cytoplasmic membrane.

In addition, it was shown that FITC-labeled short peptides AEDG, Glu-Asp-Arg (EDR), Lys-Glu-Asp-Gly (KEDG) penetrate the nucleus and nucleolus of HeLa cells. HeLa cells were incubated with FITC-labeled peptides for 12 h (Fedoreeva et al. 2011). It is known that the nucleus of euariotic cells has a system of transport pores (nucleopores) formed by protein complexes nucleoporins. The inner diameter of nucleopores is about 50 nm; therefore, they are permeable to freely diffusing low molecular weight substances with a molecular weight of up to 3.5 kDa, which include short peptides (Ohno et al. 1998). The results obtained make it possible to consider the possibility of direct interaction of short peptides with DNA. In recent years, the method of molecular modeling has been increasingly used to analyze nanostructures, which include short peptides (Sokolova et al. 2012).

Comparison of the spatial arrangement of functional groups on the surface of the large groove of double-stranded DNA and side groups of regulatory peptides showed that the AEDG peptide can bind to the complementary DNA site on the promoter region of the gene, causing local strand separation, and thereby initiating the process of RNA gene transcription by polymerase II (Khavinson et al. 2012a, b Anisimov and Khavinson 2010).

The ATTTS sequence complementary to the AEDG peptide was found in the promoter portions of the Ki67, P53, IL-2, MMP2 and Tram1 genes and telomerase. It was experimentally proved that the addition of the AEDG peptide to the culture of human lung fibroblasts induces the expression of the telomerase gene, telomerase activity, and promotes the elongation of telomeres by 2.4 times (Khavinson and Malinin 2005).

Activation of gene expression is accompanied by an increase in the number of cell divisions by 42.5%, which demonstrates overcoming the Hayflick limit of cell division (Anisimov and Khavinson 2010). In addition, in the pineal gland cell culture of young and old animals under the influence of the tetrapeptide, an increase in the synthesis of MMP2 and Ki67 proteins and a decrease in the synthesis of the proapoptotic protein p53 were observed (Khavinson et al. 2012a, b).

Epigenetics postulates the tissue, subcellular, and age specificity of DNA methylation and indicates that the character of DNA methylation in cancer cells is different than in normal cells (Fedoreeva et al. 2011). Taking these facts into account, it can be assumed that the same biologically active short peptide can bind to DNA depending on the nature of its methylation and will have different effects on gene functions in different tissues (cells), in the nucleus and mitochondria, in young and old cells, in normal and malignant cells. Thus, specific (complementary) peptide-DNA interactions can epigenetically control the genetic functions of a cell, and this mechanism probably played an important role at the earliest stages of life and evolution (Vanyushin and Khavinson 2016).

23.5 Elimination and Resorption of CP in the Kidneys

The question of whether short peptides are excreted in the urine or resorbed in the kidneys is still controversial (Bumbaca et al. 2019; Litvin et al. 2019; Shen et al. 2001). According to some researchers, short peptides resistant to hydrolysis by gastrointestinal tract enzymes are eliminated in the urine (Timofeeva et al. 2005). In another work, using confocal microscopy, it was shown that the introduction of a fluorescently labeled polypeptide into the intestines of rats led to its accumulation in the epithelium vesicles of the proximal renal tubule, which indicates the role of the kidneys in the metabolism of not only endogenous, but also exogenous peptides and proteins (Natochin et al. 2005). These data are consistent with the study, which indicates that the metabolism of peptides with a molecular weight of less than 60 kDa occurs in the proximal renal tubules (Diao and Meibohm 2013).

23.6 Clinical Efficiency of Oral Administration of Dietary Supplements Based on Short Peptides

Long-term experimental studies have shown that under the influence of unfavorable environmental factors, emotional stress, development of age-related pathology, the process of the body's main systems' self-regulation is disturbed, the correction of which is possible with the help of specific short peptides. In addition, the data presented in the review that short peptides may be resistant to hydrolysis in the gastrointestinal tract and blood, as well as epigenetically regulate gene expression, thereby normalizing metabolism, served as the basis for the development of drugs and biologically active food supplements based on short peptides (Khavinson 2014; Khavinson et al. 2019).

Based on the short peptides Ala-Glu-Asp-Gly (AEDG) and Lys-Glu (KE), biologically active food supplements Epimental® (Epitalon) and Regevil® (vilon) have been developed.

Epimental® (Epitalon) was developed based on the results of an experimental study of the tetrapeptide Ala-Glu-Asp-Gly. Long-term experimental studies have shown that Ala-Glu-Asp-Gly regulates metabolic processes in the cells of the neuroendocrine system, activates antioxidant defense processes, stimulates pineal and extrapineal melatonin synthesis in EC cells of the stomach and intestines, leading to optimization of the biorhythms of cortisol secretion and other hormones (Anisimov et al. 2003; Khavinson and Malinin 2005; Khavinson et al. 2012a; Khavinson 2014; Khavinson et al. 2011c).

The safety of dietary supplements Ala-Glu-Asp-Gly was confirmed by the results of a study of its general toxic effect. In the study of acute toxicity, it was found that a single administration of an Ala-Glu-Asp-Gly solution to animals at a dose 5000 times higher than the therapeutic one recommended for clinical use does not cause toxic reactions. The study of the subacute and chronic toxicity of Ala-Glu-Asp-Gly indicates the absence of side effects during its long-term use in doses exceeding the therapeutic one by 100–1000 times. When assessing the general state of animals, morphological and biochemical parameters of peripheral blood, morphological state of internal organs, state of the cardiovascular and respiratory systems, liver and kidney function, pathological changes in the body were not found. Thus, the absence of a general toxic effect indicates the safety of using Ala-Glu-Asp-Gly as a biologically active food supplement in order to maintain the function of the neuroendocrine system.

Clinical study on the effectiveness of a biologically active food supplement Epimental® (Epitalon) showed its high efficacy in subjects with conditions caused by chronic stress factors (Trofimova et al. 2021).

The study involved 560 subjects aged 35 to 68 years (260 men and 300 women) with conditions after prolonged exposure to occupational or psychoemotional stress, including those caused by frequent changes in time zones.

Stress is a pathological process, which consists in the formation of a complex of nonspecific protective, compensatory and pathological reactions of the body that arise in response to the action of extreme or pathological stimuli that threaten homeostasis: pain, hypoxia, hunger, psycho-emotional overstrain and other emergency factors that lead to changes of the same type in the lymphoid tissue, including the thymus gland, blood composition, adrenal glands, leading to a change in the biorhythm of hormone secretion. There are close links between stress and occurrence of physical illness. Modern experimental and clinical data, based on observations of people and animals, confirm the results obtained by Hans Selye, the classic of the theory of stress, and reveal the psychological processes by which emotional reactions to stress can make a person susceptible to this or another disease.

Recently, more and more often in clinical practice, the cause of disturbances in the biorhythm of hormone synthesis and related disorders on the part of the autonomic nervous system (insomnia, emotional lability, apathy, etc.) is jet lag disorder, a mismatch between the human biorhythm and the daytime rhythm, due to the frequent change of time belts.

The subjects were randomly divided into two groups: tmain and control. The subjects of the main group (270 people) received dietary supplements Epimental® (Epitalon), 1 capsule per day with meals for 20 days. The control group included 290 subjects with similar conditions who received a placebo in a similar manner. An informed consent was signed with each study participant in accordance with protocol No. 7 dated March 5, 2018, approved by the ethics committee of the St. Petersburg Institute of Bioregulation and Gerontology.

All the subjects were under occupational or psychoemotional stress for a long time, including 142 people exposed to jet lag, 430 people under occupational stress caused by extreme psychoemotional stress at work, lack of rest during long time. The subjects complained of an asthenic state: general weakness, decreased appetite, headaches, sleep disturbances, increased irritability, apathy, emotional lability, rapid fatigue, decreased performance, decreased memory and attention, dizziness, increased sweating, changes in blood pressure.

The effectiveness of the use of dietary supplements Epimental® (Epitalon) was assessed subjectively, by studying the dynamics of the subjects' complaints, and by objective indicators, including the determination of cortisol, adrenocorticotropic hormone (ACTH) and melatonin levels in the blood serum. The content of melatonin in the blood was measured twice: in the morning at 9:00 and in the evening at 21:00.

It was found that the use of Epimental® (Epitalon) improved the general condition of patients in the study group. The subjects who received Epimental® (Epitalon) noted improvements concerning all the following indicators: apathy, emotional lability, sleep disturbance, rapid fatiguability, decreased performance, and decreased attention focusing.

While apathy, fatigue, decreased performance were noted during the initial examination in all patients of both groups, then after correction with the Epimental® (Epitalon) dietary supplement, these complaints decreased in 60–70% of the cases in patients of the main group. While in the control group, there were no significant changes, in addition, complaints of emotional lability and decreased concentration of attention even increased. It should be noted that the regulation of the function of the neuroendocrine system using Epimental® (Epitalon) contributed to the restoration of sleep in more than half of the subjects. If during the initial examination 67,4% of respondents mentioned this complaint, then at the secondary examination there were 32,2% of patients complaining of the sleep disturbance. In the control group, no significant changes were recorded.

It must be mentioned that upon repeated examination, all subjective indicators in patients of the main group significantly differ from those in patients in the control group. Thus, the complaints of patients, characterizing the state after prolonged exposure to stress factors, had a positive trend only in the subjects of the main group.

As can be seen from above, the use of Epimental® (Epitalon) in subjects exposed to prolonged occupational or psycho-emotional stress, including those caused by frequent changes in time zones, contributed to an improvement in subjective indicators that significantly differed from those before the use of dietary supplements and from indicators in patients of the control group.

During the initial study of the level of melatonin in the blood of patients of both groups, it was found that the level of melatonin in the morning (9:00) was decreased by 1.3 times, the level of melatonin in the evening (21:00) was decreased by 2.3 times compared to the lower limit normal values. After application of Epimental® (Epitalon), there was a significant increase in the level of melatonin in the morning (from 6.9 ± 1.1 up to 18.2 ± 1.4 pg/ml) and in the evening (from 35.1 ± 1.3 up to 72.3 ± 3.6 pg/ml) to the lower limit of the reference values. At the same time, in the subjects of the control group, the indicators of the level of melatonin in the morning (from 6.3 ± 0.9 up to 7.8 ± 1.1 pg/ml) and in the evening (from 31.4 ± 1.2 up to 37.2 ± 1.9 pg/ml) increased slightly and did not reach the lower limit of the norm.

The use of Epimental® (Epitalon) in patients exposed to prolonged occupational or psycho-emotional stress contributed to the stabilization of the hormonal status, which indicates the leveling of maladjustment disorders and catabolic reactions.

The content of cortisol and adrenocorticotropic hormone (ACTH) before the examination was noted at the lower limit of the norm, which indicated the depletion of the reserves of the adrenal cortex. The level of cortisol in the main group was 239.1 ± 18.9 nmol/L, while in control group it was 232.7 ± 20.1 nmol/L; the level of ACTH is the main group was 12.6 ± 1.0 pg/ml, while in control group—10.9 ± 0.9 pg/ml. Significant difference between these indicators in both groups were not recorded. After complex treatment with Epimental® (Epitalon), the level of cortisol and ACTH in the blood plasma returned to normal and was detected in the middle region of the reference values of these indicators, which is extremely important for optimizing the body's response to stress factors. These changes correlated with improvements in subjective scores. The lack of positive dynamics of the content of cortisol and ACTH in the blood of the subjects of the control group is noticeable.

Thus, the results of the studies have shown that the use of Epimental® (Epitalon) dietary supplements to food contributed to the improvement of the neuroendocrine system of the body, which explains the improvement in the general condition in patients of the main group. Against the background of the use of dietary supplements Epimental® (Epitalon), a decrease in complaints of sleep disturbances, headaches, dizziness, apathy, weakness, rapid fatigueability, decreased performance, emotional lability, decreased memory and attention, increased sweating, decreased appetite was noted.

During the use of Epimental® (Epitalon), no side effects, complications and drug dependence were identified.

The results of a clinical study of Epimental® (Epitalon) dietary supplements allow us to conclude that Epimental® food supplements have a regulatory effect on the neuroendocrine system, which allows us to recommend its use in people exposed to prolonged occupational or psycho-emotional stress, including jet-lag syndrome caused by frequent change of time zones (Trofimova et al. 2021).

Regevil® (vilon) Lys-Glu (KE) was developed based on the results of an experimental study of the Lys-Glu dipeptide (KE). Long-term experimental studies have shown that peptides have a tissue-specific effect on the cells of those tissues for which they are specific. According to experimental studies, the Lys-Glu (KE) dipeptide regulates metabolic processes in the cells of the immune system, improves the indicators of cellular and tissue homeostasis in the cells of the immune system, restores impaired immunological reactivity, activates antioxidant defense processes, and stimulates tissue regeneration in the event of their suppression. These properties of the Lys-Glu (KE) dipeptide are the mechanism of its immunomodulatory and anti-stress action. (Anisimov and Khavinson 2010; Khavinson and Malinin 2005; Khavinson 2014; Khavinson et al. 2011d).

The safety of Lys-Glu (KE) dipeptide was confirmed by the results of a study of its general toxic effect. In the study of acute toxicity, it was found that a single administration of a solution of Lys-Glu (KE) dipeptide to animals in a dose 5000 times higher than the therapeutic one recommended for clinical use does not cause toxic reactions. The study of the subacute and chronic toxicity of the Lys-Glu (KE) dipeptide indicates the absence of side effects with its long-term use in doses exceeding the therapeutic one by 100–1000 times. When assessing the general state of animals, morphological and biochemical parameters of peripheral blood, morphological state of internal organs, state of the cardiovascular and respiratory systems, liver and kidney function, pathological changes in the body were not detected. Thus, the absence of a general toxic effect indicates the safety of using Lys-Glu (KE) dipeptide as a biologically active food supplement in order to maintain the function of the immune system.

A clinical study of the effectiveness of the dietary supplement Regevil® (vilon) has shown its high efficiency for the complex restoration of the functions of the immune system in pathological conditions of various origins, including for accelerating tissue regeneration after various injuries, when exposed to extreme environmental factors, as well as in elderly and senile people to maintain the functions of the immune system (Trofimova et al. 2021).

The study involved 520 people aged 43 to 76 years (268 men and 252 women) during the period of convalescence after suffering acute respiratory, bacterial or viral diseases.

It is known that various factors of a physical, chemical and biological nature, depending on the duration or intensity of their impact on the human body, can lead to depletion of adaptive and compensatory mechanisms and cause profound disturbances in various links of the immune system.

Pathological disorders in the immune system, as a rule, contribute to a protracted course of the underlying disease with a tendency to relapse, a decrease in the body's resistance to infection and the development of severe complications.

The subjects were randomly divided into two groups: the main and the control. The subjects of the main group (276 people) received dietary supplements for food Regevil® (vilon), 1 capsule per day with meals for 20 days. The control group included 244 people with similar conditions who received a placebo in a similar manner. An informed consent was signed with each study participant in accordance with Protocol No. 2 dated January 24, 2018, approved by the Ethics Committee of the St. Petersburg Institute of Bioregulation and Gerontology.

The effectiveness of the use of dietary supplements to food Regevil® (vilon) was assessed by the dynamics of complaints of the examined and by a number of objective indicators: general clinical examination of blood and urine, immunological examination of peripheral blood (the number of T- and B-lymphocytes was determined by the method of immunofluorescence with monoclonal antibodies obtained to differentiation antigens of lymphocytes CD3, CD4, CD8, CD20; the content of immunoglobulins of various classes was determined by the method of radial immunodiffusion in gel according to Mancini; functional activity of T-lymphocytes was determined by the lymphocyte migration test (LMIT) with ConA). Subjects complained of asthenic state: general weakness, loss of appetite, headaches, sleep disturbance, increased irritability, apathy, emotional lability, rapid fatiguability, reduced performance.

The studies have shown that the majority of the surveyed who are in the recovery period, regardless of the etiology of the disease (viral as a complication after influenza, acute respiratory infections or pneumococcal), have disorders in the immune status, manifested in a decrease in the number of CD3+, CD4+ cells with a slight increase in the number of lymphocytes with the CD8+ phenotype, which indicates a decrease in the level of immunoreactivity (CD4+/CD8+). The results of LMIT with ConA characterize a decrease in the functional activity of T-lymphocytes (mainly CD8+, i.e., T-suppressors/killers). The content of CD20+-cells, representing a subpopulation of B-lymphocytes, did not significantly differ from normal values, but, at the same time, an increase in the amount of immunoglobulins M and G in blood serum was observed.

It should be noted that the quantitative indicators of the content of CD3+ and CD4+ cells are characteristic of the lower limits of physiological fluctuations in their number in persons of a given age, which may indicate a depletion of their immune system. As a rule, persons with a secondary immunodeficiency state had a pronounced asthenic syndrome and significant changes in the cardiovascular system.

The results of the conducted studies indicate that Regevil® (vilon) is an effective solution for the correction of secondary immunodeficiencies developing in response to exposure to extreme factors. The use of Regevil® (vilon) allowed to normalize the impaired parameters of the immune system in 86% of cases.

The greatest effect from the use of Regevil® (vilon) was observed in relation to subpopulations of T-lymphocytes and their functional activity: there was a significant increase in the content of CD3+ and CD4+ lymphocytes to the level of normal values, with their initial significant decrease, normalization of the CD4+/CD8+ ratio, a significant decrease in the LMIT index. A less distinct reaction was observed on the part of the B-system of immunity (CD20+), probably due to its greater conservatism and insufficient duration of the drug intake, although a tendency to an increase in the content of B-lymphocytes up to the lower limit of the norm was noted. Attention is also drawn to a significant increase in the leukocyte content compared with the indicator before the use of the peptide: initially this indicator was at the lower limit of the norm ($4.6 \pm 0.6 \times 10^9$/l), after the course of using the drug, the indicator increased to optimal values ($6.5 \pm 0.2 \times 10^9$/l), which indicates a more rapid reversal of the inflammatory process in the body than in patients of the control group.

The positive dynamics of laboratory indicators was accompanied by a pronounced improvement in subjective indicators. So, after the course with the use of Regevil® (vilon), the subjects who had pneumonia noted a significant improvement in their general condition, leveling of residual effects of bronchial and pulmonary dysfunction (reduction of cough and shortness of breath) and a decrease in the severity of asthenic syndrome, always accompanying secondary immunodeficiencies caused an infectious disease. Thus, the obtained results of the study indicate not only the immunomodulatory effect of the drug, but also its ability to accelerate tissue regeneration due to the immunostimulating effect.

No side effects, complications, contraindications, drug dependence with the use of Regevil® (vilon) were identified during the clinical study.

Thus, the clinical study carried out has shown that Regevil® (vilon) promotes the normalization of cellular immunity indicators, stimulates the processes of tissue regeneration in case of their suppression, does not cause side effects, complications and drug dependence. The dietary supplement for food Regevil® (vilon) is recommended to be used to accelerate the restoration of the functions of the immune system after inflammatory diseases of infectious and non-infectious genesis (including after pneumonia of viral and bacterial etiology), it is also recommended for the older people to maintain the functional activity of the immune system (Trofimova et al. 2021).

Thus, oral administration of the dietary supplements Epimental® (epitalon) and Regevil® (vilon), created on the basis of the short peptides Ala-Glu-Asp-Gly (AEDG) and Lys-Glu (KE), turned out to be effective in clinical research aimed at maintaining the functional activity of the immune neuroendocrine systems of the body (Trofimova et al. 2021).

23.7 Conclusion

Experimentally and clinically demonstrated possibility of oral administration of drugs based on short peptides is the indication for their preventive and therapeutic uses. Some of the main characteristics of these short peptides are: their resistance to the action of enzymes of the gastrointestinal tract and blood plasma; their ability to activate absorption of various biological substances in the gastrointestinal tract; and their ability to penetrate into the cytoplasm, nucleus and nucleolus of target cells of various tissues. Furthermore, interaction of short peptides with DNA and their effects on the epigenetic regulation of gene expression provides an explanation for their high biological activity and successful use as substances with physiologically adequate and potentially geroprotective actions.

Compliance with Ethical Standards

Funding This study was not funded.

Conflict of Interest All authors declare they have no conflict of interest.

References

Addison JM, Burston D, Dalrymple JA et al (1975) A common mechanism for transport of di- and tri-peptides by hamster jejunum in vitro. Clin Sci Mol Med 49(4):313–322. https://doi.org/10.1042/cs0490313

Anisimov VN, Khavinson VKh (2010) Peptide bioregulation of aging: results and prospects. Biogerontology 11(2):139–149. https://doi.org/10.1007/s10522-009-9249-8

Anisimov VN, Khavinson VKh, Popovich IG, Zabezhinski MA et al (2003) Effect of Epitalon on biomarkers of aging, life span and spontaneous tumor incidence in female Swiss-derived SHR mice. Biogerontology 4:193–202. https://doi.org/10.1023/A:1025114230714

Bai JP, Chang LL, Guo JH (1995) Targeting of peptide and protein drugs to specific sites in the oral route. Crit Rev Ther Drug Carrier Syst 12(4):339–371. https://doi.org/10.1615/CritRevTherDrugCarrierSyst.v12.i4.30

Boldyrev AA, Aldini G, Derave W (2013) Physiology and pathophysiology of carnosine. Psychol Rev 93(4):1803–1845. https://doi.org/10.1152/physrev.00039.2012

Boyko SS, Kolyvanov GB, Zherdev VP (2007) Experimental study of the pharmacokinetics of a tryptophan-containing dipeptide GB-115. Bull Exp Biol Med 144(9):285–287

Boyko SS, Ostrovskaia RU, Zherdev VP et al (2000) Pharmakocinetics and permeability of the blood-brain barrier for a new acyl-prolyl-dipeptide with nootropic property after the peroral administration. Bull Exp Biol Med 129(4):426–429

Bumbaca B, Li Z, Shah DK (2019) Pharmacokinetics of protein and peptide conjugates. Drug Metab Pharmacokinet 34(1):42–54. https://doi.org/10.1016/j.dmpk.2018.11.001

Caputi S, Trubiani O, Sinjari B, Trofimova S et al (2019) Effect of short peptides on neuronal differentiation of stem cells. Int J Immunopathol Pharmacol 33:1–12. https://doi.org/10.1177/2058738419828613

Denisov G, Wanaski S, Luan P et al (1998) Binding of basic peptides to membranes produces lateral domains enriched in the acidic lipids phosphatidylserine and phosphatidylinositol 4.5-bisphosphate: an electrostatic model and experimental results. Biophys J 74(2):731–744. https://doi.org/10.1016/S0006-3495(98)73998-0

Derave W, De Courten B, DiBaba SP (2019) An update on carnosine and anserine research. Amino Acids 51:1–4. https://doi.org/10.1007/s00726-018-02689-9

Diao L, Meibohm B (2013) Pharmacokinetics and pharmacokinetic-pharmacodynamic correlations of therapeutic peptides. Clin Pharmacokinet 52(10):855–868. https://doi.org/10.1007/s40262-013-0079-0

Duchardt F, Fotin-Mleczek M, Schwarz H et al (2007) A comprehensive model for the cellular uptake of cationic cell-penetrating peptides. Traffic (Copenhagen Denmark) 8(7):848–866. https://doi.org/10.1111/j.1600-0854.2007.00572.x

Dyer J, Beechey RB, Gorvel JP et al (1990) Glycyl-L-proline transport in rabbit enterocyte basolateral-membrane vesicles. Biochem J 269(3):565–571. https://doi.org/10.1042/bj2690565

Fedoreeva LI, Kireev II, Khavinson VKh, Vanyushin BF, Хавинсон ВХ (2011) Penetration of short fluorescence labeled peptides into the nucleus in HeLa cells and in vitro specific interaction of peptides with deoxyribooligonucleotides and DNA. Biochemistry 76(11):1505–1516

Futaki S, Goto S, Sugiura Y (2003) Membrane permeability commonly shared among arginine-rich peptides. J Mol Recogn 16(5):260–264. https://doi.org/10.1002/jmr.635

Khavinson V, Trofimova S, Trofimov A, Solomin I (2019) Molecular-physiological aspects of regulatory effect of peptide retinoprotectors. Stem Cell Rev Rep 15(3):439–442. https://doi.org/10.1007/s12015-019-09882-7

Khavinson VKh (2014) Peptides, genome aging. Adv Gerontol 4(4):337–345. https://doi.org/10.1134/S2079057014040134

Khavinson VKh, Malinin VV (2005) Gerontological aspects of genome peptide regulation. Basel (Switzerland): Karger AG. https://doi.org/10.1159/isbn.978-3-318-01193-7

Khavinson VKh, Fedoreeva LI, Vanyushin BF (2011) Short peptides modulate the effect of endonucleases of wheat seedling. Dokl Biochem Biophys 437(1):64–67. https://doi.org/10.1134/S1607672911020025

Khavinson VKh, Timofeeva MN, Malinin VV et al (2001) Effect of the dipeptide Vilon on activity of digestive enzyme in rats of various ages. Bull Exp Biol Med 131(6):583–585. https://doi.org/10.1023/A:1012319122696

Khavinson VKh, Linkova N, Kozhevnikova E, Trofimova S (2021) EDR peptide: possible mechanism of gene expression and protein synthesis regulation involved in the pathogenesis of Alzheimer's disease. Molecules 26(1):159. Special Issue Peptide Therapeutics 2.0, 1–16. https://doi.org/10.3390/molecules26010159
Khavinson VKh, Linkova NS, Trofimov AV et al (2011) Morphofunctional fundamentals for peptide regulation of aging. Biol Bull Rev 1(4):390–394
Khavinson VKh, Linkova NS, Kvetnoy IM et al (2012a) Molecular cellular mechanisms of peptide regulation of melatonin synthesis in pinealocyte culture. Bull Exp Biol Med 153(2):255–258. https://doi.org/10.1007/s10517-012-1689-5
Khavinson VKh, Linkova NS, Chalisova NI et al (2011) Effect of short peptides on expression of signaling molecules in organotypic pineal cell culture. Bull Exp Biol Med Cell Technol Biol Med 3:138–141
Khavinson VKh, Tarnovskaya SI, Linkova NS et al (2012b) Short cell-penetrating peptides: a model of interactions with gene promoter sites. Bull Exp Biol Med 154(9):391–396
Khavinson VKh, Polyakova VO, Linkova NS et al (2011) Peptides regulate cortical thymocytes differentiation, proliferation, and apoptosis. J Amino Acids 2011:1–5. https://doi.org/10.4061/2011/517137
Khavinson VKh, Egorova VV, Timofeeva NM et al (2002) Effect of Vilon and Epitalon on Glucose and Glycine absorption in various regions of small intestine in aged rats. Bull Exp Biol Med 133(5):494–496. https://doi.org/10.1023/A:1019878224754
Khavinson V, Linkova N, Diatlova A, Trofimova S (2020) Peptide regulation of cell differentiation. Stem Cell Rev Rep 16(1):118–125. https://doi.org/10.1007/s12015-019-09938-8
Kubo T, Yanagihara K, Sato Y, Morita Y et al (2012) Enhancement of gene silencing effect and membrane permeability by peptide-conjugated 27-nucleotide small interfering RNA. Mol Nucleic Acid Analogs 17(9):11089–11102. https://doi.org/10.3390/molecules170911089
Litvin AA, Shevchenko R, Kolyvanov GB, Bochkov PO (2019) Elimination half-life of short peptide drugs in humans extrapolated from animal pharmacokinetic pharmacokinetic studies. Pharm Chem J 53(5):685–688. https://doi.org/10.1007/s11094-019-02063-3
Liu R, Xing L, Fu Q, Zhou G-h, Zhang W-g (2016) A review of antioxidant peptides derived from meat muscle and by-products. Antioxidants 5(3):32. https://doi.org/10.3390/antiox5030032
Matthews DM, Payne JW (1975) Peptide transport in protein nutrition, pp 61–146
Mendelson SD (2008) Metabolic syndrome and psychiatric illness: interactions, pathophysiology, assessment & treatment. Nutr Suppl Metab Syndr 141–186. https://doi.org/10.1016/B978-012374240-7.50012-7
Natochin IV, Seliverstova EV, Burmakin MV (2005) Kidney accumulation of yellow fluorescent protein after its absorption in the rat intestine. Russ J Physiol 91(10):1195–1204
Natochin IV, Komissarchik II, Snigirevskaia ES et al (2004) Immunocytochemical localization of vasopressin at its absorption by cells of rat small intestine. Tsitologiia 46(11):953–959
Ohno M, Fornered M, Mattaj IW (1998) Nucleocytoplasmic transport: the last 200 nanometers. Cell 92(3):327–336. https://doi.org/10.1016/S0092-8674(00)80926-5
Prutskova NP, Seliverstova EV (2012) Absorption of arginine-vasopressin and arginine-vasotocin in small intestine of the frog Rana temporaria. Zhournal Evoliutsionnoi Biokhimii Fiziologii 48(1):54–62
Ryan KK, Seeley RJ (2013) Physiology. Food as a hormone. Science 339(6122):918–919. https://doi.org/10.1126/science.1234062
Sevostianova NN, Linkova NS, Polyakova VO et al (2013) Immunomodulating effects of vilon and its analogue in the culture of human and animal thymus cells. Bull Exp Biol Med 154:562–565. https://doi.org/10.1007/s10517-013-2000-0
Shen H, Smith DE, Brosius FC III (2001) Developmental expression of PEPT1 and PEPT2 in rat small intestine, colon, and kidney. Pediatr Res 49:789–795. https://doi.org/10.1203/00006450-200106000-00013
Sinjari B, Diomede F, Khavinson V, Mironova E et al (2020) Short peptides protect oral stem cells from ageing. Stem Cells Rev Rep 16(1):159–166. https://doi.org/10.1007/s12015-019-0992-3

Sokolova OS, Shaitan KV, Grizel AV et al (2012) Three-dimensional structure of human Kv10.2 Ion Channel studied by single particle electron microscopy and molecular modeling. Russ J Bioorganic Chem 38(2):177–184

Timofeeva MN, Khavinson VKh, Malinin VV et al (2005) Effect of peptide livagen on activity of digestive enzymes in gastrointestinal tract and non-digestive organs in rats of different ages. Adv Gerontol 16:92–96

Timofeeva NM, Iezuitova NN, Gromova LV (2000) The current concepts on the absorption of monosaccharides, amino acids and peptides in the mammalian small intestine. Adv Physiol Sci 31(4):24–37

Trofimova SV, Trofimov AV, Ivko OM (2021) Ways to enhance the reserved capabilities of a one's body under the influence of negative industrial and climate factors. Doctor (3)

Tünnemann G, Martin RM, Haupt S et al (2006) Cargo-dependent mode of uptake and bioavailability of TAT-containing proteins and peptides in living cells. FASEB J 20(11):1775–1784. https://doi.org/10.1096/fj.05-5523com

Tutelyan VA, Khavinson VKh, Malinin VV (2003) Physiological role of a short peptides in nutrition. Bull Exp Biol Med 135(1):4–10

Ugolev AM, Gruzdkov AA, De Laey P, Egorova VV, Iezuitova NN et al (1975) Substrate interactions on the intestinal mucosa: a concept for the regulation of intestinal digestion. Br J Nutr 34(2):205–220

Vanyushin BF, Khavinson VKh (2016) Epigenetics - a different way of looking at genetics. Short biologically active peptides as epigenetic modulators of gene activity. In: Doerfler W, Biihm P (eds) Springer International Publishing Switzerland, pp 69–90. https://doi.org/10.1007/978-3-319-27186-6_5

Vargas-Bello-Pérez E, Márquez-Hernández RI, Hernández-Castellano LE (2019) Bioactive peptides from milk: animal determinants and their implications in human health. J Dairy Res 86(2):136–144. https://doi.org/10.1017/S0022029919000384

Zhou AW, Gou J, Du YC (1994) NLPR, an agonist of AVP 4–8, increases NGF gene expression in memory-impaired rat brain. Biomed Pept Proteins Nucleic Acids Struct Synth Biol Activity 1(1):57–58

Chapter 24
Fasting and Caloric Restriction for Healthy Aging and Longevity

Sandeep Sharma and Gurcharan Kaur

Abstract Intermittent fasting (IF) is widely practiced for health benefits among people of various societies by adopting regimens which vary in terms of dietary patterns and duration of fast. Also, sustained periods of caloric restriction (CR) without malnutrition have been shown to be a potent modulator of lifespan resulting in lower incidence of metabolic disorders like type 2 diabetes, cardiovascular diseases, cancer, and neurological disorders. IF regimens such as alternate day fasting, time restricted feeding, protein restriction etc. have recently emerged as potential alternate approaches to CR which do not involve any major changes in quality and quantity of nutritional intake. This chapter reviews the different regimens of IF and CR used in model organisms and in humans to ascertain their efficacy for metabolic fitness, resistance to age-related diseases and longevity as well as their underlying molecular and cellular mechanisms. Moreover, promoting health-oriented and disease preventive approaches are more viable options for healthy aging and longevity than continuing with disease-oriented research and therapeutic strategies.

Keywords Caloric restriction · Dietary restriction · Time restricted fasting · Metabolic syndrome · Circadian rhythms · Alternate day fasting

24.1 Introduction

Biological aging and its underlying molecular and cellular processes cannot be considered as a disease, and therefore, reorienting the focus of aging research to health-oriented and disease-preventive strategies is required (Rattan 2014). Amongst such approaches, dietary restriction (DR) has emerged to be of prime importance in maintaining and improving mental as well as physical health status in older adults (Zupo et al. 2020; Currenti et al. 2021). Caloric restriction (CR) refers to a dietary

S. Sharma
Department of Comparative Biology and Experimental Medicine, University of Calgary, Calgary, AB T2N 4N1, Canada

G. Kaur (✉)
Department of Biotechnology, Guru Nanak Dev University, Amritsar, India
e-mail: gurcharan.biotech@gndu.ac.in

S. I. S. Rattan and G. Kaur (eds.), *Nutrition, Food and Diet in Ageing and Longevity*, Healthy Ageing and Longevity 14, https://doi.org/10.1007/978-3-030-83017-5_24

intervention which recommends an overall 20–40% reduction in daily caloric intake, whereas, DR is based on a broader scope of dietary interventions that involves restrictions CR without compromising on the quality of nutrition is an effective non-pharmacological intervention which is reported to promote health span in numerous non-human species (Weindruch 1996; Ingram et al. 2007; Mattison et al. 2012) as well as in humans (Fontana et al. 2004, 2010; Fontana and Partridge 2015).

Several clinical trials conducted in the last decade have reported the benefits of short and prolonged bouts of CR in weight reduction and improvement in several physiological markers of health (Most et al. 2017, 2018; Redman and Ravussin 2011). However, long term implementation of daily traditional CR has lower success rate due to poor compliance by individuals (Barte et al. 2010; Scheen 2008). Moreover, ample availability of energy-rich food and beverages in the present-day societies baffles the individual's ability to continue with traditional CR regimen (Swinburn and Egger 2002). Keeping in view the rapid rate of population aging, some innovative and easy to implement strategies are needed to improve healthspan (Dzau et al. 2019). Intermittent fasting-dietary restriction (IF-DR) regimen based on feeding/fasting timings manipulation is emerging as an alternative and innovative intervention to promote healthy aging.

24.2 Efficacy of CR Intervention in Aging

Aging-associated changes in physiological functions affect the nutrient requirements of individuals by directly altering their appetite and body weight. Moreover sensory changes such as loss of taste and smell also reduce caloric intake in the aged people. Similarly loss of protein and lean body mass (sarcopenia) in old persons is associated with reduction in energy requirements. Although recent studies suggest that reducing calorie consumption and maintaining below-average body weight throughout life does reduce chronic disease load and increases life span, there is scant data available to confirm the potential beneficial effect of CR, specifically in the older populations. Also, it will not be appropriate to suggest at this juncture that applying lifelong CR is the only way to achieve beneficial long term health benefits such as reducing inflammatory markers, and improving metabolic functions. Therefore, additional studies as well as policy development in this direction are urgently required to establish appropriate nutritional requirements and CR regimens for older adults before making firm recommendations for this population.

After the initial report by McCay showing that CR intervention can extend lifespan in rats, several studies have reported robust potential of CR in delaying age-related impairments and lifespan extension in humans (McDonald and Ramsey 2010; Anderson and Weindruch, 2012). The applicability and efficacy of CR as an aging intervention and lifespan extension have also been studied in detail in non-human primates. A significant improvement in morbidity and mortality was reported with 25% daily CR in adult monkeys (Colman et al. 2009). The efficacy of CR in primates has been found to be dependent on the age of CR onset as in the young onset animals, CR failed to show any survival advantage. Moderate CR onset in adult primates delayed the onset of age-associated pathologies and significant lifespan extension. Moreover, timing of onset of CR also requires consideration of its negative impact on reproductive health in young animals in addition to the advantage of adult-onset CR in lifespan extension (Mattison et al. 2012).

Several recent studies on animals and humans have reported beneficial effects of CR on different health markers, thus providing new approaches for prevention of lifestyle diseases and healthy aging. Some well-known examples are population of Okinawa Island (Willcox and Willcox 2014), Calorie Restriction in Biosphere 2 (Walford et al. 2002), and members enrolled in a clinical trial by Calorie Restriction Society International, who self-impose CR and believed that it will enhance their lifespan (Fontana et al. 2004; Holloszy and Fontana 2007; Kraus et al. 2019). CALERIE (Comprehensive Assessment of Long-term Effects of Reducing Intake of Energy) study in humans resulted in improved insulin sensitivity and cardiovascular health (Das et al. 2007; Redman et al. 2011). Initial reports from short-term CR studies of 6- or 12-months part of CALERIE-I in overweight individuals showed reduction in body weight, improved glucose regulation and cardiovascular health (Most et al. 2017, 2018). CALERIE-II conducted in lean individuals with 25% CR at 12 and 24 months showed lower resting metabolism, energy expenditure and sustained metabolic fitness (Ravussinet al. 2015; Kraus et al. 2019). Although both CALERIE studies were unable to completely match many physiological effects of CR in rodents, these results are consistent with some beneficial effects of CR earlier reported in monkeys (Edwards et al. 1998) showing promise of CR as a practical tool for healthy lifespan in humans. There is not adequate evidence available in literature regarding nutritional requirements of old people and their healthy body weight.

24.3 Cellular and Molecular Basis of Potential Beneficial Effects of CR

Over the years extensive research in a wide variety of species has uncovered several pathways for the beneficial effect of CR including lifespan extension. This section deals with cellular and molecular effectors of the proposed mechanisms of CR. The lifespan extension effect of CR has been attributed to multiple neural, systemic, tissue-specific, and cell autonomous mechanisms (Fontana and Partridge 2015; Fontana 2017). At cellular level, the lifespan extension effect of CR involves increased stress resistance (Hine et al. 2015a; b), autophagy (Singh and Cuervo 2012) and chromatin remodeling (Dang et al. 2014). CR targets molecular effectors involved in energy sensing and utilization to improve cellular energetics and metabolic homeostasis. CR also promotes anti-inflammatory intestinal microbiota, and reduces obesity and metabolic dysfunctions (Tilg and Kaser 2011).

Molecular effectors of CR-related lifespan extension include a variety of kinases, deacetylase enzymes, transcription factors and co-activators involved in cellular energetics pathways. FOXO, a member of forkhead family of transcription factors, in mammals and its invertebrate homologue DAF-16, both have been implicated in increased lifespan (Seo et al. 2015). FOXO has a very interesting role in both stress resistance and apoptosis under variety of environmental conditions and its activity therefore may influence target gene expression relevant for energy homeostasis, glucose metabolism, ROS, oxidative stress, stress resistance, autophagy and cell cycle (Webb and Brunet 2014; Wang et al. 2014). AMP-activated protein kinase (AMPK) is involved in the adaptive response to cellular energy deficit or changes in cellular energetic demand. Reduction in AMPK activation has been reported with aging, whereas, activation of AMPK pathways in multiple tissues is reported with CR (Reznick et al. 2007; Canto and Auwerx 2011). The lifespan extension effect of CR is at least in part dependent on mTOR signaling to regulate metabolism, insulin sensitivity, autophagy, immunity and stress response (Kapahi et al. 2010; Kennedy and Lamming 2016). Metabolic sensors like SIRT1 and AMPK directly regulate PGC1a (peroxisome proliferator activated receptor gamma coactivator 1alpha, a family of nuclear receptor transcription factors) activity through deacetylation and phosphorylation, respectively, and improve metabolic fitness. The mechanistic Target of Rapamycin (mTOR) is a protein kinase implicated in nutrient and energy sensing pathways and mTOR is negatively regulated by AMPK. Overexpression of sirtuins including SIRT1 involved in histone deacetylation improves cellular energetics and metabolic homeostasis in addition to reducing NF-kB signaling (Guarente 2013). All of these studies suggest that regulation of nutrient and fuel sensitive pathways by CR is a shared mechanism to increase metabolic health and lifespan extension. Future studies using genomic, proteomic, and metabolomic approaches may help to understand the tissue-specific effects of CR in both animals and humans, and to elucidate the complex underlying biological processes involved in the anti-aging and life-prolonging effects of CR.

24.4 IF-DR: Novel Strategies to Improve Metabolic Health and Longevity

Pioneer research to explore the potential of CR-stimulated longevity was spearheaded by Mark Mattson (Mattson 2005; Mattson and Wan 2005). However, recent advances in this area of research have provided much deeper insights into the impact of novel dietary restriction approaches on longevity and healthspan in animal models as well as in humans (Harvie et al. 2011, 2013; Mattson et al. 2017; Anton et al. 2018). Several alternative approaches to traditional CR have acclaimed prominence as novel dietary regimens which may be more efficient to stimulate positive adaptive processes without energy restriction and weight loss (Dorling et al. 2020). IF is the most acclaimed of these novel approaches that requires either adjustment of timings for nutrient intake or the frequency of eating to enforce periodic bouts of fasting i.e. 100% energy restriction, generally recommended for $\geq$12 h (Anton et al. 2018; Patterson and Sears 2017). IF based approaches are hypothesized to enhance physiological functions and slow down disease progression attributed to prolonged gaps of daily energy restriction (Anton et al. 2018). The potential benefits of IF observed in animal studies have challenged the dogmatic viewpoint that CR is a prerequisite of longevity-promoting diets, and have encouraged scientists in the aging field to test the efficacy of these newer dietary strategies in humans (Anton et al. 2018; Fontana and Partridge 2015). Some emerging IF strategies to improve health such as alternate day fasting (ADF), alternate-day modified fasting (ADMF), 5:2IF, time restricted fasting/feeding (TRF), and protein restriction (PR) in diet have been reported to improve markers of aging in both pre-clinical and clinical set up. Therefore, this area of aging research is gaining momentum to explore whether these novel strategies offer superiority compared to the traditional CR to stimulate improvements in health and longevity (Dorling et al. 2020; Hoddy et al. 2020).

24.4.1 Alternate-Day Fasting

Alternate-day fasting (ADF) is one of the widely studied IF regimens in animals that involves food withdrawal for 24 h on every other day with ad libitum access to water (Varady and Hellerstein 2007). The lifespan extension efficacy of the ADF regimen in rodents varies with species and age of onset (Goodrick et al. 1990; Arum et al. 2009). Several studies performed with rodents, including those from our lab, have shown promising effects of alternate day IF regimen on stress response, neural and synaptic plasticity and cognition (Duanet al. 2001, 2003; Lee et al. 2002a, b; Sharma and Kaur 2005, 2007, 2008; Kumar et al. 2009). Interestingly, we observed that the beneficial effects of early onset of IF in rats negatively influenced hypothalamo-hypophysial-gonadal axis and compromised their reproductive health (Kumar and Kaur 2013). On the other hand, IF regimen started either in middle-age (Singh et al. 2015, 2017) or in late-age in rats have shown health promoting effects in reversal of age-related impairments in stress, neuronal plasticity, inflammation and cognition (Kaur et al. 2008; Sharma et al. 2010; Singh et al. 2012).

Keeping in view the human's limitation to constantly maintain a certain level of CR, Stekovic et al. (2019) carried out a clinical trial of ADF in 30 healthy non-obese and 60 controls on conventional western diet, and observed striking reduction in overall calorie intake for a period of more than 6 months. Moreover, ADF regimen was more easily tolerated than chronicCR, and showed similar beneficial effects on the cardiovascular health and fat mass. Further proteome and metabolome of subjects categorized as long-term adopters of ADF showed a significant increase in circulating levels of lipids and a decrease in amino acids like methionine on fasting days. Low systemic levels of methionine and other amino acids have been reported in model organisms to extend lifespan by reducing mTOR pathway activity and corresponding upregulation in cell autophagy (deCabo and Mattson 2019). Safety and tolerability of ADF was evaluated in another study by Catenacci et al. (2016), which reported that alterations in body weight and composition, lipids, and insulin sensitivity index were comparable with moderate daily CR regimen. Similarly, another recent study observed comparable effects of ADF to CR on health parameters over eight weeks in women with obesity and reported that higher energy intake on feeding days could offset hunger pangs on calorie restriction days and assisted in compliance to ADF regimen (Hutchison et al. 2019). Apart from potential negative consequence on reproductive health related to early initiated CR, practical implications in terms of continuously practicing ADF in daily life is another limitation for human population.

24.4.2 Alternate-Day Modified Fasting

Keeping in view the difficulties of compliance due to 100% CR during ADF regimen on the fasting day, its modified approach (ADMF) has examined IF strategies that permits ≤25% consumption of habitual daily calories intake during fasting days and ad libitum feeding on alternate days (Johnson et al. 2007). Subsequent study by Wegman et al. (2015) reported that ADMF was well tolerated and decreased plasma insulin of healthy and lean subjects. Similar studies in obese but healthy individuals reported that 2–3 months of ADMF lowered their adiposity and improved CVDs as well as inflammatory markers (Bhutani et al. 2013; Varady et al. 2013; Hoddy et al. 2014), irrespective of macronutrientcomposition (Klempel et al. 2013) and meal timings (Hoddy et al. 2014) on fasting days. Moreover, the health benefits of ADMF, and traditional CR in terms of weight loss, weight maintenance, or cardioprotection were found comparable over 12 months in 100 obese participants in the age group of 18–65 years (Trepanowski et al. 2017, 2018; Gabel et al. 2019). Although no foolproof evidence is available till date to demonstrate that ADMF offers significant advantage to markers of aging as compared to traditional CR, but relatively easy compliance to ADMF regimen as compared to traditional CR certainly suggests its superiority for implementation of ADMF as an effective lifestyle intervention in the current obesogenic environment.

24.4.3 Intermittent Fasting Regimen 5:2

Another novel dietary approach which has been tested recently (Anton et al. 2018) is categorized as 5:2IF, which allows ad libitum normal diet eating for 5 days in a week and severe/complete energy restriction on 2 days per week. Fasting is recommended on either consecutive or non-consecutive days. The advantage of 5:2IF regimen is its flexibility of fasting bouts which makes it easy to adopt by individuals having inconsistent work schedules and social commitments. In a pilot study, Harvie et al. (2011) selected 107 premenopausal overweight or obese women for either six months of traditional CR or 5:2IF and restricted the overall energy intake of subjects by 25% from baseline energy requirements. The main benefit of 5:2IF over CR was greater improvements in insulin resistance and fasting insulin levels, whereas no difference was observed in markers of energy metabolism, inflammation, and quality of life between these two regimens. The data from such studies may help to assess the importance of fasting independent of energy balance between different regimens. Similar findings have been reported by some other recent studies assessing 5:2IF in relation to traditional CR with no between-group changes in glycemic control, weight loss, quality of life and attrition (Carter et al. 2016; Conley et al. 2018; Headland et al. 2018). Taken together, the current evidence does not clearly explain the superiority of 5:2IF over traditional CR in improving markers of aging and longevity and demands future long-term studies in different population groups.

24.4.4 Time Restricted Feeding/Fasting as an Emerging IF Strategy

Eating behaviors are mostly evaluated based on nutritional quality and quantity of food, but little attention is paid to the temporal patterns of eating and their role in the etiology of diseases. Food overeating behaviors as well as excessive consumption of processed foods with high salt, sugars and fats are the major factors in the development of chronic lifestyle-associated pathologies (Zollner 1990; Zarrinpar et al. 2016; Mozaffarian 2016; Micha et al. 2017). Period between start of first meal to the end of the last meal of a day is considered as the daily feeding time window. Recently, Kant (2018) collected the data of daily feeding time from an American cohort of 15,000 adults and reported that for most of the individuals the estimated feeding time was 12 h, which even reached 15 h for more than half of them. Similarly, another study from India found that erratic eating pattern and prolonging daily feeding time may be a risk factor in the development of metabolic disorders (Gupta et al. 2017). These recent studies suggest that the onset of non-communicable diseases may be prevented/slowed down by time-restricted feeding (TRF), a regimen of IF in which everyday's nutrient intake is restricted to few hours (usually to 12 h during the day), without any consideration given to alter nutrient quality or calories intake. TRF regimen suggests that the daily food consumption be limited within a period of 4–12 h, which introduces a fasting window of 12–20 h per day (Chaix et al. 2014). The major difference between IF and TRF regimens is that although caloric restriction is not required during feeding time in TRF, but a daily eating window must be consistently maintained (Moon et al. 2020). Studies on experimental animals have reported that TRF dietary regimen attenuates the onset/progression of metabolic diseases against pre-existing obesity, T2D, hyperinsulinemia, hepatic steatosis, inflammation (Hatori et al. 2012; Rothschild et al. 2014).

The concept of TRF was developed keeping in view its relevance to circadian rhythms, which are daily 24 h rhythms of body in physiology, metabolism and behavior sustained under constant light or dark conditions (Xie et al. 2019). Several recent pilot studies on humans have reported potential beneficial effects of TRF regimen on metabolic health parameters (Jamshed et al. 2019; Sutton et al. 2018; Tinsley et al. 2017). Jamshed et al (2019) studied the effects of early TRF (skipping dinner) on eleven overweight adults and observed that only 4 days of early TRF altered the expression of 6 circadian clock genes as well as upregulated the expression of both SIRT1 and LC3A which play important role in autophagy.

Autophagy is well reported to play a key role in protecting the body against multiple chronic diseases like diabetes, CVDs, cancers, and neurodegenerative diseases, by recycling used and damaged proteins and organelles. Sutton et al. (2018) reported that 8 h of early time restricted feeding without reducing the food intake for 5 weeks although did not cause weight loss but enhanced insulin sensitivity in prediabetic men. Similarly, a recent study carried out on subjects undergoing orthodox religious fasting reported that time restricted eating may be beneficial to provide better metabolic and glycemic profile (Karras et al. 2021). Due to variations in protocols used and nature of study samples, TRF studies have so far produced mixed data on the superiority of this regimen compared to other IF paradigms. However, Sutton et al. (2018) hypothesized that TRF implemented during earlier periods of the waking phase may elicit more noticeable benefits due to added advantage of utilizing circadian rhythms. Specifically, they postulated that the food consumed during the early waking hours may give greater insulin sensitivity and thermic effect thus providing a more suitable time window for food consumption to enhance metabolic endpoints. Their study on pre-diabetic men reported that early hours TRF regimen improved secondary outcome measures such as insulin sensitivity, hypertension, and oxidative stress in these subjects but without any weight loss (Sutton et al. 2018).

24.5 Cellular and Molecular Basis of Potential Beneficial Effects of IF-DR and Its Modified Versions

Fasting involving meal skipping on designated days of the week or certain times during a calendar year is a feature of traditional rituals among many religious groups including Hindus, Muslims, Christians, Jews, Buddhists and others. In economically rich societies, there is ample supply and access to food for most of the people; but during evolution both humans and animals had intermittent access to the food, and this is still true for wild animals which face extended bouts of fasting in nature. Organisms tend to adapt their physiology and adjust cellular energetics to achieve metabolic homeostasis while facing this fundamental challenge of uncertain and extended fasting bouts to enhance their survival under such adverse conditions. Quiescence is an example of one such adaptive survival mechanisms evolved in response to ecological constraints like prolonged fasting. It is not surprising that quiescence-related genes are also important in the control of lifespan (Baugh 2013). Food deprivation has been shown to extend lifespan in a variety of organisms including *E. coli* (Gonidakis et al. 2010), yeast (Longo et al. 2012), nematodes (Kaeberlein et al. 2006; Lee et al. 2006) and flies (Partridge et al. 2005). Mammals respond to acute fasting bouts lasting 12–14 h by lowering blood glucose by more than 20% and increasing levels of fat-derived ketone bodies, free fatty acids and gluconeogenesis. Interestingly a similar metabolic adaptation resulting in acetic acid accumulation in response to food deprivation was reported in both bacteria and yeast (Gonidakis et al. 2010; Longo et al. 2012). These metabolic adaptations involving switching to alternate fuel may have originated first in microorganism and later evolved and appeared in mammals which can respond to glucose deprivation by switching to fatty acids and ketone bodies as an alternate fuel (Cahill 2006).

Fasting increases metabolic efficiency at the mitochondrial level, increases levels of chaperones, inducesautophagy and genomic stability (Mattson et al. 2014). Several studies have reported that in alternate day feeding regimen, IF-DR improves sensory, motor, and cognitive functions including learning and memory in rodents (Singh et al. 2012; Fontan-Lozano et al. 2007) attributed to increased neurogenesis from neural stem cells and synaptic plasticity (Lee et al. 2002a; b). These results can be explained with evolutionary perspective where hunger can engage sensory and cognitive functions accompanied with neuroendocrine changes leading to increased motivation that enables foraging and food seeking behaviors. Additionally, where hungry animals are hypervigilant and active during foraging and on the contrary food consumption and satiety supports a relatively sedentary state. Fasting and aerobic exercise share many similarities in the physiological responses involved in glucose metabolism, cardiovascular and autonomic regulation (Anson et al. 2003). Brain derived neurotrophic factor (BDNF) signaling has been shown to be sensitive to physical activity and exercise. Several studies have shown increased BDNF signaling in response to IF-DR in rodents (Duan et al. 2001, 2003). Thus, BDNF signaling seems to be a shared mechanism explaining common responses to both fasting and activity involving appetite, metabolism, cardiovascular and autonomic regulation (Rothman et al. 2012).

From evolutionary perspective, individuals whose bodily functions were maintained under the fasted state were more successful in acquiring food as well as had better fertility rates (Mattson et al. 2018). IF in combination with physical activity results in depletion of liver glycogen stores on one hand and simultaneously enhances ketone bodies production from adipose-cell-derived fatty acids. Mild stressful conditions such as IF and exercise that initially have a physiological burden but bring about potentially beneficial effects are collectively called 'hormetins' (Rattan 2008). Pioneer work by Mattson's team reports that intermittent metabolic switching, i.e. repeating cycles of fasting and feeding on daily basis may optimize brain functions and provide resilience throughout the lifespan and further explain that the phenomenon of hormesis represents a key concept for such beneficial effects of IF (Mattson 2008). Santoro et al. (2020) propose that aging may be the integrated result of the adaptive responses such as flexibility of energy metabolism, inflammaging and stress response, and consider IF has hormetic effects to promote healthy aging.

24.6 Cross Talk Between Circadian Rhythms, and Time-Restricted Feeding for Healthy Aging

The role of orexigenic and anorexigenic factors controlling short term homeostatic feeding is well studied. Daily feeding patterns are controlled by circadian clocks, including the resetting of master clock in the suprachiasmatic nuclei by ambient light and other brain clocks by feeding time, via hormonal, nutrient and visceral cues (Challet 2019).The plasticity of the circadian system to accommodate change in ambient light or food availability although is an advantage in nature to adapt to different seasons, but such plasticity may also become a liability in modern societal setup where light and food are both available *ad libitum* round the clock. As a result, almost all human beings in modern society are voluntarily overriding this natural cycle of diurnal circadian rhythm by self-selecting sleep–wake timings as per their convenience and suitability to work schedule. Night-shift workers have been observed to experience such chronodisruption with hormonal imbalance, metabolic disorders, and increased incidence of cancer (Davis and Mirick 2006). Furthermore, there is bidirectional link between feeding and sleep patterns. In the modern human lifestyle, extended wakefulness allows food ingestion behavior to continue late into the night, which significantly contributes to increased caloric intake that often correlates with modern human lifestyle. Moreover, eating at a sub-optimal time of the 24 h circadian rhythm promotes excessive energy storage instead of energy expenditure and often results in overweight/obesity. Chronic disruptions of diurnal rhythms of feeding, fasting and sleep-wake cycles i.e. chrono-disruption, can lead to circadian desynchronization with deleterious health consequences (Manoogian and Panda 2017). Transgenic knockout mouse models of circadian rhythm genes like CLOCK, BMAL1 and PER 1,2 have also been shown to result in premature aging, increased age-related health impairments and reduced lifespan (Dubrovsky et al. 2010; Kondratov et al. 2006; Lee 2005). Mice subjected to long-term daytime restricted feeding showed increased amplitude of CLOCK gene expression, increased expression of catabolic factors and reduced levels of disease markers (Sherman et al. 2011). Daytime TRF does not entrain SCN, whereas, on the contrary CR is shown to entrain SCN, indicating that energy reduction can affect central oscillator (Froy 2018). Interestingly, the timing of food presentation in IF determines its effect on circadian rhythms such that when food is introduced in the light period, mice exhibited arrhythmicity in clock gene expression and in contrast, night-time feeding yielded rhythms like those generated during *ad libitum* feeding (Froy et al. 2009). Based on these findings, it may be suggested that adherence to timed feeding schedules especially restricting feeding to daytime may help to prevent circadian desynchronization and improve energy homeostasis and metabolic fitness during aging.

24.7 Conclusion

CR and different regimens of IF when integrated into the standard medical care may hold great potential for the prevention and treatment of aging-associated chronic metabolic diseases. The regimens such as ADF, 5:2IF, TRF, and protein restriction have recently emerged as potential alternate approaches to CR, which do not involve any major changes in quality and quantity of nutritional intake. These novel approaches have been shown in several pilot studies to achieve strong effects on several disease markers which constitute the underlying basis of metabolic syndrome, cardiovascular diseases, cancer, and neurodegenerative diseases. Although the mechanisms of action of these regimens is still poorly understood, their major impact appears to promote coordinated beneficial effects on the process of aging as well as aging-associated diseases, unlike conventional pharmacological therapies which target inhibition of specific molecules/enzymes. Moreover, TRF regimen which recommends restricting the daily feeding time window in alignment with circadian rhythms may prove to be more beneficial in prevention and/or slow down of aging process and promote healthspan. Taking lessons from Okinawa island people, balanced diet consumed in moderate quantities during time-restricted window may be the mantra to achieve wholesome nutrition as it does not require much conscious effort and is also easy to compliant for lifelong. These practices, which are already gaining popularity as lifestyle interventions may eventually accommodate modern healthcare in various settings. Further future interventional studies on basic and translational research are warranted on a large number of human subjects to elucidate the impact of TRF on different parameters of physical and mental health.

Compliance with Ethical Standards

Conflict of Interest All authors declare they have no conflict of interest.

References

Anderson RM, Weindruch R (2012) The caloric restriction paradigm: implications for healthy human aging. Am J Hum Biol 24:101–106

Anson RM, Guo Z, de Cabo R, Iyun T, Rios M, Hagepanos A, Ingram DK, Lane MA, Mattson MP (2003) Intermittent fasting dissociates beneficial effects of dietary restriction on glucose metabolism and neuronal resistance to injury from calorie intake. Proc Natl Acad Sci U S A 100:6216–6220

Anton SD, Moehl K, Donahoo WT, Marosi K, Lee SA, Mainous AG III, Leeuwenburgh C, Mattson MP (2018) Flipping the metabolic switch: understanding and applying the health benefits of fasting. Obesity (silver Spring) 26:254–268

Arum O, Bonkowski MS, Rocha JS, Bartke A (2009) The growth hormone receptor gene-disrupted mouse fails to respond to an intermittent fasting diet. Aging Cell 8:756–760

Barte JCM, Ter Bogt NCW, Bogers RP, Teixeira PJ, Blissmer B, Mori TA, Bemelmans WJE (2010) Maintenance of weight loss after lifestyle interventions for overweight and obesity, a systematic review. Obes Rev 11:899–906

Baugh LR (2013) To grow or not to grow: nutritional control of development during Caenorhabditis elegans L1 arrest. Genetics 194:539–555

Bhutani S, Klempel MC, Kroeger CM, Trepanowski JF, Varady KA (2013) Alternate day fasting and endurance exercise combine to reduce body weight and favorably alter plasma lipids in obese humans. Obesity 21:1370–1379

Cahill GF Jr (2006) Fuel metabolism in starvation. Annu Rev Nutr 26:1–22

Canto C, Auwerx J (2011) Calorie restriction: is AMPK a key sensor and effector? Physiology 26:214–224

Carter S, Clifton PM, Keogh JB (2016) Intermittent energy restriction in type 2 diabetes: a short discussion of medication management. World J Diabetes 7:627–630

Catenacci VA, Pan Z, Ostendorf D, Brannon S, Gozansky WS, Mattson MP, Martin B, MacLean PS, Melanson EL, Troy Donahoo W (2016) A randomized pilot study comparing zero-calorie alternate-day fasting to daily caloric restriction in adults with obesity. Obesity 24:1874–1883

Chaix A, Zarrinpar A, Miu P, Panda S (2014) Time-restricted feeding is a preventative and therapeutic intervention against diverse nutritional challenges. Cell Metab 20:991–1005

Challet E (2019) The circadian regulation of food intake. Nat Rev Endocrinol. 15:393–405

Colman RJ, Anderson RM, Johnson SC, Kastman EK, Kosmatka KJ, Beasley TM, Allison DB, Cruzen C, Simmons HA, Kemnitz JW, Weindruch R (2009) Caloric restriction delays disease onset and mortality in rhesus monkeys. Science 325:201–204

Conley M, LeFevre L, Haywood C, Proietto J (2018) Is two days of intermittent energy restriction per week a feasible weight loss approach in obese males? Randomised Pilot Study. Nutr Diet. 75:65–72

Currenti, W., Godos, J., Castellana, S., Caruso, G., Ferri, R., Caraci, F., Grosso, G., Galvano, F.: Association between time restricted feeding and cognitive status in older Italian adults. Nutrients 13:191 (2021)

Dang W, Sutphin GL, Dorsey JA, Otte GL, Cao K, Perry RM, Wanat JJ, Saviolaki D, Murakami CJ, Tsuchiyama S, Robison B (2014) Inactivation of yeast Isw2 chromatin remodeling enzyme mimics longevity effect of calorie restriction via induction of genotoxic stress response. Cell Metab 19:952–966

Das SK, Gilhooly CH, Golden JK, Pittas AG, Fuss PJ, Cheatham RA, Tyler S, Tsay M, McCrory MA, Lichtenstein AH, Dallal GE (2007) Long-term effects of 2 energy-restricted diets differing in glycemic load on dietary adherence, body composition, and metabolism in CALERIE: a 1-y randomized controlled trial. Am J Clin Nutr 85:1023–1030

Davis S, Mirick DK (2006) Circadian disruption, shift work and the risk of cancer: a summary of the evidence and studies in Seattle. Cancer Causes Control 17:539–545

de Cabo R, Mattson MP (2019) Effects of intermittent fasting on health, aging, and disease. Nengl J Med 381(26):2541–2551

Dorling JL, Martin CK, Redman LM (2020) Calorie restriction for enhanced longevity: the role of novel dietary strategies in the present obesogenic environment. Aging Res Rev 64:101038

Duan W, Guo Z, Jiang H, Ware M, Li XJ, Mattson MP (2003) Dietary restriction normalizes glucose metabolism and BDNF levels, slows disease progression, and increases survival in huntingtin mutant mice. Proc Natl Acad Sci U S A 100:2911–2916

Duan W, Lee J, Guo Z, Mattson MP (2001) Dietary restriction stimulates BDNF production in the brain and thereby protects neurons against excitotoxic injury. J MolNeurosci 16:1–12

Dubrovsky YV, Samsa WE, Kondratov RV (2010) Deficiency of circadian protein CLOCK reduces lifespan and increases age-related cataract development in mice. Aging (Albany NY) 2:936–944

Dzau VJ, Inouye SK, Rowe JW, Finkelman E, Yamada T (2019) Enabling healthful aging for all-the national academy of medicine grand challenge in healthy longevity. N Engl J Med 381:1699–1701

Edwards IJ, Rudel LL, Terry JG, Kemnitz JW, Weindruch R, Cefalu WT (1998) Caloric restriction in rhesus monkeys reduces low density lipoprotein interaction with arterial proteoglycans. J Gerontol A Biol Sci Med Sci 53:443–448

Fernandes L, Paúl C (2017) Editorial: aging and mental health. Front Aging Neurosci 1–3

Flatt T (2012) A new definition of aging? Front Genet 3:148

Fontana L, Partridge L (2015) Promoting health and longevity through diet: from model organisms to humans. Cell 161:106–118
Fontana L (2017) The science of nutritional modulation of aging. Aging Res Rev 39:1–2
Fontana L, Meyer TE, Klein S, Holloszy JO (2004) Long-term calorie restriction is highly effective in reducing the risk for atherosclerosis in humans. PNAS 101:6659–6663
Fontana L, Partridge L, Longo VD (2010) Extending healthy life span–from yeast to humans. Science 328:321–326
Fontán-Lozano Á, Sáez-Cassanelli JL, Inda MC, de los Santos-Arteaga M, Sierra-Domínguez SA, López-Lluch G, Delgado-García JM, Carrión ÁM (2007) Caloric restriction increases learning consolidation and facilitates synaptic plasticity through mechanisms dependent on NR2B subunits of the NMDA receptor. J Neurosci 27:10185–10195
Froy O (2018) Circadian rhythms, nutrition and implications for longevity in urban environments. Proc Nutr Soc 77:216–222
Froy O, Chapnik N, Miskin R (2009) Effect of intermittent fasting on circadian rhythms in mice depends on feeding time. Mech Aging Dev 130:154–160
Gabel K, Kroeger CM, Trepanowski JF, Hoddy KK, Cienfuegos S, Kalam F, Varady KA (2019) Differential effects of alternate-day fasting versus daily calorie restriction on insulin resistance. Obesity 27:1443–1450
Gonidakis S, Finkel SE, Longo VD (2010) Genome-wide screen identifies Escherichia coli TCA-cycle-related mutants with extended chronological lifespan dependent on acetate metabolism and the hypoxia-inducible transcription factor ArcA. Aging Cell 9:868–881
Goodrick CL, Ingram DK, Reynolds MA, Freeman JR, Cider N (1990) Effects of intermittent feeding upon body weight and lifespan in inbred mice: interaction of genotype and age. Mech Ageing Dev 55:69–87
Guarente L (2013) Calorie restriction and sirtuins revisited. Genes Dev 27:2072–2085
Gupta NJ, Kumar V, Panda S (2017) A camera-phone based study reveals erratic eating pattern and disrupted daily eating-fasting cycle among adults in India. PLoS ONE 12:0172852
Harvie M, Wright C, Pegington M, McMullan D, Mitchell E, Martin B, Cutler RG, Evans G, Whiteside S, Maudsley S, Camandola S (2013) The effect of intermittent energy and carbohydrate restriction v. daily energy restriction on weight loss and metabolic disease risk markers in overweight women. Br J Nutr 110:1534–1547
Harvie MN, Pegington M, Mattson MP, Frystyk J, Dillon B, Evans G, Cuzick J, Jebb SA, Martin B, Cutler RG, Son TG (2011) The effects of intermittent or continuous energy restriction on weight loss and metabolic disease risk markers: a randomized trial in young overweight women. Int J Obes (lond) 35:714–727
Hatori M, Vollmers C, Zarrinpar A, DiTacchio L, Bushong EA, Gill S, Leblanc M, Chaix A, Joens M, Fitzpatrick JA, Ellisman MH (2012) Time-restricted feeding without reducing caloric intake prevents metabolic diseases in mice fed a high-fat diet. Cell Metab 15:848–860
Headland ML, Clifton PM, Keogh JB (2018) Effect of intermittent energy restriction on flow mediated dilatation, a measure of endothelial function: a short report. Int J Environ Res Public Health 15:1166
Hine C, Harputlugil E, Zhang Y, Ruckenstuhl C, Lee BC, Brace L, Longchamp A, Treviño-Villarreal JH, Mejia P, Ozaki CK, Wang R (2015) Endogenous hydrogen sulfide production is essential for dietary restriction benefits. Cell 160:132–144
Hine C, Harputlugil E, Zhang Y, Ruckenstuhl C, Lee BC, Brace L, Longchamp A, Treviño-Villarreal JH, Mejia P, Ozaki CK, Wang R (2015) Endogenous hydrogen sulfide production is essential for dietary restriction benefits. Cell 160(1–2):132–144
Hoddy KK, Kroeger CM, Trepanowski JF, Barnosky A, Bhutani S, Varady KA (2014) Meal timing during alternate day fasting: impact on body weight and cardiovascular disease risk in obese adults. Obesity (Silver Spring) 22(12):2524–2531
Hoddy KK, Marlatt KL, Çetinkaya H, Ravussin E (2020) Intermittent fasting and metabolic health: from religious fast to time-restricted feeding. Obesity 28:S29–S37
Holloszy JO, Fontana L (2007) Caloric restriction in humans. Exp Gerontol 42:709–712

Hutchison AT, Liu B, Wood RE, Vincent AD, Thompson CH, O'Callaghan NJ, Wittert GA, Heilbronn LK (2019) Effects of intermittent versus continuous energy intakes on insulin sensitivity and metabolic risk in women with overweight. Obesity 27:50–58

Ingram DK, Young J, Mattison JA (2007) Calorie restriction in nonhuman primates: assessing effects on brain and behavioral aging. Neuroscience 145:1359–1364

Jamshed H, Beyl RA, Della Manna DL, Yang ES, Ravussin E, Peterson CM (2019) Early time-restricted feeding improves 24-hour glucose levels and affects markers of the circadian clock, aging, and autophagy in humans. Nutrients 11:1234

Johnson JB, Summer W, Cutler RG, Martin B, Hyun DH, Dixit VD, Pearson M, Nassar M, Tellejohan R, Maudsley S, Carlson O (2007) Alternate day calorie restriction improves clinical findings and reduces markers of oxidative stress and inflammation in overweight adults with moderate asthma. Free Radical Biol Med 42:665–674

Kaeberlein TL, Smith ED, Tsuchiya M, Welton KL, Thomas JH, Fields S, Kennedy BK, Kaeberlein M (2006) Lifespan extension in Caenorhabditis elegans by complete removal of food. Aging Cell 5:487–494

Kant AK (2018) Eating patterns of US adults: meals, snacks, and time of eating. Physiol Behav 193:270–278

Kapahi P, Chen D, Rogers AN, Katewa SD, Li PWL, Thomas EL, Kockel L (2010) With TOR, less is more: a key role for the conserved nutrient-sensing TOR pathway in aging. Cell Metab 11:453–465

Karras SN, Koufakis T, Adamidou L, Antonopoulou V, Karalazou P, Thisiadou K, Mitrofanova E, Mulrooney H, Petróczi A, Zebekakis P, Makedou K (2021) Effects of orthodox religious fasting versus combined energy and time restricted eating on body weight, lipid concentrations and glycaemic profile. Int J Food Sci Nutr 72:82–92

Kaur M, Sharma S, Kaur G (2008) Age-related impairments in neuronal plasticity markers and astrocytic GFAP and their reversal by late-onset short term dietary restriction. Biogerontology 9:441–454

Kennedy BK, Lamming DW (2016) The mechanistic target of rapamycin: the grand conducTOR of metabolism and aging. Cell Metab 23:990–1003

Klempel MC, Kroeger CM, Varady KA (2013) Alternate day fasting (ADF) with a high-fat diet produces similar weight loss and cardio-protection as ADF with a low-fat diet. Metabolism 62:137–143

Kondratov RV, Kondratova AA, Gorbacheva VY, Vykhovanets OV, Antoch MP (2006) Early aging and age-related pathologies in mice deficient in BMAL1, the core componentof the circadian clock. Genes Dev 20:1868–1873

Kraus WE, Bhapkar M, Huffman KM, Pieper CF, Das SK, Redman LM, Villareal DT, Rochon J, Roberts SB, Ravussin E, Holloszy JO (2019) 2 years of calorie restriction and cardiometabolic risk (CALERIE): exploratory outcomes of a multicentre, phase 2, randomised controlled trial. Lancet Diabetes Endocrinol 7:673–683

Kumar S, Kaur G (2013) Intermittent fasting dietary restriction regimen negatively influences reproduction in young rats: a study of hypothalamo-hypophysial-gonadal axis. PLoS ONE 8, e52416

Kumar S, Parkash J, Kataria H, Kaur G (2009) Interactive effect of excitotoxic injury and dietary restriction on neurogenesis and neurotrophic factors in adult male rat brain. Neurosci Res 65:367–374

Lee CC (2005) The circadian clock and tumor suppression by mammalian period genes. Methods Enzymol 393:852–861

Lee GD, Wilson MA, Zhu M, Wolkow CA, De Cabo R, Ingram DK, Zou S (2006) Dietary deprivation extends lifespan in Caenorhabditis elegans. Aging Cell 5:515–524

Lee J, Duan W, Mattson MP (2002) Evidence that brain-derived neurotrophic factor is required for basal neurogenesis and mediates, in part, the enhancement of neurogenesis by dietary restriction in the hippocampus of adult mice. J Neurochem 82:1367–1375

Lee J, Seroogy KB, Mattson MP (2002) Dietary restriction enhances neurotrophin expression and neurogenesis in the hippocampus of adult mice. J Neurochem 80:539–547
Longo VD, Mattson MP (2014) Fasting: molecular mechanisms and clinical applications. Cell Metab 19:181–192
Longo VD, Shadel GS, Kaeberlein M, Kennedy B (2012) Replicative and chronological aging in Saccharomyces cerevisiae. Cell Metab 16:18–31
López-Otín C, Blasco MA, Partridge L, Serrano M, Kroemer G (2013) The hallmarks of aging. Cell 153:1194–1217
Manoogian EN, Panda S (2017) Circadian rhythms, time-restricted feeding, and healthy aging. Aging Res Rev 39:59–67
Mattison JA, Roth GS, Beasley TM, Tilmont EM, Handy AM, Herbert RL, Longo DL, Allison DB, Young JE, Bryant M, Barnard D (2012) Impact of caloric restriction on health and survival in rhesus monkeys from the NIA study. Nature 489:318–321
Mattson MP, Wan R (2005) Beneficial effects of intermittent fasting and caloric restriction on the cardiovascular and cerebrovascular systems. J NutrBiochem 16:129–137
Mattson MP (2005) Energy intake, meal frequency, and health: a neurobiological perspective. Annu Rev Nutr 25:237–260
Mattson MP (2008) Dietary factors, hormesis and health. Aging Res Rev 7:43–48
Mattson MP, Allison DB, Fontana L, Harvie M, Longo VD, Malaisse WJ, Mosley M, Notterpek L, Ravussin E, Scheer FA, Seyfried TN (2014) Meal frequency and timing in health and disease. Proc Natl Acad Sci U S A 111:16647–16653
Mattson MP, Longo VD, Harvie M (2017) Impact of intermittent fasting on health and disease processes. Aging Res Rev 39:46–58
Mattson MP, Moehl K, Ghena N, Schmaedick M, Cheng A (2018) Intermittent metabolic switching, neuroplasticity and brain health. Nat Rev Neurosci 19:63–80
McDonald RB, Ramsey JJ (2010) Honoring Clive McCay and 75 years of calorie restriction research. J Nutr 140:1205–1210
Micha R, Peñalvo JL, Cudhea F, Imamura F, Rehm CD, Mozaffarian D (2017) Association between dietary factors and mortality from heart disease, stroke, and type 2 diabetes in the United States. JAMA 317:912–924
Moon S, Kang J, Kim SH, Chung HS, Kim YJ, Yu JM, Cho ST, Oh CM, Kim T (2020) Beneficial effects of time-restricted eating on metabolic diseases: a systemic review and meta-analysis. Nutrients 12:1267
Most J, Gilmore LA, Smith SR, Han H, Ravussin E, Redman LM (2018) Significant improvement in cardiometabolic health in healthy nonobese individuals during caloric restriction-induced weight loss and weight loss maintenance. Am. J. Physiol-Endocrinol Metab 314:E396–E405
Most J, Tosti V, Redman LM, Fontana L (2017) Calorie restriction in humans: an update. Aging Res Rev 39:36–45
Mozaffarian D (2016) Dietary and policy priorities for cardiovascular disease, diabetes, and obesity: a comprehensive review. Circulation 133:187–225
Partridge L, Piper MD, Mair W (2005) Dietary restriction in Drosophila. Mech Ageing Dev 126:938–950
Patterson RE, Sears DD (2017) Metabolic effects of intermittent fasting. Annu Rev Nutr 37:371–393
Rattan SI (2008) Hormesis in aging. Aging Res Rev 7:63–78
Rattan SI (2014) Aging is not a disease: implications for intervention. Aging Dis. 5:196–202
Ravussin E, Redman LM, Rochon J, Das SK, Fontana L, Kraus WE, Romashkan S, Williamson DA, Meydani SN, Villareal DT, Smith SR (2015) A 2-year randomized controlled trial of human caloric restriction: feasibility and effects on predictors of health span and longevity. J Gerontol A Biol Sci Med Sci 70:1097–1104
Redman LM, Ravussin E (2011) Caloric restriction in humans: impact on physiological, psychological, and behavioral outcomes. Antioxid Redox Signal 14:275–287
Redman LM, Huffman KM, Landerman LR, Pieper CF, Bain JR, Muehlbauer MJ, Stevens RD, Wenner BR, Kraus VB, Newgard CB, Kraus WE (2011) Effect of caloric restriction with and

without exercise on metabolic intermediates in nonobese men and women. J Clin Endocrinol Metab 96(2):E312–E321

Reznick RM, Zong H, Li J, Morino K, Moore IK, Hannah JY, Liu ZX, Dong J, Mustard KJ, Hawley SA, Befroy D (2007) Aging-associated reductions in AMP-activated protein kinase activity and mitochondrial biogenesis. Cell Metab 5:151–156

Rothman SM, Griffioen KJ, Wan R, Mattson MP (2012) Brain-derived neurotrophic factor as a regulator of systemic and brain energy metabolism and cardiovascular health. Ann N Y Acad Sci 1264:49–63

Rothschild J, Hoddy KK, Jambazian P, Varady KA (2014) Time-restricted feeding and risk of metabolic disease: a review of human and animal studies. Nutr Rev 72:308–318

Rowe GC, Arany Z (2014) Genetic models of PGC-1 and glucose metabolism and homeostasis. Rev Endocr Metab Disord 15:21–29

Santoro A, Zhao J, Wu L, Carru C, Biagi E, Franceschi C (2020) Microbiomes other than the gut: inflammaging and age-related diseases. Semin Immunopathol 42(5):589–605

Scheen AJ (2008) CB1 receptor blockade and its impact on cardiometabolic risk factors: overview of the RIO programme with rimonabant. J Neuroendocrinol 20:139–146

Seo M, Seo K, Hwang W, Koo HE, Hahm J, Yang J, Han SK, Hwang D, Kim S, Jang SK, Lee Y, Nam HG, Lee SV (2015) RNA helicase HEL-1 promotes longevity by specifically activating DAF-16/FOXO transcription factor signaling in Caenorhabditis elegans. Proc Natl Acad Sci USA 112(31):E4246–4255

Sharma S, Kaur G (2005) Neuroprotective potential of dietary restriction against kainate-induced excitotoxicity in adult male Wistar rats. Brain Res Bull 67:482–491

Sharma S, Kaur G (2007) Intermittent dietary restriction as a practical intervention in aging. Ann N Y Acad Sci 1114:419–427

Sharma S, Kaur G (2008) Dietary restriction enhances kainate-induced increase in NCAM while blocking the glial activation in adult rat brain. Neurochem Res 33:1178–1188

Sharma S, Singh R, Kaur M, Kaur G (2010) Late-onset dietary restriction compensates for age-related increase in oxidative stress and alterations of HSP 70 and synapsin 1 protein levels in male Wistar rats. Biogerontology 11:197–209

Sherman H, Frumin I, Gutman R, Chapnik N, Lorentz A, Meylan J, le Coutre J, Froy O (2011) Long-term restricted feeding alters circadian expression and reduces the level of inflammatory and disease markers. J Cell Mol Med 15:2745–2759

Singh R, Cuervom AM (2012) Lipophagy: connecting autophagy and lipid metabolism. Int J Cell Biol 2012:2820–2041

Singh R, Lakhanpal D, Kumar S, Sharma S, Kataria H, Kaur M, Kaur G (2012) Late-onset intermittent fasting dietary restriction as a potential intervention to retard age-associated brain function impairments in male rats. Age (dordr) 34:917–933

Singh R, Manchanda S, Kaur T, Kumar S, Lakhanpal D, Lakhman SS, Kaur G (2015) Middle age onset short-term intermittent fasting dietary restriction prevents brain function impairments in male Wistar rats. Biogerontology 16:775–788

Singh H, Kaur T, Manchanda S, Kaur G (2017) Intermittent fasting combined with supplementation with Ayurvedic herbs reduces anxiety in middle aged female rats by anti-inflammatory pathways. Biogerontology 18:601–614

Stekovic S, Hofer SJ, Tripolt N, Aon MA, Royer P, Pein L, Stadler JT, Pendl T, Prietl B, Url J, Schroeder S (2019) Alternate day fasting improves physiological and molecular markers of aging in healthy, non-obese humans. Cell Metab 30:462–476

Sutton EF, Beyl R, Early KS, Cefalu WT, Ravussin E, Peterson CM (2018) Early time-restricted feeding improves insulin sensitivity, blood pressure, and oxidative stress even without weight loss in men with prediabetes. Cell Metab 27:1212–1221

Swinburn B, Egger G (2002) Preventive strategies against weight gain and obesity. Obes Rev 3:289–301

Tilg H, Kaser A (2011) Gut microbiome, obesity, and metabolic dysfunction. J Clin Invest 121:2126–2132

Tinsley GM, Forsse JS, Butler NK, Paoli A, Bane AA, La Bounty PM, Morgan GB, Grandjean PW (2017) Time-restricted feeding in young men performing resistance training: a randomized controlled trial. Eur J Sport Sci 17:200–207

Trepanowski JF, Kroeger CM, Barnosky A, Klempel M, Bhutani S, Hoddy KK, Rood J, Ravussin E, Varady KA (2018) Effects of alternate-day fasting or daily calorie restriction on body composition, fat distribution, and circulating adipokines: secondary analysis of a randomized controlled trial. Clin Nutr 37:1871–1878

Trepanowski JF, Kroeger CM, Barnosky A, Klempel MC, Bhutani S, Hoddy KK, Gabel K, Freels S, Rigdon J, Rood J, Ravussin E (2017) Effect of alternate-day fasting on weight loss, weight maintenance, and cardioprotection among metabolically healthy obese adults: a randomized clinical trial. JAMA Internal Med 177:930–938

Varady KA, Hellerstein MK (2007) Alternate-day fasting and chronic disease prevention: a review of human and animal trials. Am J Clin Nutr 86:7–13

Varady KA, Bhutani S, Klempel MC, Kroeger CM, Trepanowski JF, Haus JM, Hoddy KK, Calvo Y (2013) Alternate day fasting for weight loss in normal weight and overweight subjects: a randomized controlled trial. Nutr J 12:1–8

Walford RL, Mock D, Verdery R, MacCallum T (2002) Calorie restriction in biosphere 2: alterations in physiologic, hematologic, hormonal, and biochemical parameters in humans restricted for a 2-year period. J Gerontol A Biol Sci Med Sci 57:B211–B224

Wang L, Karpac J, Jasper H (2014) Promoting longevity by maintaining metabolic and proliferative homeostasis. J Exp Biol 217:109–118

Wegman MP, Guo MH, Bennion DM, Shankar MN, Chrzanowski SM, Goldberg LA, Xu J, Williams TA, Lu X, Hsu SI, Anton SD (2015) Practicality of intermittent fasting in humans and its effect on oxidative stress and genes related to aging and metabolism. Rejuvenation Res 18:162–172

Weindruch R (1996) The retardation of aging by caloric restriction: studies in rodents and primates. Toxicol Pathol 24:742–745

Webb A, E., Brunet, A., (2014) FOXO transcription factors: key regulators of cellular quality control. Trends Biochem Sci 39:159–169

Willcox B, J., Willcox, D, C., (2014) Caloric restriction, CR mimetics, and healthy aging in Okinawa: controversies and clinical implications. Curr Opin Clin Nutr Metab Care 17:51–58

Xie Y, Tang Q, Chen G, Xie M, Yu S, Zhao J, Chen L (2019) New insights into the circadian rhythm and its related diseases. Front Physiol 10:682–701

Zarrinpar A, Chaix A, Panda S (2016) Daily eating patterns and their impact on health and disease. Trends Endocrinol Metab 27:69–83

Zöllner N (1990) The relevance of diet for civilization diseases, especially atherosclerosis. Wien Med Wochenschr Supplement 106, suppl-11

Zupo R, Castellana F, Bortone I, Griseta C, Sardone R, Lampignano L, Lozupone M, Solfrizzi V, Castellana M, Giannelli G, De Pergola G (2020) Nutritional domains in frailty tools: working towards an operational definition of nutritional frailty. Aging Res Rev 101148

Chapter 25
Calorie Restriction Mimetics and Adult Stem Cells

Bauyrzhan Umbayev, Yuliya Safarova, Aislu Yermekova, and Timur Saliev

Abstract The regulation of nutrition and metabolism plays a pivotal role in regulating the aging process. The nutrition is a critical external factor influencing the development of aging and associated disorders. One of the well-known dietary interventions to slow aging and reduce mortality is calorie restriction. The results of the range of studies indicate that calorie restriction facilitates the increase in life expectancy and a decrease in the aging processes. The effect of calorie restriction on the aging process has been associated with a wide range of the reactions in different types of cells, particularly the stem cells. This Chapter discusses the role of the stem cells in aging processes and associated disorders in the context of calorie restriction strategies. It encompasses the analysis of the results of preclinical and clinical studies on the relationship between the calorie restriction and adult stem cell function in order to understand the effects of calorie restriction on the health and longevity. The Chapter highlights the role of stem cells in neurogenesis, alterations in stem cell function under the influence of calorie restriction. In addition, the potential of the calorie restriction mimetics as aging modulators has been discussed too.

Keywords Adult stem cells · Aging · Calorie restriction · Calorie restriction mimetics · Autophagy · MTOR

25.1 Introduction

The results of numerous studies have not yet led to the universal concept of aging. During the last decades, several distinct, interrelated and interdependent, biological mechanisms underlying the process of aging and age-related diseases have been

---

B. Umbayev · Y. Safarova · A. Yermekova
Center for Life Sciences, National Laboratory Astana, Nazarbayev University, Nur-Sultan, Kazakhstan

Y. Safarova
School of Engineering and Digital Sciences, Nazarbayev University, Nur-Sultan, Kazakhstan

T. Saliev (✉)
S.D. Asfendiyarov Kazakh National Medical University, Almaty, Kazakhstan

S. I. S. Rattan and G. Kaur (eds.), *Nutrition, Food and Diet in Ageing and Longevity*, Healthy Ageing and Longevity 14, https://doi.org/10.1007/978-3-030-83017-5_25

identified (Kennedy et al. 2014). These mechanisms encompass macromolecular damage, metabolism, stem cells and regeneration, proteostasis, adaptation to stress, inflammation and epigenetics (Kennedy et al. 2014). In the context of the current understanding of aging, it has been assumed that human body has developed adaptive strategies to recognize and neutralize the combination of the stress factors that affect the biological mechanisms described above and adaptive consequences of protective responses formed during the evolution process (Epel and Lithgow 2014; Santoro et al. 2020). In case the human body fails to adapt to the stressor, it leads to the disruption of the homeostasis and eventually may cause the development of age-related changes (Santoro et al. 2020).

One of the most important environmental factors associated with the vital activity of the body is the availability of nutrients (Speakman 2013). The lack of the nutrients is a possibly predictable event and in the course of evolution adaptive mechanisms for starvation have been formed, but excess of nutrients (especially some in particular) and associated obesity is non-adaptive, and facilitates metabolic dysfunction (Santoro et al. 2020).

The importance of the regulation of nutrition and metabolism is demonstrated in recent advances in proteomics and metabolomics indicate that metabolic signalling pathways play a pivotal role in regulating the aging process and nutrition can be a critical external factor influencing the development of aging (Schüler et al. 2020). Moreover, according to the hypothesis "metabolic age score" metabolic changes accumulate over time, and an estimate of metabolic age score is an informative measurement of the biological age (López-Otín et al. 2016; Hertel et al. 2016). In compliance with the numerous studies congenital defects in the metabolic signalling pathways accelerate aging, and the maintenance of longevity is closely related to the metabolic processes (López-Otín et al. 2016; Barzilai et al. 2012, Catic 2018).

The regulation of aging by the metabolic processes can be considered within the concept of hormesis, where a certain level of the metabolic disorders weakens the effects of aging due to the activation and upregulation of the protective cellular pathways. However, over-activation of the metabolism leads to an acceleration of the aging processes (Gems and Partridge 2008; Santoro et al. 2020). This concept agrees with the numerous data that indicate that calorie restriction (CR) from a normal diet, and not just a lack of overeating, facilitates the increase in life expectancy and a decrease in the aging processes of animals and humans (Heilbronn and Ravussin 2003; Flanagan et al. 2020; Franceschi et al. 2018). Along with the organismal level, CR promotes changes in the aging phenotypes at the tissue, cellular and molecular levels. (Ma et al. 2020).

The selection of the optimal CR program is a very urgent task in gerontology. Considering CR in the prism of hormesis, the use of the moderate (usually intermittent) stress is necessary to obtain a beneficial effect as it was noted in the pioneering works of Mattson (Mattson 2008). In addition, it was assumed that adaptive responses arising under the influence of the food stress in the aging organisms lose the intensity of the response and so-called "metabolic flexibility", i.e. the ability to balance between the energy consumption and energy storage (Storlien et al. 2004). Perhaps this is one of the reasons for the change in the body's sensitivity to CR with age. In addition, other negative aspects of CR should be noted: acceleration of the sarcopenia in old organisms, possible malnutrition, difficulty in maintaining the diet in the elderly, etc. (Xie et al. 2020; Locher et al. 2016, Madeo et al. 2019). In this regard, a possible alternative is a chemical compound capable of imitating CR – calorie restriction mimetics (Shintani et al. 2018, Madeo et al. 2019).

The difficulty in assessing the effect of CR on the aging process is associated with a wide range of the reactions in different types of cells, tissues and organs caused by aging and the subsequent various adaptive and non-adaptive consequences of the calorie restriction (Ma et al. 2020). To improve the understanding of this processes we should focus on the impact of CR on the adult stem cells, which are the key in maintaining tissue homeostasis. Adult stem cells (ASC) (specific tissue stem cells, somatic stem cells) have been found in many human and animal tissues. ASCs are undifferentiated cells (relative to the functional tissue) capable of proliferation, self-renewal and differentiation into tissue-specific precursors to maintain the tissue homeostasis and tissue regeneration upon the illness or injury (Loeffler and Roeder 2002). The most studied adult stem cells are hematopoietic stem cells and mesenchymal stem cells (Clevers 2015, Gonzalez and Bernad 2012). Like normal somatic cells, somatic stem cells are exposed to various stress factors throughout the life, which leads to aging, (Jones and Rando 2011, Sharpless and DePinho 2007, Alt et al. 2012, Liu and Rando 2011, Rando 2006) and this can be one of the reasons of the overall organism aging (Fukada et al. 2014; Mimeault and Batra 2009). Taking into the account the key role of the stem cells in maintaining a dynamic balance in tissue and organ homeostasis, it is easy to assume that they play a central role in aging and the pathophysiology of the various age-associated diseases such as cardiovascular and cerebrovascular diseases, malignant tumors, diabetes, autoimmune diseases, recurrent infections, impaired wound healing and other diseases (Schultz and Sinclair 2016, Goodell and Rando 2015, Smith and Daniel 2012, Sharpless and DePinho 2007, Boyette and Tuan 2014).

One of the main mechanisms in aging of the adult stem cells is a disruption of proteostasis and autophagy (Chang 2020). At the same time, CR contributes to the restoration of these processes inside the cell (Chang 2020, Chung and Chung 2019) and therefore the CR mimetics should have a similar effect on the aging of the adult stem cells.

In recent years, much attention has been paid to the development of anti-aging strategies aimed at preventing and/or slowing down the aging process based on the use of the body's internal abilities for self-healing (Santoro et al. 2020), it is important to consider the potential of the CR mimetics in maintaining normal tissue homeostasis through the activation of the adult stem cells.

25.2 Calorie Restriction and Adult Stem Cells

The best known dietary intervention to slow aging and reduce mortality is calorie restriction (CR). Food intake is reduced below the energy requirements without malnutrition or deprivation of the essential nutrients. (Heilbronn and Ravussin 2003; Flanagan et al. 2020). In various preclinical models CR strikingly increase life expectancy (Heilbronn and Ravussin 2003; Flanagan et al. 2020). Data from controlled clinical trials in humans associates CR with prolonged life spanonaver age by 1–5 years (Flanagan et al. 2020). The specific mechanisms underlying CR are not clear, but existing evidence suggests that CR stimulates cellular defense mechanisms such as autophagy, mitochondrial efficacy, decreases ROS production, and decreases inflammatory cytokines (Flanagan et al. 2020). According to the preclinical studies CR is strongly dependent on the genotype and sex, and the growth hormone (GH)/insulin-like growth factor-1 (IGF-1) axis (Komatsu et al. 2019). Given the importance of adult stem cells in aging, many researchers have studied the relationship between the calorie restriction and adult stem cell function to understand the effects of CR on the health and longevity. In a number of studies caloric restriction modulates the functionality of adult stem cells and may have positive effects on the biology of stem cells in various tissues (Mazzoccoli et al. 2014, Maharajan et al. 2020).

The most sensitive cells to caloric restriction are hematopoietic stem cells (Wilkinson and Yamazaki 2018). For example, restricting diet to 75% food intake for about 2 years in BALB/cByJ mice (BALB)inhibits haematopoiesis and prevents HSC senescence. Interestingly, HSC function in 25 month old CR mice was better than in 3 month old mice without CR (Chen et al. 2003). In addition, compared to the long-term exposure, 5 months calorie restriction meals improved function of HSC to a lesser extent. (Chen et al. 2003). In turn, Tangetal found that a long-term diet of 70% of food intake (30% CR) for 6 and 12 months positively affects the phenotypes of aging HSC (increased number of HSCs and bias towards myeloid HSC during aging.) and the resting state of HSC in C57BL/6 J mice, but at the same time negatively affects B-lymphopoiesis in mice, disrupting the differentiation of HSC down the lymphoid lineage. The negative effect of 30% CR on B-lymphopoaesis was mediated by a decrease in the level of IL-6, IL-7, while the effect of CR on the resting state of cell was associated with a decrease in the concentration of IGF1 (Tang et al. 2016).

On the other hand, a lifelong caloric restriction facilitated stabilization of the bone marrow cellularity, but the number of HSCs increased compared to the animals with aging without calorie restriction. Also, the lifelong calorie restriction did not improve the reduced HSC functionality (Lazare et al. 2017). The authors suggest that the inconsistency of the results is associated with different experimental design (different duration of the diet - fasting kinetics), and also point to the importance of the composition of the standard diet, especially the content of valine (Lazare et al. 2017).

Intestinal stem cells are another attractive object of the research to assess the effect of CR on the adult stem cells (Bruens et al. 2020; Igarashi and Guarente 2016; Yilmaz et al. 2012). Early studies showed that a 60–70% CR-diet increased the number of stem cells in the intestinal crypts (Igarashi and Guarente 2016; Yilmaz et al. 2012) and reduced the incidence of the intestinal polyps by almost 60% (Mai et al. 2003). In addition, Yilmaz et al. revealed that Paneth cells in the intestinal stem cell nichemediate the effects of CRonISC. An important finding in the study of the effect of CR on ISC was made recently, where at 40% CR for 8 weeks in mice, the length and width of the small intestine did not change, but there was a reversible increase in crypt diameter, which was associated with an increase in the number of Lgr5 + stem cells and lysozyme + Paneth cells (niche expansion) (Bruens et al. 2020). Another significant results revealed that the increase in the number of stem cells associated with the CR diet leads to the displacement of the stem cells carrying pathological mutations (oncogenic APC mutations) from the niche as a result of competition of the stem cells on the basis of fitness and also leads to a lower retention of mutations in the intestine (Bruens et al. 2020).

In addition, CR affects not only Lgr5 + stem cells, but also reserve ISCs. CR expands the pool of the reserve ISCs and increases the regenerative capacity of the intestinal epithelium (Yousefi et al. 2018).

The mechanism of the intestinal stem cells regulation is based on the kinase-regulated signalling pathway, mammalian target of the rapamycin complex 1 (mTORC1) (Igarashi and Guarente, 2016; Yilmaz et al. 2012; Yousefi et al. 2018). This signalling pathway shows the opposite responses in the ISC compared to the niche cells (Maharajan et al. 2020). Thus, CR favours a decrease in mTOR signalling in Paneth cells and induce the bone stromal antigen 1 (Bst-1), which converts NAD1 into secreted cyclic ribose ADP (cADPR) and activates calcium signalling and this, in turn, stimulates the proliferation of neighbouring stem cells (Yilmaz et al. 2012). Suppression of mTORC1 by exposure to CR was also noted in reserve ISCs (Yousefi et al. 2018). In contrast, a diet associated with caloric restriction activates the mTORC1-p70 ribosomal S6 kinase (S6K1) axis and increases the number of Lgr5 + stem cells through the NAD-dependent protein deacetylase SIRT1(Igarashi and Guarente, 2016).

Interesting data was obtained in a recent study, when in a short-term (9 days) 60% CR diet the number of the stem cells increased, but the proliferation and size of the organelles obtained from the crypts diminished. At the same time, a similar diet in the germ-free (GF) mice did not cause changes in the size of organelles, which designate the role of the microbiome in the effect of CR on the intestinal stem cells (Glenny et al. 2020).

Adipose tissue is very sensitive to a calorie restriction, mitigating age-related adipocyte size increase and stimulating the production of the functional beige fat in the subcutaneous and visceral adipose tissue, and prevent aging of the white adipose tissue (Sheng et al. 2020, Fabbiano et al. 2016). However, the effect of CR on the adipose tissue derived mesenchymal stem cells is poorly understood. Keeping C57BL/6 mice at the age of 4 months or 21–29 months on a 60% CR diet for 9 months ensued a decrease in the age-related increase in the number of ASCs, but it also reduced cell clonality during aging and CR (Schmuck et al. 2011).

Thermoregulation studies in CR animals showed the importance of the ambient temperature and coat and skin for regulating the energy balance (Ravussin et al. 2012). 60% CR diet for 6 months in Swiss mice results in a decrease in subcutaneous fat reserves and thickening of the epidermis and an increase in the pool of the hair follicle stem cells. The authors of this work suggest that the thermoregulatory adaptive evolutionary mechanism compensates for the heat loss (Forni et al. 2017).

One of the most dangerous conditions associated with aging is dementia and other cognitive impairments that dramatically reduce the productivity and quality of life in elderly (Desai et al. 2010). Therefore, the regulation of the brain aging processes with diet, and in particular calorie restriction, draws in much consideration of the researchers. (Dias et al. 2020). Mild to moderate CR ($\leq$ 40%) CR launched at an early stage improved neurovascular functions of the brain, cognitive function and memory in aging animals compared to the controls (Parikh et al. 2016; Kaptan et al. 2015). Considering the role of stem cells in neurogenesis, alterations in stem cell function under the influence of CR have called the increased attention in studying the effects of CR on cognitive function and brain neuroplasticity. (Apple et al. 2019; Dias et al. 2020). Several preclinical studies show that CR stimulates cell proliferation in the dentate gyrus in the hippocampus and subventricular zone (SVZ), where neurogenesis occur throughout life and slows down with aging (Park et al. 2013; Kaptan et al. 2015; Apple et al. 2019). On the other hand, long-term CR at the age of 3 to 11 months had no effect on neurogenesis in the granular cell layer (GCL), but stimulated the survival of the newly formed glial cells in the hilus of the dentate gyrus (Bondolfi et al. 2004).

A recent study by Apple et al. (2019) found out that calorie restriction has a differential effect on the proliferation and production of neurons by the neuronal stem cells in the SVZ, depending on the age at which the diet was initiated (Apple et al. 2019). If 60% of CR was launched at an early age (6 to 7 months) within 16 weeks, it supported the increase of proliferation of stem and progenitor cells in the SVZ. Moreover, if the CR diet was started at an old age (from 12 to 18 months), these animals did not have an increase in proliferation, but the neurogenesis was equivalent to that in the young mice. The authors suggest that perhaps caloric restriction may improve neonatal survival of neurons. In addition, in this study, CR was shown to contribute to the ability of the neural stem cells in aged mice to differentiate into neurons *in vivo*. A possible effect of CR on the preservation of the neural stem cell function in the old brain has been associated with the amelioration of the progressive inflammation (decreased number of the activated microglia and cytokine expression) (Apple et al. 2019).

One of the mechanisms involved in increasing the adaptiveness and resistance of the neurons in CR may be the induction of the neurotrophic factors, such as brain neurotrophic factor (BDNF) (Fontán-Lozano et al. 2008; Kishi et al. 2014). It should be noted that there is conflicting evidence regarding BDNF expression in CR, if short-term exposure to 15% CR for 4 weeks in young females resulted in increased levels of BDNF in the hippocampus and prefrontal cortex (PFC) in adulthood (Kaptan et al. 2015), than long-term exposure to 30% CR in male mice starting at 3 months of age for 9 months or over 17 months reduced BDNF levels compared to the animals fed ad libitum (Yang et al. 2014). Most studies suggest that one of the key mechanisms of regulation induced by CR might be the inhibition of the decrease in age-related autophagy, in particular through the mTOR signalling pathway (Maiese et al. 2013; Raman et al. 2013). Several studies revealed that aging-related neurodegeneration is accompanied by the increased activation of mTOR signalling (Maiese et al. 2013; Troca-Marín et al. 2011). On the other hand according to Yang et al. there is a parallel decrease in mTOR signalling and activity with age in the hippocampus of mice (Yang et al. 2014). Therefore, it is assumed that the mTOR signalling cascade play different roles in the hippocampus of young and old mice (Yang et al. 2014). However, the main mechanisms mediating the relationship between CR and neurogenesis are poorly understood. Interesting data were obtained in a recent study where short-term CR (overnight) activates new neurons of the olfactory bulb born in adult mice in a ghrelin-dependent manner (Ratcliff et al. 2019).

One of the main reasons of frailty in older people and their loss of independence is age-related degenerative muscle loss, deterioration of the muscle fibres, and a decrease in the strength of muscle contraction, which is called sarcopenia (Xie et al. 2020). Several animal studies have been known to show that CR leads to a reduction or delay in the age-related defects that occur in the skeletal muscle (Xie et al. 2020; Boldrin et al. 2017). Muscle stem cells are also sensitive to the effects of CR. Thus, an early study conducted in young C57BL/6 mice (2 months), and on old mice (18 months), which were exposed to CR (1 week at 20% restriction and 11 weeks at 40%) showed that CR promoted an increase in frequency of skeletal muscle stem cells in young and old mice. These cells had an increased number of the mitochondria and muscle after CR showed increased regeneration rate (Cerletti et al. 2012). Later studies also showed that 25 weeks exposure to 30% CR increased the skeletal muscle stem cells in aged mice (60 weeks old) (Sato et al. 2017). Also in this study, mice on CR diet showed the reorganization of the transcriptome in muscle stem cells, manifested in the regular transcription of genes associated with the self-renewal and decreased activity of the genes involved in inflammation or repair of the mitochondrial DNA (mtDNA). This might be an indirect evidence of the restoration of the stem cells and the slowing down of their aging.

More recently, a 40% CR short-term (3 months) or long-term (6 months) diet in 12-week-old adult male Sprague–Dawley rats sustained the clonal myogenic activity in muscle stem cells (Abreu et al. 2020). At the same time, interestingly, calorie restriction did not affect mitochondrial (oxygen consumption rates, including basal and physiological respiration, ATP-associated and dependent on proton leakage respiration, maximum and reserve respiration) or glycolytic function. (Abreu et al. 2020). Overall, these results showed that caloric restriction did not result in overt changes in the muscle stem cell metabolism.

In addition, the effects of CR on the muscle stem cells depend on many factors, such as lineage, sex and age of animals according to the findings of Boldrin et al. in 2017. (Boldrin et al. 2017). Researchers showed that short-term (3 months) and longer (9 and 19 months) CR have different effects on the skeletal muscle in male and female C57Bl/6 and DBA/2 mice (shorter-lived strain)[36]. The exposure to CR was shown to increase the number of the satellite cells by 6 months, but their number decreases by 12 months. In addition, CR increased fibrosis by increasing the collagen VI content in mouse muscle and decreased muscle regenerative response. Also CR makes satellite cells less proliferative *in vitro* (Boldrin et al. 2017).

Reducing oxidative stress, increasing mitochondrial function, suppressing inflammation and apoptosis, and activating autophagy are thought to play an important role in CR-mediated regulation of the sarcopenia, but the exact mechanism by which CR affects stem cells in the skeletal muscle is still a matter of debate.

It should be noted that restricting calories can have both positive and negative consequences. (Xie et al. 2020). Age has been shown to play an important role in the response to calorie restriction. For example, it was shown that short-term CR (4 weeks) led to the accelerated muscle degradation in mice in old SD rats (25 months) (Park et al. 2017). In this aspect, it is the effect of calorie restriction on sarcopenia in the elderly that is one of the main obstacles in calorie restriction therapy (Xie et al. 2020). Thus, based on the above results of the various studies, it could be concluded that calorie restriction has an impact on the functioning of the adult stem cells. At the same time, the degree of the impact depends on the duration of the calorie restriction diet, and in the safest and most effective diet option is a short-term diet. In addition, the age at which the diet is launched is an important parameter. Thus, the best effects were achieved with a diet started at a young age, whereas a diet in old age may not have positive results or even lead to the negative consequences. It should also be noted that the heterogeneity of the adult stem cells is responsible for the variability of the signalling pathways underlying the mechanisms of CR effect.

In addition, there is evidence that often the calorie restricted diet used in various models is not complete and often leads to malnutrition in both mice and rats (Cerqueira and Kowaltowski 2010). Another limitation is the loss of the muscle mass and a decrease in the body mass index in CR in the elderly, which carries potential risks associated with an increased risk of disability and mortality. (Xie et al. 2020). Taken together, these potential risks of CR are of concern that must be considered before it could be widely used. The establishment of standardized preclinical complete diet models is needed to study the mechanisms of CR in more detail.

Considering the risks of CR for the elderly, it is necessary to develop alternative approaches, one of which may be the use of strategies based on simulating the effects of CR, for example, the use of calorie restriction mimetics (Ingram et al. 2006).

25.3 Calorie Restriction Mimetics as Aging Modulators

However, despite the accumulated data on the effect of calorie restriction on slowing aging, the body responses in older adults differ from those in younger people. Thus the health benefits of calorie restriction in elderly remain uncertain (Locher et al. 2016). Changing the eating behaviour of older adults and, moreover, maintaining long-term calorie restriction is challenging (Madeo et al. 2019). Therefore, the idea to develop techniques or compounds that can reproduce the effect of calorie restriction without restricting the food intake seems very attractive (appealing). In 1998, Lane et al. found that feeding rats with 2-deoxy-d-glucose (2DG) for 24 weeks could mimic the metabolic effects of long-term CR intervention without significant toxicity or sustained alteration in food intake (Lane et al. 1998). Based on the obtained results, they proposed the concept of calorie restriction mimetics (CRMs), i.e. compounds that demonstrate the systemic effects of CR (Madeo et al. 2019).

Currently, in the broadest terms, CRMs are viewed as any intervention that demonstrates the systemic effects of CR and increases life expectancy and prevents aging. These include anorectic or nutrient absorption inhibiting agents such as caloric restriction mimetics, calorie-reducing drugs such as sodium glucose co-transporter 2 inhibitors (SGLT2i), and even bariatric surgery (Ingram et al. 2006).

An important debatable issue in this concept is the precise definition of the concept of CR mimetics as pharmacological agents (Shintani et al. 2018, Madeo et al. 2019). Over the past two decades, there has been no consensus on the biochemical and functional determination of CR mimetics (Ingram and Roth 2015). Pioneers of the mimetic concept, Ingram et al. suggested descriptors specific to CRM: (1) mimics the metabolic, hormonal, and physiological effects of the CR; (2) activates stress response pathways observed in CR and enhances the stress protection; (3) produces CR-like effects on longevity, reduces age-related disease, and maintains more youthful function; and (4) does not significantly reduce food intake, at least over the short-term (Ingram and Roth, 2015; Ingram et al. 2006, 2004). Based on these descriptors, the concept of mimetics is a broad view of the CR mimetics.

In a narrower sense, the basic properties of CR mimetics are thought to be associated with the regulation of autophagy and glucose metabolism as key mechanisms of calorie restriction within the cell (Shintani et al. 2018, Madeo et al. 2019). According to Madeo et al. CRms are compounds that activate autophagy by promoting the deacetylation of the cellular proteins, by (1) depleting acetyl coenzyme A (AcCoA), (2) inhibiting acetyltransferases, and/or (3) stimulating deacetylases (Madeo et al. 2019).

The question of the ability of CRMs to increase life expectancy is unclear. It is assumed that the overall cumulative effect of the CRM exposure is an increase in life expectancy and a decrease in age-related disorders (Ingram et al. 2006). However, the effectiveness of CRMs in this aspect is unclear and, for example, so far, only rapamycin has shown a steady effect on the increase in life expectancy as in rodents regardless of their sex (Miller et al. 2011; Harrison et al. 2009). But at the same time, in another study, rapamycin has a limited effect on aging in mammals and increases lifespan, possibly activating mechanisms of suppression of carcinogenesis (Neff et al. 2013). The lack of CRMs effectiveness to increase lifespan may be due to the fact, that the positive effects of the CR are mediated through the regulation of many intracellular signaling pathways, whereas CRMs are apparently involved in the activation of only some of the pathways and there is no single CR mimetic capable of mimicking CR alone. Therefore, some studies suggest the use of a combination of several CR mimetics (Ingram and Roth 2015). Thus, most likely CRM is a group of molecularly unrelated compounds capable of partially causing effects similar to CR on the cells, tissues, and organs.

However, despite the absence of the mimetics that completely mimick the effects of the calorie restriction, the currently known CRMs are able to positively influence the aging processes of the cardiovascular system, and also reduce acute ischemia in preclinical models (Sciarretta et al. 2020). Slowing down the aging of the neuromuscular junctions and muscle fibers has also been observed with the exposure to CRMs (Stockinger et al. 2018). What's more, CRMs have beneficial effects on the brain aging and prevent Alzheimer's disease (Van Cauwenberghe et al. 2016; Chiba et al. 2010). In addition, numerous studies indicate a protective role for CRMs in diabetes and obesity (Chiba et al. 2010).

25.4 Calorie Restriction Mimetics and Adult Stem Cell Aging

The previous section of the chapter described the relationship between CR and adult stem cell function, and reported studies showing that CR has a generally positive effect on the adult stem cells and inhibits aging. Accordingly, it is logical to assume that when exposed to CRMs, one can expect their influence on the functioning of the stem cells through the activation of the molecular pathways involved in the response to CR. Therefore, in this section of the chapter, we summarize the impact of CRMs on the adult stem cells in aging.

The intestinal tract is the first organ in contact with food, and at the same time is very sensitive to calorie restriction (Peña-Villalobos et al. 2019), therefore, first of all, it is necessary to pay attention to the reactions of the intestinal stem cells to the action of CRMs. It has been shown that the antidiabetic drug metformin, which is classified as CRMs (Martel et al. 2021), improve aging phenotypes (hyperproliferation, centrosome amplification, and accumulation of DNA damage) in the Drosophila intestinal stem cells through the down-regulation of the AKT / TOR signaling pathway (Na et al. 2015, 2018). It was recently found that alpha-lipoic acid (ALA; 1,2-dithiolane-3-pentanoic acid), also referred to as CRM, also had a positive effect on the aging phenotypes of the Drosophila intestinal stem cells, resulting in the suppression of the age-related hyperproliferation of the intestinal stem cells (Du et al. 2020). Interestingly, the effect of ALA on stem cells was achieved not due to the antioxidant capacity, but due to the activation of the expression of specific genes associated with the autophagy and endocytosis in old cells (Du et al. 2020).

Nicotinamide riboside (NR), a precursor of nicotinamide adenine dinucleotide (NAD^+), is a potent anti-inflammatory agent and an aging modulator (Mehmel et al. 2020) with CRMs properties (Madeo et al. 2019). A study by Igarashi et al. (2019) found that oral administration of NR at a concentration of 500 mg/kg/day in drinking water for 6 weeks resulted in ISC recovery in aged mice *in vivo*, while ex vivo studies showed improved colony formation in aged mice and the effect of NR was blocked by the mTORC1 inhibitor rapamycin or the SIRT1 inhibitor EX527 (Igarashi et al. 2019). At the same time, data on the effect of the mTORC1 inhibitor rapamycin, one of the most studied CRMs, on the intestinal stem cells is controversial (He et al. 2020; Igarashi and Guarente 2016; Yilmaz et al. 2012). On the one hand, a recent study showed that mTORC1 is strongly upregulated in the ISC cells in aged mice and mTORC1 inhibition resulted in a partial improvement in aging phenotypes in 16 diet-restriction-induced ISC expansion of one-month-old mice (He et al. 2020). In contrast, Igarashi et al. demonstrated that caloric restriction induces an increase in mTORC1 activity in ISCs leading to the increased cell proliferation, whereas rapamycin suppressed dietary restriction-induced ISC proliferation (Igarashi and Guarente 2016).

The functioning of the hematopoietic stem cells can also be altered by CRMs, especially by the compounds associated with the regulation of mTOR. The importance of mTOR in HSC aging is evidenced by the results of a study that showed that age-related functional decline in HSC is improved in long-lived mTOR mutant mice (Selman et al. 2016). And also in another study, mTOR activity increases in the HSC of mice with aging, and rapamycin at a dose of 4 mg/kg every other day for 6 weeks increased lifespan, restored self-renewal and hematopoiesis of HSC in old mice (22 months) (Chen et al. 2009). An ex vivo study showed that rapamycin treatment inhibited the cellular senescence, possibly through Bmi1 activation and p16 inhibition, and led to the stimulation of ex vivo expansion and long-term hematopoietic repair of HSCs (Luo et al. 2014).

Resveratrol also affects hematopoiesis *in vitro* and *in vivo* (Rimmelé et al. 2014; Zhang et al. 2010; Matsui et al. 2012). Rimmelé et al. demonstrated that daily administration of resveratrol (5 mg/kg) for three weeks increased the frequency and the total number of the Lin - Sca1 + c-Kit + (LSK) cells in the bone marrow of the C57BL/6 mice (Rimmelé et al. 2014). In addition, the treatment with the resveratrol improved the state of the LSK-HSC, positively influenced the microenvironment of the bone marrow, and partially corrected the abnormal status of the cell cycle in a mouse model of the Fanconi anemia (Zhang et al. 2010). A similar effect on the same Fanconi model of anemia was demonstrated by metformin, a therapy that led to an improvement in the hematopoiesis, increased the size of the hematopoietic stem cell compartment, and enhanced the rest of hematopoietic stem cells and progenitor cells. Moreover, metformin reduced the DNA damage and improved the spontaneous chromosome breakage in cells (Zhang et al. 2016).

The importance of the autophagy activation when exposed to CRMs on the skeletal mouse stem cells has been shown in a number of studies (Ramos et al. 2012). It was shown that in prematurely senescent mice, the muscle-derived stem/progenitor cells (MDSPCs) exhibit abnormalities in proliferation, chondrogenic, osteogenic, and myogenic differentiation and regenerative potential, and cultivation in the rapamycin-containing media resulted in the improved differentiation and decreased apoptosis and aging (Kawakami et al. 2019).

One of the key regulators of the autophagy is Sirtuin 1 (Sirt1), also known as a NAD-dependent deacetylase sirtuin-1, the nutrient sensor, so resveratrol as an activator of the Sirt1 can influence the functioning of the satellite cells (Alway et al. 2014). Resveratrol, for example, caused the restoration of muscle mass in the plantar muscle of old rats after hanging the hind limbs, which was associated with the improvement of the satellite cell proliferation in the hind limb muscles (Bennett et al. 2013). On the other hand, the 10-month consumption of the resveratrol had a protective effect against the aging-induced oxidative stress in the skeletal muscles, but did not attenuate sarcopenia in mice (Jackson et al. 2011). This is possible because the effect of resveratrol may depend on its concentration. Resveratrol at low doses (10 μM) caused the myoblast cell cycle stop, migration and promoted the muscle regeneration *in vitro* by attenuating ROS exposure, while higher doses (40–60 μM) suppressed these effects (Bosutti and Degens, 2015).

Metformin, as a regulator of autophagy, also caused the metabolic reprogramming in the fate of the skeletal muscle stem cells. Interestingly, in a recent study by Pavlidou et al. it was shown that unlike resveratrol, metformin did not lead to the activation of Sirt1 in the satellite cells, and contributed to the maintenance of the satellite cells at rest, caused a delay in their differentiation *in vitro* and slowed down the process of muscle regeneration after the cardiotoxin injury *in vivo* (Pavlidou et al. 2019). On the other hand, the metformin improved *in vivo* the regeneration after a burn injury by activating the proliferation of the Pax7-positive satellite cells via the AMPK signalling pathway (Yousuf et al. 2020). The authors suggest that the differences between the two studies are due to the fact, that the burn injury causes systemic inflammation that lasts 2 weeks, as opposed to the shorter exposure to cardiotoxin. In addition, after a burn, animals are more mobile and this affects the muscle regeneration (Yousuf et al. 2020). Moreover, conflicting results have been obtained with the myogenic cell differentiation. On the one hand, it was found that in myoblasts, the metformin-treated MyoD and p21cip1 are not activated, which negatively affects the myogenic differentiation (Pavlidou et al. 2017). In contrast, another study showed that metformin did not affect the proliferation but enhanced the myogenic cell differentiation (Senesi et al. 2016). These results demonstrate the complex nature of the effects of metformin on the muscle stem cells and the need for more detailed studies.

Several studies have shown an association between the changes in polyamine levels in the muscle fibers and the skeletal muscle atrophy and hypertrophy (Lee and MacLean, 2011). It was shown that *in vitro* polyamines (putrescine, spermidine and spermine) cause activation of the satellite cells and the expression of myogenic regulatory factors (Thornton et al. 2013). *in vivo* studies have shown that intraperitoneal injections of spermidine at a dose of 100 mg/kg of body weight every other day for 32 days induced the autophagy in the satellite cells and activated the resting satellite cells in mice (Zhang et al. 2018).

Simulating the effect of calorie restriction on the cardiovascular system is one of the most important properties of mimetics (Sciarretta et al. 2020). An interesting question is how much adult heart stem cells are involved in this process, which can play a central role in the age-related remodeling of the heart (Carresi et al. 2021).

Among the known mimetics, the most effective heart stem cell stimulant (CSC) is the resveratrol. The main results were obtained from the studies related to the transplantation of the CSCs for the treatment of myocardial infarction. A study in Sprague–Dawley rats demonstrated that preliminary systemic administration of resveratrol at a concentration of 2.5 mg/kg/day for 2 weeks improved the myocardial tissue environment and increased the survival and proliferation and the differentiation of transplanted CSCs in the area of the myocardial infarction (Gurusamy et al. 2010).

Pre-treatment with resveratrol for 60 min with the CSC before transplantation resulted in the improved cardiac function and enhanced engraftment of the implanted stem cells that had increased the expression of Nrf2, Ref-1 and NFoB (Gorbunov et al. 2012). Resveratrol at a concentration of 2.5 mg/kg per day for 4 weeks was also shown to activate the transplanted Sca-1 + CSC cells in the heart. The authors showed that the upregulation of the VEGF and SDF-1α is the likely mechanism underlying the resveratrol action (Ling et al. 2017). Resveratrol has been shown in several studies to promote the differentiation of the Human Induced Pluripotent Stem Cells into cardiomyocytes through inhibition of the canonical Wnt signaling pathway and the SRF-miR-1 (Liu et al. 2016). It was also shown that the resveratrol concentration of 10 μmol/L was optimal for stimulating the differentiation of the mouse embryonic stem cells into cardiomyocytes (Ding et al. 2016).

Rapamycin also alters the functional properties of the CSCs, which is associated with the importance of mTOR activity for the vital activity of the heart stem cells (Zheng et al. 2017). Several studies have shown that rapamycin effectively accelerates the differentiation of the embryonic stem cells into the cardiomyocytes (Qiu et al. 2017; Lu et al. 2017). Investigations that demonstrate the role of rapamycin regulation of the mTOR activity in the senescent CSCs are important. Studies of the effect of rapamycin on the CSC showed that *in vitro* the treatment of the cells with the rapamycin (10 nM) and resveratrol (0.5 μM) reduced the cellular senescence and IL1β secretion and increased the rate of cell proliferation, due to the mechanisms that were activated by the increased phosphorylation of AMPK. Moreover, this preliminary pharmacological treatment of the old CSCs *in vitro* led to an improvement in their reparative potential *in vivo* (Avolio et al. 2014).

A recent study showed that a treatment with rapamycin (100 nM) markedly improves the cellular functions, attenuates the replicative senescence, and promotes the proliferation of the senescent human cardiac progenitor cells (hCPCs) (Park et al. 2020). In addition, the long-term treatment with rapamycin improves the clonogenic potential and maintains the migration/differentiation capacity of the old hCPCs (Park et al. 2020). It was also shown that the rapamycin pretreatment of mesenchymal stem cells isolated from the rat bone marrow enhances the cardiogenesis and differentiation of the transplanted cells into the cardiomyocytes in a model of myocardial infarction (Li et al. 2020). However, metformin impaired the homing and survival of the MSCs in the heart in the diabetic streptozotocin-induced cardiomyopathy in rats (Ammar et al. 2021).

25.5 Calorie Restriction Mimetics for Neural Stem Cells and Neurogenesis

The previous section of the chapter summarized the results of the studies showing that the CR enhances the neurogenesis. Therefore, it is necessary to consider the effects of the mimetics on the neural stem cells and neurogenesis. The available data indicate that the resveratrol has different effects on neurogenesis of a young and healthy organism (Park et al. 2012) under various stress conditions (Madhyastha et al. 2013; Moriya et al. 2011; Shen et al. 2016) and aging (Kumar et al. 2016). Thus, the administration of the resveratrol (1–10 mg/kg) for 14 days inhibited the proliferation and survival of the neural progenitor cells (NPCs) in the dentate gyrus of the hippocampus of the young C57BL/6 mice and reduced the levels of the phosphorylated form of the cyclic AMP response element-binding protein (pCREB) and the BDNF in the hippocampus (Park et al. 2012). While in healthy animals resveratrol suppressed the neurogenesis, in rats exposed to embryonic stress, the resveratrol treatment increased the number of neonatal neurons in the hippocampus and the BDNF expression (Madhyastha et al. 2013).

Resveratrol was also shown to stimulate the neurogenesis and mRNA expression of the brain neurotrophic factor in the hippocampus of mice with chronic fatigue (Pavlidou et al. 2019). *in vitro*, the exposure to resveratrol on the neural stem cells after the oxygen–glucose deprivation/reoxygenation leads to the increased cell survival and proliferation, reduces apoptosis and the MDA levels, and also increases the SOD activity, GSH content, and the expression of Nrf2, HO- 1, and NQO1 proteins (Li et al. 2020). Resveratrol also exhibited a two-phase effect on NPCs under the culture conditions (Kumar et al. 2016). While low concentrations of resveratrol (10 μM) induced cell proliferation through the signal pathways associated with the signal-regulated extracellular kinases (ERK) and p38 kinases phosphorylation, the high concentrations (>20 μM) leveled these effects. The administration of resveratrol (20 mg/kg body weight) to old rats (15 months) enhanced the neurogenesis in the region of the dentate gyrus of the hippocampus (Kumar et al. 2016).

Numerous *in vitro* and *in vivo* studies have confirmed that metformin may be an important regulator of the metabolism and functioning of the brain cells, including the stem cells, and has a great potential in the treatment of the neurodegenerative diseases (Markowicz-Piasecka et al. 2017; Jiang and Liu, 2020). Several studies have shown the activation of the neural stem cells by metformin in the various injuries. It has been shown that metformin causes an activation of the neural stem cells in the spinal cord only in males, and the progenitor cells in both sexes. Exposure to metformin resulted in the improved functional outcomes after trauma in the thoracic spinal cord (Gilbert and Morshead, 2019).

On the other hand, in a study conducted by Ruddy et al. where it was shown that the effect of metformin on the NPCs depended on sensitivity to sex hormones, especially the estradiol, and only females increased the proliferation of progenitor cells. At the same time, the neural stem cells and the neurogenesis were activated to the same extent in both sexes in a model of neonatal stroke (Ruddy et al. 2019). In an another study, it was demonstrated that administration of metformin for 7 days activates the endogenous NPCs, expands the NPC pool, and aids migration and differentiation of the NPCs in the damaged neonatal brain in a hypoxia-ischemic injury model (Dadwal et al. 2015). Using the advanced glycosylation end products (AGEs) model of the neuronal damage, metformin was shown to increase the survival of the human neural stem cells (hNSC) and enhance the expression of the AMPK, PGC1α, NRF-1, and Tfam (Chung et al. 2015).

Metformin preconditioning of the human induced pluripotent stem cells of the neural stem cells prior to the brain transplantation damaged by stroke has also been shown to improve their engraftment and regeneration and recovery after (Ould-Brahim et al. 2018). Activation of the atypical PKC-CBP pathway, which enhances the neurogenesis in the brain of the adult mice, is assumed to be signalling pathways involved in the regulation of the response to the metformin action on NPCs (Wang et al. 2012). In addition, Fatt et al. revealed that the neuronal differentiation of the NPCs extracted from the subventricular zone was enhanced by metformin and depended on the activation of the AMPK-aPKC-CBP pathway (Fatt et al. 2015).

Also, the metformin treatment has recently been shown to result in rejuvenation of the oligodendrocyte progenitor cells (OPCs) and the improvement of remyelination in old animals after focal demyelination (Neumann et al. 2019).

The mTOR signaling pathway is a key in the neuronal and glial differentiation processes and the maintenance of the neural stem cell stemness (LiCausi and Hartman 2018). Intra-cerebro-ventricular infusion of rapamycin (0.5 mM) into the left ventricle of mice was shown to reduce the number of proliferating neural stem cells for almost two times within 7 days (Paliouras et al. 2012). In addition, this study showed that rapamycin suppresses the division and differentiation of the neural stem cells *in vitro* and reduces both the size and the number of the neural stem cells grown as neurospheres (Paliouras et al. 2012).

An interesting comparative study of the effect of rapamycin or metformin on the proliferating neural precursor SVZ and DG cells was carried out in mice (Kusne et al. 2014). A 9-week regimen of daily intraperitoneal (ip) injection of rapamycin at a dose of 75 μg/kg (low dose) or 2.5 mg/kg (high dose) has been shown to reduce the proliferation of the neural stem cells in the SVZ region of the brain of adult C57BL/6 J., while a similar regimen of metformin administration at a dose of 200 mg/kg per day did not cause such an effect. A decrease in the proliferation and differentiation was also found under the influence of rapamycin, but not metformin (Kusne et al. 2014).

The study of the mTOR system in the brain during aging showed that the total number of the neural stem cells and their proliferation in the hippocampus sharply decreases with age and this correlates with a decrease in the activity of the mTOR signalling in the old brain, including the NSC (Romine et al. 2015). It was shown that a single injection of rapamycin (10 mg/kg) led to a significant decrease in the number of proliferating cells in the hippocampus in young animals. Also, if activation of the mTOR system with ketamine improved the neurogenesis in the hippocampus, then the joint intraperitoneal injections of ketamine (10 mg/kg) and rapamycin (10 mg/kg) to 12 month old mice reduced the level of proliferation (Romine et al. 2015). Collectively, these data indicate that the use of rapamycin may exacerbate brain aging, and in this aspect the compound is not a calorie restriction mimetic.

Endogenous polyamines play an important role in the neural differentiation, learning and memory (Guerra et al. 2016) and the levels of polyamines in the aging brain gradually decrease as they age (Sigrist et al. 2014). Spermidine and spermin have recently been shown to induce autophagy and inhibit brain aging in the SAMP8 mice (Xu et al. 2020). Perhaps this deceleration is also associated with the effect of spermidine on the neural stem cells, as evidenced by another study, the cultivation of the NPC with spermidine, the cell migration was facilitated, the number of neurites increased and the BDNF expression increased 7 days after the addition of SPD (Signor et al. 2017).

Thus, some of the chemical compounds that are classified as the CRMs undoubtedly have the properties that allow them to mimic calorie restriction, including through the mechanisms associated with the activation of the adult stem cells. However, the impact of the CRMs on stem cells can be both positive and negative depending on the type of stem cells, which is associated with the heterogeneity of the adult stem cells in the body. Moreover, the effects of the CRMs on stem cells can be considered within the concept of hormesis, when the small doses of the CRMs induced the activation of endogenous stem cells, while the large doses suppressed. It should also be noted that the CRMs often have the antioxidant properties, and in high doses they presumably can cause the so-called antioxidant stress in healthy cells and disrupt a normal redox homeostasis, and this, on the contrary, can lead to an accelerated aging (Kornienko et al. 2019).

Apart from that, the responses of the adult stem cells to the mimetics are age and gender dependent, so that they vary greatly from positive to negative. In addition, the state of a niche or tissue microenvironment has a strong influence on the functioning of the adult stem cells, which, under the various stresses and diseases, can modulate the effects of mimetics. The role of the microbiome, which may be involved in mediating the effects of the CRMs on stem cells, remains unexplored. We believe that this direction is the very promising in the development of a further strategy for the use of mimetics. Therefore, further studies are required to determine the role of metabolic reprogramming in the fate of stem cells with the help of mimetics. It has to include more accurate and detailed characterization and understanding of the molecular and cellular mechanisms of the aging process. Moreover, there is a need to elucidate the interaction of the microbiome axis - stem cells and determine the optimal concentrations of mimetics and their combinations. Finally, the search for the new compounds with the most complete properties mimicking calorie restrictions are needed.

Acknowledgements This research was funded by the Science Committee of the Ministry of Education and Science of the Republic of Kazakhstan (Grant No. AP09058312) and the CRP grant of Nazarbayev University No. 091019CRP2113.

Compliance with Ethical Standards

Conflict of Interest All authors declare they have no conflict of interest.

References

Abreu P, Serna JDC, Munhoz AC, Kowaltowski AJ (2020) Calorie restriction changes mu scle satellite cell proliferation in a manner independent of metabolic modulation. Mech Ageing Dev 192:111362

Alt EU, Senst C, Murthy SN, Slakey DP, Dupin CL, Chaffin AE, Kadowitz PJ, Izadpanah R (2012) Aging alters tissue resident mesenchymal stem cell properties. Stem Cell Res 8:215–225

Alway SE, Myers MJ, Mohamed JS (2014) Regulation of satellite cell function in sarcopenia. Front Aging Neurosci 6:246

Ammar HI, Shamseldeen AM, Shoukry HS, Ashour H, Kamar SS, Rashed LA, Fadel M, Srivastava A, Dhingra Sa (2021) Metformin impairs homing ability and efficacy of mesenchymal stem cells for cardiac repair in streptozotocin-induced diabetic cardiomyopathy in rats. Am J Physiol-Heart Circ Physiol 320(4):H1290–H1302. https://doi.org/10.1152/ajpheart.00317.2020

Apple DM, Mahesula S, Fonseca RS, Zhu C, Kokovay E (2019) Calorie restriction protects neural stem cells from age-related deficits in the subventricular zone. Aging 11:115–126

Avolio E, Gianfranceschi G, Cesselli D, Caragnano A, Athanasakis E, Katare R, Meloni M, Palma A, Barchiesi A, Vascotto C, Toffoletto B, Mazzega E, Finato N, Aresu G, Livi U, Emanueli C, Scoles G, Beltrami CA, Madeddu P, Beltrami AP (2014) Ex vivo molecular rejuvenation improves the therapeutic activity of senescent human cardiac stem cells in a mouse model of myocardial infarction. Stem Cells 32:2373–2385

Barzilai N, Huffman DM, Muzumdar RH, Bartke A (2012) The critical role of metabolic pathways in aging. Diabetes 61:1315–1322

Bennett BT, Mohamed JS, Alway SE (2013) Effects of resveratrol on the recovery of muscle mass following disuse in the plantaris muscle of aged rats. PLOS ONE 8:e83518

Boldrin L, Ross JA, Whitmore C, Doreste B, Beaver C, Eddaoudi A, Pearce DJ, Morgan JE (2017) The effect of calorie restriction on mouse skeletal muscle is sex, strain and time-dependent. Sci Rep 7:5160

Bondolfi L, Ermini F, Long JM, Ingram DK, Jucker M (2004) Impact of age and caloric restriction on neurogenesis in the dentate gyrus of C57BL/6 mice. Neurobiol Aging 25:333–340

Bosutti A, Degens H (2015) The impact of resveratrol and hydrogen peroxide on muscle cell plasticity shows a dose-dependent interaction. Sci Rep 5:8093

Boyette LB, Tuan RS (2014) Adult stem cells and diseases of aging. J Clin Med 3

Bruens L, Ellenbroek SIJ, Suijkerbuijk SJE, Azkanaz M, Hale AJ, Toonen P, Flanagan DJ, Sansom OJ, Snippert HJ, van Rheenen J (2020) Calorie restriction increases the number of competing stem cells and decreases mutation retention in the intestine. Cell Rep 32:107

Carresi C, Scicchitano M, Scarano F, Macrì R, Bosco F, Nucera S, Ruga S, Zito MC, Mollace R, Guarnieri L, Coppoletta AR, Gliozzi M, Musolino V, Maiuolo J, Palma E, Mollace V (2021) The potential properties of natural compounds in cardiac stem cell activation: their role in myocardial regeneration. Nutrients 13:275

Catic A (2018) Cellular metabolism and aging. Prog Mol Biol Transl Sci 155:85–107

Cerletti M, Jang YC, Finley LWS, Haigis MC, Wagers AJ (2012) Short-term calorie restriction enhances skeletal muscle stem cell function. Cell Stem Cell 10:515–519

Cerqueira FM, Kowaltowski AJ (2010) Commonly adopted caloric restriction protocols often involve malnutrition. Ageing Res Rev 9:424–430

Chang NC (2020) Autophagy and stem cells: self-eating for self-renewal. Front Cell Dev Biol 8

Chen C, Liu Y, Liu Y, Zheng P (2009) mTOR regulation and therapeutic rejuvenation of aging hematopoietic stem cells. Sci Sign 2:ra75-ra75.

Chen J, Astle CM, Harrison DE (2003) Hematopoietic senescence is postponed and hematopoietic stem cell function is enhanced by dietary restriction. Exp Hematol 31:1097–1103

Chiba T, Tsuchiya T, Komatsu T, Mori R, Hayashi H, Shimokawa I (2010) Development of calorie restriction mimetics as therapeutics for obesity, diabetes, inflammatory and neurodegenerative diseases. Curr Genomics 11:562–567

Chung KW, Chung HY (2019) The effects of calorie restriction on autophagy: role on aging intervention. Nutrients 11:2923

Chung MM, Chen YL, Pei D, Cheng YC, Sun B, Nicol CJ, Yen CH, Chen HM, Liang YJ, Chiang MC (2015) The neuroprotective role of metformin in advanced glycation end product treated human neural stem cells is AMPK-dependent. Biochimica et Biophysica Acta (BBA) Molecular Basis of Disease 1852:720–731

Clevers H (2015) What is an adult stem cell? Science 350:1319–1320

Dadwal P, Mahmud N, Sinai L, Azimi A, Fatt M, Wondisford FE, Miller FD, Morshead CM (2015) Activating endogenous neural precursor cells using metformin leads to neural repair and functional recovery in a model of childhood brain injury. Stem Cell Rep 5:166–173

Desai AK, Grossberg GT, Chibnall JT (2010) Healthy brain aging: a road map. Clin Geriatr Med 26:1–16

Dias IR, Santos CDS, Magalhães CODE, de Oliveira LRS, Peixoto MFD, de Sousa RAL, Cassilhas RC (2020) Does calorie restriction improve cognition? IBRO Rep 9:37–45

Ding H, Xu X, Qin X, Yang C, Feng Q (2016) Resveratrol promotes differentiation of mouse embryonic stem cells to cardiomyocytes. Cardiovasc Ther 34:283–289

Du G, Qiao Y, Zhuo Z, Zhou J, Li X, Liu Z, Li Y, Chen H (2020) Lipoic acid rejuvenates aged intestinal stem cells by preventing age-associated endosome reduction. EMBO Rep 21:e49583–e49583

Epel ES, Lithgow GJ (2014) Stress biology and aging mechanisms: toward understanding the deep connection between adaptation to stress and longevity. J Gerontol Ser A 69:S10–S16

Fabbiano S, Suárez-Zamorano N, Rigo D, Veyrat-Durebex C, Dokic AS, Didier JC, Trajkovski M (2016) Caloric restriction leads to browning of white adipose tissue through type 2 immune signaling. Cell Metab 24:434–446

Fatt M, Hsu K, He L, Wondisford F, Miller FD, Kaplan DR, Wang J (2015) Metformin acts on two different molecular pathways to enhance adult neural precursor proliferation/self-renewal and differentiation. Stem Cell Rep 5:988–995

Flanagan EW, Most J, Mey JT, Redman LM (2020) Calorie restriction and aging in humans. Annu Rev Nutr 40:105–133

Fontán-Lozano A, López-Lluch G, Delgado-García JM, Navas P, Carrión AM (2008) Molecular bases of caloric restriction regulation of neuronal synaptic plasticity. Mol Neurobiol 38:167–177

Forni MF, Peloggia J, Braga TT, Chinchilla JEO, Shinohara J, Navas CA, Camara NOS, Kowaltowski AJ (2017) Caloric restriction promotes structural and metabolic changes in the skin. Cell Rep 20:2678–2692

Franceschi C, Ostan R, Santoro A (2018) Nutrition and inflammation: are centenarians similar to individuals on calorie-restricted diets? Annu Rev Nutr 38:329–356

Fukada S, Ma Y, Uezumi A (2014) Adult stem cell and mesenchymal progenitor theories of aging. Front Cell Dev Biol 2:10

Gems D, Partridge L (2008) Stress-response hormesis and aging: "that which does not kill us makes us stronger." Cell Metab 7:200–203

Gilbert EAB, Morshead CM (2019) Metformin activates neural stem and progenitor cells in the spinal cord and improves functional outcomes following injury. FASEB J 33:208.3

Glenny E, Liu J, Touvron M, Vance N, Magness S, Bulik C, Van LL, Carroll I (2020) Severe calorie restriction induces gut microbiota-dependent intestinal stem cell dysfunction. Curr Dev Nutr 4:1558–1558

Gonzalez MA, Bernad A (2012) Characteristics of adult stem cells. Adv Exp Med Biol 741:103–120

Goodell MA, Rando TA (2015) Stem cells and healthy aging. Science 350:1199–1204

Gorbunov N, Petrovski G, Gurusamy N, Ray D, Kim DH, Das DK (2012) Regeneration of infarcted myocardium with resveratrol-modified cardiac stem cells. J Cell Mol Med 16:174–184

Guerra GP, Rubin MA, Mello CF (2016) Modulation of learning and memory by natural polyamines. Pharmacol Res 112:99–118

Gurusamy N, Ray D, Lekli I, Das DK (2010) Red wine antioxidant resveratrol-modified cardiac stem cells regenerate infarcted myocardium. J Cell Mol Med 14:2235–2239

Harrison DE, Strong R, Sharp ZD, Nelson JF, Astle CM, Flurkey K, Nadon NL, Wilkinson JE, Frenkel K, Carter CS, Pahor M, Javors MA, Fernandez E, Miller RA (2009) Rapamycin fed late in life extends lifespan in genetically heterogeneous mice. Nature 460:392–395

He D, Wu H, Xiang J, Ruan X, Peng P, Ruan Y, Chen Y-G, Wang Y, Yu Q, Zhang H, Habib SL, de Pinho RA, Liu H, Li B (2020) Gut stem cell aging is driven by mTORC1 via a p38 MAPK-p53 pathway. Nat Commun 11:37

Heilbronn LK, Ravussin E (2003) Calorie restriction and aging: review of the literature and implications for studies in humans. Am J Clin Nutr 78:361–369

Hertel J, Friedrich N, Wittfeld K, Pietzner M, Budde K, van der Auwera S, Lohmann T, Teumer A, Völzke H, Nauck M, Grabe HJ (2016) Measuring biological age via metabonomics: the metabolic age score. J Proteome Res 15:400–410

Igarashi M, Guarente L (2016) mTORC1 and SIRT1 cooperate to foster expansion of gut adult stem cells during calorie restriction. Cell 166:436–450

Igarashi M, Miura M, Williams E, Jaksch F, Kadowaki T, Yamauchi T, Guarente L (2019) NAD+ supplementation rejuvenates aged gut adult stem cells. Aging Cell 18:12935

Ingram DK, Anson RM, de Cabo R, Mamczarz J, Zhu M, Mattison J, Lane MA, Roth GS (2004) Development of calorie restriction mimetics as a prolongevity strategy. Ann N Y Acad Sci 1019:412–423

Ingram DK, Roth GS (2015) Calorie restriction mimetics: can you have your cake and eat it, too? Ageing Res Rev 20:46–62

Ingram DK, Zhu M, Mamczarz J, Zou S, Lane MA, Roth GS, Decabo R (2006) Calorie restriction mimetics: an emerging research field. Aging Cell 5:97–108

Jackson JR, Ryan MJ, Alway SE (2011) Long-term supplementation with resveratrol alleviates oxidative stress but does not attenuate sarcopenia in aged mice. J Gerontol Ser A 66A:751–764

Jiang LL, Liu L (2020) Effect of metformin on stem cells: molecular mechanism and clinical prospect. World J Stem Cells 12:1455–1473

Jones DL, Rando TA (2011) Emerging models and paradigms for stem cell ageing. Nat Cell Biol 13:506–512

Kaptan Z, Akgün-Dar K, Kapucu A, Dedeakayoğulları H, Batu Ş, Üzüm G (2015) Long term consequences on spatial learning-memory of low-calorie diet during adolescence in female rats; hippocampal and prefrontal cortex BDNF level, expression of NeuN and cell proliferation in dentate gyrus. Brain Res 1618:194–204

Kawakami Y, Hambright WS, Takayama K, Mu X, Lu A, Cummins JH, Matsumoto T, Yurube T, Kuroda R, Kurosaka M, Fu FH, Robbins PD, Niedernhofer LJ, Huard J (2019) Rapamycin rescues age-related changes in muscle-derived stem/progenitor cells from progeroid mice. Mol Therapy Methods Clin Dev 14:64–76

Kennedy BK, Berger SL, Brunet A, Campisi J, Cuervo AM, Epel ES, Franceschi C, Lithgow GJ, Morimoto RI, Pessin JE, Rando TA, Richardson A, Schadt EE, Wyss-Coray T, Sierra F (2014) Geroscience: linking aging to chronic disease. Cell 159:709–713

Kishi T, Hirooka Y, Nagayama T, Isegawa K, Katsuki M, Takesue K, Sunagawa K (2014) Calorie restriction improves cognitive decline via up-regulation of brain-derived neurotrophic factor. Int Heart J 56:14–168

Komatsu T, Park S, Hayashi H, Mori R, Yamaza H, Shimokawa I (2019) Mechanisms of calorie restriction: a review of genes required for the life-extending and tumor-inhibiting effects of calorie restriction. Nutrients 11:3068

Kornienko JS, Smirnova IS, Pugovkina NA, Ivanova JS, Shilina MA, Grinchuk TM, Shatrova AN, Aksenov ND, Zenin VV, Nikolsky NN, Lyublinskaya OG (2019) High doses of synthetic antioxidants induce premature senescence in cultivated mesenchymal stem cells. Sci Rep 9:1296

Kumar V, Pandey A, Jahan S, Shukla RK, Kumar D, Srivastava A, Singh S, Rajpurohit CS, Yadav S, Khanna VK, Pant AB (2016) Differential responses of Trans-Resveratrol on proliferation of neural progenitor cells and aged rat hippocampal neurogenesis. Sci Rep 6:28142

Kusne Y, Goldberg EL, Parker SS, Hapak SM, Maskaykina IY, Chew WM, Limesand KH, Brooks HL, Price TJ, Sanai N, Nikolich-Zugich J, Ghosh S (2014) Contrasting effects of chronic, systemic

treatment with mTOR inhibitors rapamycin and metformin on adult neural progenitors in mice. Age (dordr) 36:199–212
Lane MA, Ingram DK, Roth GS (1998) 2-Deoxy-D-glucose feeding in rats mimics physiologic effects of calorie restriction. J Anti Aging Med 1:327–337
Lazare S, Ausema A, Reijne AC, van Dijk G, van Os R, de Haan G (2017) Lifelong dietary intervention does not affect hematopoietic stem cell function. Exp Hematol 53:26–30
Lee NK, Maclean HE (2011) Polyamines, androgens, and skeletal muscle hypertrophy. J Cell Physiol 226:1453–1460
Li Z-H, Wang Y-L, Wang H-J, Wu J-H, Tan Y-Z (2020) Rapamycin-preactivated autophagy enhances survival and differentiation of mesenchymal stem cells after transplantation into infarcted myocardium. Stem Cell Rev Rep 16:344–356
Licausi F, Hartman NW (2018) Role of mTOR complexes in neurogenesis. Int J Mol Sci 19:1544
Ling L, Gu S, Cheng Y (2017) Resveratrol activates endogenous cardiac stem cells and improves myocardial regeneration following acute myocardial infarction. Mol Med Rep 15:1188–1194
Liu H, Zhang S, Zhao L, Zhang Y, Li Q, Chai X, Zhang Y (2016) Resveratrol enhances cardiomyocyte differentiation of human induced pluripotent stem cells through inhibiting canonical WNT signal pathway and enhancing serum response factor-miR-1 axis. Stem Cells Int 2016:2524092–2524092
Liu L, Rando TA (2011) Manifestations and mechanisms of stem cell aging. J Cell Biol 193:257–266
Locher JL, Goldsby TU, Goss AM, Kilgore ML, Gower B, Ard JD (2016) Calorie restriction in overweight older adults: do benefits exceed potential risks? Exp Gerontol 86:4–13
Loeffler M, Roeder I (2002) Tissue stem cells: definition, plasticity, heterogeneity, self-organization and models–a conceptual approach. Cells Tissues Organs 171:8–26
López-Otín C, Galluzzi L, Freije JMP, Madeo F, Kroemer G (2016) Metabolic control of longevity. Cell 166:802–821
Lu Q, Liu Y, Wang Y, Wang W, Yang Z, Li T, Tian Y, Chen P, Ma K, Jia Z, Zhou C (2017) Rapamycin efficiently promotes cardiac differentiation of mouse embryonic stem cells. Biosci Rep 37:BSR20160552
Luo Y, Li L, Zou P, Wang J, Shao L, Zhou D, Liu L (2014) Rapamycin enhances long-term hematopoietic reconstitution of ex vivo expanded mouse hematopoietic stem cells by inhibiting senescence. Transplantation 97:20–29
Ma S, Sun S, Geng L, Song M, Wang W, Ye Y, Ji Q, Zou Z, Wang S, He X, Li W, Esteban CR, Long X, Guo G, Chan P, Zhou Q, Belmonte JCI, Zhang W, Qu J, Liu G-H (2020) Caloric restriction reprograms the single-cell transcriptional landscape of rattus norvegicus Aging. Cell 180:984-1001.e22
Madeo F, Carmona-Gutierrez D, Hofer SJ, Kroemer G (2019) Caloric restriction mimetics against age-associated disease: targets, mechanisms, and therapeutic potential. Cell Metab 29:592–610
Madhyastha S, Sekhar S, Rao G (2013) Resveratrol improves postnatal hippocampal neurogenesis and brain derived neurotrophic factor in prenatally stressed rats. Int J Dev Neurosci 31:580–585
Maharajan N, Vijayakumar K, Jang C, Cho G-W (2020) Caloric restriction maintains stem cells through niche and regulates stem cell aging. J Mol Med 98:25–37
Mai V, Colbert LH, Berrigan D, Perkins SN, Pfeiffer R, Lavigne JA, Lanza E, Haines DC, Schatzkin A, Hursting SD (2003) Calorie restriction and diet composition modulate spontaneous intestinal tumorigenesis in ApcMin mice through different mechanisms. Cancer Res 63:1752
Maiese K, Chong ZZ, Shang YC, Wang S (2013) mTOR: on target for novel therapeutic strategies in the nervous system. Trends Mol Med 19:51–60
Markowicz-Piasecka M, Sikora J, Szydłowska A, Skupień A, Mikiciuk-Olasik E, Huttunen KM (2017) Metformin - a future therapy for neurodegenerative diseases : theme: drug discovery, development and delivery in alzheimer's disease guest editor: davide brambilla. Pharm Res 34:2614–2627
Martel J, Chang S-H, Wu C-Y, Peng H-H, Hwang T-L, Ko Y-F, Young JD, Ojcius DM (2021) Recent advances in the field of caloric restriction mimetics and anti-aging molecules. Ageing Res Rev 66:101240

Matsui K, Ezoe S, Oritani K, Shibata M, Tokunaga M, Fujita N, Tanimura A, Sudo T, Tanaka H, McBurney MW, Matsumura I, Kanakura Y (2012) NAD-dependent histone deacetylase, SIRT1, plays essential roles in the maintenance of hematopoietic stem cells. Biochem Biophys Res Commun 418:811–817

Mattson MP (2008) Hormesis defined. Ageing Res Rev 7:1–7

Mazzoccoli G, Tevy MF, Borghesan M, Vergini MRD, Vinciguerra M (2014) Caloric restriction and aging stem cells: the stick and the carrot? Exp Gerontol 50:137–148

Mehmel M, Jovanović N, Spitz U (2020) Nicotinamide riboside-the current state of research and therapeutic uses. Nutrients 12:1616

Miller RA, Harrison DE, Astle CM, Baur JA, Boyd AR, de Cabo R, Fernandez E, Flurkey K, Javors MA, Nelson JF, Orihuela CJ, Pletcher S, Sharp ZD, Sinclair D, Starnes JW, Wilkinson JE, Nadon NL, Strong R (2011) Rapamycin, but not resveratrol or simvastatin, extends life span of genetically heterogeneous mice. J Gerontol A Biol Sci Med Sci 66:191–201

Mimeault M, Batra SK (2009) Aging of tissue-resident adult stem/progenitor cells and their pathological consequences. Panminerva Med 51:57–79

Moriya J, Chen R, Yamakawa J-I, Sasaki K, Ishigaki Y, Takahashi T (2011) Resveratrol improves hippocampal atrophy in chronic fatigue mice by enhancing neurogenesis and inhibiting apoptosis of granular cells. Biol Pharm Bull 34:354–359

Na HJ, Park JS, Pyo JH, Jeon HJ, Kim YS, Arking R, Yoo MA (2015) Metformin inhibits age-related centrosome amplification in Drosophila midgut stem cells through AKT/TOR pathway. Mech Ageing Dev 149:8–18

Na HJ, Pyo JH, Jeon HJ, Park JS, Chung HY, Yoo MA (2018) Deficiency of Atg6 impairs beneficial effect of metformin on intestinal stem cell aging in Drosophila. Biochem Biophys Res Commun 498:18–24

Neff F, Flores-Dominguez D, Ryan DP, Horsch M, Schröder S, Adler T, Afonso LC, Aguilar-Pimentel JA, Becker L, Garrett L, Hans W, Hettich MM, Holtmeier R, Hölter SM, Moreth K, Prehn C, Puk O, Rácz I, Rathkolb B, Rozman J, Naton B, Ordemann R, Adamski J, Beckers J, Bekeredjian R, Busch DH, Ehninger G, Graw J, Höfler H, Klingenspor M, Klopstock T, Ollert M, Stypmann J, Wolf E, Wurst W, Zimmer A, Fuchs H, Gailus-Durner V, Hrabe de Angelis M, Ehninger D (2013) Rapamycin extends murine lifespan but has limited effects on aging. J Clin Invest 123:3272–3291

Neumann B, Baror R, Zhao C, Segel M, Dietmann S, Rawji KS, Foerster S, McClain CR, Chalut K, van Wijngaarden P, Franklin RJM (2019) Metformin restores CNS remyelination capacity by rejuvenating aged stem cells. Cell stem cell 25:473-485 e8

Ould-Brahim F, Sarma SN, Syal C, Lu KJ, Seegobin M, Carter A, Jeffers MS, Doré C, Stanford WL, Corbett D, Wang J (2018) Metformin preconditioning of human induced pluripotent stem cell-derived neural stem cells promotes their engraftment and improves post-stroke regeneration and recovery. Stem Cells Dev 27:1085–1096

Paliouras GN, Hamilton LK, Aumont A, Joppé SE, Barnabé-Heider F, Fernandes KJL (2012) Mammalian target of rapamycin signaling is a key regulator of the transit-amplifying progenitor pool in the adult and aging forebrain. J Neurosci 32:15012

Parikh I, Guo J, Chuang K-H, Zhong Y, Rempe RG, Hoffman JD, Armstrong R, Bauer B, Hartz AMS, Lin A-L (2016) Caloric restriction preserves memory and reduces anxiety of aging mice with early enhancement of neurovascular functions. Aging 8:2814–2826

Park HR, Kong KH, Yu BP, Mattson MP, Lee J (2012) Resveratrol inhibits the proliferation of neural progenitor cells and hippocampal neurogenesis. J Biol Chem 287:42588–42600

Park J-H, Glass Z, Sayed K, Michurina TV, Lazutkin A, Mineyeva O, Velmeshev D, Ward WF, Richardson A, Enikolopov G (2013) Calorie restriction alleviates the age-related decrease in neural progenitor cell division in the aging brain. Eur J Neurosci 37:1987–1993

Park JH, Lee NK, Lim HJ, Ji ST, Kim Y-J, Jang WB, Kim DY, Kang S, Yun J, Ha JS, Kim H, Lee D, Baek SH, Kwon S-M (2020) Pharmacological inhibition of mTOR attenuates replicative cell senescence and improves cellular function via regulating the STAT3-PIM1 axis in human cardiac progenitor cells. Exp Mol Med 52:615–628

Park SE, Chung HY, Park YJ (2017) Calorie Restriction Facilitates Aging-Related Muscle Loss in the Aged Mice. The FASEB Journal 31:1b306

Pavlidou T, Marinkovic M, Rosina M, Fuoco C, Vumbaca S, Gargioli C, Castagnoli L, Cesareni G (2019) Metformin delays satellite cell activation and maintains quiescence. Stem Cells Int 2019:5980465

Pavlidou T, Rosina M, Fuoco C, Gerini G, Gargioli C, Castagnoli L, Cesareni G (2017) Regulation of myoblast differentiation by metabolic perturbations induced by metformin. PLOS ONE 12:e0182475

Peña-Villalobos I, Casanova-Maldonado I, Lois P, Sabat P, Palma V (2019) Adaptive physiological and morphological adjustments mediated by intestinal stem cells in response to food availability in mice. Front Physiol 9:1821

Qiu XX, Liu Y, Zhang YF, Guan YN, Jia QQ, Wang C, Liang H, Li YQ, Yang HT, Qin YW, Huang S, Zhao XX, Jing Q (2017) Rapamycin and CHIR99021 coordinate robust cardiomyocyte differentiation from human pluripotent stem cells via reducing p53-dependent apoptosis. J Am Heart Assoc 6:e005295

Raman L, Kong X, Kernie SG (2013) Pharmacological inhibition of the mTOR pathway impairs hippocampal development in mice. Neurosci Lett 541:9–14

Ramos FJ, Chen SC, Garelick MG, Dai DF, Liao CY, Schreiber KH, Mackay VL, An EH, Strong R, Ladiges WC, Rabinovitch PS, Kaeberlein M, Kennedy BK (2012) Rapamycin reverses elevated mTORC1 signaling in lamin A/C-deficient mice, rescues cardiac and skeletal muscle function, and extends survival. Sci Transl Med 4:144ra103

Rando TA (2006) Stem cells, ageing and the quest for immortality. Nature 441:1080–1086

Ratcliff M, Rees D, McGrady S, Buntwal L, Hornsby AKE, Bayliss J, Kent BA, Bussey T, Saksida L, Beynon AL, Howell OW, Morgan AH, Sun Y, Andrews ZB, Wells T, Davies JS (2019) Calorie restriction activates new adult born olfactory-bulb neurones in a ghrelin-dependent manner but acyl-ghrelin does not enhance subventricular zone neurogenesis. J Neuroendocrinol 31:e12755

Ravussin Y, Leduc CA, Watanabe K, Leibel RL (2012) Effects of ambient temperature on adaptive thermogenesis during maintenance of reduced body weight in mice. Am J Physiol Regul Integr Comp Physiol 303:R438–R448

Rimmelé P, Lofek-Czubek S, Ghaffari S (2014) Resveratrol increases the bone marrow hematopoietic stem and progenitor cell capacity. Am J Hematol 89:E235–E238

Romine J, Gao X, Xu X-M, So KF, Chen J (2015) The proliferation of amplifying neural progenitor cells is impaired in the aging brain and restored by the mTOR pathway activation. Neurobiol Aging 36:1716–1726

Ruddy RM, Adams KV, Morshead CM (2019) Age- and sex-dependent effects of metformin on neural precursor cells and cognitive recovery in a model of neonatal stroke. Sci Adv 5:eaax1912

Santoro A, Martucci M, Conte M, Capri M, Franceschi C, Salvioli S (2020) Inflammaging, hormesis and the rationale for anti-aging strategies. Ageing Res Reviews 64:101142

Sato S, Solanas G, Peixoto FO, Bee L, Symeonidi A, Schmidt MS, Brenner C, Masri S, Benitah SA, Sassone-Corsi P (2017) Circadian reprogramming in the liver identifies metabolic pathways of aging. Cell 170:664-677.e11

Schmuck EG, Mulligan JD, Saupe KW (2011) Caloric restriction attenuates the age-associated increase of adipose-derived stem cells but further reduces their proliferative capacity. Age (dordr) 33:107–118

Schüler SC, Gebert N, Ori A (2020) Stem cell aging: the upcoming era of proteins and metabolites. Mechanisms of Ageing and Development 190:111288

Schultz MB, Sinclair DA (2016) When stem cells grow old: phenotypes and mechanisms of stem cell aging. Development 143:3

Sciarretta Sebastiano, Forte Maurizio, Castoldi Francesca, Frati Giacomo, Versaci Francesco, Sadoshima Junichi, Kroemer Guido, Maiuri Maria Chiara (2020) Caloric restriction mimetics for the treatment of cardiovascular diseases. Cardiovasc Res 117(6):1434–1449. https://doi.org/10.1093/cvr/cvaa297

Selman C, Sinclair A, Pedroni SMA, Irvine EE, Michie AM, Withers DJ (2016) Evidence that hematopoietic stem cell function is preserved during aging in long-lived S6K1 mutant mice. Oncotarget 7:29937–29943

Senesi P, Montesano A, Luzi L, Codella R, Benedini S, Terruzzi I (2016) Metformin treatment prevents sedentariness related damages in mice. J Diabetes Res 2016:8274689

Sharpless NE, Depinho RA (2007) How stem cells age and why this makes us grow old. Nat Rev Mol Cell Biol 8:703–713

Shen C, Cheng W, Yu P, Wang L, Zhou L, Zeng L, Yang Q (2016) Resveratrol pretreatment attenuates injury and promotes proliferation of neural stem cells following oxygen-glucose deprivation/reoxygenation by upregulating the expression of Nrf2, HO-1 and NQO1 in vitro. Mol Med Rep 14:3646–3654

Sheng Y, Xia F, Chen L, Lv Y, Lv S, Yu J, Liu J, Ding G (2020) Differential responses of white adipose tissue and brown adipose tissue to calorie restriction during aging. J Gerontol Ser A 76:393–399

Shintani H, Shintani T, Ashida H, Sato M (2018) Calorie Restriction mimetics: upstream-type compounds for modulating glucose metabolism. Nutrients 10:1821

Signor C, Girardi BA, Wendel AL, Frühauf PKS, Pillat MM, Ulrich H, Mello CF, Rubin MA (2017) Spermidine improves the persistence of reconsolidated fear memory and neural differentiation in vitro: Involvement of BDNF. Neurobiol Learn Memory 140:82–91

Sigrist SJ, Carmona-Gutierrez D, Gupta VK, Bhukel A, Mertel S, Eisenberg T, Madeo F (2014) Spermidine-triggered autophagy ameliorates memory during aging. Autophagy 10:178–179

Smith JA, Daniel R (2012) Stem cells and aging: a chicken-or-the-egg issue? Aging Dis 3:260–268

Speakman JR (2013) Evolutionary perspectives on the obesity epidemic: adaptive, maladaptive, and neutral viewpoints. Annu Rev Nutr 33:289–317

Stockinger J, Maxwell N, Shapiro D, Decabo R, Valdez G (2018) Caloric restriction mimetics slow aging of neuromuscular synapses and muscle fibers. J Gerontol Ser A 73:21–28

Storlien L, Oakes ND, Kelley DE (2004) Metabolic flexibility. Proc Nutr Soc 63:363–368

Tang D, Tao S, Chen Z, Koliesnik IO, Calmes PG, Hoerr V, Han B, Gebert N, Zörnig M, Löffler B, Morita Y, Rudolph KL (2016) Dietary restriction improves repopulation but impairs lymphoid differentiation capacity of hematopoietic stem cells in early aging. J Exp Med 213:535–553

Thornton KJ, Chapalamadugu KC, Doumit ME, Murdoch GK (2013) Polyamines enhance satellite cell activation and expression of myogenic regulatory factors. FASEB J 27:1146.7

Troca-Marín JA, Alves-Sampaio A, Montesinos ML (2011) An increase in basal BDNF provokes hyperactivation of the akt-mammalian target of rapamycin pathway and deregulation of local dendritic translation in a mouse model of down's syndrome. J Neurosci 31:9445

van Cauwenberghe C, Vandendriessche C, Libert C, Vandenbroucke RE (2016) Caloric restriction: beneficial effects on brain aging and Alzheimer's disease. Mamm Genome 27:300–319

Wang J, Gallagher D, Devito LM, Cancino GI, Tsui D, He L, Keller GM, Frankland PW, Kaplan DR, Miller FD (2012) Metformin activates an atypical pkc-cbp pathway to promote neurogenesis and enhance spatial memory formation. Cell Stem Cell 11(1):23–35

Wilkinson AC, Yamazaki S (2018) The hematopoietic stem cell diet. Int J Hematol 107:634–641

Xie W-Q, Xiao W-F, Tang K, Wu Y-X, Hu P-W, Li Y-S, Duan Y, Lv S (2020) Caloric restriction: implications for sarcopenia and potential mechanisms. Aging 12:24441–24452

Xu T-T, Li H, Dai Z, Lau GK, Li B-Y, Zhu W-L, Liu X-Q, Liu H-F, Cai W-W, Huang S-Q, Wang Q, Zhang S-J (2020) Spermidine and spermine delay brain aging by inducing autophagy in SAMP8 mice. Aging 12:6401–6414

Yang F, Chu X, Yin M, Liu X, Yuan H, Niu Y, Fu L (2014) mTOR and autophagy in normal brain aging and caloric restriction ameliorating age-related cognition deficits. Behav Brain Res 264:82–90

Yilmaz ÖH, Katajisto P, Lamming DW, Gültekin Y, Bauer-Rowe KE, Sengupta S, Birsoy K, Dursun A, Yilmaz VO, Selig M, Nielsen GP, Mino-Kenudson M, Zukerberg LR, Bhan AK, Deshpande V, Sabatini DM (2012) mTORC1 in the Paneth cell niche couples intestinal stem-cell function to calorie intake. Nature 486:490–495

Yousefi M, Nakauka-Ddamba A, Berry CT, Li N, Schoenberger J, Simeonov KP, Cedeno RJ, Yu Z, Lengner CJ (2018) Calorie restriction governs intestinal epithelial regeneration through cell-autonomous regulation of mTORC1 in reserve stem cells. Stem Cell Rep 10:703–711

Yousuf Y, Datu A, Barnes B, Amini-Nik S, Jeschke MG (2020) Metformin alleviates muscle wasting post-thermal injury by increasing Pax7-positive muscle progenitor cells. Stem Cell Res Ther 11:18

Zhang L, Gong H, Sun Q, Zhao R, Jia Y (2018) Spermidine-activated satellite cells are associated with hypoacetylation in ACVR2B and Smad3 binding to myogenic genes in mice. J Agric Food Chem 66:540–550

Zhang Q-S, Tang W, Deater M, Phan N, Marcogliese AN, Li H, Al-Dhalimy M, Major A, Olson S, Monnat RJ, Grompe M (2016) Metformin improves defective hematopoiesis and delays tumor formation in Fanconi anemia mice. Blood 128(24):2774–2784

Zhang QS, Marquez-Loza L, Eaton L, Duncan AW, Goldman DC, Anur P, Watanabe-Smith K, Rathbun RK, Fleming WH, Bagby GC, Grompe M (2010) Fancd2-/- mice have hematopoietic defects that can be partially corrected by resveratrol. Blood 116:5140–5148

Zheng B, Wang J, Tang L, Shi J, Zhu D (2017) mTORC1 and mTORC2 play different roles in regulating cardiomyocyte differentiation from embryonic stem cells. Int J Dev Biol 61:65–72

Chapter 26
Healthy Nutrition for Older People

Marios Kyriazis

Abstract The role of nutrition in health has fascinated humans for millennia. Currently, there is a substantial body of research guiding better practices to prevent or treat age-related conditions through nutritional interventions. Chronic degenerative conditions such as cardiovascular disease, sarcopenia and frailty, diabetes, cancer, osteoporosis, prostatic hyperplasia, menopause, age-related macular degeneration and many others, are amenable to nutritional interventions which, in association with specific treatments, may help alleviate the burden of the disease. Nutritional principles can also be applied for the prevention of each condition. In this chapter, there is a presentation regarding some of these principles, mainly focusing on the latest findings in each case. The aim is not to discuss all diets and nutrients suitable for each condition, but rather to use nutrition as a pretext to elucidate some biological mechanisms and processes involved in age-related degeneration. Certain intricacies of chronic degeneration, inflammation, oxidative damage and other age-related processes are at the heart of this discussion, which also examines the action of a number of nutrients or nutritional supplements. It is important to highlight that due to the fact that age-related diseases are multifactorial, it is necessary to employ a multi-pronged approach, tailored to the needs of each patient.

Keywords Health · Ageing · Age-related diseases · Inflammation · Cancer

26.1 Introduction

Chronic degeneration leads to age-related diseases which, in turn, affect the normal function of each one of us. From a clinical perspective, it is important for each individual to be able to live and operate in an environment within the limits imposed by physical, psychological and sociocultural challenges. If one is able to successfully overcome these challenges, then this individual may 'age successfully' and function well, to achieve their life goals. Although there are several elements which may

M. Kyriazis (✉)
National Gerontology Centre, Larnaca, Cyprus

S. I. S. Rattan and G. Kaur (eds.), *Nutrition, Food and Diet in Ageing and Longevity*, Healthy Ageing and Longevity 14, https://doi.org/10.1007/978-3-030-83017-5_26

help us overcome successfully these life challenges, this chapter will focus on the nutritional aspects.

Ageing, in other words, 'time-related dysfunction' (Kyriazis 2020), is associated with an increased risk of certain conditions and diseases. Here we will discuss some common ones: osteoarthritis, cardiovascular disease, cancer, age related macular degeneration, menopause, osteoporosis, Parkinson's disease, dementia, sarcopenia and frailty, and benign prostatic hyperplasia. On some occasions, the nutrients discussed here have been used since antiquity for their healthy-sustaining properties. Modern science has attempted to elucidate the exact mechanisms of action, and how these mechanisms may translate into concrete clinical benefits. In addition, many nutrients exhibit hormetic, dose–response benefits, and this presents an opportunity to discuss the increasingly important phenomenon of hormesis (Rattan and Kyriazis 2019). Below is a presentation highlighting certain advantages of nutrients, as these may be used against age-related conditions. The presentation is in no particular order, but aims to be balanced, provides a stimulus for further exploration, and covers most such conditions.

26.2 Osteoarthritis

At the heart of osteoarthritic (OA) changes is chronic inflammation. Therefore, nutritional products which regulate inflammation should provide benefit, both with regards to anatomical changes and clinical symptoms. One example is the case of catechins, chiefly found in green tea. Green tea catechins downregulate several inflammation processes and upregulate anabolic factors (Reddy et al. 2020). It was also shown that catechins can modulate miRNA expression and improve collagen protection (Luk et al. 2020).

In addition to the catechins, the role of dietary fatty acids is increasingly being recognized. In a 16-week randomized, double-blind, placebo-controlled study examining the effects of fish oil (2000 mg/day docosahexaenoic acid plus 400 mg/day eicosapentaenoic acid), it was found that these oils improve the pain of OA in overweight patients (Kuszewski et al. 2020). Supplementation with omega-3 fatty acids is associated with decreased pain and improvement of function, through modulation of apoptosis, reduction of oxidative stress and a decrease in prostaglandin production (Loef et al. 2019). One of the actions of docosahexaenoic acid, i.e. modulation of apoptosis, has been observed in OA cell models (Xu et al. 2019).

Another nutritional agent, which can have some benefit in OA is resveratrol, acting as an anti-inflammatory, anti-oxidative factor. Resveratrol modulates the function of the TLR4/Akt/FoxO1 axis which is involved in inflammation. Specifically, it upregulates phosphoinositide-3-kinase-Akt (PI3K/Akt) phosphorylation, inactivates FoxO1, and reduces Toll-like receptor 4 (TLR4) and inflammation activity (Xu et al. 2020). In a model of temporomandibular osteoarthritis, treatment with resveratrol was found to be associated with prevention of inflammation, reduction of chondrocyte apoptosis, and overall reduction of cartilage destruction (Yuce et al.

2020). Resveratrol is widely considered as having several beneficial actions in many conditions, and it will be discussed again later.

Other nutritional agents which were found to benefit OA:

- Ginger (through inhibition of prostaglandins, acting as an agonist of vanilloid nociceptor, and as antioxidation agent) (Rondanelli et al. 2020).
- Garcinia mangostana (a sweet, juicy tropical fruit) by targeting and suppressing pro-inflammatory cytokines (such as IL-6, TNF-α, and INF- γ) and other factors (NF-κB, STAT3, and COX-2) (Chiu et al. 2020). Extracts of this fruit could also be beneficial in other conditions where inflammation plays a role (Tsai et al. 2020).
- Vitamin K. There are two types of this vitamin, namely vitamin K1 (phylloquinone) and vitamin K2 (a series of menaquinones) with some differences in properties. Vitamin K is known to be associated with modulation of chronic inflammation and it can thus be of benefit, as a dietary supplement, in OA as well as in several other age-related conditions (Simes et al. 2020). Vitamin K1 is found mostly in green leafy vegetables (kale, spinach, lettuce, Swiss chard). Vitamin K2 is in meat, dairy products, liver, eggs, and Japanese "natto," (fermented soy beans).

26.3 Cardiovascular Disease

The nutritional approach into prevention and treatment of cardiovascular disease in older people is wide-ranging. Here, the discussion will revolve around only certain aspects of this approach, presenting the latest research with regards to only a few nutritional factors.

One relevant such factor is vitamin E, which reduces oxidative stress and inflammation during cardiovascular events. Studies show that vitamin E deficiency is associated with an increased risk of cardiovascular events, although studies of vitamin E supplementation in such events have not been definitive (Ziegler et al. 2020). It may be the case that acute supplementation during myocardial infarction will yield better results than chronic preventative consumption through the diet.

Another nutrient which plays a huge role in cardiovascular disease is the group of omega-3 fatty acids already mentioned above. The role of omega-3 fatty acids (such as eicosapentaenoic acid, and docosahexaenoic acid) is well recognized in reducing inflammatory mediators (cytokines and leukotrienes) and atherosclerosis. One of the mechanisms involved is reprogramming of triglyceride-rich lipoproteins (TRLs) (Shibabaw 2020).

Studies of marine omega-3 fatty acids (1 g/d) and vitamin D3 (2000 IU/d) in 25 871 men aged over 50 and in women aged over 55 years for five years, did not find significant benefits of vitamin D with regards to cardiovascular events, although there was a reduction of total coronary risk from the marine fatty acids (Manson et al. 2020). But, as vitamin D can be of benefit in other age-related conditions (for instance osteoporosis), its role remains important. Nevertheless, omega-3 fatty acids are being proven consistently beneficial in reducing the risk of cardiovascular disease.

An important subject that should be mentioned here is that of Calorie Restriction, including the practice of Intermittent Fasting. This notion is crucial in ageing, and can provide information regarding all age-related degenerative conditions. However, it is something that can be practically difficult or impossible to apply on humans as a daily practice. During a calorie restricted diet, the subject is fed approximately 30% (or other proportions) of the normal *ad libidum* diet, and this has been shown to have a host of positive (as well as negative) effects on the organism. The effects are evident when the total amount of calories is restricted, regardless of the diet composition (Brandhorst and Longo 2019). Due to the fact that a routine calorie restricted diet cannot be applied on an average human (due to practical restrictions, hunger and other adverse effects), another pattern that can be of use is intermittent fasting. A short-term 'fasting-mimicking' dietary approach is a periodic, low-calorie, and low-protein dietary approach which can have the beneficial effects without the adverse effects of a chronic and continuing calorie restricted diet (Crupi et al. 2020).

In more general terms, and leaving the subject of calorie restriction aside, adherence to a healthy dietary pattern is as significant as the contents of such a healthy diet. Healthy dietary patterns are those generally low in red meat, salt, and refined sugars, and high in oily fish, deep coloured fruit and vegetables, fibre, pulses, nuts, and whole grains. However, importance should not be given on single nutrient approaches. A more significant aspect is the wider dietary habits which involve consumption of several food groups, over long periods of time, rather than episodically. For instance, a Western type dietary pattern has a higher cardiovascular risk, compared to a consistent Mediterranean type diet (Najafi et al. 2020).

26.4 Cancer

It is beyond the scope of this chapter to discuss nutritional support for all cancers. However, it may be interesting to give a glimpse into the role of certain nutrients which, alone or in combination, could be used to improve cancer-related symptoms or risks. The following is a short list:

- Coriolus versicolor, an edible medicinal mushroom, consumed mainly in China. Two of its polysaccharides, namely CVPn and CVPa were shown to induce nitric oxide production, phagocytosis and reduction of Tumour Necrosis Factor (TNF) (Zhang et al. 2020). In other words, coriolus may exhibit increased immunomodulation and thus be useful in inhibiting the growth of cancerous cells.
- Resveratrol in association with curcumin. In a recent study, the combination of these two nutrients was found to enhance autophagy, modulate apoptosis and provide protection against cancer both in vitro and in vivo (Patra et al. 2020). Resveratrol is abundant in red wine, blueberries and dark chocolate, while curcumin is found turmeric. Both are also available in oral supplement form.
- L-carnosine (β-alanine, l-histidine). Found in red meat and game meat, carnosine is a pluripotent agent with a host of effects. Apart from acting as an antioxidant and

anti-glycator, it is a heavy metal chelator, and apoptosis modulator (Chmielewska et al. 2020). It has been used against several degenerative conditions, such as cataract, diabetes, cancer, neurodegeneration, and even schizophrenia (Banerjee and Poddar 2020). Carnosine will be mentioned again below.

- Chrysin (5,7-dihydroxyflavone) is found in honey, propolis, and passion flower. It exhibits antiproliferative activities and it can be used in association with conventional chemotherapy with cisplatin (Sherif et al. 2020). Its antineoplastic effect has been studied with regards to cancers of the lung, breast, colon, cervix, stomach, melanoma, and liver (Ganai et al. 2020), so it remains a promising agent.

The concept of calorie restriction mentioned above, can also be useful in the general case of cancer. Through a restrictive dietary pattern, it may be possible to activate autophagy, and this may have positive outcomes during chemotherapy. An increased dependency of cancer cells on autophagy may be exploited through nutritional interventions, although research on humans is still unsatisfactory (Cozzo et al. 2020). Autophagy is a process whereby dysfunctional or unnecessary cellular components are eliminated by the cell, and this mechanism has gained increased attention by researchers with regards to ageing. In any case, calorie restriction and/or intermittent fasting are known to exert benefits on insulin, IGF, cortisol, sexual hormones, oxidation, inflammation, as well as on markers such as FOXO, AMPK and SIRT-1 (Longo and Fontana 2010).

It is clear from the above short discussion, that nutrition plays an important role here and that research is confirming the beneficial action of several nutrients. However, it is also important to mention that it is unlikely that any single nutrient will have a noticeable effect. Best results may be obtained through a combinational nutritional approach, in association with other lifestyle measures, tailored specifically to the individual.

26.5 Age-Related Macular Degeneration

The subject of nutrition features high in the list of prevention measures recommended for Age-Related Macular Degeneration (ARMD). It is worth remembering that ARMD, together with glaucoma, are neurodegenerative conditions and are the leading causes of blindness in older populations.

Research in ARMD over the past several decades has examined the action of nutrients (vitamins/minerals/factors) such as vitamins C and E, beta-carotene, zinc, lutein, zeaxanthin, copper and the polyunsaturated fatty acid docosahexaenoic acid (DHA) among others (Walchuk and Suh 2020). Some of the recommended nutrients may exhibit hormetic (Rattan and Demirovic 2009) benefits, being beneficial in low doses and detrimental in higher doses. The typical example is resveratrol which inhibits neovascularization, reduces reperfusion damage and improves vascular serum biomarkers (Richer et al. 2013). Resveratrol is available mostly in red wine (hormetic doses!) and as an oral supplement.

Exogenous supplementation with NAD (nicotinamide adenine dinucleotide) has been found to play an important role both in therapy and in prevention (Cimaglia et al. 2020). NAD is a factor which regulates retinal cell metabolism and homeostasis, and mechanisms involved in this respect are the induction of mitophagy and the regulation of oxidative stress (Wei et al. 2019).

Several other nutritional manipulations have been associated with a decreased risk of ARMD. For example, a decreased omega-6/omega-3 ratio protects against neovascular ARMD (Mance et al. 2011). It has been suggested that a omega-6/omega-3 ratio of 1 is the normal evolutionary ideal for health, whereas in Western diets this ratio is as high as 15 to 1 (Simopoulos 2006). This increased ratio is associated with several other chronic conditions, including cardiovascular diseases, cancer and chronic inflammation, as mentioned above. In this respect, increasing the intake of omega-3 fatty acids should be the aim of dietary interventions. Therefor a diet rich in cold-water fatty fish (herring, tuna, sardines, salmon, and mackerel), in seeds/nuts (walnuts, flaxseed, soya bean, canola) or in the form of oral supplements, has been recommended (National Institute of Health, Office of Dietary Supplements https://ods.od.nih.gov/factsheets/Omega3FattyAcids-Consumer/, retrieved 20 January 2021).

Certain alternative interventions may also be of some use. A retrospective single-group study using a treatment modality based on intravenous nutrition in association with microcurrent stimulation and light therapy was found useful (Kondrot 2015). The authors of this study found significant improvements in visual acuity contrast and visual fields.

There are foods that are best avoided in ARMD. Examples include processed foods that contain trans fats, palm oil, lard, vegetable shortening, and margarine. High-fat dairy foods, fatty beef, pork and lamb are also to be avoided.

26.6 Menopausal Symptoms

Nutritional support for menopausal problems has been the subject of a large number of studies, for many decades. Nutrients or supplements studied include black cohosh, flaxseed, calcium, red clover, vitamin D, and, of course, soya. The following is a short comment on some of these, concentrating on the latest published research at the time of writing.

- Black cohosh *(Actaea racemosa, Cimicifuga racemosa)*. In a systematic study it was found that black cohosh extract was significantly better than placebo in the treatment of menopausal symptoms, with no significant adverse effects (Castelo-Branco et al. 2020).
- Flaxseed. A very interesting study explored the relationship between lignan-rich oilseeds such as flaxseed oil, and the gut microbiota. The study showed that such oilseeds interact with faecal microbiota in premenopausal women and, through modulation of enterolignans, act as a substitute for human oestrogen (Corona et al. 2020). This leads to a better clinical profile, reducing menopausal symptoms.

- Soya genistein (Thangavel et al. 2019). One of the best studied elements of soya is the group of isoflavones (phytoestrogens), and specifically genistein. This forms a large part of the isoflavone content of soya, perhaps as high as 60%. Genistein improves glucose metabolism, induces apoptosis in cancer, has antioxidant effects, and modulates postmenopausal symptoms such as hot flushes, anxiety and depression.

Polyphenols (isoflavones, genistein etc.) exhibit a hormetic, biphasic dose–response activity. In other words, they are beneficial in low dose and detrimental in higher doses, having only a defined window of positive result, beyond which they become detrimental (Leri et al. 2020).

Apart from nutrients used to alleviate symptoms of the menopause itself, it is also necessary to consider nutrients which are of benefit in conditions associated with the menopause, such as osteoporosis. Vitamin D status can affect both menopausal symptoms and osteoporosis, as mentioned below. Dietary macronutrients such as vitamin D consumption may have a positive effect in this respect. It is known that serum 25-hydroxyvitamin D concentration is correlated with menopausal status, and that oral consumption of vitamin D may be able to modify any adverse effects associated with the menopause (Chun et al. 2020). Sources of dietary vitamin D include fish such as salmon, sardines, herring and mackerel, red meat and liver, egg yolks and specially fortified foods, and of course, as oral dietary supplements. It is worth mentioning at this point that vitamin D deficiency is widespread, including (paradoxically) in countries where there are long periods of sunlight. Levels of 25(OH)D below 30 nmol/L are encountered in over 20% of the population in India, Tunisia, Pakistan, and Afghanistan. Specifically, in India, it is estimated that half a billion people may have low vitamin D levels (Cashman 2020).

26.7 Osteoporosis

The case of osteoporosis is an example of the complexity of ageing, in the sense that it is not just one disease related to degeneration, but its presence is associated with other conditions affecting the elderly, such as frailty, falls, mobility problems, pain etc. In addition, it is an example of how a multifactorial approach to treatment is necessary. It is not enough to employ only a dietary or a pharmacological approach, but several other interventions need to be used in association, e.g. physical exercise, hormonal manipulation, and appropriate lifestyle measures. This underlines the requirement to deal with ageing in a multidisciplinary fashion, where treatment of one condition may have secondary beneficial effects on another. In any case, nutrition does play a significant role here, as obesity and osteoporosis are interrelated, and as it is known that the Western diet with its saturated fat content does influence osteoporosis (Martyniak et al. 2020). In this respect, polyunsaturated fats in the diet are beneficial because these reduce bone loss and enhance osteogenesis (Bao et al. 2020). Certain other nutrients do have a proven benefit, and these include minerals (calcium, phosphorus and magnesium), vitamin D, vitamin B12 (Kalimeri et al. 2020) protein rich foods, fibre, fruits and vegetables, and prebiotic foods (Ilesanmi-Oyelere and Kruger 2020).

Other studies have shown the benefit of tomato (green tomato extract) which improves bone formation through Bmp2-Smad 1/5/8-Runx2 signaling, and modulate the nuclear factor kappa-B (RANKL)/osteoprogeterin (OPG) pathway, which in turn improves bone resorption (Nirmala et al. 2020). It is an opportunity here to mention that BMPs (Bone Morphogenetic Proteins), are members of the TGF-β superfamily, which have strong osteogenic activity and stimulate mesenchymal osteoprogenitor cells to become mature osteoblasts. This shows that nutritional factors may indeed modulate bone metabolism, therefore a wise dietary choice should have a positive therapeutic result.

A relevant vitamin not usually considered by nutritionists in osteoporosis, is vitamin K. This vitamin plays a role in cardiovascular health and in bone development, as already mentioned above in the case of osteoarthritis. The National Academy of Science (Food and Nutrition Board), has determined that the dietary requirements of healthy people are around 100 ug/day, but the requirements in older people with chronic degenerative conditions could be different. There is an intricate relationship between vitamin K, gut microbiota and osteoporosis. Certain intestinal bacteria produce vitamin K and thus affect bone metabolism, bone composition and, ultimately, fracture risk. The most relevant bacteria in this respect are those of the genus Bacteroides (fracture risk increases as their concentration decreases), and those of the Rikenellaceae family (Ozaki et al. 2020). Therefore, it is important to maintain good intestinal flora through use of probiotics (BOX 1). This will not only have an effect on bowel health but can also affect bone and even neuronal health, as mentioned below in the case of dementia.

26.8 Alzheimer's Disease

Among other factors, inflammation, and one of its markers, Tumor Necrosis Factor alpha (TNFα), have been implicated in Alzheimer's Disease (AD). A high-fat diet increases expression of TNFα in the tissues but this does not necessarily translate into a worsening cognition score (Jackson et al. 2020). There are multiple factors and processes contributing to AD and it would be simplistic to claim that a suitable nutrition will inevitably lead to a reduced risk of the disease. Nevertheless, it is possible to discuss some dietary factors that could possibly be useful, both in prevention and in patients with established AD. For instance, resveratrol could be of benefit, particularly if combined with physical exercise. The mechanism here could be due to resveratrol's effects on apoptosis, neuroinflammation and reduction of Aβ oligomers (brain amyloid-beta) (Broderick et al. 2020).

Enhancing mitophagy will be mentioned below, in the case of Parkinson's disease. However, mitophagy is also implicated in other neurodegenerative diseases including

Alzheimer's Disease. During mitophagy there is elimination of age-damaged mitochondria. Nutrients such as resveratrol, curcumin, astaxanthin and spermidine (which are found in the Mediterranean and Okinawan diets), are increasingly being promoted as effective in enhancing mitophagy (Varghese et al. 2020).

The relationship between the gut microbiota and chronic neurodegenerative conditions is increasingly being recognized, and has been briefly mentioned above. Dysregulation of the function of gut microbiota, affects the gut-brain axis and has repercussions on neurological function (Zhu et al. 2020). Therefore, maintaining an effective balance and function of the gut microbiota, usually through nutritional manipulations, should result in a decreased risk of neurodegeneration, including Alzheimer's dementia (Kowalski and Mulak 2019) (BOX 1). For instance, ketogenic diets and intermittent fasting were found to have an effect on Alzheimer's disease, at least in a rat model (Park et al. 2020). In this study it was found that intermittent fasting and a diet high in starch may decrease the progress of dementia. It is, however, suggested that because there is a great variation in the gut microbiota between people, it is necessary to tailor-made each intervention to optimally suit each individual.

****

BOX 1

Ways to improve gut microbiota include (Valdes et al. 2018):

* Probiotics (Bifidobacterium and Lactobacillus species)

* Dietary fibre

* Avoidance of antibiotics, pesticides, and food additives, such as emulsifiers

* Fermented foods such as kimchi (fermented cabbage), sauerkraut, кombucha (a fermented beverage of black tea and sugar), kefir (fermented milk product), miso (fermented soybean, barley or brown rice), tempeh (fermented soybean product), raw cheeses made from unpasteurized milk, and Greek yogurt.

****

The potential of carnosine against chronic degeneration has already been mentioned above, but it is worth returning to this nutrient as it appears to have a confirmed anti-neurodegenerative action (Schön et al. 2019). Carnosine is available in red meat and also as an over-the-counter food supplement. It can cross the blood-brain barrier, reduces amyloid beta polymerization, and decreases aggregates of amyloid. In addition, carnosine protects brain vascular endothelial cells (RBE4) against toxicity induced by beta amyloid, and modulates brain-derived neurotrophic factor (BDNF) and nerve growth factor (NGF) in some animal models (Schön et al. 2019). Overall, it optimizes energy metabolism, improves the function of mitochondria in neurons and acts as a suppressor of chronic neurodegeneration, making it an ideal nutritional supplement in age-related brain diseases (Banerjee and Poddar 2020).

Also, at this point it is worth mentioning that a patient suffering from dementia is more likely to be frail, weak, and sarcopenic, with other concomitant conditions. It is therefore important to consider all aspects of health in this type of patient and establish a nutritional program suited to the specific (not the general) needs of each such patient.

26.9 Parkinson's Disease

Conditions which can contribute to the pathology of Parkinson's disease (PD) may include dysbiosis due to dietary habits, and increased intestinal permeability (Lister 2020). It may thus be possible to alter the progression of the disease by suitable manipulation of these nutrition-related conditions. Principal factors in this respect could be the status of vitamin D and B complex, omega fatty acids, probiotics and coenzyme Q10.

An interesting approach in Parkinson's disease is to modulate mitophagy, in other words, the elimination of damaged mitochondria, through nutritional factors. We know that there is mitochondrial dysfunction in dopaminergic neurons in PD, therefore any measures which may result in improved mitophagy may lead to an improvement in clinical symptoms. Promotion of mitophagy can be achieved through boosting of Parkin, a ubiquitin ligase, by consuming a low protein, plant-based diet (McCarty and Lerner 2020). This could be one of the reasons why East Asian and sub-Saharan Africa diets (mostly vegan) have a decreased risk of PD compared to the high protein diets of Western countries.

A diet rich in spermidine may also enhance mitophagy (Yang et al. 2020). Spermidine can be found in the components of the Mediterranean diet such as mushrooms, legumes, and whole grains, as well as in soy and corn. The role of spermidine in protecting against neurodegeneration in general (not only PD but also Alzheimer's disease, and ageing degeneration) is important and is slowly but increasingly being elucidated.

Hydrogen sulfide (HS) is a signaling molecule with multiple actions, and it exhibits typical hormetic properties. It provides protection against oxidative damage and it is also a cytoprotector. It plays a positive role in neurogenesis (Sun et al. 2020). N-acetylcysteine and methionine are precursors of HS, and are available in oral supplements. However, HS acts in a hormetic manner, meaning that while low doses can be of benefit, higher doses may be detrimental (Calabrese et al. 2010).

In addition, in PD there is generation of cytotoxic factors (such as peroxynitrite) by microglia, which then cause dysfunction of dopaminergic neurons. It has been suggested that modulation of peroxynitrite through nutritional factors may translate into clinical benefit in PD (McCarty and Lerner 2020). Compounds which may play a role in this respect include spirulina, vitamin D, caffeine, probiotics, promotion of hydrogen sulfide (taurine, N-acetylcysteine) and generally low-protein diets which may also help in improving response to levodopa drug therapy (Guebila and Thiele 2016).

Certain plant alkaloids, such as palmatine, have exhibited inhibition activities against anti-acetylcholinesterase, and are thus being studied further with regards to their phytotherapeutic effects against Parkinson's disease (Chaves et al. 2020). Palmatine is found in extracts of the Coscinium fenestratum plant, for example. It has several other health-improving properties, and it exhibits a hormetic, dose–response activity – essentially acting as a hormetin (Long et al. 2019).

Studies of diet patterns show that 'healthy' diets (such as the Mediterranean diet) are associated with a reduced incidence of prodromal symptoms of Parkinson's disease (Mosberry et al. 2020). Specifically, it was shown that increased adherence to a Mediterranean diet is associated with a corresponding decrease in the risk of non-motor prodromal symptoms of Parkinson's disease in older people, such as depression, constipation, urinary dysfunction and daytime somnolence (Maraki et al. 2019).

26.10 Sarcopenia

A common condition which contributes to physical frailty and falls in later life is sarcopenia. It is a progressive condition where there is loss of skeletal muscle mass and reduction of physical strength. Sarcopenia has been associated with low-grade chronic inflammation, with abnormalities in protein metabolism in the muscle tissues, and mitochondrial dysfunction. Clinically, it has been associated with several detrimental effects common in ageing, such as falls, hospitalization, chronic disability, and institutionalization.

An interesting view is that dietary interventions may influence miRNAs in muscle (myomiRNAs) which regulate development and maintenance of muscle tissue (Barbiera et al. 2020). Dysregulation of myomiRNAs is found in ageing (Drummond et al. 2008), and it has been shown that nutritional factors may play a positive role here (Iannone et al. 2020). See BOX 2.

****

BOX 2

Nutrient-dependent miRNA regulation may be achieved with:

* Essential amino acids (EAAs), which increase miRNA function (Drummond et al. 2009)

* Glucose and amino acids (e.g. Leucine) (Chen et al. 2013)

* Albumin (Soeters et al. 2019)

****

There are several studies confirming that certain nutrients may impact positively on sarcopenic patients. For instance, daily consumption of yogurt fortified with vitamins D (1000 IU) and C (500 mg) together with beta-Hydroxy beta-Methyl Butyrate (3 g) for 12 weeks in a randomized, double-blind controlled manner, was found to improve muscle strength and anabolic functions in older patients with sarcopenia (Nasimi et al. 2020). The role of beta-Hydroxy beta-Methyl Butyrate (HMB) in building muscle mass in sarcopenia is intriguing. It is a metabolite of leucine which regulates muscular protein synthesis and has a host of other actions (Cruz-Jentoft 2018). For instance, it:

- Decreases muscle cell apoptosis
- Stabilizes cell membranes
- Inhibits the ubiquitin-proteasome pathway, thus reducing proteolysis

- Enhances proliferation and differentiation of stem cells in muscle
- May prevent muscle loss in bed-ridden patients who cannot exercise

Therefore, although more studies are needed in order to confirm solidly the effects of HMB, it is fair to suggest that its use can help older people with sarcopenia, by improving lean muscle mass, muscle strength and physical function (Oktaviana et al. 2019). This nutrient is usually taken by mouth in powder form.

Royal jelly is another nutrient which may be helpful in sarcopenia. It delays muscular functional decline during ageing and prevents muscle fiber atrophy (Shirakawa et al. 2020). Royal jelly, and other bee products such as propolis and bee pollen, are rich in antioxidants with strong action, including flavonoids and phenols, as well as being rich in several amino acids. These can modulate inflammatory response and decrease oxidative and glycation stress in muscular tissues, reduce catabolic activity, improve stem-cell function, improve AMPK metabolism, and enhance muscle protein synthesis (Ali and Kunugi 2020).

The activity of AMPK (5'-adenosine monophosphate-activated protein kinase) is one of the most interesting subjects in the field of ageing. AMPK controls muscle metabolism and modulates anabolic and catabolic events. It improves muscle mass and muscle tissue regeneration (Thomson 2018). It is known that obesity is associated with a reduced AMPK function, although chronic activation of AMPK, is also found in obesity and in impaired pancreatic function. We see once again that hormetic, dose-response principles come into play, when long-term, indiscriminate activation of AMPK is not necessarily beneficial and should be avoided (Lyons and Roche 2018).

26.11 Benign Prostatic Hyperplasia (BPH)

Peanut sprouts (germinated peanuts) containing high amounts of resveratrol are eaten in salads or in stir-fries usually in a vegetarian diet context. Extracts of the Yesan sprout variety were found to affect molecular markers of BPH, such as 5α-reductase, fibroblast growth factor, and the apoptotic markers Bcl-2, and Bax. In model rats, the extract also reduced the size, weight and thickness of prostate (Song et al. 2020).

Nutritional habits which may affect zinc in men over 50 years old, can also affect the risk of BPH. Zinc has consistently been shown to play an active role in BPH, although it is not a simple matter of increasing the daily dietary amounts of zinc and expecting to see clinical result. Zinc metabolism depends on factors that may promote or hinder zinc absorption. These factors can be endogenous and/or exogenous, such as drugs or a diet low in zinc (Sauer et al. 2020). Zinc exhibits a concentration-dependent effect on inflammation, oxidation and apoptosis, through modulation of TNF-α and IL-6 (Hacioglu et al. 2020). Foods rich in zinc include oysters, red meat and poultry, beans, nuts, and whole grains.

Although it is known than cranberry extracts may have an effect on urinary tract infections, little is known about its effect on BPH. In a model using male Sprague-Dawley rats treated with cranberry powder, it was found that the extract has a significant decreasing effect on prostate weight, dihydrotestosterone (DHT), and 5-alpha reductase (An et al. 2020).

An interesting nutrient and hormetin in the case of prostate, is ginger. Hormetins are natural or synthetic compounds which can produce hormesis (Rattan and Kyriazis 2019). Ginger (*Zingiber officinale Roscoe)* has numerous effects including antidiabetic, lipid-lowering and anti-inflammatory actions. Its activities include apoptosis modulation, improvement of autophagy, and cell regulation. In addition, ginger has oestrogen-modulating effects which means that it can be beneficial in prostate hyperplasia, as well as in other conditions where oestrogen is involved (osteoporosis, menopause and certain cancers) (Kiyama 2020).

26.12 Conclusion

It is important to advise the patient that there is no universal diet or a nutritional pattern that can protect against all age-related diseases. Each disease, while sharing certain common characteristics of ageing (for instance oxidation, inflammation, degeneration, and glycation), has to be seen separately and focused on the needs of each particular patient. The principles of personalised medicine should be applied in all aspects of nutritional interventions, both for prevention and for treatment. It is known that diets which are generally considered ‘healthy’ in some countries (e.g. Mediterranean diet) may not be suited to the cultural, social or biological makeup of other countries (Asia, for instance) (Moffat and Morell-Hart 2020). Therefore, a dietary pattern has to be tailored individually to each patient, and be considered with other lifestyle measures, specific to that patient. Nevertheless, there are certain pluripotent nutrients that can be recommended in most conditions. Examples include resveratrol, carnosine, omega-3 acids, and vitamins D, E and C.

Nutrition may help in the prevention and treatment of age-related degenerative conditions through modulation of oxidation, glycation, inflammation and cell metabolism. It is not, however, the only important factor in these conditions, neither it is an approach that can work on its own, separate from other lifestyle interventions. The interaction between nutrients and physical exercise, smoking and excess alcohol, as well as genetic factors, play an intricate role in achieving optimal health for the ageing patient. Research has clarified a great amount of basic biological mechanisms of how this can be achieved. The study of these mechanisms provides useful insights regarding the mechanisms of ageing, hormesis, and optimal clinical function in ageing.

Compliance with Ethical Standards

Conflict of Interest All authors declare they have no conflict of interest.

References

Ali AM, Kunugi H (2020) Apitherapy for age-related skeletal muscle dysfunction (Sarcopenia): a review on the effects of Royal Jelly, Propolis, and Bee Pollen. Foods 9:1362

An YJ, Lee JY, Kim Y, Jun W, Lee YH (2020) Cranberry powder attenuates benign prostatic hyperplasia in rats. J Med Food. https://doi.org/10.1089/jmf.2020.4779

Banerjee S, Poddar MK (2020) Carnosine research in relation to aging brain and neurodegeneration: A blessing for geriatrics and their neuronal disorders. Arch Gerontol Geriatr 91:104239. https://doi.org/10.1016/j.archger.2020.104239. Epub ahead of print

Bao M, Zhang K, Wei Y, Hua W, Gao Y, Li X, Ye L (2020) Therapeutic potentials and modulatory mechanisms of fatty acids in bone. Cell Prolif 3(2):e12735

Barbiera A, Pelosi L, Sica G, Scicchitano BM (2020) Nutrition and microRNAs: Novel Insights to Fight Sarcopenia. Antioxidants (Basel) 2 9(10):E951

Brandhorst S, Longo VD (2019) Dietary restrictions and nutrition in the prevention and treatment of cardiovascular disease. Circulation Res 124:952–965

Broderick TL, Rasool S, Li R, Zhang Y, Anderson M, Al-Nakkash L, Plochocki JH, Geetha T, Babu JR (2020) Neuroprotective effects of chronic resveratrol treatment and exercise training in the 3xTg-AD mouse model of alzheimer's disease. Int J Mol Sci 21(19):7337

Calabrese V, Cornelius C, Dinkova-Kostova AT, Calabrese EJ, Mattson MP (2010) Cellular stress responses, the hormesis paradigm, and vitagenes: novel targets for therapeutic intervention in neurodegenerative disorders. Antioxid Redox Signal 13(11):1763–1811

Cashman KD (2020) Vitamin D deficiency: defining, prevalence, causes, and strategies of addressing. Calcif Tissue Int 106:14–29

Castelo-Branco C, Gambacciani M, Cano A, Minkin MJ, Rachoń D, Ruan X, Beer AM, Schnitker J, Henneicke-von Zepelin HH, Pickartz S (2020) Review & meta-analysis: isopropanolic black cohosh extract iCR for menopausal symptoms - an update on the evidence. Climacteric Oct 6:1–11. https://doi.org/10.1080/13697137.2020.1820477. Epub ahead of print

Chaves SKM, Afzal MI, Islam MT, Hameed A, Da Mata AMOF, Da Silva Araújo L, Ali SW, Rolim HML, De Medeiros MDGF, Costa EV, Salehi B, Martins N, Arif AM, Imran M, Sharifi-Rad J, Melo-Cavalcante AAC, Feitosa CM (2020) Palmatine antioxidant and anti-acetylcholinesterase activities: a pre-clinical assessment. Cell Mol Biol (Noisy-le-grand) 66(4):54–59

Chen X, Huang Z, Chen D, Yang T, Liu G (2013) MicroRNA-27a is induced by leucine and contributes to leucine-induced proliferation promotion in C2C12 cells. J Mol Sci 14:14076–14084

Chiu YS, Wu JL, Yeh CT, Yadav VK, Huang HS, Wang LS (2020) γ-Mangostin isolated from Garcinia mangostana L. suppresses inflammation and alleviates symptoms of osteoarthritis via modulating miR-124–3p/IL-6/NF-κB signaling. Aging (Albany NY) 12(8):6630–6643

Chmielewska K, Dzierzbicka K, Inkielewicz-Stępniak I, Przybyłowska M (2020) Therapeutic potential of carnosine and its derivatives in the treatment of human diseases. Chem Res Toxicol 33(7):1561–1578

Chun H, Kim GD, Doo M (2020) Differences in the association among the vitamin D concentration, dietary macronutrient consumption, and metabolic syndrome depending on pre- and postmenopausal status in Korean women: a cross-sectional study. Diabetes Metab Syndr Obes 13:3601–3609

Cimaglia G, Votruba M, Morgan JE, André H, Williams PA (2020) Potential therapeutic benefit of NAD+ supplementation for glaucoma and age-related macular degeneration. Nutrients 12(9):2871

Corona G, Kreimes A, Barone M, Turroni S, Brigidi P, Keleszade E, Costabile A (2020) Impact of lignans in oilseed mix on gut microbiome composition and enterolignan production in younger healthy and premenopausal women: an in vitro pilot study. Microb Cell Fact 19(1):82

Cozzo AJ, Coleman MF, Pearce JB, Pfeil AJ, Etigunta SK, Hursting SD (2020) Dietary energy modulation and autophagy: exploiting metabolic vulnerabilities to starve cancer. Front Cell Dev Biol 8:590192

Crupi AN, Haase J, Brandhorst S, Longo VD (2020) Periodic and intermittent fasting in diabetes and cardiovascular disease. Curr Diab Rep 20(12):83

Cruz-Jentoft AJ (2018) Beta-hydroxy-beta-methyl butyrate (HMB): from experimental data to clinical evidence in sarcopenia. Curr Protein Pept Sci 19(7):668–672

Drummond MJ, McCarthy JJ, Fry CS, Esser KA, Rasmussen BB (2008) Aging differentially affects human skeletal muscle microRNA expression at rest and after an anabolic stimulus of resistance exercise and essential amino acids. Am J Physiol Endocrinol Metab 295(6):E1333–E1340

Drummond MJ, Glynn EL, Fry CS, Dhanani S, Volpi E, Rasmussen BB (2009) Essential amino acids increase MicroRNA-499, -208b, and -23a and downregulate myostatin and myocyte enhancer factor 2C mRNA expression in human skeletal muscle. J Nutr 139:2279–2284

Ganai SA, Sheikh FA, Baba ZA (2020) Plant flavone Chrysin as an emerging histone deacetylase inhibitor for prosperous epigenetic-based anticancer therapy. Phytother Res. Sep 15. https://doi.org/10.1002/ptr.6869. Epub ahead of print

Guebila MB, Thiele I (2016) Model-based dietary optimization for late-stage, levodopa-treated, Parkinson's disease patients. NPJ Syst Biol Appl 2:16013

Hacioglu C, Kacar S, Kar F, Kanbak G, Sahinturk V (2020) Concentration-dependent effects of zinc sulfate on DU-145 human prostate cancer cell line: oxidative, apoptotic, inflammatory, and morphological analyzes. Biol Trace Elem Res 195(2):436–444

Iannone F, Montesanto A, Cione E, Crocco P, Caroleo MC, Dato S, Rose G, Passarino G (2020) Expression patterns of muscle-specific miR-133b and miR-206 correlate with nutritional status and sarcopenia. Nutrients 12(2):297

Ilesanmi-Oyelere BL, Kruger MC (2020) Nutrient and dietary patterns in relation to the pathogenesis of postmenopausal osteoporosis-a literature review. Life (basel) 10(10):220

Jackson JM, Bay AA, Barter JD, Ni L, Caudle WM, Serra MC, Wharton W, Hackney ME (2020) The role of nutrition and inflammation on cognition in a high-risk group for alzheimer's disease. J Alzheimers Dis Rep 4(1):345–352

Kalimeri M, Leek F, Wang NX, Koh HR, Roy NC, Cameron-Smith D, Kruger MC, Henry CJ, Totman JJ (2020) Folate and vitamin B-12 status is associated with bone mineral density and hip strength of postmenopausal Chinese-Singaporean women. JBMR plus 4(10):e10399

Kiyama R (2020) Nutritional implications of ginger: chemistry, biological activities and signaling pathways. J Nutr Biochem 86:108486

Kondrot EC (2015) Improvement in vision parameters for participants treated with alternative therapies in a 3-day program. Altern Ther Health Med 21(6):22–35

Kowalski K, Mulak A (2019) Brain-gut-microbiota axis in alzheimer's disease. J Neurogastroenterol Motil 25(1):48–60

Kuszewski JC, Wong RHX, Howe PRC (2020) Fish oil supplementation reduces osteoarthritis-specific pain in older adults with overweight/obesity. Rheumatol Adv Pract 4(2):rkaa036

Kyriazis M (2020) Aging as "Time-Related Dysfunction": A Perspective. Frontiers in Medicine 7, 371. https://www.frontiersin.org/article/10.3389/fmed.2020.00371

Leri M, Scuto M, Ontario ML, Calabrese V, Calabrese EJ, Bucciantini M, Stefani M (2020) Healthy effects of plant polyphenols: molecular mechanisms. Int J Mol Sci 21(4):1250

Lister T (2020) Nutrition and lifestyle interventions for managing Parkinson's disease: a narrative review. J Mov Disord 13(2):97–104

Loef M, Schoones JW, Kloppenburg M, Ioan-Facsinay A (2019) Fatty acids and osteoarthritis: different types, different effects. Joint Bone Spine 86(4):451–458

Long J, Song J, Zhong L, Liao Y, Liu L, Li X (2019) Palmatine: a review of its pharmacology, toxicity and pharmacokinetics. Biochimie 162:176–184

Longo VD, Fontana L (2010) Calorie restriction and cancer prevention: metabolic and molecular mechanisms. Trends Pharmacol Sci 31:89–98

Luk HY, Appell C, Chyu MC, Chen CH, Wang CY, Yang RS, Shen CL (2020) Impacts of green tea on joint and skeletal muscle health: prospects of translational nutrition. Antioxidants (basel) 9(11):E1050

Lyons CL, Roche HM (2018) Nutritional Modulation of AMPK-Impact upon metabolic-inflammation. Int J Mol Sc 19(10):3092

Mance TC, Kovacević D, Alpeza-Dunato Z, Stroligo MN, Brumini G (2011) The role of omega6 to omega3 ratio in development and progression of age-related macular degeneration. Coll Antropol 35(Suppl 2):307–310

Manson JE, Bassuk SS, Cook NR, Lee IM, Mora S, Albert CM, Buring JE; VITAL Research Group (2020) Vitamin D, Marine n-3 Fatty Acids, and Primary Prevention of Cardiovascular Disease Current Evidence. Circ Res 126(1):112-128

Maraki MI, Yannakoulia M, Stamelou M, Stefanis L, Xiromerisiou G, Kosmidis MH, Dardiotis E, Hadjigeorgiou GM, Sakka P, Anastasiou CA, Simopoulou E, Scarmeas N (2019) Mediterranean diet adherence is related to reduced probability of prodromal Parkinson's disease. Mov Disord 34(1):48–57

Martyniak K, Wei F, Ballesteros A, Meckmongkol T, Calder A, Gilbertson T, Orlovskaya N, Coathup MJ (2020) Do polyunsaturated fatty acids protect against bone loss in our aging and osteoporotic population? Bone 7:115736. https://doi.org/10.1016/j.bone.2020.115736. Epub ahead of print

McCarty MF, Lerner A (2020) Nutraceuticals targeting generation and oxidant activity of peroxynitrite may aid prevention and control of Parkinson's disease. Int J Mol Sci 21(10):3624

McCarty MF, Lerner A (2020) Low risk of Parkinson's disease in quasi-vegan cultures may reflect GCN2-mediated upregulation of Parkin. Adv Nutr nmaa112. https://doi.org/10.1093/advances/nmaa112

Moffat T, Morell-Hart S (2020) Promoting the value of all food heritage. Department of Anthropology, Mc Master University. https://anthropology.mcmaster.ca/news/dr-tina-moffat-and-dr-shanti-morell-hart-weigh-in-on-the-mediterranean-diet. Accessed 2 Jan 2021

Molsberry S, Bjornevik K, Hughes KC, Healy B, Schwarzschild M, Ascherio A (2020) Diet pattern and prodromal features of Parkinson disease. Neurology 95(15):e2095–e2108

Najafi M, Mozaffari H, Jalilpiran Y, Mokhtari P, Teymouri M, Faghih S (2020) The associations between dietary patterns and cardiovascular risk factors among adults: a cross-sectional study. Clin Nutr ESPEN 40:300–308

Nasimi N, Sohrabi Z, Dabbaghmanesh MH, Eskandari MH, Bedeltavana A, Famouri M, Talezadeh P (2020) A novel fortified dairy product and sarcopenia measures in sarcopenic older adults: a double-blind randomized controlled trial. J Am Med Dir Assoc S1525–8610(20):30741–30746

Nirmala FS, Lee H, Kim JS, Ha T, Jung CH, Ahn J (2020) Green tomato extract prevents bone loss in ovariectomized rats, a model of osteoporosis. Nutrients 12(10):3210

Oktaviana J, Zanker J, Vogrin S, Duque G (2019) The effect of β-hydroxy-β-methylbutyrate (HMB) on sarcopenia and functional frailty in older persons: a systematic review. J Nutr Health Aging 23(2):145–150

Ozaki D, Kubota R, Maeno T, Abdelhakim M, Hitosugi N (2020) Association between gut microbiota, bone metabolism, and fracture risk in postmenopausal Japanese women. Osteoporos Int Nov 25. https://doi.org/10.1007/s00198-020-05728-y. Epub ahead of print

Park S, Zhang T, Wu X, Yi Qiu J (2020) Ketone production by ketogenic diet and by intermittent fasting has different effects on the gut microbiota and disease progression in an alzheimer's disease rat model. J Clin Biochem Nutr 67(2):188–198

Patra S, Pradhan B, Nayak R, Behera C, Rout L, Jena M, Efferth T, Bhutia SK (2020) Chemotherapeutic efficacy of curcumin and resveratrol against cancer: Chemoprevention, chemoprotection, drug synergism and clinical pharmacokinetics. Semin Cancer Biol 2:S1044-579X(20)30215-7

Rattan SIS, Kyriazis M (eds) (2019) The Science of Hormesis in Health and Longevity. Academic Press Elsevier, Amsterdam

Rattan SIS, Demirovic D (2009) Hormesis can and does work in humans. Dose Response 8(1):58–63

Reddy AT, Lakshmi SP, Maruthi Prasad E, Varadacharyulu NC, Kodidhela LD (2020) Epigallocatechin gallate suppresses inflammation in human coronary artery endothelial cells by inhibiting NF-κB. Life Sci 258:118136

Richer S, Stiles W, Ulanski L, Carroll D, Podella C (2013) Observation of human retinal remodeling in octogenarians with a resveratrol based nutritional supplement. Nutrients 5(6):1989–2005

Rondanelli M, Fossari F, Vecchio V, Gasparri C, Peroni G, Spadaccini D, Riva A, Petrangolini G, Iannello G, Nichetti M, Infantino V, Perna S (2020) Clinical trials on pain lowering effect of ginger: a narrative review. Phytother Res https://doi.org/10.1002/ptr.6730. Epub ahead of print

Sauer AK, Vela H, Vela G, Stark P, Barrera-Juarez E, Grabrucker AM (2020) Zinc deficiency in men over 50 and its implications in prostate disorders. Front Oncol 10:1293

Schön M, Mousa A, Berk M, Chia WL, Ukropec J, Majid A, Ukropcová B, de Courten B (2019) The potential of carnosine in brain-related disorders: a comprehensive review of current evidence. Nutrients 11(6):1196

Sherif IO, Al-Mutabagani LA, Sabry D, Elsherbiny NM (2020) Antineoplastic activity of chrysin against human hepatocellular carcinoma: new insight on GPC3/SULF2 axis and lncRNA-AF085935 expression. Int J Mol Sci 21(20):7642

Shibabaw T (2020) Omega-3 polyunsaturated fatty acids: anti-inflammatory and anti-hypertriglyceridemia mechanisms in cardiovascular disease. Mol Cell Biochem. Nov 11. https://doi.org/10.1007/s11010-020-03965-7. Epub ahead of print

Shirakawa T, Miyawaki A, Matsubara T, Okumura N, Okamoto H, Nakai N, Rojasawasthien T, Morikawa K, Inoue A, Goto A, Washio A, Tsujisawa T, Kawamoto T, Kokabu S (2020) Daily oral administration of protease-treated royal jelly protects against denervation-induced skeletal muscle atrophy. Nutrients 12(10):3089

Simes DC, Viegas CSB, Araújo N, Marreiros C (2020) Vitamin K as a diet supplement with impact in human health: current evidence in age-related diseases. Nutrients 12(1):138

Simopoulos AP (2006) Evolutionary aspects of diet, the omega-6/omega-3 ratio and genetic variation: nutritional implications for chronic diseases. Biomed Pharmacother 60(9):502–507

Soeters PB, Wolfe RR, Shenkin A (2019) Hypoalbuminemia: pathogenesis and clinical significance. J Parenter Enter Nutr 43:181–193

Song JH, Hwang B, Chung HJ, Moon B, Kim JW, Ko K, Kim BW, Kim WR, Kim WJ, Myung SC, Moon SK (2020) Peanut sprout extracts cultivated with fermented sawdust medium inhibits benign prostatic hyperplasia in vitro and in vivo. World J Mens Health 38(3):385–396

Sun HJ, Wu ZY, Nie XW, Bian JS (2020) Role of hydrogen sulfide and polysulfides in neurological diseases: focus on protein S-Persulfidation. Curr Neuropharmacol. 2020 Sep 5. https://doi.org/10.2174/1570159X18666200905143550. Epub ahead of print

Thangavel P, Puga-Olguín A, Rodríguez-Landa JF, Zepeda RC (2019) Genistein as potential therapeutic candidate for menopausal symptoms and other related diseases. Molecules 24(21):3892

Thomson DM (2018) The role of AMPK in the regulation of skeletal muscle size, hypertrophy, and regeneration. Int J Mol Sc 19(10):3125

Tsai HH, Chen CW, Yu PL, Lin YL, Hsieh RH (2020) Mangosteen pericarp components alleviate progression of prostatic hyperplasia and mitochondrial dysfunction in rats. Sci Rep 10(1):322

Valdes AM, Walter J, Segal E, Spector TD (2018) Role of the gut microbiota in nutrition and health BMJ 361:k2179

Varghese N, Werner S, Grimm A, Eckert A (2020) Dietary mitophagy enhancer: a strategy for healthy brain aging? Antioxidants (basel) 9(10):E932

Walchuk C, Suh M (2020) Nutrition and the aging retina: a comprehensive review of the relationship between nutrients and their role in age-related macular degeneration and retina disease prevention. Adv Food Nutr Res 93:293–332

Wei Q, Hu W, Lou Q, Yu J (2019) NAD+ inhibits the metabolic reprogramming of RPE cells in early AMD by upregulating mitophagy. Discov Med 27(149):189–196

Xu F, Song Y, Guo A (2019) Anti-apoptotic effects of docosahexaenoic acid in IL-1β-induced human chondrosarcoma cell death through involvement of the MAPK signaling pathway. Cytogenet Genome Res 158(1):17–24

Xu X, Liu X, Yang Y, He J, Jiang M, Huang Y, Liu X, Liu L, Gu H (2020) Resveratrol exerts anti-osteoarthritic effect by inhibiting TLR4/NF-κB signaling pathway via the TLR4/Akt/FoxO1 axis in IL-1β-stimulated SW1353 cells. Drug Des Devel Ther 14:2079–2090

Yang X, Zhang M, Dai Y, Sun Y, Aman Y, Xu Y, Yu P, Zheng Y, Yang J, Zhu X (2020) Spermidine inhibits neurodegeneration and delays aging via the PINK1-PDR1-dependent mitophagy pathway in C. elegans. Aging (Albany NY) 12(17):16852–16866

Yuce P, Hosgor H, Rencber SF, Yazir Y (2020) Effects of intra-articular resveratrol injections on cartilage destruction and synovial inflammation in experimental temporomandibular joint osteoarthritis. J Oral Maxillofac Surg S0278–2391(20):31133–31142

Zhang X, Cai Z, Mao H, Hu P, Li X (2020) Isolation and structure elucidation of polysaccharides from fruiting bodies of mushroom Coriolus versicolor and evaluation of their immunomodulatory effects. Int J Biol Macromol 5:S0141-8130(20)34943-6

Zhu F, Li C, Chu F, Tian X, Zhu J (2020) Target dysbiosis of gut microbes as a future therapeutic manipulation in alzheimer's disease. Front Aging Neurosci 12:544235

Ziegler M, Wallert M, Lorkowski S, Peter K (2020) Cardiovascular and metabolic protection by vitamin E: a matter of treatment strategy? Antioxidants (basel) 9(10):935

Chapter 27
Nutritional Regulation of Metabesity

Ebenezer I. O. Ajayi, Olorunfemi R. Molehin, Stephen A. Adefegha, Aderonke E. Fakayode, Johnson O. Oladele, and Olawumi A. Samuel

Abstract Prolonged inactivity, sedentary lifestyle and unbalanced nutrition predispose to weight gain and obesity, also termed as metabesity (metabolic obesity). Deficient metabolic sugar processing, impaired brain energy generation, dysfunctional important pathways are risk factors in the pathogenesis of diabetes. Other degenerative diseases, including neurodegenerative, and cardiovascular diseases can make metabesity more complex. These may become aggravated over time owing to several epigenetic factors, including fatty diet, lack of exercise, lack of adequate sleep, sedentary lifestyle, alcohol, and drug use. Metabesity is underlined by vascular- and neuro-inflammation, and encompasses a wide physiological and clinical manifestation. It is imperative to note that the existing interventions are only for the management of diabetes, they are not curative. Secondary complications still occur in many cases of diabetes. While continuous, concerted efforts go on to mitigate this phenomenon, dietary manipulations and modulation remain a promising therapeutic and lifestyle approach. The interrelationship between the central nervous system (mainly the hypothalamus and the brain stem) and diabetes has been elucidated. The

E. I. O. Ajayi (✉)
Diabesity Complications and Other Neglected Infectious Diseases Group, Department of Biochemistry, Osun State University, Osogbo, Nigeria
e-mail: ebenezer.ajayi@uniosun.edu.ng

E. I. O. Ajayi · O. A. Samuel
Membrane Biophysics and Nanotechnology Laboratories, Mercedes and Martin Ferreyra Institute of Medicine, IMMF-INIMEC-CONICET-UNC, Cordoba, Argentina

O. R. Molehin
Department of Biochemistry, Faculty of Science, Ekiti State University, Ado-Ekiti, P.M.B. 5363, Ado-Ekiti 360001, Nigeria

S. A. Adefegha
Functional Foods and Nutraceuticals Unit, Department of Biochemistry, Federal University of Technology, Akure, P.M.B. 704, Akure 340001, Nigeria

A. E. Fakayode
Department of Biochemistry and Molecular Biology, Faculty of Science, Obafemi Awolowo University, Ile-Ife, P.M.B. 13, Ile-Ife 220282, Nigeria

J. O. Oladele
Biochemistry Unit, Department of Chemical Sciences, Kings University, Ode Omu, Osun, Nigeria

S. I. S. Rattan and G. Kaur (eds.), *Nutrition, Food and Diet in Ageing and Longevity*, Healthy Ageing and Longevity 14, https://doi.org/10.1007/978-3-030-83017-5_27

hypothalamic neuronal and glial populations have been found to systemically regulate blood glucose levels in such a manner that the hypothalamic astrocytes capacity for glucose sensing affect the overall glucose homeostasis. Inflammatory responses rising from diabetes and obesity affects the astrocytes and the regulation of metabolic hormones. Cancers are notorious for being voracious, and aggressively depicting angiogenesis in order to amass available cell nutrients to themselves by all means and at any cost. As a member of the inflammatory metabesity concept, especially when metastatic, cancer angiogenesis is always to the detriment of normal apparently healthy cell populations within the systems of an organism. Should there be one, or a comorbidity of any two or more of the illnesses consisting metabesity, how may nutrition, nutrigenomics, nutraceuticals and/or functional foods be deployed to regulate, manage, mitigate, treat and cure them. In this chapter, we attempt to answer this question.

Keywords Metabesity · Nutrition · Inflammation · Gut · Brain

27.1 Introduction

The role of a balanced and regulated nutrition is crucial for health and longevity. Imbalanced nutrition combined with prolonged inactivity and a sedentary lifestyle predispose one to excessive weight gain and obesity, often termed as metabesity within the general concept of metabolic syndrome. The balance between energy intake (food consumption) and energy expenditure (basal metabolic rate, i.e. biochemical processes required to maintain cellular viability, physical activity and adaptive thermogenesis) is tightly regulated (Fall et al. 2013). A homeostatic network maintains energy stores through a complex interplay between the feeding regulatory centres in the central nervous system (CNS), particularly in the hypothalamus and the regulated storage and mobilization of fat stores that maintains the body energy stores (Calle et al. 2003). Thus, genes that encode the molecular components of this system may underlie obesity and related disorders. (Nielsen et al. 2014; Calle et al. 2003; Hu 2007).

Metabolic syndrome is generally perceived as an important health problem regarding cardiovascular health, and with the concept of metabesity more comprehensive definition has been proposed (Harville et al. 2012). It is a constellation of risk factors that predisposes to an enhanced risk of Type 2 diabetes and cardiovascular disease. The clustering of at least three of five risk factors, elevated blood pressure, elevated fasting plasma glucose, high triglycerides, low HDL-cholesterol and increased body weight, have been variously described to constitute the metabolic syndrome. Obesity and diabetes mellitus remain the leading causes of reduced health span and life span throughout the world. Hence, it is not surprising that these areas are at the center of highly active areas of research. The identification of novel mechanisms underlying these metabolic disorders sets the stage for uncovering new potential therapeutic strategies. Obesity is a complex and chronic non-communicable disease that affects more than a third of the world's population (Hruby and Hu 2015). It has been shown that obesity in middle age can shorten life expectancy by 4–7 years (Peeters et al. 2003).

Metabesity refers to metabolic aberrations associated with obesity. These include low-grade inflammation, mitochondrial dysfunction, and changes in gut microbiome. Along with a genetic component, the phenotypes in metabesity are largely the result of sedentary lifestyles and unhealthy eating habits (Nielsen et al. 2014). Metabesity is associated with several co-morbidities including an increased risk for cardiovascular conditions like hypertension, heart failure, myocardial infarction, stroke, and sudden death. Insulin resistance, high blood pressure and glucose levels, visceral adiposity, progressive atherosclerosis, dyslipidaemia and fatty liver are common in obese individuals (Ha et al. 2018). Metabesity adversely impacts endocrine balances in the body and increases the risk of degenerative conditions like dementia.

Metabesity is an impending epidemic of huge public health implications with enormous clinical, socioeconomic, and humanistic burdens. Interventions to combat sedentary lifestyle and unhealthy eating should be introduced early in life to prevent the onset and progression of metabesity (Cuffe et al. 2018).

27.2 Pathophysiology of Metabesity

Aging itself is largely a metabolic condition. As we get older, the day-to-day operations of metabolism inflict damage on human cells and organs. Moreover, as this damage accumulates, metabolism itself is thrown into disarray and these things are no longer coordinated with each other, causing metabolism to malfunction further and inflict even the more damage (Davies et al. 2019). Among the many signs of metabolic discoordination is a buildup of visceral fat, which may be partly a symptom and partly a cause of aging. It is well known that carrying excess visceral fat tissue in various ways increases risk of age-related diseases, shortens life expectancy, and raises lifetime medical expenditure (Kay et al. 2016). Aging, therefore, along with diabetes, cardiovascular and neurodegenerative diseases, and cancer, is itself an additional component of metabesity (Kay et al. 2016). It is also well known that the western diet and lifestyle similarly contribute to the same metabolic dysfunction and contribute to signs of premature aging. So given how intimately connected aging, metabesity and disease are, seeking to address the metabolic roots of various diseases might also lead to the discovery of the methods for improving the aging process itself, with positive ramifications for everything from obesity to arthritis. Of all the major health threats to emerge, none has challenged the very foundations of public health so profoundly as the rise of chronic non-communicable diseases (Salehi et al. 2017). Heart disease, cancer, diabetes, and chronic respiratory diseases, once linked only to affluent societies, are now global, and the poor suffer the most. These diseases share four risk factors: tobacco use, excessive use of alcohol, unhealthy diets, and physical inactivity thus leading to metabesity (Davies et al. 2019). The action of insulin in lowering blood glucose levels results from the suppression of hepatic glucose production and the increased glucose uptake into muscle and adipose tissue via GLUT4 contribute greatly to metabesity (Hruby and Hu 2015). Muscle has long been considered the major site of insulin-stimulated glucose uptake in vivo, with adipose tissue contributing relatively little to total body glucose disposal. On the other hand, various transgenic studies have raised the possibility of a greater role for glucose uptake into fat in systemic glucose homeostasis. Over-expression of GLUT4 selectively in fat tissue enhances whole body insulin sensitivity and glucose tolerance (Shepherd and Kahn 1999) and knocking out GLUT4 selectively from fat tissue results in a degree of insulin resistance similar to that seen with muscle-specific knockout of GLUT4. In all forms of obesity, there is downregulation of GLUT4, a major factor contributing to the impaired insulin-stimulated glucose transport in adipocytes leading to the onset of metabesity (Fig. 27.1) (Shepherd and Kahn 1999). However, in the skeletal muscle of obese humans, GLUT4 expression is normal. It has also been suggested that defective glucose transport may be due to impaired translocation, docking, or fusion of GLUT4-containing vesicles with the plasma membrane (Tozzo et al. 1997). "With obesity, there is reduced glucose disposal in adipose tissue. It has been suggested that obesity leads to the development of hyperglycemia, hyperlipemia, hyperinsulinemia, and insulin resistance. Molecules like FFA, leptin, or TNF-α, all of which are released from adipose tissue, are known to affect glucose homeostasis indirectly (Fig. 27.1).

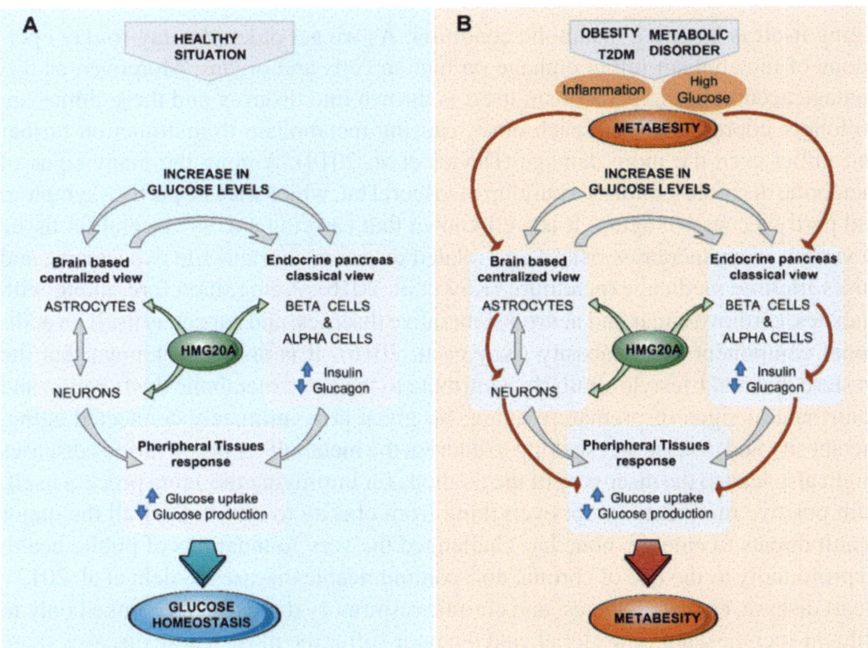

Fig. 27.1 Pathogenesis of metabesity (Benoit and Francisco 2019)

Undoubtedly there are other, as yet undiscovered, molecules from adipose tissue that influence systemic metabolism. Obesity is associated with increased basal lipolysis in adipose tissue, and elevated circulating free fatty acids". Acute-phase serum amyloid A (SAA), a lipolytic adipokine in humans, stimulates basal lipolysis (Cuffe et al. 2018). The lipolysis has been postulated to be an autocrine feedback mechanism by which increased SAA production from enlarged adipocytes A into the circulation may contribute to insulin resistance. The SAA act through the CLA-1 and the extracellular signal-regulated kinase signaling pathway to stimulate lipolysis directly. Alternatively, increased lipolysis by SAA may be indirect, i.e. through the stimulation of the lipolytic cytokines viz IL-6 and TNF-α. (Hinney et al. 2010). Plasma triglyceride (TG) concentration is another metabolic variable, most affected in obesity. It has been suggested that there is tissue resistance to insulin-mediated glucose uptake, which in turn accelerates the very low-density lipoprotein (VLDL), TG production rate and leads to endogenous hypertriglyceridemia. In obesity there is decreased Lipoprotein lipase-mediated lipolysis of chylomicron-TG and ineffective inhibition of hormone sensitive lipase-mediated lipolysis in adipose tissue (Cuffe et al. 2018). SAA has also a direct effect on cholesterol metabolism. Being an apolipoprotein by nature, it is the apoprotein of high-density lipoprotein (HDL). The inter-action of SAA with HDL may impair the function of HDL as an anti-atherogenic molecule and facilitate its degradation contributing to the occurrence of metabesity (Harville et al. 2012). Adipose tissue-derived proteins have been defined as adipokines, and have been implicated in the pathogenesis of chronic inflammation in obesity (Fig. 27.1). The study of adipose tissue on inflammation was considerably impacted by the demonstration of resident macrophages in adipose tissue. The possible mechanisms underlying the infiltration of macrophages into adipose tissue may be the chemokines by adipocytes, which would then attract resident macrophages. Recent studies have suggested that macrophages infiltrate adipose tissue as part of a scavenger function in response to adipocyte necrosis (Heymsfield and Wadden 2017; la Fleur and Serlie 2014). Obesity is associated with elevated levels of circulating proinflammatory cytokines such as plasminogen activator inhibitor-1 (PAI-1), C-reactive protein (CRP), TNF-α, and IL-6 and monocyte chemoattractant protein-1 (MCP-1). PAI-1 is elevated in subjects with metabolic complications of obesity, and is expressed in the stromal fraction of adipose tissue, including endothelial cells. The cell types involved in the inflammatory response in obesity are not fully delineated. Recent attention has focused on adipose tissue macrophages (ATMs) as a mediator of inflammatory responses in adipose tissue. In addition to the production of pro-inflammatory cytokines that promote metabolic complications, adipose tissue is the sole source of adiponectin, which is anti-inflammatory and associated with protection from atherosclerosis (Shepherd et al. 1993).

27.3 Risk Factors of Metabesity

27.3.1 Lipotoxicity

Excessive accumulation of lipids (intracellular lipid accumulation) in tissues other than adipose tissue beyond the ability of the tissues to oxidize them, leads to cellular dysfunction which is commonly known as lipotoxicity, a crucial condition in the pathogenesis of diabetes as one of the underlining factors in insulin resistance and favours gradual atrophy of beta (β) cells in the pancreas. Amplified plasma concentrations of free fatty acids associated with obesity occurred as a result of excessive accumulation by tissues (liver, pancreas, skeletal muscles) connected with glucose homeostasis regulation. Visceral and subcutaneous adipose tissues are the main origins of free fatty acids (Belfort et al. 2005), but the latter release free fatty acids faster than the former (Bjorntorp 1991). Triacylglycerols have been assumed to be the only lipid that causes insulin resistance; however, recent approaches have focused on bioactive lipids that can alter the enzymatic activities in insulin metabolism. Diacylglycerols are examples of lipids that can alter the insulin pathway and they formed in a cell via de novo synthesis or through the breakdown of phospholipids (Yu et al. 2002; Schmitz-Peir 2002). They are also vital second messengers implicated in the regulation of intracellular activities which include signal transduction, differentiation, and proliferation. Despite their usefulness, undue accumulation of diacylglycerols enhances the activation of protein kinase C (PKC) that causes aberrant insulin signaling which leads to impaired GLUT4 translocation to the cell membrane. Overexpression of glycerol-3-phosphate acyltransferase 1 which catalyzed the acylation of glycerol-3-phosphate during the synthesis of diacylglycerols, has been reported to cause steatosis and increased insulin resistance (Bednarski et al. 2016). Long-chain acyl-CoA are examples of lipids that can alter the insulin pathway (Hulver et al. 2003; Ellis et al. 2000). The previous report documented that a marked increase in long-chain acyl-CoA levels were observed in the liver, skeletal and adipose tissues of obese people (Blachnio-Zabielska et al. 2012). Activated long-chain acyl-CoAs serve as substrates in β-oxidation and in the synthesis of other lipids such as ceramides, diacylglycerols, and triacylglycerols (Cronan, 1997) and has been reported to activates protein kinase C which enhances serine phosphorylation at 307 (Ser307) in IRS-1. This further inhibited phosphorylation of tyrosine in the insulin receptor and enhance phosphoinositide 3-kinases (PI3K) activation required for insulin action (Tomas et al. 2002; Itani et al. 2002).

27.3.2 Lipid

Excess energy in the form of triglycerides is stored mainly in the adipose tissue. These lipids are stored during energy balance as lipid droplets of adipocytes through hyperplasia (increase in number of adipocyte) or hypertrophy (enlarge the size of adipocytes) (Hausman et al. 2001). Obesity has been linked with adipose tissue dysfunction induced by hypertrophy and it has been proposed to be crucial in the pathogenesis of metabolic diseases such as insulin resistance (Wang et al. 2005; Kim et al. 2007; Weyer et al. 2000). During physical exercise and shortage of energy in the body, free fatty acids are produced through lipolysis of the triglycerides stored in adipocytes. These free fatty acids are then transported through the circulatory system to tissues where they are used as source of energy. Free fatty acids have been implicated in the pathogenesis of obesity-associated metabolic disturbances particularly insulin resistance. Free fatty acids through the portal circulation can directly enter the liver in obesity and increase hepatic free fatty acids level which in turns enhance gluconeogenesis and lipid synthesis with insulin resistance in the liver (Boden 1997). Peripheral insulin resistance in humans can be induced by elevated level of circulating free fatty acids (Boden 1997; Kelley et al. 1993). Furthermore, free fatty acids serve as ligands for the toll-like receptor 4 (TLR4) complex (Shi et al. 2006) and activate cytokine secretion of macrophages (Suganami et al. 2005), thereby influencing inflammation of adipose tissue which enhance obesity-induced metabolic complications.

27.3.3 Adipokines

Adipokines are referred to as the various kinds of adipose tissue-derived secretory factors which are products of excess adiposity and adipocyte dysfunction. They also play significant role in the pathogenesis and development of many metabolic diseases through modification of inflammatory responses and alteration of lipid and glucose and lipid homeostasis (Hauner 2005; Halberg et al. 2008). Adipokines can be classified based on their activities as insulin-sensitizing adipokines with anti-inflammatory properties such as secreted frizzled-related protein 5 (SFRP5) and adiponectin; and insulin resistance adipokines with pro-inflammatory activities such as angiopoietin-like protein 2, visfatin, retinol binding protein 4 (RBP4), plasminogen activator inhibitor (PAI)-1, resistin, leptin, IL-18, IL-6, TNF-α, and MCP-1 (Hotamisligil et al. 1993; Amrani et al. 1996; Sartipy and Loskutoff 2003; Rotter et al. 2003).

27.3.3.1 Leptin

Leptin, a hormone produced in the adipose cells, is essential in the regulation of energy homeostasis and is richly found in adipose tissue especially in adipocytes (Friedman and Hallas 1998). It suppresses appetite and activates energy expenditure (Friedman and Hallas 1998). In obese subjects, the level of Leptin mRNA expression in the adipose tissue and its level in the circulation are increased (Considine et al. 1996; Kouidhi et al 2010). This increase may be due to the presence of leptin resistance (Friedman and Hallas 1998). Leptin, independent of actions on energy expenditure, food intake, or body weight, also performs crucial function in the regulation of glucose homeostasis. Leptin regulates pancreatic β-cell activities and enhances sensitivity of insulin in the skeletal muscle and liver (Marroqui et al. 2012). Furthermore, due to cytokine-like structure of leptin and its receptor being a member of the class I cytokine receptor (gp130) superfamily, leptin has been proposed to have pro-inflammatory activities (Paz-Filho et al. 2012). Leptin binds to its receptor to activates the Janus kinase (JAK)/signal transducers and activators of transcription (STAT) signal transduction pathway which lead to its numerous functions. Leptin has been reported to suppress the secretion of the anti-inflammatory cytokine IL-4 and stimulates the secretion of IL-2 and IFN-γ which are pro-inflammatory cytokines (Lord et al. 1998). Endotoxin (lipopolysaccharide, LPS) and pro-inflammatory cytokines (TNF-α, IL-1) has been documented to increase the level of Leptin mRNA expression in the adipose tissue and circulating leptin level (Grunfeld et al. 1996).

27.3.3.2 MCP-1/CCL2 and Other Chemokines

Chemokines and their receptors are significantly expressed in subcutaneous and visceral adipose tissue in patients suffering obesity (Huber et al. 2008). They are essential in selective recruitment of lymphocytes and monocytes. Migration and infiltration of monocytes/macrophages are mainly regulated by macrophage chemoattractant protein-1/C–C motif chemokine ligand 2 (MCP-1/CCL2). MCP-1/CCL2 induces inflammation in the adipose tissue by pulling in inflammatory cells from the blood stream and its expression in the adipocytes related to adiposity (Deshmane et al. 2009). Over-secretion of MCP-1/CCL2 enhances recruiting process of macrophage and exacerbates the metabolic phenotype in adipose tissue (Kanda et al. 2006; Kamei et al. 2006). However, absence of MCP-1/CCL2 or its receptor lessens accumulation of macrophages and prevent against insulin resistance (Kanda et al. 2006; Lumeng et al. 2008). Adipocyte-derived MCP-1/CCL2 has been reported to induce inflammation in human adipose tissue without the involvement of leukocytes or macrophages, even though many cells such as macrophages/leukocytes and adipocyte in adipose tissue secrete MCP-1/CCL2 (Meijer et al. 2011). CXCL14, C-X-C motif chemokine ligand 5 (CXCL5) and CCL5 are other examples of chemokines that have been implicated in obesity-induced insulin resistance and in the migration and infiltration of monocytes/macrophages (Keophiphath et al. 2009; Chavey et al. 2009). In adipose tissue of obese subjects, multiple chemokine ligands and receptors with inflammation activities have reported (Tourniaire et al. 2013).

27.3.3.3 Tumour Necrosis Factor Alpha (TNF-α)

Tumour necrosis factor alpha (TNF-α) is another pro-inflammatory cytokine which has been implicated in the development of insulin resistance and obesity (Hotamisligil et al. 1993). TNF α is produced by different kind of cells and tissues adipose tissue (Hotamisligil et al. 1993), lymphocytes, macrophages, monocytes, and skeletal muscles (Saghizadeh et al. 1996) and is involved in various physiological processes including inflammation. TNFα performs its activities through its two key receptors (p50 and p75) with concomitant activation of serine kinases such as p38 MAPK, IκBα kinase β (IKKβ), and c-Jun NH2-terminal kinase (JNK), which enhance transcription and expression of different pro-inflammatory mediators (Wellen and Hotamisligil 2005). TNFα is also involved in the induction of insulin resistance via its influence on the regulation of lipid and carbohydrate metabolism when its operating in an auto- or paracrine ways (Hotamisligil and Spiegelman 1994). Oversecretion of TNFα due to increase in number of fatty acids in the cell was first established in adipose tissue of obese and diabetic rodents (Hotamisligil et al. 1993). Although adipocytes produce TNF-α in rodents but the secretion is weak in humans, a study demonstrated positive correlation between insulin levels, obesity, and insulin resistance, and increased TNFα secretion in human adipocytes (Kem et al. 2001). In human, despite improving the inflammatory status of healthy overweight subjects with metabolic syndrome and insulin resistance, and eliminating the TNF-α, there is no significant effect on the insulin resistance (Wascher et al. 2011). Similarly, treatment of human subjects with TNF-α antagonist did not enhance insulin sensitivity in humans (Bernstein et al. 2006). These observations may be due to compensatory effects of other pro-inflammatory cytokines. Meanwhile, many pro-inflammatory cytokines secreted by adipose tissue such as IL-6, TNF-α, and IL-1 have been implicated in in disrupting insulin signaling and cause metabolic dysregulation (McArdle et al. 2013). TNF-α which is a component of inflammation pathway can activate cytokine cascades (both antagonistic and synergistic reactions) which regulates the production and expression of other hormones, cytokines and their receptors (Illei and Lipsky 2000).

27.4 The Significance of Metabesity in Other Degenerative Diseases

27.4.1 Metabesity and Diabetes

Globally, the incidence of type 2 diabetes is estimated to be about 9.3% (463 million people) in 2019, it is estimated to rise to 10.2% (578 million) by 2030 and 10.9% (700 million) by 2045 (Saeedi et al. 2019). The link between metabesity and diabetes can be understood by considering the link between obesity and diabetes in relation to metabolism.

Obesity is a major risk factor for diabetes. Majority of patients who are obese have an increased level of free fatty acids. It has been hypothesized that free fatty acids may be an important link between obesity, and type 2 diabetes mellitus (Hussain et al. 2010). Wilkin et al. proposed one of the widely accepted theories that links body mass index and obesity, the "accelerator hypothesis".

Increasing waist line, low chronic inflammation, systemic or tissue-specific increased mitochondrial damage, accumulated and circulating advanced glycation end products, atherosclerotic plaques, dysfunctional unfolded protein response mechanism are common in obesity, insulin resistance and type 2 diabetes mellitus. A cascade of events ensues that causes remodeling of protein structure and loss/gain of function which over time imposes progressive acquired phenotypic changes on the cells within the system. Once the extracellular matrix becomes gradually compromised, ageing may set in.

27.4.2 Metabesity and Cardiovascular Diseases

Obesity is one of the risk factors for cardiovascular diseases. It has been associated with diseases such as atherosclerosis, myocardial infraction, heart failure, cardiac arrhythmias amongst others (Poirier et al. 2006). Metabolic alterations in lipid storage has been identified in both obesity and atherosclerosis due to accumulation of triglycerides and cholesterol in both disease conditions (Csige et al. 2018).

Cardiovascular events are manifestations of complications arising from underlying disease conditions such as an atherosclerosis, which in turn is characterized by increased carotid plaques, among other cardiovascular challenges (Perez et al. 2019). Carotid plaques are more common and advanced in patients that have higher basal metabolic indices than in those having apparently normal body weight (McGill et al. 2002). Inflammation is responsible for major cascades and events that precipitate into atherosclerosis including endothelial dysfunction, atherosclerotic plaques etc. (Shoelson et al. 2007).

Heart failure, another major cardiovascular event, seems to be increasing worldwide in terms of prevalence (Mozaffarian et al. 2015). Increase in body mass index is correlated with increase in cardiovascular events in both men and women although at different rates (Kenchaiah et al. 2002). Excessive weight potentiates increases in aldosterone levels and the mineralocorticoid receptor expression, promoting interstitial cardiac fibrosiss, platelet aggregation, carotid plaques, blood pressure anomalies, and endothelial dysfunction (Olivier et al. 2017).

Once diagnosed, close monitoring of the heart condition is essential in order to avoid cardiovascular events rising from associated illnesses including obesity, diabetes, hypertension, atherosclerosis (and palques), agimh and cancer. Different models of atherosclerosis plaque measurement exist for ascertaining the extent of the disease progression that may lead to local obstruction to or from the heart, vascular inflammation, oedema, among others.

27.4.3 Metabesity and Neurodegenerative Diseases

Neurodegenerative diseases are increasingly becoming a cause of concern globally owing majorly to morbidity, disability and a decrease in the quality of life resulting for their incidence and prevalence (Ashrafian et al. 2013). They include Alzheimer's disease, Hungtinton's disease, Parkinson's disease, amylotipic lateral sclerosis (ALS) amongst others. Protein glycation and protein glycation can affect protein folding leading defects. Once misfolded, the proteins can aggregate, causing diseases such as Alzheimer's disease, Parkinson's disease, Huntington's disease.

There is a growing body of evidence that increase in adiposity is correlated with the risk of developing Alzheimer's disease as well as dementia. The risk of incidence of Alzheimer's disease and dementia has been correlated with midlife on set of obesity (Whitmer et al. 2008; Gustafson 2006). Increased adiposity affects dysfunctional glucose sensing, hyperinsulinaemia, and insulin resistance (Ashrafian et al. 2013). Obesity has also been linked to dyslipidaemia, disordered fatty acid and cholesterol metabolism. Taken together there is an underlying distortion in most biomolecules in the body as a result of obesity especially in neurodegenerative disease conditions (Ashrafian et al. 2013). It has also been observed that neurodegeneration was alleviated via amyloid clearance following the direct central administration of insulin to the brain (Reger et al. 2006; Holscher 2012).

Parkinson's disease, which is usually present in old age, is characterized by tremor, dementia, muscle rigidity amongst other factors. Owing to the loss of dopaminergic neurons in the substantial nigra. Obesity leads to the depletion of the striatal dopamine receptors (Wang et al. 2001). An epidemiological study has also correlated obesity and Parkinson's disease as lean patients have a reduced risk of developing Parkinson disease and most obese patients have a higher risk of developing Parkinson's disease (Abblott et al. 2002).

27.5 Nutritional Intervention Against Metabesity

27.5.1 The Function of Macronutrients and Other Dietary Components

Excess accumulation and storage of adipose tissue which is a condition of obesity can be successfully ameliorated through special combinations of macronutrient. The classes of food (lipids, carbohydrates, and proteins) have different energy metabolism thus, total calories contents of macronutrients in diets that will cause changes in body composition and enhance weight loss must be considered. Moreover, altering the macronutrient composition of isocaloric diets may cause changes in their energy density. The carbohydrate and protein balance must be maintained in the body due to its limited ability to store protein and carbohydrate, and its immeasurable ability to store fat (Flatt 1988). Dietary macronutrient can influence the body's energy balance base on the current energy status of the body (ie, in a positive, negative, or neutral energy balance). Dietary intervention that improves weight loss and post–weight-loss regain has been reported to be through employing dietary protein levels (25–35% of energy as protein). High protein diets have been documented to reduce intake of energy via modification of satiety hormones and enhance diet-induced thermogenesis which support a negative energy balance (Halton and Hu 2004; Paddon-Jones et al. 2008; Abete et al. 2010; Pesta and Samuel 2014). High protein diets also enhance weight management results in a randomized controlled trials and ecological studies (Due et al. 2004; Larsen et al. 2010; Ankarfeldt et al. 2014; Wycherley et al. 2012). In clinical study, weight loss has been found across a wide range of varying proportions of macronutrient compositions especially, modification of most dietary recommended macronutrients (carbohydrates and fats) (Sacks et al. 2009). Similarly, variation in response to dietary intervention has been attributed to the wide interindividual differences in weight loss. For instance, insulin sensitivity is an important in the weight loss response on low-carbohydrate or high-carbohydrate diets. An improved response is observed in a low-carbohydrate against a high-carbohydrate diet in insulin-resistant, but not in insulin-sensitive individuals (Pittas et al. 2005; Cornier et al. 2005; Ebbeling et al. 2007). In spite of these variations, recent evidence indicates that the average weight loss responses to a wide range of dietary macronutrient outlines and other dietary alterations are alike and largely a function of compliance and the energy restriction achieved (Jensen et al. 2014). Currently, managing energy balance is the most promising nutrition intervention for obesity. The future direction will be to formulate and recommend personalized nutrient patterns to address the specific needs of individual patients.

27.5.2 The Function of Bioactive Compounds

Due to the upsurge in the prevalence of metabesity and associated complications, plant derived diets have been employed in the treatment and management of obesity owing to their abundant health benefits. Some medicinal plants contained bioactive compounds that can be used to inhibit the pathogenesis of metabesity.

27.5.2.1 Effects of Bioactive Compounds in Tea

Galactoatechin gallate (GCG), epicatechin (EC), epigallocatechin (EGC), galactatechin (GC), catechin (C), gallate (EGCG), and epigallocatechin are the substantial bioactive components of green tea. Administration of EGCG to rats caused a decrease on food intake and has been suggested to be helpful in weight management. Green tea catechins (GTC) mitigated excess increase of liver and body weights through the modulation of serum and liver triglyceride levels. Rats fed with GTC supplementation displayed a significant reduction in body weight (Yan et al. 2013). The potential mechanisms of weight management effects attributed to green tea polyphenols include the ability of tea polyphenols to stimulate AMP-activated protein kinase which is bioavailable in adipose tissue, skeletal muscle, and liver; and the ability of components of tea to reduce calorie intake via decrease of the proteins' and lipids' absorption into the intestine.

27.5.2.2 Effects of Bioactive Compounds in Coffee and Green Coffee

Chlorogenic acids, caffeolyl-feruloylquinic acid, p-coumaroylquinic acid, feruloylquinic acid, dicaffeoylquinic acid, caffeoylquinic acid, and methyl-xanthine alkaloid caffeine are some of the significant bioactive components of coffee (Pan et al. 2016; Burwell and Vilsack 2015). In a randomized controlled trial, weight control was observed in 100 patients with obesity (25 males and 95 females) who consumed of 3–5 cups/day of coffee for three weeks (Pan et al. 2016; Burwell and Vilsack 2015). In another study using tablets of decaffeinated green coffee extract, consumption coffee inhibited excess weight and accumulation (Martinez-Saez et al. 2014). Recently, green coffee bean extract markedly reduced fat mass and body weight. It was noted that the antioxidant properties of coffee may have contributed to the observed decrease in body fat and body weight (Tamara et al. 2011). Drinking of coffee and green coffee significantly reduced excess weight and also markedly reduced BMI, however, further studies are needed to elucidate the mechanism via which coffee elicit this health benefit.

27.5.2.3 Effects of Resveratrol

Resveratrol is a polyphenolic phytoalexin isolated in many than 70 plant species and is found in high proportion in the skin of red grapes. Dark chocolate, blueberries, nuts, pomegranates, berries, and tea comprise of resveratrol in varying concentrations (Diepvens et al. 2007). Resveratrol displays difference kinds of therapeutic benefits, including antioxidant and anti-inflammatory (Xia et al. 2017). Resveratrol has been reported to be an activator of sirtuin 1, a crucial molecular target regulating cellular energy metabolism and mitochondrial homeostasis. Adenosine monophosphate-activated protein kinase (AMPK) is another important target of resveratrol, indicating that it can perform a critical function in regulation of energy homeostasis; by stimulating AMPK, resveratrol demonstrates a lipid-lowering effect. Resveratrol has potential anti-obesity effects by promoting lipolysis and β-oxidation, reducing lipogenesis, decreasing proliferation of adipocytes, and suppressing differentiation (Tomé-Carneiro et al. 2012).

Although the effects of resveratrol have been extensively explored in animal models (Pereira et al. 2015; Wang et al. 2015), only few clinical studies have been carried out, and the results are inconsistent. Furthermore, contrary to the substantial preclinical discoveries of beneficial metabolic effects of resveratrol in an inflammation setting or obesity, the findings of human clinical trials of resveratrol effects on obesity-related morbidities have been inconclusive. For instance, Timmers et al. (2011) in a cross-over study reported that 150 mg/day of resveratrol treatment reduced serum inflammatory markers, increased energy expenditure, and decreased adipose tissue lipolysis and plasma fatty acid and glycerol levels of obese men (Timmers et al. 2011). In another study, Resveratrol treatment did not improve hepatic lipid content, blood pressure, glucose homeostasis, or inflammatory status in middle-aged men with metabolic syndrome. On the contrary, it significantly increased total cholesterol and LDL cholesterol (Kjær et al. 2017). In another clinical investigation, Konings et al. (2014) examined the effects of 30 days resveratrol treatment (150 mg/day) on the adipocyte size and gene expression patterns in obese men. It was documented that resveratrol treatment decreased the size of abdominal subcutaneous adipocytes (Konings et al. 2014).

Although, many beneficial effects have been reported in some clinical trials, but with some conflicting information and discrepancies (Novelle et al. 2015). It has been observed that resveratrol presents a dose–response hormesis in the biological models in which it has been tested, affecting several outcomes with medical and therapeutic significance (Scapagnini et al. 2014). Arzola-Paniagua et al. (2016) observed not significant decreases in leptin/adiponectin ratio, leptin, triglycerides, fat mass, waist circumference, and BMI in obese individual's treated with resveratrol therapy. Resveratrol is used combined with other phytochemicals, too. For example, twelve weeks of combined epigallocatechin-3-gallate and resveratrol supplementation increased mitochondrial capacity and stimulated fat oxidation in obese humans (Most et al. 2016). However, another report reported that high levels of resveratrol supplementation treatment had no effect on metabolic events, adipose tissue content, and energy expenditure. Findings from a clinical study conducted by Poulsen et al. (2013) were contrary to the anti-obesity potential of resveratrol in obese men; The report documented that no effect was seen on inflammatory or metabolic biomarkers, oxidation rates of lipid, blood pressure, ectopic or visceral fat content, or resting energy expenditure (Poulsen et al. 2013).

Evidence from these limited clinical studies combined with the results from in vitro and animal studies indicate that potential anti-obesity effects of resveratrol may be achieved through dietary supplementation. The differing results could be due to the variable doses selected in the assays and to the different clinical backgrounds of the study subjects. However, the optimal doses and study period for the anti-obesity potential of resveratrol remain to be determined (Kjær et al. 2017; Zhao et al. 2017).

27.5.2.4 Effects of Quercetin

Quercetin is a flavonoid and another active compound found in foods, fruit, wine, apples, ginger, garlic, onions, and vegetables. There are several in vitro and in vivo investigations aimed at elucidating the beneficial effects of quercetin in obesity, but only little clinical findings were documented on human studies and clinical trials that have been carried out to examine the effects of quercetin on obesity treatment (Chen et al. 2016; Nabavi et al. 2015). A study assessed the effects of administration of quercetin in overweight obese subjects with various apolipoprotein E genotypes; the quercetin (150 mg/day/subject) decreased the waist circumference and triacylglycerol concentration (Pfeuffer et al. 2013). Quercetin supplementation offered beneficial effect against some risk factors of cardiovascular disease, though exerted slightly pro-inflammatory effects (Pfeuffer et al. 2013; Brüll et al. 2015). Quercetin mitigated oxidative stress in obese animal models, but Shanely et al. documented that quercetin

(500 or 1000 mg/day/subject for 12 weeks) has no effect on oxidative stress and antioxidant in obese subjects (Shanely et al. 2010). Presently, a clinical trial that is still under phase II stage investigation, is exploring whether quercetin changes the absorption of glucose by the body in obese subjects (Trails, 2003). Further studies are needed to further examine the bioactive effects and bioavailability of quercetin.

27.5.2.5 Effects of Bioactive Compounds in *Curcuma longa*

The most bioactive compound in the *Curcuma longa* is curcumin. *Curcuma longa* is one of the most used spice in India and other Asian states. It is the first records about the turmeric as a useful medicine and has been consumed for many centuries as medicine of Ayurveda, which means "science of long life". Curcumin exhibits many therapeutic and biological functions such as antiangiogenesis, antioxidation, anticancer, immunomodulatory, anti-inflammation in different organs in the body (Strimpakos and Sharma 2008; Aggarwal 2010). There are significant evidences about potency of curcumin in enhancing and promoting β-oxidation, suppressing fatty acid synthesis, and reducing fat storage (Bradford 2013; Jayarathne et al. 2017). Scientific and clinical studies have demonstrated the health benefit effect of curcumin in enhancing insulin sensitivity, modulating the bodyweight change (gain), and mitigating pathogenesis of diabetes (Ramirez-Bosca et al. 2000; Ramirez Boscáa et al. 2000; Panahi et al. 2016).

Unlike the studies on the effects of curcumin in cells or animals, studies on obese subjects are limited. Mohammadi et al. (2013) documented the first clinical trial using curcumin in obesity treatment. In the study, obese subjects were treated with a commercial formulation of curcumin (1 g/day) supplemented with a bioavailability enhancer, piperine, for a month. Even though there were no changes in body fat, weight, or BMI, there is a marked reduction in serum triglyceride levels following treatment with curcumin. this indicate that curcumin can influence and improve insulin actions (Mohammadi et al. 2013). Furthermore, oral treatment of curcumin supplementation (1 g/day) for 30 days substantially attenuated oxidative stress burden in obese individuals (Sahebkar et al. 2013). Curcumin supplementation had beneficial effects on hip circumference, high-density lipoprotein levels, waist circumference, body mass index, and triglyceride/high-density lipoprotein ratio in overweight and obese female adolescents (Saraf-Bank et al. 2019).

In another study, 30-day treatment of curcumin (500 mg/day) decreased serum levels of inflammatory cytokines IL-1β and IL-4 of obese patients (Ganjali et al. 2014). This suggests the anti-inflammatory activity of curcumin in obesity therapy. In a randomized control trial carried out by Di Pierro et al. (2015), curcumin was reported to decrease body weight and omental adipose tissue among overweight people affected by metabolic syndrome. Taken together, studies have demonstrated that curcumin exhibited anti-obesity and anti-inflammatory properties via adipose tissue by reducing lipid storage, adiposity, and stimulating lipid oxidation (Zhao et al. 2017). Although curcumin has been used in clinical trials, its multidimensional pharmacological properties, pharmacokinetics, and adverse effects in obesity therapy need to be cautiously examined. A joint report of the Food and Agriculture Organization (FAO) and the World Health Organization (WHO) has recommended 1 mg/kg body weight of curcumin as maximum daily usage (WHO 2000). Investigations into curcumin continues to provide novel insights into metabolic regulation that may ultimately translate into effective therapy (Jin et al. 2018).

27.6 Conclusion

While taking care to apply nutritional interventions to prevent, mitigate, delay or reduce metabesity, care must be taken to avoid aggravating the situation. Metabesity may get out of control if nutrition becomes a source of dietary advanced glycation end products (AGEs) of which Western diets, and modern-day foods may be replete of, owing to mode the of cooking and the length of heat exposure, especially with dry heat (including frying, grilling, toasting, baking, and barbecuing). Excessive consumption of AGEs has been implicated in obesity, diabetes, inflammation, aging, neurodegenerative and cardiovascular diseases. Rather, low heat cooking (boiling, steaming) with water, raw consumption of organic plant-based foods or those exposed to adequate cooking time, vegetarian or vegan diet may help improve gut function and health, while ensuring the health of the brain, heart, kidney and other organs affected by obesity and diabetes (diabesity) complications, and the interconnected diseases with metabolic roots. Overall, this will serve to extend healthspan, and healthy lifespan.

Compliance with Ethical Standards

Conflict of Interest All authors declare they have no conflict of interest.

References

Abbott RD et al (2002) Midlife adiposity and the future risk of Parkinson's disease. Neurology 59(7):1051–1057

Abete I, Astrup A, Martinez JA et al (2010) Obesity and the metabolic syndrome: role of different dietary macronutrient distribution patterns and specific nutritional components on weight loss and maintenance. Nutr Rev 2010(68):214–231

Aggarwal BB (2010) Targeting inflammation-induced obesity and metabolic diseases by curcumin and other nutraceuticals. Annu Rev Nutr 30:173–199

Amrani A et al (1996) Interleukin-1 effect on glycemia in the non-obese diabetic mouse at the pre-diabetic stage. J Endocrinol 148:139–148

Ankarfeldt MZ et al (2014) Body characteristics, [corrected] dietary protein and body weight regulation. Reconciling conflicting results from intervention and observational studies? PLoS One 2014(9):e101134

Arzola-Paniagua MA, García-Salgado López ER, Calvo-Vargas CG, Guevara-Cruz M (2016). Efficacy of an orlistat-resveratrol combination for weight loss in subjects with obesity: a randomized controlled trial. Obesity 24:1454–1463

Bednarski T, Pyrkowska A, Opasinska A, Dobrzyn P (2016) Regulation of cardiac metabolism and function by lipogenic factors. Postepy Hig Med Dosw (online) 70:644–653

Belfort R et al (2005) Dose-response e_ect of elevated plasma free fatty acid on insulin signaling. Diabetes 54:1640–1648

Bernstein LE, Berry J, Kim S, Canavan B, Grinspoon SK (2006) Effects of Etanercept in patients with the metabolic syndrome. Arch Intern Med 166:902–908

Bjorntorp P (1991) Metabolic implications of body fat distribution. Diab. Care 14:1132–1143

Blachnio-Zabielska AU et al (2012) Increased bioactive lipids content in human subcutaneous and epicardial fat tissue correlates with insulin resistance. Lipids 47:1131–1141

Boden G (1997) Role of fatty acids in the pathogenesis of insulin resistance and NIDDM. Diabetes 46:3–10

Bradford PG (2013) Curcumin and obesity. BioFactors 39:78–87

Brüll V et al (2015) Effects of a quercetin-rich onion skin extract on 24 h ambulatory blood pressure and endothelial function in overweight-to-obese patients with (pre-) hypertension: a randomised double-blinded placebo-controlled cross-over trial. Br J Nutr 2015(114):1263–1277

Burwell H, Vilsack H (2015) Dietary Guidelines Advisory Committee USDA. DGAC 2015, Chapter 1–7

Calle EE, Rodriguez C, Walker-Thurmond K, Thun MJ (2003) Overweight, obesity, and mortality from cancer in a prospectively studied cohort of U.S. adults. N Engl J Med 348:1625–1638

Chavey C et al (2009) CXC ligand 5 is an adipose-tissue derived factor that links obesity to insulin resistance. Cell Metab 9:339–349

Chen S, Jiang H, Wu X, Fang J (2016) Therapeutic effects of Quercetin on inflammation, obesity, and type 2 diabetes. Mediators Inflamm 2016:9340637

Considine RV et al (1996) Serum immunoreactive-leptin concentrations in normal-weight and obese humans. N Engl J Med 334:292

Cornier M-A, Donahoo WT, Pereira R et al (2005) Insulin sensitivity determines the effectiveness of dietary macronutrient composition on weight loss in obese women. Obes Res 13:703–709

Cronan JE, Jr (1997) In vivo evidence that acyl coenzyme a regulates DNA binding by the Escherichia coli FadR global transcription factor. J Bacteriol 179:1819–1823

Cuffe H et al (2018) Targeted deletion of adipocyte Abca1 (ATP-binding cassette transporter A1) impairs diet-induced obesity. Arterioscler Thromb Vasc Biol 38:733–743

Davies LC, Rice CM, McVicar DW, Weiss JM (2019) Diversity and environmental adaptation of phagocytic cell metabolism. J Leukoc Biol 105:37–48

Deshmane SL, Kremlev S, Amini S, Sawaya BE (2009) Monocyte chemoattractant protein-1 (MCP-1): an overview. J Interferon Cytokine Res 29:313–326

Di Pierro F, Bressan A, Ranaldi D, Rapacioli G, Giacomelli L, Bertuccioli A (2015) Potential role of bioavailable curcumin in weight loss and omental adipose tissue decrease: Preliminary data of a randomized, controlled trial in overweight people with metabolic syndrome. Preliminary study. Eur Rev Med Pharmacol Sci 19:4195–4202

Diepvens K, Westerterp KR, Westerterp-Plantenga MS (2007) Obesity and thermogenesis related to the consumption of caffeine, ephedrine, capsaicin, and green tea. Am J Physiol Regul Integr Comp Physiol 292:R77–R85

Due A, Toubro S, Skov AR et al (2004) Effect of normal-fat diets, either medium or high in protein, on body weight in overweight subjects: a randomised 1-year trial. Int J Obes Relat Metab Disord 28:1283–1290

Ebbeling CB, Leidig MM, Feldman HA et al (2007) Effects of a low-glycemic load vs low-fat diet in obese young adults: a randomized trial. JAMA 297:2092–2102

Ellis BA et al (2000) Long-chain acyl-CoA esters as indicators of lipid metabolism and insulin sensitivity in rat and human muscle. Am J Physiol Endocrinol Metab 279:E554–E560

Fall C, Osmond C (2013) Commentary: the developmental origins of health and disease: an appreciation of the life and work of Professor David J.P. Barker, 1938–2013. Int J Epidemiol. 42:1231–1232 Oxford University Press
Flatt JP (1988) Importance of nutrient balance in body weight regulation. Diabetes Metab Rev 4:571–581
Friedman JM, Halaas JL (1998) Leptin and the regulation of body weight in mammals. Nature 395:763–770
Ganjali S et al (2014) Investigation of the e_ects of curcumin on serum cytokines in obese individuals: A randomized controlled trial. Sci. World J. 2014:898361
Gauthier BR, Bermúdez-Silva FJ (2019) Advances in genetics of regeneration in metabesity. Gene 10(5):383
Grunfeld C et al (1996) Endotoxin and cytokines induce expression of leptin, the ob gene product, in hamsters. J Clin Investig 97:2152–2157
Gustafson D (2006) Adiposity indices and dementia. Lancet Neurol 5(8):713–720
Ha EE, Bauer RC (2018) Emerging roles for adipose tissue in cardiovascular disease. Arterioscler Thromb Vasc Biol 38:e137–e144
Halberg N, Wernstedt-Asterholm I, Scherer PE (2008) The adipocyte as an endocrine cell. Endocrinol Metab Clin N Am 37:753–768
Halton TL, Hu FB (2004) The effects of high protein diets on thermogenesis, satiety and weight loss: a critical review. J Am Coll Nutr 23:373–385
Harville EW, Srinivasan S, Chen W, Berenson GS (2012) Is the metabolic syndrome a small baby syndrome? The bogalusa heart study. Metab Syndr Relat Disord 10:413–421 Mary Ann Liebert, Inc
Hauner H (2005) Secretory factors from human adipose tissue and their functional role. Proc Nutr Soc 64:163–169
Hausman DB, DiGirolamo M, Bartness TJ, Hausman GJ, Martin RJ (2001) The biology of white adipocyte proliferation. Obes Rev 2:239–254
Heymsfield SB, Wadden TA (2017) Mechanisms, pathophysiology, and management of obesity. N Engl J Med 376(3):254–266
Hinney A, Vogel CI, Hebebrand J (2010) From monogenic to polygenic obesity: recent advances. Eur Child Adolesc Psychiatry 19:297–310
Hotamisligil GS, Spiegelman BM (1994) Tumor necrosis factor alpha: a key component of the obesity-diabetes link. Diabetes 43:1271–1278
Hotamisligil GS, Shargill NS, Spiegelman BM (1993) Adipose expression of tumor necrosis factor-alpha: direct role in obesity-linked insulin resistance. Science 259:87–91
Hruby A, Hu FB (2015) The epidemiology of obesity: a big picture. Pharmacoeconomics 33:673–689
Hu FB (2007) Obesity and mortality: watch your waist, not just your weight. Arch Intern Med 167:875–876
Huber J et al (2008) CC chemokine and CC chemokine receptor profiles in visceral and subcutaneous adipose tissue are altered in human obesity. J Clin Endocrinol Metab 93:3215–3221
Hulver MW et al (2003) Skeletal muscle lipid metabolism with obesity. Am J Physiol Endocrinol Metab 84:E741–E747
Hussain A, Hydrie MZI, Claussen B, Asghar S (2010) Type 2 Diabetes and obesity: a review. J Diabetol 2(1):1–7
Illei GG, Lipsky PE (2000) Novel, antigen-specific therapeutic approaches to autoimmune inflammatory diseases. Curr Opin Immunol 12:712–718
Itani SI, Ruderman NB, Schmieder F, Boden G (2002) Lipid-induced insulin resistance in human muscle is associated with changes in diacylglycerol, protein kinase C, and IkappaB-alpha. Diabetes 51:2005–2011
Jayarathne S, Koboziev I, Park OH, Oldewage-Theron W, Shen CL, Moustaid-Moussa N (2017) Anti-inflammatory and anti obesity properties of food bioactive components: effects on adipose tissue. Prev Nutr Food Sci 22:251–262

Jensen MD, Ryan DH, Apovian CM et al (2014) 2013 AHA/ACC/TOS guideline for the management of overweight and obesity in adults: a report of the American College of Cardiology/American Heart Association task force on practice guidelines and the obesity society. Circulation 129(25 Suppl 2):S102–S138

Jin T, Song Z, Weng J, Fantus IG (2018) Curcumin and other dietary polyphenols: Potential mechanisms of metabolic actions and therapy for diabetes and obesity. Am J Physiol Endocrinol Metab 314:E201–E205

Kamei N et al (2006) Overexpression of monocyte chemoattractant protein-1 in adipose tissues causes macrophage recruitment and insulin resistance. J Biol Chem 281:26602–26614

Kanda H et al (2006) MCP-1 contributes to macrophage infiltration into adipose tissue, insulin resistance, and hepatic steatosis in obesity. J Clin Investig 116:1494–1505

Kasuga M (2006) Insulin resistance and pancreatic β-cell failure. J Clin Invest 116(7):1756–1760

Kay AM, Simpson CL, Stewart JA Jr (2016) The role of AGE/RAGE signaling in Diabetes-mediated vascular calcification. J Diab Res 2016:6809703

Kelley DE, Mokan M, Simoneau JA, Mandarino LJ (1993) Interaction between glucose and free fatty acid metabolism in human skeletal muscle. J Clin Investig 92:91–98

Keophiphath M, Rouault C, Divoux A, Clement K, Lacasa D (2009) CCL5 promotes macrophage recruitment and survival in human adipose tissue. Arterioscler Thromb Vasc Biol 30:39–45

Kern PA, Ranganathan S, Li C, Wood L, Ranganathan G (2001) Adipose tissue tumor necrosis factor and interleukin-6 expression in human obesity and insulin resistance. Am J Physiol Endocrinol Metab 280:E745–E751

Kim JY et al (2007) Obesity-associated improvements in metabolic profile through expansion of adipose tissue. J Clin Investig 117:2621–2637

Kjær TN et al (2017) No beneficial effects of resveratrol on the metabolic syndrome: a randomized placebo-controlled clinical trial. J Clin Endocrinol Metab 102:1642–1651

Konings E et al (2014) The effects of 30 days resveratrol supplementation on adipose tissue morphology and gene expression patterns in obese men. Int J Obes 38:470–473

Kouidhi S et al (2010) Relationship between subcutaneous adipose tissue expression of leptin and obesity in Tunisian patients. Tunis Med 88:569–572

la Fleur SE, Serlie MJ (2014) The interaction between nutrition and the brain and its consequences for body weight gain and metabolism; studies in rodents and men. Best Pract Res Clin Endocrinol Metab 28(5):649–659

Larsen TM, Dalskov S-M, van Baak M et al (2010) Diets with high or low protein content and glycemic index for weight-loss maintenance. N Engl J Med 363:2102–2113

Lord GM, Matarese G, Howard JK, Baker RJ, Bloom SR, Lechler RI (1998) Leptin modulates the T-cell immune response and reverses starvation-induced immunosuppression. Nature 394:897–901

Lumeng CN, DelProposto JB, Westcott DJ, Saltiel AR (2008) Phenotypic switching of adipose tissue macrophages with obesity is generated by spatiotemporal differences in macrophage subtypes. Diabetes 57:3239–3246

Marroquí L et al (2012) Role of leptin in the pancreatic β-cell: Effects and signaling pathways. J Mol Endocrinol 49:R9–R17

Martinez-Saez N et al (2014) A novel antioxidant beverage for body weight control based on co_ee Silverskin. Food Chem 150:227–234

McArdle MA, Finucane OM, Connaughton RM, McMorrow AM, Roche HM (2013) Mechanisms of obesity-induced inflammation and insulin resistance: Insights into the emerging role of nutritional strategies. Front Endocrinol (lausanne) 4:52

McGill HC, McMahan CA, Herderick EE (2002) Obesity accelerates the progression of coronary atherosclerosis in young men. Circulation 105(23):2712–2718

Meijer K et al (2011) Human primary adipocytes exhibit immune cell function: Adipocytes prime inflammation independent of macrophages. PLoS One 6:e17154

Mohammadi A et al (2013) Effects of supplementation with curcuminoids on dyslipidemia in obese patients: a randomized crossover trial. Phytother Res 27:374–379

Most J et al (2016) Combined epigallocatechin-3-gallate and resveratrol supplementation for 12 wk increases mitochondrial capacity and fat oxidation, but not insulin sensitivity, in obese humans: a randomized controlled trial. Am J Clin Nutr 104:215–227
Mozaffarian D, Benjamin EJ, Go AS (2015) Heart disease and stroke statistics–2015 update: a report from the American Heart Association. Circulation 131(4):29–322
Nabavi SF, Russo GL, Daglia M, Nabavi SM (2015) Role of quercetin as an alternative for obesity treatment: you are what you eat! Food Chem 179:305–310
Nielsen JH, Haase TN, Jaksch C, Nalla A, Søstrup B, Nalla AA et al (2014) Impact of fetal and neonatal environment on beta cell function and development of diabetes. Acta Obstet Gynecol Scand 93:1109–1122
Novelle MG, Wahl D, Dieguez C, Bernier M, de Cabo R (2015) Resveratrol supplementation: where are we now and where should we go? Ageing Res Rev 21:1–15
Olivier A, Pitt B, Girerd N (2017) Effect of eplerenone in patients with heart failure and reduced ejection fraction: potential effect modification by abdominal obesity Insight from the EMPHASIS-HF Trial. Eur J Heart Failure 19(9):1186–1197
Paddon-Jones D, Westman E, Mattes RD et al (2008) Protein, weight management, and satiety. Am J Clin Nutr 87:1558S-S1561
Pan MH, Tung YC, Yang G, Li S, Ho CT (2016) Molecular mechanisms of the anti-obesity e_ect of bioactive compounds in tea and co_ee. Food Funct 7:4481–4491
Panahi Y et al (2016) Effects of supplementation with curcumin on serum adipokine concentrations: a randomized controlled trial. Nutrition 32:1116–1122
Paz-Filho G, Mastronardi C, Franco CB, Wang KB, Wong ML, Licinio J (2012) Leptin: molecular mechanisms, systemic pro-inflammatory effects, and clinical implications. Arq Bras Endocrinol Metabol 56:597–607
Peeters A, Barendregt JJ, Willekens F, Mackenbach JP, Al MA, Bonneux L et al (2003) Obesity in adulthood and its consequences for life expectancy: a life-table analysis. Ann Intern Med 138:24–32
Pereira S et al (2015) Resveratrol prevents insulin resistance caused by short-term elevation of free fatty acids in vivo. Appl Physiol Nutr Metab 2015(40):1129–1136
Perez H, Flores Allende GA, Ajayi EIO, Porta DJ, Garcia NH (2019) Atherosclerlsis plaque area reduction: working hypothesis to prevent cardiovascular event. Comp Clin Pathol 28:907-914
Pesta DH, Samuel VT (2014) A high-protein diet for reducing body fat: mechanisms and possible caveats. Nutr Metab 11:53
Pfeuffer M et al (2013) Effect of quercetin on traits of the metabolic syndrome, endothelial function and inflammation in men with di_erent APOE isoforms. Nutr Metab Cardiovasc Dis 23:403–409
Pittas AG, Das SK, Hajduk CL et al (2005) A low-glycemic load diet facilitates greater weight loss in overweight adults with high insulin secretion but not in overweight adults with low insulin secretion in the CALERIE trial. Diab Care 28:2939–2941
Poirier P, Giles TD, Bray GA (2006) Obesity and cardiovascular disease: pathophysiology, evaluation, and effect of weight loss: an update of the 1997 American Heart Association Scientific Statement on Obesity and Heart Disease from the Obesity Committee of the Council on Nutrition, physical activity, and metabolism. Circulation 113(6):898–918
Poulsen MM et al (2013) High-dose resveratrol supplementation in obese men: an investigator-initiated, randomized, placebo-controlled clinical trial of substrate metabolism, insulin sensitivity, and body composition. Diabetes 62:1186–1195
Ramirez Boscáa A et al (2000) An hydroalcoholic extract of Curcuma longa lowers the abnormally high values of human-plasma fibrinogen. Mech Ageing Dev 114:207–210
Ramirez-Bosca A et al (2000) An hydroalcoholic extract of curcuma longa lowers the apo B/apo A ratio. Implications for atherogenesis prevention. Mech Ageing Dev 119:41–47
Rotter V, Nagaev I, Smith U (2003) Interleukin-6 (IL-6) induces insulin resistance in 3T3-L1 adipocytes and is, like IL-8 and tumor necrosis factor-alpha, overexpressed in human fat cells from insulin-resistant subjects. J Biol Chem 278:45777–45784

Sacks FM, Bray GA, Carey VJ et al (2009) Comparison of weight-loss diets with different compositions of fat, protein, and carbohydrates. N Engl J Med 2009(360):859–873

Saghizadeh M, Ong JM, Garvey WT, Henry RR, Kern PA (1996) The expression of TNF alpha by human muscle. Relationship to insulin resistance. J Clin Investig 97:1111–1116

Sahebkar A et al (2013) Curcuminoids modulate pro-oxidant-antioxidant balance but not the immune response to heat shock protein 27 and oxidized LDL in obese individuals. Phytother Res 27:1883–1888

Salehi P, Hsu I, Azen CG, Mittelman SD, Geffner ME, Jeandron D (2017) Effects of exenatide on weight and appetite in overweight adolescents and young adults with Prader-Willi syndrome. Pediatr Obes 12:221–228

Saraf-Bank S, Ahmadi A, Paknahad Z, Maracy M, Nourian M (2019) Effects of curcumin on cardiovascular risk factors in obese and overweight adolescent girls: A randomized clinical trial. Sao Paulo Med J 137:414–422

Sartipy P, Loskutoff DJ (2003) Monocyte chemoattractant protein 1 in obesity and insulin resistance. Proc Natl Acad Sci USA 100:7265–7270

Scapagnini G et al (2014) Dose response biology of resveratrol in obesity. J. Cell Commun Signal. 8:385–391

Schmitz-Peier C (2002) Protein kinase C and lipid-induced insulin resistance in skeletal muscle. Ann N Y Acad Sci 967:146–157

Shanely RA, Knab AM, Nieman DC, Jin F, McAnulty SR, Landram MJ (2010) Quercetin supplementation does not alter antioxidant status in humans. Free Radic Res 44:224–231

Shepherd PR, Kahn BB (1999) Glucose transporters and insulin action–implications for insulin resistance and diabetes mellitus. N Engl J Med 341:248–257

Shepherd PR, Gnudi L, Tozzo E, Yang H, Leach F, Kahn BB (1993) Adipose cell hyperplasia and enhanced glucose disposal in transgenic mice overexpressing GLUT4 selectively in adipose tissue. J Biol Chem 268:22243–22246

Shi H, Kokoeva MV, Inouye K, Tzameli I, Yin H, Flier JS (2006) TLR4 links innate immunity and fatty acid-induced insulin resistance. J Clin Investig 116:3015–3025

Shoelson SE, Herrero L, Naaz A (2007) Obesity, inflammation, and insulin resistance. Gastroenterology 132(6):2169–2180

Strimpakos AS, Sharma RA (2008) Curcumin: preventive and therapeutic properties in laboratory studies and clinical trials. Antioxid Redox Sign 10:511–545

Suganami T, Nishida J, Ogawa Y (2005) A paracrine loop between adipocytes and macrophages aggravates inflammatory changes: role of free fatty acids and tumor necrosis factor alpha. Arterioscler Thromb Vasc Biol 25:2062–2068

Tamara B et al (2011) Antioxidant-rich co_ee reduces DNA damage, elevates glutathione status and contributes to weight control: Results from an intervention study. Mol Nutr Food Res 55:793–797

Timmers S et al (2011) Calorie restriction-like e_ects of 30 days of resveratrol supplementation on energy metabolism and metabolic profile in obese humans. Cell Metab 14:612–622

Tomas E et al (2002) Enhanced muscle fat oxidation and glucose transport by ACRP30 globular domain: Acetyl-CoA carboxylase inhibition and AMP-activated protein kinase activation. Proc Natl Acad Sci USA 99:16309–16313

Tomé-Carneiro J et al (2012) Consumption of a grape extract supplement containing resveratrol decreases oxidized LDL and ApoB in patients undergoing primary prevention of cardiovascular disease: a triple-blind, 6-month follow-up, placebo-controlled, randomized trial. Mol Nutr Food Res 56:810–821

Tourniaire F et al (2013) Chemokine Expression in Inflamed Adipose Tissue Is Mainly Mediated by NF-κB. PLoS One 8:e66515

Tozzo E, Gnudi L, Kahn BB (1997) Amelioration of insulin resistance in streptozotocin diabetic mice by transgenic overexpression of GLUT4 driven by an adipose-specific promoter. Endocrinology 138:1604–1611

Trials.gov C (2003–2020) Investigating the use of Quercetin on Glucose Absorption in Obesity, and Obesity with Type 2 Diabetes. 2003–2020. http://clinicaltrials.gov/show/NCT00065676. Accessed 31 July 2003
Wang GJ et al (2001) Brain dopamine and obesity. Lancet 357(9253):354–357
Wang Y, Rimm EB, Stampfer MJ, Willett WC, Hu FB (2005) Comparison of abdominal adiposity and overall obesity in predicting risk of type 2 diabetes among men. Am J Clin Nutr 81:555–563
Wang S et al (2015) Resveratrol induces brown-like adipocyte formation in white fat through activation of AMP-activated protein kinase (AMPK) alpha1. Int J Obes 39:967–976
Wascher TC, Lindeman JH, Sourij H, Kooistra T, Pacini G, Roden M (2011) Chronic TNF-α neutralization does not improve insulin resistance or endothelial function in "healthy" men with metabolic syndrome. Mol Med 17:189–193
Wellen KE, Hotamisligil GS (2005) Inflammation, stress, and diabetes. J Clin Investig 115(5):1111–1119
Weyer C, Foley JE, Bogardus C, Tataranni PA, Pratley RE (2000) Enlarged subcutaneous abdominal adipocyte size, but not obesity itself, predicts type II diabetes independent of insulin resistance. Diabetologia 43:1498–1506
Whitmer RA, Gustafson DR, Barrett-Connor E, Haan MN, Gunderson EP, Yaffe K (2008) Central obesity and increased risk of dementia more than three decades later. Neurology 71(14):1057–1064
WHO (2000) Evaluation of certain food additives. WHO Technical report Series, 891. WHO, Geneva, Switzerland
Wycherley TP, Moran LJ, Clifton PM et al (2012) Effects of energy-restricted high-protein, low-fat compared with standard-protein, low-fat diets: a meta-analysis of randomized controlled trials. Am J Clin Nutr 96:1281–1298
Xia N, Daiber A, Förstermann U, Li H (2017) Antioxidant effects of resveratrol in the cardiovascular system. Br J Pharmacol 174:1633–1646
Yan JQ, Yan Z, BaoLu Z (2013) Green tea catechins prevent obesity through modulation of peroxisome proliferator-activated receptors. Sci. China Life 56:804–810
Yu C et al (2002) Mechanism by which fatty acids inhibit insulin activation of insulin receptor substrate-1 (IRS-1)-associated phosphatidylinositol 3-kinase activity in muscle. J Biol Chem 277:50230–50236
Zhao Y et al (2017) The beneficial effects of Quercetin, curcumin, and resveratrol in obesity. Oxid Med Cell Longev 2017:1459497

Chapter 28
Why Ashwagandha for Healthy Ageing? Evidence from Cultured Human Cells

Renu Wadhwa, Sukant Garg, Mallika Khurana, and Sunil C. Kaul

Abstract Ageing is an innate indispensable physiological process largely conceived as general decline in body functions and defense mechanisms. While ageing per se is not a disease, there are many age-related pathologies, modulation of which is considered as anti-ageing in several ways. Chronic stress often triggers senescence-inducing mechanisms manifested as premature/rapid ageing. The latter is associated with a high incidence of stress-related disorders such as cancer, neurodegeneration, metabolic disorders and muscle/bone dysfunctions. Cell culture system provides an easy and convenient experimental system to study mechanisms of natural (replicative) and stress-induced ageing. We, over the years, have researched molecular mechanisms of ageing and age-related pathologies, and their modulation with natural compounds using cell culture as a model system. Among several others, bioactives from Ashwagandha (*Withania somnifera*) have emerged as useful natural compounds with a variety of activities and are hence predicted to assist in health care in stress and disease states. In this chapter, we describe highlights of our research work demonstrating the therapeutic potential of Ashwagandha leaves that offer advantage over roots in terms of availability, processing and being enriched with active compounds.

Keywords Ashwagandha · Stress · Protection · Cancer · Inhibition · QOL

28.1 Introduction

The word "ageing" possesses an inherent meaning of decline, at large. Globally increase in ageing population and environmental stress are the two major concerns in the modern society. It is largely seen as increasing incidence of a variety of pathologies that severly affect quality of life (QOL) at all ages, but more apparent in elderly populations. With parallel advancements in medicine and drug deveopl-ment, although most pathologies have new treatment opportunities, maintenance of

R. Wadhwa (✉) · S. Garg · M. Khurana · S. C. Kaul
AIST-INDIA DAILAB, DBT-AIST International Center for Translational and Environmental Research (DAICENTER), National Institute of Advanced Industrial Science and Technology (AIST), Tsukuba 305 8565, Japan
e-mail: renu-wadhwa@aist.go.jp

S. I. S. Rattan and G. Kaur (eds.), *Nutrition, Food and Diet in Ageing and Longevity*, Healthy Ageing and Longevity 14, https://doi.org/10.1007/978-3-030-83017-5_28

QOL and healthy ageing remains relatively less explored aspect as yet. It has been well documented that a good diet, supplements enriched with natural compounds with antioxidant, anti-inflammatory and antistress ingredients help in maintaining the body homeostasis and brain health, collectively maintaining QOL through extended lifespan. Often, these aspects are tracked back to traditional home medicine, oral wisdom and experiences that have not been validated enough by laboratory studies. Only from last one decade, several laboratories worldwide have engaged modern biotechnology tools to validate activities in herbal extracts and their active components. One such herb is Ashwagandha, also called "Queen of Ayurveda" for its popular use in Indian traditional home medicine. Multiple activities including anticancer, antistress and remedial potential for old age-related neurodegenerative pathologies, their active principles and underlying mechanism(s) have been investigated using cell culture and mice models. Here we provide a review on our findings on the use of cell culture and molecular technologies as a valid approach to test the effects of Ashwagandha extracts and bioactives, and their relevance to healthy ageing.

28.2 Use of Cell Culture Systems For Understanding Ageing, Stress and Disease Mechanisms

Cell culture technologies that have established methods to grow human cells (as attached to substratum or in suspended in culture medium) provide an extremely useful experimental system to investigate the molecular mechanisms of life processes including ageing, stress and a variety of diseases. Furthermore, they provide a convenient platform for pharmacological studies including drug screenings, pharmacodynamics and pharmacokinetics. Cell culture started with the hanging drop technique in which a viable tissue was hung in growth favoring culture medium. It has progressed to most complex techniques not only for cells from a wide variety of lineages but also for cells in stressed, diseased and rejuvenated state including the induced pluripotent and artificial cell systems (Lovitt et al. 2014; Yao and Asayama 2017; Kokubu et al. 2015; Kwakkenbos et al. 2016; Ostrow et al. 2015). The principles behind all of these are, however, to count the cells, and inspect their morphology and functional characteristics by molecular/biochemical/imaging assays. In a typical drug discovery regimen, the response of living cells against the effect of a drug is tested in terms of viability, morphology, growth patterns, nucleic acid rearrangements and modulation of the central dogma in target Vs control cells (Ghanemi 2015). Human normal somatic cells have a limited life span and enter a state of permanent growth arrest after dividing for a limited number of times, also known as the Hayflick limit (Hayflick and Moorhead 1961). Cancer cells, on the other hand, often possess robust and infinite proliferation capability, and are established into immortal cell lines that offer convenient material for research on its molecular mechanisms and drug discovery (Wadhwa et al. 1991; Reddel 2000; Daniele et al. 2002; Kumar et al. 2021).

Cellular senescence is an indispensable outcome of life processes such as, mitochondrial function and oxidative stress, cell division and telomere shortening, and other intra- and extra-cellular stresses that constitute the normal byproduct of various biological processes. DNA-protein and protein-protein interactions are essential part of normal molecular signaling (Yang et al. 2020). Errors in these interactions lead to malfunctions of proteins, hypo- or hyper-activation of signaling often lead to accumulation of aggregated, misfolded macromolecules and molecular garbage. Inefficiency in clearance of misfolding and aggregated macromolecules due to decline in repair and recycle signaling including DNA repair, protein degradation and autophagy often connect to pathological conditions including premature ageing. In these premises, natural/synthetic compounds that have the potential to prevent or revert stress-induced molecular changes and the pathogenesis are often refer to as anti-ageing compounds. Cell based assays provide a convenient way to investigate the anti-stress and anti-ageing potential/efficacy of the test compounds. A large number of compounds (~250) have been discovered by such cell-based assays (Bakula et al. 2019). On the other hand, cytotoxicity of the test compounds, time and dose dependent response assays are the established first steps of anticancer drug discovery regimen. Furthermore, cell morphology, attachment and migration characteristics often provide hints to their mechanism of action. We earlier developed Quantitative and Qualitative Cell Viability assay (QCV assay) that provides an easy and reliable method to identify short- and long-term response of cells to a variety of cytotoxic, cytostatic and cytoprotective compounds (Garg et al. 2018a). Using high throughput screening of several compounds, we had identified that a natural withanolide – Withanone, and Withanone-rich extract from Ashwagandha leaves, as well as caffeic acid phenethyl ester (CAPE) and CAPE-rich extract from honeybee propolis cause toxicity to cancer cells selectively (Widodo et al. 2008, 2009, 2010, Priyandoko et al. 2011, Grover et al. 2012, Wadhwa et al. 2016b, Bhargava et al. 2018). Using cell-based loss of function screening assays, in which cells are treated with either gene specific siRNA/shRNA or randomized ribozymes before exposing to the test reagents, have revealed mechanism of action of these compounds, at least in part.

Stress-induced premature senescence in cultured cells is also commonly used as a model system to test the stress activities of various environmental factors in which the stressed phenotype is validated by monitoring the expression of stress-induced proteins and occurrence of structural and functional alterations. In order to test the anti-ageing mechanisms, effects of the test compounds on replicative or stress-induced senescence are determined by *in vitro* serial passaging. Molecular mechanism(s) of action are determined by analysis of cell characteristics, changes in the expression level of proteins and mRNAs that determine these characteristics. For example, replicative and premature senescence can be determined by senescence-associated β-galactosidase assay, increase in the expression level of H2AX, p53, $p21^{WAF1}$, $p14^{ARF}$, $p16^{INK4A}$ proteins that signify increased level of DNA damage and induction of growth arrest by p53 and pRb tumor suppressor pathways (Kameda et al. 2021; von Zglinicki et al. 2021; Zhang et al. 2021). Phosphorylation of HP1γ on serine 93, required for incorporation of HP1γ into senescence-associated heterochromatin foci (SAHF), has been established as a specific marker for senescence.

We had earlier reported that mortalin, a member of Hsp70 family of proteins, is frequently enriched in cancer cells and decreases with replicative senescence in normal cells (Deocaris et al. 2013; Na et al. 2016). Cancer cells compromised for mortalin entered senescence like growth arrest. (Deocaris et al. 2009; Gao et al. 2015; Yoo et al. 2010; Ryu et al. 2014; Wadhwa et al. 2000). In cancer cells, mortalin sequesters the tumor suppressor p53 protein in the cytoplasm and thereby inactivates it (Kaul et al. 2005; Wadhwa et al. 2006). Consistently, cells in which mortalin protein has been compromised or its interactions with p53 is abrogated by natural/synthetic compounds showed activation of p53 function and induction of either growth arrest or apoptosis (Wadhwa et al. 2000; 2002a; 2002b; Widodo et al. 2008; Sari et al. 2020). Furthermore, subcellular distribution (staining pattern) of mortalin was seen to shift from perinuclear (typical in cancer cells) to pancytoplasmic (typical in normal cells). In line with these, the natural and synthetic compounds that caused senescence in cancer cells also showed shift in mortalin staining pattern from perinuclear to pancytoplasmic type (Wadhwa et al. 1993; Widodo et al. 2007).

Many cellular models of cancer have been introduced for molecular investigations at the gene, transcript and protein levels. The mutated versions of tumor suppressor proteins or their functional activation by various cellular factors including the binding proteins have been identified and assigned to drive oncogenic signaling. Tumor suppressor proteins, such as p53, pRB, $p14^{ARF}$ and $p1^{6INK4A}$ have emerged as key regulators of senescence-related growth arrest (Duncan et al. 2000; Itahana et al. 2003; Larsen 2004). On the other hand, a large variety of cancers have been shown to possess either mutated or inactivated forms of these proteins. It has been well established that mutations in p53 can activate many signaling pathways other than those activated by the loss of wild type p53 activity (Fig. 28.1). Cancer cells possessing specific p53 mutants have served as convenient models to understand their role in carcinogenesis and metastasis (Shay et al. 1995; Duncan and Reddel 1997). Wild type p53 acts as a transcription factor for a large number of proteins involved in regulation of cell cycle progression, apoptosis and stress response. In turn, it is regulated by numerous proteins that regulate its subcellular localization, phosphorylation, tetramerization that are essential factors for its activity as transcriptional factor. pRb, known as the product of the retinoblastoma tumor susceptibility gene (*RB1*) is also a major tumor suppressor protein. Cells containing mutated pRB possess chromosomal instability and aneuploidy, contributing to tumorigenesis. Hypo-phosphorylated pRB binds and inactivates E2F family of transcriptional factors and inhibit the progression of cell cycle. Phosphorylation of pRB by cyclin-CDK complexes leads to the release of E2F from pRB-E2F complex (Fig. 28.1). The latter is an essential factor for entry of cells to S-phase. In this way, cyclin-CDK complexes play a key role in cell cycle proliferation or arrest (Roupakia et al. 2021). Activity of cyclin-CDK complexes is inhibited by two classes of inhibitors, $p21^{WAF1}$ (a major effector of p53) and $p16^{INK4A}$. The latter is coded from INK4a (MTS1, CDKN2) locus that codes for another unrelated tumor suppressor protein Alternative Reading Frame protein ($p14^{ARF}$) (Roupakia et al. 2021). It acts upstream of p53 and enhances its transcriptional activation function required for cell cycle arrest. Both $p16^{INK4A}$ and $p14^{ARF}$ have been found to inactivated in cancers by gene silencing mediated by either point

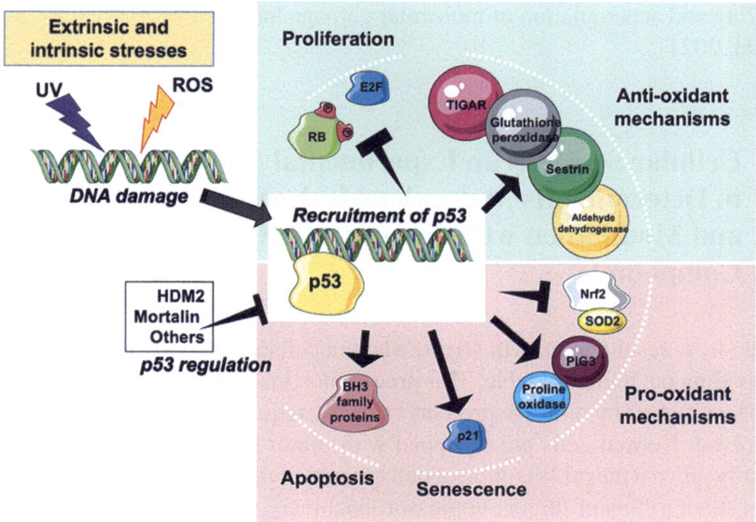

Fig. 28.1 Schematic diagram showing multiple activities of the tumor suppressor protein, p53

mutations or DNA methylation. Cell culture models have played key role not only to understand these mechanisms but also as drug discovery platforms (Larsen 2004).

Besides cancer, cell culture models are commonly used to understand the mechanisms of stress, ageing, age-related pathologies and their intervention (D'Souza et al. 2021; Ali et al. 2021). Most of these are governed by altered repair capacity and defense mechanisms leading to accumulative errors in replication and metabolic processes. Alzheimer's disease (AD) is a classic example of age-related pathologies involving neuro-degeneration due to protein misfolding, accumulation of oxidatively damaged proteins (accumulation of beta-amyloid fibrils and hyperphosphorylated tau proteins). These deposits are neurotoxic and have been found to be correlated with cognitive impairment. Parkinson's disease (PD), another age-related neurodegenerative disorder, involves the loss of dopaminergic neurons and causes symptoms like tremors, bradykinesia and rigid muscles. Age-related Macular Degeneration (AMD) is another old age disease, where the vision is impaired due to neovascular or atrophic degradation of the macula of the retina in the eye. Cardiovascular diseases such as atherosclerosis and myocardial infraction occur due to dyslipidemic states in the body wherein the blood vessels accumulate cholesterol leading to plaque development and vascular remodeling, eventually leading to blocked arteries-veins and then infraction of a portion of the heart. Stroke-ischemic, hemorrhagic and stroke that are highly correlated with incidence of high blood pressure, show age-related trends. Stroke patients usually loose a part of the brain function and thus wholly or partly impairing the bodily function. Cell culture has played a major role so far to help in deciphering the pathophysiology of these diseases and drug designing. Besides the specific gene mutations and functions, contributions of oxidative stress, protein misfolding and accumulation of molecular garbage has been acknowledged at large (Ali et al. 2021).

28.3 Cellular Stress as an Experimental Model to Determine Its Molecular Mechanisms and Modulation with Natural and Synthetic Compounds

Cellular stress results in growth arrest, where a cell enters a cytostatic phase i.e., it stops dividing but remains viable. This process hosts a complex change in the protein homeostasis, whereby multiple proteins critically required for cell cycle progression are involved. Normal cells are equipped with tumor suppression mechanisms that essentially prevent the proliferation of defective or stressed cells. p53 and pRB constitute two main pillars of tumor suppressor machinery that regulate the cell proliferation in normal and stressed conditions. These in turn are regulated by multitude of factors, intra and extra-cellular conditions. Inefficient tumor suppressor pathways result in escape of cells from growth arrest and continue proliferation even in stressed conditions leading to activation of carcinogenesis. On the other hand, reactivation of tumor suppressor mechanisms in cancer cells put them into growth arrest/apoptosis and are considered as a viable anti-cancer strategy (Amellem et al. 1996; Itahana et al. 2002; Cheung et al. 2010; Mowla et al. 2014).

A variety of stress models are used in cell culture to evaluate the response of cells to stress in terms of their growth characteristics, biochemical and molecular profiles. Some of the commonly used stress reagents are listed in Table 28.1 (Kalra et al. 2020). It has been established that the stress-induced changes in expression level of cell cycle regulating proteins put cells to arrest in a particular cell cycle phase. This could be due to one or more simultaneously orchestrating processes viz., DNA damage, activation of tumor suppressors, downregulation of proliferation pathways. Each of these processes are characterized not only by specific proteins but also by their threshold level that may be monitored by either biochemical or imaging assays. Replicative senescence is associated with telomeric attrition and DNA damage stress in an accumulative manner. These along with the decline in repair mechanisms yield the final arrest of cells in nondividing, but viable, state that is marked by increase in the level of p53, $p21^{WAF1}$, $p14^{ARF}$ and $p16^{INK4A}$ proteins (Itahana et al. 2002; von Zglinicki et al. 2021). At the same time, E2F, Cyclins and CDK family proteins that are essential for cell cycle progression show decline. A variety of stresses have been shown to mimic these changes and hence cause premature senescence in normal human cells (Mohamad Kamal et al. 2020). Reversal of these changes and induction of repair pathways are considered as anti-ageing effects. For example, mechanisms that have been shown to play key role in preventing oxidative insults such as Nrf2-dependent-pathway and antioxidant proteins are monitored to evaluate the anti-stress and anti-ageing potential of natural/synthetic compounds (Yuan et al. 2021). Natural

Table 28.1 Frequently used chemical models of stresses in cell culture studies

Stress types	Chemical models for cell culture studies	Concentration range	Biochemical activities
Environmental stresses			
Heavy metal	Sodium (meta) arsenite	0.5–100 μM	Protein misfolding, aggregation, induction of stress proteins
	Cadmium nitrate tetrahydrate	0.5–5 μM	Cytotoxicity
Smoke	Nicotine	0.2–2 mM	Potent parasympathomimetic stimulant
	Benzo[a]pyrene	20–200 μM	Mutagenic stressor metabolite
Diesel/fuel	2-Nitro-9-fluorenone	0.1–1 μM	Mutagenic diesel-exhaust photoproduct
Plastics	Vanadium (V) oxide	0.4–10 μM	Amphoteric oxide, oxidizing agent
	Titanium (IV) oxide	0.5–5 mM	Inflammatory/lysosomal activity defects
	Bisphenol A bis (chloroformate)	0.02–200 μM	Synthetic xenoestrogen/ endocrine disruptor
Aldehydes	Formaldehyde	150 μM–1.5 mM	Organic aldehydes/potent carcinogen
Polycyclic aromatic HCs	1-Hydroxypyrene	10–100 μM	Potent excretory metabolite
Radiation	Ultraviolet	25 mJ/Cm^2	DNA damage
Biological stresses			
Chemotherapeutic drugs	Doxorubicin	0.125–1.25 μM	DNA intercalation/DNA damage response
Hypoxia	Cobalt chloride	100–1000 μM	Erythropoiesis
Oxidative damage	Hydrogen peroxide	100–1000 μM	Oxidation of proteins, membrane lipids and DNA
Alcohol	Ethanol	1–10% (v/v)	Dehydration and hepatic damage
Physiological stresses (Anger, emotions)	Epinephrine	5–100 μM	Hormonic neurotransmitter peptide/steroid

compounds have largely been explored *in vitro* so as to elucidate their antioxidant, anti-stress and hence anti-ageing properties. Various stressors have been studied with respect to their protective activities on cells and key cell metabolic processes. Here, we present the details of a few such common stresses with a summary of protective activities by natural compounds.

In a conventional cell culture protocol, cells are subjected to oxidative stress by H_2O_2 or other chemical reagents and recovered in either the normal cell culture medium or the one supplemented with test compounds. Using brain derived cells, it has been established that the oxidative stress leads to upregulation of reactive oxygen species (ROS) and reduction in several key molecules involved in neurogeneration and memory related functions (Wadhwa et al. 2016a). On the other hand, some natural compounds including withanone, astaxanthin, fucoxanthin and resveratrol were shown to protect against oxidative damage and promote neurodifferentiation (Priyandoko et al. 2011, Kirtonia et al. 2020, Shah et al. 2015, Garg et al. 2018b, Afzal et al. 2019; 2020). Molecular mechanisms of oxidative stress-induced damage and protection by natural compounds can also be investigated using cell culture system. UV radiation is another common stressor that produces ROS, oxidative stress, and most importantly DNA damage leading to ageing, age-related pathologies and active oncogenesis. Cell culture system has been used to discover UV-protective compounds. For example, withanone (one of the withanolies from Ashwagandha) was shown to protect cells against UV-induced oxidative stress (Priyandoko et al. 2011). A combination of Withanone and Cucurbitacin B (a bioactive compound present in the plants of Cucurbitaceae family) was shown to possess anti-stress capability (Garg et al. 2018b). Interestingly, several findings derived from cell culture system has been proved in animal models. In mice model, UV radiation-induced wrinkles were corrected when treated with resveratrol by induction of anti-inflammatory and antioxidant pathways (Arora et al. 2020). Fucoxanthin, a marine carotenoid, has been shown to combat UV by downregulation of filaggrin against sunburn, by regulation of the NLRP3 inflammasome in UVB exposed keratinocytes *in vitro* (Matsui et al. 2016). Similarly, astaxanthin, another carotenoid pigment present in marine organisms, has been shown to cause protective effects *in vitro*, in neuronal cancers and against UV on ageing skin in clinical trials (Ng et al. 2021; Eren et al. 2019). Many other chemical models representing the environmental stresses including benzo[a]pyrene for smoke, sodium arsenite for heavy metal, 2-nitro-9-fluorenone for diesel, bisphenol A bis(chloroformate) for plastics, cobalt chloride for hypoxia, etc. have been studied using cell culture systems, and provide an easy and affordable high throughput screening system to identify anti-stress compounds (Kalra et al. 2020).

28.4 Ashwagandha - Bioactivities Discovered in Cell Culture Models

Scientifically known as *Withania somnifera*, Ashwagandha, Winter cherry or Indian ginseng, belongs to Solanaceae family and readily grows in drier parts of India, Pakistan, Afghanistan, Sri Lanka, Congo and South Africa. It is characterized by erect branching, a height of about 1.50 m, minutely stellate tomentose, fleshy, tapering and whitish brown roots, ovate leaves and green flowers. The plant grows well between 600–1200 m altitude in sandy loam or light red soil with a pH of 7.5–8.0, and temperature range of 15 to 39 °C. The semi-tropical areas receiving 500–750 mm rainfall are suitable for its cultivation, however, it requires a dry season during its growing period. Widely prated and boasted in Ayurveda, Ashwagandha is known as the remedy of all diseases. It has been used since the ancient times as a rejuvenizer, mood stabilizer, aphrodisiac, anti-diabetic, anticancer, antihypertensive, and immunomodulator. Scientific studies in the past few decades have validated its anti-stress, anticancer, anti-inflammatory, adaptogenic, cardiovascular, immunomodulatory and radiosensitizer activities. Ashwagandha is particularly rich in a group of phytochemicals called withanolides. These are steroidal lactones and their derivatives (Wadhwa et al. 2016a). Several laboratory studies have reported anticancer activities in the extracts derived from roots and leaves of Ashwagandha and the actives were identified as Withaferin-A (Wi-A) and Withanone (Wi-N).

28.5 Anticancer Activity—Experimental Evidence

We initially prepared the alcoholic extract from Ashwagandha leaves and called it i-Extract. Bioactivity in i-Extract was examined by *in vitro* assays, using a variety of human cancer cells. i-Extract caused toxicity to cancer cells; however it was nontoxic to normal cells at the equivalent doses (Kaur et al. 2004; Widodo et al. 2007). Analysis of i-Extract by High Pressure Liquid Chromatography (HPLC) showed that it contained Wi-A and Wi-N as predominant constituents (Widodo et al. 2007). Cell culture-based assays showed that the extracts with high ratio of Wi-N/Wi-A exhibited selective toxicity to cancer cells. Molecular studies were undertaken on the identified cellular targets for anticancer activities of i-Extract. We adopted loss-of-function screening using shRNA and randomized ribozyme libraries. shRNAs that caused resistance to the cytotoxic effect of i-Extract were selected and their gene targets were anticipated to be crucial for i-Extract induced death in cancer cells (Widodo et al. 2007, 2008). Interestingly, computational analysis of these gene targets identified p53 tumor suppressor as one of the key elements of this effect. We then examined the activities of p53 protein by biochemical and imaging analyses. Interestingly, we found that the i-Extract and Wi-N caused activation of p53 tumor suppressor pathway in cancer, but not normal, cells. Wi-A, on the other hand, was seen to activate p53 in both the cell types. It was consistent that Wi-A was cytotoxic to both cancer and normal cells, and confirmed that the activation of p53 is one of the major mechanisms of anticancer activity in Ashwagandha bio-actives. It has been previously reported that mortalin binds to p53 in cancer cells. The interaction of mortalin and p53 in cell cytoplasm was shown to cause retention of p53 in the cytoplasm and thus inhibiting its transcriptional activation function in the nucleus (Wadhwa et al. 2016b, 2006; Yun et al. 2017). Ashwagandha bio-actives were shown to inhibit mortalin-p53 interaction, causing retranslocation of p53 to the nucleus and reactivation of its transcriptional activation function that results either in growth arrest (allowing DNA repair) or apoptosis. Interestingly, i-Extract treated cancer cells exhibited a shift in staining pattern of mortalin from perinuclear (typical of cancer cells) to pancytoplasmic (typical of normal cells) pattern (Widodo et al. 2008; Grover et al. 2012; Kataria et al. 2012; 2013). In light of the reports that mortain-p53 interaction occurs in cancer cells whereas normal cells lack this interaction, the above molecular data provided justification for the selective toxicity role of i-Extract to cancer cells. Taken together, it was suggested that i-Extract blocked mortalin-p53 interactions resulting in nuclear translocation and transcriptional activation of p53 (endorsed by increase in $p21^{WAF1}$ and induction of senescence-like growth arrest in cancer cells). Furthermore, i-Extract treated mutant p53 harbouring cells showed decrease in (i) its expression level and (ii) expression of $p21^{WAF1}$ indicative of wild type p53 activity. These data suggested that i-Extract reactivates tumor suppressor function of p53 in cancer cells and could be utilised for cancer treatment (Widodo et al. 2008). Multimode anticancer activity of i-Extract was confirmed by selection of other proteins in loss-of-function screenings. These included (i) TPX2, a microtubule-associated protein that regulates Aurora-A function in an allosteric manner. Aurora-A is enriched in cancers and plays a key role in mitosis (centrosome maturation and chromosome segregation in particular). Due to such essential role of Aurora A in cell division, its inhibitors have been proposed as anticancer drugs. Most of these inhibitors work by blocking its ATP binding site, common to other kinases and hence limited by their specificity. Wi-A was shown to inhibit TPX2 and block TPX2-Aurora A complex. (ii) TFAP2A plays an important role in tumor growth by regulation of several proteins including insulin like growth factor receptor-1, MMP-2, E-cadherin, c-kit, Bcl-2, HER-2, and Smad signaling. (iii) ING1, an ING family protein, involved in human cellular senescence, tumor suppression and apoptosis. (iv) LHX3, a homeodomain transcription factor that is involved in embryonic development, cell fate determination, and oncogenesis. These data suggested that the cancer cell killing by i-Extract might involve several proteins and multimodal action. Analyses with computational and molecular docking tools further revealed the involvement of several kinds of biological processes including cell cycle, nucleic acid metabolism and cellular stress and oncogenesis at two levels (i) mitochondrial stress that results in change in membrane potential and reactive oxygen species (ROS) (ii) DNA damage stress leading to activation of DNA damage and repair machinery. It has been well established that whereas a moderate increase in ROS promotes cell proliferation and differentiation, its excessive amount causes irreversible oxidative damage to DNA, proteins and lipids leading to cell death (Widodo et al. 2007, 2008,

2010). Cancer cells frequently exhibit high oxidative stress and increased generation of reactive oxygen species (ROS) as compared to the normal counterparts. The higher level of ROS has been considered as a selective therapeutic target for cancer treatment. Consistent with the cytotoxicity of Wi-A, Wi-N and i-Extract, induction of ROS was detected. Furthermore, i-Extract and Wi-N lead to an induction of ROS only in cancer cells and hence was interpreted to be responsible for selective cancer cell killing. Indeed, analysis of mitochondrial membrane potential (MMP) and ultrastructure revealed decrease in MMP and structure deterioration (swollen and small fragmented mitochondria with reduced cristae) in i-Extract treated cancer cells (Widodo et al. 2010).

Several independent studies have supported multimodal anticancer activity of Wi-A. These include (i) inactivation of NFκB, a multifunctional transcriptional factor that regulates cytokine production and stress response (Dubey et al. 2018), (ii) collapse of intermediate filament protein Vimentin, an important regulator of cell shape and migration (Thaiparambil et al. 2011; Ivaska et al. 2007), (iii) oxidative stress in endoplasmic reticulum and mitochondria and mitc-FOS-FLIP-TRAIL induced apoptosis (Nishikawa et al. 2015), (iv) inhibition of proteasome-mediated degradation leading to upregulation of p21, BAX, IκBα and apoptotic cell death, (v) inhibition of STAT3 phosphorylation (an anti-apoptotic, pro-migratory factor in various types of cancers, (vi) inhibition of Hedgehog signaling pathway, an important contributor to proliferation, differentiation, and maintenance of tissue patterning, (vii) inhibition of c-MET, AKT and RAF-1 signaling, (viii) inhibition of snail and e-Cadherin mediating EMT signaling (Hsu et al. 2019), (ix) downregulation of BRCA1 and HSF1 proteins and (x) downregulation of HPV oncoproteins in human cervical cancer cells and restoration of p53 mediated growth arrest or apoptosis. Furthermore, Wi-A has been shown to sensitize cancer cells to radiotherapy and work in a synergistic manner with other standard clinical anticancer compounds including sorafenib (Cohen et al. 2012), oxaliplatin (Li et al. 2015) and cisplatin (Kakar et al. 2012, 2014, 2016). It caused inhibition of MRN complex and inhibition of telomere maintenance in ALT cancer cells (Yu et al. 2017).

With the experimental strategies described above, effect of water extract of Ashwagandha leaves (WEX) that contained low levels of Wi-N or Wi-A but moderate amount of triethylene glycol (TEG) was investigated (Wadhwa et al. 2013). TEG is a member of dihydroxy alcohol family, known to have high ability to hold water molecules. It has been established as a low toxicity mild disinfectant towards a variety of airborne, solution and surface bound microbes including bacteria and viruses. A similar compound, triethylene tetramine (TETA) has been reported to be a novel ligand for γ-quadruplex and possess many kinds of biological activities, including telomerase inhibition and induction of senescence in tumor cells. Low dose of TETA caused growth arrest of tumor cells in culture. Biochemical, molecular and imaging analyses revealed that the cytotoxicity of TEG is mediated by activation of p53 and pRb tumor suppressor pathways. However, in contrast to the selective activation of p53 in response to i-Extract treatment as discussed above, WEX caused activation of p53 both in cancer as well as normal cells. On the other hand, pRb phosphorylation and cyclin B1 decreased only in cancer cells. RB-phosphorylation causes release of E2F from RB-E2F complex, and transcription of genes required for cell cycle progression. Decrease in pRB phosphorylation in WEX-treated cancer cells exhibited inhibition of E2F and cell cycle arrest. These cell-based studies provided evidence that (i) ashwagandha possess multiple bioactives with multimodal mechanism of action and (ii) consumption of leaves as such may be more effective than the partial alcohol or water extracts. In view of this, we investigated the anticancer potential of Ashwagandha leaf powder and reported that mice fed with Ashwagandha (500 mg/Kg body weight) showed no toxicity, but strong suppression of tumor progression. Furthermore, a mixture of leaf powder with gamma cyclodextrin (γCD) showed higher efficacy, may be due to increased stability and bioavailability of active ingredients (Wadhwa et al. 2013; Kaul et al. 2016).

Several studies have suggested that using combinations of bioactives could not only have synergistic therapeutic efficacy but also offer side benefits such as healthy ageing and lifespan extension of the healthy cells. Some of the examples are Wi-N+Wi-A (Gao et al. 2014), Cucurbitacin B+Wi-N (Garg et al. 2018b, 2020) and CAPE+Wi-A (Sari et al. 2020). An analogue of Wi-A (2, 3-dihydro-3β-methoxy Withaferin A) was shown to lack cytotoxic effect, instead showed senescence inducing potential in a variety of cancer cells even at a relatively low-therapeutic index dose (Chaudhary et al. 2017, 2019). It was shown to play a protective role in healthy cells against various intracellular stresses. Whereas cucurbitacin was found to be toxic to both cancer and healthy cells (Li et al. 2016), its combination with withanone (CucWi-N) was selectively toxic to cancer cells (Garg et al. 2020). At differential doses, it also offered anti-metastatic, anti-stress, senescence-inducing and differentiation-inducing potencies in cancer cells without causing any harm to the healthy tissue. It offered health benefits in terms of protection to healthy cells against harmful environmental toxins and even promoted healthy differentiation of the brain cancer cells (Garg et al. 2018b). Some reports on investigations of bioactivities of Ashwagandha in animal models are listed here. In a rat model, 250–750 mg/kg daily feeding of Ashwagandha root extract showed reduced incidence of gentamycin-induced nephrotoxicity (Jeyanthi and Subramanian 2009). This feature was characteristically attributed to the antioxidant potential in the Ashwagandha root constituents. Additionally, Ashwagandha leaves, fresh and dry tubers showed strong antioxidant activity in the DPPH and lipid peroxidation models (Sumathi et al. 2007), where the leaves showed the highest potential. Alcoholic root extract provided protection against lead toxicity in a mouse model (Chaurasia et al. 2000).

In view of the above findings, hydroponic cultivation of Ashwagandha that could offer stable and defined resource of its bioactive components was established. By altering light and geography, hydroponically cultivated Ashwagandha leaves were engineered to be enriched in Wi-A and Wi-N (Kaul et al. 2016). Of note, Wi-N and Wi-A in leaves at the concentration of 3 and 1.8 mg/g in the usual setting increased to about 8.6 and 3.4 mg/g upon such manipulations, respectively. Furthermore, it was shown that pre-soaking of seeds and cultivation of Ashwagandha plant in Vermicompost, resulted in higher germination, denser leaf and root mass, and higher concentrations of active withanolides per se (Kaur et al. 2018).

28.6 Antistress Bioactives and their Bioactivities

Ageing is governed by multiple genetics and environmental factors. Oxidized proteins buildup with age as a normal byproduct of life-driving processes and increasing errors and inefficiencies in repair and recycle processes. Oxidative damage has been established as a major contributory factor to the age-related pathologies such as, Parkinson's and Alzheimer's disease. Proteasome plays an important role in degradation of damaged, mutant and misfolded proteins. Progressive failure in maintenance of protein quality and garbage management capacities by regular clearing of the damaged proteins lead to an accumulation of misfolded/modified proteins. Strengthening of proteasomal function has been predicted as an antiageing strategy to manage the accumulation of garbage and provide rejuvenation. As signified by its species name, *somnifera*, Ashwagandha contains high amounts of withanolides and has sleep-inducing activity. It is often described as Aphrodite herb (after the Greek goddess of beauty and sexual attractiveness) and believed to increase sexual desire. Intake of Ashwagandha works in various ways to enrich the body's physical and mental wellbeing, which likely connects to pleasurable sexual experiences. Inability to enjoy sexual pleasure is associated with fatigue and nervous exhaustion; Ashwagandha promotes calmness and restfulness, which may account for its aphrodisiac activity. Ashwagandha affects the hypothalamic-pituitary-adrenal hormonal and signaling axis as well as the neuroendocrine system, which controls how our bodies react to stress. Thus, Ashwagandha induces a state of harmony among various physiological functions. The adaptogenic activities of Ashwagandha—its ability to promote balance in the body and help deal with stress—are the outcome of combined positive functions of several potent withanolides, alkaloids, and other compounds. Pre-clinical research and clinical studies support therapeutic use of this plant to improve cardiovascular function, liver function, and cognitive and memory-related ailments (Wadhwa et al. 2016a; Gannon et al. 2019; Kulkarni and Dhir 2008; Modak et al. 2007; Ven Murthy et al. 2010).

28.6.1 Bioactive: Withanone

In our efforts to characterize Ashwagandha activities and their molecular mechanisms, we initially prepared the leaf extract of Ashwagandha (i-Extract) that showed anticancer activity as described in the above section. In contrast to up-regulation of 21^{WAF1} in cancer cells, it caused down-regulation in normal cells. Since $p21^{WAF1}$ is tightly involved in initiation and establishment of cellular senescence, we anticipated that the Wi-N treated cells may have anti-ageing effect. In long term serial passaging of normal human fibroblasts, indeed Wi-N treated cells underwent further 10–12 population doublings (Widodo et al. 2009). When examined at parallel population doublings, Wi-N treated normal fibroblasts showed about 20% less auto-fluorescence (accumulated molecular damage) and senescence-associated β-gal activity (marker for senescence) supporting that Wi-N-induced increase in population doublings of cells may be due to the delayed senescence in normal human fibroblasts (Widodo et al. 2009). Increase in oxidative stress and decrease in proteasomal function and accumulation of molecular garbage have been established as the main causes of senescence-associated functional decline both in cellular and organismal models. In our experiments, normal cells were exposed to either H_2O_2 or UV for oxidative stress and followed by their recovery in either normal or Wi-N supplemented medium (Widodo et al. 2009). Biochemical and imaging analyses supported that the presence of Wi-N protected cells against UV as well as oxidative stress (Widodo et al. 2009). Furthermore, premature ageing was induced by proteasomal inhibitor, epoxomicin. The effect of Wi-N in this premature ageing model system revealed protection against the premature ageing. Biochemical analysis revealed that Wi-N has a strong proteasomal inducing activity in treated cells. Thus Wi-N was seen to be not only selectively toxic to cancer cells, but also protect normal human fibroblasts against the toxicity caused by Wi-A (Widodo et al. 2009).

28.6.2 Bioactive: 3β-Methoxy Withaferin A

Anticancer activity of Wi-A was supported by induction of wild type p53 activities. Interestingly, cells when treated with 3β-methoxy withaferin A (mWi-A) showed neither cytotoxicity nor any effect on the p53 signaling (Chaudhary et al. 2017, 2019). NRF-2 is an important regulator of cellular oxidative stress response. It is upregulated in many cancers and provides an advantage of survival against oxidative stress. NRF-2 was identified as an important target of Wi-A. The level of NRF-2 was examined in Wi-A and mWi-A treated cells. When cells were treated with Wi-A there was a decrease in NRF2 expression, but no change was observed in mWi-A treated cells. Similar phenomenon was observed with ROS induction in response to Wi-A treatment. Increase in CARF and GADD45, DNA damage regulatory proteins, was also recorded in Wi-A, but not in mWi-A treated cells. 3βmWi-A lacked cytotoxic effect and was well tolerated at higher concentrations (Chaudhary et al. 2017). Rather it protected normal cells against oxidative, chemical and UV stress by activating anti-stress and pro-survival signaling.

Wi-A, and 3βmWi-A are natural co-occurring withanolides in Ashwagandha leaves. 3βmWi-A, with the substitution of a β-methoxy group at Wi-A parental ring is suggested to be a natural Wi-A analogue. Wi-A is potent anticancer compound, shown to inhibit cancer cell proliferation, metastases and EMT processes, 3βmWi-A lacked these activities. Consistently, Wi-A, but not 3βmWi-A, led to vimentin aggregation. Wi-A, but not 3βmWi-A, treated cells showed decrease in Vimentin, N-cadherin and E-cadherin proteins, signifying compromised cell migration and invasion. Wi-A was earlier shown to target hnRNP-K, a key regulator of cell migration and angiogenesis. Reduced levels of MMPs, uPA proteases and VEGF targets viz. phospho-p38MAPK, heat shock protein 27 (HSP27) that promote cancer cell migration were indeed found in Wi-A treated cells. 3βmWi-A treated cells lacked these effects. Contrary to Wi-A,

3βmWi-A (i) was found to be safe and possess stress-relieving activity (ii) when given subsequent to a variety of stress factors including Wi-A, protects normal cells against their toxicity, and (iii) is a vital compound that may guard normal cells against the toxicity associated with various targeted therapeutic regimes in clinical practice. Normal cells exposed to oxidative, UV and genotoxic (doxorubicin, cisplatin and etoposide) drug induced stress, when subsequently recovered in 3βmWi-A supplemented culture medium showed better survival (Chaudhary et al. 2019). Furthermore, an increase in 3βmWi-A induced recovery was found to be concentration and time dependent, supporting the pro-survival effect of 3βmWi-A. The data was confirmed by expression analysis of stress marker proteins that provided molecular evidence to the differential regulation of apoptosis and survival pathways by Wi-A and 3βmWi-A, respectively. Of note, 3βmWi-A protected normal cells against Wi-A cytotoxicity to some extent and provided molecular evidence to the long-trusted belief in Ayurveda's holistic approach of herbal extracts over the purified components.

28.6.3 Bioactive: Triethylene Glycol

Triethylene glycol (TEG) was detected in the water extract of Ashwagandha leaves by NMR analysis. It inhibited cancer cell proliferation by activation of pRB tumor suppressor pathway. TEG is primarily used for industrial purpose and very little is known about its use in the biological systems. TEG has been found to have little or no toxicity in various animal studies suggesting that it is relatively safe in biological system. In Indian system of traditional home medicine, Ayurveda, the root or whole plant extract of Ashwagandha (*Withania somnifera*) has been used to assist in sleep. In search of sleep-inducing active components in Ashwagandha, we investigated the effect of various components on sleep regulation by oral administration in mice. We found that the alcoholic extract that contained high amount of active withanolides was ineffective to induce sleep in mice. However, the water extract which contains TEG as a major component induced significant amount of non-rapid eye movement (NREM) sleep with slight change in rapid eye movement (REM) sleep (Kaushik et al. 2017). Commercially available TEG also increased NREM sleep in mice in a dose-dependent manner. The study had demonstrated that the TEG is an active sleep-inducing component in Ashwagandha leaves and could potentially be useful for insomnia therapy, a common age-related disorder.

28.6.4 Bioactivity: Anti-neurodegenerative

Stress has a deteriorative impact on the brain function. Prolonged exposure to stressful stimuli has been known to cause neuro-inflammation and neurodegeneration, cumulating in several diseases that show age-related trend. At the molecular level, these are the result of either intracellular oxidative stresses or malfunctioning of protein recycling. In cell culture-based studies, brain-derived cells when treated with Ashwagandha extract showed protections against oxidative stresses and even showed differentiation to their functional states. For example, these treated cells were protected against H_2O_2-mediated ROS production and DNA damage. They showed upregulation of GFAP (a marker for glial-differentiation), and NF200 and MAP2 (markers for neuro-differentiation). The low dose cocktail of withanone with cucurbitacin B (CucWi-N) also showed similar effects with an upregulation of GAP43, MAP2 and GFAP proteins (Shah et al. 2009, 2015; Kataria et al. 2011, 2012, 2013).

Scopolamine is a potent acetylcholine receptor antagonist that inhibits central cholinergic neuronal activity and impairs memory as evident by several behavioral parameters like water maze and passive avoidance latency test. Scopolamine-induced amnesic mouse model is one of the well-recognized pharmacological animal models of memory dysfunction. Ashwagandha leaf extract protected the glioma and neuroblastoma cells against scopolamine-induced stress. In mice model, scopolamine supplementation led to dose- and time-dependent suppression of brain-derived neurotrophic factor (BDNF) and GFAP genes in brain tissue, the effect was suppressed by co-supplementation of the Ashwagandha extract or Wi-N. Biochemical analysis showed that Wi-N caused suppression of ROS and upregulation of MAP2, NF200, PSD-95 and GAP43 proteins advocating for its neuro-protective and differentiation-inducing potential. We also identified the muscarinic receptors (acetylcholine-responding post-synaptic responders) to be some of the targets of Wi-N. Whereas scopolamine drastically reduced actin polymerization and Arc (Activity-regulated cytoskeletal associated protein that plays a key role in synaptic plasticity and memory consolidation), Ashwagandha extract and Withanone treated mice showed increase in ARC expression and dendrite spine density (Konar et al. 2011, 2019, 2020; Gautam et al. 2013).

Traumatic events (ischemia, injury and epilepsy) and neurodegeneration (Huntington's disease, Alzheimer's disease and Parkinson's diseases) which often show age-related trend involve glutamate toxicity and severe neurological deficits. Ashwagandha water extract was shown to promote repair and regenerative mechanisms in glutamate (excitatory neurotransmitter acting on the NMDA and AMPA receptors)-induced toxicity in retinoic acid-differentiated brain cancer cells. Using the brain derived cell culture systems, it was shown that Ashwagandha extract not only attenuated glutamate excitotoxicity, but also showed the potential to improve the neural plasticity, learning and memory functions. Neurodegenerative and cognitive disorders as well as loss of memory impair daily aspects of personal, work and social that increases with age. Ashwagandha has been recommended as a potential treatment option for these ailments due to its well- recognized anti-stress, anti-inflammatory, immunomodulatory, antioxidant, cortisol lowering, pro-cholinergic, anti-glutamatergic and neuroprotective properties. Steroidal lactones such as glycowithanolides, sitoindosides, and Withaferin A, have been validated as potent brain antioxidant, neuroprotective and memory enhancing factors in cell and animal studies. In animal models of stress (stress measuring assays including swimming, sleep deprivation as well as chemical and physical stresses), pretreatment with Ashwagandha extract normalized the levels of antioxidative enzymes and caused reduction in lipid peroxidation products like malonidialdehyde and thiobarbituric acid reactive substances. Chronic administration of Ashwagandha enhanced memory in Morris Maze Test and in immobilization stress rat models.

As also stated above, Ashwagandha is the most recognized nootropic or natural memory and cognitive enhancer-herb in Indian traditional home medicine system. It is recommended to recover memory decline in ageing and neurodegenerative pathologies as well as enhance basal memory function of healthy individuals. The data has been experimentally supported by its effects such as (i) activation of antioxidant defense system, (ii) induction of cholinergic system, (iii) up-regulation of neural plasticity, and (iv) regeneration of damaged neurons and synapses. Stressed cells have been shown to possess significant disruption of neural network involving both dendrite and axonal atrophy and premature neurodegeneration. Ashwagandha extract and its bioactives including Withanolide A, Withanoside IV, and Withanoside VI have been shown to stimulate dendrite and axonal growth and reconstructed damaged synapses in stressed neuroblastoma cell lines (Wadhwa et al. 2016a). Oxidative stress contributes to majority of brain dysfunctions including memory impairment. It has been tightly connected to ageing and neurodegenerative diseases such as Parkinson's, Alzheimer and Huntington's. Rotenone is a neurotoxic compound that mimics Parkinson's associated memory and cognitive impairment. Ashwagandha leaf extract was shown to normalize the level of oxidative stress markers and restores GSH level. Amongst physiological models, hypoxia- the lack of oxygen supply has been shown to cause oxidative stress, degenerate neural cells and affect memory function. Ashwagandha extracts with strong antioxidant potential has been shown to successfully ameliorating stress and restoring memory. In an experimental animal model of epilepsy, Ashwagandha extract and withanolide-A were shown to be neuro-protective and recover spatial memory loss by reduction in oxidative stress (Manjunath and Muralidhara 2013, 2015).

In order to support the use of Ashwagandha extracts for medicinal use relating directly or indirectly to healthy ageing, a number of clinical studies have been reported. Consumption of Ashwagandha extract appear to reduce chronic stress, anxiety and insomnia and improve memory and cognition in a dose-dependent manner (Raut et al. 2012; Choudhary et al. 2017). At just 120 mg/day dose, it reduced the severity of obsessive–compulsive disorder (Jahanbakhsh et al. 2016). At 400 mg/day dose, it significantly decreased the fasting blood glucose and serum triglycerides (Agnihotri et al. 2013). Similar effects were also seen at 500 mg/day dose (Pingali et al. 2014). Ziegenfuss et al. (2018) showed that the use of 500 mg/day dose may enhance the physical strength and performance. Against subclinical hypothyroidism, 600 mg/day Ashwagandha root extract effectively normalized the serum thyroid indices (Sharma et al. 2018). At 675 mg/day dose, the extract significantly reduced the severity of male infertility, and increase the sperm count, semen volume, and sperm motility (Ambiye et al. 2013). Similarly, the dose of 1000 mg/day given to Schizophrenia patients, a medium effect on depression and anxiety in cases of schizophrenia were observed (Gannon et al. 2019).

28.6.5 Anti-diabetic

Diabetes, characterized by hyperglycemia and hyperinsulinemia, is one of the major health problems in the twenty-first century. One or more symptoms of diabetes (glycosuria, dysbiosis, polyphagia, polyuria, polydipsia, sleep issues, weakness and tiredness) appear at mid age and often show age-related worsening trend. Main treatments of diabetes include the use of metformin (example of Biguanides) and recombinant insulin which are regarded as safe, well-tolerated and highly effective. Other classes of drugs include Sulfonylureas—glimepiride, glipizide, glyburide, Thiazolidinediones—pioglitazone, Actos generic, Alpha-glucosidase inhibitors—Acarbose, and Meglitinides—naateglinide. Combinatorial therapy is usually employed to ensure insulin spikes in the patient. Although, the use of Western medicine is highly affective but turns out to be expensive and adverse effect laden, herbal medication is hence being researched upon as alternatives.

Ashwagandha withanolides have been shown to possess anti-hyperglycemic activity in several experimental models. Studies have suggested that low oral doses of Ashwagandha cause anti-hyperglycemic action which is due to their ability to stimulate insulin release from pancreatic cells, glucose transporters and inhibition of enzymatic activities of α-amylase and α-glucosidase involved in starch degradation and digestive process. Furthermore, Wi-A and Wi-N have been shown to modulate the functions of cellular chaperones responsible for tissue repair and other nuclear processes regulating oxidative stress and cellular stress responses. A clinical study indicated that the regular intake of low dose of Ashwagandha root powder could induce potassium sparing diuretic effects in diabetic patients and also reduce serum cholesterol and triglycerides in hyperlipidemic patients. Report of another 30-day exploratory clinical study observed an improvement in sleep quality of the patients. Significant improvements of endothelial function, as well as in biomarkers of oxidative stress, systemic inflammation, and HbA1C levels were also observed. Animal studies on anti-hyperlipidemic and other beneficial effects of Ashwagandha withanolides against altered lipid metabolism in diabetic or hyperlipidemic animals have revealed increase in bile acid synthesis, reduction in HMG-CoA activity and improvements in antioxidative status of the animals. Furthermore, psychiatric conditions that often accompany or caused by metabolic disorders like diabetes have been shown to be benefited by Ashwagandha (Udayakumar et al. 2009; Parihar et al. 2016; Kaur and Kaur 2017; Saleem et al. 2020, Durg et al. 2020).

Most recently, in scenario of global COVID-19 pandemic by Severe Acute Respiratory Syndrome Coronavirus 2 (SARS-coronavirus 2; SARS-CoV-2) that emerged in December 2019 in Wuhan, China, all aspects of normal life are affected. High infected numbers and deaths world-wide have evoked urgent initiation of new line of drug and vaccine development on one-hand and repurposing of the existing drugs on the other. Because of the immunity enhancing properties of several kinds of traditional home medicine, they have been recommended for the prevention and treatment of COVID-19. Ashwagandha bio-actives, Wi- and Wi-N, were explored for their potential for antiCOVID-19 activity using bioinformatics and experimental technologies.

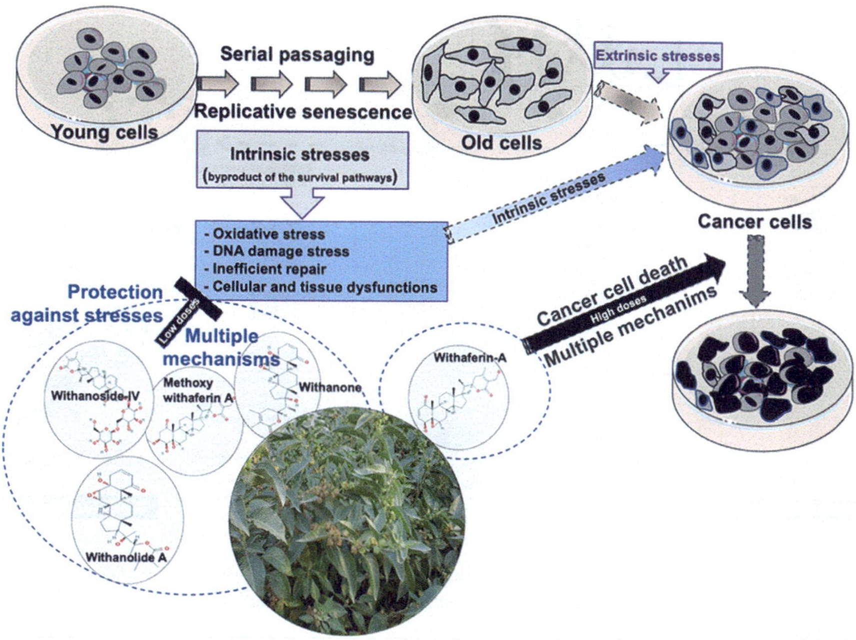

Fig. 28.2 Schematic representation of intrinsic and extrinsic stresses involved in replicative senescence and carcinogenesis, and their modulation by treatment with Ashwagandha bioactives

These were tested for their binding to a highly conserved protein (M^{pro}, an essential protein for virus replication and survival) of SARS-CoV-2. It was found that Wi-N, but not Wi-A, binds strongly to M^{pro} with the binding energy equivalent to its known inhibitor (N3) predicting that Wi-N may serve as a natural drug for COVID-19. Binding potential of Wi-A and Wi-N to cell membrane protein (TPMRSS2) that acts as a gate for entry of virus to the host cell was also investigated. Both Wi-A and Wi-N have been predicted to bind stably with TMPRSS2; Wi-N showed stronger interactions. Furthermore, human cells treated with Wi-N showed low level of expression of TMPRSS2 predicting three-way action of Wi-N to deal with SARS-CoV-2 (blocking its entry to the host cells by interaction with TMPRSS2, downregulation of TMPRSS2 expression and diminishing the viral survival through inhibition of viral M^{pro} protein) (Kumar et al. 2020a, b). Taken together, cell culture-based studies have provided evidence that whereas Wi-A has strong anticancer potential, Wi-N offers antistress, neurogenerative and antivirus activities that may contribute to healthy ageing. Therefore, purposeful consumption of Ashwagandha warrants attention to the amount and ratio of active withanolides that have been shown to vary in different parts of the plant (Kumar et al. 2020a).

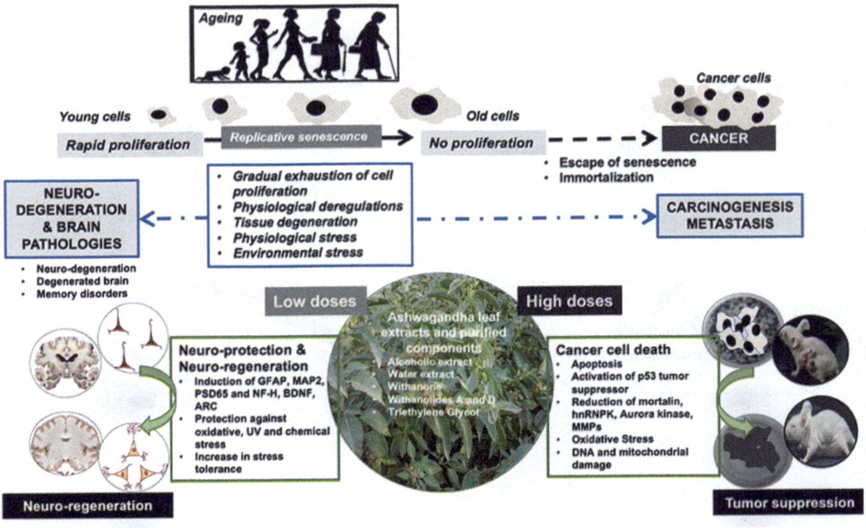

Fig. 28.3 Schematic representation of ageing, old age-related pathologies at cell, tissue and organismal levels, and their modulation by treatment with Ashwagandha bioactives

28.7 Conclusion

Ageing is an extremely complex, unavoidable byproduct of life processes. Defined as a time-related normal process, it is complicated by a variety of tissue dysfunctions resulting in pathological conditions. Intrinsic and extrinsic stresses impact the ageing process by several ways, often leading to premature ageing. Recent laboratory studies have demonstrated the strength of cell culture and molecular technologies to investigate the mechanisms of these processes and their manipulation by natural and synthetic compounds. Ashwagandha-derived withanolides have been shown to possess a variety of bioactivites including anticancer (at high doses) and anti-stress (at low doses) in cell culture (Fig. 28.2) as well as mouse models (Fig. 28.3) accounting for its value for disease therapeutics on one hand and healthy ageing on the other, respectively.

Compliance with Ethical Standards

Funding This study was funded by DBT (India) and AIST (Japan). Mallika Khurana is a recipient of the MEXT (Monbukagakusho), Japan scholarship.

Conflict of Interest All authors declare they have no conflict of interest.

References

Afzal S, Garg S, Ishida Y, Terao K, Kaul SC, Wadhwa R (2019) Rat glioma cell-based functional characterization of anti-stress and protein deaggregation activities in the marine carotenoids, astaxanthin and fucoxanthin. Mar Drugs 17:189

Afzal S, Garg S, Adiga D, Ishida Y, Terao K, Kaul SC, Wadhwa R (2020) Anti-stress, glial- and neuro-differentiation potential of resveratrol: characterization by cellular, biochemical and imaging assays. Nutrients 12:671

Agnihotri AP, Sontakke SD, Thawani VR, Saoji A, Goswami VS (2013) Effects of withania somnifera in patients of schizophrenia: a randomized, double blind, placebo controlled pilot trial study. Indian J Pharmacol 45:417–418

Ali M, Saleem U, Anwar F, Imran M, Nadeem H, Ahmad B, Ali T, Atta URR, Ismail T (2021) Screening of synthetic isoxazolone derivative role in alzheimer's disease: computational and pharmacological approach. Neurochem Res 46:905–920

Ambiye VR, Langade D, Dongre S, Aptikar P, Kulkarni M, Dongre A (2013) Clinical evaluation of the spermatogenic activity of the root extract of ashwagandha (Withania somnifera) in oligospermic males: a pilot study. Evid Based Complement Alternat Med 2013:571420

Amellem O, Stokke T, Sandvik JA, Pettersen EO (1996) The retinoblastoma gene product is reversibly dephosphorylated and bound in the nucleus in S and G2 phases during hypoxic stress. Exp Cell Res 227:106–115

Arora D, Khurana B, Nanada S (2020) Statistical development and in vivo evaluation of resveratrol-loaded topical gel containing deformable vesicles for a significant reduction in photo-induced skin ageing and oxidative stress. Drug Dev Ind Pharm 46:1898–1910

Bakula D, Ablasser A, Aguzzi A, Antebi A, Barzilai N, Bittner MI, Jensen MB, Calkhoven CF, Chen D, Grey A, Feige JN, Georgieviskaya A, Gladyshev VN, Golato T, Gudkov AV, Hoppe T, Kaeberlein M, Katajisto P, Kennedy BK, Lal U, Martin-Villalba A, Moskalev AA, Ozerov I, Petr MA, Reason RDC, Tyshkovskiy A, Vanhaelen Q, Zhavoronkov A, Scheibye-Knudsen M (2019) Latest advances in ageing research and drug discovery. Ageing (albany NY) 11:9971–9981

Bhargava P, Grover A, Nigam N, Kaul A, Doi M, Ishida Y, Kakuta H, Kaul SC, Terao K, Wadhwa R (2018) Anticancer activity of the supercritical extract of Brazilian green propolis and its active component, artepillin C: bioinformatics and experimental analyses of its mechanisms of action. Int J Oncol 52:925–932

Chaudhary A, Klra RS, Huang C, Prakash J, Kaul SC, Wadhwa R (2017) 2,3-Dihydro-3beta-methoxy withaferin-A protects normal cells against stress: molecular evidence of its potent cytoprotective activity. J Nat Prod 80:2756–2760

Chaudhary A, Kalra RS, Malik V, Katiyar SP, Sundar D, Kaul SC, Wadhwa R (2019) 2, 3-Dihydro-3beta-methoxy withaferin-A lacks anti-metastasis potency: bioinformatics and experimental evidences. Sci Rep 9:17344

Chaurasia SS, Panda S, Kar A (2000) Withania somnifera root extract in the regulation of lead-induced oxidative damage in male mouse. Pharmacol Res 41:663–666

Cheung CT, Kaul SC, Wadhwa R (2010) Molecular bridging of ageing and cancer: A CARF link. Ann N Y Acad Sci 1197:129–133

Choudhary D, Bhattacharyya S, Bose S (2017) Efficacy and safety of ashwagandha (Withania somnifera (L.) Dunal) root extract in improving memory and cognitive functions. J Diet Suppl. 14:599–612

Cohen SM, Mukerji R, Timmermann BN, Samadi AK, Cohen MS (2012) A novel combination of withaferin A and sorafenib shows synergistic efficacy against both papillary and anaplastic thyroid cancers. Am J Surg 204:895–900, discussion 900–901

Daniele N, Halse R, Grinyo E, Yeaman SJ, Shepered PR (2002) Conditionally immortalized cell lines as model systems for high-throughput biology in drug discovery. Biochem Soc Trans 30:800–802

Deocaris CC, Kaul SC, Wadhwa R (2009) The versatile stress protein mortalin as a chaperone therapeutic agent. Protein Pept Lett 16:517–529

Deocaris CC, Lu WJ, Kaul SC, Wadhwa R (2013) Druggability of mortalin for cancer and neuro-degenerative disorders. Curr Pharm Des 19:418–429

D'souza GX, Rose SE, Knupp A, Nicholson DA, Keene CD, Young JE (2021) The application of in vitro-derived human neurons in neurodegenerative disease modeling. J Neurosci Res 99:124–140

Dubey S, Kallubai M, Sarkar A, Subramanyam R (2018) Elucidating the active interaction mechanism of phytochemicals withanolide and withanoside derivatives with human serum albumin. PLoS ONE 13:e0200053

Dug S, Bavage S, Shivaram SB (2020) Withania somnifera (Indian ginseng) in diabetes mellitus: a systematic review and meta-analysis of scientific evidence from experimental research to clinical application. Phytother Res 34:1041–1059

Duncan EL, Reddel RR (1997) Genetic changes associated with immortalization a review. Biochemistry (mosc) 62:1263–1274

Duncan EL, Wadhwa R, Kaul SC (2000) Senescence and immortalization of human cells. Biogerontology 1:103–121

Eren B, Tuncay Tanriverdi S, Aydin Kose F, Ozer O (2019) Antioxidant properties evaluation of topical astaxanthin formulations as anti-ageing products. J Cosmet Dermatol 18:242–250

Gannon JM, Brar J, Rai A, Chengappa KNR (2019) Effects of a standardized extract of Withania somnifera (Ashwagandha) on depression and anxiety symptoms in persons with schizophrenia participating in a randomized, placebo-controlled clinical trial. Ann Clin Psychiatry 31:123–129

Gao R, Shah N, Lee JS, Katiyar SP, Li L, Oh E, Sundar D, Yun CO, Kaul SC, Wadhwa R (2014) Withanone rich combination of ashwagandha withanolides restricts metastasis and angiogenesis through hnRNP-K. Mol Cancer Ther 13:2930–2940

Gao R, Singh R, Kaul Z, Kaul SC, Wadhwa R (2015) Targeting of DNA damage signaling pathway induced senescence and reduced migration of cancer cells. J Gerontol A Biol Sci Med Sci 70:701–713

Garg S, Kaul SC, Wadhwa R (2018) Anti-Stress and glial differentiation effects of a novel combination of cucurbitacin B and withanone (CucWi-N): experimental evidence. Ann Neurosci 25:201–209

Garg S, Huifu H, Kaul SC, Wadhwa R (2018) Integration of conventional cell viability assays for reliable and reproducible read-outs: experimental evidence. BMC Res Notes 11:403–408

Garg S, Huifu H, Kumari A, Sundar D, Kaul SC, Wadhwa R (2020) Induction of senescence in cancer cells by a novel combination of cucurbitacin B and withanone: molecular mechanism and therapeutic potential. J Gerontol A Biol Sci Med Sci 75:1031–1041

Gautam A, Wadhwa R, Thakur MK (2013) Involvement of hippocampal arc in amnesia and its recovery by alcoholic extract of ashwagandha leaves. Neurobiol Learn Mem 106:177–184

Ghnemi A (2015) Cell cultures in drug development: applications, challenges and limitations. Saudi Pharm J 23:453–454

Grover A, Singh R, Shandiyla A, Priyandoko D, Agarwal V, Bisaria VS, Wadhwa R, Kaul SC, Sundar D (2012) Ashwagandha derived withanone targets TPX2-Aurora A complex: computational and experimental evidence to its anticancer activity. PLoS ONE 7:e30890

Hayflick L, Moorhead PS (1961) The serial cultivation of human diploid cell strains. Exp Cell Res 25:585–621

Hsu JH, Chang PM, Cheng TS, Kuo YL, Wu AT, Tran TH, Yang YH, Chen M, Tsai YC, Chu YS, Huang TH, Huang F, Lai JM (2019) Identification of withaferin A as a potential candidate for anti-cancer therapy in non-small cell lung cancer. Cancers (basel). 11:1003

Itahana K, Dimri GP, Hara E, Itahana Y, Zou Y, Desprez PY, Campisi J (2002) A role for p53 in maintaining and establishing the quiescence growth arrest in human cells. J Biol Chem 277:18206–18214

Itahana K, Zou Y, Itahana Y, Martinez JL, Beausejour C, Jaocbs JJ, Van Lohuizen M, Band V, Campisis J, Dimri GP (2003) Control of the replicative life span of human fibroblasts by p16 and the polycomb protein Bmi-1. Mol Cell Biol 23:389–401

Ivaska J, Pallari HM, Nevo J, Eriksson JE (2007) Novel functions of vimentin in cell adhesion, migration, and signaling. Exp Cell Res 313:2050–2062

Jahanbaksh SP, Manteghi AA, Emami SA, Mahyari S, Gholampour B, Mohammadpour AH, Sahebkar A (2016) Evaluation of the efficacy of Withania somnifera (Ashwagandha) root extract in patients with obsessive-compulsive disorder: a randomized double-blind placebo-controlled trial. Complement Ther Med 27:25–29

Jeyanthi T, Subramanian P (2009) Nephroprotective effect of Withania somnifera: a dose-dependent study. Ren Fail 31:814–821

Kakar SS, Jala VR, Fong MY (2012) Synergistic cytotoxic action of cisplatin and withaferin A on ovarian cancer cell lines. Biochem Biophys Res Commun 423:819–825

Kakar SS, Ratajczak M, Powell KS, Moghadamafalahi M, Miller DM, Batra SK, Singh SK (2014) Withaferin a alone and in combination with cisplatin suppresses growth and metastasis of ovarian cancer by targeting putative cancer stem cells. PLoS ONE 9:e107596

Kakar SS, Worth CA, Wang Z, Carter K, Ratajczak M, Gunjal P (2016) DOXIL when combined with Withaferin A (WFA) targets ALDH1 positive cancer stem cells in ovarian cancer. J Cancer Stem Cell Res. 4:e1002

Kalra RS, Chaudhary A, Omar A, Cheung CT, Garg S, Kaul SC, Wadhwa R (2020) Stress-induced changes in CARF expression determine cell fate to death, survival, or malignant transformation. Cell Stress Chaperones 25:481–494

Kameda M, Mikawa T, Yokode M, Inagaki N, Kondoh H (2021) Senescence research from historical theory to future clinical application. Geriatr Gerontol Int 21:125–130

Kataria H, Shah N, Kaul SC, Wadhwa R, Kaur G (2011) Water extract of ashwagandha leaves limits proliferation and migration, and induces differentiation in glioma cells. Evid Based Complement Alternat Med. 2011:267614

Kataria H, Wadhwa R, Kaul SC, Kaur G (2012) Water extract from the leaves of Withania somnifera protect RA differentiated C6 and IMR-32 cells against glutamate-induced excitotoxicity. PLoS ONE 7:e37080

Kataria H, Wadhwa R, Kaul SC, Kaur G (2013) Withania somnifera water extract as a potential candidate for differentiation based therapy of human neuroblastomas. PLoS ONE 8:e55316

Kaul SC, Aida S, Yaguchi T, Kaur K, Wadhwa R (2005) Activation of wild type p53 function by its mortalin-binding, cytoplasmically localizing carboxyl terminus peptides. J Biol Chem 280:39373–39379

Kaul SC, Ishida Y, Tamura K, Wada T, Iitsuka T, Garg S, Kim M, Gao R, Nakai S, Okamoto Y, Terao K, Wadhwa R (2016) Novel methods to generate active ingredients-enriched ashwagandha leaves and extracts. PLoS ONE 11:e0166945

Kaur T, Kaur G (2017) Withania somnifera as a potential candidate to ameliorate high fat diet-induced anxiety and neuroinflammation. J Neuroinflammation 14:201–206

Kaur K, Rani G, Widodo N, Nagpal A, Taira K, Kaul SC, Wadhwa R (2004) Evaluation of the anti-proliferative and anti-oxidative activities of leaf extract from in vivo and in vitro raised ashwagandha. Food Chem Toxicol 42:2015–2020

Kaur A, Singh B, Ohri P, Wang J, Wadhwa R, Kaul SC, Pati PK, Kaur A (2018) Organic cultivation of ashwagandha with improved biomass and high content of active withanolides: use of vermicompost. PLoS ONE 13:e0194314

Kaushik MK, Kaul SC, Wadhwa R, Yanagisawa M, Urade Y (2017) Triethylene glycol, an active component of Ashwagandha (Withania somnifera) leaves, is responsible for sleep induction. PLoS ONE 12:e0172508

Kirtonia A, Sethi G, Garg M (2020) The multifaceted role of reactive oxygen species in tumorigenesis. Cell Mol Life Sci 77:4459–4483

Kokubu Y, Asashima M, Kurisaki A (2015) Establishment and culture optimization of a new type of pituitary immortalized cell line. Biochem Biophys Res Commun 463:1218–1224

Konar A, Shah N, Singh R, Saxena N, Kaul SC, Wadhwa R, Thakur MK (2011) Protective role of ashwagandha leaf extract and its component withanone on scopolamine-induced changes in the brain and brain-derived cells. PLoS ONE 6:e27265

Konar A, Gupta R, Shukla RK, Maloney B, Khanna VK, Wadhwa R, Lahiri DK, Thakur MK (2019) M1 muscarinic receptor is a key target of neuroprotection, neuroregeneration and memory recovery by i-Extract from Withania somnifera. Sci Rep 9:13990

Konar A, Kakra RS, Chaudhary A, Nayak A, Guruprasad KP, Satyamoorthi K, Ishida Y, Terao K, Kaul SC, Wadhwa R (2020) Identification of caffeic acid phenethyl ester (CAPE) as a potent neurodifferentiating natural compound that improves cognitive and physiological functions in animal models of neurodegenerative diseases. Front Ageing Neurosci 12:561925

Kulkarni SK, Dhir A (2008) Withania somnifera: an Indian ginseng. Prog Neuropsychopharmacol Biol Psychiatry 32:1093–1105

Kumar V, Dhanjal JK, Kaul SC, Wadhwa R, Sundar D (2021) Withanone and caffeic acid phenethyl ester are predicted to interact with main protease (M(pro)) of SARS-CoV-2 and inhibit its activity. J Biomol Struct Dyn 39(11):3842–3854

Kumar V, Dhanjal JK, Bhargava P, Kaul A, Wang J, Zhang H, Kaul SC, Wadhwa R, Sundar D (2020) Withanone and withaferin-A are predicted to interact with transmembrane protease serine 2 (TMPRSS2) and block entry of SARS-CoV-2 into cells. J Biomol Struct Dyn 1–27. https://doi.org/10.1080/07391102.2020.1775704

Kumar S, Sarma P, Kaur H, Prajapat M, Bhattacharyya A, Avti P, Sehkhar N, Kaur H, Bansal S, Mahendiratta S, Mahalmani VM, Singh H, Prakash A, Kuhad A, Medhi B (2021) Clinically relevant cell culture models and their significance in isolation, pathogenesis, vaccine development, repurposing and screening of new drugs for SARS-CoV-2: a systematic review. Tissue Cell 70:101497

Kwakkenbos MJ, Van Helden PM, Beaumont T, Spits H (2016) Stable long-term cultures of self-renewing B cells and their applications. Immunol Rev 270:65–77

Larsen CJ (2004) pRB, p53, p16INK4a, senescence and malignant transformation. Bull Cancer 91:399–402

Li X, Zhu F, Jiang J, Sun C, Wang X, Shen M, Tian R, Shi C, Xu M, Peng F, Guo X, Wang M, Qin R (2015) Synergistic antitumor activity of withaferin A combined with oxaliplatin triggers reactive oxygen species-mediated inactivation of the PI3K/AKT pathway in human pancreatic cancer cells. Cancer Lett 357:219–230

Li K, Yu Y, Sun S, Liu Y, Garg S, Kaul SC, Lei Z, Gao R, Wadhwa R, Zhang Z (2016) Functional characterisation of anticancer activity in the aqueous extract of helicteres angustifolia L. roots. PLoS ONE 11:e0152017

Lovitt CJ, Shelper TB, Avery VM (2014) Advanced cell culture techniques for cancer drug discovery. Biology (basel) 3:345–367

Manjunath MJ, Muralidhara (2013) Effect of Withania somnifera supplementation on rotenone-induced oxidative damage in cerebellum and striatum of the male mice brain. Cent Nerv Syst Agents Med Chem 13:43–56

Manjunath MJ, Muralidhara (2015) Standardized extract of Withania somnifera (Ashwagandha) markedly offsets rotenone-induced locomotor deficits, oxidative impairments and neurotoxicity in Drosophila melanogaster. J Food Sci Technol 52:1971–1981

Matsui M, Tanaka K, Higashiguchi N, Okawa H, Yamada Y, Tanaka K, Taira S, Aoyama T, Takanishi M, Natsume C, Takakura Y, Fujita N, Hashimoto T, Fujita T (2016) Protective and therapeutic effects of fucoxanthin against sunburn caused by UV irradiation. J Pharmacol Sci 132:55–64

Modak M, Dixit P, Londhe J, Ghaskhadbi S, Devasagayam TP (2007) Indian herbs and herbal drugs used for the treatment of diabetes. J Clin Biochem Nutr 40:163–173

Mohomad Kamal NS, Safuan S, Shamsuddin S, Foroohzandeh P (2020) Ageing of the cells: Insight into cellular senescence and detection Methods. Eur J Cell Biol 99:151108

Mowla SN, Lam EW, Jat PS (2014) Cellular senescence and ageing: the role of B-MYB. Ageing Cell 13:773–779

Na Y, Kaul SC, Ryu J, Lee JS, Ahn HM, Kaul Z, Kalra RS, Li L, Widodo N, Yun CO, Wadhwa R (2016) Stress chaperone mortalin contributes to epithelial-mesenchymal transition and cancer metastasis. Cancer Res 76:2754–2765

Ng QX, De Deyn M, Lke W, Foo NX, Chan HW, Yeo WS (2021) Effects of astaxanthin supplementation on skin health: a systematic review of clinical studies. J Diet Suppl 18:169–182

Nishikawa Y, Okuzaki D, Fukushima K, Mukai S, Ohno S, Ozaki Y, Yabuta N, Nojima H (2015) Withaferin A induces cell death selectively in androgen-independent prostate cancer cells but not in normal fibroblast cells. PLoS ONE 10:e0134137

Ostrow KL, Donaldson K, Blakeley J, Belberg A, Hoke A (2015) Immortalized human schwann cell lines derived from tumors of schwannomatosis patients. PLoS ONE 10:e0144620

Parihar P, Shetty R, Ghafourifar P, Parihar MS (2016) Increase in oxidative stress and mitochondrial impairment in hypothalamus of streptozotocin treated diabetic rat: antioxidative effect of Withania somnifera. Cell Mol Biol (Noisy-le-grand) 62:73–83

Pingali U, Pilli R, Fatima N (2014) Effect of standardized aqueous extract of Withania somnifera on tests of cognitive and psychomotor performance in healthy human participants. Pharmacognosy Res 6:12–18

Priyandoko D, Ishii T, Kaul SC, Wadhwa R (2011) Ashwagandha leaf derived withanone protects normal human cells against the toxicity of methoxyacetic acid, a major industrial metabolite. PLoS ONE 6:e19552

Raut AA, Rege NN, Tadvi FM, Solanki PV, Kene KR, Shiolkar SG, Pandey SN, Vaidya RA, Vaidya AB (2012) Exploratory study to evaluate tolerability, safety, and activity of ashwagandha (withania somnifera) in healthy volunteers. J Ayurveda Integr Med. 3:111–114

Reddel RR (2000) The role of senescence and immortalization in carcinogenesis. Carcinogenesis 21:477–484

Roupakia E, Markopolous GS, Kolettas E (2021) Genes and pathways involved in senescence bypass identified by functional genetic screens. Mech Ageing Dev 194:111432

Ryu J, Kaul Z, Yoon AR, Liu Y, Yaguchi T, Na Y, Ahn HM, Gao R, Choi IK, Yun CO, Kaul SC, Wadhwa R (2014) Identification and functional characterization of nuclear mortalin in human carcinogenesis. J Biol Chem 289:24832–24844

Saleem S, Muhammad G, Hussain MA, Altaf M, Bukhari SNA (2020) Withania somnifera L.: Insights into the phytochemical profile, therapeutic potential, clinical trials, and future prospective. Iran J Basic Med Sci 23:1501–1526

Sari AN, Bhargava P, Dhanjal JK, Putri JF, Radhakrishnan N, Shefrin S, Ishida Y, Terao K, Sundar D, Kaul SC, Wadhwa R (2020) Combination of withaferin-A and CAPE provides superior anticancer potency: bioinformatics and experimental evidence to their molecular targets and mechanism of action. Cancers (basel) 12:1160

Shah N, Kataria H, Kaul SC, Ishii T, Kaur G, Wadhwa R (2009) Effect of the alcoholic extract of ashwagandha leaves and its components on proliferation, migration, and differentiation of glioblastoma cells: combinational approach for enhanced differentiation. Cancer Sci 100:1740–1747

Shah N, Singh R, Sarangi U, Saxena N, Chaudhary A, Kaur G, Kaul SC, Wadhwa R (2015) Combinations of ashwagandha leaf extracts protect brain-derived cells against oxidative stress and induce differentiation. PLoS ONE 10:e0120554

Sharma AK, Basu I, Singh S (2018) Efficacy and safety of ashwagandha root extract in subclinical hypothyroid patients: a double-blind, randomized placebo-controlled trial. J Altern Complement Med 24:243–248

Shay JW, Tomlinson G, Piatyszek MA, Gollahon LS (1995) Spontaneous in vitro immortalization of breast epithelial cells from a patient with Li-Fraumeni syndrome. Mol Cell Biol 15:425–432

Sumathi S, Padma PR, Gathampari S, Vidhya S (2007) Free radical scavenging activity of different parts of withania somnifera. Anc Sci Life 26:30–34

Thaiparambil JT, Bender L, Ganesh T, Kline E, Patel P, Liu Y, Tighouart M, Vertino PM, Harvey RD, Garcia A, Marcus AI (2011) Withaferin A inhibits breast cancer invasion and metastasis at sub-cytotoxic doses by inducing vimentin disassembly and serine 56 phosphorylation. Int J Cancer 129:2744–2755

Udayakumar R, Kasthurirengan S, Mariashibu TS, Rajesh M, Anbazhagan VR, Kim SC, Ganpathi A, Choi CW (2009) Hypoglycaemic and hypolipidaemic effects of Withania somnifera root and leaf extracts on alloxan-induced diabetic rats. Int J Mol Sci 10:2367–2382

Ven Murthy MR, Ranjekar PK, Ramassamy C, Deshpande M (2010) Scientific basis for the use of Indian ayurvedic medicinal plants in the treatment of neurodegenerative disorders: ashwagandha. Cent Nerv Syst Agents Med Chem 10:238–246

Von Zglinicki T, Wan T, Miwa S (2021) Senescence in post-mitotic cells: a driver of ageing? Antioxid Redox Signal 34:308–323

Wadhwa R, Kaul SC, Ikawa Y, Sugimoto Y (1991) Protein markers for cellular mortality and immortality. Mutat Res 256:243–254

Wadhwa R, Kaul SC, Mitsui Y, Sugomoto Y (1993) Differential subcellular distribution of mortalin in mortal and immortal mouse and human fibroblasts. Exp Cell Res 207:442–448

Wadhwa R, Sugihara T, Yoshida A, Nomura H, Reddel RR, Simpson R, Maruta H, Kaul SC (2000) Selective toxicity of MKT-077 to cancer cells is mediated by its binding to the hsp70 family protein mot-2 and reactivation of p53 function. Cancer Res 60:6818–6821

Wadhwa R, Taira K, Kaul SC (2002) An Hsp70 family chaperone, mortalin/mthsp70/PBP74/Grp75: what, when, and where? Cell Stress Chaperones 7:309–316

Wadhwa R, Colgin L, Yaguchi T, Taira K, Reddel RR, Kaul SC (2002) Rhodacyanine dye MKT-077 inhibits in vitro telomerase assay but has no detectable effects on telomerase activity in vivo. Cancer Res 62:4434–4438

Wadhwa R, Takano S, Kaur K, Deocaris CC, Perieia-Smith OM, Reddel RR, Kaul SC (2006) Upregulation of mortalin/mthsp70/Grp75 contributes to human carcinogenesis. Int J Cancer 118:2973–2980

Wadhwa R, Singh R, Gao R, Shah N, Widodo N, Nakamoto T, Ishida Y, Terao K, Kaul SC (2013) Water extract of Ashwagandha leaves has anticancer activity: identification of an active component and its mechanism of action. PLoS ONE 8:e77189

Wadhwa R, Konar A, Kaul SC (2016) Nootropic potential of Ashwagandha leaves: beyond traditional root extracts. Neurochem Int 95:109–118

Wadhwa R, Nigam N, Bhargava P, Dhanjal JK, Goyal S, Grover A, Sundar D, Ishida Y, Terao K, Kaul SC (2016) Molecular characterization and enhancement of anticancer activity of caffeic acid phenethyl ester by gamma cyclodextrin. J Cancer 7:1755–1771

Widodo N, Kaur K, Shreshta BG, Takagi Y, Ishii T, Wadhwa R, Kaul SC (2007) Selective killing of cancer cells by leaf extract of ashwagandha: identification of a tumor-inhibitory factor and the first molecular insights to its effect. Clin Cancer Res 13:2298–2306

Widodo N, Takagi Y, Shreshtha BG, Ishii T, Kaul SC, Wadhwa R (2008) Selective killing of cancer cells by leaf extract of ashwagandha: components, activity and pathway analyses. Cancer Lett 262:37–47

Widodo N, Shah N, Priyandoko D, Ishii T, Kaul SC, Wadhwa R (2009) Deceleration of senescence in normal human fibroblasts by withanone extracted from ashwagandha leaves. J Gerontol A Biol Sci Med Sci 64:1031–1038

Widodo N, Priyandoko D, Shah N, Wadhwa R, Kaul SC (2010) Selective killing of cancer cells by ashwagandha leaf extract and its component withanone involves ROS signaling. PLoS ONE 5:e13536

Yang L, Chen Y, Liu Y, Xing Y, Miao C, Zhao Y, Chang X, Zhang Q (2020) The role of oxidative stress and natural antioxidants in ovarian ageing. Front Pharmacol 11:617843

Yao T, Asayama Y (2017) Animal-cell culture media: history, characteristics, and current issues. Reprod Med Biol 16:99–117

Yoo JY, Ryu J, Gao R, Yaguchi T, Kaul SC, Wadhwa R, Yun CO (2010) Tumor suppression by apoptotic and anti-angiogenic effects of mortalin-targeting adeno-oncolytic virus. J Gene Med 12:586–595

Yu Y., Katiyar SP, Sundar D, Kaul Z, Miyako E, Zhang Z, Kaul SC, Reddel RR, Wadhwa R (2017) Withaferin-A kills cancer cells with and without telomerase: chemical, computational and experimental evidences. Cell Death Dis 8:e2755

Yuan H, Xu Y, Luo Y, Wang NX, Xiao JH (2021) Role of Nrf2 in cell senescence regulation. Mol Cell Biochem 476:247–259

Yun CO, Bhargava P, Na Y, Lee JS, Ryu J, Kaul SC, Wadhwa R (2017) Relevance of mortalin to cancer cell stemness and cancer therapy. Sci Rep 7:42016

Zhang K, Wang L, Chen H, Shi Y, Liu Y, Liu J, Hong X, Liu JP (2021) Pulmonary alveolar stem cells undergo senescence, apoptosis and differentiation by p53-dependent and p53-independent mechanisms in telomerase deficient mice. Clin Exp Pharmacol Physiol 48:651

Ziegenfuss TN, Kedia AW, Sandrock JE, Raub BJ, Kerksick CM, Lopez HL (2018) Effects of an Aqueous Extract of Withania somnifera on Strength Training Adaptations and Recovery: The STAR Trial. Nutrients 10:1807

Chapter 29
The Fact and Fiction of Nutritional Claims About Health and Longevity

Éric Le Bourg

Abstract For centuries, various feeding regimens have been proposed to live longer with a better health. However, in many cases, they have not been proven to have any positive effect, and they can even have negative ones. This chapter reviews the claims of various regimens (e.g. detox diets, fasting). It is doubtful that they could increase human lifespan and improve health, provided one adopts a balanced regime with no excessive intake and avoids junk food.

Keywords Claims · Fact · Fiction · Food restriction · Anti-aging · Healthspan

29.1 Introduction

For centuries, various regimes have been proposed to live longer and provide a better health (see the chapter 25 in this book: Stambler 2020a; Haber 2004; Stambler 2020b). Modulating food intake appears as an obvious strategy to improve health because everybody has observed that gluttony or famine bring with them a bad health. Going a step farther, it could be expected that modulating both the quantity and quality of food has beneficial or deleterious effects on healthspan and, ultimately, on lifespan.

One of the first modern claims that a specific food could have such effects was that of the 1908 Nobel laureate Elie Metchnikoff, who wrote (translated from French) that "the introduction of microbes cultivated from sour milk can bring about a considerable change in the intestinal flora in a direction favourable to the preservation of health … (and) making old age more physiological than it is at present and probably also of prolonging human lifespan" (Metchnikoff 1903). This was the beginning of the yoghurt industry (Fisberg and Machado 2015) and the good news is that eating yoghurt does not provoke any harm. This is however not the case with all regimens claiming to increase lifespan. One of the issues with such regimens is that fear of aging can lead many people to support them despite any counter-argument. Because this behaviour is not only that of adepts but can also be that of scientists, the net result

É. Le Bourg (✉)
Centre de Recherches sur la Cognition Animale (CRCA), Centre de Biologie Intégrative (CBI Toulouse), Université de Toulouse, CNRS, UPS, Toulouse, France
e-mail: eric.le-bourg@univ-tlse3.fr

S. I. S. Rattan and G. Kaur (eds.), *Nutrition, Food and Diet in Ageing and Longevity*, Healthy Ageing and Longevity 14, https://doi.org/10.1007/978-3-030-83017-5_29

is that some people are even more convinced that they adopt the best way to live long if they eat a specific food, ingest antioxidants, or fast. Partly connected to this issue is that of fashion. From time to time, a new long-life recipe shoots to the fame and many people are immediately convinced of it, before giving up when a new recipe is proposed on the TV, magazines or social networks. This is an issue because, in some cases, these recipes are dangerous and it has been warned that "some dietary supplements have been associated with potentially serious health consequences for senior citizens" (United States General Accounting Office 2001).

Therefore, fear of ageing, fashion, or over-enthusiasm of the lay public and scientists are obstacles to know the merits and pitfalls of regimens and/or individual nutritional components claiming to improve life at old age, or even to live longer, and this should be taken into account when reviewing the effects of these regimens. The purpose of this chapter is to separate the fact and fiction of nutritional claims about health and longevity.

29.2 Vitamins and Other Micro-nutrients

Vitamins are often considered by the lay public to be "good" because of their name, vitamin meaning life and amine, and of their association with fruits and vegetables. Most of people know that a lack of vitamins can provoke severe health issues and some think that the higher the dose, the best is the effect on health. While the first part of this sentence is correct, the second one is not. However, relying on this wrong rationale, many people, including some scientists, believe that one should supplement food with vitamins, even if it already contains a large part of fruits and vegetables. The first issue is that high doses of some vitamins can be toxic, as it is the case with vitamin E (Miller et al. 2005; Bjelakovic et al. 2007). The second issue, less tragic, is that high doses can be not assimilated and are excreted, like vitamin C in urine. Nevertheless, the double Nobel laureate Linus Pauling promoted high doses of vitamin C because "vitamin C will have great value in controlling the problems associated with advancing age" (Pauling 1983) and also "improve the general health of nearly every cancer patient (and) increase survival time somewhat" (Pauling and Moertel 1986). Regarding this claim of Pauling on cancer, Moertel, in the same two-handed article, severely wrote: "Whether one is dealing with treatment of the common cold or of cancer, and whether one is dealing with a benign vitamin or a highly toxic chemotherapeutic drug, it would seem to best serve the interest of the patient if public advocacy of a proposed treatment is withheld until that treatment has been proved effective by definitive studies of sound scientific design".

In any case, the opinion of a Nobel laureate can lead to major consumption changes and the main result is that, nowadays, the internet is replete with sites promoting vitamin C. The Nobel laureate Elie Metchnikoff provoked the birth of a new industry, that of yoghurt, and the Nobel laureate Linus Pauling made that many people take 1 g or more of vitamin C daily and produce an expensive urine, for the best of pharmaceutical industry. The difference is that yoghurt is a food, and thus useful, and that taking 1 g of vitamin C daily is useless, particularly if people eat fruits.

Beside vitamins, other micro-nutrients, such as mineral ones, are advertised and sold as recipes to live longer. For instance, selenium (Se), which is involved in antioxidant enzymes such as glutathione peroxidase, is sold on the internet for "optimized health and longevity support". However, while selenium deficiency can be an issue, for instance in Sweden (Alehagen et al. 2016), and "may suggest the value of modest Se supplementation in order to improve the health of the Swedish population", it is not the case in the USA, a result which pleas against supplementation because the window for therapeutic effects is narrow and toxic effects are soon observed (references in Bleys et al. 2007). Yet, dozens of selenium supplements are for sale in the USA.

29.3 Antioxidants

The free radical theory of ageing (Harman 1956) posits that ageing "seems to be a more or less direct function of the metabolic rate", as it was postulated by Pearl (1928) in his rate of living theory, and that, because "the concentration of free radicals increased with increasing metabolic activity", antioxidants could "slow down the aging process and thereby put off the appearance of the diseases associated with it". This theory gained early a very large popularity because it was easy to understand: a higher metabolism means more radicals, more attacks, and thus a shorter life. In fact, Harman (1956) provided the biochemical mechanism of the postulated negative effect of metabolism on lifespan. However, it seems clear for many years, among species (e.g. Austad and Fischer 1991) or in the same species (e.g. Lints 1989), that a higher metabolism is not linked to a lower lifespan. Furthermore, the free radicals theory of ageing does not seem to explain lifespan. For instance, among 13 primate species, Csiszar et al. (2012) concluded that "increased longevity in this sample of primates is not causally associated with low cellular reactive oxygen species generation". Within species, Le Bourg's reviews (2001,2003,2005) concluded that it is doubtful that antioxidants could be of help to increase lifespan in animal models or in human beings.

Finally, Speakman and Selman (2011) wrote that there are "fundamental questions over whether the free-radical damage theory remains useful for understanding the ageing process". These authors emphasised that "it has been recently suggested that perhaps the lifespan effect is too harsh a criterion by which the theory should be evaluated", but one can bet that if clear effects of antioxidants had been observed on lifespan, it would not be considered "too harsh a criterion". The authors argued that focusing on health is more sound because oxidative stress is linked with various diseases. For instance, cataract is linked with a depletion of glutathione but, for the time being, "antioxidant supplementation has been shown to be ineffective in preventing or delaying cataract" (Lim et al. 2020). Beyond cataract, Luo et al. (2020) put the last nail in the coffin, concluding that "while observational studies mainly found detrimental effects of high oxidative stress levels on disease status, randomized clinical trials examining the effect of antioxidant supplementation on disease status generally showed null effects".

In such conditions, why taking antioxidants to delay ageing and increase lifespan? There is simply no reason to do that because, after decades of research, supplementing food with antioxidants has not been shown to have any effect. Yet, it remains that there is a plethora of antioxidants on the internet, for the best of companies selling them.

29.4 Other Dietary Supplements

Companies have an innumerable imagination when they intend to sell dietary supplements, as it can be easily seen on the internet when typing "dietary supplements aging" and selecting "shopping". In addition to vitamins and antioxidants, many other supplements are for sale, and it would be a waste of time to enumerate all of them. Just focus on a famous one, because advertised, again, by a Nobel laureate.

Fermented papaya was promoted by the 2008 Nobel laureate Luc Montagnier as having anti-ageing effects and boosting antioxidant defences. It was particularly proposed in 2002 to the pope John Paul-2 as a remedy against the Parkinson's disease. In 2004, the French food safety agency concluded that there was not any evidence for an anti-ageing effect and that (translated from French) "the use of recommendations by scientific personalities, which is decried in the context of the promotion of pharmaceutical products, is not acceptable for foodstuffs either" (AFSSA 2004). Since then, a few articles have been published, not always in well-known journals, and it seems that fermented papaya has not shown a significant amount of positive effects on the ageing process, which could induce its use as a classical drug.

This is probably the final word about dietary supplements: if some scientists think that the supplement X can be of use to improve or delay ageing or increase lifespan, it is incumbent upon them to work with pharmaceutical companies to test their expected new anti-ageing drug according to the usual rules. If X eventually appears to be of some use, more than a placebo, it will be incorporated in the list of authorised and prescribed drugs. If it is not the case, X will be one of the numerous disappointing hopes, a common fate when testing new drugs. By contrast, promoting X in the mass-media without sticking to this process is an abuse of the public credibility.

29.5 Specific Regimens (Mediterranean, Unbalanced, and Okinawa Diets)

Beyond the various supplements, several diets have been proposed to live long with a good health. The most famous one is probably the Mediterranean diet with many fruits and vegetables rather than excessive meat and fats. This diet seems to favour a better healthspan (e.g. Féart et al. 2010) and reduce all-cause mortality (e.g. Eleftheriou et al. 2018). It is not clear whether individual components of the Mediterranean diet explain its positive effects, as each component may have a slight positive effect, if any (Eleftheriou et al. 2018). However, it seems clear that people eating fruits and vegetables are less prone to eat junk food and thus eating more fruits is also eating less junk food. This way, adopting widely the Mediterranean diet could reduce the global threat of obesity and cardio-vascular diseases (GBD 2017). It is not so surprising that this diet has positive effects on health or, rather, not-deleterious effects. Indeed, eating fruits and vegetables, proteins coming from fish and other animals, is not really different from the diet of the human kind for millennia. It is highly different from the consumption of excessive sugar, soft drinks, junk food and other modern foodstuffs not conceived to provide appropriate calories and a balanced regimen with lipids, proteins and carbohydrates, but to be massively sold with the lowest possible price and the highest profit. This Mediterranean diet thus appears to be a traditional feeding regimen but it should not be confused with the so-called paleolithic diet, which promotes 34% of calories coming from proteins (Eaton and Konner 1985). This paleolithic diet does not seem to be linked with beneficial effects on health (Dinu et al. 2020), which is maybe not surprising because a 35% level of proteins is dangerous as it exceeds the capacity of the liver to convert nitrogen in urea (Bilsborough and Mann 2006).

Beside the non-balanced paleolithic diet, other regimens have been proposed with either a low or high level of proteins, carbohydrates or lipids. Using a non-balanced regimen appears to be an error from the beginning, as it artificially alters the spontaneous feeding choice, because subjects forced to ingest an unbalanced diet could tend to reestablish their nutritional target as soon as possible. For instance, Simpson et al. (2003) showed that people fed for two days with a diet containing much more than the usual 15% of calories coming from proteins decrease their food intake by nearly 50%, with the result that the total consumption of proteins is hardly modified. Conversely, subjects forced to eat less proteins for two days increase their food intake by nearly 35%, the total consumption of proteins being also hardly modified. Thus, subjects eat less in the first case and more in the second one, because they strictly regulate the intake of proteins rather than the number of calories. Could it be that one of the causes of the epidemic of obesity in the USA is the decrease of the protein content, from 17% of calories in 1971–1975 in the US diet to a nadir of 15.1% in 2001–2002 (Ford and Dietz 2013), and the necessity of an increased food intake to eat enough proteins?

Similarly, it has been claimed that the Okinawa diet could improve health and increase lifespan (Willcox et al. 2006). The feeding regimen observed after World War 2 and up to the 1960s was depleted in proteins (9% of calories), which makes that, combined with a low number of calories (1785 kcal/day), Okinawans ate ca 57% of the proteins eaten by other Japanese people (Le Bourg 2012). This covers respectively, only 55, 70 and 84% of the daily protein needs of pregnant women, adult men and women (Committee on dietary references intakes 2005). In 1950, ca 20% of Okinawans had therefore nutritional deficiency symptoms, such as a delayed menarche for 10% of women and a deficient lactation for 18% of them (Willcox et al. 2007). Regarding lifespan, Okinawans outlived other Japanese people when the infant mortality was twice higher in mainland Japan (e.g. 90‰ in 1940) than in Okinawa (45‰), which obviously decreased mean lifespan on mainland, but this higher lifespan in Okinawa does not longer exist today (Le Bourg 2012). Today, the malnutrition linked to World War 2 is over, and there is no doubt that the traditional Okinawan diet, with vegetables, sweet potatoes, and soy (Willcox et al. 2006), is of a better quality than the modern occidental food with soft drinks, junk food and so on. Yet, it seems there are no results showing that the Okinawan diet can increase lifespan (Le Bourg 2012), as claimed by some authors (e.g. Willcox et al. 2006).

29.6 Dietary Restriction

Restricting the food intake is called dietary restriction (DR), but also calorie restriction or food restriction. Whatever the term, it means a decrease of food intake without malnutrition. It has been claimed that DR "will extend the life of all species in the Animalia kingdom, including the human species" (Masoro 2006) and that "it would be surprising if appropriately applied, chronic DR would not significantly increase the average lifespan of people" (Weindruch 2006). These very optimistic claims are however maybe not warranted because, even if it is often claimed that DR increases lifespan in many species, if not all, it is not the case (Le Bourg 2010), and, among mice, the results are dependent on the genetic background (Liao et al. 2010). In non-human primates, it has been reported that DR increased lifespan in the mouse lemur *Microcebus murinus* (Pifferi et al. 2018) and one of the two studies in the macaque *Macaca mulatta* also reported a higher lifespan in food-restricted animals (Colman et al. 2014; Mattison et al. 2012). However, these results are better explained by a low lifespan in the control group because of a bad diet and/or overfeeding (Le Bourg 2018).

Regarding humans, results on people self-practicing DR or in controlled studies, such as CALERIE (e.g. Ravussin et al. 2015) show that DR can decrease weight, body-mass index, body fat, cholesterol level, triglycerides, blood pressure, fasting glucose and insulin level (Fontana et al. 2004). All these indices are predictive of a better cardio-vascular health. However, body temperature (−0.2 °C, Heilbronn et al. 2006), lean mass (−5%, Racette et al. 2006), muscular strength (−7%, Weiss et al. 2007), mineral bone density, particularly on hip (Villareal et al. 2006), can also decrease. A more recent study concluded that while DR decreased absolute muscle strength, it increased the strength when expressed relative to body weight (Racette et al. 2017), which could suggest that DR had no real negative effect. However, one can oppose that a DR subject would have more difficulty to carry a 20 kg concrete block than before beginning DR, and that having a higher strength when expressed on the body weight would be of no help to carry the load.

It is not to say that DR protocols are useless as they could provide clues to fight obesity but it is not sure that DR could improve health of people of a normal weight, with an appropriate diet and making some exercise, thus, to people already applying well-known recipes to avoid a premature death. Furthermore, "people with a low body-mass index (BMI, say below 21 kg/m^2) should also be cautioned against practicing CR, as they could soon become underweight, which is a risk factor for various health concerns" (Le Bourg and Redman 2018). It could even be said that DR could be risky for lean people, as adverse events linked to DR (infections, gastrointestinal disorders, and so on) are more probable in subjects with a normal body mass index (BMI: 22–25 kg/m^2) than in overweight ones (BMI: 25–28 kg/m^2) (Romashkan et al. 2016; Le Bourg 2018). Le Bourg and Redman (2018) have warned that DR "with the aim of delaying aging and increasing longevity is not an approved treatment for self-administration", and particularly that "the risk for elderly people is often not to eating too much, but eating too little", because there is a severe risk of undernutrition in elderly people (Kagansky et al. 2005).

However, other authors do not hesitate to write that people should consume starch and fat blockers in addition to a moderate DR (Kurzweill and Grossman 2010), despite the fact that starch and fat blockers are drugs with side-effects. They are prescribed against diabetes and obesity and certainly not in the absence of any disease, even if the internet is replete with advertisements for these products. Similarly, other authors claim that DR could be useful to people, like Most et al. (2017) who write that a forced 20% DR without malnutrition during World War 2 in Norway made that mortality from cardio-vascular diseases dropped by 30%. Because the cited study of Strøm and Jensen (1951) reports a very "close correlation" with the up to 40% decrease of fat intake, it could be concluded that DR decreased mortality but tobacco consumption also decreased by ca 50% during the war (Lund et al. 2009). It is not a risky hypothesis to state that this decreased consumption of tobacco had probably a huge effect on cardio-vascular mortality.

DR is a procedure used in laboratory and clinical studies to better understand, particularly, how to decrease the toll of obesity or cardio-vascular diseases. However, it is a risky business if not done in a controlled study in a laboratory under strict supervision of physicians. Thus, people involved in, for instance, the CALERIE study are not at risk, but those who decide to restrict their food intake on their own are at risk, particularly if they are rather old. It is maybe not surprising that DR is so fashionable in the USA, because obesity is an important problem for a large part of the US population. One may argue that people wishing to live long and not to become obese should first adopt safe behaviours: discarding junk food and snacking, eating more moderately than the current US food intake, making exercise. It is often a hard challenge to modify a behaviour, particularly when the social environment does not encourage to do it, but it is the price to pay to enjoy a better healthspan.

29.7 Detox Diets

For many decades, not to say centuries or millennia, some people have claimed that the organism could be auto-intoxicated and that it was necessary to clean the intestine, as narrated by Frexinos (1997). This way, various recipes have been proposed to clean the body and, more precisely, the intestine with purges (e.g. Guelpa 1910). Today, colon cleansing, or colon hydrotherapy, is still in favour but the so-called detox diet, which is in essence very similar to these old ideas, is more fashionable. While a very few people adopt colon cleansing to detoxify their body, there are numerous proposals of detox programs combining fasting, specific foods to clean the body, and physical exercise. However, "commercial detox diets rarely identify the specific toxins they aim to remove or the mechanisms by which they eliminate them" (Klein and Kiạt 2015). These authors, reviewing the very few results on detox diets, concluded that "considering the financial costs to consumers, unsubstantiated claims and potential health risks of detox products, they should be discouraged by health professionals" and, very wisely, that "there is no doubt that sustained healthy habits are of greater long-term value than the quick fixes offered by commercial detox diets". There is not a single word to add.

29.8 Fasting

DR is not an easy way of life. Indeed, only a very few people are ready to restrict their food intake for years or decades in the hope to live longer. In a way, it is accepting a boring life and hoping it will last for longer than for other people, which is not very enjoyable. It would be much better if a short DR, or repeated short bouts of DR (intermittent fasting), could be used with the same results.

Fasting may have positive effects on health and survival to some threats. In *D. melanogaster* flies, fasting increases survival to anoxia, a model of ischemia–reperfusion in flies (Vigne et al. 2009), or to cold (Le Bourg 2013). A 3-days fast with water increases survival in mice subjected to renal ischemia-reperfusion, similarly to a 4-weeks DR (Mitchell et al. 2010). One week after reperfusion the 10 fasted and 10 DR mice were alive but only 4 out of 10 of control ones. In addition, fasting with chemotherapy helps mice to survive an artificial cancer more than chemotherapy alone (Lee et al. 2012). In women with breast cancer, a randomised clinical trial (de Groot et al. 2020) with 131 patients suggested that fasting before chemotherapy could reduce the tumour more than chemotherapy alone in a per-protocol analysis (comparing patients that really complied with the protocol), the results being less clear in intention to treat (groups are compared even if some subjects did not respect the protocol, as in real life). Fasting is thus studied by scientists and, for instance, some authors wonder whether a preoperative fasting lasting more than the usual night could improve surgical outcomes (Longchamp et al. 2017).

Therefore, fasting seems promising in therapy and some authors highly promote it in books or media, as for instance Valter Longo (2018). Intermittent fasting is also fashionable and, at a first sight, it could appear as a nice idea to preserve health and, again, there are many sites promoting fasting cures on internet (e.g. in the US, https://thefastingmethod.com/, in France https://www.ffjr.com/) or in books. One of the issues with intermittent fasting is that some authors are over-enthusiast to the point they write problematic sentences. For instance, de Cabo and Mattson (2020) write that "on the island of Okinawa, the traditional population typically maintains a regimen of intermittent fasting and has low rates of obesity and diabetes mellitus, as well as extreme longevity". However, such "extreme longevity" does not exist (see above) and the Okinawa diet is not intermittent fasting: there is no fast for one or two days from time to time, but an avoidance of eating at satiety. In addition, writing that "clinical studies have focused mainly on overweight young and middle-age adults, and we cannot generalize to other age groups the benefits and safety of intermittent fasting", is hazardous, as the main issue in elderly people is under-nutrition: it would be risky for elderly people to fast more than they already and non-intentionally do.

People practicing intermittent fasting often say they feel better, with more energy. This is true and can be easily explained. When a subject or an animal is fasting, the nutritional needs are no longer satisfied, and the organism sounds the alarm to implement various means to restore the previous physiological state. Thus, more energy is diverted to seek food, at the expense of body reserves, cognition is in alert to locate food, and so on. As said by Longo and Mattson (2014), "hunger is an adaptive response to food deprivation that involves sensory, cognitive and neuroendocrine changes that motivate and enable food-seeking behaviors". The net result is that animals or subjects are better at walk and people have the feeling they have a better mood and cognition. A part of this process has been observed in *D. melanogaster* flies by Connolly (1966). Flies deprived of food put in a novel environment maintain a high activity level for at least 7 h, while ad libitum-fed flies show a regular time-linked decrease.

This is what happens… at the beginning. After that, if fasting lasts for too long, the available energy is decreasing and subjects can become sluggish, physically and mentally. In other words, not too old people with a good health are not at risk if they decide to fast for one day from time to time. Concluding that because they feel better, more awake, and even smarter, intermittent fasting will improve healthspan and even increase lifespan is an illusion linked to the physiological state occurring during a short fasting. It has been shown that regular fasting during the whole life (4 days every 2 weeks) could increase lifespan in mice (Sogawa and Cubo 2000), but it also paradoxically increased their weight. This is probably not the result hoped by people prone to fast, but it is also commonly observed in people under a restricted regimen. They lose weight, before often to regain more, probably because the organism is preparing to a next fast by making more reserves than before the fast: once bitten, twice shy!

29.9 Conclusions

For centuries, if not millennia, people have hoped to enjoy a better health by modifying their diet. It is a sound idea and there is no debate on the necessity to favour a regimen covering nutritional needs, but not more. However, it is not sensible to think that miraculous regimens do exist: if it were the case, humans would have discovered them since time immemorial. In fact, humans have discovered the miraculous regimen for a long life: a balanced regimen providing proteins, lipids and carbohydrates, as well as essential nutriments such as vitamins. Thus, a wise strategy is to discard modern junk food known to favour cardio-vascular diseases and, eventually, a shorter lifespan. Yet, it is also wise to resist ads promoting risky regimens with no proteins, or only proteins, dietary supplements daily, detox diets, and all new avatars of the modern quacks wearing a white blouse on a computer screen.

The global obesity epidemic is a real issue and the Covid-19 pandemic has shown that, among the younger patients, those transferred in intensive care units were often obese (e.g. Hajifathalian et al. 2020) and also that obesity increased the fatality risk (de Larochelambert et al. 2020). It is maybe a sufficient reason to adopt an appropriate food regimen and to discard other ones.

Compliance with Ethical Standards

Conflict of Interest The author has no conflict of interest.

References

AFSSA (2004) Avis de l'agence française de sécurité sanitaire des aliments relatif aux justificatifs concernant la composition d'un complément alimentaire à base d'extrait de papaye fermentée et les différentes allégations de santé mettant en avant l'action anti-âge. (Advice of the French Food Safety Agency on the justifications concerning the composition of a food supplement based on fermented papaya extract and the various health claims highlighting the anti-ageing action). Afssa – Saisine n° 2004-SA-0228

Alehagen U, Johansson P, Björnstedt M, Rosén A, Post C, Aaseth J (2016) Relatively high mortality risk in elderly Swedish subjects with low selenium status. Eur J Clin Nutr 70:91–96

Austad SN, Fischer KE (1991) Mammalian aging, metabolism, and ecology: evidence from the bats and marsupials. J Gerontol 46:B47–B53

Bilsborough S, Mann N (2006) A review of issues of dietary protein intake in humans. Int J Sport Nutr Exerc Metab 16:129–152

Bjelakovic G, Nikolova D, Gluud LL, Simonetti RG, Gluud C (2007) Mortality in randomized trials of antioxidant supplements for primary and secondary prevention. Systematic review and meta-analysis. J Amer Med Assoc 297:842–857

Bleys J, Navas-Acien A, Guallar E (2007) Selenium and diabetes: more bad news for supplements. Ann Intern Med 147:271–272

Colman RJ, Beasley TM, Kemnitz JW, Johnson SC, Weindruch R, Anderson RM (2014) Caloric restriction reduces age-related and all-cause mortality in rhesus monkeys. Nature Commun 5:3557

Committee on dietary references intakes (2005) Dietary reference intakes for energy, carbohydrate, fiber, fat, fatty acids, cholesterol, protein, and amino acids (2002/2005). National Academies Press 2002/2005, Washington

Connolly KJ (1966) Locomotor activity in *Drosophila* as a function of food deprivation. Nature 209:224

Csiszar A, Podlutsky A, Podlutskaya N, We S, Merlin SZ, Philipp EE, Doyle K, Davila A, Recchia FA, Ballabh P, Jt P, Ungvari Z (2012) Testing the oxidative stress hypothesis of aging in primate fibroblasts: is there a correlation between species longevity and cellular ROS production? J Gerontol A Biol Sci Med Sci 67:841–852 de Cabo R, Mattson MP (2020) Effects of intermittent fasting on health, aging, and disease. N Engl J Med 381:2541–2551

de Larochelambert Q, Marc A, Antero J, Le Bourg E, Toussaint JF (2020) Covid-19 mortality: a matter of vulnerability among nations facing limited margins of adaptation. Front Public Health. https://doi.org/10.3389/fpubh.2020.604339

de Groot S, Lugtenberg RT, Cohen D, Welters MJP, Ehsan I, Vreeswijk MPG, Smit VTHBM, de Graaf H, Heijns JB, Portielje JEA, van de Wouw AJ, Imholz ALT, Kessels LW, Vrijaldenhoven S, Baars A, Kranenbarg EM, Carpentier MD, Putter H, van der Hoeven JJM, Nortier JWR, Longo VD, Pijl H, Kroep JR, Dutch Breast Cancer Research Group (BOOG) (2020) Fasting mimicking diet as an adjunct to neoadjuvant chemotherapy for breast cancer in the multicentre randomized phase 2 DIRECT trial. Nat Commun 11:3083

Dinu M, Pagliai G, Angelino D, Rosi A, Dall'Asta M, Bresciani L, Ferraris C, Guglielmetti M, Godos J, Del Bo' C, Nucci D, Meroni E, Landini L, Martini D, Sofi F (2020) Effects of popular diets on anthropometric and cardiometabolic parameters: an umbrella review of meta-analyses of randomized controlled trials. Adv Nutr 11:815–833

Eaton SB, Konner M (1985) Paleolithic nutrition. A consideration of its nature and current implications. N Eng J Med 312:283–289

Eleftheriou D, Benetou V, Trichopoulou A, La Vecchia C, Bamia C (2018) Mediterranean diet and its components in relation to all-cause mortality: meta-analysis. Br J Nutr 120:1081–1097

Féart C, Samieri C, Barberger-Gateau P (2010) Mediterranean diet and cognitive function in older adults. Curr Opin Clin Nutr Metab Care 13:14–18

Fisberg M, Machado R (2015) History of yogurt and current patterns of consumption. Nutr Rev 73(suppl. 1):4–7

Fontana L, Meyer TE, Klein S, Holloszy JO (2004) Long-term caloric restriction is highly effective in reducing the risk for atherosclerosis in humans. Proc Natl Acad Sci USA 101:6659–6663

Ford ES, Dietz WH (2013) Trends in energy intake among adults in the United States: findings from NHANES. Am J Clin Nutr 97:848–853

Frexinos J (1997) L'art de purger (The art of purging). Editions Louis Pariente, Paris

GBD 2015 Obesity Collaborators, Afshin A, Forouzanfar MH, Reitsma MB, Sur P, Estep K, Lee A, Marczak L, Mokdad AH, Moradi-Lakeh M, Naghavi M, Salama JS, Vos T, Abate KH, Abbafati C, Ahmed MB, Al-Aly Z, Alkerwi A, Al-Raddadi R, Amare AT, Amberbir A, Amegah AK, Amini E, Amrock SM, Anjana RM, Ärnlöv J, Asayesh H, Banerjee A, Barac A, Baye E, Bennett DA, Beyene AS, Biadgilign S, Biryukov S, Bjertness E, Boneya DJ, Campos-Nonato I, Carrero JJ, Cecilio P, Cercy K, Ciobanu LG, Cornaby L, Damtew SA, Dandona L, Dandona R, Dharmaratne SD, Duncan BB, Eshrati B, Esteghamati A, Feigin VL, Fernandes JC, Fürst T, Gebrehiwot TT, Gold A, Gona PN, Goto A, Habtewold TD, Hadush KT, Hafezi-Nejad N, Hay SI, Horino M, Islami F, Kamal R, Kasaeian A, Katikireddi SV, Kengne AP, Kesavachandran CN, Khader YS, Khang YH, Khubchandani J, Kim D, Kim YJ, Kinfu Y, Kosen S, Ku T, Defo BK, Kumar GA, Larson HJ, Leinsalu M, Liang X, Lim SS, Liu P, Lopez AD, Lozano R, Majeed A, Malekzadeh R, Malta DC, Mazidi M, McAlinden C, McGarvey ST, Mengistu DT, Mensah GA, Mensink GBM, Mezgebe HB, Mirrakhimov EM, Mueller UO, Noubiap JJ, Obermeyer CM, Ogbo FA, Owolabi MO, Patton GC, Pourmalek F, Qorbani M, Rafay A, Rai RK, Ranabhat CL, Reinig N, Safiri S, Salomon JA, Sanabria JR, Santos IS, Sartorius B, Sawhney M, Schmidhuber J, Schutte AE, Schmidt MI, Sepanlou SG, Shamsizadeh M, Sheikhbahaei S, Shin MJ, Shiri R, Shiue I, Roba HS, Silva DAS, Silverberg JI, Singh JA, Stranges S, Swaminathan S, Tabarés-Seisdedos R, Tadese F, Tedla BA, Tegegne BS, Terkawi AS, Thakur JS, Tonelli M, Topor-Madry R, Tyrovolas S, Ukwaja KN, Uthman OA, Vaezghasemi M, Vasankari T, Vlassov VV, Vollset SE, Weiderpass E, Werdecker A, Wesana J, Westerman R, Yano Y, Yonemoto N, Yonga G, Zaidi Z, Zenebe ZM, Zipkin B, Murray CJL (2017) Health effects of overweight and obesity in 195 countries over 25 years. N Engl J Med 377:13–27

Guelpa G (1910) Autointoxication et désintoxication (Autointoxication and detoxification). Doin et fils, Paris

Haber C (2004) Life extension and history: the continual search for the fountain of youth. J Geront Biol Sci 59:B515–B522
Hajifathalian K, Kumar S, Newberry C, Shah S, Fortune B, Krisko T, Ortiz-Pujols S, Zhou XK, Dannenberg AJ, Kumar R, Sharaiha RZ (2020) Obesity is associated with worse outcomes in COVID-19: analysis of early data from New York City. Obesity 28:1606–1612
Harman D (1956) Aging: a theory based on free radical and radiation chemistry. J Geront 11:298–300
Heilbronn LK, De Jonge L, Frisard MI, Delany JP, Larson-Meyer DE, Rood J, Nguyen T, Martin CK, Volaufova J, Most MM, Greenway FL, Smith DA Ravussin E (2006) Effect of 6-month calorie restriction on biomarkers of longevity, metabolic adaptation, and oxidative stress in overweight individuals. J Am Med Assoc 295:1539–1548
Kagansky N, Berner Y, Koren-Morag N, Perelman L, Knobler H, Levy S (2005) Poor nutritional habits are predictors of poor outcome in very old hospitalized patients. Am J Clin Nutr 82:784–791
Klein AV, Kiat H (2015) Detox diets for toxin elimination and weight management: a critical review of the evidence. J Human Nutr Diet 28:675–686
Kurzweil R, Grossman T (2010) Bridges to life. In: Fahy GM, West MD, Coles LS, Harris SB (eds.) The future of aging. Pathways to human life extension, pp 3–22. Springer
Le Bourg E (2001) Oxidative stress, aging and longevity in *Drosophila melanogaster*. FEBS Lett 498:183–186
Le Bourg E (2003) Antioxidants as modulators. In: Rattan SIS (ed) Modulating aging and longevity. Kluwer Academic Publishers, Dordrecht, pp 183–203
Le Bourg E (2005) Antioxidants and aging in human beings. In: Rattan SIS (ed) Aging interventions and therapies. World Scientific Publishers, Singapore, pp 85–107
Le Bourg E (2010) Predicting whether dietary restriction would increase longevity in species not tested so far. Ageing Res Rev 9:289–297
Le Bourg E (2012) Dietary restriction studies in humans: focusing on obesity, forgetting longevity. Gerontology 58:126–128
Le Bourg E (2013) Fasting can protect young and middle-aged *Drosophila melanogaster* flies against a severe cold stress. Biogerontology 14:513–529
Le Bourg E (2018) Does calorie restriction in primates increase lifespan? Revisiting studies on macaques (*Macaca mulatta*) and mouse lemurs (*Microcebus murinus*). BioEssays 40:1800111
Le Bourg E, Redman LM (2018) Do-it-yourself calorie restriction: the risks of simplistically translating findings in animal models to humans. BioEssays 40:1800087
Lee C, Raffaghello L, Brandhorst S, Safdie FM, Bianchi G, Martin-Montalvo A, Pistoia V, Wei M, Hwang S, Merlino A, Emionite L, De Cabo R, Longo VD (2012) Fasting cycles retard growth of tumors and sensitize a range of cancer cell types to chemotherapy. Sci Transl Med 4:124ra27
Liao CY, Rikke BA, Johnson TE, Diaz V, Nelson JF (2010) Genetic variation in the murine lifespan response to dietary restriction: from life extension to life shortening. Aging Cell 9:92–95
Lim JC, Grey AC, Zahraei A, Donaldson PJ (2020) Age-dependent changes in glutathione metabolism pathways in the lens: new insights into therapeutic strategies to prevent cataract formation—a review. Clin Exp Ophtalmology 48:1031–1042
Lints FA (1989) The rate of living theory revisited. Gerontology 35:36–57
Longchamp A, Harputlugil E, Corpataux JM, Ozaki CK, Mitchell JR (2017) Is overnight fasting before surgery too much or not enough? How basic aging research can guide preoperative nutritional recommendations to improve surgical outcomes: a mini-review. Gerontology 63:228–237
Longo VD (2018) The longevity diet. Penguin Random House, New York
Longo VD, Mattson MP (2014) Fasting: molecular mechanisms and clinical applications. Cell Metab 19:181–192
Lund KE, Lund M, Bryhni A (2009) Tobacco consumption among men and women 1927–2007. Tidsskr nor Legeforen 129:1871–1874
Luo J, Mills K, le Cessie S, Noordam R, van Heemst D (2020) Ageing, age-related diseases and oxidative stress: what to do next? Ageing Res Rev 57:100982

Masoro EJ (2006) Dietary restriction-induced life extension: a broadly based biological phenomenon. Biogerontology 7:153–155
Mattison JA, Roth GS, Beasley TM, Tilmont EM, Handy AM, Herbert RL, Longo DL, Allison DB, Young JE, Bryant M, Barnard D, Ward WF, Qi W, Ingram DK, de Cabo R (2012) Impact of caloric restriction on health and survival in rhesus monkeys from the NIA study. Nature 489:318–321
Metchnikoff E (1903) Études sur la nature humaine: essai de philosophie optimiste. Masson, Paris. (English translation: The nature of Man: studies in optimistic philosophy, Putnam's GP sons, New York, 1907)
Miller ER III, Pastor-Barriuso R, Dalal D, Riemersma RA, Appel LJ, Guallar E (2005) Meta-analysis: high-dosage vitamin E supplementation may increase all-cause mortality. Ann Intern Med 142:37–46
Mitchell JR, Verweij M, Brand K, van de Ven M, Goemaere N, van den Engel S, Chu T, Forrer F, Müller C, de Jong M, van IJcken W, IJzermans JN, Hoeijmakers JH, de Bruin RW (2010) Short-term dietary restriction and fasting pre-condition against ischemia reperfusion injury in mice. Aging Cell 9:40–53
Most J, Tosti V, Redman LM, Fontana L (2017) Calorie restriction in humans: an update. Ageing Res Rev 39:36–45
Pauling L (1983) Vitamin C and longevity. Agressologie 24:317–319
Pauling L, Moertel C (1986) A proposition: megadoses of vitamin C are valuable in the treatment of cancer. Nutr Rev 44:28–32
Pearl R (1928) The rate of living. Knopf, London
Pifferi F, Terrien J, Marchal J, Dal-Pan A, Djelti F, Hardy I, Chahory S, Cordonnier N, Desquilbet L, Hurion M, Zahariev A, Chery I, Zizzari P, Perret M, Epelbaum J, Blanc S, Picq JL, Dhenain M, Aujard F (2018) Caloric restriction increases lifespan but affects brain integrity in grey mouse lemur primates. Comm Biol 1:30
Racette SB, Weiss EP, Villareal DT, Arif H, Steger-May K, Schechtman KB, Fontana L, Klein S, Holloszy JO (2006) One year of caloric restriction in humans: feasibility and effects on body composition and abdominal adipose tissues. J Geront Biol Sci 61A:943–950
Racette SB, Rochon J, Uhrich ML, Villareal DT, Das SK, Fontana L, Bhapkar M, Martin CK, Redman LM, Fuss PJ, Roberts SB, Kraus W (2017) Effects of two years of calorie restriction on aerobic capacity and muscle strength. Med Sci Sports Exerc 49:2240–2249
Ravussin E, Redman LM, Rochon J, Das SK, Fontana L, Kraus WE, Romashkan S, Williamson DA, Meydani SN, Villareal DT, Smith SR, Stein RI, Scott TM, Stewart TM, Saltzman E, Klein S, Bhapkar M, Martin CK, Gilhooly CH, Holloszy JO, Hadley EC, Roberts SB; Calerie Study Group (2015) A 2-year randomized controlled trial of human caloric restriction: feasibility and effects on predictors of health span and longevity. J Gerontol A Biol Sci Med Sci 70:1097–1104
Romashkan SV, Das SK, Villareal DT, Ravussin E, Redman LM, Rochon J, Bhapkar M, Kraus WE, CALERIE Study Group (2016) Safety of two-year caloric restriction in non-obese healthy individuals. Oncotarget 7:19124–19133
Simpson SJ, Batley R, Raubenheimer D (2003) Geometric analysis of macronutrient intake in humans: the power of protein? Appetite 41:123–140
Sogawa H, Kubo C (2000) Influence of short-term repeated fasting on the longevity of female (NZB × NZW) F1 mice. Mech Ageing Dev 115:61–71
Speakman JR, Selman C (2011) The free-radical damage theory: accumulating evidence against a simple link of oxidative stress. BioEssays 33:255–259
Stambler I (2020a) Cultural myths and legends about food for eternal life. This book
Stambler I (2020b) History of life-extensionism. Encyclopedia of Biomedical Gerontology 2:228–237
Strøm A, Jensen RA (1951) Mortality from circulatory diseases in Norway 1940–1945. Lancet 1:126–129
United States General Accounting Office (2001) Health products for seniors: "anti-aging" products pose potential for physical and economic harm. Report to chairman, special committee on aging, US senate, GAO-01-1129

Vigne P, Tauc M, Frelin C (2009) Strong dietary restrictions protect *Drosophila* against anoxia/reoxygenation injuries. PLoS ONE 4:e5422
Villareal DT, Fontana L, Weiss EP, Racette SB, Steger-May K, Schechtman KB, Klein S, Holloszy JO (2006) Bone mineral density response to caloric restriction-induced weight loss or exercise-induced weight loss: a randomized controlled trial. Arch Intern Med 166:2502–2510
Weindruch R (2006) Will dietary restriction work in primates? Biogerontology 7:169–171
Weiss EP, Racette SB, Villareal DT, Fontana L, Steger-May K, Schechtman KB, Klein S, Ehsani AA, Holloszy JO, Washington University School Of Medicine Calerie Group (2007) Lower extremity muscle size and strength and aerobic capacity decrease with caloric restriction but not with exercise-induced weight loss. J Appl Physiol 102:634–640
Willcox DC, Willcox BJ, Todoriki H, Curb JD, Suzuki M (2006) Caloric restriction and human longevity: what can we learn from the Okinawans? Biogerontology 7:173–177
Willcox BJ, Willcox DC, Todoriki H, Fujiyoshi A, Yano K, He Q, Curb JD, Suzuki M (2007) Caloric restriction, the traditional Okinawan diet, and healthy aging: the diet of the world's longest-lived people and its potential impact on morbidity and life span. Ann NY Acad Sci 1114:434–455

Part V
Conclusions and Suggestions

Chapter 30
Nutrition, Food and Diet: Recapitulation, Conclusions and Suggestions

Suresh I. S. Rattan and Gurcharan Kaur

Abstract Here we recapitulate the main themes of the issues related to nutrition, food and diet discussed in the twenty-nine chapters in this book. The insights from various contributions have been extracted to conceptualize how nutritional components are not only the source of building blocks and energy for the body, but also serve numerous critical functions in maintaining health. This wealth of information can be useful for making recommendations and suggestions for food for health and longevity, especially in old age. A central message derived from this collection of articles is that optimizing our nutrition by making the right choice of food, and aligning our eating behaviors with our bio-social rhythms, are the most effective general strategies to maintain, recover and enhance individual- and public-health.

Keywords Nutrition · Longevity · Healthy ageing · Macro- and micronutrients · Diet and culture · Nutritional interventions

30.1 Introduction

In this volume of the bookseries "Healthy Ageing and Longevity", we have succeeded in bringing together a diversified group of researchers and academicians reviewing the published literature and presenting their personal views about nutritional components, foods, dietary patterns, and nutrition- and diet-based interventions for health and longevity.

It is worth repeating from our preface to this volume that nutrition, food and diet are the terms often used casually and uncritically as overlapping ones. However, while "nutrition" generally refers to the macro- and micro-nutrients essential for health and survival, nutrition is normally obtained from animal- or plant-based foods. We do not eat nutrition, we eat food, without always being aware or conscious of the nutritious

S. I. S. Rattan (✉)
Department of Molecular Biology and Genetics, Aarhus University, Aarhus, Denmark
e-mail: rattan@mbg.au.dk

G. Kaur
Department of Biotechnology, Guru Nanak Dev University, Amritsar, India

S. I. S. Rattan and G. Kaur (eds.), *Nutrition, Food and Diet in Ageing and Longevity*, Healthy Ageing and Longevity 14, https://doi.org/10.1007/978-3-030-83017-5_30

value of the food. Moreover, these nutritional values may or may not match our bodily requirements which depend upon various factors, such as age, general health status, physiological demands and accessibility. Even more importantly, there are crucial socio-cultural dimensions which give shape to our taste, preferences and taboos towards food (Samaddar et al. 2020). Furthermore, the three commonly identified meals of the day – breakfast, lunch and dinner – are as varied globally as different populations, societies and cultures. And above all of this lurks in the background the evolutionary history of *Homo sapiens* from being hunter-gatherers to becoming agriculturists (Finch 2010), and to the present state of becoming the consumers of highly processed food products with overeating behaviors wherever food is available in abundance.

In this concluding chapter, we try to recapitulate the main messages which can be derived from various chapters in this book (Table 30.1), and how that information can be useful for making recommendations and suggestions for practical use and for future research. Perhaps the central carry home message is obvious that optimizing our nutrition by making the "right" choice of food and aligning our eating behaviors with our bio-social rhythms are the most effective general strategies to maintain, recover and even enhance individual- and public-health.

30.2 Nutrition

Proteins, carbohydrates and lipids are the three essential nutritional components which provide us the basic material for building our biological structures and for producing the energy required for all physiological and biochemical processes. Additionally, minerals and micronutrients, including vitamins, facilitate the optimal utilisation of the primary nutritional components by catalyzing numerous biochemical processes, by increasing their bioavailability and absorption, and by balancing the microbiome crucial for maintaining health *(see Chapters 1 to 5)*. Another category of nutrients, termed nutritional hormetins, are various compounds in the food, which by virtue of their toxic nature, induce and stimulate body's intrinsic repair and maintenance systems. This apparently paradoxic effects of toxic compounds in the food as health-beneficial nutritional hormetins is in line with the basic understanding of the biological mechanisms of health and survival through adaptive responses and hormesis (Rattan and Kyriazis 2019) *(see chapter 6)*. Furthermore, it is in this context that we need to understand and appreciate the evolutionary basis of human nutritional requirements *(see chapter 7)*. This is also the basis of understanding human nutrigenomics (how nutritional components affect gene expression) and nutrigenetics (how our genetic heterogeneity affects our response to various nutrients) (Mutch et al. 2005; Wittwer et al. 2011).

Table 30.1 Chapter sequence, titles, authors and page numbers

Chapter number, title and authors	Page numbers
Chapter-01: Dietary proteins: functions and health benefits – KhetanShevkani and Shivani Chourasia	3–37
Chapter-02: Carbohydrates as nutritional components for health and longevity – GözdeOkburan and Ceren Gezer	39–52
Chapter-03: Fats and oils in health and longevity Kaustuv Bhattacharya and Suresh I.S. Rattan	53–62
Chapter-04: Micronutrients in ageing and longevity – BoženaĆurko-Cofek	63–83
Chapter-05: Probiotics and prebiotics in healthy aging – Aayushi Kadam, Deepak Kadam, KanchanlataTungare, Heena Shah	85–108
– Chapter-06: Nutritional hormetins in ageing and longevity – Suresh Rattan	109–122
Chapter-07: Notes towards an evolutionary biology of nutrition – Donovon German and Michael Rose,	123–151
Chapter-08: Animal- and plant-based food for health and longevity – AzzaSilotry Naik, Majeed Jamakhani, Madhavi R. Verneker and S. S. Lele,	155–177
Chapter-09: Fermented foods in ageing and longevity – S. Heeba and P. Nisha	179–192
Chapter-10: Milk and other dairy product trends in health and longevity – Dino Demirovic Holmquist	193–198
Chapter-11: Anti-inflammatory foods in ageing and longevity – Ceren Gezer	199–219
Chapter-12: Nutraceuticals and functional foods in aging and aging-associated diseases – Jatinder Pal Singh, Balwinder Singh, and Amritpal Kaur	221–238
Chapter-13: Food for brain health – Vittorio Calabrese, Angela Trovato, Maria Scuto, Maria Laura Ontario, Francesco Rampulla, Vincenzo Zimbone, Ignazio La Mantia, Valentina Greco, Sebastiano Sciuto, Roberto Crea, Edward J. Calabrese, and Giuseppe Dionisio	239–274
Chapter-14: Nutrition and exercise to maintain physical functioning during ageing – Pol Grootswagers, Marco Mensink, Lisette de Groot	275–298
Chapter-15: Dietary patterns and healthy ageing – Ligia Dominguez and Mario Barbagallo	301–314
Chapter-16: Ketogenic diet, circadian rhythm and aging – Anita Jagota and Sushree Abhidhatri Sharma	315–330
Chapter-17: Diet according to the traditional Chinese medicine for health and longevity – Yifang Zhang	331–356
Chapter-18: Indian traditional foods and diets: combining traditional wisdom with modern science of nutraceuticals and functional foods – Jyoti S. Gokhle, S. S. Lele and Laxmi Ananthanarayan	357–392

(continued)

Table 30.1 (continued)

30.3 Food

Almost all the food we eat is of the animal and plant origin, although continuing advancements in nutrition research and innovation of food industry may open up new vistas including man-made synthetic foods. Besides the inherent nutritional composition of plant- and animal-based foods, farming and agricultural practices, methods of food preparation and preservation, and the amounts consumed are equally important factors for their eventual health benefits (see *chapter 8 and 9*). Moreover, prevalent climatic conditions and the local production of raw materials have always challenged human-beings across civilizations and cultures to explore and develop different processes of preservation and modification, such as fermentation. The claimed health benefits of fermented foods are attributed to a combination of bioactive components released as by-products of the fermentation process and the presence of probiotics in these foods, which support the gut microbiota (see *chapter 9*).

Various foods consumed traditionally in different cultures, and some of the claims made for their potential modes of action, have been scientifically validated to some extent. Such foods are often touted as complete foods, anti-inflammatory foods, food for brain, food for physical endurance, functional foods and so on (see *chapter 11–13*). Traditional foods enriched with a variety of phytoconstituents and other minerals, vitamins and natural or synthetic hormetins are generally promoted under the banner of "functional foods". A similar trend can be seen for milk and dairy products with novel and innovative formulationswith a focus on functionality, health, and easy availability (see *chapter 10*).

30.4 Diet and Culture

The next stage after having access to food is the manner in which that food is consumed, and that is what gives it the status of becoming diet or a meal. This is also where the socio-cultural aspects of dietary habits and traditions become more important than the biochemistry and molecular biology of the nutritional components of food (Samaddar et al. 2020). The so-called Mediterranean diet, paleo diet, ketogenic diet, Chinese diet, Ayurvedic diet, vegetarian diet, and more recently, vegan diet, are some of the varied expressions of such social and cultural practices (*see chapter 15-18*). The health outcomes of these variety of dietary and mealing patterns prevalent in different cultures have been helpful in cross-cultural adaptation to both healthy eating patterns on one hand and bringing awareness about the health-risk associations of certain food habits on the other. It is in this context that the subject of chrononutrition, that is how our daily rhythms affect and regulate our nutritional needs, has become an important aspect of healthy eating *(see chapters 17, 18)*. Similarly, the so-called "nutrient timing" involves eating foods at strategic times in order to achieve certain outcomes, such as muscle growth, sports performance and fat loss (Kessler and Pivovarova-Ramich 2019). Looking back into the cultural history of longevity foods, one realizes that it is not just the right combination of nutrients which is important, but the elaborate social practices, rituals and normative behaviors for obtaining, preparing and consuming food are equally, if not more, important aspects of health-promoting eating (*see chapter 20*).

Thus, dietary patterns and meal-timing seem to be more predictive of health-related quality of life than foods or nutrients on their own (Fontana and Partridge 2015; Kessler and Pivovarova-Ramich 2019). Encouraging a shift to healthy dietary patterns, however, critically depends not only on the food availability, accessibility and affordability, but also on the behavioral, intentional and socio-cultural conditions of consumers for making their food choices. An old saying that "we are what we eat" may be corrected to that "we eat what we are"!

30.5 Nutritional and Dietary Interventions

Food is often considered as one of the three pillars of human health and survival – the other two being physical activity, and social and mental engagement. Therefore, manipulating and modifying food components and food habits have always been attractive targets for intervention. This is exemplified by extensive research done on the effects of individual nutrients, short peptides, and plant extracts in stimulating, inhibiting or regulating various molecular pathways in relation to ageing, and age-related diseases, including cancer, in experimental animal model systems and cells in culture *(see chapters 21, 23, 28)*. More recently, single-nutrient-based or multiple-nutrients-based selective killing of senescent cells, and calorie restriction mimetics as anti-ageing therapies are being tested and developed *(see chapters 22, 25)*. These therapeutic strategies follow the same "one target, one missile" pharmaceutical rationale, and consider ageing as a treatable disease. However, often the promises made, based on the results obtained from such experimental studies, are either naïve extrapolations from experimental model systems or overhyped claims for human applications, which deliver very little in reality (Rattan 2020) *(see chapter 29)*.

Some other innovative, holistic and relatively easy to implement food- and diet-based interventions strategies are caloric/dietary restriction and time restricted eating (TRE) regimens. For example, intermittent fasting (IF), based on feeding/fasting timing manipulation, is emerging as a promising intervention to promote health and longevity. Meta-analyses of several pilot studies in human subjects show that TRE, with eating restricted to a few hours in the day time, has beneficial effects on several health parameters (Mattson et al. 2014) (*see chapter 24*). Similarly, several other practical recommendations are made for the optimal nutrition in old age (*see chapter 26*), and for the regulation of metabolic syndromes, especially metabesity (*see chapter, 27*).

30.6 Conclusions and Suggestions

Food is more than just being one of the three pillars of health. Food is both the foundation and the scaffolding for the building and survival of an organism on a daily basis. Scientific research on the macro- and micro-nutrient components of food has developed deep understanding of their molecular, biochemical and physiological roles. Various recommendations are repeatedly made and modified for some optimal daily requirements of nutrients for maintaining and enhancing health, and for the prevention and treatment of diseases. Can one envisage developing a "nutrition pill" for perfect health, which could be used globally, across cultures, and at all ages? We don't think so.

So far, most of our knowledge about nutrients comes from the experimental studies made with individual active components, or, at best, with the combination of one or two compounds. However, in reality, we obtain these nutrients from the food where they exist together with hundreds of other interacting compounds, and become modified through the process of cooking and preservation. Scientifically, there is very little known about the health-related effects of innumerable combinations of food sources, cooking methods, dietary combinations and other factors. However, there is a lot of folk knowledge available in all cultures about the dos and don'ts of food, for which we need scientific verification and validation. For this purpose, we must rethink and modify our presently established scientific protocols of double blinded randomized control trials which are practically impossible to apply on food and dietary research at the population level.

Finally, nutrition, food and diet are the basis of life, health and longevity. We have gathered a huge amount of information with respect to what nutrition is required for health and survival, which foods can provide what nutrition and to what extent, and which dietary and eating practices seem to be more or less health beneficial. Whereas in some parts of the world, abundant availability and accessibility of food has led to its overuse, abuse and consequent life-style diseases, in other parts of the world economic disparity and food scarcity keep perpetuating starvation, malnutrition, poor health and shortened lifespan. It is usually not the lack of knowledge about the optimal nutrition, food and diet that forces us to make bad choices; rather it is either our inability to access such foods, or our gullibility to fall prey to the exaggerated claims in the commercial interests of the food companies, which determine what and how much we consume. Science will and must continue to gather more information about the biochemical, physiological and other aspects of food, which we should apply wisely and globally for the health and longevity of all.

Compliance with Ethical Standards

Conflict of Interest All authors declare they have no conflict of interest.

References

Finch CE (2010) Evolution of the human lifespan and diseases of aging: Roles of infection, inflammation, and nutrition. Proc Natl Acad Sci U S A 107:1718–1724

Fontana L, Partridge L (2015) Promoting health and longevity through diet: from model organisms to humans. Cell 161:106–118

Kessler K, Pivovarova-Ramich O (2019) Meal timing, aging, and metabolic health. Int J Mol Sci 20:1911

Mattson MP et al (2014) Meal frequency and timing in health and disease. Proc Natl Acad Sci U S A 111:16647–16653

Mutch DM, Wahli W, Williamson G (2005) Nutrigenomics and nutrigenetics: the emerging faces of nutrition. FASEB J 19:1602–1616

Rattan SIS (2020) Naive extrapolations, overhyped claims and empty promises in ageing research and interventions need avoidance. Biogerontology 21:415–421

Rattan SIS, Kyriazis M (eds) (2019) The science of hormesis in health and longevity. Academic Press, UK

Samaddar A (2020) Capturing diversity and cultural drivers of food choice in eastern India. Int J Gastron Food Sci 22:100249

Wittwer J, Rubio-Aliaga I, Hoeft B, Bendik I, Weber P, Daniel H (2011) Nutrigenomics in human intervention studies: current status, lessons learned and future perspectives. Mol Nutr Food Res 55:341–358

Author Index

S. I. S. Rattan and G. Kaur (eds.), *Nutrition, Food and Diet in Ageing and Longevity*, Healthy Ageing and Longevity 14, https://doi.org/10.1007/978-3-030-83017-5

O

P

R

S

T

Tagliaferri, Federica, 301
Trofimova, Svetlana V., 491
Trovato, Angela, 239
Tungare, Kanchanlata, 85

U

V

W

Y

Z

The manufacturer's authorised representative in the EU is Springer Nature Customer Service Centre GmbH, Europaplatz 3, 69115 Heidelberg, Germany. If you have any concerns regarding our products, please contact ProductSafety@springernature.com

Printed and bound by CPI Group (UK) Ltd, Croydon, CR0 4YY
30/09/2026
02225504-0018